T0190010

Lecture Notes in Computer Science 10193

Commenced Publication in 1973
Founding and Former Series Editors:
Gerhard Goos, Juris Hartmanis, and Jan van Leeuwen

More information about this series at http://www.springer.com/series/7409

Joemon M. Jose · Claudia Hauff
Ismail Sengor Altıngovde · Dawei Song
Dyaa Albakour · Stuart Watt
John Tait (Eds.)

Advances in Information Retrieval

39th European Conference on IR Research, ECIR 2017
Aberdeen, UK, April 8–13, 2017
Proceedings

 Springer

Editors
Joemon M. Jose (ID)
University of Glasgow
Glasgow
UK

Dyaa Albakour
Signal Media
London
UK

Claudia Hauff
TU Delft - EWI/ST/WIS
Delft
The Netherlands

Stuart Watt
Turalt/University Health Network
Toronto
Canada

Ismail Sengor Altingovde
Middle East Technical University
Ankara
Turkey

John Tait
JohnTait.net Ltd. and BCS IRSG
Sunderland
UK

Dawei Song
Open University
Milton Keynes
UK

ISSN 0302-9743   ISSN 1611-3349   (electronic)
Lecture Notes in Computer Science
ISBN 978-3-319-56607-8   ISBN 978-3-319-56608-5   (eBook)
DOI 10.1007/978-3-319-56608-5

Library of Congress Control Number: 2017936705

LNCS Sublibrary: SL3 – Information Systems and Applications, incl. Internet/Web, and HCI

Printed on acid-free paper

This Springer imprint is published by Springer Nature
The registered company is Springer International Publishing AG
The registered company address is: Gewerbestrasse 11, 6330 Cham, Switzerland

Preface

These proceedings contain the full papers, short papers, and demonstrations selected for presentation at the 39th European Conference on Information Retrieval (ECIR 2017). The event was organized by the School of Computing Science and Digital Media, Robert Gordon University, Aberdeen, Scotland. The conference was held during April 8–13, 2017, in Aberdeen, Scotland, UK.

ECIR 2017 received a total of 248 submissions in three categories: 135 full papers, 101 short papers, and 12 demonstrations. The geographical distribution of the authors was as follows: 53% were from Europe, 33% from Asia, 11% from North and South America, 1% from Australia, and 2% from Africa. All submissions were reviewed by at least three members of an international two-tier Program Committee. Of the full papers submitted to the conference, 36 were accepted for oral presentation (27% of the submitted ones). Of the short papers submitted to the conference, 35 were accepted for poster presentation (35% of the submitted ones). In addition, seven demonstrations (58% of the submitted ones) were accepted. The accepted contributions represent the state of the art in information retrieval, cover a diverse range of topics, propose novel applications, and indicate promising directions for future research. We thank all Program Committee members for their time and effort in ensuring the high quality of the ECIR 2017 program.

ECIR 2017 included a panel on "Information Retrieval" organized by Maarten DeRijke. The panel stems from the fact that information retrieval (IR) has always been concerned with retrieving the most relevant information from huge amounts of data including user interaction data. IR is in the midst of a radical paradigm shift, common also to many other research fields, in becoming an increasing data-driven science due to, for example, recent developments in deep learning, crowdsourcing, user interaction analysis, and so on. The goal of the panel was to discuss the emergent trends in this area, their advantages, their pitfalls, and their implications for the future of our field.

ECIR 2017 hosted one half-day tutorial: "Efficient Query Processing Infrastructure" by Nicola Tonellotto (Istituto di Scienza e Tecnologie dell'Informazione, Italy) and Craig Macdonald (University of Glasgow, UK). In addition, ECIR 2017 hosted four workshops covering a range of IR topics. These workshops were:

- The Fifth International Workshop on Bibliometric-Enhanced Information Retrieval (BIR2017)
- Exploitation of Social Media for Emergency Relief and Preparedness (SMERP)
- The Second International Workshop on Online Safety, Trust and Fraud Prevention (OnST'2017)
- Social Media for Personalization and Search (SoMePeAS)

Short descriptions of these workshops and tutorial are included in the proceedings. In addition, a doctoral consortium was organized on the first day of the conference.

We would like to thank our invited speakers for their contributions to the program: Laura Dietz (University of New Hampshire, USA), Jaime Teevan (Microsoft Research), and Alexander Hauptmann (CMU). We are grateful to the panel led by Stefan Rüger for selecting the recipient of the 2016 Microsoft BCS/BCS IRSG Karen Spärck Jones Award, and we congratulate Jaime Teevan on receiving this award. In addition, we are continuing with the Test of Time Award, to recognize research that had long-standing influence. We thank Norbert Fuhr for charing this committee.

The final day of the conference was dedicated to the Industry Day; it brought together academic researchers and industry by offering a mix of invited talks by industry leaders and presentations of novel and innovative ideas from research. Finally, ECIR 2017 would not have been possible without the generous financial support from our sponsors. We thank all of these sponsors.

February 2017

Joemon Jose
Claudia Hauff
Ismail Altıngövde
Dawei Song
Pia Borlund
Jaana Kekalainen
Emine Yimaz
Mohamed Gaber
Udo Kruschwitz
Tony Russell-Rose
Dyaa Albakour
Stuart Watt
Daqing He
Fazli Can
Eero Sorumunen
John Tait
Ayse Goker

Organization

General Chairs

Ayse Goker   Robert Gordon University, UK
John Tait   johntait.net Ltd. and BCS IRSG, UK
  (Acting General Chair)

Program Chairs

Claudia Hauff   TU Deft, The Netherlands
Joemon M. Jose   University of Glasgow, UK

Short Paper Chairs

Ismail Altıngövde   Middle East Technical University, Turkey
Dawei Song   Open University, UK

Workshop Chairs

Pia Borlund   University of Copenhagen, Denmark
Jaana Kekalainen   University of Tampere, Finland

Industry Day Chairs

Udo Kruschwitz   University of Essex, UK
Tony Russell-Rose   UXLabs, UK

Demo Chairs

Dyaa Albakour   Signal Media, UK
Stuart Watt   Turalt/University Health Network, Toronto, Canada

Tutorial Chairs

Fazli Can   Bilkent University, Turkey
Daqing He   University of Pittsburgh, USA
Eero Sorumunen   University of Tampere, Finland

Proceedings Chairs

Long Chen   University of Glasgow, UK
Haitao Yu   University of Tsukuba, Japan

Doctoral Consortium Chairs

Mohamed Gaber   Robert Gordon University, UK
Emine Yilmaz   University College London, UK

Local Organization

Virginia Dawod   Robert Gordon University, UK
  (Local Chair)
Michael Heron (Web Chair)   Robert Gordon University, UK
John Isaacs (Support)   Robert Gordon University, UK
Leszek Kaliciak (Support)   Robert Gordon University, UK
Stewart Massie (Support)   Robert Gordon University, UK

Publicity Chairs

Martin Halvey   University of Strathclyde, UK
Adam Wyner   University of Aberdeen, UK

Sponsorship Chairs

Susan Craw   Robert Gordon University, UK
Andrei Petrovski   Robert Gordon University, UK

Test of Time Award Chair

Norbert Fuhr   University of Duisberg-Essen, Germany

Best Paper Award Chair

Hideo Joho   University of Tsukuba, Japan

Program Committee

Full-Paper Meta-Reviewers

Ioannis Arapakis   Yahoo Labs, Spain
Krisztian Balog   University of Stavanger, Norway
Pablo Castells   Universidad Autonoma, Madrid, Spain
Fabio Crestani   University of Lugano (USI), Switzerland
Bruce Croft   University of Massachusetts, Amherst, USA
Maarten de Rijke   University of Amsterdam, The Netherlands
Nicola Ferro   University of Padua, Italy
Norbert Fuhr   University of Duisberg-Essen, Germany
Eric Gaussier   IMAG, France
Cathal Gurrin   Dublin City University, Ireland
Morgan Harvey   Northumbria University, UK

Gareth Jones	Dublin City University, Ireland
Jaap Kamps	University of Amsterdam, The Netherlands
Evangelos Kanoulas	University of Amsterdam, The Netherlands
Gabriella Kazai	Lumi, UK
Udo Kruschwitz	University of Essex, UK
Oren Kurland	Technion, Israel
David Losada	University of Santiago de Compostela, Spain
Ilya Markov	University of Amsterdam, The Netherlands
Wolfgang Neijdl	I3S, Germany
Iadh Ounis	University of Glasgow, UK
Gabriela Pasi	University of Milan, Italy
Benjamin Piwowarski	CNRS, France
Stefan Rueger	Knowledge Media Institute, UK
Pavel Serdyukov	Yandex, Russia
Hui Yang	Georgetown University, USA
Ke Zhou	University of Nottingham, UK

Full-Paper Program Committee

Mikhail Ageev	Moscow State University, Russia
Dirk Ahlers	NTNU, Norway
Ahmet Aker	University of Sheffield, UK
Dyaa Albakour	Signal Media, UK
Rami Alkhawaldeh	University of Glasgow, UK
Omar Alonso	Microsoft, USA
Giambattista Amati	Fondazione Ugo Bordoni, Italy
Linda Andersson	Vienna University of Technology, Austria
Avi Arampatzis	Democritus University of Thrace, Greece
Jaime Arguello	University of North Carolina at Chapel Hill, USA
Leif Azzopardi	University of Strathclyde, UK
Alvaro Barreiro	University of A Coruna, Spain
Roberto Basili	University of Rome Tor Vergata, Italy
Srikanta Bedathur	IBM Research, India
Alejandro Bellogin	Universidad Autonoma de Madrid, Spain
Patrice Bellot	University - CNRS (LSIS), France
Pablo Bermejo	Universidad de Castilla-La Mancha, Spain
Catherine Berrut	LIG, University Joseph Fourier of Grenoble I, France
Alessandro Bozzon	Delft University of Technology, The Netherlands
Pavel Braslavski	Ural Federal University, Russia
Paul Buitelaar	National University of Ireland, Galway, Ireland
Fidel Cacheda	Universidad de A Coruna, Spain
Pável Calado	Universidade de Lisboa, Portugal
Fazli Can	Bilkent University, Turkey
Ivan Cantador	Universidad Autonoma de Madrid, Spain
Claudio Carpineto	Fondazione Ugo Bordoni, Italy
Lu'sa Coheur	IST/INESC-ID Lisboa, Portugal

Ronan Cummins	University of Cambridge, UK
Alfredo Cuzzocrea	ICAR-CNR and University of Calabria, Italy
Eva D'Hondt	LIMSI, CNRS, Université Paris-Saclay, France
Martine De Cock	University of Washington Tacoma, USA
Pablo de La Fuente	Universidad de Valladolid, Spain
Arjen de Vries	Radboud University, The Netherlands
Emanuele Di Buccio	University of Padua, Italy
Giorgio Maria Di Nunzio	University of Padua, Italy
Laura Dietz	University of New Hampshire, USA
Huizhong Duan	WalmartLabs, USA
Carsten Eickhoff	ETH Zurich, Switzerland
Liana Ermakova	PSU, IRIT, Russia
Hui Fang	University of Delaware, USA
Juan M. Fernandez-Luna	University of Granada, Spain
Ingo Frommholz	University of Bedfordshire, UK
Patrick Gallinari	LIP6, University of Paris 6, France
Kavita Ganesan	University of Illinois at Urbana Champaign, USA
Anastasia Giachanou	University of Lugano, Switzerland
Giorgos Giannopoulos	Imis Institute, "Athena" R.C., Greece
Lorraine Goeuriot	Laboratoire d'Informatique de Grenoble, France
Julio Gonzalo	UNED, Spain
Michael Granitzer	University of Passau, Germany
Guillaume Gravier	CNRS, IRISA, France
Matthias Hagen	Bauhaus-Universität Weimar, Germany
Allan Hanbury	Vienna University of Technology, Austria
Preben Hansen	Stockholm University, Sweden
Donna Harman	NIST, USA
Jer Hayes	IBM, Ireland
Ben He	University of Chinese Academy of Sciences, China
Daqing He	University of Pittsburgh, USA
Nathalie Hernandez	IRIT, France
Frank Hopfgartner	University of Glasgow, UK
Andreas Hotho	University of Würzburg, Germany
Gilles Hubert	IRIT, France
Dmitry Ignatov	National Research University, Russia
Shen Jialie	Northumbria University, UK
Jiepu Jiang	University of Massachusetts Amherst, USA
Hideo Joho	University of Tsukuba, Japan
Nattiya Kanhabua	Aalborg University, Denmark
Diane Kelly	University of Tennessee, USA
Liadh Kelly	Trinity College Dublin, Ireland
Yiannis Kompatsiaris	CERTH – ITI, Greece
Kyumin Lee	Utah State University, USA
Wang-Chien Lee	Pennsylvania State University, USA
Teerapong Leelanupab	King Mongkut's Institute of Technology, Thailand
Liz Liddy	Syracuse University, USA

Alan Said   University of Skövde, Sweden
Rodrygo Santos   Universidade Federal de Minas Gerais, Brazil
Florence Sedes   IRIT P. Sabatier, France
Giovanni Semeraro   University of Bari, Italy
Azadeh Shakery   University of Tehran, Iran
Parikshit Sondhi   University of Illinois at Urbana Champaign, USA
Yang Song   Google, USA
Eero Sormunen   University of Tampere, Finland
Venkat Subramaniam   IBM Research – India, India
Pascale Sžbillot   IRISA, France
Lynda Tamine   IRIT, France
Marko Tkalcic   Free University of Bolzano, Italy
Tassos Tombros   Queen Mary University of London, UK
Ming-Feng Tsai   National Chengchi University, Taiwan
Theodora Tsikrika   Information Technologies Institute, CERTH, Greece
Denis Turdakov   Institute for System Programming RAS, Russia
Yannis Tzitzikas   University of Crete and FORTH-ICS, Greece
Thierry Urruty   XLIM-SIC, France
David Vallet   Google, Australia
Marieke Van Erp   VU University Amsterdam, The Netherlands
Sumithra Velupillai   KTH Royal Institute of Technology, Sweden
Suzan Verberne   Radboud University Nijmegen, The Netherlands
Stefanos Vrochidis   Information Technologies Institute, Greece
Jun Wang   University College London, UK
Colin Wilkie   University of Glasgow, UK
Christa Womser-Hacker   Universität Hildesheim, Germany
Tao Yang   Ask.com and UCSB, USA
Fajie Yuan   University of Glasgow, UK
Dan Zhang   Facebook, USA
Duo Zhang   University of Illinois, Urbana-Champaign, USA
Guido Zuccon   Queensland University of Technology, Australia

Short-Paper Program Committee Members

Mikhail Ageev   Moscow State University, Russia
Dirk Ahlers   NTNU, Norway
Ahmet Aker   University of Sheffield, UK
Dyaa Albakour   Signal Media Ltd., UK
Rami Alkhawaldeh   University of Glasgow, UK
Omar Alonso   Microsoft, USA
Giambattista Amati   Fondazione Ugo Bordoni, Italy
Avishek Anand   L3S Research Center, Germany
Linda Andersson   Vienna University of Technology, Austria
Ioannis Arapakis   Yahoo Labs, Spain
Jaime Arguello   University of North Carolina at Chapel Hill, USA
Leif Azzopardi   University of Strathclyde, UK

Frank Hopfgartner	University of Glasgow, UK
Andreas Hotho	University of Würzburg, Germany
Gilles Hubert	IRIT, France
Dmitry Ignatov	National Research University, Russia
Shen Jialie	Northumbria University, UK
Jiepu Jiang	University of Massachusetts Amherst, USA
Hideo Joho	University of Tsukuba, Japan
Gareth Jones	Dublin City University, Ireland
Jaap Kamps	University of Amsterdam, The Netherlands
Nattiya Kanhabua	Aalborg University, Denmark
Evangelos Kanoulas	University of Amsterdam, The Netherlands
Pinar Karagoz	Middle East Technical University (METU), Turkey
Gabriella Kazai	Lumi, UK
Diane Kelly	University of Tennessee, USA
Liadh Kelly	Trinity College Dublin, Ireland
Yiannis Kompatsiaris	CERTH, ITI, Greece
Ralf Krestel	University of Potsdam, Germany
Udo Kruschwitz	University of Essex, UK
Tayfun Kucukyilmaz	TED University, Turkey
Oren Kurland	Technion, Israel
Kyumin Lee	Utah State University, USA
Teerapong Leelanupab	King Mongkut's Institute of Technology, Thailand
Shangsong Liang	University College London, UK
Christina Lioma	University of Copenhagen, Denmark
Fernando Loizides	University of Wolverhampton, UK
David Losada	University of Santiago de Compostela, Spain
Bernd Ludwig	University Regensburg, Germany
Mihai Lupu	Vienna University of Technology, Austria
Craig MacDonald	University of Glasgow, UK
Andrew Macfarlane	City University London, UK
Walid Magdy	University of Edinburgh, UK
Marco Maggini	University of Siena, Italy
Thomas Mandl	University of Hildesheim, Germany
Stephane Marchand-Maillet	University of Geneva, Switzerland
Miguel Martinez-Alvarez	Signal Media, UK
Bruno Martins	IST, Instituto Superior Tecnico, Portugal
Yosi Mass	IBM Haifa Research Lab, Israel
Richard McCreadie	University of Glasgow, UK
Edgar Meij	Bloomberg, UK
Marcelo Mendoza	Universidad Tecnica Federico Santa Maria, Chile
Boughanem Mohand	IRIT University, Paul Sabatier Toulouse, France
Josiane Mothe	Institut de Recherche en Informatique de Toulouse, France
Samir Moussa	Signal Media, UK
Hannes Mühleisen	Centrum Wiskunde & Informatica (CWI), The Netherlands

Philippe Mulhem	LIG-CNRS, France
Henning Müller	HES-SO, Switzerland
Franco Maria Nardini	ISTI-CNR, Italy
Wolfgang Neijdl	I3S, Germany
Dong Nguyen	University of Twente, The Netherlands
Jian-Yun Nie	University of Montreal, Canada
Boris Novikov	St.-Petersburg University, Russia
Andreas Nuernberger	University of Magdeburg, Germany
Neil O'Hare	Yahoo! Research, Spain
Michael Oakes	University of Wolverhampton, UK
Iadh Ounis	University of Glasgow, UK
Rifat Ozcan	Independent Researcher, Turkey
Deepak P.	Queen's University Belfast, UK
Gabriela Pasi	University of Milan, Italy
Virgil Pavlu	Northeastern University, USA
Pavel Pecina	Charles University in Prague, Czech Republic
Raffaele Perego	ISTI-CNR, Italy
Vivien Petras	HU Berlin, Germany
Karen Pinel-Sauvagnat	IRIT, France
Florina Piroi	Vienna University of Technology, ISIS, IFS, Austria
Benjamin Piwowarski	CNRS, France
Vassilis Plachouras	Thomson Reuters, UK
Georges Quenot	Laboratoire d'Informatique de Grenoble, France
Naeem Ramzan	University of West of Scotland, UK
Jesus Alberto Rodriguez Perez	University of Glasgow, UK
Dmitri Roussinov	University of Strathclyde, UK
Stefan Rueger	Knowledge Media Institute, UK
Tony Russell-Rose	UXLabs, UK
Alan Said	University of Skövde, Sweden
Rodrygo Santos	Universidade Federal de Minas Gerais, Brazil
Pascale Sebillot	IRISA, France
Florence Sedes	IRIT, University of P. Sabatier, France
Pavel Serdyukov	Yandex, Russia
Azadeh Shakery	University of Tehran, Iran
Parikshit Sondhi	University of Illinois at Urbana Champaign, USA
Yang Song	Google, USA
Eero Sormunen	University of Tampere, Finland
Venkat Subramaniam	IBM Research, India
Lynda Tamine	IRIT, France
Tassos Tombros	Queen Mary University of London, UK
Nicola Tonellotto	ISTI-CNR, Italy
Ming-Feng Tsai	National Chengchi University, Taiwan
Theodora Tsikrika	Information Technologies Institute, CERTH, Greece
Denis Turdakov	Institute for System Programming RAS, Russia
Yannis Tzitzikas	University of Crete and FORTH-ICS, Greece

Thierry Urruty	XLIM-SIC, France
David Vallet	Google, Australia
Marieke Van Erp	VU University Amsterdam, The Netherlands
Sumithra Velupillai	KTH Royal Institute of Technology, Sweden
Suzan Verberne	Radboud University Nijmegen, The Netherlands
Stefanos Vrochidis	Information Technologies Institute, Greece
Jun Wang	University College London, UK
Stuart Watt	turalt, Canada
Colin Wilkie	University of Glasgow, UK
Christa Womser-Hacker	Universität Hildesheim, Germany
Hui Yang	Georgetown University, USA
Tao Yang	Ask.com and UCSB, USA
Fajie Yuan	University of Glasgow, UK
Dan Zhang	Facebook, USA
Duo Zhang	University of Illinois, Urbana-Champaign, USA
Peng Zhang	Tianjin University, China
Guido Zuccon	Queensland University of Technology, Australia

Demonstration Reviewers

Omar Alonso	Microsoft, USA
Krisztian Balog	University of Stavanger, Norway
Matthias Hagen Bauhaus	Universität Weimar, Germany
Alessandro Bozzon	Delft University of Technology, The Netherlands
David Corney	Signal Media, UK
Laura Dietz	University of New Hampshire, USA
Michael Granitzer	University of Passau, Germany
Allan Hanbury	Vienna University of Technology, Austria
Jussi Karlgren	Gavagai and KTH, Sweden
Julia Kiseleva	Eindhoven University of Technology, The Netherlands
Udo Kruschwitz	University of Essex, UK
Craig MacDonald	University of Glasgow, UK
Richard McCreadie	University of Glasgow, UK
Samir Moussa	Signal Media, UK
Andreas Nuernberger	Otto von Guericke University of Magdeburg, Germany
Vassilis Plachouras	Thomson Reuters, UK
Eero Sormunen	University of Tampere, Finland
Sacel Vargas	Mendeley Ltd., UK
Guido Zuccon	Queensland University of Technology, Australia

Doctoral Consortium Reviewers

Eugene Agichtein	Emory University, USA
Jaime Arguello	University of North Carolina at Chapel Hill, USA
Ben Carterette	University of Delaware, USA
Fernando Diaz	Microsoft Research, USA

Evangelos Kanoulas   University of Amsterdam, The Netherlands
Rishabh Mehrotra   University College London, UK
Filip Radlinski   Google, UK
Milad Shokouhi   Microsoft, UK

Test of Time Award Committee

Maristella Agosti   University of Padua, Italy
Fabio Crestani   University of Lugano, Switzerland
Eric Gaussier   IMAG, France
Djoerd Hiemstra   University of Twente, The Netherlands
Iadh Ounis   University of Glasgow, UK
Fabrizio Sebastiani   Consiglio Nazionale delle Ricerche, Italy

Best Paper Award Committee

Luanne Freund   The University of British Columbia, Canada
Nobert Fuhr   University of Duisburg-Essen, Germany
Hideo Joho   University of Tsukuba, Japan
Yiqun Liu   Tsinghua University, China

Student Grant Committee

Haiming Liu   University of Bedfordshire, UK
John Tait   johntait.net Ltd. and BCS IRSG, UK
Nirmalie Wiratunga   Robert Gordon University, UK

Abstracts of Doctoral Consortium
Papers and Tutorial

Interactive Technology Trend Detection

Noushin Fadaei

University of Hildesheim, Universitätspl. 1, 31141 Hildesheim, Germany
fadaei@uni-hildesheim.de

Industrial companies and research institutes constantly seek trends and analyze them in order to find new technologies or detect transforming or improving industries. Patent databases are natural candidates to extract this information from, as they collect recent activities of major research and development departments. However, semantic analysis of patents can be quite challenging for NLP tools using available resources, as they are mostly built upon newspaper corpora [3]; while containing professional terms, patents do not follow a standard form of wordings and instead of the usual concepts, an uncommon abbreviation, explanation or unregulated breaking of the components might appear in the text. General and commercial titles also influence the task of assigning the patents to categories. Defining the keywords of the patents as well as their possible paraphrases demands professional knowledge which can be provided by an expert user in an interactive approach. The interactive system can also benefit from the experts knowledge in order to adapt the quantities of a defined uptrend.

In this work, we propose a relevance feedback system which aims to provide users with suggestions on new trends and critical topics of the desired domain or theme, while allowing them to adjust the procedure for better results. Clustering and trend detection are of main components of this system. Patent clustering demands soft clustering techniques as patents share various technologies, uses or materials [2]. Using the clusters, a time-series chart is produced for each topic by means of the application dates of the patents included. The system is then detects the uptrends through these charts. In this work, we also build a benchmark using the World Intellectual Property Organization (WIPO) reports [1] which contain expert queries for categories of a theme. The same queries can be used on our database to achieve gold standard classes.

References

1. The World Intellectual Property Organization (WIPO). http://www.wipo.int/portal/en/index.html
2. Fadaei, N., Mandl, T., Schwantner, M., Sofean, M., Werner, K., Struß, J., Womser-hacker, C.: Patent analysis and patent clustering for technology trend mining. In: HIER Workshop, Hildesheim (2015)
3. Marcus, M., Marcinkiewicz, M.A., Santorini, B.: Building a large annotated corpus of English: the Penn Treebank. In: Computational Linguistics, pp. 313–330. MIT Press, Cambridge (1993)

Right to Information Query Analysis for Predicting Amendments

Nayantara Kotoky

Indian Institute of Technology Guwahati, Guwahati, India
nayantara@iitg.ernet.in

The constitution of India has undergone 101 amendments from 1950 to September 2016 (almost two per year). Compared to the other democracies, amendments in India are a frequent process. There are no specific rules that states what exactly qualifies for an amendment. As of today, the amendment process in India is completely manual. The process is triggered by observing the effects/execution of a particular law of our country and taking feedback from the citizens. One way to observe citizens' emotions and collect feedback is to look at the interactions between the citizens and the government. The Right to Information Act (RTI) 2005 allows Indian citizens to access governmental information, records etc. by posting RTI queries/application. Such queries are present in every public institution, and these queries are laden with the citizens' concerns, issues and emotions. The aim of this work is to collect RTI queries from institutions across India and analyse them. The objective of such analysis is to uncover underlying patterns in the RTI query-reply process that are indicators of potential amendments to Indian laws. We want to find latent patterns such as 'transparency' of institutions and 'effectiveness of implementation' of the RTI act across India. Both of these parameters can identify issues in the working of public institutions, and are suggestive of amendments. In this regard, the following research questions are addressed:

1. What constitutes a potential amendment?
2. How to model transparency of an institution?
3. How to model effectiveness of implementation of an act?
4. Can transparency and/or implementation effectiveness be used as cues for tentative amendments?
5. What learning algorithms can be used for finding such patterns?

We choose psychometric modelling to identify the above two parameters. A synthetic matrix of reply statistics that resembles our collected RTI data has been constructed, and analysed via Graded Response Model. The experimental outcomes are two-fold:

1. Each institution is assigned a transparency value, and indicates that not all institutions are equally transparent in replying to RTI queries.
2. There is variation in the difficulties in acquiring RTI replies across India based on the query category. This is a characteristic of the department/section which contains the information required in the RTI queries, indicating that the RTI rules are non-uniform and ineffective in its implementation across India.

Efficient Query Processing Infrastructures

Nicola Tonellotto[1] and Craig Macdonald[2]

[1] ISTI, National Research Council of Italy, Pisa, Italy
nicola.tonellotto@isti.cnr.it
[2] School of Computing Science, University of Glasgow, Glasgow, G12 8QQ, UK
craig.macdonald@glasgow.ac.uk

Abstract. Typically, techniques that benefit effectiveness of information retrieval (IR) systems have a negative impact on efficiency. Yet, with the large scale of Web search engines, there is a need to deploy efficient query processing techniques to reduce the cost of the infrastructure required. The proposed tutorial aims to provide a detailed overview of the infrastructure of an IR system devoted to the *efficient yet effective processing of user queries*. This tutorial will guide the attendees through the main ideas, approaches and algorithms developed in the last 30 years in query processing. In particular, we will illustrate, with detailed examples and simplified pseudo-code, the most important dynamic pruning techniques adopted in major search engines, as well as the state-of-the-art innovations in query processing, such as impact-sorted and blockmax indexes. We will also describe how modern search engines exploit such algorithms with learning-to-rank (LtR) models to produce effective results, exploiting new approaches in LtR query processing. Finally, this tutorial will introduce query efficiency predictors for dynamic pruning, and discuss their main applications to scheduling, routing, selective processing and parallelisation of query processing, as deployed by a major search engine.

Contents

Abstracts of Doctoral Consortium Papers and Workshops

Entity Linking to One Thousand Knowledge Bases

Ning Gao[1]([⊠]) and Silviu Cucerzan[2]

[1] University of Maryland, College Park, USA
ninggao@umd.edu
[2] Microsoft Research, Redmond, USA
silviu@microsoft.com

Abstract. We address the task of entity linking to multiple knowledge bases (KB). In particular, we investigate the use of over one thousand domain-specific KBs derived from Wikia.com collections in conjunction with the Wikipedia collection as a background-knowledge repository. Our system employs a two-step approach: for each document, a supervised model with a large set of features detects whether there exists a Wikia collection whose domain matches the document; when such a collection is available, the system extracts and resolves the entity mentions in the document to the KB obtained by merging the Wikipedia KB and the KB corresponding to the matched Wikia collection. Otherwise, the system employs only the background KB for analysis, in a standard entity-detection-and-linking framework. On a Web news articles dataset, our system achieves 90% precision in detecting domain-accurate Wikia collections while providing also high linking accuracy (93%) to the KB of the matched Wikia collection.

1 Introduction

Entity detection and linking (EDL), also known as entity recognition and disambiguation (ERD), is the task of identifying mentions of entities in text (*detection/recognition*) and assigning the detected mentions to entities in a large knowledge base (*linking/disam-biguation*). The establishment of large encyclopedic collections such as Wikipedia and Freebase has drawn considerable attention towards this task [2,3,6,10,14,16]. However, focusing exclusively on Wikipedia or similar general knowledge repositories is not sufficient in many real-world scenarios, such as entity-based indexing of corporate data and entity-based indexing of news because numerous domain-specific entities and concepts are absent from general-use encyclopedic KBs (e.g., finance, food, fiction).

Our study finds that more than 80% of the entities in the domain-specific collections used in this paper are likely not to have entries in Wikipedia. In particular, while there exists a comprehensive Wikia collection for the Pokémon world with over 14,000 entity pages, Wikipedia contains only a few tens of pages for Pokémon entities. Similarly, there is a Wikia collection for the "Survivor"

J.M. Jose et al. (Eds.): ECIR 2017, LNCS 10193, pp. 1–14, 2017.
DOI: 10.1007/978-3-319-56608-5_1

television franchise with over 3,000 entity pages, of which only very few have pages in Wikipedia.

Most of the previous work on entity linking employs only one single KB as linking target. To the best of our knowledge, there has been very little done to address the problem of entity linking to a large number of KBs automatically. We propose a framework in which one large encyclopedic KB is employed as a comprehensive repository of general knowledge in conjunction with over one thousand independent domain-specific KBs. For the former, we employ Wikipedia; for the latter, we employ a large set of Wikia collections from http://www.wikia.com/Wikia.

Rather than attempting to merge all KBs into one single knowledge base, we employ a paradigm in which the domain-specific KBs are kept separate, and they *get activated on the fly* and used in conjunction with the general KB when they are needed for analyzing a document pertaining to those respective domains. This paradigm is interesting because of several aspects: First, merging a large collection of diverse ontologies/KBs is very challenging. Second, in an enterprise context, many domain-specific KBs are designed to be protected with authority access and are not allowed to be merged into a general, externally-maintained repository. For example, aeronautical companies such as Boeing and Airbus deal with domain/company-specific terminologies as well as numerous technologies that need be kept private (from the ouside world or from each other) in addition to many entities and concepts that are part of common knowledge. Conflating such terminologies and knowledge repositories into one KB or with each other would be both daunting and undesirable. Third, facts and relations from some domain-specific collections, such as those that target fictional work, may be valid only with respect to that particular domain. For example, the city of London has entries in *Harry Potter* Wikia, *Baker Street* Wikia and hundreds of other Wikia collections. However, the *British Ministry of Magic* being located in *London* is a "fact" only in the *Harry Potter* domain. Merging directly the knowledge for entities in different domains into one canonical entity entry can lead to inaccuracies, conflicting information, and noise.

2   Related Work

The tasks of linking to multiple KBs and merging KBs for entity linking have been tackled only to a small degree previously. Most entity linking work, starting with the works of Bunescu and Paşca [1], Cucerzan [3], and Mihalcea and Csomai [10], has employed Wikipedia for deriving a reference KB for the task. The Text Analysis Conference track on Knowledge Base Population (TAC-KBP) established the evaluation framework for entity linking [7,8], in which a target KB with over 800,000 entities was derived from the Wikipedia collection as of October 2008, and thousands of documents from a large corpus of news and Web text were annotated with entity mentions. Interestingly, 57% of the evaluated entities were not in the targeted KB [9].

Ruiz-Casado et al. [13] and Niemann et al. [11] studied the task of automatically assigning Wikipedia entries to WordNet synsets, which can be considered as simple one-direction merging of *two* KBs. However, extending those approaches to bi-directional merging over thousands of KBs seems extremely difficult. Sil et al. [15] proposed an open-database named entity disambiguation system that is able to resolve entity mentions detected in text to an arbitrary KB provided in Boyce-Codd normal form. However, this work focuses on distant supervision and domain-adaption, and relies on *manually* identifying a KB that matches the analyzed documents, without addressing the tasks of detecting domain-specific KBs or maintaining a multi-KB structure automatically. Demartini et al. [5] used probabilistic reasoning and crowdsourcing techniques for the task of entity linking over *four* KBs (DBpedia, Freebase, Geonames and New York Times). However, the KBs are simply "merged", and then the candidate entities are triaged by TF-IDF methods. Pereira [12] proposed an idea of resolving the task of entity linking to multiple KBs by using different textual and KB features, and ontology modularization to select entities in the same semantic context, although the detailed structure is not discussed in the paper.

These approaches have dealt with a small number of KBs, for either the purpose of entity linking or similar tasks, none of which has the ability to deal with the complexity caused by a very large number of KBs. Employing such a large repository of KBs and automatically mapping documents to domain-specific KBs from this repository to perform entity linking has not been reported until now.

3 Datasets

We employed the English Wikipedia collection from August 11, 2014, and we crawled all 1,163 available Wikia collections in English within the top 5,000 Wamranked list[1] as of June 13, 2014. Table 1 compares page and linkage statistics between Wikipedia and the employed Wikia collections. *Number of pages* denotes the total number of entity pages in Wikipedia and all Wikia collections. *Page length* shows the mean and standard deviation (SD) for the length (in characters) of Wikipedia and Wikia pages. *Micro* shows the average over all the Wikia pages, while *Macro* shows the average over all the Wikia collections. While the Wikipedia pages are on average 1.7 times longer than Wikia pages, the difference between the average page length of the two sources is not significant given the high standard deviations inside these collections.

Table 1 also shows statistics for the existing linkage between collections as created by the wiki contributors. *Inner links* is the average number of links on a page from one collection (whether Wikipedia or a Wikia collection) to other pages in the same collection. *Out links* for Wikipedia is the average number of links from a Wikipedia page to pages in any Wikia collection. *Out links* for Wikias is the average number of links from a Wikia page to Wikipedia pages. *Cross links* is the average number of links from pages in one Wikia to pages

---

[1] Wamrank is the official ranking from the Wikia website, which evaluates the health and vitality of collections.

Table 1. Page and linkage statistics for Wikipedia and Wikia.

		Wikipedia	Wikia set	
			Macro	Micro
Number of pages		4,591,935	3,059,412	
Page length	Mean	4,284	2,573	2,454
	Stdev	7,901	2,992	8,801
Inner links	Mean	31.6	8.4	11.8
	Stdev	77.8	6.9	36.7
Out links	Mean	0	0.14	0.09
	Stdev	0.09	0.65	2.18
Cross links	Mean		0.03	0.03
	Stdev		0.1	4.19

in other Wikias. Compared to the rich *inner links*, both *out links* and *cross links* for Wikipedia and Wikia are sparse, suggesting that the Wikipedia and the domain-specific Wikias are relatively isolated from each other.

4   Linking Wikia Collections to Wikipedia

Because editorial links between Wikipedia and the Wikias are quite rare, we attempt to connect more strongly the Wikia collections in our study to Wikipedia by employing NEMO, a state-of-the-art Wikipedia-based EDL system [4]. The text of each Wikia page is analyzed with NEMO to identify entity mentions and automatically link them to Wikipedia when possible. Table 2 shows more statistics for both editorial links and automatically generated links from Wikia collections to Wikipedia. When accounting only for editorial links, 243 out of the 1,163 Wikias are completely isolated from Wikipedia (i.e., they do not contain any links to Wikipedia). As expected, the automatic linking process is able to connect all Wikia collections to Wikipedia. We split both types of links from Wikia to Wikipedia into three mutually exclusive sets, as follows:

Wikia-to-entity are the links for which the entity mention appears on the homepage of the Wikia collection and is string-wise identical with both the name

Table 2. Statistics for *editorial links* (contributor-created) and *automatic links* (as generated by NEMO) from Wikia to Wikipedia.

	Editorial links	Automatic links
Wikia collections with links to Wikipedia	920	1,163
Wikia-to-entity linkage	58	616
Entity-to-entity linkage	4,456	518,335
Mention-to-entity linkage	196,928	1,801,203

Table 3. Performance of entity detection (ED) and linking (EL) from Wikia to Wikipedia.

ED recall for overlapping mentions	0.74
ED recall for exact boundaries	0.63
EL Wikia-to-entity accuracy	1.00
EL entity-to-entity accuracy	0.86
EL mention-to-entity accuracy	0.84

of the Wikia collection and the mapped Wikipedia entity. For example, in the text of the *Harry Potter* Wikia's homepage, there is a mention of "Harry Potter" that gets identified and linked to the Wikipedia page *Harry Potter*. These types of links are likely to capture cases in which there exists a whole Wikia collection dedicated to one entity in Wikipedia.

Entity-to-entity are the links for which the entity mention in the text of a Wikia page is identical to the title of the Wikia page. For example, on the page *J.K. Rowling* in the *Harry Potter* Wikia, there is a mention "J.K. Rowling", which gets linked to the Wikipedia entity *J.K. Rowling*. These links are likely to indicate duplicate coverage of an entity in both Wikipedia and the analyzed Wikia collection.

Mention-to-entity are the links that are not included in the former two categories. For example, on the page *Zubeida Khan* in the *Harry Potter* Wikia, there is a mention "Pakistan" linked to the Wikipedia entity *Pakistan*. These links are likely to indicate the case when a Wikipedia entity being mentioned in the analyzed Wikia collection.

Using all editorial links as ground-truth, we evaluate the recall and accuracy for the automatic entity detection (ED) and entity linking (EL) processes. The results are shown in Table 3. Because of the large number of Wikia contributors, who are not trained for NLP style annotations, the linkage is inconsistent (for example, some links include determiners, possessive particles, or titles/occupations, while others do not), which we refer to as boundary inconsistency. In our error analysis of the missed links, we found out that 4.2% of those are numbers and 33% are lowercase words (e.g., blurb, film, novel). For the detected links with identical boundaries, we further evaluate the disambiguation accuracy of the employed linking system for each of the three types of links. We obtain 100% accuracy for Wikia-to-entity links, and close to 85% for the other two types of links. We also measured the linking accuracy for the detected overlapping mentions with different boundaries and obtained that over half of those were resolved to the same Wikipedia entities as those chosen by the Wikia contributors.

5 System Architecture

We propose an approach in which we use Wikipedia as a background encyclopedic KB and Wikia collections as domain-specific KBs. As shown in Fig. 1, for

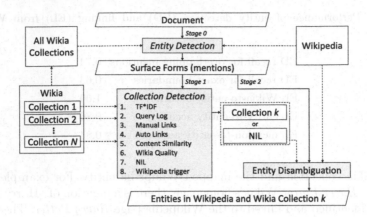

Fig. 1. System architecture.

each input document, our multi-KB entity linking system first employs in *Stage 0* the information from all Wikia collections and Wikipedia to detect candidate entity mentions, referred to as *surface forms* henceforth. It then detects in *Stage 1* the best matching Wikia collection (or returns NIL if no matching Wikia collection is found). In *Stage 2*, it links the surface forms in the document to either Wikipedia or the selected Wikia collection. This final stage employs a modified version of a state-of-the-art EDL system [4], which targets Wikipedia and the selected Wikia collection as one single KB, but with two strongly connected subcomponents. When a mention can be resolved to both Wikipedia and Wikia pages, we give preference to the disambiguation in Wikia, based on the intuition that pages from the domain-specific collections are more relevant in the context of a document that belongs to the respective domain. For example, in an article about the "Sherlock and Holmes" movie, linking the mention "London" to the entity *London* in the *Baker Street* Wikia should be preferable to linking it to *London* in Wikipedia.

The detection of an appropriate Wikia collection is vital in this framework because incorrect collection choice can lead to wrong linked entities. We frame the collection detection task as a supervised ranking problem, in which the Wikia collections are scored and ranked based on how well they match the input document. We employ a boosted-tree learning-to-rank framework in which we investigate eight groups of features that attempt to measure the topical matching between a document and each Wikia collection, as described further. Several feature groups are novel by task design (i.e., Wikia to Wikipedia Linkage, Wikia Collection Quality, Wikipedia Triggers). Other feature groups (e.g., TFIDF, Content Matching) have been widely used in other tasks such as document classification and document retrieval.

TF*IDF. We employ a group of 13 features that are variants of similarities between a document and a Wikia collection, as inspired by the TF*IDF framework. Formally, for an input document d with m detected surface forms

$S_d = \{s_d^1, \cdots, s_d^m\}$ and for a candidate Wikia collection c, let tf_i denote the frequency of s_d^i in the input document, $|c|$ the number of pages in c, w_i the term frequency of s_d^i in collection c, and idf_i the inverse document frequency of s_d^i in all the Wikia collections, computed as $idf_i = \log_2\left(1 + \frac{N}{N(s_d^i)}\right)$, where $N=1{,}163$ is the number of Wikia collections in our repository and $N(s_d^i)$ is the number of those collections that contain s_d^i as a linked mention (either editorial or automatic). The formulas for the features employed are as following (all sums are over all mentions from s_d^1 to s_d^m):

$$\sum_{i=1}^{m} tf_i * idf_i \qquad \sum_{i=1}^{m} \sqrt{tf_i} * idf_i \qquad \sum_{i=1}^{m} \frac{tf_i * idf_i * w_i}{\sqrt{|c|}}$$

$$\sum_{i=1}^{m} \frac{tf_i * idf_i}{\sqrt{|c|}} \qquad \sum_{i=1}^{m} \frac{\log_2{(tf_i)} * idf_i}{\sqrt{|c|}} \qquad \sum_{i=1}^{m} \frac{\sqrt{tf_i * w_i} * idf_i}{\sqrt{|c|}}$$

$$\sum_{i=1}^{m} tf_i * idf_i * w_i \qquad \sum_{i=1}^{m} \sqrt{tf_i} * w_i * idf_i \qquad \sum_{i=1}^{m} \log_2{(tf_i)} * idf_i$$

$$\sum_{i=1}^{m} \frac{\log_2{(tf_i * w_i)} * idf_i}{\sqrt{|c|}} \qquad \sum_{i=1}^{m} \frac{\sqrt{tf_i} * idf_i}{\sqrt{|c|}} \qquad |c|$$

$$\sum_{i=1}^{m} \log_2{(tf_i * w_i)} * idf_i$$

Web Search Logs. We attempt to capture the relatedness between Wikipedia pages and Wikia collections by mining the query logs of a major Web search engine to identify queries for which a user visited both a Wikipedia page and a Wikia page for more than 30 s each after retrieving them as search results. In such cases, we create a connection between the respective Wikipedia page and Wikia collection. For example, if we detect that a user submitted the query "jeff moss", then visited the Wikipedia *Jeff Moss* page and also the *Jeff Moss* page in the *Muppet* Wikia, which were returned by the search engine in the top ranked results, we create a connection between the Wikipedia entity *Jeff Moss* and the Wikia collection *Muppet*. In this way, we associate to each Wikia collection c a set of Wikipedia entities $E_c = \{e_c^1, \cdots, e_c^{m_c}\}$. We were able to extract in total 72,030 such connections. Given an input document, we analyze it first with the Wikipedia-based EDL system and obtain a set $E_d = \{c_d^1, \cdots, e_d^m\}$ of extracted Wikipedia entities. We compute the relatedness of the input document and a Wikia collection c as the cardinality of the intersection $|E_d \cap E_c|$.

Wikia to Wikipedia Linkage. The editorial and automatic links from Wikia collections to Wikipedia pages are also used to calculate the similarity between an input document and each Wikia collection. Let e_d^i be a referenced Wikipedia entity in the analyzed document, and l_c^i the number of links from Wikia collection c to the Wikipedia entity e_d^i. Then $\sum_{i=1}^{m} l_c^i$ is the total number of links from the Wikia collection c to the entities $E_d = \{e_d^1, \cdots, e_d^m\}$ referenced in the input document. This number can be employed as a similarity score between the document and the collection c. Since we have both editorial links and automatic links from the Wikia collections to Wikipedia, further organized into three categories (Wikia-to-entity, entity-to-entity, and mention-to-entity), we can compute in this manner a total of six features.

Content Matching. We devise a group of 5 features to measure the importance of the surface forms detected in a document d with respect to a candidate Wikia collection c. As previously, let S_d denote the set of surface forms in the input document and $w_c(s)$ denote the frequency of a surface form s as a linked mention (either editorially or automatically) in collection c. We denote with L_c the set of all surface forms employed in linked entity mentions in collection c. We calculate as features the number of surface forms from the document that are in the collection $|S_d \cap L_c|$, the total frequency of the matched surface forms in the collection $\sum_{s \in S_d \cap L_c} w_c(s)$, the coverage of the matched surface forms in the Wikia collection $\frac{|S_d \cap L_c|}{|L_c|}$, the frequency-based coverage $\frac{\sum_{s \in S_d \cap L_c} w_c(s)}{\sum_{l \in L_c} w_c(l)}$, as well as a binary indicator whether the title/first line of the document contains the Wikia collection name as a substring.

Wikia Collection Quality. We employ 5 features to measure the quality of Wikia collections, as well as the novelty provided by each Wikia collection with respect to Wikipedia. We employ the Wamrank of Wikia collections as a measure of quality. For novelty, we use the number of surface forms in the collection that are novel with respect to Wikipedia, and the percentage of novel surface forms in the target Wikia. Additionally, we compute the number and the percentage of surface forms from the input document that are in the candidate Wikia but not in Wikipedia.

NIL. We employ a binary feature as a NIL indicator, which is set to 1 for NIL and 0 for any candidate collection c.

Wikipedia Triggers. After we train Stage 1 by using only the previously discussed 31 features on the training set employed in the experiments on news articles (as discussed in the next section), we employ it to pre-analyze all Wikipedia pages and to detect for each page the best matching Wikia collection. We obtain that 31% of the Wikipedia pages *trigger* a Wikia collection, while 69% get assigned the NIL class. For example, the Wikipedia page *Albus Dumbledore* triggers the *Harry Potter* Wikia, while the page *Piotr Kuncewicz* does not trigger any Wikia collection. We can devise now an extra binary feature for the matching of a Wikia collection c to an input document d, by using the triggers of the entities extracted by the Wikipedia-based EDL system from the d. We assign to this feature the value 1 if there exists a Wikipedia entity extracted from d that triggers c, and 0 otherwise.

6   Experiments

As noted in Sect. 2, there are no existing systems or data collections designed for the task of entity linking to a large number of KBs. To evaluate our work, we employ two sets of documents, consisting of Wikia pages and news stories.[2]

---

[2] The annotations can be downloaded at http://www.umiacs.umd.edu/~ninggao/
publications.

6.1   Wikia Pages

From all Wikia pages with more than three linked entity mentions in the 1,163 collections employed in our study, we randomly select a quarter for training the boosted tree ranking system of Stage 1, a quarter for dev-test, and the remaining half (1,568,325 pages) for final testing. Each Wikia page is employed as an input document to our system. We use the collection to which a page belongs as ground-truth for Wikia detection (Stage 1), and the editorial links in the text of each Wikia page to other Wikia pages or Wikipedia pages as ground-truth for linking (Stage 2). Note that because any Wikia page in these sets belongs to one of the candidate Wikias, we do not have NIL triggers in this setting.

Wikia Collection Detection. Table 4 shows the performance of Wikia detection for the Wikia pages in the test set. The *average number of trigger candidates* to Wikia pages is 206. The *recall* and *precision* @1 are 0.95 and 0.84 respectively. The *accuracy* for detecting the ground-truth Wikia collection is 0.79, and the Mean Reciprocal Rank (MRR) is 0.85. In our error analysis, we found that there are 316 Wikia collection pairs that have more than 100 pages wrongly detected to each other as the target Wikia collection, which account for 43.4% of the detection errors. Table 5 shows the top five such pairs. For example, there are 5,302 pages in the *starwars* Wikia and the *swfanon* Wikia wrongly detected as being from the other collection. The former contains information about the "Star Wars" universe, including movies, characters, video games, etc., while the latter contains information about "Star Wars" gathered or written by fans. We manually browsed the home pages of these detected Wikia pairs and judged which pairs are *comparable* (cover the same domain). We found that 89.6% of them are comparable collections, and thus, linking entities to one another is actually informative.

Table 4. Wikia collection detection for Wikia pages.

Average number of candidates	206
Recall	0.95
Precision (accuracy/recall)	0.84
Accuracy	0.79
MRR	0.85

Table 5. Top five Wikia pairs that get confused to one other

Wikia pairs	Confusion count
swfanon, starwars	5,302
stexpanded, memory-beta	2,427
pokemonfanon, pokemon	2,387
classiccars, automobile	1,737
doctor-who-collectors, tardis	1,728

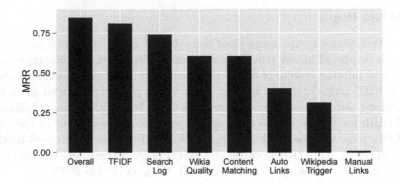

Fig. 2. Performance of single feature groups

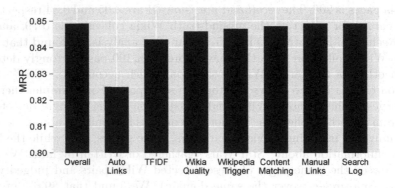

Fig. 3. Feature group ablation study

Feature Study. To further analyze the performance of each feature group, we randomly select 20 documents with at least three linked entity mentions from each Wikia collection. We split this into two groups of 11,630 Wikia pages for training and testing. Figure 2 shows the MRR obtained overall (0.85) and by using individual feature groups. The TF*IDF group performs the best for Wikia detection (0.82), followed by the Web search logs group (0.74), Wikia quality (0.60), and the content matching features (0.60). A rather surprising finding is the extremely poor performance of the editorial (manual) linkage subgroup (0.01) in comparison to the automatic linkage subgroup (0.40). While the features based on automatic links do not achieve very high MRR by themselves in our feature ablation study shown in Fig. 3, we note that that the overall accuracy gets the largest drop when the auto links features are suppressed (from 0.85 to 0.82).

Entity Linking Accuracy. For the Wikia pages with correctly triggered Wikia in Stage 1, we evaluate the linking of the detected surface forms to Wikipedia and the triggered Wikia. As previously, we use the links created by the Wikia contributors as the linkage ground truth. The results are shown in Table 6. The EL evaluation is done only for exact boundary matching. As expected, the

Table 6. Entity detection and linking results

Total number of links to Wikia	8,305,241	Total number of links to Wikipedia	271,744
ED recall for overlapping mentions	0.77	ED recall for overlapping mentions	0.66
ED recall for exact boundaries	0.67	ED recall for exact boundaries	0.53
EL Wikia accuracy	0.93	EL Wikipedia accuracy	0.74
Wikipedia predicted links/Wikia ground-truth	0.01	Wikia predicted links/Wikipedia ground-truth	0.12

performance of *ED recall* and *EL accuracy* is substantially better for Wikia than for Wikipedia, partly because the input documents belong to the domain of the triggered Wikia, and partly because the ambiguity of surface forms in each Wikia is much lower than the ambiguity in Wikipedia.

Table 6 also shows that 1% of the surface forms linked by Wikia contributors to other pages in Wikia get linked by our system to Wikipedia. Conversely, 12% of surface forms that are linked by contributors to Wikipedia are resolved by the employed EDL system to Wikia. While we did not perform an exhaustive analysis of those errors, we noticed that a large number of them are outer links to Wikipedia created for the purpose of interlinking the collections, as in the sentence "Reference Wikipedia Harry Potter page", in which "Harry Potter" is linked to *Harry Potter* in Wikipedia, while our systems links it to the page with the same name in the triggered *Harry Potter* Wikia.

6.2 News Articles

Wikia pages provide a large pre-annotated dataset for training and evaluating. However, there are two shortcomings with the strategy of employing them for training and testing. First, there are no NIL examples in these data. Second, the Wikia pages in the test are likely to be more similar content-wise to the pages on which the training is done. Therefore, we also evaluate the performance of our multi-KB entity linking system on a set of news articles crawled from the Web.

Wikia Detection for News Articles. For training, we use: (1) 1,000 local news articles from a local news station in a major city (names not disclosed to preserve anonymity of the review process). We automatically assign NIL as triggers by adding "NIL" as a trigger candidate; (2) 3,000 randomly selected Wikia pages, each annotated with the collection to which it belongs as trigger; (3) 246 news articles from the same news source, with the property that their title contains one of the names of the Wikia collections employed in our study. For example "Batman star Christian Bale visits shooting victims" contains the Wikia name *Batman*, which we assign to this article as its Wikia trigger. (4) 300 randomly selected Wikipedia pages with manually assigned Wikia triggers by human annotators as ground truth. For testing, we use 260 news articles selected from 15 popular news sites (e.g. CNN), selected through a process that uses the query logs of a large Web search engine as follows: after submitting a query, the user must open a Wikipedia page, a Wikia page, and a news article from the selected websites in any order, and spend at least 30 s on each of those pages.

Table 7. News article Wikia triggering.

Non-NIL (65%)	Correct	Correct + interesting
precision@1	0.90	0.95
NIL (35%)	Correct	
precision@1	0.77	

Table 7 shows the trigger Wikia detection results for the tested news articles. To obtain the ground-truth, two judges read the news article and judged post-hoc the top 3 returned trigger Wikia collections (possibly including NIL), by employing three labels: *correct*, *interesting* or *wrong*. For example, for a news article with title "Game Of Thrones author George RR Martin talks season 4", the Wikia collection of *Game of Thrones* was the correct answer. While the Wikia collection of *Ice and Fire* is not correct, since this Wikia focuses on the book series rather than the TV series, it might still be labeled as an interesting collection for linking. Note that for a news article, there could be more than one correct trigger Wikia collection. For example, both collections *marvel-movies* and *marvel* are judged as correct for a news article titled "Agent Coulsons' secret is out". The annotation inter-agreement is 91.5%. Table 7 shows the results of this evaluation, with a relatively high precision @1 of 0.90 for Wikia triggers and 0.77 for NIL when employing the strict definition of correctness. Moreover, for the news articles with Non-NIL trigger Wikia collections, 83% of the top 2 Non-NIL Wikia triggers are correct, with an additional 6% judged as interesting.

Entity Detection and Linking Accuracy. We further evaluate the effectiveness of ED and EL by using the triggered *Wikia* (the first prediction) or *Wikipedia* as linking targets, shown in Table 8. By employing all the information from Wikia collections and Wikipedia, 2882 surface forms from news articles are detected, in which 500 are randomly selected as evaluation mentions. In Table 8, *Wikia* employs the information from the triggered Wikia and Wikipedia in the step of entity detection, and linking the detected surface forms to the triggered Wikia. On the other hand, *Wikipedia* employs only the information from Wikipedia for entity detection, and linking the mentions to Wikipedia. Two independent annotators evaluate the correctness of the ED and EL steps. As shown in Table 8, by including all the information from the triggered Wikia, the system is able to find additional 6.2% of mentions comparing with using only

Table 8. Evaluation for news article entity detection and linking.

	Wikia	Wikipedia
ED accuracy	1	0.942
EL accuracy	0.928	0.805

Wikipedia. More importantly, the system achieves a 15.3% relative improvement (63.1% error reduction) on linking accuracy by using the triggered Wikia as linking target rather than Wikipedia.

7 Conclusion

We investigated a multi-KB entity linking framework that employs one general knowledge KB (Wikipedia) and a large set of domain-specific KBs (Wikia collections) as linking targets. We developed a supervised model with a large and diverse set of features to detect when a domain-specific KB matches a document targeted for entity analysis. The system obtained high performance for both Wikia detection and entity linking to Wikipedia and Wikia. The performance of both entity detection and entity disambiguation improved by targeting in conjunction the triggered Wikia and Wikipedia as opposed to only the Wikipedia collection.

References

1. Bunescu, R., Paşca, M.: Using encyclopedic knowledge for named entity disambiguation. In: NACL, pp. 9–16 (2006)
2. Cassidy, T., Ji, H., Ratinov, L.A., Zubiaga, A., Huang, H.: Analysis and enhancement of Wikification for microblogs with context expansion. In: COLING, pp. 441–456 (2012)
3. Cucerzan, S.: Large-scale named entity disambiguation based on Wikipedia data. In: EMNLP-CoNLL, pp. 708–716 (2007)
4. Cucerzan, S.: Named entities made obvious: the participation in the ERD 2014 evaluation. In: ERD@SIGIR, pp. 95–100 (2014). http://dblp.uni-trier.de/db/conf/sigir/erd2014
5. Demartini, G., Difallah, D.E., Cudré-Mauroux, P.: Zencrowd: leveraging probabilistic reasoning and crowdsourcing techniques for large-scale entity linking. In: WWW, pp. 469–478 (2012)
6. Hoffart, J., Yosef, M.A., Bordino, I., Fürstenau, H., Pinkal, M., Spaniol, M., Taneva, B., Thater, S., Weikum, G.: Robust disambiguation of named entities in text. In: EMNLP, pp. 782–792 (2011)
7. Ji, H., Grishman, R., Dang, H.T., Griffitt, K., Ellis, J.: Overview of the TAC 2010 knowledge base population track. In: TAC (2010)
8. McNamee, P., Dang, H.T.: Overview of the TAC 2009 knowledge base population track. In: TAC, vol. 17, pp. 111–113 (2009)
9. Mcnamee, P., Dredze, M., Gerber, A., Garera, N., Finin, T., Mayfield, J., Piatko, C., Rao, D., Yarowsky, D., Dreyer, M.: HLTCOE approaches to knowledge base population at TAC 2009. In: TAC (2009)
10. Mihalcea, R., Csomai, A.: Wikify!: linking documents to encyclopedic knowledge. In: CIKM, pp. 233–242 (2007)
11. Niemann, E., Gurevych, I.: The people's web meets linguistic knowledge. automatic sense alignment of Wikipedia and wordnet. In: IWCS, pp. 205–214 (2011)
12. Pereira, B.: Entity linking with multiple knowledge bases: an ontology modularization approach. In: Mika, P., et al. (eds.) ISWC 2014. LNCS, vol. 8797, pp. 513–520. Springer, Cham (2014). doi:10.1007/978-3-319-11915-1_33

13. Ruiz-Casado, M., Alfonseca, E., Castells, P.: Automatic assignment of wikipedia encyclopedic entries to wordnet synsets. In: Szczepaniak, P.S., Kacprzyk, J., Niewiadomski, A. (eds.) AWIC 2005. LNCS (LNAI), vol. 3528, pp. 380–386. Springer, Heidelberg (2005). doi:10.1007/11495772_59
14. Shen, W., Wang, J., Luo, P., Wang, M.: Linking named entities in Tweets with knowledge base via user interest modeling. In: SIGKDD, pp. 68–76 (2013)
15. Sil, A., Cronin, E., Nie, P., Yang, Y., Popescu, A.M., Yates, A.: Linking named entities to any database. In: EMNLP-CoNLL, pp. 116–127 (2012)
16. Zheng, Z., Si, X., Li, F., Chang, E.Y., Zhu, X.: Entity disambiguation with freebase. In: IEEE/WIC/ACM, pp. 82–89 (2012)

Where Do All These Search Terms Come From? – Two Experiments in Domain-Specific Search

Daniel Hienert[(⊠)] and Maria Lusky

GESIS – Leibniz Institute for the Social Sciences, Cologne, Germany
daniel.hienert@gesis.org, maria.lusky@gmail.com

Abstract. Within a search session users often apply different search terms, as well as different variations and combinations of them. This way, they want to make sure that they find relevant information for different stages and aspects of their information task. Research questions which arise from this search approach are: Where do users get all the ideas, hints and suggestions for new search terms or their variations from? How many ideas come from the user? How many from outside the IR system? What is the role of the used search system? To investigate these questions we used data from two experiments: first, from a user study with eye tracking data; second, from a large-scale log analysis. We found that in both experiments a large part of the search terms has been explicitly seen or shown before on the interface of the search system.

Keywords: Search terms · Search process · Session · Social sciences · Digital library · Interactive information retrieval

1 Introduction

For simple information needs users can enter some keywords into the search bar and most of the times receive the right answer. However, for more complex information needs users tend to vary their search terms, add new terms or use combinations of them in order to achieve better results and to uncover new aspects of an advanced information problem. This scenario of searching information is a rather complex one as we have an interplay between the user, the search system and information outside the search system, e.g. in other online or offline sources. Input for new search attempts can therefore be derived from several sources and may additionally be subject to cognitive processes by the user.

A first set of research questions therefore is: What are the sources of new search terms? What is the share of input coming from the user, the search system or other sources? Where and when in the search process are potential new search terms recognized? Further research questions are: How long does it take until a potential term is used in a search? And which cognitive processes are applied on it? The answers to these questions have implications for the design of our search systems. They tell us where, when and how in the search process users are getting ideas for new search

© Springer International Publishing AG 2017
J.M. Jose et al. (Eds.): ECIR 2017, LNCS 10193, pp. 15–26, 2017.
DOI: 10.1007/978-3-319-56608-5_2

terms. This can be a basis for designing new supporting services within a search system that help users in the right place at the right time of the search session.

To answer the basic question where and when users get ideas for new search terms from, we use data from two related experiments in the field of social science literature search: (1) a task-based user study with 32 subjects and recorded eye tracking data, and (2) a large scale log analysis with log data of nine years. The first experiment will tell us explicitly if users have seen new search terms in their search process before they use them. The second experiment can tell us on a large scale if new search terms have been shown on the system before being used by the user.

2   Related Work

In this section we will present related work on interactive search models, evaluation models and the analysis of search terms used in a search session.

2.1   Models for Information Search

The classical Cranfield paradigm is a rather technical model with the goal to optimize search results for a given query. Interactive Information Retrieval (IIR), in contrast, tries to incorporate the user into the search process and explicitly take into account the interactivity between the user, the system and the content. The IIR evaluation model of Cole et al. [2] for example models the search process by starting at a problematic situation a user is facing, which triggers the overall goal and the task to seek information with different seeking strategies to solve the issue. Another framework for IIR is the IPRP model [4] which sketches the search process as transitions between situations, where the user can choose in each situation from a list of choices. Another search model is exploratory search [12] which explicitly addresses the case of a user who is not only looking up a simple information fact, but who is engaging in a more complex problem or unknown area and who is learning and investigating, trying to understand the problem a bit better step by step in his search process.

2.2   Evaluation Methods

For the evaluation of IIR systems and situations, different methods can be used. IR evaluation for a long time has focused primarily on the system view. However, user studies can give valuable insights on how users interact with IR systems. Kelly [11] gives a good overview of user-oriented evaluation methods for IIR. Advantages of these kinds of studies are that real users are observed (maybe within a given task) and the way they interact with the system. These methods enable us to investigate the information seeking behavior of users on the one hand and how an IR system can support users (or hinder them) to gain new insights on the other hand. Disadvantages of user studies are that they are often costly, small-scaled and their significance can therefore be limited.

Eye tracking as a method in IIR evaluation can be used for various purposes. First, it shows the user's attention to different parts of the IIR system's interface, e.g. the search bar or an item on the result page. For example, the F pattern is known as a regularity of how users read web pages [13]. Second, it shows which kinds of texts (title, abstract etc.) users are scanning and how they do it. Longer dwell times can e.g. indicate the user's interest in an item. Third, eye movement patterns can reveal cognitive representation of information acquisition and were used to derive user groups of different domain knowledge and working on different search tasks [3]. The E-Z Reader model [15] assumes that text reading is a serial process with the user's attention to one word after the other. Each of these attention spots is called a fixation. A jump from one fixation to the next one is called a saccade. Within a fixation the E-Z Reader model divides the process of understanding the word meaning (lexical processing) in two stages L1 and L2. The first stage L1 describes the "familiarity check" – the basic word identification – which can be processed with a maximum mean time of 104 ms [16]. With the end of this stage the programming of the saccade to the next word is initiated. The second stage L2 ends with the full understanding of the word. Both stages take an overall time from 151 ms to 233 ms on average [15]. The time for lexical processing depends on a number of variables such as the word length, the word frequency in a language corpus and the word/text difficulty [15].

Log analysis as an evaluation methodology in IIR stands in the middle between user- and system-oriented studies. Log analysis can capture user interaction with the system on a large scale, however, it cannot anticipate the user's information need, the task, the overall problem, the situation and context of the search [9]. It is important to distinguish between web search engine log analysis and digital library (DL) log analysis [1]: in web search retrieved documents are web pages; in DL search documents are maintained by information professionals and are often organized by knowledge organization systems. Also, DL search is often specific for a certain domain, community or topic.

2.3 Analysis of Search Term Usage

The focus in IIR on interactivity also suggests having a deeper look at the whole search process. Thereby the event(s) of a user entering keywords into the search bar is certainly important. Transaction log analysis (TLA) has already dealt with different *statistical measures* of search term usage for a long time [14]: How many search terms were used? How long are search terms on average? In this sense a lot of studies were conducted in different domains (e.g. for Pubmed users [7]). Along that, users of *different domains* search differently: for example for the domains of history and psychology see [19]. On the one hand the *effectiveness* of different sources of search terms had been investigated, especially the use of a controlled vocabulary from a thesaurus vs. free uncontrolled terms [17]. Another aspect are the *patterns of query reformulation*: In which way do users add, delete and replace query terms? For example, Jansen et al. [8] found that generalization and specialization are main transition patterns in web search. Jiang and Ni [10] recently studied what affects word changes in query reformulation based on word-, query- and task-level.

So far, in research only little attention has been given to the *sources of search terms*. Spink and Saracevic [18] conducted a "real-life" study with academic users from several domains and identified five sources of search terms: (1) the question statement the subjects had to fill out with their own information problem, (2) user interaction, (3) a thesaurus, (4) an intermediary and (5) the retrieved items. Yue et al. [20] did a smaller work investigating where query terms come from in collaborative web search. We build up on this research and investigate if users have explicitly seen search terms before applying them in a free search. In a large-scale experiment we check if search terms have been shown on the system before being used.

3   Evaluation Context

In this section we first briefly describe the evaluation system, a real-world digital library for social science literature information. Then we report on the typical search processes in the search system to understand what users' possibilities are for getting search term suggestions.

3.1   System Description

Sowiport [5] is a digital library for social science information with more than nine million bibliographic records, full texts and research projects. The portal gives an integrated search access to twenty German and English-language databases. About 25,000 unique visitors per week are visiting the portal, mainly from German-speaking countries. One of the services for supporting users in their search process is the Combined Term Suggestion Service (CTS) [5]. When the user enters characters into the search bar, the service proposes different term suggestions: (1) auto completion terms from the thesaurus for the social sciences, (2) related, broader and narrower terms from the thesaurus, (3) statistically related terms from a co-occurrence analysis based on titles and abstracts, and (4) author names based on auto completion.

3.2   Search Process

The search process in Sowiport normally follows regular patterns which already were visualized and analyzed with the WHOSE toolkit [6] and which are comparable to the ones in other literature information systems. A first possibility is that users enter Sowiport via the homepage. They can then directly initiate a search via the search bar, where term recommendations from the CTS are shown. The user can also switch to the advanced search form and start there. The next step is the result page which shows a list of twenty documents with title, authors, source and a highlighting text fragment that shows the textual context where the user terms were found in the document. Each document has (where available) links to Google Scholar, Google Books and to the full text (via DOI or URL directly to the journal, proceedings, archive, university or personal websites). Users can follow these links and read (parts of) the full text outside the Sowiport system. On the result page, users can continue and refine their search by paging, choosing from the facets, entering new search terms, or starting a new search for persons, proceedings or journals from the metadata of each record. If one of the records seems to be relevant, the user can enter the detailed view with a click on the title. Then, all metadata entries such as title, source, categories, topics, abstract, references and citations are shown. From here, the users can continue by choosing from similar or related records on the left page section, by choosing a document from references or citations, by entering new search terms in the search bar above or by initiating a new search by clicking on the metadata entries. A large part of users enters Sowiport through a detailed view of a record coming directly from a search engine. These users can then continue their search process with the options of the detailed view.

We can distinguish between two possibilities of how users can initiate a new search process: (1) by simply clicking on a link. This can be done in the result list for authors, proceedings, journals and from the facet section and in the detailed view for all metadata of the record (authors, keywords, categories, journal, proceeding) or (2) by manually entering new search terms into the search bar. This can be done in the search bar on the home page, in the advanced search form and always in the search bar above the result list and the detailed view. *In this paper we will focus on where users get ideas and suggestions for new search terms from when entering them freely in a search form* (for brevity we call it in the following a "*free search*") as here users explicitly enter new search terms which come from the user's mind (and are not readily prepared by the system).

Suggestions for new search terms can come on the *system side*: (1) from the search term recommender when entering terms in one of the search forms, (2) on the result page from titles, authors, sources and highlighted fragments of each search result, (3) from the facet section shown on the result page on the left, (4) from the detailed view which shows all fields such as title, source, categories, topics, abstract, references and citations. Additionally, search terms can derive from (5) the full text which is checked typically *outside the retrieval system* and finally (6) from the *user side* who may have some keywords on his mind, a list of references printed out on his desk or printed text with markers here and there.

4 Experiment I: User Study

For a first investigation we used data from a user study. For each free search we investigated if the search term was seen by the user on the search system by using eye-tracking data.

4.1 Description

We used data from a lab study with two groups of 16 subjects each (20 female, 12 male) that took place in single sessions with a duration of 30 min. While one group consisted of bachelor and master students, the other group comprised only postdoctoral researchers. All subjects worked in different fields of the social sciences. The students were between 22 and 35 years old (m = 26.38, sd = 3.76), while the age of the postdocs ranged between 30 and 62 (m = 40.19, sd = 9.23). On a 5-point Likert scale (1 = "very rarely", 5 = "very often"), the subjects rated their frequency of use of digital libraries on average with 2.78 (sd = 1.02) and of Sowiport with 2.22 (sd = 1.14). They also considered their search experience in digital libraries as moderate (m = 2.91, sd = 0.91).

All subjects were given the same document about the topic "education inequality", opened in Sowiport, and were asked to find similar documents using our digital library. To do so, they had a total time of 10 min. During the task their eye movements as well as the screen were recorded. We made sure the conditions were the same in each session: The subjects used a mouse, a keyboard and a 22″-monitor connected to a laptop. The laptop display served as an observation screen. All subjects worked with Mozilla Firefox. For tracking their eye movements we used the remote eye tracking device *SMI iView RED 250* that was attached to the bottom side of the stimulus monitor. We calibrated the eye tracker with each subject using a 9-point calibration with a sampling frequency of 250 Hz and only then started the experiment. For creating the eye tracking experiment as well as analyzing the gaze data, we used the corresponding software *SMI Experiment Suite 360°*.

4.2 Methodology

For analyzing the subjects' eye movements we created a gaze replay video for each subject, showing their scan paths during the whole session in order to determine the individual words the subjects looked at. The eye tracking software enabled us to make full screen records that also captured the navigation bar of the web browser and dynamic elements like the search term recommender. We used a fixation time threshold of 104 ms as the beginning of the L2 period when the user starts to semantically understand the word. Since the user study was limited to the interaction between the user and our search system, these are the only two sources where search terms could be derived from. Therefore, we first detected each time a subject conducted a free search during the experiment and captured the search terms that were used. In a second step, we carefully observed the subject's scan paths of the session and checked if they had read the search terms before.

4.3 Results

The analysis of the gaze replay videos shows that for this task users are scanning through the result lists and detailed views looking for information that can help to solve the task. As a starting point they especially scan the metadata of the seed document, its references, citations and related entries. They use the title, keywords, abstract, references and citations to browse to related documents and conduct new searches. Terms for free searches were seen explicitly on the result list, in the detailed view or in other parts of the system. Figure 1 shows the detailed results. The users conducted 82 free searches. About 78% of user search terms were seen explicitly on the system before being used for a free search. The largest part comes from the detailed view (51.22%), then from the CTS (9.76%), the result list (4.88%), the references (4.88%), from related entries (4.88%) and from the thesaurus (2.44%). Metadata fields from which search terms were taken are the title (58.93%), keywords (28.57%), abstract (7.14%), authors (3.57%), and categories (1.79%). In 21.95% of the cases the used search term had not been seen by the user prior to the search, which means that the search term was formed by the user. The diagrams in Fig. 1 also show that the student and the postdoc group have very similar results.

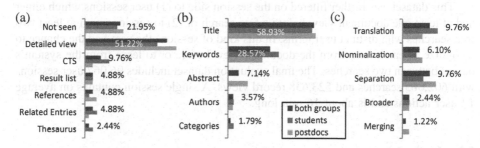

Fig. 1. User study: (a) sources and (b) metadata fields where the search terms were seen and (c) the distribution of cognitive operations.

In a lot of cases the terms later used for a free search query were seen by the user several times during the session. We measured an average time of 3:44 min from first sight to search and an average time of 1:27 min from last sight to search.

One third (29.27%) of the participants conducted cognitive operations of the terms seen. We identified the following categories: (1) *translation* (e.g. from German to English), (2) *separation* of compound terms and then taking only one part of the term for searching, (3) *nominalization* of terms from e.g. personifications to substantives, (4) *merging* of two terms seen and (5) *broadening* of terms.

5   Experiment II: Log Analysis

In this second experiment we used the insight from the first experiment and wanted to find out on a large scale if applied search terms in a free search were shown before on the system. We used a log-based approach and computed for every free search if the used search term had been shown before in the session. Here, the investigation of search term sources was limited to the system side.

5.1   Dataset

For this experiment we used nine years of Sowiport's log data from between November 2007 and July 2016. The data derives from two different technical systems underlying Sowiport and from different sources, such as log files and logs in database tables. The dataset was cleaned from bots and search engines.

We extracted two user actions from the log data to a user action table: (1) A search action (*"search"*) with the database fields session-id, timestamp, search form type (simple, advanced, URL), search field type (all, author, keyword, title, location, date, institution journal/proceedings, topic-feed), the user search terms and result list ids. A *free search* based on keywords (not persons, numbers, locations etc.) can then be identified from the action table by having the search form type set to "simple" or "advanced" and the search field type set to "all" or "keyword" and the user search terms not being empty. (2) A view record action (*"view_record"*) with the fields session-id, timestamp and the doc-id of the viewed record.

This dataset was further filtered on the session side to (1) user sessions which either had at least one document view before a free search or (2) to sessions with at least two free searches with distinct user terms. In this kind of sessions the user had the chance to recognize a search term from the document view before or to learn from the system's output between two searches. The final evaluation dataset includes 96,067 user sessions with 602,065 searches and 523,638 record views. A single session contains on average 12 user actions and is about 16 min long.

5.2   Methodology

We built an algorithm that takes each individual user session and goes through each action, step by step in temporal order. For each session step we collected the metadata of the records which had been shown on the system in a collector. The metadata was cleaned from German and English stop words and stemmed to facilitate the comparison to user search terms later on. For a search action we collected the metadata of the result list entries (title, persons, keywords, categories). For a view_record action we collected the metadata of the viewed record (title, persons, keywords, categories, abstract). References and citations for that record would only be added to the collector if the user had clicked the appropriate tab in the user interface. Some information shown on the system were not collected, because it would have been too costly to compute them for each single search and record view. This affects namely the facet section on the result list and the highlighting fragments for each record that show in which context the user's search terms were found. For the detailed view we left the similar and related documents out of computation.

For each search action, the algorithm first checked if the search terms were taken from the term recommender. If not, it checked if the (stemmed and stop-word cleaned) search terms were shown in a previous session step by comparing them to the collected metadata. Therefore, it went backwards through the session, starting from the search event. Then each search term was compared to the metadata fields in the collector. The ordering of different metadata fields (title, keywords etc.) in the collector had an influence on the field in which the user term is found, because the user term was first checked against the first entry, then the second and so on. We chose the order of the user study (see Fig. 1) as an empirical basis. For each hit, the session step, the source, document and metadata field where the term was found and the search term itself were recorded.

5.3 Results

Figure 2 shows the results of the log experiment. A share of *38.29%* (215,376 of 562,426) user search terms were shown by the system before being used in a free search. The source was in most cases (25.02%) the result list, then the detailed view (13.27%), followed by the term recommender CTS (2.9%) and marginally the references (19 times - ~0%). Metadata fields, where search terms were derived from are keywords (57.13%), title (18.45%), persons (10.38%), abstract (8.45%) and categories (5.58%). We also measured the distance between the search action and the step in which the search term was shown on the system. Figure 2(c) shows that a large part (29.59% of 38.29% maximum) was shown within three steps, which is quite near the search action. Within 10 steps almost all search terms that were used were shown on the system (35.79% of 38.29% maximum). There are on average 2:30 min/9.35 session steps between first occurrence and the search and 2:04 min/3.64 session steps between last occurrence and the search step. On average, a term was shown 8.76 times within a session before being used in a free search.

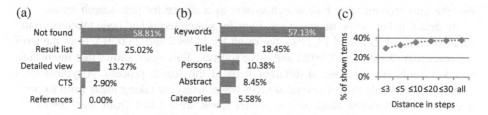

Fig. 2. Log analysis: (a) sources and (b) metadata fields where the search terms were shown on the system and (c) the distance to the search action in session steps.

6 Discussion

The two different experimental approaches in our case have well completed each other. The user experiment visualized the process that users are explicitly scanning the user interface for information and in particular showed that in their free searches users apply terms they have seen before on the search system. Here, two different sources – system and user – were examined as possible sources of search terms. The log experiment then concentrated on the system side as a source for search terms and checked if there is a regularity.

In the user study a large part (78%) came from the system and was seen; the rest came from the user and other sources. This really high value can be surely ascribed to the specific evaluation task. We additionally experimented with lower and higher fixation times. With a fixation time of 50 ms some more search terms had been recognized before the search, with 151 ms some less, but the core of search terms which were seen was stable.

In the log analysis we found a value of about 38% of terms that were shown before being used in a free search. This is still a high value, but surely based on a different kind of user population with a diversity of tasks and topics. In the log analysis we can only assume that the users have explicitly seen the terms. However, the identified scan process in the user study, the number of search terms occurrence in the session prior search and the scale of the experiment in the log analysis indicate a high probability for this being true.

In both experiments a considerable amount of free search terms originated from different parts of the system, which should give system designers a higher responsibility to support users in finding the right terms. Support has to be given not only via a typical term recommender (which has been long-time acknowledged in our field), but also in all steps of the search process, as well as while viewing the entries in the result list and checking a record in detail.

In terms of system and user sources, Spink and Saracevic [18] in their experiment found that user interaction was responsible for 23% of the search terms, while 11% came from Term Relevance Feedback [the rest came from the question statement (38%), thesaurus (19%) and intermediary (9%)]. Certainly, our and their results are hard to compare, because of the different settings of the experiment. However, on the system side they have focused on a relevance feedback loop, in which users chose terms from documents they found relevant. This is in contrast to our experiment, where we take into account the *whole* search system as a source for new search terms.

In detail, in both experiments suggestions for search terms had been taken from the detailed view (51.22% and 13.27%), the result list (4.88% and 25.02%), from the term recommender (9.76% and 2.90%) and other sources. This again shows that interesting new keywords are extracted at different steps of the search process. A typical term recommender is only one of several sources where users are taking ideas from for new free searches. Metadata fields where search terms were taken from were relatively similar in both experiments. Most came from the keyword Section (28.07% and 57.13%) and the title (59.65% and 18.45%), from the abstract (7.14% and 8.45%), persons (3.57% and 10.38%), and categories (1.79% and 5.58%).

Following the search processes in the user experiment showed that search terms were shown several times in the system before users applied them in a free search. In the log analysis, applied search terms had been shown in the system up to eight times before being used. Although both experiments had different kinds of tasks (exploratory search in the user experiment; a diversity of tasks in the log analysis), the time spans from first sight and last sight until search are comparable. It took about 3:44/2:30 min from first sight and 1:27/2:04 min from last sight to the search event. Additionally, the log experiment shows us that the largest share of terms were shown within three session steps – thus from an interaction perspective really near the search action.

All in all, by taking into account the whole search system, we can see that steps in the session beforehand influence the actual step, which is a strong argument for the whole session or interactive information retrieval discussion.

7 Conclusion and Future Work

In this paper we conducted two experiments to investigate where users are taking ideas and suggestions for new search terms in free searches from. The user experiment showed well the process of scanning information and taking term suggestions from the system that have been shown at different sources, such as the result list, the detailed view or the term recommender. The log analysis showed on a large scale that one third of search terms had been shown on the system before the users conducted a search query with these terms. Answering our research question from the beginning, we can say that a good share of search terms comes from the system. The other parts are information from outside the system, but from online sources (e.g. reading full texts or articles in another tab) and from the user side with printed texts, ideas from discussions etc.

Search terms were seen and shown up to eight times in the search session and it could take some minutes until they were used in a free search. This again shows that the segmentation of the search process to query-response is too short-sighted, but user perception in the process minutes before querying can massively influence the actual action step. This also somehow negates user models with the assumption that the actual step is only influenced by the action before. The user experiment also showed that users are conducting cognitive processing of seen terms such as translation or separation.

We can conclude that finding new search terms is a process: (I) A good share of new free search terms comes from the system. (II) Search terms are shown and seen several times on the system before being used. (III) Terms can come from different parts of the system and from different metadata fields. (IV) Search terms are seen at different points in time within the session and it can take some time until they are used. (V) New search terms partly underlie cognitive operations from the user.

This research shows that searching and especially finding new free search terms is a complex process with interaction between the user, the system, the content and other entities online and offline. The user's state is influenced by all parts of the system and the user influences the system's state. In future work we want to concentrate even more on examining which interaction processes happen within a whole search session and how we can develop more suitable user models that capture these processes.

Acknowledgements. This work was partly funded by the DFG, grant no. MA 3964/5-1; the AMUR project at GESIS. The authors thank the focus group IIR at GESIS for fruitful discussions and suggestions.

References

1. Agosti, M., et al.: Web log analysis: a review of a decade of studies about information acquisition, inspection and interpretation of user interaction. Data Min. Knowl. Discov. **24** (3), 663–696 (2011)
2. Cole, M., et al.: Usefulness as the criterion for evaluation of interactive information retrieval. In: Proceedings of the Workshop on Human-Computer Interaction and Information Retrieval, pp. 1–4 (2009)
3. Cole, M.J., et al.: User activity patterns during information search. ACM Trans. Inf. Syst. **33** (1), 1:1 1:39 (2015)
4. Fuhr, N.: A probability ranking principle for interactive information retrieval. Inf. Retrieval **11**(3), 251–265 (2008)
5. Hienert, D., et al.: Digital library research in action: supporting information retrieval in Sowiport. D-Lib Mag. **21**(3/4) (2015). http://dx.doi.org/10.1045/march2015-hienert

6. Hienert, D., van Hoek, W., Weber, A., Kern, D.: WHOSE – a tool for whole-session analysis in IIR. In: Hanbury, A., Kazai, G., Rauber, A., Fuhr, N. (eds.) ECIR 2015. LNCS, vol. 9022, pp. 172–183. Springer, Cham (2015). doi:10.1007/978-3-319-16354-3_18
7. Dogan, R.I., et al.: Understanding PubMed user search behavior through log analysis. Database J. Biol. Databases Curation **2009**, bap018 (2009)
8. Jansen, B.J., et al.: Patterns and transitions of query reformulation during web searching. Int. J. Web Inf. Syst. **3**(4), 328–340 (2007)
9. Jansen, B.J.: Search log analysis: what it is, what's been done, how to do it. Libr. Inf. Sci. Res. **28**(3), 407–432 (2006)
10. Jiang, J., Ni, C.: What affects word changes in query reformulation during a task-based search session? In: Proceedings of the 2016 ACM Conference on Human Information Interaction and Retrieval, CHIIR 2016, Carrboro, North Carolina, USA, 13–17 March 2016, pp. 111–120 (2016)
11. Kelly, D.: Methods for evaluating interactive information retrieval systems with users. Found Trends Inf. Retrieval **3**(1–2), 1–224 (2009)
12. Marchionini, G.: Exploratory search: from finding to understanding. Commun. ACM **49**(4), 41–46 (2006)
13. Nielsen, J.: F-shaped pattern for reading web content. https://www.nngroup.com/articles/f-shaped-pattern-reading-web-content/
14. Peters, T.A.: The history and development of transaction log analysis. Libr. Hi Tech **11**(2), 41–66 (1993)
15. Reichle, E.D., et al.: E-Z Reader: a cognitive-control, serial-attention model of eye-movement behavior during reading. Cogn. Syst. Res. **7**(1), 4–22 (2006)
16. Reichle, E.D., et al.: Using E-Z Reader to simulate eye movements in nonreading tasks: a unified framework for understanding the eye-mind link. Psychol. Rev. **119**(1), 155–185 (2012)
17. Rowley, J.: The controlled versus natural indexing languages debate revisited: a perspective on information retrieval practice and research. J. Inf. Sci. **20**(2), 108–119 (1994)
18. Spink, A., Saracevic, T.: Interaction in information retrieval: selection and effectiveness of search terms. JASIS **48**(8), 741–761 (1997)
19. Yi, K., et al.: User search behavior of domain-specific information retrieval systems: an analysis of the query logs from PsycINFO and ABC-Clio's Historical Abstracts/America: History and Life. J. Am. Soc. Inf. Sci. Technol. **57**(9), 1208–1220 (2006)
20. Yue, Z., et al.: Where do the query terms come from? An analysis of query reformulation in collaborative web search. In: Proceedings of the 21st ACM International Conference on Information and Knowledge Management, pp. 2595–2598. ACM, New York (2012)

Using Query Performance Predictors
to Reduce Spoken Queries

Jaime Arguello[1]([⊠]), Sandeep Avula[1], and Fernando Diaz[2]

[1] University of North Carolina at Chapel Hill, Chapel Hill, USA
jarguell@email.unc.edu
[2] Microsoft Research, Cambridge, UK

Abstract. The goal of query performance prediction is to estimate a
query's retrieval effectiveness without user feedback. Past research has
investigated the usefulness of query performance predictors for the task of
reducing verbose textual queries. The basic idea is to automatically find
a shortened version of the original query that yields a better retrieval. To
date, such techniques have been applied to TREC topic descriptions (as
surrogates for verbose queries) and to long textual queries issued to a web
search engine. In this paper, we build upon an existing query reduction
approach that was applied to TREC topic descriptions and evaluate its
generalizability to the new task of reducing spoken query transcriptions.
Our results show that we are able to outperform the original spoken
query by a small, but significant margin. Furthermore, we show that
the terms that are omitted from better-performing sub-queries include
extraneous terms not central to the query topic, disfluencies, and speech
recognition errors.

1 Introduction

Speech-enabled search allows users to formulate queries using spoken language.
The search engine transcribes the spoken query using an automatic speech recognition (ASR) system and then runs the textual query against the collection.
While speech is a natural means of communicating an information need, spoken queries pose a challenge for speech-enabled search engines, for two reasons:
(1) spoken queries are longer than textual queries and may include terms that
are not central to the query topic [5], and (2) spoken queries may have speech
recognition errors that can cause a significant drop in retrieval performance [10].

In this paper, we focus on the task of automatically reducing spoken query
transcriptions in order to improve retrieval performance. We evaluate an approach that extends the algorithm proposed by Kumaran and Carvalho [11], which
was originally evaluated using TREC topic descriptions as surrogates for verbose
textual queries. Our approach proceeds in three steps. First, given a spoken query
transcription, we generate a set of candidate sub-queries to consider (including
the original spoken query). Second, we use a regression model to predict each
sub-query's retrieval performance compared to the original. Finally, we use a

© Springer International Publishing AG 2017
J.M. Jose et al. (Eds.): ECIR 2017, LNCS 10193, pp. 27–39, 2017.
DOI: 10.1007/978-3-319-56608-5_3

weighted rank fusion method to combine the rankings from the top-k sub-queries with the greatest predicted performance.

The regression model is trained to predict the difference in performance between a candidate sub-query and the original query as a function of a set of features. Following prior work, we experimented with three types of features: *pre-retrieval* query performance features, *post-retrieval* query performance features, and drift features. Our query performance features estimate the candidate sub-query's effectiveness. On the other hand, to avoid drifting too far from the original query topic, our drift features capture the relatedness between a candidate sub-query and the original. We present an evaluation on 5,000 spoken queries that were obtained using a crowdsourced study and transcribed using three freely available ASR systems provided by AT&T, IBM, and WIT.AI.

This paper makes the following contributions. First, we propose an extension of an existing query reduction approach and achieve comparable results on the task of reducing TREC topic descriptions. Second, we evaluate the *generalizablity* of our approach to the new task of reducing spoken queries. Third, we describe the types of spoken query terms that are dropped in order to improve retrieval performance, which suggest unique challenges and opportunities for improving spoken query retrieval. Finally, we describe our collection of 5,000 spoken queries which are based on the 250 TREC 2004 Robust Track topics and are therefore associated with a reusable IR test collection. Our spoken query transcriptions are available for others to extend our research.[1]

2 Related Work

Our work builds on three areas of prior research: (1) query performance prediction, (2) automatically reducing verbose queries, and (3) using query performance predictors to improve spoken query retrieval.

Query Performance Prediction: Query performance predictors estimate a query's effectiveness without user feedback. Pre-retrieval measures capture evidence such as the query's specificity, topical coherence, and estimated rank stability [8]. Query specificity measures consider the query terms' inverse document frequency (IDF) and inverse collection term frequency (ICTF) values [6,9,22]. Other specificity measures include the *query-scope*—proportional to the number of documents with at least one query term—and *simplified clarity*—equal to the KL-divergence between the query and collection language models [9]. Topical coherence can be measured using the degree of co-occurrence between query terms [8]. Finally, the rank stability can be estimated using the query terms' variance of TF.IDF weights across documents in the collection [22].

Post-retrieval measures capture evidence such as the topical coherence of the top results, the actual rank stability, and the extent to which similar documents obtain similar retrieval scores. The *clarity* score measures the KL-divergence between the language model of the top documents and a background model of

---

[1] https://ils.unc.edu/~jarguell/ecir2017/.

the collection [6]. Rank stability methods perturb the query [20,24], the documents [23], or the retrieval system [2], and measure the degree of change in the output ranking. Finally, the auto-correlation score from Diaz [7] considers the extent to which documents with a high text similarity obtain similar retrieval scores.

Reducing Verbose Queries: Kumaran and Carvalho [11] focused on automatically reducing TREC topic descriptions. They used learning-to-rank (LTR) to predict the sub-query with the best performance and used query performance predictors as features. The authors focused on a heuristically-chosen sample of all possible sub-queries and found a 6.8% improvement in average precision on the TREC 2004 Robust Track collection. Balasubramanian et al. [3] evaluated a similar technique on verbose queries issued to a commercial web search engine and considered only sub-queries with $n-1$ terms. Xue et al. [19] focused on reducing TREC topic descriptions and trained a sequential model to label each query-term as 'keep' or 'do not keep' using query performance predictors as features. The authors found greater improvements by combining the predicted sub-query with the original. Xue and Croft extended this idea by combining sub-queries in a weighted fashion, setting the mixing parameters based on the LTR output [18]. Zhao and Callan trained a classifier to predict a query term's importance by combining performance predictors with features such as the query-term's rareness, abstractness, and ambiguity [21]. Their results found greater improvements for more verbose queries (i.e., TREC descriptions vs. titles).

Improving Spoken Query Retrieval: Prior work has also considered improving spoken query recognition using evidence similar to some of the query performance predictors mentioned above. Mamou et al. [14] focused on re-ranking the ASR system's n-best list using term co-occurrence statistics in order to favor transcribed queries with semantically related terms. Li et al. [13] combined language models generated from different query-click logs to bias the ASR output in favor of previously run queries with clicks. Peng et al. [15] focused on re-ranking the n-best list using post-retrieval evidence such as the number of search results and the number of exact matches in the top results. Arguello et al. [1] used a wide-range of pre- and post-retrieval query performance predictors to re-rank the ASR system's n-best list.

3 Data Collection

Our spoken queries were collected as part of a user study reported in a previous paper. We provide a general description of the study and the ASR systems used, and refer the reader to Arguello et al. [1] for additional details.

User Study: Spoken queries were collected using Amazon Mechanical Turk (MTurk). Participants were given a search task description and were asked to produce a recording of how they would request the information from a speech-enabled search engine. Each MTurk Human Intelligence Task (HIT) proceeded as follows. First, participants were given a set of instructions and a link to a video explaining the HIT. Participants were then asked to click a "start" button to open the main voice recording page in a new browser tab. While loading, the main voice recording page asked participants to grant access to their computer's microphone. Participants were required to grant access in order to continue. The main voice-recording page provided participants with three items: (1) a "view task" button that displayed the search task description in a pop-up window, (2) Javascript widgets to record the spoken query and save the recording as a WAV file on their computer, and (3) an HTML form to upload the saved WAV file to our server. The search task was displayed in a pop-up window to prevent participants from reading the search task description while producing their recording.

Each MTurk HIT was priced at $0.15 USD. We restricted our HITs to workers with a 95% acceptance rate or greater and to workers within the U.S. Finally, in order to gather spoken queries from a wide range of participants, each worker was allowed to complete a maximum of 100 HITs (2% of all HITs). In total, we collected spoken queries from 167 participants.

Search Tasks: We developed 250 search tasks based on the 250 topics from the TREC 2004 Robust Track. We used the TREC description and narrative as guidelines and situated each task in a background scenario that gave rise to the information need. We collected 20 spoken queries per search task for a total of 5,000 spoken queries. An example search task and spoken query are provided below.

> **TREC Topic ID and Title:** 303 - Hubble Telescope Achievements
> **TREC Description:** Identify positive accomplishments of the Hubble telescope since it was launched in 1991.
> **Search Task Description:** You recently saw a picture of space taken by the Hubble telescope and now you are curious about the scientific advances made possible by the Hubble telescope since its launch in 1991. Find information about the positive accomplishments of the Hubble telescope, which include the ability to gather new and better-quality data that has led to new discoveries, theories, and areas of inquiry.
> **Example Spoken Query:** "What scientific advances have been made as a result of the Hubble telescope?"

ASR Systems: In this work, we treated the ASR system as a "black box" and used three freely available speech-to-text APIs provided by AT&T, IBM, and WIT.AI. All three APIs accept a WAV file as input and return the most confident transcription in JSON format.

Spoken Queries vs. TREC Topic Descriptions: In this work, we test the *generalizability* of a query reduction approach on TREC topic descriptions and spoken query transcriptions. Thus, we were interested in the differences between TREC topic descriptions and the spoken queries produced by our participants. We focus on the query transcriptions produced by the AT&T API.

Our spoken queries are different than the 250 TREC topic descriptions from the 2004 Robust Track in two important ways. First, our spoken queries are shorter. Including stopwords, our spoken queries have an average of 10.11 ± 4.81 words, while TREC topic descriptions have an average of 16.76 ± 8.89 words. Excluding stopwords, our spoken queries have an average of 5.055 ± 2.22 words, while TREC topic descriptions have an average of 9.12 ± 5.32.[2] Both TREC topic descriptions and our spoken queries were about 45% stopwords.

Second, when issued as queries against the TREC 2004 Robust Track collection, TREC topic descriptions produced better retrievals than our spoken queries. TREC topic descriptions achieved an average precision of 0.240, while our spoken queries achieved an average precision of 0.113.

Taken together, these two trends suggest that reducing spoken queries may be more difficult than reducing TREC topic descriptions. Our spoken queries had fewer topical terms and a lower baseline performance.

4   Algorithm

The goal of our algorithm is to select sub-queries that perform better than the original query transcription. Our approach is similar to the one proposed by Kumaran and Carvalho [11], and proceeds in three steps: (1) generate a candidate set of sub-queries to consider (including the original), (2) predict the retrieval performance of each candidate sub-query, and (3) combine the retrievals from the top-k sub-queries with the highest predicted performance.

Step 1: A query with n terms has $2^n - 1$ sub-queries (excluding the null query). We considered a much smaller subset of sub-queries using the following heuristics. First, we only considered sub-queries with 3–6 terms. Second, we only considered sub-queries with at least one noun. Third, to favor topically coherent queries, we only considered the 25 sub-queries with the highest average mutual information between query-term pairs. Finally, we included the original query in the candidate set. Similar heuristics were used in prior work [11].

Step 2: To perform the second step, we trained a regression model to predict each candidate sub-query's absolute increase or decrease in retrieval performance compared to the original query. We trained support vector regression models using the LibLinear toolkit.[3] At test time, we simply selected the candidate sub-queries with the greatest predicted performance. As described in more detail below, we measured retrieval performance in terms of P@10, NDCG@30, and average precision (AP). We trained different regression models for different metrics. Each sub-query was represented as a vector of features (Sect. 5), and feature values were normalized to zero-min and and unit-max separately for each candidate set of sub-queries. In other words, we used each feature's min and max values from the set of sub-queries associated with the *same* spoken query.

---

[2] We used the SMART stopword list.
[3] https://www.csie.ntu.edu.tw/~cjlin/liblinear/.

Step 3: Finally, to perform the third step, we used a *weighted* version of the Reciprocal Rank Fusion (RRF) method [4] to combine the document rankings from the top-k sub-queries with the *greatest* predicted performance. Let $\mathcal{R}_i$ denote the document ranking from the ith sub-query with the greatest predicted performance, and let $\mathcal{R}_i(d)$ denote the rank of document d in $\mathcal{R}_i$. Documents retrieved by the top-k sub-queries were scored as follows:

$$\text{score}(d) = \sum_{i=1}^{k} \frac{1}{i} \times \frac{1}{t + \mathcal{R}_i(d)}. \tag{1}$$

Parameter t mitigates the impact of highly ranked documents that are outliers, and we set it to $t = 60$ based on prior work [4].

5 Features

We used three types of features: pre-retrieval query performance predictors, post-retrieval query performance predictors, and drift features. The numbers in parentheses indicate the number of features in each group.

Pre-retrieval Features (5): Prior work shows that well-performing queries tend to contain discriminative terms that appear in only a few documents. We included four features aimed to capture this type of evidence. Following prior work, we included the average inverse document frequency (IDF) value across query terms [6,9,22]. The *query-collection similarity* (QCS) score measures the extent to which the query terms appear many times in only a few documents [22]. The *query scope* score is inversely proportional to the number of documents with at least one query term [9]. Finally, the *simplified clarity* score measures the KL-divergence between the query and collection language models [9].

Prior work also shows that a query is more likely to perform well if the query terms describe a coherent topic. We included one feature to capture this type of evidence. Our point-wise mutual information (PMI) feature measures the average degree of co-occurrence between query-term pairs [8].

Post-Retrieval Features (6): A query is more likely to perform well if the top-ranked documents describe a coherent topic. We included five features aimed to capture this type of evidence. The *clarity* score measures that KL-divergence between the language model of the top results and the collection [6]. The *query feedback* score measures the degree of overlap between the top results before and after query-expansion [24]. A greater overlap suggests that the original query is more on-topic. Finally, we considered the *normalized query commitment* (NQC) score, which measures the standard deviation of the top document scores. We included three NQC scores: the standard deviation of the top document scores, the standard deviation of the scores *above* the mean top-document score, and the standard deviation of the scores *below* the mean top-document score [16].

Prior work also shows that in an effective retrieval, similar documents have similar retrieval scores [7]. We included one feature to model this type of evidence. The autocorrelation score from Diaz [7] measures to extent to which results with a high cosine similarity have similar scores.

Drift Features (2): The aim of our drift features was to favor sub-queries that do not drift too far from the original. We included two features to measure this type of evidence. Our *relevance model similarity* feature computes the similarity between the language model of the top results from the original query and the candidate sub-query. Following Lavrenko *et al.* [12], relevance models were estimated by combining the language models of the top-10 results weighted by their retrieval scores. The relevance model similarity was computed using the Bhattacharyya correlation. Finally, we measured the Jaccard coefficient between the top-10 results from the original and candidate sub-query. All drift feature values were 1.0 for the original query, which was included in the candidate set.

6   Evaluation Methodology

Retrieval performance was measured by issuing queries against the TREC 2004 Robust Track collection. We used Lucene's implementation of the query-likelihood model with Dirichlet smoothing ($\mu = 1000$), and used the Krovetz stemmer and the SMART stopword list. We evaluated in terms of P@10, NDCG@30, and average precision (AP).

Models were evaluated using 20-fold cross-validation. In order to train and test using spoken queries from *different* TREC topics, all 20 spoken queries for the same topic were assigned to the same fold. We report average performance across held-out folds and measured statistical significance using the approximation of Fisher's randomization test described in Smucker *et al.* [17]. We used the same cross-validation folds in all our experiments. Thus, when measuring statistical significance, the randomization was applied to the 20 pairs of performance values for the two models being compared.

We compare against two baseline approaches: (1) selecting the best-performing candidate sub-query (oracle) and (2) running the original spoken query transcription (original). Parameter k and SVM regression parameter c were tuned by doing a second level of cross-validation.

7   Results

Our evaluation results are presented in Table 1. We present results using the ASR output from our three speech-to-text APIs: AT&T (Table 1a), IBM (Table 1b), and WIT.AI (Table 1c). Additionally, we applied our approach to the task of reducing TREC topic descriptions (Table 1d). We present results in terms of average precision (AP), P@10 and NDCG@30.

The rows labeled original show the performance of the original spoken query in Table 1a–c and the original TREC topic description in Table 1d. The rows labeled all show the performance of our models using *all* features. The rows labeled no.x show the performance of our models using all features *except* for pre-retrieval query performance features (no.pre), post-retrieval query performance features (no.post) and drift features (no.drift). The rows labeled oracle show the performance of the best candidate sub-query. This is not a "true" oracle experiment because we did not consider every possible sub-query. However, it determines whether Step 1 in our approach was able to select sub-queries that perform better than the original.

In Step 3 of our approach, we combined the rankings from the top-k sub-queries with the greatest predicted performance using Eq. 1. We were interested in evaluating the contribution of this step to retrieval performance. To this end, we considered three additional alternatives: (1) selecting the *single* sub-query with the greatest predicted performance ($k = 1$), (2) combining the rankings from *all* candidate sub-queries in a weighted fashion as described in Eq. 1 ($k = $ max), and (3) combining the rankings from all candidate sub-queries in an *unweighted* fashion by omitting factor $1/i$ from Eq. 1 ($k = $ max, unweighted).

The results in Table 1 show seven important trends. First, overall retrieval performance was better for the IBM and WIT.AI APIs than the AT&T API. As it turns out, the AT&T API had more ASR errors, possibly because it uses a language model less well-suited for queries or for the topics associated with the 2004 Robust Track collection. Our goal was not to compare speech-to-text APIs. However, as described below, our results suggest that we can improve retrieval performance for spoken queries with varying degrees of ASR error.

Second, across all APIs and evaluation metrics, our models using all features (all) performed at the same level or significantly better than the baseline of running the original query (original). Improvements were higher in terms of AP than P@10 and NDCG@30, suggesting that our approach was able to retrieve more relevant documents beyond the top-10 results. In terms of AP, performance improvements compared to the original query were in the 4–5% range. We observed similar trends on TREC topic descriptions. On the task of reducing TREC topic descriptions, Kumaran and Carvalho [11] reported a 6.8% improvement in AP on the same TREC 2004 Robust Track collection. In our case, we observed a 5.0% improvement when using all features (all) and a 6.25% improvement when ignoring pre-retrieval features (no.pre).

Third, our approach (all) outperformed the alternative of selecting the *single* sub-query with the greatest predicted performance ($k = 1$). In all cases, setting $k = 1$ resulted in a drop in retrieval performance. This result suggests that combining the rankings from the top sub-queries yields a more robust solution.

Fourth, our results show that combining the rankings from all candidate sub-queries in a weighted fashion ($k = $ max) is a reasonable alternative. In most cases, setting $k = $ max resulted in only a slight drop in performance. This result shows that our approach is not very sensitive to parameter k. In retrospect, this makes sense, as factor $1/i$ in Eq. 1 places much more emphasis on the top sub-queries than the bottom ones.

Fifth, combining rankings from all sub-queries in an *unweighted* fashion ($k = $ max, unweighted) resulted in a large drop in performance. In fact, in all cases, the drop in performance compared to the original query was statistically significant. This result shows that effectively reducing spoken queries (and TREC topic descriptions) is not simply a matter of combining sub-queries without first estimating their retrieval performance. In other words, this result validates Steps 2 and 3 of our approach.

Table 1. Results using TREC topic descriptions and the spoken query transcriptions generated using the AT&T, IBM, and WIT.AI APIs. The percentages indicate percent improvement over the original query (original). A ▲ and ▼ denotes a significant increase and decrease in performance compared to original, respectively. We report significance at the $p < .05$ level using Bonferroni correction.

	AP	P@10	NDCG@30
original	0.113	0.206	0.197
all	0.119 (5.31%)▲	0.210 (2.25%)▲	0.203 (3.12%)▲
all (k=1)	0.116 (2.65%)▲	0.207 (0.67%)	0.200 (1.63%)
all (k=max)	0.118 (4.42%)▲	0.208 (1.27%)	0.202 (2.37%)
all (k=max, unweighted)	0.109 (−3.54%)▼	0.199 (−2.99%)▼	0.191 (−3.13%)▼
no.pre	0.118 (4.42%)▲	0.208 (1.16%)	0.202 (2.32%)▲
no.post	0.115 (1.77%)	0.207 (0.79%)	0.200 (1.31%)
no.drift	0.118 (4.42%)▲	0.206 (0.01%)	0.199 (1.15%)
oracle	0.146 (29.20%)▲	0.285 (38.56%)▲	0.258 (31.20%)▲

(a) AT&T spoken query transcriptions

	AP	P@10	NDCG@30
original	0.165	0.293	0.282
all	0.173 (4.85%)▲	0.300 (2.16%)▲	0.290 (2.77%)▲
all (k=1)	0.170 (3.03%)▲	0.296 (0.82%)	0.286 (1.35%)
all (k=max)	0.173 (4.85%)▲	0.300 (2.32%)▲	0.290 (2.75%)▲
all (k=max, unweighted)	0.160 (−3.03%)▼	0.288 (−2.04%)▼	0.276 (−2.23%)▼
no.pre	0.173 (4.85%)▲	0.299 (2.00%)▲	0.290 (2.72%)▲
no.post	0.168 (1.82%)	0.296 (0.86%)	0.286 (1.20%)
no.drift	0.171 (3.64%) ▲	0.295 (0.65%)	0.285 (0.77%)
oracle	0.211 (27.88%)▲	0.395 (34.60%)▲	0.361 (27.74%)▲

(b) IBM spoken query transcriptions

	AP	P@10	NDCG@30
original	0.183	0.321	0.308
all	0.191 (4.37%)▲	0.327 (1.77%)	0.317 (2.70%)▲
all (k=1)	0.188 (2.73%)▲	0.324 (0.77%)	0.312 (1.16%)
all (k=max)	0.190 (3.83%)▲	0.326 (1.50%)	0.316 (2.42%)▲
all (k=max, unweighted)	0.177 (−3.28%)▼	0.314 (−2.26%)▼	0.302 (−2.15%)▼
no.pre	0.192 (4.92%)▲	0.326 (1.44%)	0.316 (2.52%)▲
no.post	0.185 (1.09%)	0.323 (0.42%)	0.312 (1.14%)
no.drift	0.190 (3.83%)▲	0.323 (0.61%)	0.311 (1.02%)
oracle	0.228 (24.59%)▲	0.422 (31.26%)▲	0.385 (24.99%)▲

(c) WIT.AI spoken query transcriptions

(continued)

Table 1. (*continued*)

	AP	P@10	NDCG@30
original	0.240	0.403	0.384
all	0.252 (5.00%)▲	0.417 (3.49%)	0.393 (2.37%)
all (k=1)	0.245 (2.08%)	0.403 (−0.05%)	0.387 (0.69%)
all (k=max)	0.245 (2.08%)	0.403 (0.02%)	0.384 (0.00%)
all (k=max, unweighted)	0.225 (−6.25%)▼	0.380 (−5.67%)▼	0.361 (−6.04%)▼
no.pre	0.255 (6.25%)▲	0.411 (2.00%)	0.395 (2.70%)
no.post	0.240 (0.00%)	0.397 (−1.42%)	0.379 (−1.38%)
no.drift	0.251 (4.58%)▲	0.403 (−0.11%)	0.388 (1.08%)
oracle	0.301 (25.42%)▲	0.544 (34.87%)▲	0.491 (27.68%)▲

(d) TREC topic descriptions

Sixth, our feature ablation results suggest that pre-retrieval query performance features were the least predictive and that post-retrieval features were the most predictive. Omitting pre-retrieval features (no.pre) resulted in the lowest drop in performance. In most cases, no.pre still performed significantly better than the original baseline. In contrast, omitting post-retrieval features (no.post) resulted in the largest drop in performance. In all cases, no.post was statistically equal to the original baseline. This result is consistent with prior work that shows that, while post-retrieval features are more computationally expensive, they provide valuable evidence [11,19].

The final trend worth noting is that there is still room for improvement. Across all APIs and metrics, the oracle significantly outperformed the original query (original) and all our models by a large margin.

8   Discussion

Sub-query Effectiveness: Based on our results, it is clear that some candidate sub-queries perform better than others. For example, combining the rankings from all candidate sub-queries in a *weighted* fashion (based on their predicted performance) outperformed combining the rankings in an *unweighted* fashion. A natural follow-up question is: On average, what percentage of the candidate sub-queries outperformed the original query? For our spoken queries, the average percentage of candidate sub-queries that outperformed the original query were: AT&T = 29.22% ± 22.22%, IBM = 31.74% ± 21.38%, and WIT.AI = 31.58% ± 20.80%. Similarly, for TREC topic descriptions, the average percentage of better-performing sub-queries was 30.65% ± 22.22%. Across all datasets, most candidate sub-queries did not outperform the original. Thus, any method that uses sub-queries to reduce verbose queries needs to be selective about which sub-queries to focus on.

Reducing Spoken Queries: Our results in Table 1 show that we can improve retrieval performance by dropping terms from the original spoken query. We were interested in better understanding what are the types of original query terms that are omitted from a better-performing sub-query. To answer this question, we counted the number of times each term was omitted from a candidate sub-query that outperformed the original query in terms of AP. For this and the next analysis, we focus on the recognition output from the AT&T API.

The following are the top-50 most frequently dropped terms: information (2189), find (934), country (660), show (592), states (535), united (510), people (471), current (343), affect (313), list (275), negative (274), um (270), america (263), world (238), government (237), company (234), effects (229), con (226), recent (226), place (222), pro (221), type (218), call (216), industry (210), work (209), history (209), case (202), conditions (190), tax (189), international (184), worldwide (176), activity (172), treatment (170), human (163), news (159), project (158), happen (158), instance (156), law (156), impact (156), involve (154), nineteen (148), made (147), side (146), system (145), increase (142), group (142), number (139), document (138), and search (138).

Interestingly, we see three types of terms. First, we see several imperative verbs and nouns associated with 'requesting information' (e.g., find, show, list, search, information, document). Second, we see at least one disfluency (e.g., um). Third, we see terms describing *extra-topical* dimensions of the information need. For example, we see terms that suggest the desire for information about a specific time frame (e.g., history, recent, current, news), as well as terms that suggest the desire for information about a particular perspective (e.g., negative, pro, con). This last category is particularly interesting. Such terms may be problematic for search systems because they may not frequently appear in relevant documents. For instance, a document discussing historic or recent events may not actually contain the terms 'history' or 'recent'. Future work might consider whether such extra-topical terms are more popular in spoken versus textual queries.

Finally, we expected that dropping speech recognition errors would yield better-performing sub-queries. Indeed, we found evidence of this in our results. For example, we found cases where the spoken term 'lyme' was misrecognized as 'line' and omitted from better-performing sub-queries. Other example pairs (x,y) where the spoken term x was misrecognized as y and subsequently omitted from a better-performing sub-query include: (apirin, aprin); (beatify, beautify), (cult, colt); (export, expert); (fatal, foetal); (france, francis); (czech, check); (melanoma, melonoma); (nobel, noble); (pisa, pizza); (vegetation, visitation); (role, roll); and (soil, swell).

9 Conclusion

We presented an approach for reducing spoken queries. Our approach is an extension of the algorithm proposed by Kumaran and Carvalho [11], which was applied to the task of reducing TREC topic descriptions. We were able to closely approximate the level of performance reported in Kumaran and Carvalho [11] and tested the generalizability of our approach on the new task of reducing spoken queries.

Our results suggest three major trends. First, our approach yielded small, but significant improvements over the baseline of running the original transcription as the query. Second, combining the rankings from the top-k sub-queries in a weighted fashion yielded the best performance—it performed better than simply selecting the single sub-query with the greatest predicted performance and better than combining all candidate sub-queries in an *unweighted* fashion. Finally, post-retrieval query performance features were more predictive than pre-retrieval query performance features and drift features.

A post-hoc analysis found that the types of terms that are omitted from a better-performing sub-query include a combination of: (1) terms that are not central to the query topic (e.g., find, information), (2) disfluencies (e.g., um, eh), (3) terms that describe extra-topical dimensions of the information need, and (4) speech recognition errors.

Our findings point to several directions for future work. First, our results suggest several additional features that might be useful for predicting sub-query performance. For instance, non-topical terms such as 'find' and 'information' might tend to appear towards the beginning of a spoken query. Thus, features that characterize the relative positions of the dropped query terms might improve sub-query prediction performance. Also, ASR systems sometimes include term confidence values in the output transcription. Features that characterize the ASR confidence values of the dropped terms might also be useful. Finally, future work should consider whether terms associated with extra-topical dimensions of the information need, such as terms that convey temporal constraints ('historic', 'recent') or perspective constraints ('pros', 'cons'), are more common in spoken versus textual queries.

Acknowledgments. This work was supported in part by NSF grant IIS-1451668. Any opinions, findings, conclusions, and recommendations expressed in this paper are the authors' and do not necessarily reflect those of the sponsors.

References

1. Arguello, J., Avula, S., Diaz, F.: Using query performance predictors to improve spoken queries. In: Ferro, N., Crestani, F., Moens, M.-F., Mothe, J., Silvestri, F., Nunzio, G.M., Hauff, C., Silvello, G. (eds.) ECIR 2016. LNCS, vol. 9626, pp. 309–321. Springer, Heidelberg (2016). doi:10.1007/978-3-319-30671-1_23
2. Aslam, J.A., Pavlu, V.: Query hardness estimation using jensen-shannon divergence among multiple scoring functions. In: Amati, G., Carpineto, C., Romano, G. (eds.) ECIR 2007. LNCS, vol. 4425, pp. 198–209. Springer, Heidelberg (2007). doi:10.1007/978-3-540-71496-5_20
3. Balasubramanian, N., Kumaran, G., Carvalho, V.R.: Exploring reductions for long web queries. In: SIGIR (2010)
4. Cormack, G.V., Clarke, C.L.A., Buettcher, S.: Reciprocal rank fusion outperforms condorcet and individual rank learning methods. In: SIGIR (2009)

5. Crestani, F., Du, H.: Written versus spoken queries: a qualitative and quantitative comparative analysis. JASIST **57**(7), 881–890 (2006)
6. Cronen-Townsend, S., Zhou, Y., Croft, W.B.: Predicting query performance. In: SIGIR (2002)
7. Diaz, F.: Performance prediction using spatial autocorrelation. In: SIGIR (2007)
8. Hauff, C.: Predicting the effectiveness of queries and retrieval systems. Dissertation, Univeristy of Twente (2010)
9. He, B., Ounis, I.: Inferring query performance using pre-retrieval predictors. In: Apostolico, A., Melucci, M. (eds.) SPIRE 2004. LNCS, vol. 3246, pp. 43–54. Springer, Heidelberg (2004). doi:10.1007/978-3-540-30213-1_5
10. Jiang, J., Jeng, W., He, D.: How do users respond to voice input errors? Lexical and phonetic query reformulation in voice search. In: SIGIR (2013)
11. Kumaran, G., Carvalho, V.R.: Reducing long queries using query quality predictors. In: SIGIR (2009)
12. Lavrenko, V., Croft, W.B.: Relevance based language models. In: SIGIR (2001)
13. Li, X., Nguyen, P., Zweig, G., Bohus, D.: Leveraging multiple query logs to improve language models for spoken query recognition. In: ICASSP (2009)
14. Mamou, J., Sethy, A., Ramabhadran, B., Hoory, R., Vozila, P.: Improved spoken query transcription using co-occurrence information. In: INTERSPEECH (2011)
15. Peng, F., Roy, S., Shahshahani, B., Beaufays, F.: Search results based n-best hypothesis rescoring with maximum entropy classification. In: IEEE Workshop on Automatic Speech Recognition and Understanding (2013)
16. Shtok, A., Kurland, O., Carmel, D., Raiber, F., Markovits, G.: Predicting query performance by query-drift estimation. TOIS **30**(2), 11 (2012)
17. Smucker, M.D., Allan, J., Carterette, B.: A comparison of statistical significance tests for information retrieval evaluation. In: CIKM (2007)
18. Xue, X., Croft, W.B.: Modeling subset distributions for verbose queries. In: SIGIR (2011)
19. Xue, X., Huston, S., Croft, W.B.: Improving verbose queries using subset distribution. In: CIKM (2010)
20. Yom-Tov, E., Fine, S., Carmel, D., Darlow, A.: Learning to estimate query difficulty: including applications to missing content detection and distributed information retrieval. In: SIGIR (2005)
21. Zhao, L., Callan, J.: Term necessity prediction. In: CIKM (2010)
22. Zhao, Y., Scholer, F., Tsegay, Y.: Effective pre-retrieval query performance prediction using similarity and variability evidence. In: Macdonald, C., Ounis, I., Plachouras, V., Ruthven, I., White, R.W. (eds.) ECIR 2008. LNCS, vol. 4956, pp. 52–64. Springer, Heidelberg (2008). doi:10.1007/978-3-540-78646-7_8
23. Zhou, Y., Croft, W.B.: Ranking robustness: a novel framework to predict query performance. In: CIKM (2006)
24. Zhou, Y., Croft, W.B.: Query performance prediction in web search environments. In: SIGIR (2007)

Entity Linking in Queries:
Efficiency vs. Effectiveness

Faegheh Hasibi[1(✉)], Krisztian Balog[2], and Svein Erik Bratsberg[1]

[1] Norwegian University of Science and Technology, Trondheim, Norway
{faegheh.hasibi,sveinbra}@idi.ntnu.no
[2] University of Stavanger, Stavanger, Norway
krisztian.balog@uis.no

Abstract. Identifying and disambiguating entity references in queries is one of the core enabling components for semantic search. While there is a large body of work on entity linking in documents, entity linking in queries poses new challenges due to the limited context the query provides coupled with the efficiency requirements of an online setting. Our goal is to gain a deeper understanding of how to approach entity linking in queries, with a special focus on how to strike a balance between effectiveness and efficiency. We divide the task of entity linking in queries to two main steps: candidate entity ranking and disambiguation, and explore both unsupervised and supervised alternatives for each step. Our main finding is that best overall performance (in terms of efficiency and effectiveness) can be achieved by employing supervised learning for the entity ranking step, while tackling disambiguation with a simple unsupervised algorithm. Using the Entity Recognition and Disambiguation Challenge platform, we further demonstrate that our recommended method achieves state-of-the-art performance.

1 Introduction

The aim of semantic search is to deliver more relevant and focused responses, and in general an improved user experience, by understanding the searcher's intent and context behind the query provided. Identifying entity mentions in text and subsequently linking them to the corresponding entries in a reference knowledge base (KB) is known as the task of *entity linking*. It can be performed on long texts (i.e., documents), or very short texts such as web search queries; the latter is referred to as *entity linking in queries* (ELQ). It has been shown that leveraging entity annotations of queries is beneficial for various information retrieval tasks including document retrieval [8,31], entity retrieval [17,28], and task understanding [32].

Entity linking has been extensively studied for long texts [7,14,15,21,24,25]. Despite the large variety of approaches, there are two main components that are present in all entity linking systems: (i) *candidate entity ranking*, i.e., identifying entities that can be possibly linked to a mention, and (ii) *disambiguation*, i.e., selecting the best entity (or none) for each detected mention. There is also a

© Springer International Publishing AG 2017
J.M. Jose et al. (Eds.): ECIR 2017, LNCS 10193, pp. 40–53, 2017.
DOI: 10.1007/978-3-319-56608-5_4

general consensus on the two main categories of features that are needed for effective entity linking: (i) contextual similarity between a candidate entity and the surrounding text of the entity mention, and (ii) interdependence between all entity linking decisions in the text (extracted from the underlying KB). Previous studies [4,14] have investigated these aspects in a unified framework and derived general recommendations for entity linking in documents. Entity linking in queries, however, has only recently started to draw attention [3,6,16] and such systematic evaluation of the different components has not been conducted until now. With this study, we aim to fill that gap.

What is special about entity linking in queries? First, queries are short, noisy text fragments where the ambiguity of a mention may not be resolved because of the limited context. That is, a mention can possibly be linked to more than one entity (see Table 1 for examples). This is unlike entity linking in documents, where it is assumed that there is enough context for disambiguation. Second, ELQ is an online process that happens during query-time, meaning that it should be performed under serious time constraints (in contrast with traditional entity linking which is offline). The ideal solution is not necessarily the most effective one, but the one that represents the best trade-off between effectiveness and efficiency. Therefore, the same techniques that have been used for entity linking in documents may not be suitable for queries. We formulate the following two research questions:

- **RQ1.** Given the response time requirements of an online setting, what is the relative importance of candidate entity ranking vs. disambiguation? In other words, if we are to allocate the available processing time between the two, which one would yield the highest gain?
- **RQ2.** Given the limited context provided by queries, which group of features is needed the most for effective entity disambiguation: contextual similarity, interdependence between entities, or both?

To answer the above research questions, we set up a framework where different candidate entity ranking and disambiguation methods can be plugged in. For each of these components, we experiment with both unsupervised and supervised alternatives, resulting in a total of four different ELQ systems. Our candidate entity ranking and disambiguation methods draw on, and extend further, ideas from the existing literature. Supervised methods are expected to yield high effectiveness coupled with lower efficiency, while for unsupervised approaches it is the other way around. Our results reveal that it is more beneficial to use supervised learning for the candidate entity ranking step. If this step provides high-quality results, then disambiguation can be successfully tackled with a simple and elegant greedy algorithm. Moreover, our analysis shows that entity interdependencies provide little help for disambiguation. This is an interesting finding as it stands in contrast to the established postulation for entity linking in documents. Consequently, we identify a clearly preferred approach that uses supervised learning for candidate entity ranking and an unsupervised algorithm for disambiguation. Using the evaluation platform of the Entity Recognition and

Disambiguation (ERD) challenge [3], we show that our preferred approach performs on a par with the current state of the art.

The main contribution of this paper is to present the first systematic investigation of the ELQ task by bringing together the latest entity linking techniques and practices in a unified framework. In addition, we develop a novel supervised approach for entity disambiguation in ELQ, which encompasses various textual and KB-based relatedness features. Finally, we make a best practice recommendation for ELQ and demonstrate that our recommended approach achieves state-of-the-art performance. The resources developed with this paper are made available at http://bit.ly/ecir2017-elq.

Table 1. Example queries with their linked entities. Each set represents an interpretation of the query; ambiguous queries have multiple interpretations (i.e., multiple table rows).

Query	Entity linking interpretation(s)
Nashville thrift stores	{NASHVILLE TENNESSEE, CHARITY SHOP}
Obama's wife	{BARACK OBAMA}
Cambridge population	{CAMBRIDGE}
	{CAMBRIDGE (MASSACHUSETTS)}
New york pizza manhattan	{NEW YORK-STYLE PIZZA, MANHATTAN}
	{NEW YORK, MANHATTAN}

2   Related Work

Early work on entity linking relied on the contextual similarity between the document and the candidate referent entities [7,24]. Milne and Witten [25] introduced the concepts of commonness and relatedness, which are generally regarded as two of the most important features for entity linking. In contrast to early systems that disambiguate one mention at a time, collective entity linking systems exploit the relatedness between entities jointly and disambiguate all entity mentions in the text simultaneously [15,19,21,29]. Since entity linking is a complex process, several attempts have been made to break it down into standard components and compare systems in a single framework [4,14,30]. Particularly, Hachey et al. [14] reimplemented three prominent entity linking systems in a single framework and found that much of the performance variation between these systems stems from the candidate entity ranking step (called *searcher* in their framework). We follow the final recommendation of their study and divide the entity linking task into two main steps, candidate entity ranking and disambiguation, to perform a systematic investigation of entity linking in queries.

Recognizing and disambiguating entities in short texts, such as tweets and search snippets, has only recently gained attention [11,13,23]. Entity linking in queries (ELQ) is particularly challenging because of the inherent ambiguity (see Table 1). Deepak et al. [9] addressed ELQ by assigning a single entity to a mention. The Entity Recognition and Disambiguation (ERD) [3] challenge framed ELQ as the task of finding multiple interpretations of the query, and this was followed in subsequent studies [6,16,18]. Hasibi et al. [16] proposed generative models for ranking and disambiguating entities. The SMAPH system [6], on the other hand, "piggybacks" on a web search engine to rank entities, and then disambiguates them using a supervised collective approach. We consider the key features of these previous studies in a single system in order to perform a comprehensive comparison of the two main ELQ components (candidate entity ranking and disambiguation) with respect to both efficiency and effectiveness. We, however, do not include the piggybacking technique as its reliance on an external search service would seriously hinder the efficiency of the entity linking process in our setup.

3   Entity Linking in Queries

The task of entity linking in queries (ELQ) is to identify, given an input query q, a set of *entity linking interpretations* $I = \{E_1, \ldots, E_m\}$, where each interpretation $E_i = \{(m_1, e_1), \ldots, (m_k, e_k)\}$ consists of a set of mention-entity pairs. Mentions within E_i are non-overlapping and each mention m_j is linked to an entity e_j in a reference knowledge base. By way of illustration, the output of ELQ for the query "new york pizza manhattan" would be $I = \{E_1, E_2\}$, where $E_1 = \{$(new york pizza, NEW YORK-STYLE PIZZA), (manhattan, MANHATTAN)$\}$ and $E_2 = \{$(new york, NEW YORK), (manhattan, MANHATTAN)$\}$. Following [3,16], we restrict ourselves to detecting proper noun entities and do not link general concepts (e.g., "PIZZA").

We frame the ELQ problem as a sequence of the following two subtasks: *candidate entity ranking* (CER) and *disambiguation*. The first subtask takes the query q and outputs a ranked list of mention-entity pairs along with the corresponding scores. The second subtask takes this list as input and forms the set of entity linking interpretations I. For each subtask, we present two alternatives: unsupervised and supervised. The resulting four possible combinations are compared experimentally in Sect. 5.1.

3.1   Candidate Entity Ranking

This subtask is responsible for (i) identifying all possible entities that can be linked in the query and (ii) ranking them based on how likely they are link targets (in any interpretation of the query). The objective is to achieve both high recall and high precision at early ranks, as the top-ranked entity-mention pairs obtained here will be used directly in the subsequent disambiguation step. Using lexical matching of query n-grams against a rich dictionary of entity name variants allows for the identification of candidate entities with close to perfect recall [16]. We follow this approach to obtain a list of candidate entities together with their corresponding mentions in the query. Our focus of attention below is on ranking these candidate (m, e) pairs with respect to the query, i.e., estimating $score(m, e, q)$.

Unsupervised. For the unsupervised ranking approach, we take a state-of-the-art generative model, specifically, the MLMcg model proposed by Hasibi et al. [16]. This model considers both the likelihood of the given mention and the similarity between the query and the entity: $score(m, e, q) = P(e|m)P(q|e)$, where $P(e|m)$ is the probability of a mention being linked to an entity (a.k.a. *commonness* [22]), computed from the FACC collection [12]. The query likelihood $P(q|e)$ is estimated using the query length normalized language model similarity [20]:

$$P(q|e) = \frac{\prod_{t \in q} P(t|\theta_e)^{P(t|q)}}{\prod_{t \in q} P(t|C)^{P(t|q)}}, \tag{1}$$

where $P(t|q)$ is the term's relative frequency in the query (i.e., $n(t, q)/|q|$). The entity and collection language models, $P(t|\theta_e)$ and $P(t|C)$, are computed using the Mixture of Language Models (MLM) approach [27].

Supervised. Our supervised approach employs learning-to-rank (LTR), where each (query, mention, entity) triple is described using a set of features. The ranking function is trained on a set of mention-entity pairs with binary labels, with positive labels denoting the correctly annotated entities for the given query. We use a total of 28 features from the literature [6,23], which are summarized in Table 2.

3.2   Disambiguation

The disambiguation step is concerned with the formation of entity linking interpretations $\{E_1, \ldots, E_m\}$. Similar to the previous step, we examine both unsupervised and supervised alternatives, by adapting existing methods from the literature. We further extend the supervised approach with novel elements.

Unsupervised. We employ the greedy algorithm introduced in [16], which forms interpretations in three consecutive steps: (i) pruning, (ii) containment mention filtering, and (iii) set generation. In the first step, the algorithm takes the ranked list of mention-entity pairs and discards the ones with ranking score below the threshold τ_s. This threshold is a free parameter that controls the balance between precision and recall. The second step removes containment mentions (e.g., "kansas city mo" vs. "kansas city") by keeping only the highest scoring one. Finally, interpretations are built iteratively by processing mention-entity pairs in decreasing order of score and adding them to an existing interpretation E_i, where the mention does not overlap with other mentions already in E_i and i is minimal; if no such interpretation exists then a new interpretation $E_{|E|+1}$ is created.

Supervised. The overall idea is to generate all possible interpretations from a ranked list of mention-entity pairs, then employ a binary classifier to collectively select the most pertinent interpretations. Our approach is similar in spirit

Table 2. Feature set used for ranking entities, categorized to mention (M), entity (E), mention-entity (ME), and query (Q) features.

Feature	Description	Type
$Len(m)$	Number of terms in the entity mention	M
$NTEM(m)^{\ddagger}$	Number of entities whose title equals the mention	M
$SMIL(m)^{\ddagger}$	Number of entities whose title equals part of the mention	M
$Matches(m)$	Number of entities whose surface form matches the mention	M
$Redirects(e)$	Number of redirect pages linking to the entity	E
$Links(e)$	Number of entity out-links in DBpedia	E
$Commonness(e,m)$	Likelihood of entity e being the target link of mention m	ME
$MCT(e,m)^{\ddagger}$	True if the mention contains the title of the entity	ME
$TCM(e,m)^{\ddagger}$	True if title of the entity contains the mention	ME
$TEM(e,m)^{\ddagger}$	True if title of the entity equals the mention	ME
$Pos_1(e,m)$	Position of the 1st occurrence of the mention in entity abstract	ME
$SimM_f(e,m)^{\dagger}$	Similarity between mention and field f of entity; Eq. (1)	ME
$LenRatio(m,q)$	Mention to query length ratio: $\frac{|m|}{|q|}$	Q
$QCT(e,q)$	True if the query contains the title of the entity	Q
$TCQ(e,q)$	True if the title of entity contains the query	Q
$TEQ(e,q)$	True if the title of entity is equal query	Q
$Sim(e,q)$	Similarity between query and entity; Eq. (1)	Q
$SimQ_f(e,q)^{\dagger}$	LM similarity between query and field f of entity; Eq. (1)	Q

‡ Entity title refers to the `rdfs:label` predicate of the entity in DBpedia

†Computed for all individual DBpedia fields $f \in \mathcal{F}$ and also for field *content* (cf. Sect. 4.1)

to the top performing contender in the ERD challenge [6], as they also select interpretations using a collective supervised approach. However, we generate the interpretations only from the top-K mention-entity pairs (obtained from the CER step) and generate all possible interpretations out of those. We further require that mentions within the same interpretation do not overlap with each other. The value of K is set empirically, and it largely depends on the effectiveness of the CER step. If CER has high precision then K can be low, while less effective approaches can be compensated for with higher K values.

Once the candidate sets are generated, each is represented by a feature vector. We devise two main families of features: (i) set-based features are computed for the entire interpretation set, and (ii) entity-based features are calculated for

Table 3. Feature set used in the supervised disambiguation approach. Type is either query dependent (QD) or query independent (QI).

Set-based features		Type
$CommonLinks(E)$	Number of common links in DBpedia: $\bigcap_{e \in E} out(e)$.	QI
$TotalLinks(E)$	Number of distinct links in DBpedia: $\bigcup_{e \in E} out(e)$	QI
$J_{KB}(E)$	Jaccard similarity based on DBpedia: $\frac{CommonLinks(E)}{TotalLink(E)}$	QI
$J_{corpora}(E)^{\ddagger}$	Jaccard similarity based on FACC: $\frac{\|\bigcap_{e \in E} doc(e)\|}{\|\bigcup_{e \in E} doc(e)\|}$	QI
$Rel_{MW}(E)^{\ddagger}$	Relatedness similarity [25] according to FACC	QI
$P(E)$	Co-occurrence probability based on FACC: $\frac{\|\bigcap_{e \in E} doc(e)\|}{TotalDocs}$	QI
$H(E)$	Entropy of E: $-P(E)log(P(E))-(1-P(E))log(1-P(E))$	QI
$Completeness(E)^{\dagger}$	Completeness of set E as a graph: $\frac{\|edges(G_E)\|}{\|edges(K_{\|E\|})\|}$	QI
$LenRatioSet(E,q)^{\S}$	Ratio of mentions length to the query length: $\frac{\sum_{e \in E} \|m_e\|}{\|q\|}$	QD
$SetSim(E,q)$	Similarity between query and the entities in the set; Eq. (2)	QD
Entity-based features		
$Links(e)$	Number of entity out-links in DBpedia	QI
$Commonness(e,m)$	Likelihood of entity e being the target link of mention m	QD
$Score(e,q)$	Entity ranking score, obtained from the CER step	QD
$iRank(e,q)$	Inverse of rank, obtained from the CER step: $\frac{1}{rank(e,q)}$	QD
$Sim(e,q)$	Similarity between query and the entity; Eq. (1)	QD
$ContextSim(e,q)$	Contextual similarity between query and entity; Eq. (3)	QD

$\ddagger$ $doc(e)$ represents all documents that have a link to entity e
$\dagger$ G_E is a DBpedia subgraph containing only entities from E; and $K_{\|E\|}$ is a complete graph of $\|E\|$ vertices
$\S$ m_e denotes the mention that corresponds to entity e

individual entities. Features in the first group are computed collectively on all entities of the set and measured as a single value, while the members of the second group need to be aggregated (we use *min, max, avg* as aggregators). It is worth noting that each interpretation typically consists of very few entities. Therefore, considering all entities for computing set-based features is feasible; it also captures more information than one could get from aggregated pair-wise similarity features. Table 3 summarizes our feature set.

We highlight two novel and important features. $SetSim(E,q)$ measures the similarity between all entities in the interpretation E and the query q:

$$SetSim(E,q) = P(q|\theta_E) = \frac{\prod_{t \in q} P(t|\theta_E)^{P(t|q)}}{\prod_{t \in q} P(t|C)^{P(t|q)}}. \tag{2}$$

It is calculated similar to Eq. (1), the main difference being that the probability of each term is estimated based on the interpretation's language model $P(t|\theta_E)$:

$$P(t|\theta_E) = \sum_{e \in E} \sum_{f \in F} \mu_f P(t|\theta_{e_f}).$$

In similar vein, $ContextSim(e, q)$ measures the similarity between the entity and the query context, where query context is the "rest" of the query, i.e., without the mention m_e that corresponds to entity e. Formally:

$$ContextSim(e, q) = P(q - m_e|e), \qquad (3)$$

where $P(q - m_e|e)$ is computed using Eq. (1).

4   Experimental Setup

In this section we describe our data sources, settings of methods, and evaluation metrics.

4.1   Data

Knowledge base. We employ DBpedia 3.9 as our reference knowledge base and build an index of all entities that have both `rdfs:label` and `dbo:comment` predicates. The index includes the following set of fields: $\mathcal{F}$ ={*title, content,* `rdfs:label`, `dbo:wikiPageWikiLink`, `rdfs:comment`, `dbo:abstract`}, where *title* is the concatenation of `rdfs:label`, `foaf:name` and `dbo:wikiPage-Redirects` predicates, and *content* holds the content of all predicates of the entity; the remaining fields correspond to individual predicates.

Surface form dictionary. To recognize candidate entities in queries, we employ a rich surface form dictionary, which maps surface forms to entities. We utilize the FACC entity-annotated web corpora [12] and include surface forms above a commonness threshold of 0.1 [16]. Additionally, we add DBpedia name variants as surface forms; i.e., entity names from `rdfs:label`, `foaf:name`, and `dbo:wikiPageRedirects` predicates [7,11,16]. We confine our dictionary to entities present in the Freebase snapshot of proper named entities, provided by the ERD challenge [3].

Test Collections. We evaluate our methods on two publicly available test collections: Y-ERD [16] and ERD-dev [3]. The former is based on the Yahoo Search Query Log to Entities (YSQLE) dataset[1] and consists of 2, 398 queries. All results on this collection are obtained by performing 5-fold cross validation.[2]

---

[1] http://webscope.sandbox.yahoo.com/.

[2] It is important to note that Y-ERD contains queries that have been reformulated (often only slightly so) during the course of a search session; we ensure that queries from the same session are assigned to the same fold when using cross-validation.

The ERD-dev collection contains 91 queries and is released as part of the ERD challenge [3]. We apply the trained models (on the whole Y-ERD collection) to ERD-dev queries and report on the results. In addition, ERD also provides an online evaluation platform which is based on a set of 500 queries (referred to as ERD-test); the corresponding annotations are not released. We evaluate the effectiveness[3] of our recommended system using ERD-test to evaluate how it performs against the current state of the art.

4.2 Methods

Candidate Entity Ranking. For the unsupervised method (**MLMcg**), we follow [26] and use title and content fields, with weights 0.2 and 0.8, respectively. For the supervised method (**LTR**), we employ the Random Forest (RF) [2] ranking algorithm and set the number of trees to 1000 and the maximum features to 10% of size of the feature set [23]. We further include two baseline methods for reference comparison: (i) **MLM** is similar to MLMcg, but without considering the commonness score; i.e., computed based on the Eq. (1); (ii) **CMNS** ranks entities based on the commonness score, while prioritizing longer mentions, and is shown to be a strong baseline [1,16,23].

Disambiguation. The unsupervised disambiguation method (**Greedy**) involves a score threshold parameter, which is set (using a parameter sweep) depending on the CER method used: 20 for MLMcg and 0.3 in case of LTR. For the supervised disambiguation method (**LTR**), we set the number of top ranked entities K to 5 (based on a parameter sweep) and use a RF classifier with similar setting to supervised CER. For baseline comparison, we consider the top-3 performing systems from the ERD challenge: SMAPH [6], NTUNLP [5], and Seznam [10].

4.3 Evaluation

As both precision and recall matter for the candidate entity ranking step, we evaluate our methods using Mean Average Precision (MAP), recall at rank 5 (R@5), and precision at position 1 (P@1). When evaluating CER, we are only concerned about the ranking of entities; therefore, we consider each entity only once with its highest scoring mention: $score(e, q) = \arg\max_{m \in q} score(m, e, q)$. For the disambiguation step, we measure the end-to-end performance using set-based metrics (precision, recall, and F-measure), according to the strict evaluation metrics in [16]. As for efficiency, we report on the average processing time for each query, measured in seconds. The experiments were conducted on a machine with an Intel Xeon E5 2.3GHz 12-core processor, running Ubuntu Linux v14.04. Statistical significance is tested using a two-tailed paired t-test. We mark improvements with $^\triangle(p < 0.05)$ or $^\blacktriangle(p < 0.01)$, detoriations with $^\triangledown(p < 0.05)$ or $^\blacktriangledown(p < 0.01)$, and no significance by $^\circ$.

---

[3] Carmel et al. [3] do not report on the efficiency of the approaches and the online leaderboard is no longer available, hence we present only effectiveness results from Cornolti et al. [6].

5   Results and Analysis

In this section we report on our experimental results and answer our research questions.

5.1   Results

We start by evaluating the *candidate entity ranking* and *disambiguation* steps and then answer our first research question: "Given the response time requirements of an online setting, what is the relative importance of candidate entity ranking vs. disambiguation?"

Table 4. Candidate entity ranking results on the Y-ERD and ERD-dev datasets. Best scores for each metric are in boldface. Significance for line $i > 1$ is tested against lines $1..i - 1$.

Method	Y-ERD			ERD-dev		
	MAP	R@5	P@1	MAP	R@5	P@1
MLM	0.7507	0.8556	0.6839	0.7675	0.8622	0.7333
CMNS	0.7831▲	0.8230▲	0.7779▲	0.7037°	0.7222▽	0.7556°
MLMcg	0.8536▲▲	0.8997▲▲	0.8280▲▲	0.8543△▲	0.9015°▲	**0.8444**°°
LTR	**0.8667**▲▲▲	**0.9022**▲▲°	**0.8479**▲▲▲	**0.8606**△▲°	**0.9289**△▲°	0.8222°°°

Candidate Entity Ranking. Table 4 presents the results for CER on the Y-ERD and ERD-dev datasets. We find that commonness is a strong performer (this is in line with the findings of [1,16]). Combining commonness with MLM in a generative model (MLMcg) delivers excellent performance, with MAP above 0.85 and R@5 around 0.9. The LTR approach can bring in further slight, but for Y-ERD significant, improvements. This means that both of our CER methods (MLMcg and LTR) are able to find the vast majority of the relevant entities and return them at the top ranks.

Disambiguation. Table 5 reports on the disambiguation results. We use the naming convention X-Y, where X refers to the CER method (MLMcg or LTR) and Y refers to the disambiguation method (Greedy or LTR) that is applied on top. Our observations are as follows. The MLM-Greedy approach is clearly the most efficient but also the least effective one. Learning is more expensive for disambiguation than for CER, see LTR-Greedy vs. MLMcg-LTR; yet, it is also clear from this comparison that more performance can be gained when learning is done for CER than when it is done for disambiguation. The most effective method is LTR-Greedy, outperforming other approaches significantly on both test sets. It is also the second most efficient one. Interestingly, even though the MLMcg and LTR entity ranking methods perform equally well according to CER evaluation (cf. Table 4), we observe a large difference in their performance when the Greedy disambiguation approach is applied on top of them. The reason is that the absolute scores produced by LTR are more meaningful than those of MLMcg (despite the query length normalization efforts for the latter; cf. Eq. (1)). This plays a direct role in Greedy disambiguation, where score thresholding is used. We note that the reported efficiency results are meant for comparison across different approaches. For practical applications, further optimizations to our basic implementation would be needed (cf. [1]).

Table 5. End-to-end performance of ELQ systems on the Y-ERD and ERD-dev query sets. Significance for line $i > 1$ is tested against lines $1..i-1$.

Method	Y-ERD				ERD-dev			
	Prec	Recall	F1	Time	Prec	Recall	F1	Time
MLMcg-Greedy	0.709	0.709	0.709	**0.058**	0.724	0.712	0.713	**0.085**
MLMcg-LTR	0.725°	0.724°	0.724°	0.893	0.725°	0.731°	0.728°	1.185
LTR-LTR	0.731△°	0.732△°	0.731△°	0.881	0.758°°	0.748°°	0.753°°	1.185
LTR-Greedy	**0.786▲▲▲**	**0.787▲▲▲**	**0.787▲▲▲**	0.382	**0.852▲▲△**	**0.828▲△°**	**0.840▲▲△**	0.423

Based on the results, LTR-Greedy is our overall recommendation. We compare this method against the top performers of the ERD challenge (using the official challenge platform); see Table 6. For this comparison, we additionally applied spell checking, as this has also been handled in the top performing system (SMAPH-2) [6]. The results show that our LTR-Greedy approach performs on a par with the state-of-the-art systems. This is remarkable taking into account the simplicity of the Greedy disambiguation algorithm vs. the considerably more complex solutions employed by others.

Table 6. ELQ results on the official ERD test platform.

Method	F1
LTR-Greedy	0.699
SMAPH-2 [6]	0.708
NTUNLP [5]	0.680
Seznam [10]	0.669

Answer to RQ1. Our results reveal that candidate entity ranking is of higher importance than disambiguation for ELQ. Hence, it is more beneficial to perform the (expensive) supervised learning early on in the pipeline for the seemingly easier CER step; disambiguation can then be tackled successfully with an unsupervised (greedy) algorithm. (Note that selecting the top ranked entity does not yield an immediate solution; as shown in [16], disambiguation is an indispensable step in ELQ.)

5.2   Feature Analysis

We now analyze the features used in our supervised methods and answer our second research question: "Given the limited context provided by queries, which group of features is needed the most for effective entity disambiguation?" For the sake of completeness, we also report feature importance for the CER step, even though that does not directly relate to the above RQ. Figure 1(a) shows the top features used in the LTR entity ranking approach in terms of Gini score. We observe that *Matches, Commonness*, and the various query similarity features play the main role in the entity ranking function. As for the supervised disambiguation step, which is our main focus here, we selected the top 15 features independently for the MLMcg-LTR and LTR-LTR methods; interestingly, we ended up with the exact same set of features. Figure 1(b) demonstrates that nearly all influential features are query dependent; the only query independent features are P and H, capturing the co-occurrence of entities in web corpora.

Answer to RQ2. We conclude that contextual similarity features are the most effective for entity disambiguation. This is based on two observations: (i) the unsupervised (Greedy) method takes only the entity ranking scores as input, which are computed based on the contextual similarity between entity and query; (ii) the supervised (LTR) method relies the most on query-dependent features. This is an interesting finding, as it stands in contrast to the common postulation in entity linking in documents that interdependence between entities help to better disambiguate entities. Entity interdependence features (and, in general, collective disambiguation methods) are more helpful when sufficiently many entities are mentioned in the text; this is not the case for queries.

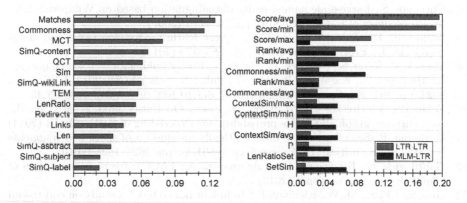

Fig. 1. Most important features used in the supervised approaches, sorted by Gini score: (Left) candidate entity ranking, (Right) disambiguation.

6   Conclusion

In this paper, we have performed the first systematic investigation of entity linking in queries (ELQ). We have developed a framework where different methods can be plugged in for two core components: *candidate entity ranking* and *disambiguation*. For each of these components, we have explored both unsupervised and supervised alternatives by employing and further extending state-of-the-art approaches. Our experiments have led to two important findings: (i) it is more rewarding to employ supervised learning for candidate entity ranking than for disambiguation, and (ii) entity interdependence features, which are the essence of collective disambiguation methods, have little benefit for ELQ. Overall, our findings have not only revealed important insights, but also provide guidance as to where future research and development in ELQ should be focused.

References

1. Blanco, R., Ottaviano, G., Meij, E.: Fast and space-efficient entity linking in queries. In: Proceedings of WSDM, pp. 179–188 (2015)
2. Breiman, L.: Random forests. Mach. Learn. **45**, 5–32 (2001)
3. Carmel, D., Chang, M.-W., Gabrilovich, E., Hsu, B.-J.P., Wang, K.: ERD 2014: entity recognition and disambiguation challenge. In: ACM SIGIR Forum, vol. 48, pp. 63–77 (2014)
4. Ceccarelli, D., Lucchese, C., Orlando, S., Perego, R., Trani, S.: Learning relatedness measures for entity linking. In: Proceedings of CIKM, pp. 139–148 (2013)
5. Chiu, Y.-P., Shih, Y.-S., Lee, Y.-Y., Shao, C.-C., Cai, M.-L., Wei, S.-L., Chen, H.-H.: NTUNLP approaches to recognizing and disambiguating entities in long and short text at the ERD challenge 2014. In: Proceedings of ERD@SIGIR (2014)
6. Cornolti, M., Ferragina, P., Ciaramita, M., Rüd, S., Schütze, H.: A piggyback system for joint entity mention detection and linking in web queries. In: Proceedings of WWW, pp. 567–578 (2016)
7. Cucerzan, S.: Large-scale named entity disambiguation based on Wikipedia data. In: Proceedings of EMNLP-CoNLL, pp. 708–716 (2007)
8. Dalton, J., Dietz, L., Allan, J.: Entity query feature expansion using knowledge base links. In: Proceedings of SIGIR, pp. 365–374 (2014)
9. Deepak, P., Ranu, S., Banerjee, P., Mehta, S.: Entity linking for web search queries. In: Hanbury, A., Kazai, G., Rauber, A., Fuhr, N. (eds.) ECIR 2015. LNCS, vol. 9022, pp. 394–399. Springer, Cham (2015). doi:10.1007/978-3-319-16354-3_43
10. Eckhardt, A., Hreško, J., Procházka, J., Smrs, O.: Entity linking based on the co-occurrence graph and entity probability. In: Proceeding of ERD@SIGIR (2014)
11. Ferragina, P., Scaiella, U.: TAGME: on-the-fly annotation of short text fragments (by Wikipedia entities). In: Proceedings of CIKM, pp. 1625–1628 (2010)
12. Gabrilovich, E., Ringgaard, M., Subramanya, A.: FACC1: freebase annotation of ClueWeb corpora, Version 1 (2013)
13. Guo, S., Chang, M.-W., Kiciman, E.: To link or not to link? A study on end-to-end tweet entity linking. In: HLT-NAACL, pp. 1020–1030 (2013)
14. Hachey, B., Radford, W., Nothman, J., Honnibal, M., Curran, J.R.: Evaluating entity linking with Wikipedia. Artif. Intell. **194**, 130–150 (2013)
15. Han, X., Sun, L., Zhao, J.: Collective entity linking in web text: a graph-based method. In: Proceedings of SIGIR, pp. 765–774 (2011)
16. Hasibi, F., Balog, K., Bratsberg, S.E.: Entity linking in queries: tasks and evaluation. In: Proceedings of ICTIR, pp. 171–180 (2015)
17. Hasibi, F., Balog, K., Bratsberg, S.E.: Exploiting entity linking in queries for entity retrieval. In: Proceedings of ICTIR, pp. 209–218 (2016)
18. Hasibi, F., Balog, K., Bratsberg, S.E.: On the reproducibility of the TAGME entity linking system. In: Ferro, N., Crestani, F., Moens, M.-F., Mothe, J., Silvestri, F., Nunzio, G.M., Hauff, C., Silvello, G. (eds.) ECIR 2016. LNCS, vol. 9626, pp. 436–449. Springer, Cham (2016). doi:10.1007/978-3-319-30671-1_32

19. Hoffart, J., Yosef, M.A., Bordino, I., Fürstenau, H., Pinkal, M., Spaniol, M., Taneva, B., Thater, S., Weikum, G.: Robust disambiguation of named entities in text. In: Proceedings of EMNLP, pp. 782–792 (2011)
20. Kraaij, W., Spitters, M.: Language models for topic tracking. In: Language Modeling for Information Retrieval, pp. 95–123 (2003)
21. Kulkarni, S., Singh, A., Ramakrishnan, G., Chakrabarti, S.: Collective annotation of Wikipedia entities in web text. In: Proceedings of SIGKDD, pp. 457–466 (2009)
22. Medelyan, O., Witten, I.H., Milne, D.: Topic indexing with Wikipedia. In: Proceedings of the Wikipedia and AI Workshop at the AAAI 2008 Conference (2008)
23. Meij, E., Weerkamp, W., de Rijke, M.: Adding semantics to microblog posts. In: Proceedings of WSDM, pp. 563–572 (2012)
24. Mihalcea, R., Csomai, A.: Wikify!: linking documents to encyclopedic knowledge. In: Proceedings of CIKM, pp. 233–242 (2007)
25. Milne, D., Witten, I.H.: Learning to link with Wikipedia. In: Proceedings of CIKM, pp. 509–518 (2008)
26. Neumayer, R., Balog, K., Nørvåg, K.: When simple is (more than) good enough: effective semantic search with (almost) no semantics. In: Baeza-Yates, R., Vries, A.P., Zaragoza, H., Cambazoglu, B.B., Murdock, V., Lempel, R., Silvestri, F. (eds.) ECIR 2012. LNCS, vol. 7224, pp. 540–543. Springer, Heidelberg (2012). doi:10.1007/978-3-642-28997-2_59
27. Ogilvie, P., Callan, J.: Combining document representations for known-item search. In: Proceedings of SIGIR, pp. 143–150 (2003)
28. Schuhmacher, M., Dietz, L., Paolo Ponzetto, S.: Ranking entities for Web queries through text and knowledge. In: Proceedings of CIKM, pp. 1461–1470 (2015)
29. Sen, P.: Collective context-aware topic models for entity disambiguation. In: Proceedings of WWW, pp. 729–738 (2012)
30. Usbeck, R. et al.: GERBIL: general entity annotator benchmarking framework. In: Proceedings of WWW, pp. 1133–1143 (2015)
31. Xiong, C., Callan, J.: EsdRank: Connecting query and documents through external semi-structured data. In: Proceedings of CIKM, pp. 951–960 (2015)
32. Yilmaz, E., Verma, M., Mehrotra, R., Kanoulas, E., Carterette, B., Craswell, N.: Overview of the TREC 2015 tasks track. In: Proceedings of TREC (2015)

Cross-Lingual Sentiment Relation Capturing for Cross-Lingual Sentiment Analysis

Qiang Chen[1], Wenjie Li[2], Yu Lei[2], Xule Liu[1], Chuwei Luo[1], and Yanxiang He[1(✉)]

[1] School of Computer Science, Wuhan University, Wuhan 430072, China
{qchen,xuleliu,luochuwei,yxhe}@whu.edu.cn
[2] Department of Computing, The Hong Kong Polytechnic University, Kowloon, Hong Kong, China
{cswjli,csylei}@comp.polyu.edu.hk

Abstract. Sentiment connection is the basis of cross-lingual sentiment analysis (**CSLA**) solutions. Most of existing work mainly focus on general semantic connection that the misleading information caused by non-sentimental semantics probably lead to relatively low efficiency. In this paper, we propose to capture the document-level sentiment connection across languages (called cross-lingual sentiment relation) for CLSA in a joint two-view convolutional neural networks (CNNs), namely Bi-View CNN (**BiVCNN**). Inspired by relation embedding learning, we first project the extracted parallel sentiments into a bilingual sentiment relation space, then capture the relation by subtracting them with an error-tolerance. The bilingual sentiment relation considered in this paper is the shared sentiment polarity between two parallel texts. Experiments conducted on public datasets demonstrate the effectiveness and efficiency of the proposed approach.

Keywords: Cross-lingual sentiment relation · Bi-View CNN · Cross-lingual sentiment analysis

1 Introduction

Aiming at addressing the imbalance issue of sentimental resources, Cross-lingual Sentiment Analysis (CLSA) has become an attractive focus in recent years. In the literature, existing work used the sentimental resources of one language, called source language, to help analyze the sentiment of the other language, called target language [10]. Like most of existing CLSA research did [4,6,7,12, 20,26], we use English as the source language and Chinese as the target language as well.

The general idea of CLSA, commonly known as one type of transfer learning, is to first build the language connections, then leverage the sentimental resources in the source language to help analyze the sentiment of target language [4,6,15,20]. One issue the existing CLSA work (as aforementioned) faced is, language connections are always built based upon superficial features

© Springer International Publishing AG 2017
J.M. Jose et al. (Eds.): ECIR 2017, LNCS 10193, pp. 54–67, 2017.
DOI: 10.1007/978-3-319-56608-5_5

(i.e., n-grams), and hence are unable to accurately express the global sentiment relation in document-level. Moreover, since superficial feature is the simple mixture of different semantics, misleading information is probably brought by the non-sentiment semantics into the learning of sentiment analysis.

Very recently, a number of latent representation learning approaches [22, 24, 25] have been proposed for CLSA problem. Most of them bridge the language gap by means of learning latent representations in the cross-lingual semantic space. Despite the success they have achieved, their focuses are mainly on the general semantic connection, which makes them not sentiment-driven. Some other work suggested to learn the sentiment embedding by incorporating the sentiment polarity information of training data into the pre-trained semantic embedding. The disadvantage of such work is that they cannot capture the complex cross-lingual sentiment relations in document-level due to the simple structures of their models [26].

In this work, we model document-level sentiment connections across languages as certain cross-lingual sentiment relations, inspired by the recent work of relation embedding learning on translation hypothesis [2,9,13,21]. More specifically, we join the sentiments from the two different languages by a novel **bi-view convolutional neural network (BiVCNN)** in order to capture the cross-lingual sentiment relation. Furthermore, since Chen et al. [4] find MT errors are inevitable to distort the pinpoint sentiment, in the proposed BiVCNN, we consider the sentiment polarity shared by a pair of parallel texts as a distortion-tolerant bilingual sentiment relation so that it can more or less offset sentiment distortions caused by MT errors.

The proposed BiVCNN has two main procedures: the bi-view sentiment extraction (BiSE) and the sentiment relation capturing (SRC). In BiSE, we independently abstract the general semantics of documents in the two languages with a convolution-pooling layer, then extract the sentiments from their general semantics with the orthogonal transformation method in [16]. In SRC, we first embed the two sentiments into the same bilingual sentiment relation space [9], then capture the sentiment relations between cross-lingual sentiments by using a translation-based relation embedding method [2,9,21]. Moreover, to offset the negative effects caused by MT errors, we introduce a distortion-tolerance for the sentiment relation capturing. Finally, we classify the sentimental relation according to the label of the pseudo-parallel documents.

The advantages of our proposed method are two-fold: extracting the sentiments in both languages and capturing their sentiment relation with a distortion-tolerance. These two advantages make our method outperform most of the state-of-the-art methods, as shown in the experiments we conducted on public datasets.

2   Related Work

The related work in this paper can be categorized to the cross-lingual sentiment analysis research and the relation embedding learning research. The cross-lingual sentiment analysis research are the direct ones which share the same tasks with our work, while the relation embedding learning research inspired our model.

2.1 Cross-Lingual Sentiment Analysis

There are two alternative solutions to cross-lingual sentiment analysis. One is traditional transfer learning. The other is latent cross-lingual representation learning.

Most research explore traditional transfer learning (TTL) and focus on knowledge adaptation. The high-quality language connection, actually the sentiment connection, is the focus of transfer learning based CLSA approaches. For example, Wan [20] applied a supervised co-training framework to iteratively adapt knowledge learned from two languages by transferring translated texts to each other. Other similar research include [4,6,7]. All these approaches rely on MT to build the sentiment connection. In addition, the unlabeled parallel data is also employed to fill the gap between two languages, such as [12,14]. The work described above all build language connections between superficial semantic features. Thereby, the non-sentimental semantics can easily brings misleading information during the learning.

Meanwhile, latent cross-lingual representation learning (CLRL) also gives some answers to CLSA. Most of CLRL approaches propose to bridge the language gap by learning cross-lingually semantic space for CLSA [19,22,24,25]. However, their language connections actually belong to general semantic relation but not sentiment relation. Based on [24], even though Zhou et al. [26] proposed to learn bilingual sentiment word embeddings by incorporating sentiment information of labeled data, their focus is not the cross-lingual sentiment relation capturing as well. The focus of this paper is to capture the document-level sentiment relation across languages in CLSA.

2.2 Relation Embedding Learning

Relation embedding learning focuses on link prediction/completion in knowledge bases. There are two assumptions on semantic relation. One is space projection, and the other is semantic translation.

Projection Based Method. The entities in the knowledge base are represented as points in the semantic space. Two entities between which a specific semantic relation exists can be projected to the same point in the semantic space with corresponding relation projection matrix/matices. Bordes et al. (2012, 2014) proposed to synchronously project two entities into the same semantic space with two projection matrices respectively [1,3]. Socher et al. [17] regarded semantic relation as a 3-way tensor. Besides, some work considered second-order correlations between entity embeddings as quadratic forms, and define bilinear score functions to catch the correlations [8,18].

Translation Based Method. Based on the directionality and arithmetic property [13], semantic relation between entities can be interpreted as translation. It means a specific relation between entities is modeled as a translation vector that points from the head entity to the tail entity. Bordes et al. [2] define relation as a translation vector which is subtraction from the tail entity to the head entity, called TransE. Wang et al. [21] proposed to project entities into specific relation hyperplanes and allowed entities to have distinct representations in different relation hyperplanes, called TranH. Lin et al. [9] improved TransH by replacing the relation hyperplane with a relation space, called TransR. In this paper, we introduce the translation based relation embedding learning to capture the cross-lingual sentiment relation for the CLSA problem.

3   Bi-View Convolutional Neural Network for Sentiment Relation Capturing

In this paper, we propose a novel bi-view convolutional neural network, namely **BiVCNN**, to capture the sentiment relations between parallel texts in CLSA. Our target is first to extract sentiments from general semantics for both of the two languages, and then to capture and identify the sentiment relations across languages. Besides, to tolerate sentiment distortions caused by MT errors, a distortion tolerance is introduced into BiVCNN when capturing the sentiment relations. In this work, the source language is English (EN), and the target language is Chinese (CN).

3.1   Problem Formation

In this work, the training data include the labeled English reviews $L_{EN} = \{(x_i^{len}, y_i)\}_{i=1}^M$, where x_i^{len} and $y_i \in \{-1, 1\}$ (negative: -1; positive: 1) represent review i and its sentiment polarity. The test data are Chinese reviews $T_{CN} = \{x_s^{tcn}\}_{s=1}^S$. To build the initial language connection, online MT service is employed to obtain the Chinese translations $L_{TrCN} = \{(x_i^{lcnTr}, y_i)\}_{i=1}^M$ of L_{EN} and also the English translations $T_{TrEN} = \{x_s^{tenTr}\}_{s=1}^S$ of T_{CN}, hence we have two pseudo-parallel datasets $[L_{EN}, L_{TrCN}]$ and $[T_{CN}, T_{TrEN}]$ for the training and testing respectively. Besides, the scales of the English lexicon and the Chinese lexicon are d^{en} and d^{cn}. The input of BiVCNN are the pseudo-parallel reviews represented as sequences of pre-trained word embeddings. Thereby, any English review or Chinese review is represented as $x_i^{en} \in \mathbb{R}^{n \times nh}$ or $x_j^{cn} \in \mathbb{R}^{m \times nh}$, where nh is the dimension of the word embedding.

Our goal is to learn the cross-lingual sentiment relations from $[L_{EN}, L_{TrCN}]$, and then to identify the sentiment polarities of $[T_{CN}, T_{TrEN}]$ according to these relations. Therefore, the CLSA problem is transformed to a cross-lingual sentiment relation capturing problem, which is driven by the initial sentiment connection between pseudo-parallel data.

3.2   The Bi-View Convolutional Neural Network

The framework of BiVCNN consists of two components, the bi-view sentiment extraction (BiSE) and the cross-lingual sentiment relation capturing (SRC).

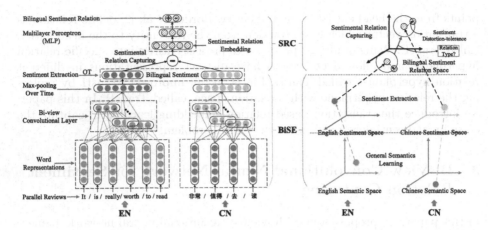

Fig. 1. The framework of Bi-View Convolutional Neural Network.

In BiSE, the monolingual sentiments of pseudo-parallel reviews are individually extracted with language-specific convolution-pooling layers and orthogonal transformations (OT) [16]. In SRC, the cross-lingual sentiment relations between parallel sentiments are captured by translating monolingual sentiments. The cross-lingual sentiment relations are specified as the sentiment polarities of the pseudo-parallel documents. The framework of BiVCNN is illustrated in Fig. 1.

3.2.1   Bi-View Sentiment Extraction
In this component, we first individually abstract the general semantics $[g_i^{l_{en}}, g_i^{l_{cn}Tr}]$ for the pseudo-parallel reviews $[x_i^{l_{en}}, x_i^{l_{cn}Tr}]$ by two language-specific convolution-pooling layers, and then extract the sentiments $[s_i^{l_{en}}, s_i^{l_{cn}Tr}]$ from $[g_i^{l_{en}}, g_i^{l_{cn}Tr}]$ by orthogonal transformations proposed in [16].

Bi-View General Semantics Extraction. For the convolution-pooling layer in the language p, the process is illustrated as the following.

Suppose that the length of the review x_i^p is n, then x_i^p can be represented as:

$$x_i^p = e_{1:n}^p = e_1^p \oplus e_2^p \oplus \cdots \oplus e_n^p \tag{1}$$

where $\oplus$ represents the concatenation operator. $e_{j+1:j+k}^p$ refers to the concatenation of words $e_{j+1}^p, e_{j+2}^p, \cdots, e_{j+k}^p$.

Convolution. The convolution layer in the language p includes h filters all of which have the same filter shape $k \times nh$, where k is the filter window. Overall, the filters are denoted as $\mathbf{W}^p = \mathbf{w}_{1:h}^p \in \mathbb{R}^{h \times k \times nh}$. For the ith filter $\mathbf{w}_i^p \in \mathbb{R}^{k \times d}$, $\mathbf{w}_i^p$ slides over the document x_i^p to generate a feature map $\mathbf{c}_i^p$ of x_i^p. Specifically, the feature $c_{i,j}^p$ in $\mathbf{c}_i^p$ is generated from k continuous words $e_{j+1:j+k}^p$, and $\mathbf{c}_i^p \in \mathbb{R}^{n-k+1}$ over $x_i^p = e_{1:k}^p, \cdots, e_{n-k+1:n}^p$ is represented by Formula (2). If the length of x_i^p

is $n < k$, x_i^p is padded as the new one of the length n with equivalent "#" respectively at the beginning and the end of x_i^p.

$$c_{i,j}^p = tanh(\mathbf{w}^p \cdot e_{j+1:j+k}^p + b^p), \mathbf{c}_i^p = [c_{i,1}^p, \cdots, c_{i,n-k+1}^p] \tag{2}$$

where $b^p \in \mathbb{R}$ is the corresponding bias term of $\mathbf{w}^p$.

Pooling. The pooling operation are performed on the feature map $\mathbf{c}_i^p$. And the maximum value $max\{\mathbf{c}_i^p\}$, the minimum value $min\{\mathbf{c}_i^p\}$ and the mean value $mean\{\mathbf{c}_i^p\}$ are concatenated as the shortcut $\hat{\mathbf{c}}_i^p$ of $\mathbf{c}_i^p$ (Formula (3), where [...] refers to the column-concatenation).

$$\hat{\mathbf{c}}_i^p = [max\{\mathbf{c}_i^p\}, min\{\mathbf{c}_i^p\}, mean\{\mathbf{c}_i^p\}] \tag{3}$$

   Generally, for simplification, we denote the convolution-pooling layer as $\mathcal{C}(x_i^p; \mathbf{w}^p, b^p)$. Hence, the general semantics of the pseudo-parallel reviews $x_i^{len} \in \mathbb{R}^{d^{en} \times n_i^{en}}$ and $x_i^{lcnTr} \in \mathbb{R}^{d^{cn} \times n_i^{cn}}$ are abstracted as:

$$g_i^{len} = \mathcal{C}(x_i^{len}; \mathbf{W}^{en}, \mathbf{b}^{en}), g_i^{lcnTr} = \mathcal{C}(x_i^{lcnTr}; \mathbf{W}^{cn}, \mathbf{b}^{cn}) \tag{4}$$

where $\mathcal{C}(x_i; \mathbf{W}^p, \mathbf{b}^p)$ is the stack of all feature maps $\{\mathcal{C}(x_i; \mathbf{w}_j^p, b_j^p) | \mathbf{w}_j^p \in \mathbf{W}^p, b_j^p \in \mathbf{b}^p\}_{j=1}^h$. Thus, $g_i^{len} \in \mathbb{R}^{3h}$ and $g_i^{lcnTr} \in \mathbb{R}^{3h}$.

Bi-View Sentiment Extraction. Inspired by sentiment extraction in [16], we also use similar orthogonal transformation methods to extract the sentiments from monolingual general semantics of the pseudo-parallel reviews. Specifically, for each language, an orthogonal matrix $Q^p \in \mathbb{R}^{3h \times 3h}$ is used to transform the monolingual semantic space into a bilingual sentiment space where non-sentimental dimensions are gradually learned to be **zeros**. Hence, the bilingual sentiment representations s_i^{len} and s_i^{lcnTr} of the reviews x_i^{len} and x_i^{lcnTr} are:

$$s_i^{len} = Q^{en} \cdot g_i^{len}, \quad s_i^{lcnTr} = Q^{cn} \cdot g_i^{lcnTr} \tag{5}$$

where $s_i^{len}, s_i^{lcnTr} \in \mathbb{R}^{3h}$. In general, the updated $Q^{p(t+1)}$ is not orthogonal without reorthogonalization after parameter updating in iterative learning.

3.2.2   Bilingual Sentiment Relation Capturing

In the literature, it is found that semantic relations can be modeled by the arithmetical operation of latent semantic representations [2,13]. Moreover, more relation embedding models for knowledge base were proposed [9,21].

   Sentiment is one constituent of semantics. Inspired by these work and basis, to capture the bilingual sentiment relation, the monolingual sentiment is first projected into the bilingual sentiment relation space by a relation-transforming matrix $\mathbf{R}^p \in \mathbb{R}^{h \times 3h}$, then the relation is captured by translating the parallel sentiments in the relation space, which is similar to TransR model proposed in [9]. The translating in this paper is defined as a weighted subtraction between vectors. Hence the bilingual sentiment relation r_i between sentiments s_i^{len} and $s_i^{l_{cn}Tr}$ is captured by Formula (6).

$$r_i = \mathbf{R}^{en} s_i^{len} \ominus_\epsilon \mathbf{R}^{cn} s_i^{l_{cn}Tr} \tag{6}$$

where $r_i \in \mathbb{R}^h$. $\ominus_\epsilon$ refers to the weighted subtraction which reduces the differences of certain dimensions between representations of different languages in which the ratio of the values of some dimensions are not ranged in $[1 - \epsilon, \frac{1}{1-\epsilon}]$. It means $\ominus_\epsilon$ in certain degree distinguishes the sentiment-related dimensions from non-sentiment ones, because non-sentiment dimensions of pseudo-parallel semantics in the bilingual semantic space, in the common sense, should be consistent after space projecting.

We then use a fully-connected multi-layer perceptron (MLP) as the classifier to identify the relation type (relation -1 or 1) of r_i according to its bilingual sentiment polarity. The process is illustrated as:

$$h_i^r = tanh(\mathbf{w}_r r_i + \mathbf{b}_r), \quad p(\hat{y}_r | h_i^r) = softmax(\mathbf{w}_h h_i^r + \mathbf{b}_h) \tag{7}$$

where $\mathbf{w}_r \in \mathbb{R}^{h \times h}$, $\mathbf{b}_r \in \mathbb{R}^h$, $\mathbf{w}_h \in \mathbb{R}^{2 \times h}$ and $\mathbf{b}_h \in \mathbb{R}^2$. Hence, $h_i^r \in \mathbb{R}^h$ and $\hat{y}_r \in \mathbb{R}^2$. We use the cross-entropy loss with the parameter regularization term as the objective function defined as:

$$\mathcal{L}_{CrsEnt}(\theta) = - \sum_{i=1}^{2} y_{i,j} \log p_{i,j}(\hat{y}_r | h_i^r) + \frac{\lambda}{2} ||\theta||^2 \tag{8}$$

To allow sentiment distortions caused by MT errors, we design a distortion-tolerance $\varepsilon_i \leqslant \varphi$ based on the hinge loss, defined by Formula (9), to alleviate the negative influence from MT errors by removing the non-sentiment dimensions while capturing the difference between sentiment dimensions.

$$\mathcal{L}_{StDt}(\theta) = max\{\varepsilon_i - \varphi, 0\}, \quad \varepsilon_i = ||mask(r_i) \odot (s_i^{len} - s_i^{l_{cn}Tr})||^2 \tag{9}$$

where $\odot$ refers to Hadamard Product. $mask(\cdot)$ refers to the mask which distinguishes non-zeros from zeros.

3.3 Training and Implementation Details

In general, to synchronously consider the classification loss and the distortion-tolerance loss, the overall objective function is defined by Formula (10). Our target is to minimize the loss to learn the optimal parameter setting for bilingual sentiment relation learning.

$$\mathcal{L}(\theta) = \alpha \mathcal{L}_{CrsEnt}(\theta) + \beta \mathcal{L}_{StDt}(\theta)$$
$$\theta^{(t+1)} := \theta^{(t)} - \eta \frac{\partial \mathcal{L}(\theta)}{\partial \theta}, \quad Q^{p(t+1)} := UV^T \tag{10}$$

where $U \Sigma V^T := Q^{p(t+1)}$.

We apply SGD with a fix learning rate η to iteratively update the parameters until the objective function is convergent. To ensure Q orthogonal, we update $Q^{p(t+1)}$ as $\bar{Q}^{p(t+1)} := UV^T$ based on SVD result $Q^{p(t+1)} = U\Sigma V^T$, where Σ is the singular value matrix, and U and V are unitary matrices. Because the matrix $\bar{Q}^{p(t+1)}$ is the orthogonal matrix and nearly equals to $Q^{p(t+1)}$.

4   Experiments

4.1   Experimental Settings

We evaluate the proposed BiVCNN model on the dataset of an open cross-lingual sentiment analysis task in NLP&CC 2013[1]. The data are Amazon product reviews in three different domains: Books, DVD and Music. For each domain, there are 4,000 labeled English reviews, 4,000 test Chinese reviews, and 500 Chinese test reviews are randomly selected as the development set. In addition, there are also some unlabeled Chinese reviews (17,814 for Books; 47,071 for DVD; 29,677 for Music). The pseudo-data sets are obtained by the online Google translator.

The ICTCLAS toolkit [23] is used to segment Chinese texts. The word embeddings of both Chinese and English are pre-trained with the Word2Vec model[2] on different datasets, respectively. More specifically, the Chinese word embeddings are trained with the unlabeled Chinese reviews in NLP&CC 2013 and 5 million unlabeled Chinese posts from Weibo[3]. For the English word embeddings, the data used are the translated unlabeled English reviews in NLP&CC 2013 and the unlabeled Amazon product reviews used in [11]. The dimensionality of the pre-trained word embeddings is 300.

The number h of the filters in the convolutional layer, the coefficients α, β and λ are finely set to 100, 1.0, 1.0e-2 and 1.0e-6 respectively by the grid search algorithm. The fixed learning rate η is set to 0.1. All matrix parameters $(\mathbf{W}^{en}, \mathbf{W}^{cn}, Q^{en}, Q^{cn}, \mathbf{R}^{en}, \mathbf{R}^{cn}, \mathbf{w}_r, \mathbf{w}_h)$ are initialized with the uniform distribution $U(-\sqrt{6/(r+c)}, \sqrt{6/(r+c)})$, where r and c denotes the number of row and column of matrices, respectively. The biases $(\mathbf{b}^{en}, \mathbf{b}^{cn}, \mathbf{b}_r, \mathbf{b}_h)$ are set to 0. The distortion-tolerance holder φ is set to 1.

---

[1] NLP&CC is an annual conference of Chinese information technology professional committee organized by Chinese Computer Federation (CCF). For more details, please refer to: http://tcci.ccf.org.cn/conference/2013/dldoc/evdata03.zip.

[2] Word2Vec is one of the models implemented in the free python library *Gensim*: http://pypi.python.org/pypi/gensim.

[3] The pre-trained word embedding models and the Weibo posts leveraged are available at: https://drive.google.com/open?id=0B0l0oLL2GUuoblNta0QyY1BkdGM.

Comparison Methods. The comparison methods are listed below:

- **Lexicon-based Method(LB)**: The standard English MPQA sentiment lexicon is translated into Chinese firstly. Then it is utilized together with a small number of Chinese turning words, negations and intensifiers to predict the sentiment polarities of the test Chinese reviews
- **Basic SVM (BSVM-CN)**: A Chinese SVM classifier which is trained on the pseudo-training data translated from the labeled English reviews
- **Co-training (CoTr)** [20]: A bidirectional transfer learning method. Labeled English training dataset is translated into Chinese, and also unlabeled Chinese is translated into English. Both of the two languages train their own SVM classifiers and jointly select samples to join in the training datasets to boost themselves. In each iteration, 10 positive and 10 negative reviews are transferred from one language to the other

- **Cross-lingual LSI (CL-LSI)**: LSI method is conducted on bilingual document-term matrix of pseudo-parallel documents to obtain their latent bilingual representations. Finally, a SVM classifier is trained with the latent bilingual representations of training data
- **Bilingual Sentiment Word Embedding (BSWE)** [26]: This method learns the bilingual sentiment word embeddings by integrating sentiment information from the labeled training data with Denoised Autoencoder

- **BIVCNN without Bilingual Sentiment Relation Capturing (BiVCNN\BSR)**: The bilingual sentiment relation capturing layer in BIVCNN is replaced by a bilingual sentiment concatenation layer
- **BiVCNN without Distortion-Tolerance (BiVCNN\DT)**: The distortion-tolerance in BIVCNN is removed
- **BIVCNN without Orthogonal Transformation (BiVCNN\OT)**: The orthogonal transformation layer in BiVCNN is removed

Those comparison methods can be categorized into three categories: the first three are the baselines methods trained with superficial features, the middle two are the state-of-the-art methods based on latent semantic representation learning, and the last three are some variants of BIVCNN which are mainly used to demonstrate the necessity of the relation capturing and distortion-tolerance of BIVCNN. The basic classifiers used in the former five methods are the SVMs with linear kernels[4] implemented in the Liblinear package [5]. For LB, BSVM-CN, CoTr and CL-LSI, in particular, the classifiers are trained based on the unigram features with TF-IDF values.

Evaluation Metrics. The evaluation metrics used in our experiments are shown below:

$$Ac(f) = \frac{p^f}{P^f}, \quad mAcc = \frac{1}{3} \cdot \sum_{f' \in \mathcal{F}} Ac(f') \tag{11}$$

where p^f is the number of correct predictions, P^f is the total number of the test data, and $\mathcal{F}$ denotes the set of domains $\{Books, DVD, Music\}$.

---

[4] The parameter setting used in this paper is '-s 7'.

Table 1. Macro performance by averaging the results of 10-times running in three domains. mAcc: mean accuracy; TTL: traditional transfer learning; CLRL: cross-lingual representation learning. * refers to the best result on average accuracy matrix.

Domain	TTL methods			CLRL methods		Our methods			
	LB	BSVM-CN	CoTr	CL-LSI	BSWE	BiVCNN\CSR	BiVCNN\DT	BiVCNN\OT	BiVCNN
Book	0.7770	0.7345	0.7980	0.7648	**0.8105**	0.7870	0.8020	0.8005	0.8040
DVD	0.7832	0.7600	0.7750	0.7878	0.8160	0.8075	0.8160	0.8122	**0.8242**
Music	0.7595	0.7388	0.7722	0.7558	0.7940	0.7952	0.7928	0.7920	**0.7962**
mAcc	0.7709	0.7444	0.7817	0.7695	0.8068	0.7932	0.8036	0.8016	**0.8081**
Average	0.7657			0.7881		**0.8016***			

4.2 Experimental Results and Analysis

Results of Comparison Experiments. The experimental results in terms of macro performance are shown in Table 1. From the results several findings can be observed.

First of all, when compared to the three state-of-the-art methods that are based on superficial features, i.e., LB, BSVM-CN and CoTr, the proposed method BiVCNN shows overwhelming advantages in all cases. This highly proves our claims in the Introduction section that the CLSA problem cannot be well solved by the learning on superficial features. This is because superficial feature is just the simple mixture of different general semantics and hence misleading information is probably brought by non-sentiment semantics into the learning process.

Second, BiVCNN also performs much better than CL-LSI, the state-of-the-art latent representation learning methods which learns latent semantic representations by projecting the two different languages into a bilingual semantic space. This is mainly due to the focus of CL-LSI is not on the sentiment relations between two languages, which will also inevitably bring non-sentimental information into the learning process. In contrast, BiVCNN captures the complex sentiment relations across different languages by introducing the relation embedding techniques into bilingual sentiment learning. Also, it avoids the negative influence of non-sentimental information by adopting a distortion-tolerance when capturing sentiment relations.

Third, BiVCNN performs better in most cases than BSWE, the best-performed bilingual embedding method reported in the CLSA literature. BSWE shows appealing performance compared to other methods that it even outperforms BiVCNN in the case of Book domain. In spite of this, BSWE incorporates sentiment information into semantic representations via an additional sentiment learning step. The sentiment classifications obtained by BSWE only take word-level information into account, which often cannot capture the global sentiment of a text. Such a two-step approach is difficult to globally manage the complex relations between document-level sentiments of different languages. However, BiVCNN synchronously learns the semantics in two languages, extracts the sentiments and captures their cross-lingual sentiment relations within a joint CNN model, and thus achieves a better overall performance than BSWE.

It is reasonable that the cross-lingual sentiment relations of different pairs of parallel documents which have different sentiments are distinguished in the bilingual sentiment space, even though each pair of parallel texts share the same bilingual sentiment point in bilingual sentiment space, which has the same hypothesis with that of semantic relation [9, 21], that is because sentiment is one constituent of semantics.

Finally, when comparing the BiVCNN architecture with three simpler variants, we find that although BiVCNN\CSR, BiVCNN\DT and BiVCNN\OT adopt the similar frameworks with BiVCNN, the performance of them degenerate significantly. This demonstrates the necessity of relation capturing, distortion-tolerance and orthogonal transformation proposed in BiVCNN. Specifically, without CSR, BiVCNN model degenerates to a joint CNN model (BiVCNN\CSR) that achieves the poorer performance compared with the other three BiVCNN methods. It suggests the significance of cross-lingual sentiment capturing in BiVCNN. By the way, the poorer experimental result of BiVCNN\DT also suggests the need of DT in BiVCNN, that is because BiVCNN\DT does not consider the MT errors which was demonstrated to change the sentiment polarity of the sentiment with a relative high probability around 0.1 [4]. Lastly, OT, as illustrated in Introduction section, aims to extract sentiments for both of the languages. It is proved to be more task-driven in sentiment analysis tasks.

Stability of The Proposed Approach. Figure 2 shows the continuous performance of BiVCNN at each iteration during the learning process. It is obvious to see that the sentiments are gradually learned with the learning epoch grows, which indicates the cross-lingual sentiment relations are captured more and more accurately. Also, the learning algorithms for all the three domains quickly converge after only 15 learning epochs, which validates the efficiency and stability of the proposed approach.

We also conduct some other experiments to study the sensitivity of the parameter ϵ in weighted substraction operation $\ominus_\epsilon$. The results show that the

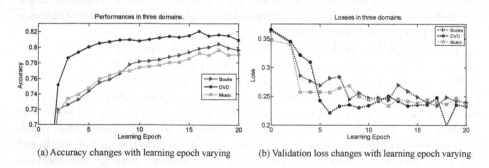

(a) Accuracy changes with learning epoch varying   (b) Validation loss changes with learning epoch varying

Fig. 2. Performance changes with learning epoch varying in three domains.

performances with different parameter settings fluctuate around the best result reported in Table 1 in a small range, which further demonstrates that the proposed method is quite stable.

5 Conclusion and Future Work

In this paper, we propose a joint CNN model, called BiVCNN, to address the CLSA problem without using direct labeled Chinese data nor direct parallel data. We introduce translation-based relation learning to capture the cross-lingual sentiment relations in the document-level. During the capturing, we extract sentiments from general semantics with an orthogonal transformation method and developed a distortion-tolerance to offset sentiment distortions caused by MT errors. The extensive experiments demonstrate the effectiveness of the proposed method. In the future, we will capture the cross-lingual sentiment relations in fine-grained linguistic units, such as the word, the aspect, and the sentence by adding soft attention.

Acknowledge. We thank all the anonymous reviewers for their detailed and insightful comments on this paper. The work described in this paper was supported by National Natural Science Foundation of China (61272291, 61672445, 61472290 and 61472291) and The Hong Kong Polytechnic University (G-YBP6, 4-BCB5 and B-Q46C).

References

1. Bordes, A., Glorot, X., Weston, J., Bengio, Y.: Joint learning of words and meaning representations for open-text semantic parsing. In: International Conference on Artificial Intelligence and Statistics, pp. 127–135 (2012)
2. Bordes, A., Usunier, N., Garcia-Duran, A., Weston, J., Yakhnenko, O.: Translating embeddings for modeling multi-relational data. In: Advances in Neural Information Processing Systems, pp. 2787–2795 (2013)
3. Bordes, A., Glorot, X., Weston, J., Bengio, Y.: A semantic matching energy function for learning with multi-relational data. Mach. Learn. **94**, 233–259 (2014)
4. Chen, Q., Li, W., Lei, Y., Liu, X., He, Y.: Learning to adapt credible knowledge in cross-lingual sentiment analysis. In: Proceedings of the 53rd Annual Meetings of the Association for Computational Linguistics, pp. 419–429. Association for Computational Linguistics (2015)
5. Fan, R.-E., Chang, K.-W., Hsieh, C.-J., Wang, X.-R., Lin, C.-J.: A library for large linear classification. J. Mach. Learn. Res. **9**, 1871–1874 (2008). JMLR.org
6. Gui, L., Xu, R., Lu, Q., Xu, J., Xu, J., Liu, B., Wang, X.: Cross-lingual opinion analysis via negative transfer detection. In: Proceedings of the 52nd Annual Meeting of the Association for Computational Linguistics, 23–25 June, Baltimore, Maryland, USA, pp. 860–865. Association for Computational Linguistics (2014)
7. He, Y.: Latent sentiment model for weakly-supervised cross-lingual sentiment classification. In: Clough, P., Foley, C., Gurrin, C., Jones, G.J.F., Kraaij, W., Lee, H., Mudoch, V. (eds.) ECIR 2011. LNCS, vol. 6611, pp. 214–225. Springer, Heidelberg (2011). doi:10.1007/978-3-642-20161-5_22

8. Jenatton, R., Roux, N.L., Bordes, A., Obozinski, G.R.: A latent factor model for highly multi-relational data. In: Advances in Neural Information Processing System, pp. 3167–3475 (2012)
9. Lin, Y., Liu, Z., Sun, M., Liu, Y., Zhu, X.: Learning entity and relation embeddings for knowledge graph completion. In: Proceedings of AAAI 2015, pp. 716–717 (2015)
10. Liu, B.: Sentiment analysis and opinion mining. In: Synthesis Lectures on Human Language Technologies, vol. 5, no. 1, pp. 1–167. Morgan & Claypool Publishers (2012)
11. McAuley, J., Targett, C., Shi, Q., van den Hengel, A.: Image-based recommendations on styles and substitutes. In: Proceedings of SIGIR 2015, 09–13 August 2015, Santiago, Chile (2015)
12. Meng, X., Wei, F., Liu, X., Zhou, M., Xu, G., Wang, H.: Cross-lingual mixture model for sentiment classification. In: Proceedings of the 50th Annual Meeting of the Association for Computational Linguistics: Long Papers-Volume 1, pp. 572–581. Association for Computational Linguistics (2012)
13. Mikolov, T., Yih, W., Zweig, G.: Linguistic regularities in continuous space word representations. In: Proceedings of the 2013 Conference of the North American Chapter of the Association for Computational Linguistics: Human Language Technologies, pp. 746–751 (2013)
14. Popat, K., Balamurali, A.R.: The haves and the have-nots: leveraging unlabelled corpora for sentiment analysis. Citeseer (2013)
15. Prettenhofer, P., Stein, B.: Cross-language text classification using structural correspondence learning. In: Proceedings of the 48th Annual Meeting of the Association for Computational Linguistics, pp. 1118–1127. Association for Computational Linguistics (2010)
16. Rothe, S., Ebert, S., Schütze, H.: Ultradense word embeddings by orthogonal transformation. arXiv preprint arXiv:1602.07572 (2016)
17. Socher, R., Chen, D., Manning, C.D., Ng, A.: Reasoning with neural tensor networks for knowledge base completion. In: Advances in Neural Information Processing Systems, pp. 926–934 (2013)
18. Sutskever, I., Tenenbaum, J.B., Salakhutdinov, R.B.: Modelling relational data using bayesian clustered tensor factorization. In: Advances in Neural Information Processing Systems, pp. 1821–1828 (2009)
19. Tang, X., Wan, X.: Learning bilingual embedding model for cross-language sentiment classification. In: Proceedings of 2014 IEEE/WIA/ACM International Joint Conferences on Web Intelligence (WI) and Intelligent Agent Technologies (IAT), pp. 134–141 (2014)
20. Wan, X.: Co-training for cross-lingual sentiment classification. In: Proceedings of the Joint Conference of the 47th Annual Meeting of the ACL and the 4th International Joint Conference on Natural Language Processing of the AFNLP: Volume 1-Volume 1, pp. 235–243. Association for Computational Linguistics (2009)
21. Wang, Z., Zhang, J., Feng, J., Chen, Z.: Knowledge graph embedding by translating on hyperplanes. In: Proceedings of the Twenty-Eighth AAAI Conference on Artificial Intelligence, pp. 1112–1119. Citeseer (2014)
22. Xiao, M., Guo, Y.: A novel two-step method for cross language representation learning. In: Advances in Neural Information Processing Systems, pp. 1259–1267 (2013)
23. Zhang, H., Yu, H., Xiong, D., Liu, Q.: HHMM-based Chinese lexical analyzer ICTCLAS. In: 2nd SIGHAN Workshop Affiliated with 42nd ACL, pp. 430–440 (2003)

24. Zhou, H., Chen, L., Huang, D.: Cross-lingual sentiment classification based on denoising autoencoder. In: Zong, C., Nie, J.Y., Zhao, D., Feng, Y. (eds.) Natural Language Processing and Chinese Computing. CCIS, vol. 496, pp. 181–192. Springer, Heidelberg (2014). doi:10.1007/978-3-662-45924-9_17
25. Zhou, G., He, T., Zhao, J., Wu, W.: A subspace learning framework for cross-lingual sentiment classification with partial parallel data. In: Proceedings of the International Joint Conference on Artificial Intelligence, Buenos Aires (2015a)
26. Zhou, H., Chen, L., Shi, F., Huang, D.: Learning bilingual sentiment word embeddings for cross-language sentiment classification. In: Proceedings of the 53rd Annual Meeting of the Association for Computational Linguistics, pp. 430–440. Association for Computational Linguistics (2015b)

Hierarchical Re-estimation of Topic Models for Measuring Topical Diversity

Hosein Azarbonyad$^{(\boxtimes)}$, Mostafa Dehghani, Tom Kenter, Maarten Marx,
Jaap Kamps, and Maarten de Rijke

University of Amsterdam, Amsterdam, The Netherlands
{h.azarbonyad,dehghani,tom.kenter,maartenmarx,kamps,derijke}@uva.nl

Abstract. A high degree of topical diversity is often considered to be
an important characteristic of interesting text documents. A recent proposal for measuring topical diversity identifies three elements for assessing diversity: words, topics, and documents as collections of words. Topic
models play a central role in this approach. Using standard topic models for measuring diversity of documents is suboptimal due to *generality*
and *impurity*. General topics only include common information from a
background corpus and are assigned to most of the documents in the
collection. Impure topics contain words that are not related to the topic;
impurity lowers the interpretability of topic models and impure topics
are likely to get assigned to documents erroneously. We propose a hierarchical re-estimation approach for topic models to combat generality and
impurity; the proposed approach operates at three levels: words, topics,
and documents. Our re-estimation approach for measuring documents'
topical diversity outperforms the state of the art on PubMed dataset
which is commonly used for diversity experiments.

1 Introduction

Quantitative notions of topical diversity in text documents are useful in several
contexts, e.g., to assess the interdisciplinarity of a research proposal [3] or to
determine the interestingness of a document [2]. An influential formalization of
diversity has been introduced in biology [17]. It decomposes diversity in terms
of *elements* that belong to *categories* within a *population* [20] and formalizes
the diversity of a population d as the expected distance between two randomly
selected elements of the population:

$$div(d) = \sum_{i=1}^{T}\sum_{j=1}^{T} p_i p_j \delta(i,j), \tag{1}$$

where p_i and p_j are the proportions of categories i and j in the population
and $\delta(i,j)$ is the distance between i and j. Bache et al. [3] have adapted this
notion of diversity to quantify the topical diversity of a text document. Words
are considered elements, topics are categories, and a document is a population.
When using topic modeling for measuring topical diversity of text document d,

© Springer International Publishing AG 2017
J.M. Jose et al. (Eds.): ECIR 2017, LNCS 10193, pp. 68–81, 2017.
DOI: 10.1007/978-3-319-56608-5_6

Bache et al. [3] model elements based on the probability of a word w given d, $P(w \mid d)$, categories based on the probability of w given topic t, $P(w \mid t)$, and populations based on the probability of t given d, $P(t \mid d)$.

In probabilistic topic modeling, at estimation time, these distributions are usually assumed to be sparse. First, the content of a document is assumed to be generated by a small subset of words from the vocabulary (i.e., $P(w \mid d)$ is sparse). Second, each topic is assumed to contain only some topic-specific related words (i.e., $P(w \mid t)$ is sparse). Finally, each document is assumed to deal with a few topics only (i.e., ($P(t \mid d)$ is sparse). When approximated using currently available methods, $P(w \mid t)$ and $P(t \mid d)$ are often dense rather than sparse [13,19,21]. Dense distributions cause two problems for the quality of topic models when used for measuring topical diversity: *generality* and *impurity*. General topics mostly contain general words and are typically assigned to most documents in a corpus. Impure topics contain words that are not related to the topic. Generality and impurity of topics both result in low quality $P(t \mid d)$ distributions.

We propose a hierarchical re-estimation process for making the distributions $P(w \mid d)$, $P(w \mid t)$ and $P(t \mid d)$ more sparse. We re-estimate the parameters of these distributions so that general, collection-wide items are removed and only salient items are kept. For the re-estimation we use the concept of *parsimony* [9] to extract only essential parameters of each distribution.

Our main contributions are: (1) We propose a hierarchical re-estimation process for topic models to address two main problems in estimating topical diversity of text documents, using a biologically inspired definition of diversity. (2) We study the efficacy of each level of re-estimation, and improve the accuracy of estimating topical diversity, outperforming the current state-of-the-art [3] on a publicly available dataset commonly used for evaluating document diversity [1].

2   Related Work

Our hierarchical re-estimation method for measuring topical diversity relates to measuring text diversity, improving the quality of topic models, model parsimonization, and evaluating topic models.

Text Diversity and Interestingness. Recent studies measure topical diversity of document [2,3,8] by means of Latent Dirichlet Allocation (LDA) [4]. The main diversity measure in this work is Rao's measure [17] (Eq. 1), in which the diversity of a text document is proportional to the number of dissimilar topics it covers. While we also use Rao's measure, we hypothesize that pure LDA is not good enough for modeling text diversity and propose a re-estimation process for adapting topic models for measuring topical diversity.

Improving the Quality of Topic Models. The two most important issues with topic models are the *generality problem* and the *impurity problem* [5,13,19, 21]. Many approaches have been proposed to address the generality problem [21–23]. The main difference with our work is that previous work does not yield sparse topic representations or topic word distributions. Soleimani and Miller [19] propose parsimonious topic models (PTM) to address the generality and impurity problems. PTM achieves state-of-the-art results compared to existing topic models. Unlike [19], we do not modify the training procedure of LDA but propose a method to refine the topic models.

Model Parsimonization. In language model parsimonization, the language model of a document is considered to be a mixture of a general background model and a document-specific language model [6,7,9,26]. The goal is to extract the document-specific part and remove the general words. We employ parsimonization for re-estimating topic models. The main assumption in [9] is that the language model of a document is a mixture of its specific language model and a general language model:

$$P(w \mid d) = \lambda P(w \mid \tilde{\theta}_d) + (1 - \lambda)P(w \mid \theta_C), \tag{2}$$

where w is a term, d a document, $\tilde{\theta}_d$ the document specific language model of d, θ_C the language model of the collection C, and λ is a mixing parameter. The main goal is to estimate $P(w \mid \tilde{\theta}_d)$ for each document. This is done in an iterative manner using EM algorithm. The initial parameters of the language model are the parameters of standard language model, estimated using maximum likelihood: $P(w \mid \tilde{\theta}_d) = \frac{tf_{w,d}}{\sum_{w'} tf_{w',d}}$, where $tf_{w,d}$ is the frequency of w in d. The following steps are computed iteratively:

E-step:

$$e_w = tf_{w,d} \cdot \frac{\lambda P(w \mid \tilde{\theta}_d)}{\lambda P(w \mid \tilde{\theta}_d) + (1 - \lambda)P(w \mid \theta_C))}, \tag{3}$$

M-step:

$$P(w \mid \tilde{\theta}_d) = \frac{e_w}{\sum_{w'} e_{w'}}, \tag{4}$$

where $\tilde{\theta}_d$ is the parsimonized language model of document d, C is the background collection, $P(w \mid \theta_C)$ is estimated using maximum likelihood estimation, and λ is a parameter that controls the level of parsimonization. A low value of λ will result in a more parsimonized model while $\lambda = 1$ yields a model without parsimonization. The EM process stops after a fixed number of iterations or after convergence.

Evaluating Topic Models. We evaluate the effectiveness of our re-estimated models by measuring the topical diversity of text documents. In addition, in Sect. 6, we analyze the effectiveness of our re-estimation approach in terms of purity in document clustering and document classification tasks. For classification, following [10,16,19], we model topics as document features with values $P(t \mid d)$. For clustering, each topic is considered a cluster and each document is assigned to its most probable topic [16,24,25].

3   Measuring Topical Diversity of Documents

To measure topical diversity of text documents, we propose HiTR (hierarchical topic model re-estimation). HiTR can be applied to any topic modeling approach that models documents as distributions over topics and topics as distributions over words.

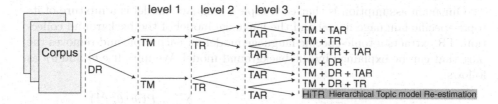

Fig. 1. Different topic re-estimation approaches. TM is a topic modeling approach like, e.g., LDA. DR is document re-estimation, TR is topic re-estimation, and TAR is topic assignment re-estimation.

The input to HiTR is a corpus of text documents. The output is a probability distribution over topics for each document in the corpus. HiTR has three levels of re-estimation: (1) **document re-estimation (DR)** re-estimates the language model per document $P(w\,|\,d)$; (2) **topic re-estimation (TR)** re-estimates the language model per topic $P(w\,|\,t)$; and (3) **topic assignment re-estimation (TAR)** re-estimates the distribution over topics per document $P(t\,|\,d)$. Based on applying or not applying re-estimation at different levels, there are seven possible re-estimation approaches; see Fig. 1. HiTR refers to the model that uses all three re-estimation techniques, i.e., TM+DR+TR+TAR. Next, we describe each of the re-estimation steps in more detail.

Document re-estimation (DR) re-estimates $P(w\,|\,d)$. Here, we remove unnecessary information from documents before training topic models. This is comparable to pre-processing steps, such as removing stopwords and high- and low-frequency words, that are typically carried out prior to applying topic models [4,11,15,16]. Proper pre-processing of documents, however, takes lots of effort and involves tuning many parameters. *Document re-estimation*, however, removes impure elements (general words) from documents automatically. If general words are absent from documents, we expect that the trained topic models will not contain general topics. After document re-estimation, we can train any standard topic model on the re-estimated documents.

Document re-estimation uses the parsimonization method described in Sect. 2. The re-estimated model $P(w\,|\,\tilde{\theta}_d)$ in (4) is used as the language model of document d, and after removing unnecessary words from d, the frequencies of the remaining words (words with $P(w\,|\,\tilde{\theta}_d) > 0$) are re-estimated for d using the following equation:

$$tf(w, d) = \left\lfloor P(w\,|\,\tilde{\theta}_d) \cdot |d| \right\rfloor,$$

where $|d|$ is the document length in words. Topic modeling is then applied on the re-estimated document-word frequency matrix.

Topic re-estimation (TR) re-estimates $P(w\,|\,t)$ by removing general words. The re-estimated distributions are used to assign topics to documents. The goal of this step is to increase the purity of topics by removing general words that have not yet been removed by DR. The two main advantages of the increased purity of topics are (1) it improves human interpretation of topics, and (2) it leads to more document-specific topic assignments, which is essential for measuring topical diversity of documents.

Our main assumption is that each topic's language model is a mixture of its topic-specific language model and the language model of the background collection. TR extracts a topic-specific language model for each topic and removes the part that can be explained by the background model. We initialize $\tilde{\theta}_t$ and θ_T as follows:

$$P(w \mid \tilde{\theta}_t) = P(w \mid \theta_t^{\mathcal{TM}}) \qquad P(w \mid \theta_T) = \frac{\sum_{t \in T} P(w \mid \theta_t^{\mathcal{TM}})}{\sum_{w' \in V} \sum_{t' \in T} P(w' \mid \theta_{t'}^{\mathcal{TM}})}$$

where t is a topic, $\tilde{\theta}_t$ is topic-specific language model of t, and θ_T is the background language model of T (the collection of all topics), $P(w \mid \theta_t^{\mathcal{TM}})$ is the probability of w belonging to topic t estimated by a topic model $\mathcal{TM}$. Having these estimations, the steps of TR are similar to the steps of parsimonization, except that in the E-step we estimate $tf_{w,t}$, the frequency of w in t, by $P(w \mid \theta_t^{\mathcal{TM}})$.

Topic assignment re-estimation (TAR) re-estimates $P(t \mid d)$. In topic modeling, most topics are usually assigned with a non-zero probability to most of documents. For documents which are in reality about a few topics, this topic assignment is incorrect and overestimates its diversity. TAR addresses the general topics problem and achieves more document specific topic assignments. To re-estimate topic assignments, a topic model is first trained on the document collection. This model is used to assign topics to documents based on the proportion of words they have in common. We then model the distribution over topics per document as a mixture of its document-specific topic distribution and the topic distribution of the entire collection.

We initialize $P(t \mid \tilde{\theta}_d)$ and $P(t \mid \theta_C)$ as follows:

$$P(t \mid \tilde{\theta}_d) = P(t \mid \theta_d^{\mathcal{TM}}) \qquad P(t \mid \theta_C) = \frac{\sum_{d \in C} P(t \mid \theta_d^{\mathcal{TM}})}{\sum_{t' \in T} \sum_{d' \in C} P(t' \mid \theta_{d'}^{\mathcal{TM}})}.$$

Here, t is a topic, d a document, $P(t \mid \tilde{\theta}_d)$ the document-specific topic distribution, and $P(t \mid \theta_C)$ the distribution of topics in the entire collection C, and $P(t \mid \theta_d^{\mathcal{TM}})$ the probability of assigning topic t to document d estimated by a topic model $\mathcal{TM}$. The remaining steps of TAR follow the ones of parsimonization, the difference being that in the E-step, we estimate $f_{t,d}$ using $P(t \mid \theta_d^{\mathcal{TM}})$.

4   Experimental Setup

Our main research question is: (RQ1) How effective is HiTR in measuring topical diversity of documents? How does it compare to the state-of-the-art in addressing the general and impure topics problem?

To address RQ1 we run our models on a binary classification task. We generate a synthetic dataset of documents with high and low topical diversity (the process is detailed below), and the task for every model is to predict whether a document belongs to the high or low diversity class. We employ HiTR to re-estimate topic models and use the re-estimated models for measuring topical diversity of documents. To gain deeper insights into how HiTR performs, we conduct a separate analysis of the last two levels of re-estimation, TR and TAR:[1] (RQ2.1) Does TR increase the purity of topics? If so, how does using the more pure topics influence the performance in topical diversity task? (RQ2.2) How does TAR affect the sparsity of document-topic assignments? And what is the effect of re-estimated document-topic assignments on the topical diversity task? To answer RQ2.1, we first evaluate the performance of TR on the topical diversity task and compare its performance to DR and TAR. To answer RQ2.2, we first evaluate TAR together with LDA in a topical diversity task and analyze its effect on the performance of LDA to study how successful TAR is in removing general topics from documents.

Dataset, Pre-processing, Evaluation Metrics, and Parameters: Following [3], we generate 500 documents with a high value of diversity and 500 documents with a low value of diversity. We select over 300,000 documents articles published between 2012 to 2015 from PubMed [1]. For generating documents with a high value of diversity, we first select 20 journals and create 10 pairs of journals. Each pair contains two journals that are relatively unrelated to each other (we use the pairs of journals selected in [3]). For each pair of journals A and B we select 50 articles to create 50 probability distributions over topics: we randomly select one article from A and one from B and generate a document by averaging the selected article's bag of topic counts. Thus, for each pair of journals we generate 50 documents with a high diversity value. Also, for each of the chosen 20 journals, we repeat the procedure but instead of choosing articles from different journals, we select them from the same journal to generate 25 non-diverse documents.

For pre-processing documents, we remove stopwords included in the standard stop word list from Python's NLTK package. In addition, we remove the 100 most frequent words in the collection and words with fewer than 5 occurrences.

Measuring Topical Diversity: After re-estimating word distributions in documents, topics, and document topic distributions using HiTR, we use the final distributions over topics per document for measuring topical diversity. Diversity of texts is computed using Rao's coefficient [3] using Eq. 1. We use the normalized angular distance δ for measuring the distance between topics, since it is a proper distance function [2].

To measure the performance of topic models on the topical diversity task, we use ROC curves and report the AUC values [3]. We also measure the *coherence* of the extracted topics; this measure indicates the purity of $P(w \mid t)$ distributions, where a high value of coherence implies high purity within topics. We estimate coherence using *normalized pointwise mutual information* between the top N words within a topic [11, 16]. As the reference corpus for computing word occurrences, we use the English Wikipedia.[2]

The topic modeling approach used in our experiments with HiTR is LDA. Following [3, 18, 19] we set the number of topics to 100. We set the two

---

[1] As the DR level of re-estimation directly employs the parsimonious language modeling techniques in [9], we omit it from our in-depth analysis.

[2] We use a dump of June 2, 2015, containing 15.6 million articles.

hyperparameters to $\alpha = 1/T$ and $\beta = 0.01$, where T is the number of topics, following [16]. In the re-estimation process, at each step of the EM algorithm, we set the threshold for removing unnecessary components from the model to 0.0001 and remove terms with an estimated probability less than this threshold from the language models, as in [9].

We perform 10-fold cross validation, using 8 folds as training data, 1 fold to tune the parameters (λ for DR, TR, and TAR), and 1 fold for testing. Our baseline for the topical diversity task is the method proposed in [3], which uses LDA. We also compare our results to PTM [19], which we use instead of LDA for measuring topical diversity. PTM is the best available topic modeling approach, and the current state of the art.

For statistical significance testing, we compare our methods to PTM using paired two-tailed t-tests with Bonferroni correction. To account for multiple testing, we consider an improvement significant if: $p \leq \alpha/m$, where m is the number of conducted comparisons and α is the desired significance. We set $\alpha = 0.05$. In Sect. 5, ▲ and ▼ indicate that the corresponding method performs significantly better and worse than PTM, respectively.

5   Results

In this section, we report on the performance of HiTR on the topical diversity task. Additionally we analyze the effectiveness of the individual re-estimation approaches.

5.1   Topical Diversity Results

Figure 2 plots the performance of our topic models across different levels of re-estimation, and the models we compare to, on the PubMed dataset. We plot ROC curves and compute AUC values. To plot the ROC curves we use the diversity scores calculated for the generated pseudo-documents with diversity labels. HiTR improves the performance of LDA by 17% and PTM by 5% in terms of AUC. From Fig. 2 two observations can be made.

First, HiTR benefits from the three re-estimation approaches it encapsulates by successfully improving the quality of estimated diversity scores. Second, the performance of LDA+TAR, which tries to address the generality problem, is higher than the performance of LDA+TR, which addresses impurity. General topics have a stronger negative effect on measuring topical diversity than impure topics. Also, LDA+DR

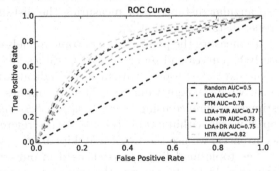

Fig. 2. Performance of topic models in topical diversity task on the PubMed dataset. The improvement of HiTR over PTM is statistically significant ($p < 0.05$) in terms of AUC.

Table 1. Topic assignments for a non-diverse document using LDA and HiTR. Only topics with $P(t \mid d) > 0.05$ are shown.

LDA			HiTR	
Topic	$P(t \mid d)$	Top 5 words	$P(t \mid d)$	Top 5 words
1	0.21	Brain, anterior, neurons, cortex, neuronal	0.68	Brain, neuronal, neurons, neurological, nerve
2	0.14	Channel, neuron, membrane, receptor, current	0.23	Channel, synaptic, neuron, receptor, membrane
3	0.10	Use, information, also, new, one	0.09	Network, nodes, cluster, community, interaction
4	0.08	Network, nodes, cluster, functional, node		
5	0.08	Using, method, used, image, algorithm		
6	0.08	Time, study, days, period, baseline		
7	0.07	Data, values, number, average, used		

outperforms LDA+TR. So, removing impurity from $P(t \mid d)$ distributions is the most effective approach in the topical diversity task, and removing impurity from $P(w \mid d)$ distributions is more effective than removing impurity from $P(w \mid t)$ distributions. Table 1 illustrates the difference between LDA and HiTR with the topics assigned by the two methods for a non-diverse document that is combined from two documents from the same journal, entitled "Molecular Neuroscience: Challenges Ahead" and "Reward Networks in the Brain as Captured by Connectivity Measures," using the procedure described in Sect. 4. As only a very basic stopword list being applied, words like *also* and *one* still appear. We expect to have a low diversity value for the combined document. However, using Rao's diversity measure, the topical diversity of this document based on the LDA topics is 0.97. This is due to the fact that there are three document-specific topics— topics 1, 2 and 4—and four general topics. Topics 1 and 2 are very similar and their δ is 0.13. The other, more general topics have high δ values; the average δ value between pairs of topics is as high as 0.38. For the same document, HiTR only assigns three document-specific topics and they are more pure and coherent. The average δ value between pairs of topics assigned by HiTR is 0.19. The diversity value of this document using HiTR is 0.16, which indicates that this document is non-diverse. Hence, HiTR is more effective than other approaches in measuring topical diversity of documents; it successfully removes generality from $P(t \mid d)$.

5.2   Topic Re-estimation Results

To answer **RQ2.1**, we focus on topic re-estimation (TR). Since TR tries to remove impurity from topics, we expect it to increase the coherence of the topics

Table 2. Topic model coherence in terms of average normalized mutual information between top 10 words in the topics on the PubMed dataset.

LDA	PTM	LDA+TR	LDA+DR+TR
8.17	9.89	9.46	10.29▲

by removing unnecessary words from topics. We measure the purity of topics based on the coherence of words in $P(w\,|\,t)$ distributions. Table 2 shows the coherence of topics according to different topic modeling approaches, in terms of average mutual information. TR significantly increases the coherence of topics by removing the impure parts from topics. The coherence of PTM is higher than of TR. However, when we first apply DR, train LDA, and finally apply TR, the coherence of the extracted topics is significantly higher than the coherence of topics extracted by PTM. We conclude that TR is effective in removing impurity from topics. Moreover, DR also contributes in making topics more pure.

5.3 Topic Assignment Re-estimation Results

To answer **RQ2.2**, we focus on TAR (topic assignment re-estimation). We are interested in seeing how HiTR deals with general topics. We sum the probability of assigning a topic to a document, over all documents: for each topic t, we compute $\sum_{d \in C} P(t\,|\,d)$, where C is the document collection. Figure 3 shows the distribution of probability mass before and after applying TAR; topics are sorted based on the topic assignment probability of LDA. LDA assigns a vast proportion of the probability mass to a relatively small number of topics, mostly general topics that are assigned to most documents. We expect that many topics are represented in some documents, while relatively few topics will be relevant to all documents. After applying TAR, the distribution is less skewed and the probability mass is more evenly distributed.

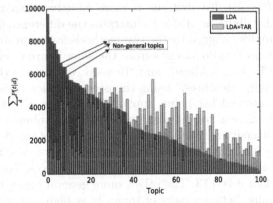

Fig. 3. The total probability of assigning topics to the documents in the PubMed dataset estimated using LDA and LDA+TAR. (The two areas are equal to the number of documents ($N \approx 300K$)).

There are topics that have a high $\sum_d P(t\,|\,d)$ value in LDA's topic assignments and a high $\sum_d P(t\,|\,d)$ value after applying TAR too; we marked them as "non-general topics" in Fig. 3. Table 3, column 2 shows the top five words for these topics. TAR is able to find these three non-general topics and their assignment probabilities to documents in the $P(t\,|\,d)$ distributions is not changed as

Table 3. Top five words for the topics detected by TAR as general topics and non-general topics.

Topic	Non-general topics	General topics
1	Health, services, public, countries, data	Use, information, also, new, one
2	Surgery, surgical, postoperative, patient, performed	Ci, study, analysis, data, variables
3	Cells, cell, treatment, experiments, used	Time, study, days, period, baseline
4		Group, control, significantly, compared, groups
5		Study, group, subject, groups, significant
6		May, also, effects, however, would
7		Data, values, number, average, used

much as the actual general topics. Thus, TAR removes general topics from documents and increases the probability of document-specific topics for each document. To further investigate whether TAR really removes general topics, Table 3, column 3 shows the top five words for the first 10 topics in Fig. 3, excluding the "non-general topics." These seven topics have the highest decrease in $\sum_d P(t \mid d)$ values due to TAR. Clearly, they contain general words and are not informative. Figure 3 shows that after applying TAR, the $\sum_d P(t \mid d)$ values have decreased dramatically for these topics, without creating new general topics.

5.4   Parameter Analysis

Next, we analyze the effect of the λ parameter on the performance of DR, TR, and TAR. Figure 4 displays the performance at different levels of re-estimation. With $\lambda = 1$, no re-estimation occurs, and all methods equal LDA. We see that DR peaks with moderate values of λ ($0.4 \leq \lambda \leq 0.45$). This reflects that documents contain a moderate amount of general information and that DR is able to successfully deal with it. For $\lambda \geq 0.8$, the performance of DR and LDA is the same and for these values of λ DR does not increase the quality of LDA.

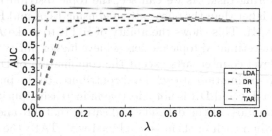

Fig. 4. The effect of the λ parameter on the performance of topics models in the topical diversity task on the PubMed dataset.

Also, the best performance of TR is achieved with high values of λ ($0.65 \leq \lambda \leq 0.75$). From this observation we conclude that topics typically need only a small amount of re-estimation. With this slight re-estimation, TR is able to improve the quality of LDA. However, for $\lambda \geq 0.75$ the accuracy of TR degrades. Lastly, TAR achieves its best performance with low values of λ ($0.02 \leq \lambda \leq 0.05$). Hence, most of the noise is in the $P(t \mid d)$ distributions and aggressive re-estimation allows TAR to remove most of it.

6   Analysis

In this section, we want to gain additional insights into HiTR and its effects on topic computation. The purity of topic assignments based on $P(t \mid d)$ distributions has the highest effect on the quality of estimated diversity scores. Thus, we investigate how pure estimated topic assignments are using HiTR. To this end, we compare document clustering and classification results, based on the topics assigned by HiTR, LDA and PTM. For clustering, following [16], we consider each topic as a cluster. Each document d is assigned to the topic that has the highest probability value in $P(t \mid d)$. For classification, we use all topics assigned to the document and consider $P(t \mid d)$ as features for a supervised classification algorithm; we use SVM. We view high accuracy in clustering or classification as an indicator of high purity of topic distributions. Our focus is not on achieving a top clustering or classification performance: these tasks are a means to assess the purity of topic distributions using different topic models.

Datasets and Metrics. We use RCV1 [12], 20-NewsGroups,[3] and Ohsumed.[4] RCV1 contains 806,791 documents with category labels for 126 categories. For clustering and classification of documents, we use 55 categories in the second level of the hierarchy. 20-NewsGroups contains ~20,000 documents (20 categories, around 1,000 documents per category). Ohsumed contains 50,216 documents grouped into 23 categories. For measuring the purity of clusters we use *purity* and *normalized mutual information* (NMI) [14]. We use 10-fold cross validation and the same pre-processing as in Sect. 4.

Purity Results. The top part of Table 4 shows results on the document clustering task. As we can see, the topic distributions extracted using HiTR score higher than the ones extracted using LDA and PTM in terms of both purity and NMI. This shows the ability of HiTR to make $P(t \mid d)$ more pure. The two-level re-estimated topic models achieve higher purity values than their respective one-level counterparts except the combination of DR and TR, which indicates that re-estimation at each level contributes to the purity of $P(t \mid d)$. The combination of TR and DR is not effective in increasing purity over its one-level counterparts on most of the datasets, indicating that TR and DR address similar issues. But when each of them is combined with TAR, the purity of the topic distributions increases, implying that DR/TR and TAR address complementary issues.

The bottom part of Table 4 shows results on the document classification task. HiTR is more accurate in estimating $P(t \mid d)$; its accuracy is higher than that of other topic models. The higher values in classification task, compared to clustering task, indicate that the most probable topic does not necessarily

---

[3] Available at http://www.ai.mit.edu/people/~jrennie/20Newsgroups/.

[4] Available at http://disi.unitn.it/moschitti/corpora.htm.

Table 4. Re-estimated topic models for document clustering (top) and document classification (bottom). For significance tests, we consider p-value $< 0.05/7$; comparisons are against PTM.

Method	RCV1		20-Newsgroups		Ohsumed	
	Purity	NMI	Purity	NMI	Purity	NMI
LDA	0.55	0.40	0.52	0.36	0.50	0.30
PTM	0.61	0.43	0.57	0.38	0.55	0.33
LDA+DR	0.57▼	0.41▼	0.56	0.39	0.53▼	0.32▼
LDA+TR	0.57▼	0.42▼	0.56	0.38	0.53▼	0.31▼
LDA+TAR	0.60	0.43	0.57	0.39	0.54	0.33
LDA+DR+TR	0.58	0.42▼	0.57	0.38	0.54	0.32
LDA+DR+TAR	0.60	0.43	0.58	0.40	0.55	0.35▲
LDA+TR+TAR	0.61	0.43	0.58	0.40▲	0.56▲	0.34▲
HiTR	**0.64▲**	**0.45▲**	**0.60▲**	**0.42▲**	**0.57▲**	**0.35▲**
	Acc	Change	Acc	Change	Acc	Change
LDA	0.76	−8%	0.81	−7%	0.50	−11%
PTM	0.82	–	0.87	–	0.56	–
LDA+DR	0.79▼		0.83▼	−5%	0.52▼	−7%
LDA+TR	0.78▼	−5%	0.83▼	−5%	0.53▼	−5%
LDA+TAR	0.82	0%	0.85▼	−2%	0.54	−4%
LDA+DR+TR	0.80▼	−2%	0.84▼	−3%	0.53▼	−5%
LDA+DR+TAR	0.83	+1%	0.86	−1%	0.56	0%
LDA+TR+TAR	0.82▲	0%	0.87	0%	0.58▲	+4%
HiTR	**0.85▲**	+4%	**0.89▲**	+2%	**0.60▲**	+7%

contain all information about the content of a document. If a document is about more than one topic, the classifier utilizes all $P(t \mid d)$ information and performs better. Therefore, the higher accuracy of HiTR in this task is an indicator of its ability to assign document-specific topics to documents.

7 Conclusions

We have proposed Hierarchical Topic model Re-estimation (HiTR), an approach for measuring topical diversity of text documents. It addresses two main issues with topic models, topic generality and topic impurity, which negatively affect measuring topical diversity scores in three ways. First, the existence of document-unspecific words within $P(w \mid d)$ (the distribution of words within documents) yields general topics and impure topics. Second, the existence of topic-unspecific words within $P(w \mid t)$ (the distribution of words within topics) yields impure topics. Third, the existence of document-unspecific topics within $P(t \mid d)$ (the distribution of topics within documents) yields general topics. We have proposed three approaches for removing unnecessary or even harmful information from probability distributions, which we combine in our method for HiTR.

Estimated diversity scores for documents using HiTR are more accurate than those obtained using the current state-of-the-art topic modeling method PTM, or a general purpose topic model such as LDA. HiTR outperforms PTM because it adapts topic models for the topical diversity task. The quality of topic models for measuring topical diversity degrades mainly because of general topics in the $P(t \mid d)$ distributions. Our topic assignment re-estimation (TAR) approach successfully removes general topics, leading to higher performance on the topical diversity task.

We analyzed the purity of topic assignments on clustering and classification tasks, where $P(t \mid d)$ distributions were directly used as features. The results confirm that HiTR is effective in removing impurity from documents; it removes impure parts from the three probability distributions mentioned, using three re-estimation approaches.

Acknowledgments. This research was supported by Ahold Delhaize, Amsterdam Data Science, Blendle, the Bloomberg Research Grant program, the Dutch national program COMMIT, Elsevier, the European Community's Seventh Framework Programme (FP7/2007-2013) under grant agreements nr 283465 (ENVRI) and 312827 (VOX-Pol), the Microsoft Research Ph.D. program, the Netherlands eScience Center under project number 027.012.105, the Netherlands Institute for Sound and Vision, the Netherlands Organisation for Scientific Research (NWO) under project nrs 314.99.108, 600.006.014, HOR-11-10, CI-14-25, 652.-002.-001, 612.-001.-551, 652.-001.-003, 314-98-071, and Yandex. All content represents the opinion of the authors, which is not necessarily shared or endorsed by their respective employers and/or sponsors.

References

1. U.S. National Library of Medicine. Pubmed Central Open Access Initiative (2010)
2. Azarbonyad, H., Saan, F., Dehghani, M., Marx, M., Kamps, J.: Are topically diverse documents also interesting? In: Mothe, J., Savoy, J., Kamps, J., Pinel-Sauvagnat, K., Jones, G.J.F., SanJuan, E., Cappellato, L., Ferro, N. (eds.) CLEF 2015. LNCS, vol. 9283, pp. 215–221. Springer, Cham (2015). doi:10.1007/978-3-319-24027-5_19
3. Bache, K., Newman, D., Smyth, P.: Text-based measures of document diversity. In KDD (2013)
4. Blei, D.M., Ng, A.Y., Jordan, M.I.: Latent dirichlet allocation. J. Mach. Learn. Res. **3**(4–5), 993–1022 (2003)
5. Boyd-Gaber, J., Mimno, D., Newman, D.: Care and feeding of topic models. In: Mixed Membership Models & Their Applic. CRC Press (2014)
6. Dehghani, M., Azarbonyad, H., Kamps, J., Marx, M.: Two-way parsimonious classification models for evolving hierarchies. In: Fuhr, N., Quaresma, P., Gonçalves, T., Larsen, B., Balog, K., Macdonald, C., Cappellato, L., Ferro, N. (eds.) CLEF 2016. LNCS, vol. 9822, pp. 69–82. Springer, Heidelberg (2016). doi:10.1007/978-3-319-44564-9_6

7. Dehghani, M., Azarbonyad, H., Kamps, J., Marx, M.: On horizontal and vertical separation in hierarchical text classification. In: ICTIR (2016)
8. Derzinski, M., Rohanimanesh, K.: An information theoretic approach to quantifying text interestingness. In: NIPS MLNLP Workshop (2014)
9. Hiemstra, D., Robertson, S., Zaragoza, H.: Parsimonious language models for information retrieval. In: SIGIR (2004)
10. Lacoste-Julien, S., Sha, F., Jordan, M.I.: DiscLDA: discriminative learning for dimensionality reduction and classification. In: NIPS (2009)
11. Lau, J.H., Newman, D., Baldwin, T.: Machine reading tea leaves: automatically evaluating topic coherence and topic model quality. In: EACL (2014)
12. Lewis, D.D., Yang, Y., Rose, T.G., Li, F.: RCV1: a new benchmark collection for text categorization research. J. Mach. Learn. Res. 5, 361–397 (2004)
13. Lin, T., Tian, W., Mei, Q., Cheng, H.: The dual-sparse topic model: Mining focused topics and focused terms in short text. In: WWW (2014)
14. Manning, C.D., Raghavan, P., Schütze, H.: Introduction to Information Retrieval. Cambridge University Press, Cambridge (2008)
15. Mehrotra, R., Sanner, S., Buntine, W., Xie, L.: Improving LDA topic models for microblogs via tweet pooling and automatic labeling. In: SIGIR (2013)
16. Nguyen, D.Q., Billingsley, R., Du, L., Johnson, M.: Improving topic models with latent feature word representations. Trans. Assoc. Comput. Linguist. 3, 299–313 (2015)
17. Rao, C.: Diversity and dissimilarity coefficients: a unified approach. Theoret. Popul. Biol. 21(1), 24–43 (1982)
18. Röder, M., Both, A., Hinneburg, A.: Exploring the space of topic coherence measures. In: WSDM (2015)
19. Soleimani, H., Miller, D.: Parsimonious topic models with salient word discovery. IEEE Trans. Knowl. Data Eng. 27(3), 824–837 (2015)
20. Solow, A., Polasky, S., Broadus, J.: On the measurement of biological diversity. J. Environ. Econ. Manag. 24(1), 60–68 (1993)
21. Wallach, H.M., Mimno, D.M., McCallum, A.: Rethinking LDA: why priors matter. In: NIPS (2009)
22. Wang, C., Blei, D.M.: Decoupling sparsity and smoothness in the discrete hierarchical dirichlet process. In: NIPS (2009)
23. Williamson, S., Wang, C., Heller, K.A., Blei, D.M.: The IBP compound Dirichlet process and its application to focused topic modeling. In: ICML (2010)
24. Xie, P., Xing, E.P.: Integrating document clustering and topic modeling. In: UAI (2013)
25. Yan, X., Guo, J., Lan, Y., Cheng, X.: A biterm topic model for short texts. In: WWW (2013)
26. Zhai, C., Lafferty, J.: Model-based feedback in the language modeling approach to information retrieval. In: CIKM (2001)

Collective Entity Linking in Tweets Over Space and Time

Wen-Haw Chong[1(✉)], Ee-Peng Lim[1], and William Cohen[2]

[1] Singapore Management University, 80 Stamford Road,
Singapore 178902, Singapore
whchong.2013@phdis.smu.edu.sg, eplim@smu.edu.sg
[2] Carnegie Mellon University, Pittsburgh, PA 15213, USA
wcohen@cs.cmu.edu

Abstract. We propose *collective entity linking* over tweets that are close in space and time. This exploits the fact that events or geographical points of interest often result in related entities being mentioned in spatio-temporal proximity. Our approach directly applies to geocoded tweets. Where geocoded tweets are overly sparse among all tweets, we use a relaxed version of spatial proximity which utilizes both geocoded and non-geocoded tweets linked by common mentions.

Entity linking is affected by noisy mentions extracted and incomplete knowledge bases. Moreover, to perform evaluation on the entity linking results, much manual annotation of mentions is often required. To mitigate these challenges, we propose *comparison-based evaluation*, which assesses the change in linking quality when one linking method modifies the output of another. With this evaluation we show that differences between collective linking and local linking, i.e. linking entities in each tweet individually, are statistically significant. In extensive experiments, collective linking consistently yields more positive changes to the linking quality, than negative changes. The ratio of positive to negative changes varies from 1.44 to 12, depending on the experiment settings.

Keywords: Concept linking · Entity linking · Entity disambiguation

1 Introduction

We explore entity linking for mentions in tweets. In entity linking, one links mentions in text, usually of named entities, to the referent entities in a given Knowledge Base (KB). Entity linking is also referred to as *entity disambiguation* or *concept linking* and is very similar to *Word Sense Disambiguation* (WSD). WSD aims to identify the correct sense of a word in a piece of text. Compared to WSD, entity linking focuses on named entity mentions. However it is unrealistic to expect detected mentions in tweets to match named entities only. This is due to social media content being written in an informal, case-insensitive manner. This increases mistakes by Named Entity Recognition (NER) tools, e.g. mistaking the term 'Merry' in the phrase 'Merry Christmas' as a named entity. Although prior work [5,11,14] do not consider such cases, in practical applications, they are

© Springer International Publishing AG 2017
J.M. Jose et al. (Eds.): ECIR 2017, LNCS 10193, pp. 82–94, 2017.
DOI: 10.1007/978-3-319-56608-5_7

impossible to exclude entirely. Hence in our entity linking work, we also cover cases of such noisy mentions. We use Wikipedia as the KB for linking. Thus our work can also be considered as Wikification [4,11].

Entity linking for social media content is challenging, as social media documents are short e.g. tweets, Foursquare shouts etc. Thus mentions arise in short documents, which may lack enough content or context for deriving features. This motivates the use of collective linking, i.e. exploiting information from multiple tweets to link mentions in a single tweet. Prior work [4,14] had considered collective linking over multiple tweets from the same user, and tweets linked by common terms or hashtags. In this work, we focus on the orthogonal aspects of space and time for collective linking. This is motivated by observations of tweeting behaviour with respect to events and geographical effects.

Event Effects. Tweets may be event related [1]. When tweet-worthy events occur, users may tweet about related entities, leading to an excess of related mentions in a space-time cube, i.e. a certain time period defined over a geographical area. Within a space-time cube, we can conduct collective linking and share linkage information across tweets. For example, the following are two actual tweets close in space and time: *<Stones>* and *<*Waiting for *@RollingStones* to come on stage so we can rock out *Singapore>*. Consider the mentions in italics. The first tweet has insufficient context for linking *Stones*. The second tweet's mentions can be linked with much less ambiguity, since *RollingStones* refers to the band entity 'The Rolling Stones' with high probability [15]. Given both tweets' space-time proximity, one can now use the second tweet's results to link the first tweet's *Stones* to the band with much more certainty.

Geographical Effects. Besides events, locations also affect tweeting behaviour. Certain entities may be more prevalent and mentioned more frequently at certain locations. Thus we can exploit geographical effects by collectively linking tweets that are close in space. For example, compare the following two tweets with mentions in italics: *<MBS* #throwback>, <Standby for SHOWTIME! @ *Marina Bay Sands>*. *MBS* in the first tweet is the surface form for many possible entities. The probability that it refers to 'Marina Bay Sands', a Singapore tourist attraction, is extremely low [15] at 0.000155. However if the second tweet with unambiguous mentions to 'Marina Bay Sands' occurs spatially near the first tweet, then it is much more plausible for the latter to be mentioning the same entity. Both event and geographical effects are often coupled due to events at Points of Interest (POI), e.g. concerts at a tourist attraction.

Challenges and Contributions. Our main contribution is a new collective entity linking method to exploit event and geographical effects. We connect tweets close in space and time to form a tweet graph, and define a novel objective function over the graph. This mitigates the challenge of entity linking for overly brief content. In addition, we introduce a comparison-based evaluation approach (see Sect. 4), which addresses the following challenges in evaluation:

– Noisy mention extraction: Automated mention extraction is noisy with mentions often being extracted in part, e.g. extracting *Garden* given *Madison Square Garden* or non-named entities being mistaken as named entities.

– Incomplete Knowledge Base: Many mentioned entities are often not in knowledge bases, even for a comprehensive KB such as Wikipedia.
– Annotation Effort: Expensive manual annotation is required to construct
ground truth linking for mentions in order to evaluate linking accuracy.

Based on comparison-based evaluation, the differences between collective linking
and local linking, i.e. linking entities in each tweet individually, are statistically
significant. Over the results of local linking, collective linking made many more
positive changes, i.e. that improves linking quality, than negative changes.

2   Related Work

The work in [9] introduces a semantic relatedness measure to quantify coherence. The measure uses only Wikipedia hyperlink structure and is inexpensive
to compute. The main idea is that semantically related entities should share
many common neighbors in Wikipedia. We use the same measure in our work.

For entity linking in tweets, Meij et al. [8] employed extensive feature engineering on content, page links and lexical word form. They then trained decision
trees for ranking entities that are related to each tweet (rather than each mention). For linking individual mentions, Liu et al.'s work [6] maximized an objective derived from coherence, mention-concept features and mention-mention features. The objective requires training of feature weights. In [14], the idea is to
exploit user interest for linking. A user's interest scores over entities are initialized and propagated over a graph of entities linked by relatedness [9]. Given a
new mention with multiple candidate entities, entities with higher interest score
are preferred. Huang et al. [4] proposed label propagation over a different form of
graph. Graph nodes are mention-entity tuples, connected based on weighted combination of various relations, e.g. coreferencing mentions, semantic-relatedness
[9] etc. After label propagation, high ranking tuples provide the linking results.

Different from the above works, we focus on orthogonal aspects such as spatial
and temporal proximity between tweets. In these aspects, the work by Fang
and Chang [2] is related. They learned entity distributions over time and large
geographical areas (smallest area considered is $100 \, km^2$) in a weakly supervised
setting. In contrast, we work in the unsupervised setting and consider much
smaller geographical areas spanning hundreds of meters. For an unsupervised
approach, TAGME [3] is applicable. Its key idea is: within the same document,
candidate entities across mentions vote for each other. For a given mention, the
entity with the highest prior is then selected from the top most voted entities.
We shall also implement TAGME as a non-collective entity linking baseline.

3   Approach

3.1   System Architecture

Our system architecture comprises of **Pre-processing**, **Local linking** and
Collective linking. Given a set of tweets for entity linking, the first pre-
processing step is mention extraction with an NER tool. The process is often

noisy with mentions being omitted or extracted partially. To mitigate this, we apply TweetNLP [10], which was specially developed for tweets. Next, for each extracted mention, we use the Google lexicon [15] to identify candidate Wikipedia entities. The lexicon lists possible mentions $\{m\}$ for each entity e along with the occurrence probability $p(e|m)$ derived from web hyperlinks.

In local linking, mentions to entities are linked individually for each tweet, without considering information from other tweets. We implemented two local linking methods: TAGME [3] and Loclink, introduced in Sect. 3.2. Local linking can be used to initialize the entity assignments for collective linking.

In collective linking, each mention in a tweet is linked using information within that tweet and from other tweets. Collective linking comprises three steps:

- **Tweet Graph Construction:** We first construct a graph that connects tweets by spatio-temporal proximity. The tweet graph is used to propagate information. Section 3.3 describes the construction process.
- **Initialization:** This means assigning an initial entity to each mention for subsequent refinement. This can be done using the results from local linking or with some other heuristics. We have opted for the former.
- **Optimization:** We define an objective function over the tweet graph and search for entity assignments to optimize it. Refer to Sect. 3.3.

3.2 LocLink: A Local Linking Method

Local linking processes each tweet individually, assigning entities that are semantically related to each other to make each tweet coherent. To quantify coherence, we adopt the semantic relatedness measure proposed in [9]. Consider entity e_a. Denote other entities with outgoing links to e_a as the set $I(e_a)$. Equivalently, regard e_a as having $|I(e_a)|$ incoming neighbors. For a pair of entities e_a, e_b with overlapping incoming neighbors, semantic-relatedness is then computed as:

$$SR(e_a, e_b) = 1 - \frac{log(max\{|I(e_a)|, |I(e_b)|\}) - log|I(e_a) \cap I(e_b)|}{log(|W|) - log(min\{|I(e_a)|, |I(e_b)|\})} \tag{1}$$

where $I(e_a) \cap I(e_b)$ are entities which link to both e_a, e_b in Wikipedia and W is the total number of Wikipedia entities. If $I(e_a) \cap I(e_b) = \emptyset$, we set $SR(e_a, e_b) = 0$.

Intra-tweet Coherence. Let d_i represent the i-th tweet containing $|m_i|$ mentions with set of linked entities $\mathbf{e}_i$. Also let m_{ia} be the a-th mention of d_i, with corresponding linked entity e_{ia}. We define the intra-tweet coherence as average semantic relatedness between its assigned entities:

$$C(d_i, \mathbf{e}_i) = \frac{1}{0.5|m_i|(|m_i| - 1)} \sum_{a=1}^{|m_i|} \sum_{b>a}^{|m_i|} SR(e_{ia}, e_{ib}) \tag{2}$$

Maximizing intra-tweet coherence makes each tweet as coherent as possible. However assigned entities can be rather obscure or rare. Hence a prior $p(e|m)$ is usually included [6,12,14] to favor more popular entities. In fact using only the prior for entity linking is a surprisingly strong baseline [5,11], while including the notion of coherence improves performance further. We use the prior from [15] and define the objective function for tweet d_i as:

$$Q_i(d_i, \mathbf{e}_i) = \xi . C(d_i, \mathbf{e}_i) + \frac{\tau}{|m_i|} \sum_{a=1}^{|m_i|} p(e_{ia}|m_{ia}) \tag{3}$$

where ξ and τ are combination weights. In the unsupervised setting, we simply let $\xi = \tau$ and assign entities to maximize Q_i. For single-mention tweets, coherence is undefined and we simply assign the entity with the highest prior to the mention. We call the above local linking method as **LocLink**.

3.3   Collective Linking in Space and Time

Inter-tweet Coherence. For collective linking, we exploit the fact that different tweets close in space and time may be related to the same event or have a common geographical effect, e.g. mentioning a common location. Therefore we expect some of the tweets to be *inter-coherent*. For computational efficiency, we shall only consider tweet pairs. Given tweets d_i and d_j with respective linked entity sets $\mathbf{e}_i$ and $\mathbf{e}_j$, we define the inter-tweet coherence as:

$$C(d_i, d_j, \mathbf{e}_i, \mathbf{e}_j) = \frac{1}{|m_i|.|m_j|} \sum_{a=1}^{|m_i|} \sum_{b=1}^{|m_j|} SR(e_{ia}, e_{jb}) \tag{4}$$

Tweet Graph Construction. Denote tweet d_i's timestamp as t_i and its location as l_i. In the simplest graph building scenario, we first retrieve geocoded tweets from a desired time interval and geographical area. For convenience, we call this a space-time cube although the geographical area need not be rectangular. For every pair of tweets d_i and d_j, we connect them if $|t_i - t_j| \leq \delta_t$ and $dist(l_i, l_j) \leq \delta_d$, where δ_t and δ_d are the respective thresholds for temporal and spatial proximities, and $dist()$ measures geographical distance.

We can relax the spatial requirement to include non-geocoded tweets. This assumes that non-geocoded tweets related to an event/POI may mention similar entities as the geocoded tweets. Thus from geocoded tweets in the initial space-time cube, we first extract mentions. We then query for more tweets with similar mentions and from same-city users (based on their profiles). We now have a mixture of tweets with and without location information. To consistently form the graph, we connect tweets based only on temporal proximity, i.e. $|t_i - t_j| \leq \delta_t$. Note that although individual edges are based on temporal proximity, the overall graph incorporates spatial-proximity since tweets are constrained to be from the initial space-time cube or users in the same city.

Objective Function. Let D and E be the set of nodes and edges respectively in the tweet graph. We define our objective function for collective linking:

$$Q(D, E, \mathbf{e}) = \frac{\alpha}{|D|} \sum_{i=1}^{|D|} C(d_i, \mathbf{e}_i) + \frac{\beta}{|E|} \sum_{(d_i, d_j) \in E} C(d_i, d_j, \mathbf{e}_i, \mathbf{e}_j)$$

$$+ \frac{\gamma}{|M|} \sum_{i=1}^{|T|} \sum_{a=1}^{|m_i|} p(e_{ia}|m_{ia}) \quad (5)$$

where $|M|$ is the total number of mentions, with set of linked entities $\mathbf{e}$; and α, β and γ are global combination weights. Essentially Q is a linear combination of intra-tweet coherence, inter-tweet coherence and the entity prior term. Thus Q encapsulates our earlier discussed intuitions about coherence and entity popularity. For a fixed set of weights, the optimization problem is to assign entities to mentions to maximize Q. For optimization, we use the decoding algorithm [6].

Parameter Settings. We consider unsupervised collective linking where labeled data is unavailable. Given that tuning/training is not possible, we consider two intuitive cases of averaging. In the first case, we use uniform weights in Q, i.e. $\alpha = \beta = \gamma$. We referred to this setting as *Uniform*. Alternatively, one can regard coherence and entity prior as very different notions and assign them equal importance. Hence in the second case, one averages over coherence and the entity prior, i.e. $\alpha = \beta, \gamma = \alpha + \beta$. We denote this setting *Avg(Coh,prior)*.

4   Comparison-Based Evaluation

Instead of heuristics/random initialization, we use local linking to initialize collective linking. This leads to a comparison-based evaluation approach. Essentially we compare initial and final linkings and determine if a change is an improvement (positive change), a degradation (negative change) or neither. As will be explained, there are several advantages in such an evaluation.

Annotation Effort. Firstly, we only need to compare linkings which are different between two linking results. This reduces the data annotation effort, compared to traditional evaluation using accuracy [13], i.e. proportion of correctly linked mentions. For example, to compute accuracy for a dataset of 100 mentions, each mention first has to be linked to the correct KB entity, typically via manual annotation [8]. In our evaluation framework, the annotation effort depends on the linkage differences between techniques and is usually less. For example, if all 100 mentions are linked by local linking and collective linking suggested 5 changes, then we only need to examine 5 changes. Clearly, more positive than negative changes is desired and implies improved performance.

Incomplete KB and Imperfect Linking. No KB can cover all mentioned entities. One can ignore unlinkable mentions or link them to the catch-all NIL entity [7,13,14]. However this discards data that may be useful for evaluation. Related to this, there is also the notion of how fine-grained a linkage needs to be, in order to be considered correct. Mentions can be linked to entities at different type or instance granularities. If one considers all coarse-grained linkages as wrong, many linkages useful for comparing techniques will be discarded.

For example, consider Table 1. The tweet was sent from the game venue during a college football match between Duke and Indiana University. Linking the mention *Duke* to Wikipedia, the most fine grained entity is e_1, i.e. Duke University's football team. However a linking technique may miss this perfect linking and choose other entities. Table 1 also lists Wikipedia entities in decreasing order of relatedness to the actual football team. Consider two techniques, one linking *Duke* to e_2, the other to e_4. Clearly the former provides useful information, even though both techniques miss out on e_1. In such cases, we still want to differentiate both techniques instead of regarding both linkings as equally wrong. If e_1 is not in the KB, but parent organizations such as e_2 and e_3 are present, it is still possible and reasonable to compare linking performance on *Duke*, instead of just discarding the mention as unlinkable. This calls for a comparison-based kind of evaluation.

Table 1. A sample tweet with mentions (in Italics). Row 2 lists candidate Wikipedia entities for the mention *Duke*, in decreasing relatedness.

Go *Duke!* #*PinstripeBowl* @ *Yankee Stadium*
• e_1: Duke_Blue_Devils_football: Duke University's football team
• e_2: Duke_Blue_Devils: Duke University's varsity sports team
• e_3: Duke_University: Duke University
• e_4: Duke: Monarch ruling over a duchy

Noisy Mention Extraction. Automated mention extraction is noisy. Often, incomplete sub-mentions are extracted. Even in cases where a mention should link to a unique entity, the notion of correct/wrong linking is less clear when sub-mentions are involved. Fortunately in comparison-based evaluation, we can compare entity assignments and pick the better one. For example, consider the tweet <Watching *Jeff Dunham* @*star* performing arts centre with the family>, where mentions (in italics) were extracted with TweetNLP [10]. The complete venue mention is *star performing arts centre*. However the sub-mention *star* was extracted, constraining entity linking to link *star*. Instead of discarding such cases, one can still compare linking results, e.g. linking to 'Movie_star' is intuitively preferred over 'Star': a luminous sphere of plasma in space. On a related note, if an extracted mention is in fact that of a non named-entity, such comparisons can also be used for evaluation.

4.1   Evaluating Changes

To evaluate changes, we define what constitutes each outcome. Firstly, we observe changes to often reduce or increase the specificity/granularity of linked entities. This leads to the consideration of parent-child relationships between entities in a type hierarchy. For brevity of discussion, we overload the term of entity types such that types can refer to semantic categories, organizations or locations. A super-type is decomposable into sub-types of finer granularities and this is applicable to semantic categories, instances, organizations and locations. For example entity e_1:'Duke_Blue_Devils_Football' is a sports team instance under the semantic category of 'American_football', and also a child organization of 'Duke_University'. For a location example 'New_York_City' (NYC) contains (and is the parent of) 'Madison_ Square_Garden', a multi-purpose indoor arena.

Clearly, we are considering more parent-child relationships beyond the semantic categories in ontologies. Hence any automated evaluation using only ontologies, e.g. the Dbpedia ontology[1] will be highly incomplete. Instead we compare type information using Wikipedia content when assessing linkage changes, e.g. e_1's Wikipedia page starts with '*The Duke Blue Devils Football team represents Duke University in the sport of American football*'.

We now discuss *positive changes* using Table 1:

- **Incorrect linking to parent entity/correct linking:** In this case, initial linking is unrelated and wrong, e.g. linking *Duke* to 'Duke', ruler of a Duchy. Changing the linking to either 'Duke_University' (a parent entity) or 'Duke_Blue_Devils_football' (the correct linking) is a positive change.
- **Parent entity to correct linking:** Eg. changing the linking for *Duke* from 'Duke_University' to 'Duke_Blue_Devils_football'. Intuitively, this provides more specific information to the system user.
- **Ancestor entity to parent entity:** In this case, the final linking is still not perfect, however the information specificity is increased, eg. changing the linking for *Duke* from 'Duke_University' to 'Duke_Blue_Devils'.
- **Incorrect sibling entity to parent entity:** We regard coarse-grained, related information as more useful than specific, but wrong information, e.g. if *Duke* is initially linked wrongly to 'Duke_Blue_Devils_men's_basketball' and changed to 'Duke_Blue_Devils', it counts as a positive change.

For the above, reversing the change direction count as *negative changes*. In addition, changes can be neither positive nor negative, e.g. replacing an incorrect entity with another. Such "neither" changes also include changing an initial unrelated entity assignment to a sibling or child entity, although this arguably improves our understanding of the tweets involved. For example, if *Duke* in Table 1 is initially linked to 'Duke' and changed to 'Duke_Blue_Devils_men's_basketball', we count it as a neither. Section 5.2 provides examples from experiments.

5   Experiments

Data. We conduct experiments on New York City (NYC) and Singapore (SG) tweets. To obtain meaningful tweets for linking (instead of trivial blabber [8]), we collect tweets near POIs or in space-time cubes covering performance events. For NYC, we obtained geocoded tweets from the CHIMPS Lab[2] that are within

---

[1] http://mappings.dbpedia.org/server/ontology/classes/.
[2] http://cmuchimps.org/.

100 meters of five popular event venues. For each venue, we consider two evenings (18:00–22:00) in Dec 2015 with the most tweets, obtaining 10 space-time cubes with an average of 24.8 tweets. For each cube, we form a spatio-temporal tweet graph for collective linking where tweets within 1 h and 100 m of each other are connected. For Singapore, we relax the spatial proximity requirement as discussed in Sect. 3.3 and obtain an average of 46.47 tweets over space-time cubes covering 17 performance events. The tweets are a mixture of geocoded and non-geocoded tweets. We connect tweets within 1 h of each other. Note that although individual edges in the tweet graph are based on temporal proximity, there is still a coarse notion of spatial proximity as most tweets are from Singapore, a small geographical area.

Following tweet graph construction, we apply both manual and automated mention extraction. For the latter, we use TweetNLP. For manual mention extraction, we process all 10 space-time cubes for NYC and 8 space-time cubes (out of 17) for SG, selected based on largest number of tweets. We link all mentions regardless of whether the parent tweets are related to the POI or event.

Local Linking Baselines. We use collective linking to modify the results of local linking. Thus the latter are equivalent to baselines. We implement *LocLink* (Sect. 3.2) with uniform weights for the objective in Eq. (3). We also implement *TAGME* [3], which is based on weighted voting among candidate entities.

5.1   Results

Results are summarized in Table 2 for New York City (NYC) tweets and Table 3 for Singapore (SG) tweets. Comparing collective linking to local linking, we see linkage improvements across all experiment settings. Consistently, collective linking makes more positive changes than negative changes, when applied on the results of local linking. In most cases, the ratio of positive to negative changes is larger than 2. The highest ratio is 12, for the experiment using NYC tweets with manually extracted mentions, TAGME for local linking and averaging over coherence and entity for Q, i.e., $Avg(coh, prior)$. The lowest ratio is 1.44, again on NYC tweets and with TweetNLP, LocLink and $Avg(coh, prior)$.

Table 2. Results on NYC tweets. Bracketed numbers are counts of unique mentions over which changes occur. (Δ: total changes, +ve: total positive, −ve: total negative, Ratio: +ve/−ve. **: significant at p-value $= 0.01$, *: sig. at p-value $= 0.05$)

Local linking method		LocLink				TAGME			
Mentions	Setting	Δ	+ve	−ve	Ratio	Δ	+ve	−ve	Ratio
Manual	*Uniform*	43	22 (14)	9 (6)	2.44**	73	37 (18)	6 (5)	6.17**
Manual	*Avg(coh, prior)*	20	13 (9)	3 (3)	4.33*	62	36 (18)	3 (3)	12.00**
TweetNLP	*Uniform*	61	23 (14)	11 (10)	2.09*	103	38 (19)	13 (12)	2.92**
TweetNLP	*Avg(coh, prior)*	50	13 (8)	9 (7)	1.44	95	35 (18)	9 (7)	3.89**

Table 3. Results on SG tweets. Notations as in Table 2.

Local linking method		LocLink				TAGME			
Mentions	Setting	Δ	+ve	−ve	Ratio	Δ	+ve	−ve	Ratio
Manual	*Uniform*	59	22 (10)	7 (4)	3.14**	93	38 (14)	8 (6)	4.75**
Manual	*Avg(coh,prior)*	28	16 (7)	2 (2)	8.00**	78	37 (16)	8 (6)	4.63**
TweetNLP	*Uniform*	83	29 (10)	9 (7)	3.22**	168	61 (21)	30 (8)	2.03**
TweetNLP	*Avg(coh,prior)*	44	23 (8)	2 (2)	11.5**	128	54 (23)	23 (6)	2.35**

Our results are statistically significant. Considering positive and negative changes, we conducted significance testing with the binomial test. The null hypothesis is that the proportion of positive and negative changes are equal. Except for one setting (TweetNLP, LocLink and $Avg(coh, prior)$), we are able to reject the null hypothesis at p-value of 0.05.

In both Tables 2 and 3, we also tabulate the number of unique mentions (in brackets) over which changes are made. This provides another view of the results accounting for mention diversity. In the trivial case, if all mentions are identical and initially wrongly linked, then it is easy to achieve many positive changes just from correcting one unique mention. However this overstates the performance advantage of collective linking due to a lack of mention diversity. From both tables, we see that the number of unique mentions for positive changes is consistently larger than that for negative changes, which is reassuring.

Collective linking exerts much of its influence through inter-tweet coherence. Recall that for *Uniform*, we use uniform weights for Q, while for $Avg(coh, prior)$, weight for the entity prior is set equal to total weights from intra and inter-tweet coherence. Thus in $Avg(coh, prior)$, inter-tweet coherence has smaller relative weight and plays a smaller role in affecting the linking results. This means that collective linking should suggest fewer changes. Indeed, we see that for a fixed mention extraction and local linking method, there are always fewer changes in $Avg(coh, prior)$ than *Uniform*.

5.2 Qualitative Analysis

Many, but not all changes are shared across experiments. Due to space constraints, we only illustrate changes for one experiment on NYC: TweetNLP for mention extraction, TAGME for local linking and uniform weighting for Q. Sample tweets are displayed in Tables 4 to 6, along with changes in the format: Initial entity → final entity. Readers can inspect Wikipedia entities by appending the entity name to the URL 'https://en.wikipedia.org/wiki/'.[3]

Positive Changes. Table 4 shows positive changes. Tweets N1 and N2 are from a college football match between Duke and Indiana University. The mention

---

[3] e.g. entity 'Duke_University' for tweet N1 (Table 4) is described in https://en.wiki pedia.org/wiki/Duke_University.

Table 4. Examples of positive changes (in bold), with affected mentions in italics.

N1	LETS GO *DUKE*!! #PinstripeBowl @Yankee Stadium
	Duke → Duke_University
N2	May be the post-season but finally getting to see the #*Hoosiers* play
	Hoosiers → Indiana_Hoosiers_football
N3	*Syracuse* game with my dad at The Garden-we're both alumni #cuse
	#cusenation #nyc
	Syracuse,_Sicily → Syracuse,_New_York

Duke in N1 is initially linked by TAGME to 'Duke': ruler of a Duchy. Collective linking then changed it to 'Duke_University'. Although this is not perfect, it is an improvement since Duke University is the parent organization of the football team involved. For N2, the final entity for *Hoosier* is correct in the strictest sense. Tweet N3 illustrates geographical effects, where surrounding tweets linked to NYC-related entities drive changes in the initial linking. For example, N3 is about a basketball game involving Syracuse University. Its final linking is a positive change, since an unrelated entity (a location in Italy) has been changed to a parent entity (university's location in NYC).

Negative Changes. Table 5 illustrates negative changes. N5's mention *World* is not from a named entity, but has been extracted by TweetNLP. It is impossible to automatically filter out all such mentions, hence linking is still conducted. The final linking in N5 is overly specific and wrong. N5 originates from NYC and surrounding tweets mentioned entities that drive the negative change. For example, mentions of NYC will drive the linking towards 'World_Wrestling_Entertainment' (WWE) since WWE's event had been held in NYC before. For N6, initial linking is to 'Yankee', which discuss usage of the word, including its usage in referring to Americans. The final linking is wrong and refers to an American baseball team.

Table 5. Examples of negative changes (in bold), with affected mentions in italics.

N5	*World*'s Most Famous Arena for my sixth sporting event in two weeks...
	World → World_Wrestling_Entertainment
N6	Incredible spread by the @*yankees*. Choice of pork,
	chicken, hot dogs and burgers. Salad bar
	Yankee → New_York_Yankees

Neither. Table 6 shows two examples where the final linking arguably improves our understanding of the tweet content. N9 is generated during a college football game. After collective linking, its mention *Bowl* is linked to a different series of football game, much better than the initial linking to 'Bowl', a container. N10's mention *WWF* is finally linked to a WWF wrestler, a more related entity than the initial linking to a nature conservation organization. Nonetheless such cases

Table 6. Sample changes (bold) for affected mentions (italics) that arguably improve tweet understanding, but are not counted as positive changes.

N9	*Bowl* Games with Famiky #CandyStripes NotPinstripes #PinstripeBowl **Bowl → Super_Bowl**
N10	I Met Former UFC Fighter & amp; *WWF* Wrestler Dan The Beast Severn At The MMA World Expo. Dan Is A... **World_Wide_Fund_for_Nature → Hulk_Hogan**

do not fall into our discussed scenarios in Sect. 4.1 and can be subjective to assess. Hence we do not count them as positive change.

6 Conclusion

Motivated by event and geographical effects, we have proposed a collective entity linking approach for tweets over space and time. In addition, we proposed a comparison-based evaluation strategy, that focuses on the linkage differences between competing entity linking techniques. This reduces manual annotation effort and mitigate challenges such as noisy mention extraction and incomplete KB. Our results show that collective linking over space and time performs much better than local linking techniques that process individual tweets. In extensive experiments, collective linking improves the linking quality of local linking.

Acknowledgements. This research is supported by DSO National Laboratories, and the National Research Foundation, Prime Ministers Office, Singapore under its International Research Centres in Singapore Funding Initiative.

We also wish to thank Dan Tasse and Jason Hong of CHIMPS Lab for providing access to the NYC tweets.

References

1. Atefeh, F., Khreich, W.: A survey of techniques for event detection in Twitter. Comput. Intell. **31**(1), 132–164 (2015)
2. Fang, Y., Chang, M.-W.: Entity linking on microblogs with spatial and temporal signals. TACL **2**, 259–272 (2014)
3. Ferragina, P., Scaiella, U.: Tagme: on-the-fly annotation of short text fragments (by wikipedia entities). In: CIKM (2010)
4. Huang, H., Cao, Y., Huang, X., Ji, H., Lin, C.-Y.: Collective Tweet Wikification based on semi-supervised graph regularization. In: ACL (2014)
5. Ling, X., Singh, S., Weld, D.S.: Design challenges for entity linking. TACL **3**, 315–328 (2015)
6. Liu, X., Li, Y., Wu, H., Zhou, M., Wei, F., Lu, Y.: Entity linking for Tweets. In: ACL (2013)
7. Mazaitis, K., Wang, R.C., Dalvi, B., Cohen, W.W.: A tale of two entity linking and discovery systems. In: TAC (2014)
8. Meij, E., Weerkamp, W., de Rijke, M.: Adding semantics to microblog posts. In: WSDM (2012)

9. Milne, D., Witten, I.: An effective, low-cost measure of semantic relatedness obtained from Wikipedia links. In: AAAI (2008)
10. Owoputi, O., O'Connor, B., Dyer, C., Gimpely, K., Schneider, N., Smith, N.A.: Improved part-of-speech tagging for online conversational text withword clusters. In: NAACL (2013)
11. Ratinov, L., Roth, D., Downey, D., Anderson, M.: Local and global algorithms for disambiguation to Wikipedia. In: ACL (2011)
12. Shen, W., Wang, J., Luo, P., Wang, M.: LIEGE: link entities in web lists with knowledge base. In: KDD (2012)
13. Shen, W., Wang, J., Luo, P., Wang, M.: Linden: linking named entities with knowledge base via semantic knowledge. In: WWW (2012)
14. Shen, W., Wang, J., Luo, P., Wang, M.: Linking named entities in Tweets with knowledge base via user interest modeling. In: KDD (2013)
15. Spitkovsky, V.I., Chang, A.X.: A cross-lingual dictionary for English Wikipedia concepts. In: LREC (2012)

Simple Personalized Search Based on Long-Term Behavioral Signals

Anna Sepliarskaia[1]([✉]), Filip Radlinski[2], and Maarten de Rijke[1]

[1] University of Amsterdam, Amsterdam, The Netherlands
{a.sepliarskaia,derijke}@uva.nl
[2] Google, London, UK
filiprad@google.com

Abstract. Extensive research has shown that content-based Web result ranking can be significantly improved by considering personal behavioral signals (such as past queries) and global behavioral signals (such as global click frequencies). In this work we present a new approach to incorporating click behavior into document ranking, using ideas of click models as well as learning to rank. We show that by training a click model with pairwise loss, as is done in ranking problems, our approach achieves personalized reranking performance comparable to the state-of-the-art while eliminating much of the complexity required by previous models. This contrasts with other approaches that rely on complex feature engineering.

1 Introduction

Search engines today combine numerous types of features when producing a ranking for a given query. They must provide ranked lists of results that are relevant (based on content), engaging (based on past user engagement), timely, and personally of interest to the user. These competing goals have led to a vast amount of work on each of them. Our focus is on personalization, which involves reranking documents on the search engine result page (SERP) so as to better satisfy a particular user's information need.

We present a novel approach to personalize search results with a model that is as effective as current state-of-the-art approaches, yet much simpler. By starting with a ranking produced by a commercial search engine, we know that the content of the top retrieved results is already likely to be of high relevance. However, we observe that usage still differentiates users and use this fact to rerank retrieval results based on implicitly collected usage. Consider, for instance, queries with only one intent but with a wide variety of relevant links such as "information retrieval conference." Links to SIGIR, ECIR, ICTIR, as well as links to general information on conferences are likely to be relevant. But each user has her own conference preference, which the system can infer from the user's past behavior—even if the user may be unable to formulate this preference directly in a query.

F. Radlinski—Work done while this author was at Microsoft, UK.

© Springer International Publishing AG 2017
J.M. Jose et al. (Eds.): ECIR 2017, LNCS 10193, pp. 95–107, 2017.
DOI: 10.1007/978-3-319-56608-5_8

Previous research on personalizing search using behavioral data has found that to improve the ranking for a given user, information from the user's short-term and long-term behavior can be used [1,4,17]. Here, short term behavior is information from the session in which the user is currently engaged; long-term behavior concerns information from all of the user's search history. We focus on the use of long-term behavior for personalizing search as long-term behavioral signals have led to larger improvements than short-term behavioral signals [1,18]. Also, short-term features cannot be used for the first query of a session, and over 40% of all sessions are of this sort [17].

At a high level, our approach calculates document scores given a query issued by a user, for each document d in the SERP. The score is a simple function combining three components: how well the document matches the query, how likely the user is to engage with documents at a given position, and how likely a user is to engage with a particular document. Perhaps surprisingly, despite not relying on handcrafted rules or sophisticated feature engineering, we show that performance is competitive with state-of-the-art models. Thus our key contribution is to show that formulating the optimization problem in this way removes the necessity for previously published complexity. We anticipate that by learning a simpler model, personalize reranking becomes more generally applicable, less complex computationally, and less error prone.

2   Related Work

There are several approaches to addressing personalized search, each with its own benefits and drawbacks. First, one needs to understand when reranking is needed. The distinction of queries in three types—navigational, informational and transactional—is well-known [2]. Users submitting navigational and transactional queries use search engines to retrieve easily findable and recognizable target results; for most navigational and transactional queries reranking is well understood [14,21]. Teevan et al. [21] show an easy and low-risk Web search personalization approach for navigational queries. Their approach achieves more than 90% accuracy. However, it works on the small segment of queries that the same user has issued at least three times. Query ambiguity is one of the indicators to inform us about changing the order of documents. Features and measures to predict it are proposed in [20]. If multiple documents have a high probability of being clicked following the query, then there is a great potential to improve the ranker.

The second type of related work concerns click models. Click models use implicit feedback to predict the probability of clicks [7]. Clicks can be a good indicator of failure or success. Features from click models are very useful for ranking documents [10,11]. However, few click models are personalized [16]. As click models use implicit feedback, manual assessment is not required nor is feature engineering. These models work well for improving the click through rate (CTR). However, to re-rank URLs the relative order of predicted relevance is more important than absolute CTR value [6]. The click model that achieves the

best performance for predicting probability of click is the User Browsing Model
(UBM) [8]. The main difference between UBM and other models is that UBM
takes into account the distance from the current document to the last clicked
document for determining the probability that the user continues browsing.

The third type of approach to behavior-based personalized search uses feature engineering to create behavior features and then learn a ranking function [13,18,22]. Work that follows this approach differs in the choice of machine
learning algorithms used. LambdaMart [3] is used in [13]. Several learning-to-
rank algorithms as well as regression models are used in [22]. Logistic regression
is used in [18]. Cai et al. [4,5] use matrix factorization and restrict themselves
to users with a sufficient volume of interactions. All of them devote significant
attention to feature engineering. For example, Masurel et al. [13] use the probability that the user skips, clicks or misses the documents. The winners of the
2014 Kaggle competition on personalized search use over 100 features [13].

3   Method

We begin by providing a general description of our personalized search method
and the intuitions behind it. At a high level, our goal is to obtain a simple yet
effective model. The simplicity is achieved by an easily interpretable function
that scores documents. The document score reflects the probability that the
document is relevant, which depends on three random variables: attractiveness
of the document to the user, attractiveness of the document to the query and
examination of the rank of the document. The uniqueness of our approach is
that, in contrast to previous models, we do not optimize the log likelihood of
click probability but explicitly fit the probability that one document is more
relevant than another in the SERP.

Our method shares traits of learning to rank methods and click models.
Inspired by approaches in non-personalized pairwise learning to rank, we explicitly model the probability that one document is more relevant than another one.
As in click models, personalized reranking involves modeling the relevance of documents using historical personal interactions with them. Further, we propose to
train our model using long-term behavioral signals, which can be compared with
classical click models [6,8,12] in its simplicity and approach, but it is as effective
as recent complex models.

In our algorithm, position bias is taken into account. We follow the position
model [7], in which it is assumed that examination of URLs on a SERP is a
function of their rank and does not depend on examinations and URLs at higher
positions. However, we assume that examination also depends on the query.
Moreover, we have a factor that reflects attractiveness of a document to a given
query. None of these parameters are personalized, therefore, we introduce new
ones that are user specific. We introduce only one type of user specific parameters
in this paper—attractiveness of a document to a specific user—but others could
easily be integrated in a similar fashion.

We first introduce some notation: (a) q denotes a query, r a rank, d a document, u a user; (b) $e_{q,r}$ denotes the examination of a document at rank r in a

SERP produced for q; (c) $a_{q,d}$ is the attractiveness of document d for query q; (d) $a_{u,d}$ is the attractiveness of d for user u. We will use the sigmoid function

$$\sigma(x) = 1/(1 + \exp(-x))$$

and the indicator function

$$I(x) = \begin{cases} 1 & \text{if } x \text{ is true} \\ 0 & \text{if } x \text{ is false.} \end{cases}$$

Given a query q submitted by user u, and a (non-personalized) SERP produced in response to q, our model re-ranks a document d that is originally placed at rank r in the SERP using the following scoring function:

$$score(q, d, u, r) = \sigma(a_{q,d}) \cdot \sigma(e_{q,r}) \cdot \sigma(a_{u,d}). \tag{1}$$

The learned parameters of the proposed model are $a_{q,d}, e_{q,r}, a_{u,d}$, which are single numbers. We use the sigmoid function to map these parameters to a probability.

We instantiate our model by training it based on implicit feedback from users. Given a query and user, we assume that the label of a given document is given by how the user interacts with it (click on it)—described specifically in Sect. 4. To achieve comparable results with the state-of-the-art model, we take inspiration from learning to rank methods and predict pairwise preferences of documents. More precisely, we map each document in the SERP to a number and the greater the difference between these numbers the higher probability that one document is more relevant than another. Specifically, for a given tuple (query q and user u) each pair of URLs d_i and d_j in a SERP with different labels is chosen. For each such pair we compute the scores $s_i = score(q, d_i, u, i)$ and $s_j = score(q, d_j, u, j)$, by using the parameters $a_{q,d_i}, e_{q,r_i}, a_{u,d_i}, a_{q,d_j}, e_{q,r_j}, a_{u,d_j}$, that were received up to that step. Let $d_i \prec d_j$ denote the event that d_i should be ranked higher than d_j. The scores are mapped to a learned probability that d_i should be ranked higher than d_j via a sigmoid function:

$$p(d_i \prec d_j) = \sigma(s_i - s_j). \tag{2}$$

We use a gradient descent formulation to minimize the cross-entropy function for each pair of documents in the SERP:

$$C(d_i, d_j) = -I(d_i \prec d_j) \cdot \log(p(d_i \prec d_j)) - (1 - I(d_i \prec d_j)) \cdot \log(p(d_j \prec d_i)). \tag{3}$$

Our method consists of three phases: first it tunes $e_{q,r}$, then $a_{q,d}$, and finally $a_{u,d}$. At each step the training procedure uses stochastic gradient descent (SGD), sequentially scanning the list of SERPs, calculating the gradient of the loss function for a SERP as

$$C_{serp} = \sum_{d_i, d_j} C(d_i, d_j), \tag{4}$$

and updating parameters $p_{uqd_1}, \ldots, p_{uqd_{10}}$ according to the following equation:

$$p_{uqd_i} \mathrel{+}= \eta \cdot \frac{\partial C_{serp}}{\partial s_i} \cdot \frac{\partial s_i}{\partial p_{uqd_i}}, \tag{5}$$

where η is a SGD-step, and p_{uqd_i} is one of $e_{q,r}$, $a_{q,d}$, $a_{u,d}$, depending on the phase.

We refer to our reranking model as specified in this section as *personalized ranked attractiveness* (PRA).

4   Experiments

In this section, we compare PRA with state-of-the-art models for personalized reranking. For this purpose we use data from the Yandex Personalized Web Search challenge [23]. We begin by noting that this dataset is the only publicly available dataset that satisfies our experimental needs. It contains information about SERPs and historical interaction with all documents shown to users: documents with their ranks and clicks on them. It also provides information on which user issued the query and interacted with the SERP.

The Yandex Personalized Web Search challenge dataset is fully anonymized. There are only numeric IDs of users, queries, query terms, sessions, URLs and their domains. The dataset comes with a full description of the SERPs contained in it: (a) the query for which the SERP was generated; (b) the ID of the user who issued the query; (c) URLs with their ranks and domains; and (d) the user's interaction with documents on the SERP, that is, indicators of clicks on documents. In case of a click, the dwell time in time units is also included. The organizers of the challenge suggest that documents with a click and dwell times not shorter than 400 time units are highly relevant to the query [23]. The following preprocessing was performed on the dataset before release: (a) queries and users are sampled from only one region (a large city); (b) sessions containing queries with a commercial intent as detected with a proprietary classifier are removed; (c) sessions with top-K most popular queries are removed; the number K is not disclosed. Some key statistics of the dataset are: (a) number of unique queries: 21,073,569; (b) number of unique urls: 703,484,26; (c) number of unique users: 5,736,333; (d) number of sessions: 34,573,630; and (e) number of clicks in the training data: 64,693,054.

Participants in the challenge are asked to rerank documents in SERPs according to the users' personal preferences.

We infer labels of URLs using a common approach [24]: (a) a 0 (irrelevant) grade corresponds to documents with no clicks or clicks whose dwell time is less than 400 time units; (b) a 1 (relevant) grade corresponds to documents that are clicked with a dwell time of more than 400 time units or clicked documents that have the lowest rank from all clicked documents in the SERP. A *satisfied* click is a click with a dwell time of at least 400 time units. We use two popular binary evaluation metrics: Precision@1 (P@1) and MAP@10.

To assess the consistency of our results, we measure the performance of our algorithms on several days. The dataset covers a period of 27 days; we use the first 20 days for training and the last 7 days (days 21–27) for testing. For each test day, we train algorithms on all days prior to the test day, and evaluate on the data collected for the test day. We do this over seven days to verify that the day of the week does not affect performance.

4.1 Training PRA

Each time the algorithm scans a SERP, we call this a "step." We use several hyper parameters to train PRA: (a) We make 5 steps for tuning each of parameters $a_{u,d}$, $a_{q,d}$, $e_{q,r}$: first, the algorithm makes 5 steps for tuning $a_{u,d}$, then 5 steps for tuning $a_{q,d}$, and finally 5 steps for tuning $e_{q,r}$. (b) We learn PRA by SGD with decreasing learning rate. In each step the learning rate is equal to the reverse square root of the number of steps $learning\ rate = 1/\sqrt{step\ number}$ (c) At the beginning we initialize all parameters $a_{u,d}$, $a_{q,d}$, $e_{q,r}$ to zero.

4.2 Baselines

We consider several experimental conditions (to be described below) and several baselines. Two baselines are considered for all experimental conditions: (a) *ranker* (ORIG) – the default order that search results were retrieved by the Yandex search engine; (b) *point-wise feature engineering* (PFE)—the winner of the Yandex Personalized Web Search challenge. The core of PFE [18] is feature engineering; it uses three types of feature. Some of the features reflect the basic ranker that feeds into the reranking: document rank, document id, query id, and so on. Another group of features describes the users' interactions with URLs: whether the user clicked, skipped or missed a document in the current session or the whole history. The third set of features are pairwise: they describe, for each pair of URLs in the SERP, which document has a higher rank. To train the PFE approach, Song [18] considers all queries and logistic regression as a classifier of satisfied clicks. (c) *User Browsing Model* (UBM) [8]—a click model that performs the best for prediction probability of click [9].

For some of our experimental conditions we consider additional baselines: (d) *past click on document* (PCLICK [21])—if the SERP contains a document that received a satisfied click from the user, then it is placed on the first rank; (e) *document click through rate* (DCTR)—rerank documents according to CTR for document-query pair.

4.3 Experimental Conditions

In the literature, multiple experimental conditions have been considered for comparing approaches to personalized reranking. We consider the following: (a) *all queries*; (b) *rerank examined documents only*, where we consider all queries but with a truncated list of documents: documents below the lowest click are

Table 1. Distribution of SERPs depending on the rank of the lowest click.

Rank of the lowest click	1	2	3	4	5	6	7	8	9	10
Percentage of SERPs	54.5	13.8	8.4	5.7	4.3	3.3	2.6	2.3	2.2	2.5

removed before running the evaluation; (c) *repeated document subset*: SERPs with documents that a person clicked on in the past; (d) *poor SERPs*; and (e) *cold start*, where we group users depending on the richness of their histories. We now describe those conditions in more detail.

All Queries. For comparability with PFE we report results on the full set of queries in the dataset and the exact same parameters as were mentioned by Song [18].

Rerank Examined Documents Only. To avoid falsely penalizing algorithms if they promote documents that are relevant but were not clicked simply because the user did not observe them, we also perform our experiments using all queries but with a truncated list of documents. Specifically, all documents below the lowest click are removed before running the evaluation. It is clear that SERPs with only the first retrieved document being clicked cannot be reranked in this condition, as all other documents are excluded for this particular analysis. To understand how the potential of algorithms to change the order of documents affects relative performance, we list the ranks of the lowest click in SERPs in the dataset in Table 1. In particular, note that after truncation, more than a half of the SERPs cannot be changed by any reranking algorithms. At the other end of the spectrum, for 2.5% of the SERPs, reranking algorithms can yield any permutation of the URLs in the originally retrieved list of results.

Repeated Document Subset. From previous studies [17,21], we know that users' behavior on repeated queries is particularly predictable. People often try to re-find documents, which they have read before [19]. Therefore, we consider a third experimental condition: the set of SERPs with documents that a person clicked on in the past. More precisely, in order for a SERP to be included in this set it should contain one and only one previously clicked document, where a past click on the document may have been for a different query. This subset of SERPs contains 13.8% of the total. For this condition, we use PCLICK [19] as an additional baseline.

Poor SERPs. From [20] we know that reranking is best applied selectively. Query ambiguity is one of the indicators to inform us about changing the order of documents and a good model should not rerank subsets of documents on which the ranker works well. For most queries, the top ranked document is clicked substantially more often than any of the other documents. However, for more ambiguous queries, or queries where the ranking is particularly poor, this is not the case. To evaluate such queries, in this subset we include queries for which the top ranked document is clicked less than twice as often as the second ranked

Table 2. Description of groups in the cold start problem.

Group number	1	2	3	4	5	6	7	8	9
Number of queries issued by users in the group	0	1–2	3–5	6–8	9–11	12–15	16–21	21–32	>32

document. A total of 48% of the SERPs in the dataset satisfy this condition. We also consider an additional baseline for this experimental condition: GCTR, the global clickthrough rate as defined in Sect. 4.1.

Cold Start. Naturally, there is the cold start problem: if a user or a query are new to the system, then it becomes more difficult to produce a proper ranking. To better understand the effectiveness of PRA we also provide information on the changes of algorithms' performance depending on the richness of users' histories. We divided users into nine groups depending on the number of sessions in their history in such a way that each group has about the same number of people, i.e., each group has roughly 11% of the users; see Table 2. The first group are the people that are new; group 2 contains users who issued one or two queries, etc. Below, we report experimental results per group.

5   Results

In this section we present our experimental results. We learned all models regardless of the experimental conditions. For each of the five experimental conditions defined above (all queries, examined documents only, repeated documents, query ambiguity and cold start problem), we report on the performance of our proposed approach, PRA, and of the baselines listed in Sect. 4.1.

5.1   All Queries

Table 3 lists the results for the "all queries" condition. We see that the performance of PRA and UBM is comparable to that of PFE, the state-of-the-art.

Table 3. Results for the "all queries" condition, on each test day: days 21–27.

		21	22	23	24	25	26	27
P@1	ORIG	0.597	0.596	0.588	0.596	0.594	0.587	0.581
	PFE	0.607	0.603	0.602	0.603	0.604	0.595	0.594
	UBM	0.603	0.600	0.596	0.600	0.600	0.591	0.587
	PRA	**0.612**	**0.610**	**0.604**	**0.611**	**0.607**	**0.600**	**0.597**
MAP	ORIG	0.719	0.718	0.713	0.718	0.714	0.712	0.709
	PFE	**0.726**	0.723	**0.723**	0.723	**0.724**	**0.718**	**0.717**
	UBM	0.724	0.724	0.719	0.724	0.722	0.717	0.713
	PRA	**0.726**	**0.725**	0.720	**0.725**	0.723	**0.718**	0.716

In terms of Precision@1 PRA always outperforms PFE and PFE outperforms UBM although the difference is not significant (t-test, p-value > 0.1). In terms of MAP, the difference between PFE, UBM and PRA is at most 0.3%, in either direction. All of these three models, PFE, UBM and PRA, significantly outperform ORIG, the production ranker (t-test, p-value < 0.01). Also, surprisingly, UBM has comparable performance with PFE (t-test, p-value > 0.1).

5.2 Rerank Examined Documents Only

We turn to the second experimental condition, where models rerank only examined documents. First, as this query set excludes documents below the lowest clicked position from reranking, all algorithms achieve higher scores, as we can see by contrasting the results in Table 4 with those in Table 3. The scores for PFE and PRA in this experimental condition are higher than in the "all queries" condition, both in terms of Precision@1 and MAP. Second, PRA outperforms PFE and UBM on both metrics. The difference in terms of Precision@1 exceeds 1.5% for each day, sometimes reaching 2.3%. Also, PRA performs significantly better than PFE, the state-of-the-art, in terms of MAP (t-test, p-value < 0.01). PFE and UBM have comparable performance.

Observing the performance differences between PRA, PFE and UBM relative to Table 3 more carefully, we note that the performance of PRA improved more due to the filtering of unobserved results. This tells us that on the complete dataset PRA promoted more documents that were not observed by the user than PFE or UBM. Thus, while the results in Table 3 are conservative (assuming all documents below the lowest actual click to be not relevant), the results in Table 4 are optimistic (restricted to documents for which we have more reliable evaluation labels). In both cases, we find that PRA outperforms PFE and UBM. We expect that results from an online evaluation would be somewhere between these two bounds.

Table 4. Results for the "rerank examined documents only" condition, on each test day: days 21–27.

		21	22	23	24	25	26	27
P@1	ORIG	0.597	0.596	0.588	0.596	0.594	0.587	0.581
	PFE	0.610	0.606	0.608	0.606	0.608	0.598	0.599
	UBM	0.610	0.600	0.606	0.613	0.610	0.600	0.597
	PRA	**0.628**	**0.627**	**0.620**	**0.629**	**0.627**	**0.621**	**0.623**
MAP	ORIG	0.719	0.718	0.713	0.718	0.717	0.712	0.709
	PFE	0.734	0.731	0.733	0.731	0.733	0.726	0.726
	UBM	0.730	0.730	0.728	0.731	0.732	0.726	0.723
	PRA	**0.741**	**0.740**	**0.735**	**0.741**	**0.740**	**0.736**	**0.737**

5.3   Repeated Document Subset

In this experimental condition we only consider SERPs that contain exactly one previously clicked document. As this segment of queries was the specific target of the method proposed by Teevan et al. [19], we consider the additional baseline PCLICK. Table 5 lists the results for this condition. PRA achieves the best overall Precision@1 scores, followed by PCLICK, PFE, UBM and ORIG. Note that the difference in performance between PRA and the other approaches is more than 1% on every single test day. Surprisingly, PCLICK significantly outperforms PFE (t-test, p-value < 0.01), even though PFE is far more complicated and includes features that reflect user interactions with documents.

Although UBM and PFE achieve a similar performance in other experimental conditions, in this one PFE achieves better results than UBM. This is a consequence of the fact that PFE is personalized and uses the whole history of a user to predict clicks. As expected, all approaches achieve better Precision@1 scores than ORIG.

Table 5. Results for the "repeated document subset" condition on each test day: days 21–27.

		21	22	23	24	25	26	27
P@1	ORIG	0.776	0.776	0.776	0.773	0.772	0.755	0.754
	PFE	0.819	0.801	0.825	0.800	0.817	0.782	0.805
	UBM	0.798	0.800	0.797	0.796	0.794	0.778	0.777
	PCLICK	0.839	0.838	0.838	0.836	0.830	0.817	0.815
	PRA	**0.851**	**0.849**	**0.848**	**0.848**	**0.842**	**0.830**	**0.831**
MAP	ORIG	0.850	0.850	0.850	0.849	0.848	0.836	0.835
	PFE	0.880	0.868	0.883	0.866	0.878	0.855	0.870
	UBM	0.866	0.867	0.866	0.865	0.864	0.853	0.853
	PCLICK	**0.894**	**0.893**	**0.893**	**0.892**	**0.888**	**0.879**	**0.880**
	PRA	0.893	0.891	0.891	0.891	0.886	0.877	**0.880**

Interestingly, the results for MAP show a different pattern. PCLICK and PRA work almost equally well: the difference between them is less than 0.3% and not statistically significant (t-test, p-value > 0.01). Both PCLICK and PRA perform significantly better than PFE (t-test, p-value < 0.01), which is better UBM, which, in turn, significantly outperforms *ORIG*.

5.4   Poor SERPs

Here we present results on ambiguous queries or queries where the ranking is particularly poor with the additional baseline DCTR; see Sect. 4.1 for a more precise definition. Table 6 shows the results on this subset for Precision@1 and MAP. For both metrics PRA outperforms other approaches, followed by PFE,

Table 6. Results for the "poor SERPs" condition on each test day: days 21–27.

		21	22	23	24	25	26	27
P@1	ORIG	0.420	0.415	0.415	0.415	0.425	0.424	0.424
	PFE	0.440	0.434	0.444	0.433	0.440	0.442	0.443
	UBM	0.440	0.435	0.437	0.437	0.440	0.444	0.443
	DCTR	0.450	0.448	0.433	0.450	0.446	0.441	0.443
	PRA	**0.460**	**0.458**	**0.458**	**0.458**	**0.457**	**0.456**	**0.458**
MAP	ORIG	0.617	0.614	0.611	0.614	0.618	0.617	0.618
	PFE	0.628	0.625	**0.630**	0.624	0.623	0.628	0.630
	UBM	0.627	0.627	0.623	0.628	0.624	0.630	0.630
	DCTR	0.628	0.626	0.610	0.627	0.621	0.617	0.620
	PRA	**0.635**	**0.633**	**0.630**	**0.634**	**0.632**	**0.629**	**0.633**

UBM, DCTR, and then ORIG. The difference between PRA and the other approaches is significant (t-test, p-value < 0.01). ORIG performs significantly worse than the other approaches, while for most test days the differences between PFE, UBM and DCTR are not significant.

Also, all algorithms work much better on the subset where the condition of *Poor SERPs* is not satisfied. The performances of ORIG, PFE, UBM and PRA are similar and the precision@1 scores are over 78%. To conclude, the PFE, UBM and PRA methods improve ambiguous queries, but do not affect non-ambiguous ones.

5.5 Cold Start Problem

In the "cold start problem" condition we provide information on the algorithms' quality depending on the richness of users' history. This experiment has several results; see Table 7 for the results for both Precision@1 and MAP.

First, despite the fact that ORIG is not personalized it performs better for users with a long history. One of the explanations of this is that people who

Table 7. Performance of algorithms depending on the number of queries issued by user.

		0	1–2	3–5	6–8	9–11	12–15	16–21	21–32	>32
P@1	ORIG	0.584	0.570	0.572	0.581	0.585	0.595	0.605	0.613	0.652
	PFE	**0.594**	**0.579**	0.581	0.592	0.597	0.608	0.620	0.631	0.684
	UBM	0.593	0.578	0.581	0.590	0.595	0.605	0.617	0.627	0.673
	PRA	0.588	0.577	**0.582**	**0.594**	**0.600**	**0.613**	**0.626**	**0.640**	**0.694**
MAP	ORIG	0.710	0.700	0.700	0.707	0.710	0.718	0.726	0.733	0.762
	PFE	0.714	0.702	0.704	0.712	0.717	0.725	0.734	0.744	0.783
	UBM	**0.715**	**0.705**	**0.706**	**0.714**	0.717	0.725	0.734	0.743	0.777
	PRA	0.710	0.700	0.703	0.713	**0.718**	**0.727**	**0.737**	**0.749**	**0.788**

use the search engine a lot learn to submit high quality queries [15]. Second, the personalized models PFE and PRA benefit more from a user's history than ORIG and UBM. For users who issued more than 32 queries the difference between ORIG and these model is more than 2% for both metrics. Also, for users with a limited history, PFE and UBM benefits more than other algorithms. However, for users with a rich history UBM performs worse than PFE, which in turn performs worse than PRA, but still much better than ORIG.

6   Conclusion

As search engines often show ten documents as a result page, most users can find a relevant item among them. However, different users have different interests. Thus for some users the first document may be relevant, but for others not. Thus we study reranking documents according to user interest. We have proposed a new simple method for personalized search based on long-term behavioral signals that matches or outperforms the state-of-the-art for this task.

We note that current state of the art solutions are effective, however they require extensive feature engineering. The most effective approaches have more than one hundred features. The second approach for this problem is manually creating rules, which is bound to work on a small segment of queries only. Another approach is click models. Click models are a very elegant solution for this problem, but in several experimental conditions work significantly worse than the state of the art. In contrast, our algorithm is applicable to all result sets, does not require feature engineering, but has comparable performance in all experimental conditions. We achieve this performance by incorporating click models with learning to rank algorithms.

We compared our proposed method with the state-of-the-art and with manually defined rules using a publicly available data set. We considered multiple experimental conditions. In all conditions we perform as least as well as the state-of-the-art and in several conditions we significantly outperform it according to both metrics used, despite the simplicity of our method.

Finally, we observe that our proposed approach only covers queries that have been seen previously, in the training data. In the future we plan to extend our approach to previously unseen queries by incorporating query similarity in our model. Also we plan to incorporate different relevance signals from query results and behavioral facets (visited pages, eye movement, etc.)

Acknowledgements. This research was supported by the Microsoft Research Ph.D. program. All content represents the opinion of the authors, which is not necessarily shared or endorsed by their respective employers and/or sponsors.

References

1. Bennett, P.N., White, R.W., Chu, W., Dumais, S.T., Bailey, P., Borisyuk, F., Cui, X.: Modeling the impact of short- and long-term behavior on search personalization. In: SIGIR, pp. 185–194 (2012)

2. Broder, A.: A taxonomy of web search. SIGIR Forum **36**(2), 3–10 (2002)
3. Burges, C., Shaked, T., Renshaw, E., Lazier, A., Deeds, M., Hamilton, N., Hullender, G.: Learning to rank using gradient descent. In: ICML, pp. 89–96 (2005)
4. Cai, F., Liang, S., de Rijke, M.: Personalized document re-ranking based on Bayesian probabilistic matrix factorization. In: SIGIR. ACM, July 2014
5. Cai, F., Wang, S., de Rijke, M.: Behavior-based personalization in web search. J. Assoc. Inform. Sci. Technol. (2017, to appear)
6. Chapelle, O., Zhang, Y.: A dynamic Bayesian network click model for web search ranking. In: WWW, pp. 1–10 (2009)
7. Chuklin, A., Markov, I., de Rijke, M.: Click Models for Web Search. Morgan & Claypool Publishers, San Rafael (2015)
8. Dupret, G., Piwowarski, B.: User browsing model to predict search engine click data from past observations. In: SIGIR 2008, pp. 331–338 (2008)
9. Grotov, A., Chuklin, A., Markov, I., Stout, L., Xumara, F., Rijke, M.: A comparative study of click models for web search. In: Mothe, J., Savoy, J., Kamps, J., Pinel-Sauvagnat, K., Jones, G.J.F., SanJuan, E., Cappellato, L., Ferro, N. (eds.) CLEF 2015. LNCS, vol. 9283, pp. 78–90. Springer, Heidelberg (2015). doi:10.1007/978-3-319-24027-5_7
10. Joachims, T.: Optimizing search engines using clickthrough data. In: KDD, pp. 133–142 (2002)
11. Joachims, T.: Evaluating retrieval peformance using clickthrough data. In: Text Mining (2003)
12. Liu, C., Guo, F., Faloutsos, C.: BBM: Bayesian browsing model from petabyte-scale data. In: KDD, pp. 537–546 (2009)
13. Masurel, P., Lefévre-Hasegawa, K., Bourguignat, C., Scordia, M.: Dataiku's solution to Yandex's personalized web search challenge. In: WSDM 2014 Workshop on Web Search Click Data (2014)
14. Matthijs, N., Radlinski, F.: Personalizing web search using long term browsing history. In: WSDM, pp. 25–34 (2011)
15. Morgan, H., Claudia, H., David, E.: Learning by example: training users with high-quality query suggestions. In: SIGIR. ACM, New York (2006)
16. Shen, S., Hu, D., Chen, W., Yang, Q.: Personalized click model through collabora tive filtering. In: WSDM, pp. 323–332 (2012)
17. Shokouhi, M., White, R.W., Bennett, P., Radlinski, F.: Fighting search engine amnesia: reranking repeated results. In: SIGIR, pp. 273–282 (2013)
18. Song, G.: Point-wise approach for Yandex personalized web search challenge. In: WSDM 2014 Workshop on Web Search Click Data (2014)
19. Teevan, J., Adar, E., Jones, R., Potts, M.: Information re-retrieval: repeat queries in Yahoo's logs. In: SIGIR, pp. 151–158 (2007)
20. Teevan, J., Dumais, S.T., Liebling, D.J.: To personalize or not to personalize: modeling queries with variation in user intent. In: SIGIR, pp. 163–170 (2008)
21. Teevan, J., Liebling, D., Geetha, G.R.: Understanding and predicting personal navigation. In: WSDM, pp. 85–94 (2011)
22. Volkovs, M.: Context models for web search personalization. In: WSDM 2014 Workshop on Web Search Click Data (2014)
23. Yandex. Personalized web search challenge (2013). http://www.kaggle.com/c/yandex-personalized-web-search-challenge/details/prizes
24. Yilmaz, E., Verma, M., Craswell, N., Radlinski, F., Bailey, P.: Relevance and effort: an analysis of document utility. In: CIKM, pp. 91–100 (2014)

The Effects of Search Task Determinability on Search Behavior

Rob Capra[✉], Jaime Arguello, and Yinglong Zhang

University of North Carolina at Chapel Hill, Chapel Hill, USA
rcapra@unc.edu

Abstract. Among the many task characteristics that influence search behaviors and outcomes, task complexity has received considerable attention. One view of task complexity is through the lens of *a priori* determinability—a measure of how much the searcher knows about the task outcomes, information requirements, and processes involved. In this paper, we explore a novel manipulation of *a priori* determinability in the context of comparative search tasks, which require comparing items (or alternatives) along different dimensions. Our manipulation involved explicitly including the *items* to be compared and/or the *dimension* by which to compare items in the search task description. We report on two user studies that investigate the effects of our manipulation on searchers' pre-task perceptions, search behaviors and post-task outcomes. Our results found that specifying the items had an effect on searchers' pre-task perceptions, but not their search behaviors and outcomes, and that specifying the dimension had no effect on perceptions, but made the task *more* difficult by possibly introducing uncertainty into the search process.

1   Introduction

A large body of prior research has investigated how search tasks vary along different dimensions. Task characteristics can relate to the search task's main activity (e.g., gathering factual information), end goal (e.g., well-defined or amorphous), task structure (e.g., its complexity), or the searcher's perceptions of the task (e.g., its expected difficulty) [11]. Studies have shown that many of these task characteristics can influence search behaviors and outcomes [12,15]. Understanding how task characteristics influence search behaviors is important to the study and design of interactive IR systems and to the development of models of how users engage in search processes.

Task complexity is one characteristic that has received considerable attention in recent work [2,3,7–9]. Task complexity is a multi-faceted concept that has been considered from different perspectives [15]. An influential approach proposed by Byström and Järvelin [5] is to view task complexity in terms of the *a priori* determinability of the task (i.e., how well a searcher is able to determine the outcomes, processes, and information requirements for a task in advance of actually performing it) [3,5,13]. A search task with low determinability is one

© Springer International Publishing AG 2017
J.M. Jose et al. (Eds.): ECIR 2017, LNCS 10193, pp. 108–121, 2017.
DOI: 10.1007/978-3-319-56608-5_9

with high *uncertainty* regarding the solution, information requirements, and the processes involved in gathering the needed information.

In this paper, we explore *a priori* determinability as a way to investigate the effects of task complexity on searchers' perceptions, search behaviors, and outcomes. Our goal was to *manipulate* the determinability of tasks while holding other task characteristics constant. To this end, we focus on *comparative* tasks. Our study participants were asked to search for information in order to compare and contrast items (or alternatives) belonging to the same category. For example, one of the tasks asked participants to compare and contrast different methods for purifying water during a hiking trip. Comparative tasks involve two important activities: (1) identifying the different *items* belonging to the given category (e.g., water filters, chemical tablets, boiling techniques) and (2) identifying the different *dimensions* by which the items can differ (e.g., the weight of the equipment, the time it takes to purify the water, the micro-organisms eliminated). We created 17 different task groups with 4 determinability levels each. Our four determinability levels were operationalized by explicitly including or excluding the items and dimensions in the task description.

We report on two crowdsourced studies (Study 1 and Study 2) that investigate the following three main research questions (RQ1-RQ3). In RQ1, we investigate whether searchers perceive differences in determinability when we include items and/or dimensions in a comparative search task description. In RQ2, we consider whether our manipulation of determinability yields differences in search behaviors and strategies. Finally, in RQ3, we investigate whether our manipulation of determinability yields differences in perceived outcome measures (e.g., difficulty, engagement, satisfaction) reported after completing the task. Study 1 investigates RQ1, and Study 2 investigates RQ2 and RQ3.

2   Related Work

Our research builds on prior work focused on understanding how task characteristics influence search behaviors and outcomes.

Tasks play an important role in understanding information seeking and searching [14]. Byström and Hansen [4] distinguish between *work tasks*, *information-seeking tasks*, and *information search tasks*. A search task is done in the context of an information-seeking task and both are done in the context of a work task. In this paper, we manipulate determinability at the information search task level.

A large body of prior work has characterized tasks along different dimensions. Li and Belkin [11] provide an extensive literature review and propose a classification scheme, including aspects of the task's activity, goal, and structure.

Different characterizations of task complexity have been proposed in prior work (see Wildemuth *et al.* [15] for a review). Campbell [6] characterized task complexity in terms of: (1) the number of required outcomes, (2) the number of paths to the outcomes, (3) the level of uncertainty about the paths, and (4) the degree of interdependence between the paths. Jansen *et al.* [8] (and later Kelly *et al.* [9]) used Anderson and Krathwohl's taxonomy of learning outcomes from educational theory [1] to create tasks with different levels of *cognitive* complexity. Cognitive complexity is associated with the amount of learning and mental effort required to complete the task. The simplest tasks (called *remember* tasks) require verifying or searching for a specific fact, while the most complex tasks (called *create* tasks) require searching in order to develop a new solution to a problem.

More closely related to our work, Byström and Järvelin [5] (and later Bell and Ruthven [3]) reduced task complexity to the *a priori* determinability of the task. Byström and Järvelin [5] defined *a priori* determinability as the extent to which a searcher is able to internalize the task at hand and deduce: (1) the task outcomes, (2) the information needed to produce the outcomes, and (3) the processes associated with gathering the required information. In later work, Bell and Ruthven [3] sought to *manipulate* the *a priori* determinability of tasks in a study. Tasks were designed to influence the *a priori* determinability of: (1) the information needed, (2) the strategy for searching, and (3) the need to synthesize information from multiple sources.

Similar to our research, past studies have investigated how different characterizations of task complexity influence participants' expectations, behaviors, and outcomes. Studies have found that complex tasks are associated with higher levels of expected difficulty [3,7,9], experienced difficulty [2,3,7,9], and search effort as indicated by measures derived from queries, clicks, bookmarks, and the task completion time [2,7–9]. Additionally, Kelly *et al.* [9] found that participants' choice of queries, query-terms, and pages visited *diverged* more from each other during complex tasks. Finally, Capra *et al.* [7] found that task complexity affected participants' engagement with a search assistance tool.

Our research adds to this body of work by investigating how a novel manipulation of task determinability in the context of comparative search tasks affects users' expectations, search behaviors, and experiences.

3   Determinability of Comparative Tasks

In this work, we manipulated the determinability of comparative tasks. Our tasks asked participants to compare and contrast items or alternatives belonging to the same category. Comparative tasks fall under the *analyze* level of cognitive complexity. According to Anderson and Krathwohl [1], *analyze* tasks require "breaking materials or concepts into parts and determining how the parts relate to each other", and may involve mental and physical activities such as "organizing and differentiating" and "creating spreadsheets". Comparative tasks involve two important activities: (1) identifying the different items associated with the given topic, and (2) identifying dimensions by which the items can differ.

Our manipulation involved making the task narrower in scope by specifying the items to be compared and/or the dimension by which to compare the items. We created 17 task topics (groups) with 4 determinability levels each, for a total of 68 task descriptions.[1] Each task included a background story that motivated the information need. The background story was consistent across all task descriptions within the same group, and the final information request was manipulated to elicit different levels of determinability. Below, we illustrate our four determinability levels for one task group. The items and dimension are shown in bold.

- **Unspecified (U):** no items or dimension specified.
  "You are planning an extended hiking trip. You heard that it can be unsafe to drink water directly from streams along the trail and that you need to purify water before drinking it. You would like to learn more about this. For this task, find out: What are different methods for purifying water to drink from streams and how do they differ?"
- **Specified Items (I):** specified two items to compare, but not the dimension.
  "You are planning... For this task, find out: How do **boiling water** and using a **charcoal filter** differ as methods for purifying water from streams?"
- **Specified Dimension (D):** specified the dimension, but not the items.
  "You are planning... For this task, find out: What are different methods for purifying water to drink from streams and how do they differ in terms of the **micro-organisms eliminated**?"
- **Both (B):** specified both items and the dimension.
  "You are planning... For this task, find out: How do **boiling water** and using a **charcoal filter** differ as methods for purifying water from streams in terms of the **micro-organisms eliminated**?"

Figure 1 illustrates a conceptual representation of comparative tasks as a grid to compare items across dimensions. Our *unspecified* tasks (Region U) left the items and dimensions completely open. Our *specified items* and *specified dimension* tasks were more narrowly focused by specifying two items to compare (Region I) or by specifying one dimension by which to compare any number of items (Region D). Finally, our *both* tasks were the most narrowly focused and limited the comparison to two items and one dimension (Region B).

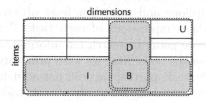

Fig. 1. Conceptual Representation of Comparative Tasks.

Our manipulation of task determinability can also be understood in light of the three factors described in Bell and Ruthven [3]. We expected that making the task more narrowly focused would produce less uncertainly in terms of the:

---

[1] Task descriptions are available at https://ils.unc.edu/searchstructures/resources/ecir2017_tasks.txt.

Table 1. Task topics, items, and dimensions used in our task descriptions.

Topic	Items	Dimension
Water purification methods	Boiling water and charcoal filter	Micro-organisms eliminated
Carpal tunnel treatments	Wrist splints and corticosteroids	Side-effects
Medicinal herbs for colds	Echinacea and St. John's Wort	Effectiveness
Motor oil for cars	Synthetic and organic motor oils	Performance
Types of rice	Rice and brown rice	Fiber content
Types of lightbulbs	Fluorescent bulbs and LEDs	Type of light
Types of ballet	Classical and neoclassical	Physical movements
Music speaker materials	Polypropylene and paper	High-frequency performance
Garden fertilizers	Organic and chemical fertilizers	Nutrient content
Types of paint thinner	Linseed and poppyseed oil	Effect on paint color
Wifi routers	Single band and dual band	Signal interference
Different types of plastic	PET and PVC	Ability to be recycled
Indoor dog breeds	Pug and Bichon Frise	Amount of exercise needed
Smoking cessation methods	Nicotine gum and nicotine patches	Success rate
Covering material for couch	Leather and microfiber	Ease of cleaning
Chinese keyboards	Pinyin and Wubi methods	Difficulty of learn
Cooking skillet materials	Aluminum and cast iron	Ability to distribute heat

(1) information needed, (2) the strategy for finding relevant content, and (3) the need to consult multiple sources. Table 1 lists the different topics, items, and dimensions associated with our 17 task groups.

4   Study 1: Search Task Evaluation

In our first research question (RQ1), we investigate whether specifying the items and/or the dimension of a comparative task might influence participants' perceptions of the *a priori* determinability and expected difficulty of the task.

To investigate this question, we conducted a crowdsourced study using the Amazon Mechnical Turk (MTurk). Participants were asked to read a series of four search task descriptions and rate their level of agreement with a set of 10 statements about their impressions of each task. Participants were asked to imagine that they were going to search for information using a web search engine in order to answer the task. Since we were primarily interested in participants' impressions of the task, participants did not actually perform the search. To gain statistical power, we designed Study 1 with task determinability as a within-subject factor. To keep the study manageable, we chose a subset of four task topics (carpal tunnel treatments, motor oil for cars, garden fertilizers, and types of plastic). Each participant did all four tasks, each with a different determinability level. Treatment combinations (n = 16) were created using a Latin square and participants were randomly assigned to one. Ultimately, we collected usable data from 63 participants.

Through a series of small-scale pilot tests, we developed a set of 10 statements (Table 2) to measure participants' perceptions about the task descriptions. These

Table 2. Study 1 questionnaire measures

Measure	Agreement Statement
PriorKnow	I already know a lot about this topic
Specificity	The task is very specific
Difficulty	I think the task will be difficult
Focused	The information requested is narrowly focused
NewInfo	The task description provides me with information that I did not already know
LackDim	There are dimensions of the task that are not specified in the description
ManyDetail	The task description has a lot of details
LookFor	Right now, I know some specific things to look for to address the task
SpecItems	The task is very specific in terms of the number of items I need to compare
SpecDim	The task is very specific in terms of the factors I need to consider when comparing the items

statements inquire about a range of concepts related to determinability and expected difficulty, including: prior knowledge, how focused the task is, whether it includes new information previously unknown to the participant, and the scope of the task in terms of the items to be compared and the dimensions by which to compare them. Participants indicated their level of agreement with the 10 statements on a 7-point scale from strongly disagree (1) to strongly agree (7). The statements were displayed below the task description. Participants were not allowed to go back after submitting responses for a task.

We recruited U.S. MTurk workers with a $\geq$ 95% acceptance rate. Participants were paid $0.75 USD and were allowed to complete the study only once.

Study 1 Results (RQ1): Table 3 summarizes the results for each of the 10 measures for each level of task determinability. We conducted one-way repeated measures ANOVAs to investigate the differences of determinability on each measure. Table 3 shows the results of this analysis, along with post-hoc comparisons using the modified Bonferroni correction outlined in Keppel [10](p.170).

Table 3 shows three important trends. First, our manipulation of determinability had a significant effect for 8 of the 10 measures—only Difficulty and LookFor

Table 3. Study 1 Questionnaire Results. The task versions with specified items are shown with gray background

	Unspecified (U)	Items (I)	Dimension (D)	Both (B)	p	post-hoc
PriorKnow	3.19 (1.62)	2.47 (1.60)	2.97 (1.70)	2.34 (1.38)	.001	U>I,B; D>B
Specificity	5.07 (1.48)	5.52 (1.27)	5.20 (1.39)	5.72 (1.07)	.000	U<I,B; D<B
Difficulty	3.52 (1.55)	3.61 (1.56)	3.66 (1.63)	3.87 (1.53)	.532	
Focused	4.51 (1.45)	5.19 (1.29)	4.79 (1.32)	4.95 (1.34)	.007	U<I,B
NewInfo	3.59 (1.70)	4.15 (1.83)	3.75 (1.62)	4.18 (1.76)	.027	U<I,B
LackDim	3.82 (1.58)	3.19 (1.50)	3.55 (1.59)	3.44 (1.48)	.057	U>I
ManyDetail	3.88 (1.44)	4.38 (1.39)	4.22 (1.55)	4.52 (1.54)	.032	U<I,B
LookFor	5.08 (1.18)	5.33 (1.17)	5.43 (1.13)	5.08 (1.52)	.131	
SpecItems	4.00 (1.65)	5.15 (1.51)	4.50 (1.76)	5.21 (1.56)	.000	U<D,I,B; D<I,B
SpecDim	4.83 (1.40)	5.02 (1.42)	4.98 (1.35)	5.35 (1.22)	.050	U<B

did not show significant differences. It is possible that participants reported similar levels of expected difficulty because all task descriptions were associated with comparative tasks at the *analyze* level of Anderson and Krathwhol's taxonomy [1].

The second important trend is that specifying the items in the task description had a strong effect on many of our measures. This can be seen by comparing task versions I and B (the two where the items were specified) with task versions U and D (the two where the items were *not* specified). The observed differences were generally in the directions we expected—the tasks with the specified items (I and B) were perceived to be more focused, had more details, and were more specific in terms of the items and dimensions to be considered. Similarly, the tasks with the specified items (I and B) were perceived to provide more information that the participant did not already know (NewInfo), and influenced participants to rate their prior knowledge as being lower (PriorKnow).

The third important trend is that specifying the dimension in the task description did *not* have a strong effect. This can be seen by comparing pairs of tasks where the only difference was the specified dimension (compare U vs. D and I vs. B). Based on our post-hoc comparisons, task versions U and D, as well as task versions I and B, were statistically equal for 9 out of 10 measures. SpecItems was the only measure for which specifying the dimension had a significant difference (U<D).

We were also interested in understanding how participants interpreted our 10 measures. To investigate this, we conducted an exploratory factor analysis with principle components analysis (using varimax rotation) and found a solution using two factors that explained 51% of the variance. All measures had factor loadings $\geq$.6 for these two factors, with no measures having a cross-loading $>$ .4. Because the measures had loadings $\geq$.6, we kept them all in our final solution.

The final factor loading matrix is shown in Table 4, and suggests that our questions measured two main concepts. Factor 1 focuses on the expected difficulty of the task. These measures were inversely related—when participants perceived the task as specifying new information, they reported having less prior

Table 4. Study 1 factor analysis

Measure	Factor 1	Factor 2
PriorKnow	−.606	
Specificity		.752
Difficulty	.693	
Focused		.741
NewInfo	.620	
LackDim		−.612
ManyDetail		.636
LookFor		.702
SpecItems		.661
SpecDims		.735

knowledge, and expected the task to be more difficult. Factor 2 focuses on the determinability of the task—the extent to which the task specified the information needed to complete it and reduced uncertainty about what to look for. Most of the measures loaded on this factor and were directly related. LackDim was negatively weighted because of its negative wording (Table 2). These results also suggest that participants did not make a strong distinction between our questions about the specification of items versus dimensions.

5   Study 2: Search Behaviors and Outcomes

In our remaining research questions, we investigate whether and how specifying items and/or dimensions of a comparative task might influence participants' search behaviors (RQ2) and perceptions about their search experience (RQ3).

To investigate these questions, we conducted a second crowdsourced study on Amazon Mechanical Turk. Participants were given a search task, were asked to search for and bookmark relevant pages, and were asked to complete a post-task questionnaire. Search tasks were presented as individual HITs on MTurk.

Each HIT presented a single task description and asked the participant to find and bookmark pages that would help them construct a response for the task. Searches were conducted using a custom-built search system that produced results using the Bing Web Search API. Participants were required to install toolbar buttons for bookmarking pages, viewing the current set of bookmarks in a pop-up window, and indicating when they were done with the task. When bookmarking a page, participants were required to provide a brief justification for why the page was useful, and the "view bookmarks" pop-up window allowed participants to delete bookmarks. Participants were required to bookmark at least 3 pages before finishing the HIT. Participants were paid $0.30 USD per HIT and were offered an $0.30 USD bonus if they bookmarked five or more pages. After finishing the task, participants completed a post-task questionnaire with questions about their level of enjoyment, engagement, interest increase, knowledge increase, perceived task difficulty, satisfaction with their solution and search strategy, and about how much time pressure they felt.

For each of our 17 task groups and 4 levels of task determinability, our goal was to collect data from 15 to 20 participants. To this end, we posted a total of 1,360 HITs on MTurk ($17 \times 4 \times 20$). Participants were randomly assigned to task-group/determinability-level combinations, but were *not* allowed to complete tasks from the same group (topic) more than once. Once all combinations of task topic and determinability level had data from at least 15 participants, we stopped the data collection. In total, we collected data for 1,317 search tasks and 348 participants. We recruited U.S. MTurk workers with a $\geq 95\%$ acceptance rate.

Study 2 Search Behavior Results (RQ2): For this and the next analysis, we conducted one-way ANOVAs to investigate the differences of determinability on each measure. Results for RQ2 are presented in Table 5. The first eight measures are associated with the level of search effort (e.g., number of queries, clicks, bookmarks, time between the query and the first SERP click (if any),

Table 5. Study 2 search behavior results. The task versions with a specified dimension are shown with gray background

	Unspecified (U)	Items (I)	Dimension (D)	Both (B)	p	post-hoc
Queries	1.91 (1.44)	1.93 (1.48)	2.32 (1.61)	2.35 (1.68)	.000	U,I <D,B
QueryLength	7.42 (4.37)	7.76 (3.34)	9.28 (6.99)	9.47 (5.81)	.000	U,I <D,B
Clicks	5.91 (3.10)	6.06 (3.11)	6.38 (3.49)	6.07 (2.96)	.300	–
ClicksPerQuery	4.04 (2.62)	4.04 (2.32)	3.58 (2.23)	3.50 (2.29)	.003	U,I > D,B
TimeToFirstClick	10.92 (24.42)	8.88 (9.55)	11.55 (22.93)	14.60 (33.38)	.028	I <B
Bookmarks	4.35 (1.10)	4.46 (1.17)	4.45 (1.18)	4.35 (1.15)	.417	
BooksPerQuery	3.12 (1.55)	3.20 (1.61)	2.71 (1.54)	2.68 (1.58)	.000	U,I > D,B
CompletionTime	330.56 (238.13)	353.94 (252.38)	374.75 (292.06)	373.21 (288.97)	.128	–
QueriesWOClicks	0.37 (0.80)	0.32 (0.79)	0.39 (0.77)	0.42 (0.81)	.440	–
QueriesWOBooks	0.50 (0.97)	0.47 (0.93)	0.61 (1.07)	0.61 (1.03)	.146	–
QueryLogLike	-46.65 (28.28)	-46.78 (20.26)	-58.85 (45.37)	-57.81 (36.95)	.000	U,I>D,B
UniqueQueries	1.31 (1.49)	1.42 (1.57)	1.82 (1.71)	1.90 (1.79)	.000	U,I <D,B
UniqueQTerms	0.84 (1.56)	0.83 (1.68)	1.02 (1.70)	0.84 (1.48)	.398	–
UniqueURLs	1.14 (1.49)	1.17 (1.51)	1.60 (1.84)	1.31 (1.47)	.001	U,I <D

and time to completion in seconds). The next two measures suggest trial-and-error (e.g., number of queries without a click and queries without a bookmark). Finally, the last four measures capture the extent to which participants' searches *diverged* from other participants who completed the same combination of task-group and determinability level. The query log-likelihood measure was computed by first generating a language model from all queries issued by the other participants who completed the same task-group/determinability combination, and then measuring the average log-likelihood of the participant's queries. A lower log-likelihood score indicates that the participant's queries contained language that was not frequently used by the other participants. Similarly, the last three measures are associated with the number of queries, query terms, and clicked URLs that were *not* observed in search sessions from the other participants.

Table 5 shows two important trends. The first main trend is that specifying the dimension had a strong effect on search behavior. This can be seen by comparing task versions D and B (the two versions where the dimension was specified) with tasks versions U and I (the two versions where the dimension was *not* specified). In terms of search effort, task versions D and B had significantly more queries, longer queries, and fewer clicks and bookmarks per query. It also took longer for participants to produce the first SERP click after issuing a query, suggesting that participants had more difficulty identifying relevant results. Task versions D and B also had more evidence of trial-and-error (more queries without clicks and bookmarks), although the differences were not significant. Finally, in terms of search strategy, while completing task versions D and B, participants issued significantly more unique queries (as evidenced by the lower query log-likelihood and greater number of unique queries), and clicked on more unique URLs.

The second important trend is that specifying the items did *not* have a strong effect. This can be seen by comparing between pairs of task versions where the only difference was the specified items (compare U vs. I and D vs. B). Both pairs of task versions were associated with similar amounts of search effort and divergence of search strategy. In fact, our post-hoc comparisons revealed no significant differences between task versions U and I and between D and B.

Study 2 Post-task Questionnaire Results (RQ3): Table 6 summarizes our post-task questionnaire results. Task determinability had a significant effect on several measures: knowledge increase, perceived difficulty, overall satisfaction, and satisfaction with the search strategy.

The trends in this data largely match the search behavior results reported in Table 5—specifying the dimension often had more impact than specifying the items. Table 6 shows that task versions D and B had lower overall satisfaction and lower satisfaction with the search strategy as compared to versions I and U, and that task version B had higher levels of difficulty than versions U and I. Tasks versions D and B also had lower ratings for enjoyment, engagement, and interest increase, but these differences did not reach statistical significance. Interestingly, knowledge increase was highest for task version I. Overall, these trends are consistent with the search behavior results and illustrate how specifying the dimension increased the effort required.

Table 6. Study 2 post-task questionnaire results. The task versions with a specified dimension are shown with gray background.

	Unspecified (U)	Items (I)	Dimension (D)	Both (B)	p	post-hoc
Enjoyment	4.69 (1.69)	4.73 (1.66)	4.47 (1.76)	4.44 (1.77)	.062	
Engagement	5.02 (1.65)	5.07 (1.71)	4.89 (1.67)	4.90 (1.74)	.448	
InterestInc	4.59 (1.86)	4.74 (1.73)	4.50 (1.89)	4.51 (1.87)	.297	
KnowledgeInc	5.03 (1.57)	5.36 (1.39)	5.09 (1.47)	5.11 (1.56)	.022	U,D,B<I
Difficulty	2.47 (1.42)	2.51 (1.48)	2.62 (1.47)	2.85 (1.58)	.005	U,I<B
OverallSat	5.64 (1.45)	5.61 (1.51)	5.38 (1.50)	5.27 (1.59)	.004	U,I>D,B
StrategySat	5.73 (1.38)	5.67 (1.40)	5.47 (1.46)	5.46 (1.52)	.029	U>D,B
TimePressure	3.25 (1.94)	3.14 (1.89)	3.30 (1.89)	3.24 (1.91)	.773	

6 Discussion

In this work, we set out to explore a novel method for manipulating the determinability of comparative search tasks. Our results reveal interesting points about how our manipulation of items and dimensions influenced participants' pre-search perceptions of a task, as well as their search behaviors and outcomes.

Study 1: Including the items in the task description influenced participants to perceive the task as being more focused and reduced their uncertainty about what to look for. Including the items also led participants to report that the tasks contained new information and that their prior knowledge of the task domain was lower. The same effects were *not* observed when the dimension was included. This was surprising to us. We expected that adding constraints of either type (items or dimensions) would increase participants' perceptions of determinability, and that there might even be an additive effect.

One possible explanation is that participants did not notice the dimension in the task description as much as they noticed the items. Another explanation is that participants did not perceive the dimension as being as strong of a constraint as the items. The items were specified as concrete noun phrases, while the dimensions were often specified as abstract concepts (e.g., "side-effects", "effectiveness", "performance", "success rate", "difficulty to learn"). These results have implications for how we define and operationalize determinability—it is not sufficient to assume that a task with more constraints is perceived to be more determinable. Based on our observations, constraints are not equal in their influence on determinability prior to working on the task.

Study 2: In Study 2, we found two interesting results: (1) specifying the items did not have an effect, and (2) specifying the dimensions did have an effect, but it was the opposite of what we expected. Based on the results of Study 1, it could be expected that specifying the items would make the task easier in Study 2. However, in Study 2, specifying the items did not yield differences in the search process or outcome measures. This may be because in the conditions where the items were *not* specified, participants were able to engage in satisficing behaviors, by bookmarking the most easily found pages or finding pages containing summaries of items.

The second interesting result from Study 2 is that specifying the dimension led to more difficult search tasks, as evidenced by greater levels of search activity, more divergent search strategies, greater levels of experienced difficulty, and lower levels of knowledge increase and satisfaction. In our initial view, we expected that adding both items and dimensions would reduce uncertainty (increasing determinability) and make the tasks easier to complete. However, adding the dimension constraint made the task more difficult, possibly because its determinability was actually *reduced*.

Task determinability involves uncertainly about different aspects of the task—the task inputs, required outcomes, and processes involved. It is possible that specifying the dimension narrowed the scope of the task and therefore reduced the uncertainly of the task outcome, but *increased* the uncertainty of the search process in different ways.

One possibility is that the dimensions of a comparative task may not be natural query-like concepts. For example, consider our "cooking skillet materials" task in Table 1. The dimension required participants to find information on how cooking materials are able to distribute heat uniformly. The language surrounding a dimension may be unknown or varied, making it more difficult to construct effective queries and identify relevant content. To gain more insight, one of the authors manually coded all queries submitted by our Study 2 participants as either containing at least one item and/or containing at least one dimension. Across all determinability levels, there were 1,441 queries with at least one item and 960 queries with at least one dimension. Indeed, this analysis suggests that it was easier for participants to explicitly search for items versus dimensions.

A second explanation is that many of our dimensions (e.g., "side-effects", "effectiveness", "performance", "success rate", "ease of cleaning", "difficulty to learn") may have introduced subjectivity into the task. Including such dimensions may have required participants to judge the credibility of information or synthesize different opinions.

Based on Bell and Ruthven's factors of determinability [3], including the dimension might have *increased* uncertainty in terms of the strategy for searching and identifying relevant content, as well as the need to integrate information from different sources. Interestingly, our results suggest that participants did not recognize this added complexity from the dimension from simply reading the task description (Study 1).

Summary: Our results provide insights into the complex relationship between task constraints and level of determinability. Our results suggest three important findings. First, task constraints that are perceived as making the task more focused may not yield differences in search behaviors and outcomes. In our case, *omitting* the items might have allowed participants to engage in satisficing behaviors when conducting the search (e.g., limiting the search to items found early on). Second, adding constraints to a task may not necessarily make it easier. In our case, specifying the dimension led to more search effort, possibly by introducing more uncertainty into the search process (e.g., constructing queries, identifying relevant content, and dealing with subjective information). Finally, while adding constraints may make a task harder, this may not be perceived before actually working on the task. This is the classic "you don't know what you don't know" paradox. In our case, participants did not perceive tasks with the dimension as being different than those without. However, the dimension led to more search activity, higher levels of difficulty, and lower levels of satisfaction. It is possible that participants experienced the added uncertainty only after starting the task (not by simply reading the description).

7   Conclusion

In this paper, we sought to create tasks with varying degrees of determinability, defined as the level of *uncertainty* regarding the task inputs, outputs, and processes involved. We focused on a specific task type (comparative tasks) and introduced a method for systematically varying task components (the items to be compared and/or the dimension by which to compare them). By including specific items or the dimension in the task description, we expected to narrow the scope of the task, increasing its determinability, and make it easier to complete.

Our results reveal a more complex situation. In Study 1, participants perceived differences in the tasks based on the items, but not the dimensions, possibly because the dimensions were more subtle in the task description. In Study 2, the items did not have an effect on search behaviors and outcomes (possibly due to satisficing behaviors in the absence of the items) and the dimensions actually made the search task harder. Interestingly, adding the dimension might have made the task less determinable by introducing uncertainly into the search process. A post-hoc analysis suggests that it was easier for participants to query for items than dimensions.

Our results have implications for experimental design, the design of search systems, and for frameworks of information seeking. From an experimental design standpoint, our results illustrate how subtle differences in task descriptions can have significant (and unexpected) influences on perceptions of tasks and on search behaviors. Wildemuth *et al.* [15] called for more research to investigate the impacts of task characteristics. Our results address this call, providing a detailed view of the effects of a specific, systematic manipulation of task determinability. From a system design perspective, our results suggest that providing recommendations or choices of dimensions in an interface (e.g., faceted search) may be especially helpful to users working on comparative tasks. Finally, our results provide additional insights into the role of *a priori* determinability in information seeking.

Acknowledgments. This work was supported in part by NSF grants IIS-1552587 and IIS-1451668. Any opinions, findings, conclusions, and recommendations expressed in this paper are those of the authors and do not necessarily reflect the views of the NSF.

References

1. Anderson, L.W., Krathwohl, D.R.: A Taxonomy for Learning, Teaching, and Assessing: A Revision of Bloom's Taxonomy of Educational Objectives. Longman, New York (2001)
2. Arguello, J.: Predicting search task difficulty. In: Rijke, M., Kenter, T., Vries, A.P., Zhai, C.X., Jong, F., Radinsky, K., Hofmann, K. (eds.) ECIR 2014. LNCS, vol. 8416, pp. 88–99. Springer, Cham (2014). doi:10.1007/978-3-319-06028-6_8
3. Bell, D.J., Ruthven, I.: Searcher's assessments of task complexity for web searching. In: McDonald, S., Tait, J. (eds.) ECIR 2004. LNCS, vol. 2997, pp. 57–71. Springer, Heidelberg (2004). doi:10.1007/978-3-540-24752-4_5
4. Byström, K., Hansen, P.: Conceptual framework for tasks in information studies. JASIST **56**(10), 1050–1061 (2005)
5. Byström, K., Järvelin, K.: Task complexity affects information seeking and use. Inf. Process. Manag. **31**(2), 191–213 (1995)
6. Campbell, D.J.: Task complexity: a review and analysis. Acad. Manag. Rev. **13**(1), 40–52 (1988)
7. Capra, R., Arguello, J., Crescenzi, A., Vardell, E.: Differences in the use of search assistance for tasks of varying complexity. In: SIGIR, pp. 23–32. ACM (2015)
8. Jansen, B.J., Booth, D., Smith, B.: Using the taxonomy of cognitive learning to model online searching. IPM **45**(6), 643–663 (2009)
9. Kelly, D., Arguello, J., Edwards, A., Wu, W.-C.: Development and evaluation of search tasks for IIR experiments using a cognitive complexity framework. In: ICTIR, pp. 101–110. ACM (2015)
10. Keppel, G., Wickens, T.D.: Design and Analysis: A Researcher's Handbook, 3rd edn. Prentice Hall, Upper Saddle River (1991)
11. Li, Y., Belkin, N.J.: A faceted approach to conceptualizing tasks in information seeking. Inf. Process. Manage. **44**(6), 1822–1837 (2008)
12. Toms, E.G.: Task-based information searching and retrieval. In: Ruthven, I., Kelly, D. (eds.) Interactive Information Seeking, Behaviour and Retrieval, pp. 43–59. Facet Publishing, London (2011). Chapter 3

13. Vakkari, P.: Task complexity, problem structure and information actions. Inf. Process. Manag. **35**(6), 819–837 (1999)
14. Vakkari, P.: Task-based information searching. Ann. Rev. Inf. Sci. Technol. **37**(1), 413–464 (2003)
15. Wildemuth, B.M., Freund, L., Toms, E.G.: Untangling search task complexity and difficulty in the context of interactive information retrieval studies. J. Doc. **70**(6), 1118–1140 (2014)

Inferring User Interests for Passive Users on Twitter by Leveraging Followee Biographies

Guangyuan Piao[✉] and John G. Breslin

Insight Centre for Data Analytics, National University of Ireland Galway,
IDA Business Park, Lower Dangan, Galway, Ireland
guangyuan.piao@insight-centre.org, john.breslin@nuigalway.ie

Abstract. User modeling based on the user-generated content of users on social networks such as Twitter has been studied widely, and has been used to provide personalized recommendations via inferred user interest profiles. Most previous studies have focused on *active users* who actively post tweets, and the corresponding inferred user interest profiles are generated by analyzing these users' tweets. However, there are also a great number of *passive users* who only consume information from Twitter but do not post any tweets. In this paper, we propose a user modeling approach using the *biographies* (i.e., self descriptions in Twitter profiles) of a user's *followees* (i.e., the accounts that they follow) to infer user interest profiles for *passive users*. We evaluate our user modeling strategy in the context of a link recommender system on Twitter. Results show that exploring the *biographies* of a user's followees improves the quality of user modeling significantly compared to two state-of-the-art approaches leveraging the *names* and *tweets* of followees.

1 Introduction

Online Social Networks (OSNs) have been growing rapidly since they first emerged in the early 2000's. A large number of users are now consuming different types of information (e.g., medical information, news) on OSNs [15] such as Twitter[1]. Therefore, inferring interests for users of these OSNs can play an important role in providing them with personalized recommendations for content. Most previous studies have inferred user interest profiles from a user's posts, such as their tweets on Twitter. The research focus in these studies has been on the user modeling of *active users* who actively generate content on Twitter. However, the percentage of *passive users* in social networks is increasing[2] (e.g., 44% of Twitter users have never sent a tweet[3]). *Passive users* are not inactive accounts, but rather users that only consume information on social networks without generating any content. In order to infer user interest profiles for passive users, some researchers have proposed linking *names* of followees (those

---

[1] https://twitter.com/.
[2] http://www.corporate-eye.com/main/facebooks-growing-problem-passive-users/.
[3] http://guardianlv.com/2014/04/twitter-users-are-not-tweeting/.

© Springer International Publishing AG 2017
J.M. Jose et al. (Eds.): ECIR 2017, LNCS 10193, pp. 122–133, 2017.
DOI: 10.1007/978-3-319-56608-5_10

whom a user is following) to Wikipedia[4] entities, and then utilizing these entities to derive abstract category-based user interests [3]. For example, if a user is following famous football players such as Cristiano_Ronaldo, they find the Wikipedia entity for Cristiano_Ronaldo, and then utilize the categories of the corresponding Wikipedia entity to infer user interests. Although this approach can extract highly accurate Wikipedia entities to boost a user's interest profile, it can only link popular Twitter accounts (e.g., the accounts of celebrities) to their corresponding Wikipedia entities. As a result, the information for a large percentage of a user's followees is often ignored.

Another piece of information that forms an important part of followees' profiles is their **biographies (bios)**. A **bio** on Twitter is a short personal description that appears in a user's profile and that serves to characterize the user's persona[5]. The length of a bio is limited to 160 characters. For example, Fig. 1 shows a user named *Bob* who has filled his bio with *"Android developer. Educator."*, which describes the user's identity.

In this paper, we investigate the bios of followees as a source of information for boosting user interest profiles. The intuition behind this is that a user might be interested in *"Android development"* if the user is following *Bob*. Our hypothesis is that, given a large number of bios of a user's followees, the entities mentioned in those bios can be leveraged for building quantified and qualified user interest profiles compared to using entities extracted based on the names of followees [3].

The contributions of our work are summarized as follows.

- We propose user modeling strategies leveraging the bios of followees for interring a user's interests by investigating two different interest propagation strategies.
- We evaluate our user modeling strategies against two state-of-the-art user modeling strategies for passive users in the context of a link recommender system on Twitter.

Fig. 1. Twitter profile.

The organization of the rest of the paper is as follows. Section 2 gives some related work, and Sect. 3 describes our proposed approaches for inferring user interest profiles. In Sect. 4, we present the Twitter dataset for our study, and Sect. 5 describes the evaluation methodology of the study. Experimental results are presented in Sect. 6. Finally, Sect. 7 concludes the paper with some future work.

2   Related Work

The largest area of work that is focused on inferring user interest profiles for *active users* is based on analyzing the tweets generated by them [1,2,9,10,13,14,16,17]. For example, Siehndel and Kawase [16] showed a prototype for generating

---

[4] https://www.wikipedia.org/.
[5] https://support.twitter.com/articles/166337.

user interest profiles based on the extracted entities from a user's tweets, and then linking these entities to 23 top-level Wikipedia categories. Kapanipathi et al. [7] extracted Wikipedia entities from a user's tweets, which were then used as activated nodes for applying various spreading activation functions based on a refined taxonomy of Wikipedia categories. As a result, a so-called weighted *Hierarchical Interest Graph* was generated for a given user. Instead of using Wikipedia categories, Piao and Breslin [14] and Orlandi et al. [11] leveraged DBpedia for propagating user interest profiles. DBpedia provides background knowledge about entities which not only includes the categories of entities, but also related entities via different properties. The authors of [14] showed that exploring some different structures of semantic information from DBpedia (i.e., categories as well as related entities) can improve the quality of user modeling in the context of a link (URL) recommender system on Twitter. Our work here is different from this line of work as we focus on inferring interests for *passive users* who do not generate tweets, but mostly just consume content from those that they follow on Twitter. In [16], the authors also suggested investigating other sources beyond tweets for user modeling. We address this research gap in our work.

Faralli et al. [5] leveraged the names of followees linked to Wikipedia entities, and then used these entities in order to infer user interest profiles for user recommendations. To the best of our knowledge, this work and the later work by [3] are the first ones exploring the use of followee profiles (in particular their names) for inferring user interest profiles, without analyzing any tweets. The authors in [5] have pointed out that leveraging followee profiles can build more stable and scalable user interest profiles than analyzing the tweets of followees. However, they also showed that only 12.7% of followees can be linked to Wikipedia entities on average. The most similar work to ours is [3]. Similar to [5], the authors in [3] first devised a method combining different heuristics for linking the followees of a user to Wikipedia entities. The linked entities were then used as activated nodes in a spreading activation function based on WiBi (Wikipedia Bitaxonomy [6]) in order to build abstracted *category-based* user interest profiles. Instead of leveraging the names of followees, we focus on the *bios* of followees for generating user interest profiles, and use the approach from [3] as one of our baseline methods (see Sect. 3.1).

3   User Modeling Approaches

In this section, we first describe two baseline methods (Sect. 3.1), and present our proposed user modeling approaches using two different propagation methods (Sect. 3.2). In this work, we define a user interest profile as follows.

Definition 1. *The interest profile of a user $u \in U$ is a set of weighted user interests (e.g., entities or categories of entities). The weight of each interest $i \in$ I: $w(u, i)$ indicates the importance of the interest i with respect to a user u.*

$$P_u = \left\{ \left(i, w(u, i) \right) \mid i \in I, u \in U \right\} \tag{1}$$

where I denotes the set of user interests, and U denotes the set of users.

3.1   Baseline Methods

SA(followees_name): Given a Twitter user u, the approach from [3] leverages the names of u's followees for user modeling. The input of this approach is a Twitter account, and the output is a *category-based* user interest profile obtained via a spreading activation method. It has three main steps for generating user interest profiles.

1. Fetch user's followees.
2. Link these to corresponding Wikipedia entities.
3. Apply a spreading activation method for the linked entities from step 2 to generate category-based profiles based on WiBi (Wikipedia Bitaxonomy[6]).

For example, if the user account *@bob* in Fig. 1 is following *@BillGates* (the Twitter account for Bill_Gates), this approach searches for the name Bill_Gates on Wikipedia in order to find the right entity for the Twitter account *@BillGates* using different heuristics. We used the author's implementation[7] [3] to link a user's followees to Wikipedia entities. The linked Wikipedia entities are activated nodes with $w(u,i) = 1$ for the next step. This approach further applies a spreading activation function from [7] (see Eq. 2) to propagate user interests from the extracted Wikipedia entities to Wikipedia categories, e.g., from Bill_Gatess to Category:Directors_of_Microsoft. The spreading activation function is defined as follows:

$$a_t(j) \leftarrow a_{t-1}(j) + d_{subnodes} \times b_j \times a_{t-1}(i) \tag{2}$$

$$d_{subnodes} = 1/\log N_{subnodes} \tag{3}$$

$$b_j = \frac{N_{e_j}}{N_{e_{cmax}}} \tag{4}$$

where j is a node (category) being activated, and i is a sub-node of j which is activating j. $d_{subnodes}$ is a decay factor based on the number of sub-nodes (sub-entities or categories) of the current category, and b_j is an *Intersect Booster* factor introduced in [7]. b_j is calculated by Eq. 4, where N_{e_i} is the total number of entities activating node j, and *cmax* is the sub-category node of j which has been activated with the maximum number of entities [7]. The weight of a node is accumulated if there are several sub-nodes activating the node.

As none of the previous studies [3,5] showed the performance of using followees' profiles (i.e., the names or bios of followees) compared to using followees' tweets, we also include a baseline method [4] using the tweets of followees for inferring user interest profiles to investigate the comparative performance of the two different approaches.

HIW(followees_tweet): This approach [4] extracts so-called *high-interest words* from each followee of a user u. The *high-interest words* consist of the top 20% of

---

[6] http://wibitaxonomy.org/.
[7] https://bitbucket.org/beselch/interest_twitter_acmsac16.

words in the ranked word list from a followee f's tweets. The latest 200 tweets from each followee are considered for our study, which results in over 13,940,000 tweets from the followees of 48 users (see Sect. 5). To construct the interest profile of u, high-interest words from all followees are aggregated by excluding the words mentioned only in a single followee's tweets. Finally, the weight of each word in u's profile is measured as $w(u, i) =$ the number of u's followees who have i as their high-interest words.

3.2   Proposed Approaches

Figure 2 presents the overview of our user modeling process, which consists of three main steps.

1. Fetch user's followees.
2. Extract Wikipedia/DBpedia [8] entities to the bios of followees.
3. Apply one of the interest propagation methods:
   (a) $SA(followees\_bio)$
   (b) $IP(followees\_bio)$.

Our approach is different from the baseline method $SA(followees\_name)$ especially in step 2. We use the Aylien API[8] to extract entities from the bios of a user's followees. The number of occurrences of each entity in the bios of followees is counted for measuring the importance of the entity with respect to a targeted user for inferring his or her interests.

SA(followees_bio): As one of our goals is investigating whether using the bio information of followees can improve the quality of user modeling compared to using the names of followees, we applied the same spreading activation algorithm (Eq. 2) for the entities extracted from the bios of followees. Therefore, the difference between this approach and $SA(followees\_name)$ is the set of activated nodes for propagation. For $SA(followees\_bio)$, the activated nodes are extracted entities from the bios of a user's followees with $w(u, i) = N_i$ which denotes the frequency of an interest i in their bios. Similar to $SA(followees\_name)$, the output of this approach is a *category-based* user interest profile.

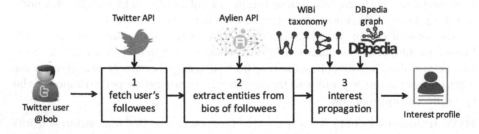

Fig. 2. Overview of our proposed approach

---
[8] http://aylien.com/.

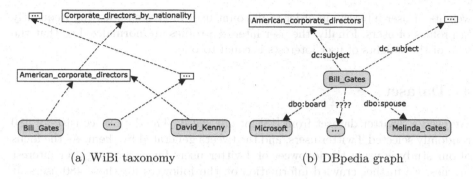

(a) WiBi taxonomy   (b) DBpedia graph

Fig. 3. Examples of WiBi taxonomy and DBpedia graph.

IP(followees_bio): Differing from the propagation of user interests using the taxonomy of Wikipedia categories, this approach uses an interest propagation method from [14]. The propagation method extends user interests using *related entities* as well as corresponding *categories* from DBpedia. DBpedia is a knowledge graph providing cross-domain knowledge extracted from Wikipedia. The difference between the WiBi taxonomy and the DBpedia graph is presented in Fig. 3. As we can see from Fig. 3(b), the DBpedia graph provides related entities in addition to the categories of an entity. For example, as well as providing categories for the entity `Bill_Gates` via the property dc[9]:`subject`, DBpedia also gives related entities such as `Microsoft` via the property dbo[10]:`board`. Therefore, as distinct from both *SA(followees_name)* and *SA(followees_bio)*, the output here is a user interest profile consisting of propagated *categories* as well as *entities*.

The authors in [14] also applied some *discounting strategies* for propagated categories, and entities via different properties. For example, a propagated category is discounted based on the log scale of the numbers of sub-pages (SP) and sub-categories (SC, see Eq. 5). A propagated entity is discounted based on the log scale of the number of occurrences of a property in the DBpedia graph (P, see Eq. 6), i.e., if the property appears frequently in the graph, the entities extended via this property should be discounted heavily. In addition, α is a decay factor for the propagation from directly extracted entities to related categories or entities ($\alpha = 2$ as in the study [14]).

$$CategoryDiscount = \frac{1}{\alpha} \times \frac{1}{\log(SP)} \times \frac{1}{\log(SC)} \tag{5}$$

$$PropertyDiscount = \frac{1}{\alpha} \times \frac{1}{\log(P)} \tag{6}$$

For all of the aforementioned user modeling approaches, after propagating user interest profiles, we further apply IDF (Inverse Document Frequency) to the

---

[9] The prefix dc denotes http://purl.org/dc/terms/.

[10] The prefix dbo denotes http://dbpedia.org/ontology/.

weights of user interests in order to discount user interests appearing frequently in profiles of users. Finally, the user interest profiles are normalized so that the sum of the weights of user interests is equal to one.

4 Dataset

We used a Twitter dataset from [13] for our study. The dataset consists of 480 randomly selected Twitter users, and the tweets generated by them. As the focus of our study is using the followees of Twitter users for generating user interest profiles, we further crawled information on the followees for those 480 users. It was possible to crawl followees for 461 of the original 480 users via the Twitter API[11] as some users did not exist anymore. As a result, the dataset consists of 461 users, and 902,544 followees of these users. Among these followees, we found that 812,483 users (around 90%) had filled out the bio field in their Twitter profiles.

Dataset for Our Experiment. As there can be a great number of followees even for a small number of users, we randomly selected 50 users with a corresponding set of 84,646 followees for our experiment. The descriptive statistics of the dataset are presented in Table 1. These 50 users have 77,825 distinct followees in total. 10% of these followees can be linked to Wikipedia entities using the approach from [3]. In contrast, 72,145 out of 77,825 (over 90%) followees have bios.

Table 1. Descriptive statistics of the dataset

# of users	50
# of followees	84,646
# of distinct followees	77,825
# of followees whose names can be linked to Wikipedia entities	7,785 (10%)
# of followees that have bios	72,145 (92.7%)

Comparison of Extracted Entities Using Names and Bios. As the entities either linked via the names or extracted from the bios of followees play a fundamental role in propagating user interests, we analyzed the number of entities that can be extracted using the two different sources. Figure 4 shows the difference between using the names and bios of followees in terms of the number of extracted entities. We can observe that using the bios of followees provides more than twice the number of entities when compared to using the names of followees. On average, 509 entities can be extracted for each user using the bios of followees, and 210 entities can be extracted for each user using the names of followees. This indicates that using the bios of followees can generate more

---

[11] https://dev.twitter.com/rest/public.

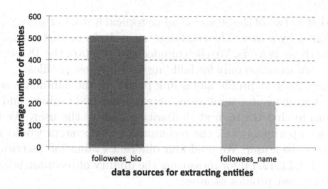

Fig. 4. Number of entities extracted via names and bios of followees.

quantified user interest profiles. We now move on to investigate whether the quantified user interest profiles generated by analyzing followees' bios have a higher quality as well, compared to those generated by linked entities based on the names of followees.

5   Evaluation Methodology

We were interested in finding out if leveraging the bios of followees for a passive user improves the quality of user modeling compared to using the names of followees. To this end, we evaluate different user interest profiles generated by different user modeling strategies in the context of a link (URL) recommender system on Twitter. Given this focus of our study, we applied a lightweight content-based recommendation algorithm for generating recommendations in the same way as previous studies [2,13,14].

Definition 2. *Recommendation Algorithm: given a user profile P_u and a set of candidate links $N = \{P_{i1}, ..., P_{in}\}$, which are represented via profiles using the same vector representation, the recommendation algorithm ranks the candidate items according to their cosine similarity to the user profile.*

Link (item) profiles were generated by applying the same propagation strategies applied for generating user interest profiles based on the content of a link. For example, given a link l, we first extract Wikipedia/DBpedia entities from the content of l, and then apply one of the aforementioned interest propagation strategies (see Sect. 3.2).

To construct a ground truth of links (URLs) for users, we assumed that links shared via a user's tweets were links representing a user's interests. Therefore, we further crawled the timelines of the 50 randomly selected users using the Twitter API, and extracted links shared in their tweets. In the same way as [14], we considered links that have at least four concepts to filter out non-topical ones which were automatically generated by third-party applications such as Swarm[12].

---

[12] https://www.swarmapp.com.

48 users were left as two of the 50 users had no topical links. On average, there were 31.46 links shared by a user. The candidate set of links consists of 1,377 distinct links shared by these 48 users. We then blinded the tweets of the 48 users, and used their followees' information only for building user interest profiles.

Given a user interest profile and a link profile in the candidate set, the recommender system measures similarities between the two profiles, and then gives the top-N links having the highest similarity scores to the user. We focused on $N = 10$ in our experiment, i.e., the recommendation system would list 10 link recommendations to a user. We used four different evaluation metrics as used in the literature [1,2,11,12,14] for measuring the quality of recommendations using different user interest profiles as input.

- **MRR.** The *MRR* (Mean Reciprocal Rank) indicates at which rank the first item *relevant* to the user occurs on average.
- **S@N.** The Success at rank N (S@N) stands for the mean probability that a relevant item occurs within the top-N ranked.
- **R@N.** The Recall at rank N (R@N) represents the mean probability that *relevant* items are successfully retrieved within the top-N recommendations.
- **P@N.** The Precision at rank N (P@N) represents the mean probability that retrieved items within the top-N recommendations are *relevant* to the user.

A significance level of alpha was set to 5% for all statistical tests. We used the *bootstrapped paired t-test*[13] to test the significance.

6   Results

Figure 5 presents the results of recommendations using different user modeling strategies in terms of four different evaluation metrics. Overall, $IP(followees\_bio)$ provides the best performance in terms of all evaluation metrics except S@10.

Comparison Between Using the Names and Bios of Followees. From Fig. 5, we observe that $IP(followees\_bio)$ as well as $SA(followees\_bio)$ which use the bios of followees for user modeling outperform $SA(followees\_name)$ which uses the names of followees. A significant improvement of $SA(followees\_bio)$ over $SA(followees\_name)$ in MRR (+63%), S@10 (+30%), P@10 (+78%), and R@10 (+84%) can be noticed ($p < 0.05$). With the same spreading activation method applied to two different sources: the names and bios of followees, the difference in terms of the four evaluation metrics clearly shows that exploring the bios of followees of passive users can infer better quality user interest profiles compared to using the names of followees.

Comparison Between Using the Bios and Tweets of Followees. Figure 5 also shows the performance of the baseline method $HIW(followees\_tweet)$, which analyzes followees' tweets for inferring *word-based* user interest profiles.

---

[13] http://www.sussex.ac.uk/its/pdfs/SPSS_Bootstrapping_22.pdf.

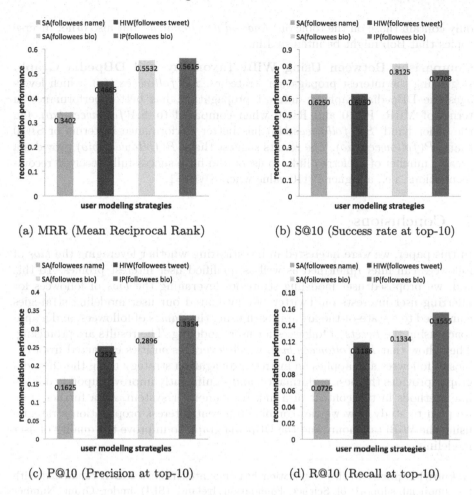

(a) MRR (Mean Reciprocal Rank)

(b) S@10 (Success rate at top-10)

(c) P@10 (Precision at top-10)

(d) R@10 (Recall at top-10)

Fig. 5. Results of the recommender system using different evaluation metrics.

The results show that our user modeling strategies using bios of followees outperform the baseline method in terms of all evaluation metrics. For instance, $IP(followees\_bio)$ outperforms $HIW(followees\_tweet)$ significantly in terms of S@10 as well as P@10 ($p < 0.05$). Considering $HIW(followees\_tweet)$ needs to analyze over 13,940,000 tweets of followees whereas $IP(followees\_bio)$ analyzes only around 77,000 bios of followees to build interest profiles for 48 users, our approach as well as $SA(followees\_name)$ [5], both of which use followees' profiles (i.e., the names or bios), are more scalable in the context of OSNs such as Twitter. On the other hand, the performance of $HIW(followees\_tweet)$ suggests that analyzing all the tweets of followees can lead to noisy information as an input for user modeling, which might decrease the quality of the inferred user interest profiles. For instance, a user who is following *Bob* (see Fig. 1) might be interested in "*Android development*", however, tweets posted by *Bob* would not only contain those on the topic of *"Android development"* but also other diverse topics that Bob might be interested in.

Comparison Between Using WiBi Taxonomy and DBpedia Graph. Regarding the interest propagation strategies, $IP(followees\_bio)$, which leverages the DBpedia graph for interest propagation, has better performance in terms of MRR, P@10 and R@10 when compared to $SA(followees\_bio)$. On the other hand, $SA(followees\_bio)$ has better performance in terms of S@10 than $IP(followees\_bio)$. The results suggest that $IP(followees\_bio)$ provides a greater number of preferred links to users who have successfully received recommendations, i.e., a higher P@10 value when S@10=1.

7   Conclusions

In this paper, we were interested in investigating whether leveraging the *bios* of followees can infer quantified as well as qualified user interest profiles. To this end, we proposed user modeling strategies leveraging the *bios* of followees for inferring user interests on Twitter. We evaluated our user modeling strategies compared to a state-of-the-art approach using the *names* of followees, and a approach using the *tweets* of followees for user modeling. The results are promising. They show that $IP(followees\_bio)$, which leverages entities extracted from the bios of followees and applies an interest propagation strategy using the DBpedia graph, provides the best performance, and significantly improves upon two baseline methods in the context of a link recommender system. As a further step, we plan to study how we can combine different interest propagation strategies using the WiBi taxonomy and the DBpedia graph to improve the quality of user modeling.

Acknowledgments. This publication has emanated from research conducted with the financial support of Science Foundation Ireland (SFI) under Grant Number SFI/12/RC/2289 (Insight Centre for Data Analytics).

References

1. Abel, F., Gao, Q., Houben, G.-J., Tao, K.: Analyzing user modeling on twitter for personalized news recommendations. In: Konstan, J.A., Conejo, R., Marzo, J.L., Oliver, N. (eds.) UMAP 2011. LNCS, vol. 6787, pp. 1–12. Springer, Heidelberg (2011). doi:10.1007/978-3-642-22362-4_1
2. Abel, F., Hauff, C., Houben, G.-J., Tao, K.: Leveraging user modeling on the social web with linked data. In: Brambilla, M., Tokuda, T., Tolksdorf, R. (eds.) ICWE 2012. LNCS, vol. 7387, pp. 378–385. Springer, Heidelberg (2012). doi:10.1007/978-3-642-31753-8_31
3. Besel, C., Schlötterer, J., Granitzer, M.: Inferring semantic interest profiles from twitter followees: does twitter know better than your friends? In: Proceedings of the 31st Annual ACM Symposium on Applied Computing, NY, USA, pp. 1152–1157. SAC 2016. ACM, New York (2016)

4. Chen, J., Nairn, R., Nelson, L., Bernstein, M., Chi, E.: Short and tweet: experiments on recommending content from information streams. In: Proceedings of the SIGCHI Conference on Human Factors in Computing Systems, pp. 1185–1194. ACM (2010)
5. Faralli, S., Stilo, G., Velardi, P.: Recommendation of microblog users based on hierarchical interest profiles. Soc. Netw. Anal. Min. 5(1), 1–23 (2015)
6. Flati, T., Vannella, D., Pasini, T., Navigli, R.: Two is bigger (and better) than one: the Wikipedia bitaxonomy project. In: ACL, vol. 1, pp. 945–955 (2014)
7. Kapanipathi, P., Jain, P., Venkataramani, C., Sheth, A.: User interests identification on twitter using a hierarchical knowledge base. In: Presutti, V., d'Amato, C., Gandon, F., d'Aquin, M., Staab, S., Tordai, A. (eds.) ESWC 2014. LNCS, vol. 8465, pp. 99–113. Springer, Heidelberg (2014). doi:10.1007/978-3-319-07443-6_8
8. Lehmann, J., Isele, R., Jakob, M., Jentzsch, A., Kontokostas, D., Mendes, P.N., Hellmann, S., Morsey, M., van Kleef, P., Auer, S.: DBpedia-a large-scale, multilingual knowledge base extracted from Wikipedia. Semant. Web J. 6(2), 167–195 (2015)
9. Michelson, M., Macskassy, S.A.: Discovering users' topics of interest on twitter: a first look. In: Proceedings of the fourth workshop on Analytics for noisy unstructured text data, pp. 73–80. ACM (2010)
10. Mislove, A., Viswanath, B., Gummadi, K.P., Druschel, P.: You are who you know: inferring user profiles in online social networks. In: Proceedings of the third ACM international conference on Web search and data mining, pp. 251–260. ACM (2010)
11. Orlandi, F., Breslin, J., Passant, A.: Aggregated, interoperable and multi-domain user profiles for the social web. In: Proceedings of the 8th International Conference on Semantic Systems, pp. 41–48. ACM (2012)
12. Piao, G.: User modeling on twitter with WordNet Synsets and DBpedia concepts for personalized recommendations. In: The 25th ACM International Conference on Information and Knowledge Management. ACM (2016)
13. Piao, G., Breslin, J.G.: Analyzing aggregated semantics-enabled user modeling on Google+ and twitter for personalized link recommendations. In: User Modeling, Adaptation, and Personalization, pp. 105–109. ACM (2016)
14. Piao, G., Breslin, J.G.: Exploring dynamics and semantics of user interests for user modeling on twitter for link recommendations. In: 12th International Conference on Semantic Systems, pp. 81–88. ACM (2016)
15. Sheth, A., Kapanipathi, P.: Semantic filtering for social data. IEEE Internet Comput. 20(4), 74–78 (2016)
16. Siehndel, P., Kawase, R.: TwikiMe!: user profiles that make sense. In: Proceedings of the 2012th International Conference on Posters and Demonstrations Track-Volume 914, pp. 61–64. CEUR-WS.org (2012)
17. Zarrinkalam, F., Fani, H., Bagheri, E., Kahani, M.: Inferring implicit topical interests on Twitter. In: Ferro, N., Crestani, F., Moens, M.-F., Mothe, J., Silvestri, F., Nunzio, G.M., Hauff, C., Silvello, G. (eds.) ECIR 2016. LNCS, vol. 9626, pp. 479–491. Springer, Heidelberg (2016). doi:10.1007/978-3-319-30671-1_35

Fusion of Bag-of-Words Models for Image Classification in the Medical Domain

Leonidas Valavanis, Spyridon Stathopoulos[✉], and Theodore Kalamboukis

Department of Informatics, Athens University of Economics and Business,
Athens, Greece
{lvalavanis,spstathop,tzk}@aueb.gr

Abstract. This paper presents a unified multimedia classification approach that integrates effectively visual and textual features. It combines
the Bag of Visual Words model (BoVW) together with a generalized
Bag of Colors (BoC) model and textual information in an early stage for
modality detection of images in the medical domain. Our contribution is
twofold: First we generalize the BoC model incorporating spatial information derived from a quad-tree decomposition of the images. Second
we propose a weighted linear combination of word embeddings for the
textual representation of the images. Experimental results conducted on
the data of the ImageCLEF contest for the years 2011, 2012, 2013 and
2016 demonstrate the effectiveness and robustness of our framework in
terms of classification accuracy outperforming all the published results
so far on the aforementioned datasets.

Keywords: Bag of Colors · Quad-tree · Image decomposition · Multimedia classification

1 Introduction

The rapid increase of multimedia information over the internet and social media
has led to the need of efficiently indexing and retrieving such information from
large scale databases. Furthermore, in certain application areas, as for example,
in the medical domain, content based image retrieval (CBIR) is an established
field of research for evidence-based diagnosis, teaching and research [17]. This
has motivated researchers to develop and continuously improve methods for
image representation, classification and retrieval. Inspired from text retrieval,
the BoVW model has shown promising results in the field of image retrieval and
classification [4,12,22,26]. This approach represents an image as a histogram of
visual words arising from intensity-based descriptors at salient points, such as
the Scale Invariant Feature Transform (SIFT) [13]. Correspondingly the BoC
model represents an image with a histogram of a predefined set of colors [25]. To
increase illumination invariance and discriminative power, of the BoVW model,
color descriptors have been proposed leading to several color variants of the
SIFT feature [1,20]. A weakness of both these models is the absence of spatial

© Springer International Publishing AG 2017
J.M. Jose et al. (Eds.): ECIR 2017, LNCS 10193, pp. 134–145, 2017.
DOI: 10.1007/978-3-319-56608-5_11

information of the visual words in the image. To address this problem, Spatial Pyramid Matching (SPM) [1, 11] was proposed, a technique that extracts features at multiple grids of different resolutions.

In this paper we propose a generalization of the BoC model based on a quad tree decomposition of images that takes into account the spatial distribution of colors. Based on the work of Wengert et al. [25], the new algorithm, referred by QBoC, generates visual words of square and orthogonal shape of homogeneous colors and different size depending on the level of the quad tree analysis of the image. In this way words are generated with high discrimination power relative to the original BoC algorithm. Experimental results of the proposed algorithm on several classification problems in the medical domain showed a substantial improvement over the original algorithm and its localized (LBoC) version. Furthermore, experimental results showed that the two models are effectively combined at an early fusion stage and produce the best results published so far on the ImageCLEF classification tasks of the past years. The remainder of the paper is organized as follows: Sect. 2, presents related research on the subject. Sections 3 and 4 describe the BoC representation and quad tree decomposition of an image respectively. Section 5 presents the evaluation framework. Section 6 discusses our benchmark datasets and presents results from the evaluation of our methods. Finally, in Sect. 7, concluding remarks are summarized with propositions for further research.

2   Related Work

Color is the simplest and most commonly used feature in CBIR. A color descriptor usually refers to the distribution of colors in the whole image (Global Color Descriptor). To incorporate spatial information images are either segmented into homogeneous regions in terms of a certain color or they are split into regular blocks and a color histogram is extracted from each region or block (Local Color Descriptor). MPEG-7 standard provides color descriptors, such as the color layout, dominant color, color autocorrelogram which are designed to capture the spatial distribution of colors. A first method to segment an image based on the color was proposed in [18]. Adjacent pixels of the same color are clustered together to form a homogeneous region and segments of a certain size and color form the vector representation of the image.

In the BoC model a color vocabulary is learned from a sample of an image collection, preferable from the train set. Using a learned color vocabulary (or palette) has shown experimentally that improves classification and retrieval performance over a flat color space quantization. The retrieval performance of the model was further increased when it was fused at an early stage with the SIFT descriptor into a compact binary signature [8]. The BoC model has been also used for classification of biomedical images in [6] showing that it is successfully combined with the BoVW-SIFT model in a late fusion manner.

One way to introduce spatial information in the representation of an image can be implemented by finding homogeneous regions based on some criterion. A common data structure which has been used for this purpose is the quad-tree. A quad-tree recursively divides a square region of an image into four equal size quadrants until a homogeneous quadrant was found or a stopping criterion is met. This structure has been used in several applications in Artificial Intelligence, robotics, games and image compression, representation and retrieval [15,21,27]. Several approaches use quad-tree decomposition to extract image features. In [23], a quad-tree decomposition based on wavelet features is proposed. The algorithm extracts texture features from homogeneous blocks at different levels of scales for each image. De Natale et al. [2] apply a quad-tree segmentation with fixed minimum and maximum region sizes to extract the distribution of dominant colors. In [19] an extended vector space based image retrieval technique was proposed which takes into account the spatial occurrence and co-occurrence of visual words inside pre-defined regions of the image obtained by quad-tree decomposition of the images up to a fixed level of resolution. In our approach we built on the BoC model and define visual words of several sizes and shape using the quad-tree decomposition of images. In the next section, for completeness, we proceed with a brief description of the BoC model [25].

3   The BoC Model

BoC creates the signature of an image from its color histogram. Unlike the traditional histogram, that uses a quantized color space, here the dominant colors are used, extracted from a sample of images taken, preferably, from the train set. The method is described in two phases: the construction of a visual vocabulary and the representation of the images by their color-histogram.

- **Visual Vocabulary:**
  1. Collect N images from the train set.
  2. Resize images to 256×256. Convert images to CIELab.
  3. Split images to 64×64 blocks of $4 \times 4 = 16$ pixels each.
  4. Select the dominant color from each block.
  5. The set of discrete colors of the blocks is clustered using k-means into m clusters.
  6. The centroids of the clusters define the visual vocabulary (V).
- **Image Representation:**
  1. for each pixel in the image
  2. replace its color with the nearest color in V.

The images are presented by vectors $x = (x_1, x_2, ..., x_n)$, where x_i denotes the feature frequency (TF, number of pixels of color i in the palette). Inspired from the vector space model in text retrieval the final weight of a feature was estimated by the TF-IDF weighting scheme defined by the relations

$$TF(x_i) = \sqrt{x_i}, \qquad IDF(x_i) = 1 + log(\frac{n}{df_i + 1}) \qquad (1)$$

where df_i is the number of images that contain color i. Finally the image-vectors are normalized with the L_1 norm which was found that provides better results compared to the Euclidean norm.

It is evident that the construction of the vocabulary and in particular the selection of its size is one of the weak points of the algorithm. A very large vocabulary can increase performance but at the cost of excessive computational requirements. On the other hand, a small size could lead to over-quantization of the color space and loss of information. We note here that sampling of images, and particularly in the medical collections, results to a vast majority of two colors, black and white. This fact influences the clustering algorithm, since these two colors are most likely to be selected as the centroids for more than one initial cluster by k-means algorithm. In the case, for example, where k-means selects more than one black centroids, then all the nearest colors to black will be assigned in one of these clusters and the rest of them may remain unaltered until the termination of the algorithm. Thus the resulting number of discrete clusters will be less than m. Furthermore, the centroids of the final clusters are biased towards the high frequency colors. Keeping only the discrete dominant colors the generated vocabulary captures a wider color space, which in turn, resulted to a more effective representation. In the next section we address the problem of introducing spatial information into the model.

4 Quad-Tree Decomposition of Images

The quad-tree decomposition, recursively subdivides an image region into quadrants if it is not homogeneous in the color. A region is considered homogeneous if all its pixels have the same color, or within some threshold of difference. In order for the algorithm to reach up to quadrants of size 1×1 the image should be represented by a square matrix $n \times n$ where n is some power of 2 ($n = 2^\rho$). An image of size $r \times c$ should first be resized into a square image $n \times n$, where n is the smallest power of 2 greater or equal to $max(r, c)$. Thus the quad tree decomposition creates a hierarchy of at most ρ levels with the root at level 1. The output of the decomposition is an array that contains the leafs of the tree defined by their upper-left corner, their size (the level of decomposition) and the color. In our experiments we have used the decomposition algorithm in [10] which has been used for image compression. The time to generate the vector representation of an image of 256×256 pixels requires in average 8.5 s on 16GB RAM computer with a 3.4MHz, i7-3770 CPU.

The boundaries between quadrants does not necessary represent quadrants of different color. Thus to further increase the discrimination power of visual words we parse the resulted quad tree and merge the siblings of the same color producing visual words of rectangular shape either vertical or horizontal as it is shown in Figs. 1 and 2.

At each level i, we check whether the quadrants form a horizontal rectangular of size $(2^{\rho-i} \times 2^{(\rho-i+1)})$ (Fig. 1), a vertical rectangular$(2^{(\rho-i+1)} \times 2^{\rho-i})$ (Fig. 2)

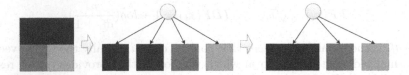

Fig. 1. Merging two neighbour siblings of the same color into a horizontal-rectangular visual word. (Color figure online)

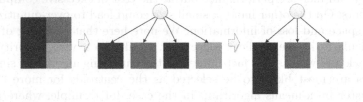

Fig. 2. Merging two siblings of the same color into a vertical-rectangular visual word. (Color figure online)

or they form double-sized squares as it is shown in Fig. 3. The visual words are described by their shape (hor, ver, sqr), size (level of the quad-tree) and color (the index of the color in the palette). In Fig. 3 the QBoC decomposition of an artificial image is shown schematically. Finally, we assign to each visual word TF-IDF weights as were defined in (1).

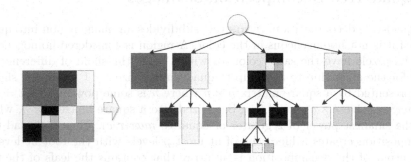

Fig. 3. Quad-tree decomposition of an artificial image. (Color figure online)

5   Evaluation Framework

In this section we present the evaluation framework for comparing our proposed algorithm versus the simple BoC and its localized version (LBoC) [25]. In our implementation of LBoC each image is split into 16×16 patches and the histogram of each patch is calculated. The histograms from a sample of images are clustered into M clusters (visual words) using k-means and each image is represented by an M-dimensional vector.

For the classification the LibLinear classifier[1] was employed, an open source library for large scale linear classification [3]. Linear SVMs are in general much faster to train and predict than the non-linear and can approximate large scale non-linear SVMs using a suitable feature map. Efficient feature mapping can be achieved using additive kernels, commonly employed in computer vision, with the homogeneous kernel map being the most common [24]. The homogeneous kernel map includes the intersection, Hellinger's, Jensen Shannon, Chi2, which allows large scale training of non-linear SVMs. The transformation of the data results into a compact linear representation which reproduces the desired kernel to a very good level of approximation. This transformation makes the use of linear SVM solvers feasible[2,3]. In our experiments, the homogeneous kernel mapping of VLFeat is used and more specifically the Chi2 kernel.

In order to train the SVM classifier, parameters must be first tuned to the best performing values while ensuring there is no over-fitting. For the SVM model, the Gamma parameter sets the homogeneity degree of the kernel, while C defines the cost parameter. Additional parameters, are the bias multiplier and the kernel type used, however, results were not greatly affected when varying their values. Tuning was performed using n-fold cross validation with random splits, for n = 10. For finding the optimal model parameters, a grid-search on parameter space (for cost and Gamma) was used using the embedded implementation of the LibLinear library. After experimentation using several parameters, results yielded better performance with cost 10, Gamma 0.5 and the L2-Regularized loss support vector.

6 Benchmark Datasets and Evaluation Results

The presented algorithms were evaluated on the classification problem with the data sets of imageCLEF contest of the years 2011, 2012, 2013 and 2016. A detailed descriptions of the data sets can be found on the contest website[4] and the overview papers [5,7,9,16]. The ImageCLEF collections contain a wide range of heterogeneous images from single to compound or multi-pane images (such as MR, x-rays or even tables) originating from various medical acquisition methods and articles. Accessing or classifying a sub-image of a multi-pane one makes the retrieval and classification a hard to solve problem. For 2011, 1,000 training and 1,000 test images classified into 18 categories were provided. In 2012 the images were classified into a hierarchy of 31 categories. In the 2013 dataset, 2,957 training and 2,582 test images were provided with the same class hierarchy as in 2012. In Table 1 we give the distribution of documents in the last three benchmarks. The full names of categories can be found in the website of the contests[5].

---

[1] https://www.csie.ntu.edu.tw/~cjlin/liblinear/.
[2] http://www.robots.ox.ac.uk/~vgg/software/homkermap/#r1.
[3] http://vision.princeton.edu/pvt/SiftFu/SiftFu/SIFTransac/vlfeat/doc/api/.
[4] http://www.imageclef.org/.
[5] http://www.imageclef.org/2016/medical.

Table 1. Collections' statistics- describe the distribution of images in the categories

Category	CLEF-2012		CLEF-2013		CLEF-2016	
	Train	Test	Train	Test	Train	Test
D3DR	25	30	46	26	201	96
DMEL	22	29	51	20	208	88
DMFL	21	13	33	33	906	284
DMLI	46	46	91	121	696	405
DMTR	29	18	46	20	300	96
DRAN	38	17	54	18	17	76
DRCO	12	13	22	1	33	17
DRCT	49	64	113	186	61	71
DRMR	43	55	97	90	139	144
DRPE	9	18	16	3	14	15
DRUS	48	13	60	85	26	129
DRXR	48	23	70	344	51	18
DSEC	5	24	29	96	10	8
DSEE	6	15	21	9	8	3
DSEM	5	14	18	1	5	6
DVDM	47	33	79	28	29	9
DVEN	32	32	64	20	16	8
DVOR	47	21	70	92	55	21
GCHE	21	50	63	19	61	14
GFIG	48	61	106	102	2954	2085
GFLO	48	50	98	20	20	31
GGEL	48	20	68	30	344	224
GGEN	47	42	89	21	179	150
GHDR	17	29	46	54	136	49
GMAT	6	14	20	5	15	3
GNCP	48	49	96	37	88	20
GPLI	10	18	28	22	1	2
GSCR	40	54	94	20	33	6
GSYS	48	47	95	16	91	75
GTAB	38	31	69	29	79	13
COMP	49	57	1105	1014		
TOTAL	1000	1000	2957	2582	6776	4166

Over the past years of the contest there was a large class of compound images that contained sub-images of several modalities something which made it difficult to train a classifier. In the 2016 contest there are no compound images and the

Table 2. Best results of classification task from imageCLEF contest.

Algorithm	CLEF2011	CLEF2012	CLEF2013	CLEF2016
textual	70.41	41.30	64.17	72.22
Visual	83.59	69.70	80.79	85.38
Mixed	86.91	66.20	81.68	88.43

dataset contains $6,776$ images in the train set and $4,166$ in the test set. However, both sets, train and test are quite unbalanced with one very large category (GFIG, $2,085$) and some other categories that contain just few images(GPLI 2) or (DSEE, 3). Thus for the 2016 data we present results from two different experiments: one with the original data of the competition and another with the training set enriched with the images of the train and test sets of the 2013 contest (2016 enriched). This was done mainly for two reasons: firstly to enrich the train set and in particular those classes with just a few images, (1–2 in some categories), and secondly for compatibility with the results in Table 2. Also, for compatibility with the performance measures used in the contest, we adopted the accuracy, defined by the proportion of successes in the test set.

As a baseline for evaluation of our proposed framework we use the results of the best runs of the contests shown in Table 2. We should note here that in our experiments we use only the data distributed by the organizers while most of the results in Table 2 were obtained using additional, visual or textual information, from external sources to train the classifiers.

6.1 QBoC Evaluation

Our first set of experiments aims to determine the impact of palette size and color space on the performance of the BoC models. For this purpose, our BoC models were tested for a grid of palette sizes (from 50 to $1,024$) and color spaces (RGB, HSV and CIELab). In all our experiments the best choice for the palette size was 512 for the BoC model and 50 for the LBoC and QBoC models and $M = 1024$ in the LBoC model. This seems reasonable because fewer colors create more visual words greater than one pixel, which have greater discriminative power, although there is a trade off between palette-size and performance. For a large enough palette size the two models coincide. Classification results across all datasets presented in Table 3 show that our proposed model substantially outperforms both BoC and LBoC models.

Table 3. BoC models evaluation

Algorithm	CLEF2011	CLEF2012	CLEF2013	CLEF2016	CLEF16 (enriched)
BoC (RGB-512)	67.97	38.70	51.39	70.3	71.09
LBoC (CIELab 50, 1024)	72.56	42.8	64.25	74.99	76.00
QBoC (RGB-50)	**77.15**	**51.7**	**67.74**	**75.78**	**78.18**

6.2 Multimodal Classification

In this section it is shown that the BoVW models, are successfully combined in an early stage with textual information and produce a very effective multi-modal classification algorithm of medical images. In early fusion, [28], image representation features extracted from different descriptors or modes are integrated into a single unified representation. Normalization techniques are applied before the integration so that features are on the same scale. In our multi-modal implementation we concatenate into one vector the image representation of the BoVW model, based on the PHOW descriptor, the representation from all the variants of BoC model together with the textual information from the image's caption.

Visual Information. In the BoVW model, small regions (local interest points) known as, salient image patches are identified that contain rich local information of the image. The extracted key-points, were expressed through the dense SIFT [1] descriptor. From those features a visual codebook was created with clustering using k-means algorithm. Each cluster (visual word) represents a different local pattern, which shares similar interest points. The histogram of an image, is created by performing a vector quantization which assigns each key-point to its closest visual word [26]. However, as it is known, the model loses the spatial information of the local descriptors due to the clustering which, severely limits their discriminative power. Pyramid Histogram of Visual Words (PHOW) addresses this problem by dividing the image into increasingly fine sub-regions of equal size, which are called pyramids. The histograms from each sub-region of the image are then concatenated into a single feature vector to form the PHOW descriptor. For our experiments, we partition the images into 1×1, 2×2 and 4×4 sub-regions and then combine the generated quantizations. Concerning the size of the codebook, a number of 1536 visual words was selected after testing of several values.

Textual Information. The textual representation of the images is defined by their caption. All the datasets, with an exception of the 2016 dataset contain a category (COMP) of compound images which may contain subfigures of different modalities. The caption of a compound image refers to all its constituent subfigures. In the 2016-dataset compound images have been split into their subfigures but the caption of the original compound image is assigned to all the subfigures. This makes difficult for the classifier to distinguish between sub-images. For the textual information we used the vector space model with TF-IDF weights of terms as a baseline. Since, captions are very short texts, traditional textual approaches, such as TF-IDF, have a difficulty to capture the semantic meaning of such texts due to the sparseness of the data. Thus we examined the use of dense word vectors, [14] known as word embeddings, as an efficient method to capture the semantic similarity of documents while reducing substantially their dimensionality that makes classification scalable. The vector representation of a caption is defined by:

$$d = \frac{1}{|d|} \sum_{t_j \in d} w_j \cdot e_j \tag{2}$$

where e_j denotes the embedding vector of the term t_j and $|d|$ the length of d. In the summation we discard terms with no embedding vectors. The coverage of the word embeddings for all our benchmarks is around 90% of the total terms. Several weights from the TF-IDF model were tested with the best results obtained with the $w_j = TF_j / \sum TF_j$. In our experiments we use the 200-dimensional word embeddings provided by the BioASK challenge[6]. These vectors were obtained by applying word2vec [14] to approximately 11 million abstracts from Pubmed. It is evident that those vectors apart of reducing the dimensionality, reduce also the train and classification time, while improving the effectiveness of classification. Table 4 presents the results of textual, visual (combining all the BoC models with the PHOW descriptor) and multi-modal classification. The results from the textual classification show a superiority of word embeddings when the train set is small (poor vocabulary) while the opposite happens with TF-IDF model. The use of our proposed QBoC model outperforms all the experiments in both: visual and multi-modal classification.

Table 4. Results of early fusion multi-modal classification

	Algorithm	CLEF2011	CLEF2012	CLEF2013	CLEF2016	CLEF2016 enriched
Textual	TF-IDF	75.98	**64.95**	66.11	66.87	**73.76**
	word2vec	**77.34**	61.4	**68.63**	**71.03**	73.48
Visual	BoC-Phow	82.91	70.70	81.88	80.74	82.74
	LBoC-Phow	84.18	71.10	81.99	82.33	84.16
	QBoC-Phow	**85.84**	**71.80**	**83.04**	**82.45**	**85.19**
Mixed	BoC-Phow-TF-IDF	86.13	71.64	84.36	81.70	84.67
	LBoC-Phow-TF-IDF	85.94	74.90	84.74	83.17	86.10
	QBoC-Phow-TF-IDF	87.11	72.30	86.17	83.70	86.61
	BoC-Phow-word2vec	87.79	**79.40**	84.36	82.36	87.21
	LBoC-Phow-word2vec	87.60	78.90	84.20	**86.25**	86.53
	QBoC-Phow-word2vec	**88.57**	77.90	**85.71**	86.10	**88.07**

7 Conclusions

We have proposed a generalization of the BoC model that incorporates spatial information into the model using a quad tree decomposition of the images. The algorithm improved performance substantially on image classification over the original BoC model and its local version. We demonstrated that the fusion of BoW models that combine low level visual features together with visual features extracted from colors and text outperforms the overall best performance from the single best classifier for each modal as even the weak performing ones

---

[6] http://participants-area.bioasq.org/.

bring complimentary information. The experiments with four benchmarks from imageCLEF contest show that the proposed framework constantly outperforms substantially other state of the art approaches and achieves the best results over all previously published on the same datasets. Our approach shows a slight lag behind the best run in the 2016 competition, which may be due to several small differences in the algorithms as it is for example, the term selection to name one.

Our results are very encouraging and open new directions for further investigation. Such a direction is the improvement on the efficiency of the proposed algorithm QBoC. The decomposition algorithm proceeds the in a top-down manner which is an expensive approach since it has been proposed for image compression and not for retrieval. Currently we are investigating a very fast implementation of QBoC based on a bottom-up decomposition of the image. Future directions of our framework include feature selection as well as the investigation of the impact of the visual vocabularies on the classification problem.

References

1. Bosch, A., Zisserman, A., Muñoz, X.: Image classification using random forests and ferns (2007)
2. De Natale, F., Granelli, F.: Structured-based image retrieval using a structured color descriptor. In: International Workshop on Content-Based Multimedia Indexing (CBMI 2001), pp. 109–115 (2001)
3. Fan, R.E., Chang, K.W., Hsieh, C.J., Wang, X.R., Lin, C.J.: Liblinear: a library for large linear classification. J. Mach. Learn. Res. **9**, 1871–1874 (2008)
4. Furuya, T., Ohbuchi, R.: Dense sampling and fast encoding for 3d model retrieval using bag-of-visual features. In: Proceedings of the ACM International Conference on image and video retrieval, p. 26. ACM (2009)
5. de Herrera, A.G.S., Kalpathy-Cramer, J., Demner-Fushman, D., Antani, S.K., Müller, H.: Overview of the imageCLEF 2013 medical tasks. In: Working Notes for CLEF 2013 Conference (2013)
6. de Herrera, A.G.S., Markonis, D., Müller, H.: Bag–of–colors for biomedical document image classification. In: Greenspan, H., Müller, H., Syeda-Mahmood, T. (eds.) MCBR-CDS 2012. LNCS, vol. 7723, pp. 110–121. Springer, Heidelberg (2013). doi:10.1007/978-3-642-36678-9_11
7. de Herrera, A.G.S., Schaer, R., Bromuri, S., Müller, H.: Overview of the image-CLEF 2016 medical task. In: Working Notes of CLEF 2016 Conference, pp. 219–232 (2016)
8. Jégou, H., Douze, M., Schmid, C.: Improving bag-of-features for large scale image search. Int. J. Comput. Vis. **87**(3), 316–336 (2010)
9. Kalpathy-Cramer, J., Müller, H., Bedrick, S., Eggel, I., de Herrera, A.G.S., Tsikrika, T.: Overview of the CLEF 2011 medical image classification and retrieval tasks. In: CLEF 2011 Labs and Workshop, Notebook Papers, 19–22 (2011)
10. Khan, M., Ohno, Y.: A hybrid image compression technique using quadtree decomposition and parametric line fitting for synthetic images. Adv. Comput. Sci. Eng. **1**(3), 263–283 (2007)
11. Lazebnik, S., Schmid, C., Ponce, J.: Beyond bags of features: Spatial pyramid matching for recognizing natural scene categories. In: 2006 IEEE Computer Society Conference on Computer Vision and Pattern Recognition, vol. 2, pp. 2169–2178. IEEE (2006)

12. Li, F.F., Perona, P.: A Bayesian hierarchical model for learning natural scene categories. In: CVPR, vol. 2, pp. 524–531 (2005)
13. Lowe, D.G.: Object recognition from local scale-invariant features. In: Proceedings of the International Conference on Computer Vision, ICCV 1999, vol. 2, p. 1150. IEEE Computer Society (1999)
14. Mikolov, T., Chen, K., Corrado, G., Dean, J.: Efficient estimation of word representations in vector space. CoRR abs/1301.3781 (2013)
15. Morvan, Y., Farin, D., De With, P.H.: Depth-image compression based on an RD optimized quadtree decomposition for the transmission of multiview images. In: IEEE International Conference on Image Processing ICIP 2007, vol. 5, pp. V-105. IEEE (2007)
16. Müller, H., de Herrera, A.G.S., Kalpathy-Cramer, J., Demner-Fushman, D., Antani, S.K., Eggel, I.: Overview of the imageCLEF 2012 medical image retrieval and classification tasks. In: Working Notes for CLEF 2012 Conference (2012)
17. Müller, H., Michoux, N., Bandon, D., Geissbühler, A.: A review of content-based image retrieval systems in medical applications - clinical benefits and future directions. I. J. Med. Inform. **73**(1), 1–23 (2004)
18. Pass, G., Zabih, R., Miller, J.: Comparing images using color coherence vectors. In: Proceedings of the Fourth ACM International Conference on Multimedia, MULTIMEDIA 1996, NY, USA, pp. 65–73. ACM, New York (1996)
19. Ramanathan, V., Mishra, S., Mitra, P.: Quadtree decomposition based extended vector space model for image retrieval. In: IEEE Workshop on Applications of Computer Vision (WACV 2011), 5–7 January 2011, Kona, HI, USA, pp. 139–144 (2011)
20. Van de Sande, K.E., Gevers, T., Snoek, C.G.: A comparison of color features for visual concept classification. In: Proceedings of the 2008 International Conference on Content-Based Image and Video Retrieval, pp. 141–150. ACM (2008)
21. Shusterman, E., Feder, M.: Image compression via improved quadtree decomposition algorithms. IEEE Trans. Image Process. **3**(2), 207–215 (1994)
22. Sivic, J., Russell, B.C., Efros, A.A., Zisserman, A., Freeman, W.T.: Discovering object categories in image collections. In: Proceedings of the International Conference on Computer Vision (2005)
23. Smith, J.M., Chang, S.F.: Quad-tree segmentation for texture-based image query. In: Blattner, M., Limb, J.O. (eds.) ACM Multimedia, pp. 279–286. ACM Press, New York (1994)
24. Vedaldi, A., Zisserman, A.: Efficient additive kernels via explicit feature maps. IEEE Trans. Pattern Anal. Mach. Intell. **34**(3), 480–492 (2012)
25. Wengert, C., Douze, M., Jégou, H.: Bag-of-colors for improved image search. In: Proceedings of the 19th International Conference on Multimedia 2011, pp. 1437–1440 (2011)
26. Yang, J., Jiang, Y.G., Hauptmann, A.G., Ngo, C.W.: Evaluating bag-of-visual-words representations in scene classification. In: Proceedings of the Internationla Workshop on Multimedia Information Retrieval, pp. 197–206. ACM (2007)
27. Yin, X., Düntsch, I., Gediga, G.: Quadtree representation and compression of spatial data. In: Peters, J.F., Skowron, A., Chan, C.-C., Grzymala-Busse, J.W., Ziarko, W.P. (eds.) Transactions on Rough Sets XIII. LNCS, vol. 6499, pp. 207–239. Springer, Heidelberg (2011). doi:10.1007/978-3-642-18302-7_12
28. Zhou, X., Depeursinge, A., Müller, H.: Information fusion for combining visual and textual image retrieval in imageCLEF@ICPR. In: Ünay, D., Çataltepe, Z., Aksoy, S. (eds.) ICPR 2010. LNCS, vol. 6388, pp. 129–137. Springer, Heidelberg (2010). doi:10.1007/978-3-642-17711-8_14

Do Topic Shift and Query Reformulation Patterns Correlate in Academic Search?

Xinyi Li$^{(\boxtimes)}$ and Maarten de Rijke

University of Amsterdam, Amsterdam, The Netherlands
{x.li,derijke}@uva.nl

Abstract. While it is known that academic searchers differ from typical web searchers, little is known about the search behavior of academic searchers over longer periods of time. In this study we take a look at academic searchers through a large-scale log analysis on a major academic search engine. We focus on two aspects: *query reformulation patterns* and *topic shifts in queries*. We first analyze how each of these aspects evolve over time. We identify important query reformulation patterns: revisiting and issuing new queries tend to happen more often over time. We also find that there are two distinct types of users: one type of users becomes increasingly focused on the topics they search for as time goes by, and the other becomes increasingly diversifying. After analyzing these two aspects separately, we investigate whether, and to which degree, there is a correlation between topic shifts and query reformulations. Surprisingly, users' preferences of query reformulations correlate little with their topic shift tendency. However, certain reformulations may help predict the magnitude of the topic shift that happens in the immediate next timespan. Our results shed light on academic searchers' information seeking behavior and may benefit search personalization.

1 Introduction

Academic search deals with the retrieval of information resources in the domain of scientific literature. Hemminger et al. [15] point out that academic search engines have become the primary portal for researchers to gain information; see also [31]. In recent years, there have been several publications focused on academic search and academic searchers. However, most are very limited in scale, and rarely reveal insights into the search behavior of academic searchers based on the analysis of large-scale transaction logs [14,23,24]. In this study we take a look at academic search through a large-scale log analysis from a major academic search engine.

Academic searchers do have a distinct search pattern that is different from the typical web searchers. For instance, in web search, the search activity becomes the least intensive on Fridays and peaks in the weekends [2]. But, as shown in Fig. 1, academic search activity peaks during weekdays, and drops in the weekends.

To study the behavior of academic searchers, we investigate two key aspects: *query reformulations* and *topic shifts*. Both have received much attention in user

© Springer International Publishing AG 2017
J.M. Jose et al. (Eds.): ECIR 2017, LNCS 10193, pp. 146–159, 2017.
DOI: 10.1007/978-3-319-56608-5_12

behavior studies of web search [4,18,26], but to the best of our knowledge, there is no previous work on revealing the query reformulation behavior and topic shifts of academic searchers that is based on a large-scale log analysis. In fact, very little is known about these two aspects of academic search.

Through this study, we provide answers to 3 research questions:

RQ1. What is the query reformulation behavior of academic searchers?

RQ2. Do academic searchers have shifts in topical interests over time?

RQ3. Is there a correlation between query reformulation behavior and topic shift?

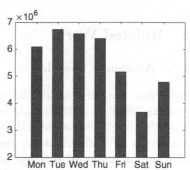

Fig. 1. Average number of queries per weekday in academic search (based on the dataset described in Sect. 3).

For the first question, we look at query reformulation behavior over time. Query reformulation happens after the user has examined the search engine result page and provides a more explicit type of feedback than clicks, which are implicit and noisy [9]. We look at five frequent types of query reformulation: *revisiting a previous query, adding terms, dropping terms, substituting part of the query,* and *issuing a completely new query.* We study how the type of reformulation behavior changes over time and find that revisiting and issuing new queries tend to happen more often as search goes on.

For the second question we take a quantitative approach to study topic shift over time. We train an LDA model [3] on all long sessions in the query log that we examine. We segment a user's queries into different timespans, and treat queries in each timespan as a bag of words. We infer a topic vector for each timespan of the user. Topic shift between successive timespans is then calculated using the Euclidean distance between the topic vectors. In this process we identify two types of user: one type increasingly focuses on topics over time and the other diversifies over time.

Finally, we conduct a correlation study to see how these two aspects—query reformulation and topical shift—are correlated with each other. We find that user's query reformulation patterns have little correlation with the tendency of topic shift, meaning that users with distinct reformulation preferences in search could be equally likely to be diversifying or focusing on topics. We also find that certain reformulations (viz. *adding terms* and *issuing new queries*) may help predict the magnitude of the next topic shift.

Contrary to previous work that studies academic searchers through surveys and user studies, this paper sheds light on the reformulation behavior and topical shifts of academic searchers through a large-scale log analysis. The insights gained help us to understand academic searchers' information seeking patterns from a much larger user base, and may be useful for personalization in academic search.

In Sect. 2 we discuss related work. In Sect. 3 we introduce the dataset characteristics. In Sect. 4 we describe our approach to study query reformulations and topic shifts. In Sect. 5 we show the result and analysis from the correlation studies. We present our conclusions in Sect. 6.

2   Related Work

2.1   Academic Search

Academic search involves the indexing and retrieval of information objects (papers, journals, authors, . . .) in the domain of academic research. The earliest academic search engine MEDLINE, which began functioning in 1971, allowed a maximum of 25 simultaneous users [28]. It was restricted to library usage and only pre-programmed searches were supported instead of online queries.

When the web became popular in the 1990s, online academic search engines started to flourish and gained popularity. Typical examples are Citeseer [11] and Aminer [38], which focus on metadata retrieval and academic network extraction respectively. There are several surveys and user studies on the search behavior of researchers on modern academic search engines [31–33], which are based on a relatively small sample of researchers. The few log analyses conducted on search engines of digital libraries are either investigating a single discipline [14,23], or limited in scale [24], as a result of which they are not representative of academic search. Moreover, they focus on basic usage statistics and lack insights on user behavior in search sessions. Recently, Li et al. [27] studied the user behavior and query failure phenomenon in academic search through a large-scale transaction log analysis.

2.2   Query Reformulations

Query reformulation is an important aspect of user behavior during search sessions. In recent years, there has been a range of studies that cover patterns and models of query reformulation [4,6,18,25,35,36], how they work in a collaborative setting [30], in voice search [20] or in mobile search [37], and their applications [5,7,21,34]. These studies show that query reformulations are the key to understanding user behavior, which will benefit retrieval tasks such as query auto completion [21] as well as topic and intent finding in users' queries [34], and which may help improve retrieval performance [13]. The findings are mostly in the domain of web search and the query reformulation behavior studied is that of the general web users.

Multiple category schemes have been used for query reformulation in the literature [4,6,18,25,35,36]. Different category schemes may correspond to (1) search engines of different designs (e.g., whether searches on multiple verticals are supported), (2) whether using search assistance is considered as a reformulation such as query suggestion, or (3) different granularities of query reformulations. Manual categorization may provide fine-grained results [6,25,35] but can not easily scale up to large query logs. On the other hand, rule-based [18,36] or learning-based [4] methods can be applied to a large query log, and are thus more suitable for analyzing long term query reformulations from a large user base.

2.3 Topic Shift in Queries

There has been a whole line of research that investigates topic mining in web search query logs [1,16,17,22], where the emphasis is on how to segment and cluster queries by topic. However, the multi-tasking nature of web searchers, which means searching and switching between multiple topics within and across sessions [29], makes it cumbersome to derive useful insights from users' topic shifts, especially over long periods.

This paper differs from previous work in academic search, by studying a large transaction log from a major academic search engine, with a focus on user behavior in search sessions. The findings are therefore better able to represent academic searchers, compared with earlier small-scale user studies and surveys. It also differs from previous work in query reformulations in web search, by revealing the academic searchers' preferences instead of those of the general web users. The paper differs from work on topic shifts in web search by looking at a different domain: academic search. Compared to the web searchers who have diverse, parallel, and fast-shifting topic interests, academic searchers are more likely to have consistent interests in a general topic. For instance, a researcher in information retrieval is more likely to stay in this general topic than diverting to biology sciences. This makes studying the long term topic shift pattern meaningful. Moreover, this study tries to link query reformulation to topic shift, and provides useful insights into their connections through a series of correlation studies.

3   Data

We study a query log from the ScienceDirect search engine,[1] containing over 39 million queries. The query log is collected from September 28, 2014 to March 5, 2015. Table 1 shows the length statistics of the query log. Two thirds of the traffic come from institution-authorized access, meaning that users in a certain IP range can access the search engine, and they may share the same session ID and user ID in the query log. Besides, many institutions use proxies or firewalls so that their IP is recorded instead of the terminal device. Therefore it is not

Table 1. Query length statistics in word count.

Category	#N	Min	Max	Mean	Median
ScienceDirect	39M	1	419	3.77	3

---

[1] http://sciencedirect.com.

possible to differentiate these IP-users. We are only confident in an ID-user one-to-one mapping when they log in or access the search engine from outside the institution. And we study these "non-IP" users only, who contribute about one third of the traffic.

With a timeout of thirty minutes as a threshold, there are a total of 4,307,889 sessions for these non-IP users, and 2,833,549 of them contain at least 3 queries which we denote as "long sessions." To obtain enough data of users, we confine the scope of users to those who have a minimum of 30 queries, and whose search behavior lasts over 30 days at least. This leaves us with 29,093 users and 1,918,334 query records.

4   Approach

In this section we describe how we study the behavior and topic change of academic searchers in a series of correlation studies.

First, we highlight the statistics of the prominent types of query reformulations from the query log. Then, we apply a time sequence-based method to make observations of how users progress in search. We break each user's queries into sequences and then align them, so that we can compare how users progress during search even if they start at a different time. Specifically, we put each user's queries into bins separated by a certain length of timespan (to be specified below). Then, we align all searchers' queries by timespan, with the first timespan of a user denoted as 0, the second as 1, in a natural number sequence. We can observe query reformulation and topic shift of users as they move from one timespan to the next. In this case, to gain enough samples from the dataset and also to ensure statistical significance in our later correlation analyses, we sample timespans of 3, 7 and 14 days long. We choose timespans of different lengths to observe whether some changes are more prominent over longer timespans. The length of timespans chosen also corresponds with the usual information seeking cycles of academic searchers, as research suggests that information-seeking happens toward a weekly basis rather than daily basis for faculty and graduate students [8,31]. Note that users may issue no query in a certain timespan; in such cases the timespan will be neglected for that user.

Query Reformulation Tendency Over Time. To uncover the reformulation preferences for the academic searchers as a whole, we examine the query reformulation preference over time for all academic searchers combined. For each timespan, we aggregate the frequency of each reformulation from all users and obtain the proportion of each reformulation type. We hypothesize that certain reformulations might happen more frequently as time goes on, for instance revisiting, because academic searchers tend to have a consistent interest in their field of study [19] and may thus need to submit a previous query repeatedly in search of new information. We try to determine if there is indeed a linear correlation of the proportion of an action over the course of time (represented as a natural number sequence of timespans). To this end, we use Pearson's correlation.

It is common for users to use a combination of the query reformulations listed in the previous section (revisiting, adding a term, dropping a term, substituting a term, new query) in order to reach their search goal. In our analysis, we calculate the proportion of each query reformulation in each time span for every user.

Topic Shift. We study the tendency of a user to shift topic over time with a quantitative approach as we aim to measure the magnitude of change in topic. We train an LDA model on long sessions that contain at least 3 queries. Each session is treated as a "document" in training because the queries within a single session mostly likely belong to the same general topic. The number of topics is set to 150, which is a reasonable value in the academic domain [12] and also ensures relatively fast convergence in Gibbs sampling. For each user, we model the queries in each timespan as a bag of words and use the trained LDA model to infer a topic vector. Then, for a given user the magnitude of topic shift between adjacent timespans is calculated using the Euclidean distance between the user's topic vectors for the two timespans.

Correlations. After studying how users' reformulation behavior and topical interest change over time, respectively, we aim to find whether there is a correlation between a user's query reformulation patterns and their topic shift tendency. Specifically, we look at two aspects of the correlations. First, the macroscopic aspect, i.e., whether a user's topic shift tendency is correlated with query reformulation preferences. For instance, suppose a user favors a specific type of reformulation, say substitution; is this user likely to be diversifying in topic shifts? Second, there is the microscopic aspect: in successive timespans, is the proportion of each reformulation type in the first timespan correlated with the topic change that happens during the next timespan? Based on the correlation findings, we consider the task of predicting the magnitude of a user's topic shift during the next timespan.

5 Results and Analysis

In this section we present the results of our analysis of users' query reformulations and topic shifts. We first analyze these two aspects separately and then perform a series of correlation studies to examine their connections.

5.1 Query Reformulation Types

To study users' query reformulation types, we apply a syntactic-based automatic categorization. Our taxonomy does not require human annotations and does not have the fine-granularity of those methods in [4,18,36]. However it is fully unsupervised and is scalable to a large query log; it contains five reformulation types that are common to the majority of taxonomies previously used for query reformulations [4,6,18,25,35,36]. The main difference is that none of these previous publications considers "revisiting queries" as a reformulation while we do

(Bruza and Dennis [6] consider "repeated query" but there is no user identifier in their query log).

Revisiting. Revisiting is issuing a query that is already in the user's search history [39]. In academic search, we find that this reformulation type is very prominent, making up 33.8% of all reformulations, which shows that academic searchers tend to have some consistent search intents and will seek information on the same topic repeatedly.

Adding terms. This type of reformulation is characterized by adding at least one term to the previous query, and corresponds to the process of refining search. This is typically seen in sessions where users start with a general query on a certain topic, then add terms to examine sub-aspects within the topic [35]. This reformulation type constitutes 8.5% of all reformulations.

Dropping terms. This is the opposite process of the adding reformulation type, constituting 5.6% of all reformulations. By dropping at least one term from the previous query, the user aims to retrieve information that is more general than the previous query [35]. This may happen when academic searchers need context information during learning.

Substituting terms. Substitution of terms is the second most prominent reformulation type that accounts for 28.0% of all reformulations. Substitution means keeping certain at least one term in the original query intact, then dropping old terms and adding new terms. Substitution behavior may happen when a user is refining a search, e.g., changing a synonym, or when the user is exploring different aspects about a certain topic [35].

New query. This reformulation concerns the situation where the user Issues a query that has no overlap of words with the previous query and that does not appear in the user's search history. Submitting a new query that is different often means a change of search intent [4]. It happens when other reformulations will not address the new intent of the users. New queries make up 24.1% of all reformulations.

Compared to web search, where substituting terms accounts for the most popular type (ranging from 22.73% to 37.5% in different datasets [4,18]), the most prominent type in academic search is revisiting and substituting terms only comes next.

5.2 Query Reformulation Tendency for All Academic Searchers Combined

Figure 2 plots all searchers' query reformulation tendency.

By definition of the correlation strength [10], there is a "very strong" positive correlation of the proportion of revisiting behavior over time, in the analyses of all timespans. This confirms our earlier hypothesis in Sect. 4, that there is an increasing trend of revisiting queries by academic searchers, which shows their consistent interests in certain topics. Interestingly, between timespans of 3 days, the tendency to submit new queries is weak, but at longer timespans (7 or 14 days),

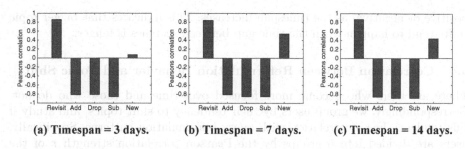

(a) Timespan = 3 days.   (b) Timespan = 7 days.   (c) Timespan = 14 days.

Fig. 2. The query reformulation preference over time for all the academic searchers, measured in correlation of the proportion of the reformulation actions (revisiting, adding terms, dropping terms, substitution and new query) over time.

we can observe a moderate positive correlation. This suggests that submitting new queries tends to happen not immediately (within a 3 day gap), but within a longer gap. The negative correlation for the other three reformulations (add, drop, and substitute) shows that users perform these reformulations less frequently in the later period of search.

5.3   Topic Change Tendency

Using the approach described in Sect. 4, we study the magnitude of the users' topic shift over time. The tendency is represented by the correlation strength: the larger the correlation, the bigger the topic shift over time for a user. Figure 3 shows the distribution of the correlation of the users, for 3 different timespans.

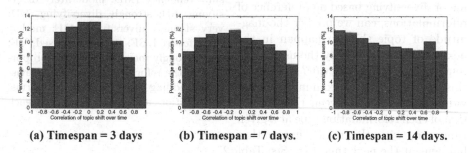

(a) Timespan = 3 days   (b) Timespan = 7 days.   (c) Timespan = 14 days.

Fig. 3. The correlation of user topic shift over time.

The correlation strength of topic shift over time indicates the evolution of user interests over time, namely whether they tend to become more focused or more diversified. In general, we find that nearly half the users tend to have increasing topic shifts over time (diversifying), and the other half have decreasing shifts (focusing). For different timespans, we see from the shape of the distribution, that there are more users showing a stronger tendency of topic shift (either positive or negative) as the timespan increases. This indicates that bigger topic shifts tend to happen when the time gap between searches is longer.

5.4   Correlation Between Reformulation Behavior and Topic Shift

There are users who become more focused over time and those who do not. Correspondingly, we group users by their tendency to shift topics, and study if this tendency has a correlation with query reformulation patterns. Specifically, users are divided into 6 groups by the Pearson correlation strength r of the topic shift tendency over time: moderately diversifying $(0.4 \leq r < 0.6)$, strongly diversifying $(0.6 \leq r < 0.8)$, very strongly diversifying $(0.8 \leq r \leq 1.0)$ and moderately focused $(-0.6 \leq r < -0.4)$, strongly focused $(-0.8 \leq r < -0.6)$, very strongly focused $(-1.0 \leq r < -0.8)$. Then we look at the correlation with the user's different reformulation type's proportions, as shown in Fig. 4.

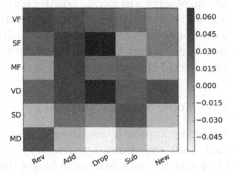

Figure 4 shows that we cannot differentiate diversifying or focused users, purely based on their query reformulation patterns. That is, the user's preference of choosing certain query reformulations is not correlated with their topic shift tendency. This is an interesting finding as it shows that even users with distinct query reformulation preferences, could be equally likely to be focusing or diversifying in search.

Taking a step back, although we cannot determine whether a user is focusing or diversifying based on preference of reformulations, can we predict the magnitude of topic shift to happen in the next timespan given only the user's current reformulation behavior? To answer this question we first examine the individual correlation between the proportion of each reformulation type at a given timespan, with the topic change that happens at the next timespan. See Table 2.

Fig. 4. The correlation of the topic shift tendency (MD: moderately diversifying, SD: strongly diversifying, VD: very strongly diversifying, MF: moderately focused, SF: strongly focused, VF: very strongly focused), with the proportion of the reformulation actions (revisiting, adding terms, dropping terms, substitution and new query) for each user.

Individually, for the majority of users there is only a weak correlation between a query reformulation type and the next topic shift. For users who show a strong correlation $(-1.00 \leq r < -0.60$ or $0.60 \leq r \leq 1.00)$, submitting new queries contributes the least to a decrease in topic shift magnitude and also the most to an increase in topic shift magnitude, respectively, compared with other reformulation types. For longer timespans, there are more users who exhibit a strong correlation. Especially when the timespan is 14 days, 21.0% of the users show a strong or very strong correlation between adding terms and topic change, and the number is even higher at 23.7% for submitting new queries. Interestingly,

Table 2. Correlation of reformulation behavior with topic shift at the next timespan. Each column shows the distribution of users (in percentage) who have different correlation strengths between a reformulation type and topic shift, in an interval of 0.2.

Correlation	Revisit	Add	Drop	Sub	New
Timespan = 3 days					
[−1.00, −0.80]	1.6%	1.4%	2.4%	3.7%	0.9%
[−0.80, −0.60]	3.7%	2.9%	4.3%	8.1%	2.3%
[−0.60, −0.40]	6.9%	6.4%	7.7%	12.9%	4.5%
[−0.40, −0.20]	12.7%	11.5%	11.7%	16.9%	7.7%
[−0.20, 0.00]	16.5%	15.5%	16.1%	17.2%	11.4%
[0.00, +0.20]	17.2%	17.5%	17.6%	14.2%	15.7%
[+0.20, +0.40]	15.8%	16.1%	15.1%	11.8%	19.3%
[+0.40, +0.60]	13.1%	14.5%	12.9%	7.6%	18.9%
[+0.60, +0.80]	8.4%	9.5%	8.0%	4.9%	13.5%
[+0.80, +1.00]	4.1%	4.6%	4.2%	2.9%	6.0%
Timespan = 7 days					
[−1.00, −0.80]	3.1%	2.4%	3.7%	5.9%	1.7%
[−0.80, −0.60]	5.3%	4.5%	6.1%	10.1%	3.4%
[−0.60, −0.40]	8.7%	7.4%	8.4%	13.1%	5.7%
[−0.40, −0.20]	11.7%	11.7%	11.2%	14.3%	8.4%
[−0.20, 0.00]	13.4%	13.9%	13.4%	14.4%	10.9%
[0.00, +0.20]	14.7%	14.4%	14.4%	12.7%	13.8%
[+0.20, +0.40]	13.7%	13.8%	14.2%	10.6%	16.4%
[+0.40, +0.60]	13.1%	14.2%	12.8%	8.6%	17.4%
[+0.60, +0.80]	9.9%	10.9%	9.3%	6.4%	13.7%
[+0.80, +1.00]	6.3%	6.9%	6.4%	4.0%	8.6%
Timespan = 14 days					
[−1.00, −0.80]	4.8%	3.9%	5.5%	8.0%	3.0%
[−0.80, −0.60]	7.0%	6.4%	7.1%	11.6%	5.1%
[−0.60, −0.40]	8.6%	8.2%	8.9%	12.6%	7.3%
[−0.40, −0.20]	11.1%	10.7%	11.1%	12.2%	9.0%
[−0.20, 0.00]	11.6%	12.3%	10.9%	11.9%	10.6%
[0.00, +0.20]	12.7%	12.3%	11.4%	11.0%	12.1%
[+0.20, +0.40]	12.7%	12.5%	12.8%	10.6%	14.5%
[+0.40, +0.60]	12.0%	12.8%	12.6%	8.5%	14.7%
[+0.60, +0.80]	10.9%	12.2%	11.1%	7.5%	13.5%
[+0.80, +1.00]	8.6%	8.8%	8.6%	6.0%	10.2%

substituting reformulations tend to correlate the least with topic change. This suggests that users tend to stay in the same general topic, or a subtopic within the general topic, while modifying only part of the original queries.

5.5 Predicting the Magnitude of the Next Topic Shift

Next, we try to utilize the observational insights that we have just gained for a prediction task: can query reformulation signals help to predict the magnitude of a user's topic shift?

More precisely, we use features from users' reformulations to predict the magnitude of topic shift at the next timespan; see Table 3. The features are the proportions and number of occurrences of query reformulations in a timespan. We cast this task as a regression task. Our training set is comprised of pairs of query reformulations and the topic shift to happen at the next timespan for all users. The test set consists of the second-last query reformulations and the next (final) topic shift for each user.

We use linear regression and three evaluation measures: correlation coefficient, mean absolute error (MAE) and root mean squared error (RMSE). The prediction results are listed in Table 4. Prediction is more accurate on shorter timespans, with the 3 day predictions reaching a medium correlation ($r = 0.4530$), while 14 day predictions being at only $r = 0.3225$. The performance difference indicates that topic shift magnitude in a shorter timespan is easier to predict than longer timespans.

Table 3. Query reformulation features for prediction of the magnitude of topic shift at the next timespan.

Name	Description
Reformulation proportions	
Revisiting_Percentage	Percentage of revisiting reformulations
Adding_Percentage	Percentage of adding term reformulations
Dropping_Percentage	Percentage of dropping term reformulations
Substitution_Percentage	Percentage of substitution reformulations
New_Query_Percentage	Percentage of new query reformulations
Reformulation occurrence numbers	
Revisiting_Number	Number of revisiting reformulations
Adding_Number	Number of adding reformulations
Dropping_Number	Number of dropping term reformulations
Substitution_Number	Number of substitution reformulations
New_Query_Number	Number of new query reformulations

Table 4. Linear regression results (correlation coefficient, mean absolute error, root mean squared error) for predicting the magnitude of a topic shift in the next timespan given query reformulation features in the current timespan.

	3 days	7 days	14 days
Correlation coefficient	0.4530	0.3906	0.3225
MAE	0.0697	0.0755	0.0805
RMSE	0.0931	0.0999	0.1057

6   Conclusion

In this study we have examined users' query reformulation behavior and their tendency of topic shift in academic search through a large-scale log analysis. We have found that over time, academic searchers as a whole tend to conduct revisiting, as well as submitting completely new queries. This pattern corresponds to the academic searcher's information needs: either seeking previous search results or new results on the same search intents, or simply pursuing new search intents. We have identified two types of topic shift patterns in users, namely the focusing type and the diversifying type.

Through a series of correlation studies, we have found that a user's preference for certain query reformulations does not correlate to their topic shift tendency. Nevertheless, users's current reformulation patterns (adding terms, submitting new queries) may help to predict the magnitude of topic change in the immediate next timespan. We further used features from query reformulations for predicting the magnitude of the next topic shift. The findings of the query reformulation behavior, topic shift type, and their connections help to improve our understanding of the behavior of academic searchers from a large user base. They may provide hints for personalized search, such as whether to provide exploratory or focusing type of search results, and recommendations of queries or papers for users.

In future work we intend to look at query reformulation patterns in the context of different search tasks, e.g., a navigational task for a single document, or a learning task for a certain research topic. And we will examine the utility of using query reformulation features to improve retrieval performance and provide better recommendations in academic search.

Acknowledgements. This research was supported by the China Scholarship Council and Elsevier. All content represents the opinion of the authors, which is not necessarily shared or endorsed by their respective employers and/or sponsors.

References

1. Aiello, L.M., Donato, D., Ozertem, U., Menczer, F.: Behavior-driven clustering of queries into topics. In: CIKM, pp. 1373–1382. ACM (2011)

2. Beitzel, S.M., Jensen, E.C., Chowdhury, A., Frieder, O., Grossman, D.: Temporal analysis of a very large topically categorized web query log. J. Assoc. Inf. Sci. Technol. **58**(2), 166–178 (2007)
3. Blei, D.M., Ng, A.Y., Jordan, M.I.: Latent Dirichlet allocation. J. Mach. Learn. Res. **3**, 993–1022 (2003)
4. Boldi, P., Bonchi, F., Castillo, C., Vigna, S.: Query reformulation mining: models, patterns, and applications. Inf. Retr. **14**(3), 257–289 (2011)
5. Bordino, I., Castillo, C., Donato, D., Gionis, A.: Query similarity by projecting the query-flow graph. In: SIGIR, pp. 515–522. ACM (2010)
6. Bruza, P.D., Dennis, S.: Query reformulation on the internet: empirical data and the hyperindex search engine. In: RIAO (1997)
7. Cai, F., de Rijke, M.: A survey of query auto completion in information retrieval. Found. Trends Inf. Retr. **10**(4), 273–363 (2016)
8. Catlow, J., Górny, M., Lewandowski, R.: Students as users of digital libraries. Qual. Quant. Methods Libr. **65**, 60–61 (2015)
9. Chuklin, A., Markov, I., de Rijke, M.: Click Models for Web Search. Morgan & Claypool Publishers, San Rafael (2015)
10. Evans, J.S.B.T., Over, D.E.: Reasoning and Rationality. Psychology Press, Hove (1996)
11. Giles, C.L., Bollacker, K.D., Lawrence, S.: Citeseer: an automatic citation indexing system. In: DL, pp. 89–98. ACM (1998)
12. Griffiths, T.L., Steyvers, M.: Finding scientific topics. Proc. Nat. Acad. Sci. **101**(Suppl. 1), 5228–5235 (2004)
13. Guan, D., Zhang, S., Yang, H.: Utilizing query change for session search. In: SIGIR, pp. 453–462. ACM (2013)
14. Han, H., Jeong, W., Wolfram, D.: Log analysis of academic digital library: user query patterns. In: iConference 2014 Proceedings, pp. 1002–1008 (2014)
15. Hemminger, B.M., Lu, D., Vaughan, K., Adams, S.J.: Information seeking behavior of academic scientists. J. Assoc. Inf. Sci. Technol. **58**(14), 2205–2225 (2007)
16. Hu, Y., Qian, Y., Li, H., Jiang, D., Pei, J., Zheng, Q.: Mining query subtopics from search log data. In: SIGIR, pp. 305–314. ACM (2012)
17. Hua, W., Song, Y., Wang, H., Zhou, X.: Identifying users' topical tasks in web search. In: WSDM, pp. 93–102. ACM (2013)
18. Jansen, B.J., Booth, D.L., Spink, A.: Patterns of query reformulation during web searching. J. Assoc. Inf. Sci. Technol. **60**(7), 1358–1371 (2009)
19. Jeng, W., He, D., Jiang, J.: User participation in an academic social networking service: a survey of open group users on mendeley. J. Assoc. Inf. Sci. Technol. **66**(5), 890–904 (2015)
20. Jeng, W., Jiang, J., He, D.: Users' perceived difficulties and corresponding reformulation strategies in Google voice search. J. Libr. Inf. Stud. **14**(1), 25–39 (2016)
21. Jiang, J.-Y., Ke, Y.-Y., Chien, P.-Y., Cheng, P.-J.: Learning user reformulation behavior for query auto-completion. In: SIGIR, pp. 445–454. ACM (2014)
22. Jones, R., Klinkner, K.L.: Beyond the session timeout: automatic hierarchical segmentation of search topics in query logs. In: CIKM, pp. 699–708. ACM (2008)
23. Jones, S., Cunningham, S.J., McNab, R., Boddie, S.: A transaction log analysis of a digital library. Int. J. Digit. Libr. **3**(2), 152–169 (2000)
24. Ke, H.-R., Kwakkelaar, R., Tai, Y.-M., Chen, L.-C.: Exploring behavior of e-journal users in science and technology: transaction log analysis of Elsevier's ScienceDirect onsite in Taiwan. Libr. Inf. Sci. Res. **24**(3), 265–291 (2002)

25. Lau, T., Horvitz, E.: Patterns of search: analyzing and modeling web query refinement. In: Kay, J. (ed.) UM99 User Modeling. CICMS, vol. 407, pp. 119–128. Springer, Heidelberg (1999). doi:10.1007/978-3-7091-2490-1_12
26. Li, L., Deng, H., He, Y., Dong, A., Chang, Y., Zha, H.: Behavior driven topic transition for search task identification. In: WWW, pp. 555–565. ACM (2016)
27. Li, X., Schijvenaars, R., de Rijke, M.: Investigating queries and search failures in academic search. Inf. Process. Manag. (2017, to appear)
28. Lindberg, D.: Internet access to the National Library of Medicine. Eff. Clin. Pract. 3(5), 256–260 (2000)
29. Mehrotra, R., Bhattacharya, P., Yilmaz, E.: Uncovering task based behavioral heterogeneities in online search behavior. In: SIGIR, pp. 1049–1052. ACM (2016)
30. Mohammad Arif, A.S., Du, J.T., Lee, I.: Examining collaborative query reformulation: a case of travel information searching. In: SIGIR, pp. 875–878. ACM (2014)
31. Niu, X., Hemminger, B.M., Lown, C., Adams, S., Brown, C., Level, A., McLure, M., Powers, A., Tennant, M.R., Cataldo, T.: National study of information seeking behavior of academic researchers in the United States. J. Assoc. Inf. Sci. Technol. 61(5), 869–890 (2010)
32. Pontis, S., Blandford, A.: Understanding "influence:" an exploratory study of academics' processes of knowledge construction through iterative and interactive information seeking. J. Assoc. Inf. Sci. Technol. 66(8), 1576–1593 (2015)
33. Pontis, S., Blandford, A., Greifeneder, E., Attalla, H., Neal, D.: Keeping up to date: an academic researcher's information journey. J. Assoc. Inf. Sci. Technol. 68(1), 22–35 (2017). Online since 22 October 2015
34. Radlinski, F., Szummer, M., Craswell, N.: Inferring query intent from reformulations and clicks. In: WWW, pp. 1171–1172. ACM (2010)
35. Rieh, S.Y., et al.: Analysis of multiple query reformulations on the web: the interactive information retrieval context. Inf. Process. Manag. 42(3), 751–768 (2006)
36. Shiri, A.: Query reformulation strategies in an interdisciplinary digital library: the case of nanoscience and technology. In: ICDIM, pp. 200–206. IEEE (2010)
37. Shokouhi, M., Jones, R., Ozertem, U., Raghunathan, K., Diaz, F.: Mobile query reformulations. In: SIGIR, pp. 1011–1014. ACM (2014)
38. Tang, J.: Aminer: toward understanding big scholar data. In: WSDM, p. 467. ACM (2016)
39. Teevan, J.: The re:search engine: simultaneous support for finding and re-finding. In: UIST, pp. 23–32. ACM (2007)

Learning to Re-rank Medical Images Using a Bayesian Network-Based Thesaurus

Hajer Ayadi[1(✉)], Mouna Torjmen Khemakhem[1], Jimmy Xiangji Huang[2],
Mariam Daoud[2], and Maher Ben Jemaa[1]

[1] ReDCAD Lab, University of Sfax, Sfax, Tunisia
hajer.ayadi@redcad.org
[2] IRLab Lab, York University, Toronto, Canada

Abstract. In this paper, we believe that representing query and images with specific medical features allows to bridge the gap between the user information need and the searched images. Queries could be classified into three categories: textual, visual and combined. We present, in this work, the list of specific medical features such as image modality and image dimensionality. We exploit these specific features in a new medical image re-ranking method based on Bayesian network. Indeed, using a learning algorithm, we construct a Bayesian network that represents the relationships among these specific features appearing in a given image collection; this network is then considered as a thesaurus (specific for that collection). The relevance of an image to a given query is obtained by means of an inference process through the Bayesian network. Finally, the images are re-ranked based on combining their initial scores and the new scores. Experiments are performed on Medical ImageCLEF datasets from 2009 to 2012 and results show that our proposed model enhances significantly the image retrieval performance compared with BM25 model.

Keywords: Bayesian network · Specific medical features · Medical image re-ranking

1 Introduction

Due to the explosive growth in medical imaging technologies, a tremendous amount of digital medical images has been produced daily in hospitals, medical research, and education. Facing such volume of images, the need for effective medical image retrieval systems to find relevant information becomes very high. Currently, text-based image retrieval (TBIR) has been shown to be successful in seeking medical images due to rich image metadata, such as filename, and surrounding text, which can be used as textual descriptions for indexing and searching. For any TBIR model, there is a great variety of methods designed to improve the effectiveness of the retrieval. Among them, we chose to use the thesauri and the image re-ranking. A typical thesaurus is composed of a set of terms and relationships among them.

© Springer International Publishing AG 2017
J.M. Jose et al. (Eds.): ECIR 2017, LNCS 10193, pp. 160–172, 2017.
DOI: 10.1007/978-3-319-56608-5_13

Previous studies [13, 15] have shown that imaging modality is an important information on the image for medical retrieval. Using the modality information, the retrieval results can often be improved significantly [15]. As imaging modalities belong to the medical-dependent (specific) features, we propose to exploit specific features for image re-ranking. The main aim of this research is to explore the possibilities of using Bayesian networks to automatically construct, for any medical image collection, a thesaurus which may be used to improve the performance of the TBIR system. The thesaurus is composed of the set of specific features defined in our previous work [2, 3]. More precisely, from the set of specific features appearing in the image collection, we build a Bayesian network that represents the strength of relationships among these features using a learning algorithm. The relevance of an image to a given query is obtained by means of an inference process through a Bayesian network. The network topology representing the dependency relationship between query features and image metadata features as well as the probability distribution encoding the strength of these relationships are calculated based on the distribution of features in image collection and queries.

Compared to studies from the literature, our research contributions present the following key new aspects: (1) This is the first attempt to exploit specific medical features to represent queries and images. Our specific features present two positive points to the medical image retrieval: on the one hand, it is specific for image search. Indeed, usually TBIR is considered as a regular text search, whence, the metadata is considered as regular text, while the main purpose of TBIR is the image search based on their metadata. The features are proposed to solve such problem. In fact, they exhibit a textual specificity for image search as they include the imaging modality (x-ray, MRI, etc.), image dimensionality (micro, gross, macro, etc.), the textual specification (histogram, analysis, finding, etc.), visual specification (color, etc.), and so on. On the other hand, the specific features are mainly specific to the medical field and more precisely the medical images, as it contains medical terminology which refers to the image modality (e.g. Ultrasound, x-ray), keywords related to the image color (e.g. brown, colored), dimensionality keywords (e.g. macro, gross). (2) We propose to use a learning algorithm for constructing a specific features-based Bayesian network, that represents the relationships among features appearing in a given image collection. This network is then used as a thesaurus (specific for that collection) to improve the performance of a TBIR by means of the re-ranking technique. So, given a particular query, we instantiate the features that compose it and propagate this information through the network until getting the image features whose posterior probabilities are combined to get the relevance score.

The remainder of this paper is organized as follows. Section 2 summarizes the related work. Section 3 describes the learning algorithm used to construct the Bayesian network representing the thesaurus. Sections 4 presents and discusses the experimental design and results. Section 5 discusses directions for future research and concludes the paper.

2   Related Work

The use of features in the information retrieval (IR) domain has received much attention from researchers, such as [4,9]. Authors in [9] used these features to predict the effectiveness of queries and information retrieval systems, authors in [2–4] used them to predict a correlation between query and retrieval function. Moreover, authors in [18] used them to re-rank documents. As a brief overview, we are going to mention the main research lines related to applying features to rank documents. Traditionally, a small number of features used as basis of retrieval model in order to rank documents [14]. Over time, this number has increased significantly. Consequently, the development of several feature categories was observed. Among these categories, we can find the following:

- Primitive textual features [14]: most of feature-based retrieval models exploit one or more of these features. The following features are among the most commonly used: Term frequency, document length, inverse document frequency, and so on.
- High-level textual features [14] represent a combination of the primitive textual features. Among these features, we can find the term weighting functions used by BM25, language modeling, and most other popular retrieval models.
- Non-textual features [14] are typically task-specific and often exploit knowledge about the domain or problem structure. Among these features, we can find: PageRank [5], URL depth [12], and query clarity [7], etc.
- Visual features [23] are extracted using visual content analysis.

Several studies [6,8,18,24] have used ranking features for ranking or re-ranking initial search results. This section briefly summarizes some of these approaches. In [24], three types of re-ranking features are proposed. They are based on visual context, initial ranking and pseudo relevance feedback. In [8], a new model for IR is proposed based on six features: query absolute frequency, query relative frequency, document absolute frequency, document relative frequency, IDF, and relative frequency in all documents. In [18], Nallapati proposed the 6 standard features in document retrieval. Those features aren't proposed based on words themselves, but on statistics of documents and queries, such as the total frequency of occurrences of all query terms in the document, or the total of the IDF-values of the query terms that occur in the document. In [6], Cao et al. proposed to construct a feature vector from each query- document pair and used it to create a ranking function that assigns scores to documents. The number of extracted features depends on dataset and varies from one data set to another. In the first one, there are about 20 features extracted from each query-document pair, including content features and hyperlink features. In the second, the standard features in document retrieval [18] are extracted for each query-document pair.

Despite the large number of existing features, there is a lack of studies defining medical-dependent (specific) features and how to use them in text based medical image retrieval purposes. Moreover, TBIR systems are usually considered as a simple text based retrieval model without taking into account the image specificity. To address these issues, we propose to integrate specific features, which are mainly specific to medical images, into a new baysian network-based thesaurus which may be used as a tool for the re-ranking process.

3   Re-Ranking Medical Images Using Bayesian Network-Based Thesaurus

In this section, we propose a new Bayesian network-based thesaurus using specific features for medical image re-ranking. Due to the great influence of the query features on the selection of the most suitable models, and the influence of image modality for medical image retrieval [2,3], we choose to integrate specific medical features and their relationships in the re-ranking process. Our approach is described in Fig. 1 as follows:

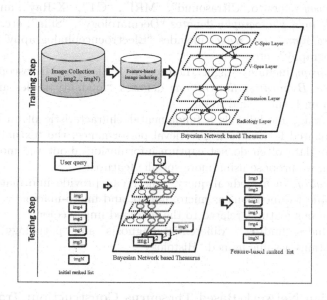

Fig. 1. Our re-ranking process using a Bayesian network based-thesaurus

1. We represent the image collection as a set of specific features, which values are predetermined and proposed in our previous work [2,3].
2. We construct a Bayesian network that represents some of the relationships among the specific features appearing in a given image collection. This network is then considered as a thesaurus.
3. We introduce a new re-ranking method based on the new thesaurus: given a particular query, we instantiate the features that compose it and propagate this information through the network until getting the features that compose the image. The image score is the sum of posterior probability of features composing this image.

3.1 Specific Medical Features

In IR, each query has a set of specific features which are heavily dependent on the searching context. As our work falls into medical image retrieval field, we propose to define medical-dependent features, which help in improving the retrieval performance. In our previous work [3], we used those features to predict the suitable retrieval model given a query. These features are defined as follows:

- *Image Modality*: is an important and fundamental image characteristic that can be used to improve retrieval performance. Image modality includes the acquisition means for acquiring and restoring images of the body, such as radiography and ultrasound. In our study, we use various medical imaging modalities proposed by the Medical Image task of the CLEF (Cross Language Evaluation Forum) evaluation campaign. The modality classes that we integrated into our specific feature set are detailed in [2, 3].
  - *Radiology* refers to "Ultrasound", "MRI", "CT", "X-Ray", and so on.
  - *Visible light photography* denotes "Dermatology", "Skin", etc.
  - *Printed signals and waves* includes "Electroencephalography", "Electrocardiography", etc.
  - *Microscopy* includes "Light Microscopy", "Electron Microscopy", etc.
  - *Generic Biomedical Illustrations* denotes "modality tables and forms", "program listing", and so on.
  Although image modality is a fundamental characteristic of an image and can be exploited for improving retrieval performance, the textual queries or image meta-data often do not capture information about the modality. For this reason, we propose using more specific features.
- *Dimensionality*: In a medical query, a user can provide information on the searched object dimensions as micro, macro, and micro-macro.
- *V-spec* contains features related to the searched image color.
- *T-spec* includes "finding", "differential diagnosis" and "pathology" as values.
- *C-spec* contains a value named "Histology".

3.2 Bayesian Network-Based Thesaurus Construction: Training Step

In the IR area, we can find two main research models based on Bayesian networks: The inference network model [20] and the belief network model [19]. The other Bayesian models have mainly been derived from these two models. However, the real challenge of these networks is the size that can exceed a few millions nodes easily. Moreover, performing exact inference in such network is dauntingly challenging and stays a real obstacle for the progress of research in this domain. In our work, as we used only specific features to represent images and queries, the experiments can be applied to large document collections without problems. We also used an approximate inference algorithm inspired from Ribeiro's [19] inference process that makes our method scalable for larger networks.

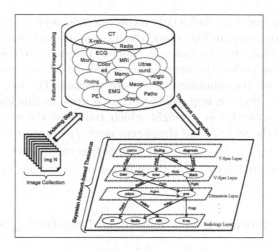

Fig. 2. The process of the thesaurus construction

In the following, we describe the steps involved in constructing the Bayesian network based thesaurus. The topology of this network represents the dependency between the features in the collection and the weight of the arcs that is, also, estimated from the distribution of features in the collections. As shown in Fig. 2, the thesaurus is constructed in the training step. Given an image collection, we have built a thesaurus based on a Bayesian network. From a feature-based image index used as a learning file, our method learns a network topology of features. The network nodes represent the specific features/value of the image collection. They are often grouped together into 9 layers called feature layers. Each layer represents an attribute (e.g., radiology, dimensionality, and microscopy), the number of nodes in each layer equals to the number of each feature/value. The feature layers are arranged in increasing order of feature frequency. The upper layer includes the less frequent features while the lower one includes frequent features. We choose this order to increase the probability of a feature belonging to lower layers given another feature belonging to upper layers.

In order to represent the relationships among the features appearing in the collection, we use directed arcs. Indeed, if two features appear often together in the collection, it means there is dependency relationships between them. Thus, the dependence between the features is strongly related to their co-occurrence. The arcs between feature nodes are directed from the less frequent features to the more frequent ones in the collection as the former highly depends on the latter (e.g., the arcs are directed from "color", as it is the less frequent feature, to "brown" in Fig. 2). Moreover, these arcs should not form any cycle and they must satisfy this condition: the arcs between layers can only be oriented from the upper layers to the lower layers because layers are arranged in increasing order of feature frequency.

In the same layer, the nodes are classified into three categories: the root nodes (includes only outgoing arcs, represented by "patho" that belongs to t-spec layer in Fig. 2), the leaf nodes (includes only incoming arcs, represented by "CT" that belongs to radiology layer in Fig. 2) and the internal nodes (includes both incoming arcs and outgoing arcs, for example "micro" belonging to dimension layer in Fig. 2). If there are cycles, the pruning steps will be carried out. The principle of pruning relies on the minimum of the arc weights. The arc is pruned if its weight is the least. So, it is impossible to have a cycle in this feature layer.

Each arc in the network has a weight which represent the weight of the relationship between the node f_i and the parent node f_j connected by the arc. This weight is defined by $P(f_i|f_j) = \frac{\#f_i \cap f_j}{\#f_j}$, with $\#f_i$ is the number of occurrence of f_i in the collection and $\#f_i \cap f_j$ is the number of occurrence of both f_i and f_j in the collection.

3.3   Bayesian Network-Based Thesaurus Use: Testing Step

Re-ranking is carried out on the basis of the generated thesaurus and the original ranking score. We propose to integrate queries and images in the Bayesian network. Hence, there will be three types of nodes that represent images, queries, and features. They are often grouped together into layers. Thus, we obtain three types of layers: *query layer, image layer and feature layer*. These nodes are connected by arcs which also will be classified into three types: The first set of arcs is *the arcs between query nodes and feature nodes* which represent the links between features and query. Their direction is from query to features. The second one is *the arcs between feature nodes and image nodes*. They model the relationship between the images and their features. They are directed from features to images. The last set of arcs is *the arcs between feature nodes*. Moreover, these arcs should not form any cycle and they must satisfy all the following conditions:

1. C1: For the image nodes, each node can only be the destination of the arcs but never the source, so it is impossible to have a cycle in this layer.
2. C2: For the query nodes, each node can only be the source of the arcs, so it is also impossible to have a cycle in this layer.
3. C3: For the feature nodes, as indicated previously, the arcs between layers can only be oriented from the upper layers to the lower layers.

It is known that exact probabilistic inference in an arbitrary Bayesian networks is NP-hard [1]. To simplify the estimation of probabilities, we propose to eliminate some nodes and arcs. This elimination is based on ignoring these nodes/arcs in the probability computation. We shall consider, img an image from the collection for which we want to calculate a relevance score and q is the given query. The elimination process as presented in Fig. 3 concerns all other image nodes and feature nodes which belong neither to q nor to img, except if they are along the path between the query features and image features. In this case, the features providing the connection between query features and image features are kept (e.g., in Fig. 3, if we eliminate the node="gros", there will no be relationship between the query feature "diagnosis" and the image feature "Radio"). In case there are more than one candidate paths, the shortest path is considered for

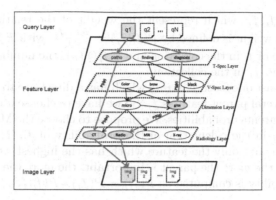

Fig. 3. The elimination process in the BayNetThes approach

connecting query features to image features. As the goal of our model is to take into account the relationships between the common features between query q and image img, but not the relationship between all features, we can eliminate the relationship between the features belonging to the image but not to the query.

Usually in Bayesian networks, each node is created with two states, "true" and "false". For each state, a conditional probability distribution is estimated. However, in our Bayesian network, we are only interested in nodes with the states are "true", as we only concern about the problem of selecting relevant images to a given query.

In order to compute the probabilities, we propose to adapt the inference's process of Ribeiro [19]. This latter includes two steps detailed as follows:

The Inference Probabilities on the Feature Nodes. $\forall f_i \in F = F_q \cup F_{img}$, where F_q is the query feature set and F_{img} is the image feature set, and f_i is a feature in the feature set and belongs to q and/or img, the probability of a node will be calculated based on its parent nodes. In our BayNetThes approach, there are two cases:

- $f_i \in F_q$: In the case where q is the parent of f_i: $P(f_i|q)$ is the probability of finding the feature f_i for a given query q. In the IR context, this corresponds to the weight of f_i relative to q. In our case, we propose to use the standardized feature frequency in query q. The probability is defined as: $P(f_i|q) = w(f_i, q)$ where $w(f_i, q)$ is the weighting function of f_i in q.

- $fi \notin F_q$: In this case, f_i is not among the query features, so, it is consequently an image feature or on the path connecting the image features to the query features. There are two cases:

(1) There is no direct arc between F_q and F_{img}. Consequently, f_i is connected to another feature f_j which is the parent node that has connection to the query features. We propose to take into consideration both $P(f_j)$ probability and

the weight (f_i, f_j) which represent the weight of the relationship between (f_i, f_j). This weight is defined by $P(f_i|f_j) = \frac{\#f_i \cap f_j}{\#f_j}$, with $\#f_i$ is the number of occurrence of f_i in the collection and $\#f_i \cap f_j$ is the number of occurrence of both f_i and f_j in the collection.

(2) There is at least one feature in img which has a direct arc to f_i: such feature may have several parent features f_j. So we have to choose a reasonable combination of parents probabilities. We propose to choose the Max-combination of parents f_j in the calculation of the probability of f_i. This combination takes into account only the feature which has the highest weight. The intuition is that, the more the parent is important, the more the child feature is. $P(f_i)$ probability is computed as $P(f_i) = P(f_j) * P(f_i|f_j)$.

The Estimation of $BayNetThes_{Score}$. The more the query features are found in the image feature layer, the more the image is relevant to the user information needs. Moreover, the query feature weights should be considered, because a feature that has significant weight must have a great impact on the query satisfaction. Hence, the number and the weight of the query features must be taken into account in the calculation of the image probability. The probability is therefore calculated as follows:

$$BayNetThes_{Score} = \frac{\sum_{f_i \in F_q} P(f_i|q) * w(f_{img}, img) * \prod_{k=1}^{n} P(f_k|f_{k-1})}{\sum_{f_i \in F_q} P(f_i|q)} \tag{1}$$

where f_i is a query feature, f_{img} is an image feature, $w(f_{img}, img)$ the frequency of feature f_{img} in the image, n is the number of feature nodes connecting the query node to the image node and $P(f_k|f_{k-1})$ is the probability of feature f_k given its parent feature f_{k-1}. We combine the Bayesian network-based thesaurus results with the initial ranking results based on new Re-Ranking scores called $(Re - rank_{BayNetThes})$. In particular, we propose modeling the $Re - rank_{BayNetThes}$ score by a simple linear combination of the normalized scores, the $initialScore$ and the $BayNetThesScore$:

$$Rerank_{BayNetThes} = \alpha * initialScore + (1 - \alpha) * BayNetThesScore \tag{2}$$

Where α is a parameter that can be tuned from 0 to 1.

4   Experiments and Results

We conduct experiments to evaluate the impact of using Bayesian network thesaurus on the re-ranking process. We compare the results of our approach to the initial results obtained by BM25 model.

4.1   Datasets and Settings

To evaluate our proposed approach, we need an annotated medical image dataset with queries and human evaluated results of them. The most available medical datasets does not meet these conditions. They are either missing the evaluation protocols (such as OHSUMED) or have a goal of textual analysis and evaluation finalities (such as TREC). On the other hand, ImageCLEFmed Evaluation Company provides some medical image collections for medical image retrieval evaluation. Since 2011, the size and scope of the collections has been involved to be closest to real world applications [10]. In this paper, we use four Image-CLEF collections in medical Image Task: two relatively small datasets (74,902 and 77,495 images present the size of the 2009 [16] and 2010 [17] datasets respectively) and two after the evolution of ImageCLEF (230,088 and 306,539 images present the size of the 2011 [11] and 2012 [15] datasets respectively). In Image-CLEFmed datasets, each image has a textual annotation that may contain a caption, a link to the full text article [22] and also the article title [16].

In order to build a strong baseline, we conduct a set of experiments to tune the parameters b in BM25 model, and we set it equals to 0.75.

4.2   Results of the Bayesian Network Based Thesaurus Re-ranking

As we mentioned in the previous section, $Re-rank_{BayNetThes}$ is a combination of initialScore and BayNetThesScore. In order to obtain the best linear combination, we conduct a set of experiments to tune the best value of α. The best value of α, which allows having the best MAP, is 0.3. Consequently, we set the value of α equals to 0.3. In Table 1, we present an analysis of the precision at top ranks. Each column presents the precision at one rank. The last one presents the MAP values on the datasets. We evaluate our approach using datasets from 2009 to 2012. Significant differences according to the Wilcoxon test ($p < 0.05$) [21] of our approach over the traditional probabilistic BM25 model are bold and starred (*). We indicate in parenthesis the percentage of improvement compared to the baseline model (BM25 model). The Re-ranking using BayNetThes approach gives significantly better results compared to BM25. Based on Table 1, we observe that improvements have been achieved on all datasets. These values increase significantly at the rank (P@5) that reached 33% compared to BM25 on 2010 dataset.

Table 1. Thesaurus-based re-rank results

Medical imageCLEF datasets	Retrieval models	P@5	P@10	P@15	P@20	P@30	P@100	MAP
2009	BM25	0.608	0.584	0.584	0.578	0.529	0.351	0.379
	Rerank BayNet-Thes	0.696* (+14.47%)	0.656* (+12.32%)	0.637* (+9.07%)	0.630* (+8.99%)	0.574* (+8.5%)	0.371 (+5.69%)	0.410* (+8.17%)
2010	BM25	0.374	0.392	0.365	0.342	0.285	0.190	0.311
	Rerank BayNet-Thes	0.499* (+33.42%)	0.431 (+9.94%)	0.416* (+13.97%)	0.393* (+14.91%)	0.324* (+13.68%)	0.199 (+4.73%)	0.341* (+9.64%)
2011	BM25	0.393	0.313	0.280	0.265	0.245	0.171	0.193
	Rerank BayNet-Thes	0.446* (+13.48%)	0.373* (+19.16%)	0.337* (+20.35%)	0.305* (+15.09%)	0.282* (+15.1%)	0.195* (+14.03%)	0.220* (+13.98%)
2012	BM25	0.418	0.313	0.266	0.243	0.190	0.098	0.193
	Rerank BayNet-Thes	0.445 (+6.45%)	0.354* (+13.09%)	0.290* (+9.02%)	0.259 (+6.58%)	0.207* (+8.94%)	0.103 (+5.1 %)	0.210* (+8.8%)

The results of the experiments vary between datasets. For 2009 dataset, Re-ranking using BayNetThes improves significantly the retrieval performance compared to BM25 on to the precision at top ranks(P@5, P@10, P@15, P@20 and p@30). This could be explained by the fact that 2009 dataset contains images and topics proposed by physicians and clinicians that beat the information needs.

Although we have only used the image caption in 2011 datasets, our model has achieved improvement of (+13.98%) according to the MAP and +20.35% according to p@15 compared to the baseline. We conclude that image captions contain relevant information.

Also, we conduct a comparative study between our approach and the existing retrieval models such as: BM25, BM25F, Dirichlet language model, and Bo1 pseudo-relevance feedback model. According to the results, our proposed model gives the best precision compared to other models.

5   Conclusions and Future Work

In this paper, we have proposed a new BayNetThes approach for re-ranking medical image by using Bayesian network based thesaurus. We proposed a new thesaurus based on Bayesian network and specific features. It is composed of three type of layers: layer of query nodes, layer of image nodes and layer of feature nodes. Those layers are connected using directed arcs outgoing from the upper layer to the lower layer. This thesaurus has been endowed with an inference mechanism which allows to obtain the probability score for each image. In particular, we have shown that the relationship between specific features has a great influence on the re-ranking process. Experiments are carried out using the medical ImageCLEF collections from 2009 to 2012. Results show that reranking using BayNetThes provides better retrieval performance than BM25 model.

In future work, we plan to include more specific features that could cover more characteristics of any user information needs in the context of image retrieval. Furthermore, medical features could be mapped to medical concepts and integrated in our feature-based Bayesian network model as a forth layer. The relationship between feature layer and concept layer may be deduced form UMLS medical terminology. Finally, we plan to conduct a comparative study between our approach and the existing retrieval models such as Language Model.

Acknowledgments. This work was supported by a discovery grant from the Natural Sciences and Engineering Research Council (NSERC) of Canada and an NSERC CREATE award. We thank all reviewers for their thorough review comments on this paper.

References

1. Acid, S., Campos, L.M.D., Fernandez-luna, J.M., Huete, J.F.: An information retrieval model based on simple bayesian networks. Int. J. Intell. Syst. **18**(2), 251–265 (2003)

2. Ayadi, H., Khemakhem, M.T., Daoud, M., Huang, J.X., Jemaa, M.B.: Mining correlations between medically dependent features and image retrieval models for query classification. In: JASIST 2016, in press
3. Ayadi, H., Khemakhem, M.T., Daoud, M., Jemaa, M.B., Huang, J.X.: Correlating medical-dependent query features with image retrieval models using association rules. In: Proceedings of the 22nd ACM CIKM, pp. 299–308 (2013)
4. Bashir, S., Rauber, A.: On the relationship between query characteristics and ir functions retrieval bias. JASIST 62(8), 1515–1532 (2011)
5. Brin, S., Page, L.: The anatomy of a large-scale hypertextual web search engine. Comput. Netw. ISDN Syst. 30(1–7), 107–117 (1998)
6. Cao, Z., Qin, T., Liu, T.-Y., Tsai, M.-F., Li, H.: Learning to rank: from pairwise approach to listwise approach. In: Proceedings of the 24th ICML, pp. 129–136 (2007)
7. Cronen-Townsend, S., Zhou, Y., Croft, W.B.: Predicting query performance. In: Proceedings of the 25th ACM SIGIR, pp. 299–306 (2002)
8. Gey, F.C.: Inferring probability of relevance using the method of logistic regression. In: Proceedings of the 17th ACM SIGIR, pp. 222–231 (1994)
9. Hauff, C., Azzopardi, L., Hiemstra, D.: The combination and evaluation of query performance prediction methods. In: Boughanem, M., Berrut, C., Mothe, J., Soule-Dupuy, C. (eds.) ECIR 2009. LNCS, vol. 5478, pp. 301–312. Springer, Heidelberg (2009). doi:10.1007/978-3-642-00958-7_28
10. Kalpathy-Cramer, J., de Herrera, A.G.S., Demner-Fushman, D., Antani, S., Bedrick, S., Müller, H.: Evaluating performance of biomedical image retrieval systems- an overview of the medical image retrieval task at ImageCLEF 2004–2014. Comput. Med. Imaging Graph. 39, 55–61 (2014)
11. Kalpathy-Cramer, J., Müller, H., Bedrick, S., Eggel, I., de Herrera, A.G.S., Tsikrika,T.: Overview of the CLEF 2011 Medical Image Classification and Retrieval Tasks. In: CLEF (Notebook Papers/Labs/Workshop) (2011)
12. Kraaij, W., Westerveld, T., Hiemstra, D.: The importance of prior probabilities for entry page search. In Proceedings of the 25th ACM SIGIR, pp. 27–34 (2002)
13. Markonis, D., Holzer, M., Dungs, S., Vargas, A., Langs, G., Kriewel, S., Müller, H.: A survey on visual information search behavior and requirements of radiologists. Methods Inf. Med. 51, 539–548 (2012)
14. Metzler, D., Croft, W.B.: Linear feature-based models for information retrieval. Inf. Retr. 10(3), 257–274 (2007)
15. Müller, H., de Herrera, A.G.S., Kalpathy-Cramer, J., Demner-Fushman, D., Antani, S., Eggel, I.: Overview of the imageCLEF 2012 Medical Image Retrieval and Classification Tasks. In: CLEF 2012 Working Notes (2012)
16. Müller, H., Kalpathy-Cramer, J., Eggel, I., Bedrick, S., Radhouani, S., Bakke, B., Kahn, C.E., Hersh, W.: Overview of the CLEF 2009 medical image retrieval track. In: Peters, C., Caputo, B., Gonzalo, J., Jones, G.J.F., Kalpathy-Cramer, J., Müller, H., Tsikrika, T. (eds.) CLEF 2009. LNCS, vol. 6242, pp. 72–84. Springer, Heidelberg (2010). doi:10.1007/978-3-642-15751-6_8
17. Müller, H., Kalpathy-Cramer, J., Eggel, I., Bedrick, S., Reisetter, J., Kahn Jr., C.E., Hersh, W.R.: Overview of the clef 2010 medical image retrieval track. In: CLEF (Notebook Papers/LABs/Workshops) (2010)
18. Nallapati, R.: Discriminative models for information retrieval. In: Proceedings of the 27th ACM SIGIR, pp. 64–71 (2004)
19. Ribeiro, B.A.N., Muntz, R.: A belief network model for IR. In: Proceedings of the 19th ACM SIGIR, pp. 253–260 (1996)

20. Turtle, H.R., Croft, W.B.: Efficient probabilistic inference for text retrieval. In: RIAO, pp. 644–662. CID (1991)
21. Wilcoxon, F.: Individual comparisons by ranking methods. Biom. Bull. 1(6), 80–83 (1945)
22. Wu, H., Sun, K., Deng, X., Zhang, Y., Che, B.: UESTC at imageCLEF 2012 medical tasks. In: Proceedings of CLEF, pp. 1–1 (2012)
23. Xu, X.-C., Xu, X.-S., Wang, Y., Wang, X.: A heterogeneous automatic feedback semi-supervised method for image reranking. In: Proceedings of the 22nd ACM CIKM, pp. 999–1008 (2013)
24. Yang, L., Hanjalic, A.: Supervised reranking for web image search. In: Proceedings of the 18th ACM MM, pp. 183–192 (2010)

A Part-of-Speech Enhanced Neural Conversation Model

Chuwei Luo[1], Wenjie Li[2(✉)], Qiang Chen[1], and Yanxiang He[1]

[1] School of Computer Science, Wuhan University, Wuhan 430072, China
{luochuwei,qchen,yxhe}@whu.edu.cn
[2] Department of Computing, The Hong Kong Polytechnic University,
Kowloon Tong, Hong Kong
cswjli@comp.polyu.edu.hk

Abstract. Modeling syntactic information of sentences is essential for neural response generation models to produce appropriate response sentences of high linguistic quality. However, no previous work in conversational responses generation using sequence-to-sequence (Seq2Seq) neural network models has reported to take the sentence syntactic information into account. In this paper, we present two part-of-speech (POS) enhanced models that incorporate the POS information into the Seq2Seq neural conversation model. When training these models, corresponding POS tag is attached to each word in the post and the response so that the word sequences and the POS tag sequences can be interrelated. By the time the word in a response is to be generated, it is constrained by the expected POS tag. The experimental results show that the POS-enhanced Seq2Seq models can generate more grammatically correct and appropriate responses in terms of both perplexity and BLEU measures when compared with the word Seq2Seq model.

Keywords: Response generation · Seq2Seq neural conversation model · Syntactic information incorporating

1 Introduction

In recent years, conversational agents that are capable of generating human-like responses are becoming more and more popular. With tremendously increased amount of conversation data on social media websites such as Twitter and Weibo developing data-driven approaches for **response generation** has received much attention. Recently, with the success of neural network based sequence-to-sequence (Seq2Seq) models [6] in natural language processing applications, researchers begin to explore the Seq2Seq encoder-decoder framework for response generation [2–5,8,9,13,14] and have achieved a certain degree of success.

However, in the existing neural conversation models, the syntactic information of the input post and the output response sentences are often overlooked. Thus, there is no guarantee that the generated responses are grammatically

© Springer International Publishing AG 2017
J.M. Jose et al. (Eds.): ECIR 2017, LNCS 10193, pp. 173–185, 2017.
DOI: 10.1007/978-3-319-56608-5_14

correct and appropriate. For example, the sentence *"it good is"* with the part-of-speech (POS) tag sequence *"PRP(personal pronoun)+JJ(adjective)+VBP(verb, non-3rd person singular present)"* may be generated by word Seq2Seq models. However, it is not difficult to see that this sentence is not grammatically well-formed. Actually, the sentence *"it is good"* with the POS tag sequence *"PRP+VBP+JJ"* is preferred. Apparently, the ability to take the order of POS tags in a sentence into account is important for a word Seq2Seq model to generate a response of high linguistic quality.

Look at another example. For the question *"How old are you?"*, word Seq2Seq neural models often tend to generate high-frequency word sequence like *"I am fine"* or *"I am ok"* as a response [8]. Obviously, they are not the appropriate responses. To better respond to the POS tag sequence of *"WRB(Wh-adverb)+JJ+VBP+PRP"*, *"PRP+VBP+CD(cardinal number)"* is more desired than *"PRP+VBP+JJ"*. Intuitively, if we can let the model know that it should produce a cardinal number after *"I am"* in this case, we may get a better response. So, the ability to generate the word conditioned on a preferred POS tag will help produce a more appropriate response.

In order to generate appropriate responses with high linguistic quality, in this work we investigate the effectiveness of utilizing the syntactic information of input posts and output responses. We introduce two POS-enhanced Seq2Seq conversation models. The first model, called POSEM-I, attaches the corresponding POS tag to each word in posts and responses to build POS-sequences that roughly represent sentence syntactic structures. It combines POS embeddings with word embeddings in the encoder, and decodes both words and their POS tags simultaneously. On the top of POSEM-I, the second model, called POSEM-II, decodes the POS tag first, and then generate the word based on the decoded POS tag. Experimental results show that our models outperform the word Seq2Seq model in terms of both perplexity and BLEU measures and can generate more appropriate responses of higher linguistic quality.

The contributions of our work can be summarized as follows. We suggest a way to integrate the syntactic information in the Seq2Seq neural conversation models, where the syntactic information is represented by the POS sequences. Since both the word and the tag orders are maintained, the grammatical errors are hence alleviated to a certain degree. We also propose an effective word generation mechanism which ensures to generate the word that belongs to the generated part-of-speech. In this way, the strong dependency that holds between the word and its POS tag is considered and can help produce a more appropriate response.

2 Related Work

The task of response generation in social medial was probably first introduced by Ritter et al. [1] who applied the phrase-based Statistical Machine Translation (SMT) techniques to *"translate"* a tweet post onto an appropriate response. They showed that SMT approaches are better suited than IR approaches on this task. Recently, with the growing interest and successful application of neural networks in machine translation, many researchers have extended these neural machine translation approaches to data-driven conversational response generation. Sordoni et al. [2] extended the work of Ritter et al. [1] to integrate prior conversational contextual information by utilizing the recurrent neural network language model (RNNLM). The consistent gains of their approaches over traditional IR and MT approaches were demonstrated. Based on a standard Seq2Seq neural network framework, Vinyals and Le [3] proposed a neural conversational model that could be trained end-to-end. Their experiment results showed the ability of this Seq2Seq model for generating simple conversations. Shang et al. [4] incorporate the attention mechanism into the Seq2Seq neural networks to generate responses for the post of the social media *Sina Weibo*. Yao et al. [11] found that in a conversation process, attention and intention play intrinsic roles. They proposed a neural network based approach that consists of three connected recurrent networks to model the attention and intention processes in conversations. Serban et al. [5] defined the generative dialogue problem as modeling the utterances and interactive structure of the dialogue. They proposed a hierarchical encoder decoder neural network framework to generate dialogue. Li et al. [8] found that Seq2Seq response generation models tend to generate safe, commonplace responses regardless of the input. They introduce a new objective function in Seq2Seq models to produce more diverse responses. To keep speaker consistency in neural response generation, Li et al. [9] represented speakers and address as distributed embeddings and incorporated these embeddings to Seq2Seq models. To model the participant role and conversational context information for two-party conversations, Luan et al. [14] incorporated both LDA topic feature and role factor as the context into RNNLM. Gu et al. [13] found a phenomenon in human-to-human conversation that humans tend to repeat entity names or even long phrases during conversations. They thus integrated the copying mechanism into Seq2Seq learning.

As we can see, most existing work on response generation adopts the Seq2Seq models. These models however are all based on the word sequences but ignore the associated syntactic information. Inevitably, they may produce the responses that contains certain grammatical errors and may be inappropriate to the post. We are thus motivated to look into the word part-of-speech and explore how to integrate the part-of-speech sequences with the word sequences.

3 Response Generation

The Seq2Seq model has been previously applied to response generation [2–5, 8,9,11]. It is essentially an encoder-decoder model. Given an input post, the encoder first transforms it into a hidden state representation. Then, the decoder generates the response based on this hidden state representation.

3.1 Standard Word Sequence-to-Sequence Model

In a typical Seq2Seq response generation model as illustrated in Fig. 1(a), given a sequence of input post $X = (x_1, x_2, ..., x_n)$ with length n and the corresponding

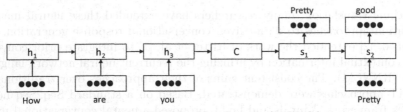

(a) Standard Word Seq2Seq Conversation Model

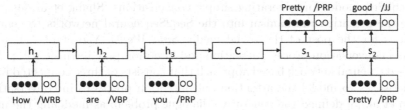

(b) Part-of-Speech Enhanced Conversation Model I(POSEM-I)

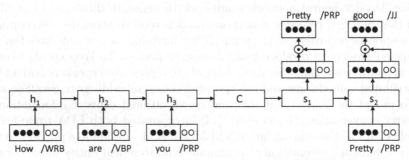

(c) Part-of-Speech Enhanced Conversation Model II(POSEM-II)

Fig. 1. Seq2Seq conversation models

response as an output sequence $Y = (y_1, y_2, ..., y_m)$ with length m, where x and y denote the word in the input and output sentences, respectively. Given a post, the likelihood of a response is estimated by:

$$P(Y|X) = \prod_{t=1}^{m} p(y_t|x_1, x_2, ..., x_n, y_1, y_2, ..., y_{t-1}) \tag{1}$$

With the recurrent neural network (RNN) encoder of the model, post X is converted into a set of high-dimensional hidden state representations $h = (h_1, h_2, ..., h_n)$. For $t = 1, ..., n$:

$$h_t = f(x_t, h_{t-1}) \tag{2}$$

where $f(\cdot)$ is a nonlinear activation function. It can be a logistic function. However, as known, the RNN may suffer from vanishing or exploding gradients for long sequences [22]. To avoid these problems, the long short-term memory

(LSTM) [15] unit or the gated recurrent unit (GRU) [16,17] could be considered. It has been found that GRU performs comparably to LSTM on sequential modeling, but GRU is computationally cheaper [16,19]. So, we choose to use GRU as the activation function $f(\cdot)$ when implementing the word Seq2Seq model.

$$z_t = \sigma(W_z e(x_t) + U_z h_{t-1})$$
$$\tilde{h} = tanh(W e(x_t) + U(r_t \circ h_{t-1}))$$
$$r_t = \sigma(W_r e(x_t) + U_r h_{t-1})$$
$$f(x_t, h_{t-1}) = z_t \circ h_{t-1} + (1 - z_t) \circ \tilde{h}$$

(3)

In the above Eq. (3), z_t is the update-gate, $\tilde{h}$ is the candidate activation and r_t is the reset gate. $\circ$ is a point-wise multiplication. σ is the sigmoid. $e(x_t)$ is word embedding of the word x_t. W_z, U_z, W, U, W_r, U_r are parameters of GRU.

Usually, a decoder will choose the last hidden state h_n as the context vector c. For $t = 1, ..., m$, it generates the response by

$$s_t = f(y_{t-1}, s_{t-1})$$
$$o_t = s_t W_{out}$$
$$p(y_t | x_1, x_2, ..., x_n, y_1, y_2, ..., y_{t-1}) = g(o_t)$$

(4)

where s_t is the hidden state in decoder at time t and $s_0 = c$. Function $g(\cdot)$ is the softmax activation function. $W_{out} \in R^{K \times V}$ is the output parameter where K is the word embedding dimensions and the vocabulary size is V. o_t is the output vector that is used to get the word probability distribution. Here, GRU is also used as function $f(\cdot)$. Note that the post encoder and the response decoder use two different RNNs with different sets of parameters.

As we know, it is crucial for any natural language generation mode to produce fluent and grammatical correct sentences. But the standard word Seq2Seq response generation model may generate a sentence that is not grammatically well-formed. To tackle this issue, a straight forward solution is to integrate the syntactic information into the word Seq2Seq model.

As is known to all, the word order in a language often follows a particular grammatical structure. For instance, there are two basic positions for adjectives in English: after *verbs* (e.g., *is good*) and before a *noun* (e.g., *good game*). This grammatical order of words is the important information that can influence the conversation generations. This explains why the sentence "*It is good.*" mentioned in the Introduction section that follows the order "*PRP+VBP+JJ*" is a good one. In order to improve the quality of the generated responses, we propose two enhanced Seq2Seq models to utilize the ordering information of part-of-speech (POS) tags (pronoun, verb, noun, etc.).

3.2 Part-of-Speech Enhanced Conversation Model I (POSEM-I)

The first model POSEM-I represents the grammatical word order in the post and response sentence by two sequences of POS tags. We design a POS tag vector to represent a POS tag. Figure 1(b) gives a brief illustration of this model. Let $P_x = (p_{x_1}, p_{x_2}, ..., p_{x_n})$ denote the POS tags sequence of input post X. $P_y = (p_{y_1}, p_{y_2}, ..., p_{y_m})$ denotes the POS tags sequence of output response Y. Each POS tag in P_x and P_y is associated with a one-hot tag vector, where p_{x_t}, $p_{y_t} \in R^L$. L is the size of POS tagset. Using the encoder RNN with GRU, both input post X and input POS tag sequences P_x are encoded into the context vector c by concatenating word embedding $e(x_i) \in R^{K_w}$ and its corresponding POS tag embedding $e(p_{x_i}) \in R^{K_p}$. For $t = 1, ..., n$:

$$H_t = [e(x_t); e(p_{x_t})]$$
$$z_t = \sigma(W_z H_t + U_z h_{t-1})$$
$$\tilde{h} = tanh(W H_t + U(r_t \circ h_{t-1}))   \quad (3')$$
$$r_t = \sigma(W_r H_t + U_r h_{t-1})$$
$$h_t = f(x_t, p_{x_t}, h_{t-1}) = z_t \circ h_{t-1} + (1 - z_t) \circ \tilde{h}$$

where h_t is the encoder hidden state. Different from the encoder of standard word Seq2Seq model formulated in Eq. (3), we use $[e(x_t); e(p_{x_i})]$ to denote the concatenation of embeddings $e(x_t)$ and $e(p_{x_t})$ to combine the word and its POS information into a single vector H_t. The first K_w dimensions of H_t correspond to words and the last K_p dimensions of H_t contain the POS tag information.

At each time step in the decoder, the hidden state that contains both word and word POS information in a response is decoded. For $t = 1, ..., m$,

$$s_t = f(y_{t-1}, p_{y_{t-1}}, s_{t-1})$$
$$ow_t = s_t[: K_w] W_{out}^1$$
$$op_t = s_t[-K_p :] W_{out}^2   \quad (4')$$
$$p(y_t | X; P_x; y_1, y_2, ..., y_{t-1}) = g(ow_t)$$
$$p(p_{y_t} | X; P_x; p_{y_1}, p_{y_2}, ..., p_{y_{t-1}}) = g(op_t)$$

where s_t represents the hidden state, $f(\cdot)$ is again the GRU (the same as in Eq. (3')). Unlike the decoder formulated in Eq. (4), $s_t[: K_w]$, i.e., the first K_w dimensions of s_t, is used to get the response words and $s_t[-K_p :]$, i.e., the last K_p dimensions of s_t is used to get the response word POS tags. Here, the output matrices are $W_{out}^1 \in R^{K_w \times V}$ and $W_{out}^2 \in R^{K_p \times L}$.

In short, with POSEM-I, we encode both post word sequence and the corresponding POS tag sequence simultaneously, and decode response word and its POS tag together. We follow the work of Dong et al. [20] and Luong et al. [21] and use the multi-task Seq2Seq learning setting to define the objective function. The objective function to be optimized is the summation of the two conditional probability terms conditioned on the context vector generated from the encoder.

$$L(\theta) = \underset{\theta}{argmax}(\prod_{t=1}^{m} p(y_t | X; P_x; y_1, y_2, ..., y_{t-1}; \theta)$$

$$+ \lambda \prod_{t=1}^{m} p(p_{y_t} | X; P_x; p_{y_1}, p_{y_2}, ..., p_{y_{t-1}}; \theta)) \quad (1')$$

where θ denotes the whole parameters of the model. λ is a hyper-parameter to control the ratio between the two conditional probability terms.

In this way, the grammatical order of words is encoded into the hidden layer and thus the context vector of this model contains not only the information of words in the sentences but also the syntactic information. At each time step during decoding, the decoder outputs both the word and the POS tag which can be used in the next time step to generate the more reasonable word and the POS tag. For example, suppose we have generated an *adjective* word "*good*" at time t. Considering that adjectives are often present before nouns in the training dataset, this *adjective* word "*good*" will help to constrain the decoder at time $t + 1$ to generate a *noun* like "*game*".

3.3   Part-of-Speech Enhanced Conversation Model II (POSEM-II)

For the second example mentioned in the Introduction section before, we can actually get a more appropriate response if we allow the decoder to produce a word that belongs to a particular favorite part-of-speech. Also, notice that a part-of-speech is a category of words that share similar grammatical properties. The words that are assigned to the same part-of-speech play the same or very similar roles within the grammatical structure of sentences. If the decoder knows to generate a word that exactly belongs to the part-of-speech generated, it will reduce the possibility for grammatical errors. Suppose at time t, the decoder has generated *noun* as a POS tag. Then, it must ensure to generate a noun word instead of an adjective word or the other non-noun word. While POSEM-I does not take the strong dependency that holds between the word and its POS tag into account, POSEM-II attempts to address this issue by generating the word not only conditioned on the hidden state but also the POS tag generated in advance.

POSEM-II has the same encoder and the same objective function as POSEM-I. As illustrated in Fig. 1(c), input post X and input POS tag sequence P_x are encoded into the context vector c together.

Different from the decoder (as formulated in Eq. (4')) of POSEM-I, the $\odot$ operation is defined in POSEM-II to ensure that the right word according to the POS tag is generated. The final word output vector ow'_t is obtained based on the original word output vector ow_t and POS tag output vector op_t by the $\odot$ operation. Formally, for $t = 1, ..., m$:

$$
\begin{aligned}
s_t &= f(y_{t-1}, p_{y_{t-1}}, s_{t-1}) \\
ow_t &= s_t[: K_w]W^1_{out} \\
op_t &= s_t[-K_p :]W^2_{out} \\
ow'_t &= ow_t \odot op_t \\
p(y_t|X; P_x; y_1, y_2, ..., y_{t-1}) &= g(ow'_t) \\
p(p_{y_t}|X; P_x; p_{y_1}, p_{y_2}, ..., p_{y_{t-1}}) &= g(op_t)
\end{aligned}
\tag{4''}
$$

In Eqs. (4''), s_t is the decoder's hidden state, $f(\cdot)$ is GRU. The details of the $\odot$ operation is as follows: A POS-word matrix $W_{pw} \in R^{L \times V}$ is built. It projects the output POS tag vector op_t to a weight vector. The weight vector has an entry to every word that belongs to the given POS tag. Then, the word is produced according to

$$ow'_t = ow_t \odot op_t = ow_t + op_t W_{pw} \tag{5}$$

In POSEM-II, the probability of the words that belong to the given part-of-speech is increased by adding ow_t to the weight vector $op_t W_{pw}$ in Eq. (5). As such, POSEM-II ensures to generate that word that belongs to the generated part-of-speech.

4   Experiments

4.1   Dataset

The training data is extracted from the Short-Text-Conversation (STC) dataset [10,23] that is a Sina Weibo[1] dataset. It contains about 196 K post and each post corresponds to an average of 28 different responses. For computational efficiency and to minimize noise, we selected the response that contains the maximum number of frequent bigram in the whole corpus. This produces a collection of 196 K Weibo post-response pairs. There are about 48 categories in Sina Weibo. To evaluate our models on different domains, we use Latent Dirichlet Allocation (LDA) [27,28] to cluster the 196K Weibo post-response pairs into 48 topics, from which we choose 5 topics to conduct the experiments, including movie, art, social news, science, and photography. The size of a topic ranges from 3,305 to 5,864 post-response pairs. For each topic, we random choose 100 posts for test and all the rest are used for training. In addition, we filter out the emoticons from the dataset.

4.2   Implementation Detail

Stanford Word Segmenter[2] and Stanford Part-Of-Speech Tagger[3] are used to segment the Chinese character sequences into words and to tag the POS information for the words. Our POS tag set is the LDC Chinese Treebank POS tag set that is used in the Stanford Chinese taggers. The words distributions of different topics are different. We construct the vocabulary for each topic using the words that appear at least twice in the dataset. All the low frequency words that do not appear in the vocabulary are replaced by a special mark "UNKnown". Following the work of Sutskever et al. [6], we trained our four-layer Seq2Seq models. The dimensions of the hidden states in all the layers are 1,000. The batch size is 128 and the learning rate is 0.5. Model parameters are initialized by randomly sampling from the uniform distribution [−0.1, 0.1]. We keep the gradients to [−5, 5] in order to avoid gradient explosion.

---

[1] http://www.weibo.com.
[2] http://nlp.stanford.edu/software/segmenter.shtml.
[3] http://nlp.stanford.edu/software/tagger.shtml.

4.3 Evaluation

How to well evaluate response generation automatically is still an open question. A goal of our models is to improve the capability of the Seq2Seq models to generate a response of better linguistic quality. Following the common practice in the work of [5, 8, 9, 11, 14], we use perplexity as the evaluation criterion. Perplexity is the most widely-used evaluation metric for language modeling. It evaluates the capability of a model for generating natural texts and can reflect the model's ability to account for the syntactic structure of the generated sentences [5]. Perplexity has also been suggested to evaluate the generative dialogue models [26], where a lower perplexity indicates better performance. Besides perplexity, following [1, 2, 8, 9] we also use multireference BLEU [24] to evaluate our models. As it has been shown that BLEU has well agreement with human judgments on response generation in the work of [9, 25].

Moreover, human evaluation is also performed. The annotators are asked to imagine themselves as the Sina Weibo users and they should treat the test posts as the Weibo post written by themselves. Like Ritter et al. [1], we perform pair-wise comparisons. We ask the annotators to judge which response is more relevant and appropriate compared against the other. We removed the pairwise responses that annotators can't decide which one is better. We present the responses to annotators in a random order.

4.4 Results

The results of perplexity are presented in Table 1. Clearly, both POSEM-I and POSEM-II outperform the standard word Seq2Seq conversation model. Compared with the standard word Seq2Seq conversation model, the POS-enhanced models achieve significant perplexity decrease. It reveals that combining POS tag sequences to the standard word Seq2Seq model helps a lot. Furthermore, by ensuring to generate the word that belongs to the generated part-of-speech, POSEM-II performs better than POSEM-I. The decoding mechanism in POSEM-II is indeed an efficient way to get better sentences with higher linguistic quality.

In Table 2, we report the BLEU results of standard word Seq2Seq model, POSEM-I and POSEM-II. As we can see, the two POS-enhanced models perform better than the standard word Seq2Seq model. By comparing POSEM-I against POSEM-II, we find that the BLEU score of POSEM-II is just a little bit better than that of POSEM-I. Actually, the BLEU criterion favors the responses that contain more high frequency and general words. They are however by no means

Table 1. Perplexity of standard Seq2Seq model and POS-enhanced models (POSEM-I and POSEM-II).

Model	Standard model	POSEM-I	POSEM-II
Perplexity	53.72	33.71	24.55

Table 2. BLEU of standard Seq2Seq model and POS-enhanced models (POSEM-I and POSEM-II). BLEU* means BLEU scores with stop words removed from generated responses.

Model	Standard model	POSEM-I	POSEM-II
BLEU	0.549	0.592	0.594
BLEU*	0.054	0.079	0.106

Table 3. Results of pairwise comparisons between various models. The column **Fraction A** lists the fraction of hits where the majority of annotators agreed Model A's response was better (more appropriate).

Model A	Model B	Fraction A	p-value
POSEM-I	Standard model	0.743	2.6e-09
POSEM-II	Standard model	0.823	2.2e-16
POSEM-II	POSEM-I	0.645	3.8e-04

the better responses. When taking a closer look at the output from the three models, we found this is the main reason why the expected significant improvement in terms of BLEU is not shown. For fair evaluation, we further report the BLEU scores with stop words removed from the generated responses, which is denoted as BLEU* in Table 2. The two proposed models now beat the word Seq2Seq conversation in BLEU scores and the significant performance gains are observed. With more strict constraints incorporated in generation mechanism, POSEM-II brings further improvement over POSEM-I. The BLEU* results also suggest that the proposed models can generate more diverse and informative responses than the word Seq2Seq model.

Table 3 presents the results of human evaluations. It is clear that the proposed two models both generate more appropriate responses than the word Seq2Seq models, while POSEM-II performs better than POSEM-I. Like Ritter et al. [1], we also show the p-value of an exact Binomial significance test. It appears that all differences are significant with above 95% confidence. The results of human evaluation further support the conclusion that incorporating the POS information helps generate more appropriate responses.

4.5   Discussion

Table 4 shows a few example responses generated by POSEM-II which is the best model in terms of both perplexity and BLEU*. In general, the output responses look fluent and reasonable. There seem rare grammatical mistakes in these responses. This is not surprising since it is our goal to improve the linguistic quality of generated responses. Statistical analysis of the average length of responses tells that the responses produced with POSEM-II are on average longer (7.25 Chinese words) than the word Seq2Seq model (5.68 Chinese words). Moreover, the word Seq2Seq model tends to generate more generic and commonplace responses [2,4,5,8].

Table 4. Sample responses generated by POSEM-II.

Input Post	Generated Response
小小年纪竟有如此身手！ He has such a good skill at this young age!	这孩子太有才了！ This guy is extremely talented!
那些奇葩的创意用品: 印章, 煎蛋器, 飞镖, 磁铁. Those wacko creative products are: stampers, egg-boilers, darts and magnets.	最后一个亮了！ The last one is amazing!
此次雅安地震是2008年汶川大地震的一次余震。 This earthquake in Ya'an was an aftershock of the 2008 disaster in Wenchuan.	我姑娘刚考上研就大地震了。 The earthquake happened as soon as my daughter was admitted to the graduate school.
我在北京, 24岁, 想去马尔代夫, 一个人。 I'm 24 years old, now in Beijing and I want to travel to Maldives alone.	我在南京, 马上下午三点半去长沙, 忘记航班号了。 I'm now in Nanjing and I'm going to Changsha at 3:30 pm. But I forget my flight number.
抹胸、露肩、缎面、蕾丝、多层A字大裙摆, 洋娃娃一般精致惊艳的美丽。 Dressed in the strapless, off the shoulder and layered A-line gown with stain and lace, she looks so beautiful!	美, 这个婚纱真是让我心动啊 This fantastic wedding dress makes me really excited!

The results shown in Table 4 and the statistics of the average length of responses make us believe that the new response generation mechanism we develop helps to produce more diverse and reasonable responses. However, the new models still can not solve the same problem discussed in [2]. As observed in our testset, if an input post is too long especially when there are many low frequency words in it, the responses produced by POSEM-II cannot often guarantee to be appropriate to the post. Nevertheless, this problem has been overcome in certain degree. As we can see in the last example, even though the input post is long and with low frequency words, POSEM-II can still output a fluent and appropriate sentence.

5 Conclusions

In this paper, we investigate how to utilize the syntactic information to improve Seq2Seq neural conversation models. We have proposed two part-of-speech enhanced neural conversation models for response generation. We have incorporated the part-of-speech tag sequences of the post and response into the standard Seq2Seq neural conversation model and our model generates the words in response conditioned on the POS tags sequence. Empirical studies confirm that our proposed part-of-speech enhanced conversation models can outperform the standard Seq2Seq neural conversation model in terms of perplexity, BLEU, and human judgments. Our study also indicates that our part-of-speech enhanced conversation models can generate more fluent, diverse and informative responses than the standard Seq2Seq neural conversation model.

Acknowledgments. The work described in this paper was supported by National Natural Science Foundation of China (61272291 and 61672445) and The Hong Kong Polytechnic University (G-YBP6, 4-BCB5 and B-Q46C).

References

1. Ritter, A., Cherry, C., Dolan, W.B.: Data-driven response generation in social media. In: Proceedings of the Conference on Empirical Methods in Natural Language Processing, pp. 583–593 (2011)
2. Sordoni, A., Galley, M., Auli, M., Brockett, C., Ji, Y., Mitchell, M., Nie, J.-Y., Gao, J., Dolan, B.: A neural network approach to context-sensitive generation of conversational responses. In: Proceedings of NAACL-HLT (2015)
3. Vinyals, O., Le, Q.: A neural conversational model. In: Proceedings of ICML Deep Learning Workshop (2015)
4. Shang, L., Zhengdong, L., Li, H.: Neural responding machine for short-text conversation. In: ACL-IJCNLP, pp. 1577–1586 (2015)
5. Serban, I.V., Sordoni, A., Bengio, Y., Courville, A., Pineau, J.: Building end-to-end dialogue systems using generative hierarchical neural network models. In: Proceedings of AAAI (2016)
6. Sutskever, I., Vinyals, O., Le, Q.V.: Sequence to sequence learning with neural networks. In: Advances in Neural Information Processing Systems (NIPS), pp. 3104–3112 (2015)
7. Wen, T.-H., Gasic, M., Mrksic, N., Pei-Hao, S., Vandyke, D., Young, S.: Semantically conditioned LSTM-based natural language generation for spoken dialogue systems. In: Proceedings of EMNLP, pp. 1711–1721 (2015)
8. Li, J., Galley, M., Brockett, C., Gao, J., Dolan, B.: A diversity-promoting objective function for neural conversation models. In: NAACL-HLT (2016a)
9. Li, J., Galley, M., Brockett, C., Gao, J., Dolan, B.: A persona-based neural conversation model. In: Proceedings of ACL (2016b)
10. Wang, H., Zhengdong, L., Li, H., Chen, E.: A dataset for research on short-text conversations. In: Proceedings of EMNLP, pp. 935–945 (2013)
11. Yao, K., Zweig, G., Peng, B.: Attention with intention for a neural network conversation model. In: NIPS Workshop on Machine Learning for Spoken Language Understanding and Interaction (2015)
12. Wen, T.-H., Vandyke, D., Mrksic, N., Gasic, M., Rojas-Barahona, L.M., Pei-Hao, S., Ultes, S., Young, S.: A network-based end-to-end trainable task-oriented dialogue system. arXiv preprint (2016). arXiv:1604.04562
13. Gu, J., Lu, Z., Li, H., Li, V.O.K.: Incorporating copying mechanism in sequence-to-sequence learning. In: Proceedings of ACL (2016)
14. Luan, Y., Ji, Y., Ostendorf, M.: LSTM based conversation models. arXiv preprint (2016). arXiv:1603.09457
15. Hochreiter, S., Schmidhuber, J.: Long short-term memory. Neural Comput. 9(8), 1735–1780 (1997)
16. Chung, J., Gulcehre, C., Cho, K., Bengio, Y.: Empirical evaluation of gated recurrent neural networks on sequence modeling. arXiv preprint (2014). arXiv:1412.3555
17. Cho, K., van Merrienboer, B., Gulcehre, C., Bougares, F., Schwenk, H., Bengio, Y.: Learning phrase representations using RNN encoder-decoder for statistical machine translation. arXiv preprint (2014). arXiv:1406.1078
18. Serban, I.V., Lowe, R., Charlin, L., Pineau, J.: A survey of available corpora for building data-driven dialogue systems. arXiv preprint (2015). arXiv:1512.05742
19. Greff, K., Srivastava, R.K., Koutník, J., Steunebrink, B.R., Schmidhuber, J.: LSTM: a search space Odyssey. CoRR, abs/1503.04069 (2015)
20. Dong, D., Wu, H., He, W., Yu, D., Wang, H.: Multi-task learning for multiple language translation. In: Proceedings of ACL (2015)

21. Luong, M.-T., Le, Q.V., Sutskever, I., Vinyals, O., Kaiser, L.: Multi-task sequence to sequence learning. In: Proceedings of ICLR (2016)
22. Bengio, Y., Simard, P., Frasconi, P.: Learning long-term dependencies with gradient descent is difficult. IEEE Trans. Neural Netw. **5**(2), 157–166 (1994)
23. Shang, L., Sakai, T., Zhengdong, L., Li, H., Higashinaka, R., Miyao, Y.: Overview of the NTCIR-12 short text conversation task. In: NTCIR-2012 (2016)
24. Papineni, K., Roukos, S., Ward, T., Zhu, W.-J.: BLEU: a method for automatic evaluation of machine translation. In: Proceedings of ACL, pp. 311–318 (2002)
25. Galley, M., Brockett, C., Sordoni, A., Ji, Y., Auli, M., Quirk, C., Mitchell, M., Gao, J., Dolan, B.: deltaBLEU: a discriminative metric for generation tasks with intrinsically diverse targets. CoRR, abs/1506.06863 (2015)
26. Pietquin, O., Hastie, H.: A survey on metrics for the evaluation of user simulations. Knowl. Eng. Rev. **28**(01), 59–73 (2013)
27. Blei, D.M., Ng, A.Y., Jordan, M.I.: Latent Dirichlet allocation. J. Mach. Learn. Res. **3**(Jan), 993–1022 (2003)
28. Hoffman, M.D., Blei, D.M., Bach, F.: Online learning for latent Dirichlet allocation. Adv. Neural Inf. Process. Syst. **23**, 856–864 (2010)

"Are Machines Better Than Humans in Image Tagging?" - A User Study Adds to the Puzzle

Ralph Ewerth[1,2(✉)], Matthias Springstein[1], Lo An Phan-Vogtmann[3], and Juliane Schütze[3]

[1] German National Library of Science and Technology (TIB), Hannover, Germany
{ralph.ewerth,matthias.springstein}@tib.eu
[2] L3S Research Center, Hannover, Germany
[3] Jena University of Applied Sciences, Jena, Germany
loan.phan-vogtmann@stud.eah-jena.de, juliane.schuetze@eah-jena.de

Abstract. "Do machines perform better than humans in visual recognition tasks?" Not so long ago, this question would have been considered even somewhat provoking and the answer would have been clear: "No". In this paper, we present a comparison of human and machine performance with respect to annotation for multimedia retrieval tasks. Going beyond recent crowdsourcing studies in this respect, we also report results of two extensive user studies. In total, 23 participants were asked to annotate more than 1000 images of a benchmark dataset, which is the most comprehensive study in the field so far. Krippendorff's α is used to measure inter-coder agreement among several coders and the results are compared with the best machine results. The study is preceded by a summary of studies which compared human and machine performance in different visual and auditory recognition tasks. We discuss the results and derive a methodology in order to compare machine performance in multimedia annotation tasks at human level. This allows us to formally answer the question whether a recognition problem can be considered as solved. Finally, we are going to answer the initial question.

1 Introduction

In the field of multimedia analysis and retrieval, human performance in recognition tasks was reported from time to time [2,9,12,13,15,16,20–23], but has not been evaluated in a consistent manner. As a consequence, the quality of human performance is not exactly known and estimates exist only for few recognition tasks. The design of the related human experiments also varies noticeably in many respects. For example, crowdsourcing is often utilized to employ annotators [9,12–16,23], which is coming along with some methodological issues. The number of human participants varies from 1 to 40 in the studies considered in this paper. The same is true for the experimental instructions and their expertise, in particular for crowdworkers. This, for example, makes it nearly unfeasible to evaluate and compare machine performance at human level across different tasks. In fact, we know little *in general* about human performance in multimedia

J.M. Jose et al. (Eds.): ECIR 2017, LNCS 10193, pp. 186–198, 2017.
DOI: 10.1007/978-3-319-56608-5_15

content analysis tasks. As a consequence, the question when such a task can be considered as solved cannot be answered easily. The related question is addressed by this paper: How can we systematically set machine performance in relation to human performance? If human ground truth data are the (only) baseline, machine performance can basically never be better than (human) ground truth data. But considering the impressive recent advances in deep learning for pattern recognition tasks, it is desirable to set machine performance in relation to human-level performance in a systematic manner.

Another issue is related to ground truth data for retrieval tasks: The relevance of multimedia documents at retrieval time for a certain user is not known in advance and it depends on the user's current search task and context. The issue of evaluating multimedia analytics systems has been also stressed recently by Zahálka et al. [25]. For example, a detective is interested in every occurrence of a suspicious object (e.g., a car) in any size. On the other hand, a TV journalist, who searches for material for re-use in order to illustrate the topic mobility, might be interested only in retrieval results showing a car in an "iconic" view, i.e., placed clearly in the foreground.

In this paper, we review a number of papers reporting human performance in visual and auditory recognition tasks. This aims at putting together some parts of the puzzle: How well do machines perform in such tasks compared to humans? To answer this question, the results are set in relation to the current state of the art of automatic pattern recognition systems. Furthermore, we present a comprehensive user study that closes the gap of comparing in detail human and machine performance in annotation tasks for realistic images, as they are used in the PASCAL VOC (Visual Object Classes) challenge [4,5], for example. More than 1000 images have been annotated by 23 participants in a non-crowdsourcing setting. The number of images also allows us to draw conclusions about rarely occuring concept categories such as "cow" or "potted plant". It is suggested to evaluate the reliability of users' annotations by Krippendorff's α [10,11], which measures the agreement among several coders. The results of the presented study are discussed and conclusions are drawn for the evaluation of computer vision and multimedia retrieval systems: A methodology is introduced that enables researchers to formally compare machine performance at human level in visual and auditory recognition tasks. To summarize, the contributions of this paper are as follows:

- Surveying and comparing human and machine performance in a number of visual and auditory pattern recognition tasks,
- presenting a comprehensive user study regarding image annotation yielding insights into the relation of human and machine performance,
- introducing the concept of inter-coder reliability in the field of multimedia retrieval evaluation for comparing human and machine performance,
- proposing an evaluation methodology that allows us to evaluate machine performance at human level in a systematic manner, and
- suggesting two indices for measuring human-level performance of systems.

The remainder of the paper is structured as follows. Section 2 surveys studies that compared human and machine performance for different visual and auditory recognition tasks. Section 3 deals with a comprehensive user study regarding image annotation and related results are presented. A methodology to evaluate machine performance at human level in a systematic way is suggested in Sect. 4. Finally, some conclusions are drawn in Sect. 5.

2   Human and Machine Performance in Visual and Auditory Recognition Tasks

In this section, we briefly survey related work which compared human and machine performance for some multimedia analysis tasks. Yet, human performance has been considered only in a small number of studies.

Some studies evaluated human performance in the task of visual concept classification. Kumar et al. [12] as well as Lin et al. [14] measured the human inter-coder agreement and compared experts against crowdsourcing annotators. Other papers reported the performance of humans and machine systems on some benchmark datasets. Jiang et al. [9] presented a dataset for consumer video understanding. For this dataset, the human annotations were significantly better than the machine results. Parikh and Zitnick [16] investigated the role of data, features, and algorithms for visual recognition tasks. An accuracy of nearly 100% is reported for humans on two PASCAL VOC datasets, whereas machine performance was around 50% on both datasets in 2008. Xiao et al. [23] presented the dataset SUN (Scene Understanding) for scene recognition consisting of 899 categories and 130,519 images. Scene categories are grouped in an overcomplete three-level tree. Human performance reached 95% accuracy at the (easy) first level and 68.5% at the third level of the hierarchy, while the machine performance of 38% accuracy was significantly below human accuracy in this study. Russakovsky et al. [18] surveyed the advances in the field of the ImageNet challenge [3] from 2010 to 2014. The best result in 2014 was submitted by Szegedy et al. [19] (GoogLeNet) and achieved an error rate of 6.66%. Russakovsky et al. compared this submission with two human annotators and discovered that the neuronal network outperformed one of them. He et al. [8] claimed their system to be the first one that surpassed human-level performance (5.1%) on ImageNet data by achieving an error rate of 4.94%.

Phillips et al. [17] conducted one of the first comparisons of face identification capabilities of humans and machines. Interestingly, at that time the top three algorithms were already able to match or to do even better face identification compared with human performance on unfamiliar faces under illumination changes. Taigman et al. [20] presented a deep learning system for face verification that improves face alignment based on explicit 3D modelling of faces. A human-level performance of 97.35% accuracy was reported for the benchmark "Faces in the Wild" (humans: 97.53% accuracy).

Other interesting comparisons between humans and machine systems include camera motion estimation (Bailer et al. [2]), music retrieval (Turnbull et al. [20]), and the geolocation estimation of images (Weyand et al. [22]). These studies also demonstrated that machines can reach or outperform human performance.

While the experiments of Parikh and Zitnick [16] with respect to re-engineering the recognition process did not provide evidence that humans are superior to machines, other reported results on PASCAL VOC and other data sets showed that humans perform significantly better on classifying natural scene categories. The reported results for visual concept classification [9,12,15,23] indicate that human performance is (far) better than the respective state of the art for automated visual concept classification at that time. Although He et al. claimed in 2015 that human-level performance has been surpassed by their approach [8], this claim remains questionable since only two human annotators were involved in the underlying study of Russakovsky et al. [18]. There are also some methodological issues in the reported experiments of the other studies, for example, experimental settings are not well defined, the employment of crowdsourcing is critical, or the number of images is too small which prevents drawing conclusions for rare classes. Hence, stronger empirical evidence still has to be provided for a meaningful comparison of human and machine performance. Therefore, in the remainder of this paper we address these issues by a comprehensive user study and derive a methodology to measure machine performance at human level.

3 User Study: Human Performance in Image Annotation

The analysis of previous work shows that the settings of the majority of studies do not allow us to compare human performance against machine learning approaches for image classification tasks. In this section, we present two user studies measuring human performance in annotating common image categories of daily life in a realistic photo collection (PASCAL VOC [4,5]). The design of this study is described in Sect. 3.1. The inter-coder agreement of the two experiments is evaluated using Krippendorff's α (Sect. 3.2). Furthermore, the results of the best machine systems submitted to PASCAL VOC's leaderboard are set in relation to the human agreement (Sect. 3.3). Finally, the results are made comparable in a systematic manner (Sect. 3.4).

3.1 Experimental Design

We have randomly selected 1,159 images from the PASCAL VOC test set. The relatively high number of images - compared to other studies - allows us to also obtain statistically relevant insights into human performance for less frequent concepts. For example, the concept cow is visible only in 34 out of 1,159 images, whereas the concept person occurs in 420 images. However, using PASCAL VOC's test set comes along with the disadvantage that the ground truth data are not available, in contrast to training and validation data. On the other hand, submission results are available only for the test set at PASCAL VOC's homepage. Therefore, we created ground truth for this test data subset by ourselves.

In total, twenty-three students (3rd and 4th year) were asked to annotate images of the test set with respect to 20 concept categories (see also Table 1), 18 students participated in the first and five other students participated in the second experiment. They were rewarded 25€ for participating in the experiment. All students were members of the Department of Electrical Engineering and Information Technology at the Jena University of Applied Sciences.

The participants were instructed to label images with respect to the presence of objects of 20 categories but without localizing them. Multiple object classes can be visible in an image, i.e., it is a multi-labeling task. This task corresponds to the classification task of the VOC challenge 2012. The study was further divided in two experiments. In the first experiment, the participants were instructed without using the PASCAL VOC annotation guidelines [1], since we aimed at measuring human performance based on common sense and existing knowledge about categories of daily life. In the second experiment, the participants were asked to annotate the images according to the PASCAL VOC guidelines.

The annotation process was divided in four batches that consisted of a slightly decreasing number of images. After each batch, the participants were allowed to make a break of 10–15 min. The annotation process had to be completed within four hours. The images were presented to all participants in the same order. They had to mark the correct object categories via corresponding checkboxes. When a user has finished annotating an image, he proceeded with the next image. The software did not allow users to return to a previously annotated image. All users completed the task within the given time limit.

3.2   Measuring Inter-Coder Agreement: Krippendorff's α

Krippendorff's α (K's α)[10,11] measures the agreement among annotators and is widely used in the social sciences to evaluate content analysis tasks. K's α is a generalization of several known reliability measures and has some desirable properties [11], it is (1) computable for more than two coders, (2) applicable to any level of measurement (ordinal, etc.) and any number of categories, (3) able to deal with incomplete and missing data, and (4) it is not affected by the number of units. In its general form K's α is equal to other agreement coefficients:

$$\alpha = 1 - \frac{D_o}{D_e}, \quad \text{where} \tag{1}$$

D_o is the observed disagreement and D_e is the expected disagreement due to chance. Krippendorff discusses differences of K's α with respect to other agreement coefficients [10] as well as explains its computation for various situations (depending on the number of coders, missing data, level of measurement, etc.) [11]. Hayes and Krippendorff provided a software that computes K's α [6].

3.3   Results for Inter-Coder Agreement

Agreement When *not* Using VOC Guidelines. The experimental results are displayed in Table 1 for the 1159 images of the PASCAL VOC test set. They show the inter-rater agreement among the 18 coders by means of K's α. Across all concept categories and users, K's α is 0.913. The largest agreement among the annotators is observable for the three categories airplane, cat, and bird, whereas the categories dining table, chair, and potted plant yield the lowest agreement.

Table 1. User agreement (K's α) on a subset of PASCAL VOC test set, number of samples per category in this *subset*, and best machine-generated results (AI-1 and AI-2, avg. precision) on the *whole* test set.

Concept	#Samples per concept	Human (K's α)	AI-1	AI-2
Airplane	77	0.980	0.986	0.998
Cat	96	0.978	0.955	0.990
Bird	101	0.976	0.934	0.976
Dog	123	0.974	0.947	0.989
Sheep	35	0.970	0.874	0.950
Cow	34	0.960	0.821	0.943
Horse	42	0.959	0.929	0.985
Bus	45	0.956	0.910	0.959
Train	55	0.953	0.960	0.987
Boat	46	0.940	0.922	0.964
Motorbike	55	0.938	0.921	0.972
Bicycle	68	0.909	0.860	0.947
TV monitor	63	0.898	0.827	0.942
Person	420	0.895	0.950	0.988
Car	126	0.848	0.836	0.948
Sofa	79	0.796	0.678	0.868
Bottle	93	0.761	0.654	0.836
Dining table	61	0.737	0.796	0.881
Chair	123	0.716	0.734	0.904
Potted plant	59	0.668	0.594	0.768
Overall/MAP	–	0.913	0.854	0.940

Some interesting observations can be made. First, larger deviations of inter-coder agreement are observable for the different categories. Applying the rule of thumb that $\alpha > 0.8$ corresponds to a "reliable" content analysis [10], it turns out that the users' annotations cannot be considered as such for five categories: sofa, bottle, dining table, chair, and potted plant (K's α even only 0.67).

Table 1 shows also the results AI-1 and AI-2, which are the best submissions at PASCAL VOC's leaderboard website[1] for the competitions comp1 and comp2,

---

[1] http://host.robots.ox.ac.uk:8080/leaderboard/main_bootstrap.php.

respectively, by means of average precision (AP). The difference between the two is that comp2 is not restricted to the training set. Please note that these results of the user study and the VOC submissions are not directly comparable at this stage due to two reasons. First, the users annotated only a subset of the original test set. Second, different evaluation measures are used. In particular, the measures differ in the way how agreement by chance is considered. A more fair comparison resolving these issues is conducted in the subsequent section.

Anyway, when we set the inter-rater agreement in relation to the performance of the currently best machine results, we find that the correlation of K's α and AP with respect to the categories is 0.88 (AI-1) and 0.89 (AI-2), respectively. In particular, it is observable that the machine learning approaches perform worst for the same five categories as humans do.

Agreement When Using VOC Guidelines. In this experiment, it is investigated whether the inter-coder agreement is improved when more precise definitions are provided to the users. For this purpose, we have asked five other students (from the same department) to annotate the same 1159 images – this time based on the PASCAL VOC annotation guidelines [1]. The guidelines give some hints how the annotator should handle occlusion, transparency, etc.

Interestingly, the inter-coder agreement was not improved by using the guidelines. The inter-coder agreement in this experiment (measured again by K's α) is 0.904, in contrast to 0.913 without guidelines. The difference between the mean values of K's α with respect to all 20 categories is not statistically relevant (paired student's t-test, two-sided, significance level of 0.05), i.e., the reliability of human annotations is equal in both experiments. Although the participants used the annotation guidelines, the annotators did not achieve a better agreement in this experiment.

3.4   Comparing Human and Machine Performance

Since ground truth data for PASCAL VOC 2012 test set are not publicly available, we have created ground truth data for the related subset on our own. The ground truth data have been created according to PASCAL VOC annotation guidelines (see above). Critical examples were discussed by three group members (experts), two of them being authors of the paper. If a consistent agreement could not be achieved, then the example was removed for the related category.

In addition, we have trained a convolutional neural network (called AI-3 from now on) and evaluated its performance on the 1159 images. This allows us to apply AI-3 on the whole PASCAL test set as well as on the subset used in the human annotation study. Finally, this link enables us to compare human and machine performance for visual concept classification on PASCAL VOC test set data. We use the convolutional neural network of He et al. [7] consisting of 152 layers, which we fine-tuned on the PASCAL VOC training dataset. The network was originally trained on the ImageNet 2012 dataset [18]. Furthermore, we have reduced the number of output neurons to the number of classes (in this case 20) and used a sigmoid transfer function to solve the multi-class labeling task of PASCAL VOC. Seven additional regions are cropped and evaluated per test image in the classification step to achieve better results. The mean average precision (MAP) for AI-3 is very similar for both datasets (0.871 vs. 0.867 on subset), although the difference of AP (for the subset and the whole test set) is larger for some classes, e.g., chair and train. The system AI-3 performs slightly better than AI-1. The results on both sets are depicted in Fig. 1.

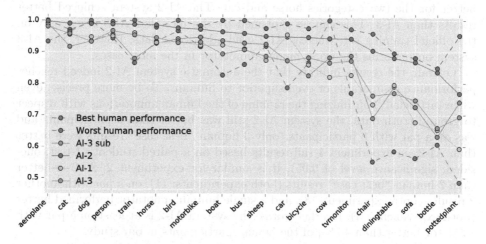

Fig. 1. Results (average precision in %) of the best and the worst human annotator, as well as of the best PASCAL VOC leaderboard submissions for comp1 and comp2.

Now, based on this analysis, the ground truth data allow us to compare human and machine performance using the same evaluation measure (average precision). But one issue still needs to be resolved. The annotators label the relevance of images only with "0" or "1" (in contrast to the real-values system scores). The (random) order of images labeled with "1" affects the measure of average precision. Due to this, we have calculated the "best case" (oracle) and the "worst case" possible retrieval results based on the human annotations.

Regarding the "best case" in the first experiment, there are three human annotators that are significantly better than AI-2 (paired student's t-test, one-sided, significance level of 0.05). These annotators achieve a mean average precision of 96.5 %, 96.3 %, and 95.3 %, respectively. However, regarding the "worst case" ordering, all human annotation results are significantly worse than AI-2, the best result (of the worst cases) achieves a mean average precision of 91.6 %. In other words, the results are sensitive to the ranking order of the images which are labeled as "relevant".

The results for the best annotator of our second experiment (using VOC guidelines) are similar. The best "best case" mean average precision is 96.9 %, this is the only human result of experiment 2 that is significantly better than

AI-2. Again, in case of "worst case" ordering, AI-2 is significantly better than all human annotation results.

Regarding both experiments, we have also estimated the best annotator (coming from experiment 2, with respect to best case ordering) and the worst annotator (coming from experiment 1, with respect to worst case ordering) with respect to our ground truth data. The results are displayed in Fig. 1. Again, the best results of the machine vision systems are presented as well. Regarding oracle results of the best human annotator, the automatic AI-2 system is only better for the two categories horse and cat. The AI-2 system achieved better results than AI-3. This can be explained by the fact that AI-3 relied only on the official training set, whereas AI-2 used additional data. However, the AI-3 system outperforms the worst human annotator in the most cases.

Overall, the results indicate that the automatic system AI-2 indeed reaches performance comparable or even superior to humans. To be more precise, even when (artificially) optimizing the ranking of the human annotations with respect to ground truth data, the system AI-2 still was better than 9 participants and was on a par with 6 participants (only 3 human "best case" results were better than AI-2) in experiment 1 (all results based on a paired student's t-test, one-sided, significance level of 0.05). It is similar for experiment 2: AI-2 is better than 2 human "best case" results (both experiments: 11), on a par with another 2 results (both experiments: 8), and a single human annotator performed better (both experiments: 4). In other words, the system AI-2 is at least on a par with 83 % (or better than 48 %) of the human participants in our study.

4   Measuring Machine Performance at Human Level

4.1   Issues of Measuring Human-Level Performance

The performance of multimedia retrieval systems is often measured by average precision (AP). However, this measure has some known drawbacks. First, the measure depends on the frequency of a concept. Considering the definition of average precision, it is clear that the lower bound is not zero but determined by the frequency of relevant documents in the collection. When randomly retrieving documents, the average precision will be equal to a concept's frequency on average. This has been stressed, e.g., by Yang and Hauptmann who suggested the measure ΔAP (delta average precision) to address this issue [24]. Second, the upper bound of 1.0 is also not reasonable. If we consider that the agreement in our user study is below 0.8 for five categories and is even only 0.67 for potted plant, the question arises how we have to interpret an average precision of, for example 67 % and 100 %, respectively. If two raters agree with K's α only with 0.67, and one rater corresponds to ground truth, it is basically possible that the 67 %-result has the same quality as the 100 %-result (from another perspective, in another context). Hence, the question remains: how can we formally measure whether a machine-based result is comparable to or even better than a human result? This question is addressed in the subsequent section.

4.2   An Experimental Methodology and Two Indices for Comparisons with Human-Level Performance

In this section, we propose an experimental methodology and two novel easy-to-use measures, called human-level performance index ($HLPI$) and human-level performance ranking index ($HLPRI$). The proposed methodology is aimed at providing a systematic guidance to measure human-level performance in multimedia retrieval tasks. It is assumed that a visual or auditory concept is either present or not in a multimedia document. Furthermore, it is assumed that a standard benchmark dataset is available. First, ground truth data G should be created by knowledgeable experts E of the related domain. If possible, the reliability of these expert annotations should be measured as well (K's α should exceed at least 0.8 as a rule of thumb) in order to ensure that the relevance of categories is well-defined. Critical examples should be subsequently discussed among the group of experts E. If no consistent decision is possible, then such examples should be appropriately marked or discarded from the dataset.

The group of human participants H should consist of at least five annotators/coders, who share a similar knowledge level regarding the target domain (e.g., experts, if performance is to be measured at expert level). The annotations of the group H are used to evaluate human performance in the given task. The annotation process should be conducted in a well-defined setting. The latter two criteria (same knowledge level, well-defined setting) normally preclude a crowd-sourcing approach. Apart from other measures, the inter-rater agreement among the annotators should be also estimated by Krippendorff's α. K's α is suggested since it can potentially also deal with other levels of measurement (than binary). Then, the agreement (accuracy) a_{human} is the median of the agreement scores (e.g., measured by K's α or AP) of the coders H with respect to G. The machine-generated result is also compared to G, yielding the agreement a_{machine}. Often, several instances of a machine system, e.g., caused by different parametrizations or training data, exist. In this case, it is also reasonable to test all these instances with respect to G and use the median as a_{machine}. In this way, it can be prevented that a system is better only due to fine-tuning or by chance. Then, the human-level performance index is defined as

$$HLPI = \frac{a_{\text{machine}}}{a_{\text{human}}}, \tag{2}$$

assuming that human inter-coder agreement is better than chance, i.e., $A_{\text{human}} > 0$. If $HLPI > 1.0$, then machine performance is possibly better. However, this has to be verified and ensured by an appropriate statistical significance test.

A second measure the human-level performance ranking index ($HLPRI$) is suggested. This index is based on a sorted list in descending order according to the agreement measurements of the n human annotators with respect to ground truth data G. Let b be the number of machine results to be evaluated that are better than human annotation results, and let w be the number of machine annotation results that are worse (in the sense of statistical relevance). Then, the human-level performance ranking index is defined by

$$HLPRI = \frac{b+1}{w+1} \tag{3}$$

$HLPRI$ of AI-2 is 2.4 and 1.5 in our experiments 1 and 2, respectively.

5   Conclusions

In this paper, we have investigated the question whether today's automatic indexing systems can achieve human-level performance in multimedia retrieval applications. First, we have presented a brief survey comparing human and machine performance in a number of visual and auditory recognition tasks. The survey has been complemented by two extensive user studies which investigated human performance in an image annotation task with respect to a realistic photo collection with 20 common categories of daily life. For this purpose, the well-knwon PASCAL VOC benchmark has been used. We have measured the human inter-coder agreement by Krippendorff's α and observed that the reliability of annotations noticeably varies for the concepts. Krippendorff's α was below 0.8 for 5 out of 20 categories, which indicates that these categories are not well-defined and are prone to inconsistent annotation. This is an issue for the creation of ground truth data and subsequent evaluation as well.

In addition, we have carefully compared human and machine performance for image annotation. It turned out that the best submission at PASCAL VOC's leaderboard is better than 11 or at least on a par with 19 out of 23 participants of our study. This indicates that the submission has indeed reached above-average human-level performance for the annotation of the considered visual concepts.

We have also addressed the issue of measuring human-level performance of multimedia analysis and retrieval systems in general. For this purpose, we have suggested an experimental methodology that integrates the assessment of human-level performance in a well-defined manner. Finally, we have derived two easy-to-use indices for measuring and differentiating human-level performance.

References

1. VOC2011 annotation guidelines. http://host.robots.ox.ac.uk/pascal/VOC/voc20 11/guidelines.html. Accessed 29 Mar 2016
2. Bailer, W., Schallauer, P., Thallinger, G.: Joanneum research at TRECVID 2005-camera motion detection. In: Proceedings of TRECVID Workshop, pp. 182–189 (2005)
3. Deng, J., Dong, W., Socher, R., Li, L.J., Li, K., Fei-Fei, L.: ImageNet: a large-scale hierarchical image database. In: Conference on Computer Vision and Pattern Recognition (CVPR), pp. 248–255. IEEE (2009)

4. Everingham, M., Eslami, S.A., Van Gool, L., Williams, C.K., Winn, J., Zisserman, A.: The PASCAL visual object classes challenge: a retrospective. Int. J. Comput. Vis. **111**(1), 98–136 (2015)
5. Everingham, M., Van Gool, L., Williams, C.K., Winn, J., Zisserman, A.: The PASCAL visual object classes (VOC) challenge. Int. J. Comput. Vis. **88**(2), 303–338 (2010)
6. Hayes, A.F., Krippendorff, K.: Answering the call for a standard reliability measure for coding data. Commun. Methods Meas. **1**(1), 77–89 (2007)
7. He, K., Zhang, X., Ren, S., Sun, J.: Deep residual learning for image recognition. arXiv preprint (2015). arXiv:1512.03385
8. He, K., Zhang, X., Ren, S., Sun, J.: Delving deep into rectifiers: surpassing human-level performance on ImageNet classification. In: Proceedings of the International Conference on Computer Vision (ICCV), pp. 1026–1034 (2015)
9. Jiang, Y.G., Ye, G., Chang, S.F., Ellis, D., Loui, A.C.: Consumer video understanding: a benchmark database and an evaluation of human and machine performance. In: Proceedings of the International Conference on Multimedia Retrieval (ICMR), p. 29. ACM (2011)
10. Krippendorff, K.: Reliability in content analysis: Some common misconceptions and recommendations. Hum. Commun. Res. **30**, 411–433 (2004)
11. Krippendorff, K.: Computing Krippendorff's alpha-reliability (2011). http://repository.upenn.edu/asc_papers/43/. Accessed 29 Mar 2016
12. Kumar, N., Berg, A.C., Belhumeur, P.N., Nayar, S.K.: Attribute and simile classifiers for face verification. In: International Conference on Computer Vision (ICCV), pp. 365–372. IEEE (2009)
13. Lake, B.M., Salakhutdinov, R., Tenenbaum, J.B.: Human-level concept learning through probabilistic program induction. Science **350**(6266), 1332–1338 (2015)
14. Lin, T.-Y., Maire, M., Belongie, S., Hays, J., Perona, P., Ramanan, D., Dollár, P., Zitnick, C.L.: Microsoft COCO: common objects in context. In: Fleet, D., Pajdla, T., Schiele, B., Tuytelaars, T. (eds.) ECCV 2014. LNCS, vol. 8693, pp. 740–755. Springer, Cham (2014). doi:10.1007/978-3-319-10602-1_48
15. Nowak, S.: Evaluation methodologies for visual information retrieval and annotation. Ph.D. thesis (2011)
16. Parikh, D., Zitnick, C.L.: The role of features, algorithms and data in visual recognition. In: Conference on Computer Vision and Pattern Recognition (CVPR), pp. 2328–2335. IEEE (2010)
17. Phillips, P.J., Scruggs, W.T., O'Toole, A.J., Flynn, P.J., Bowyer, K.W., Schott, C.L., Sharpe, M.: FRVT 2006 and ICE 2006 large-scale results. Natl. Inst. Stand. Technol. NISTIR **7408**, 1 (2007)
18. Russakovsky, O., Deng, J., Su, H., Krause, J., Satheesh, S., Ma, S., Huang, Z., Karpathy, A., Khosla, A., Bernstein, M., et al.: ImageNet large scale visual recognition challenge. Int. J. Comput. Vis. **115**(3), 211–252 (2015)
19. Szegedy, C., Liu, W., Jia, Y., Sermanet, P., Reed, S., Anguelov, D., Erhan, D., Vanhoucke, V., Rabinovich, A.: Going deeper with convolutions. In: Proceedings of the Conference on Computer Vision and Pattern Recognition (CVPR), pp. 1–9. IEEE (2015)
20. Taigman, Y., Yang, M., Ranzato, M., Wolf, L.: Deepface: closing the gap to human-level performance in face verification. In: Proceedings of the Conference on Computer Vision and Pattern Recognition (CVPR), pp. 1701–1708. IEEE (2014)
21. Turnbull, D., Barrington, L., Torres, D., Lanckriet, G.: Semantic annotation and retrieval of music and sound effects. Trans. Audio Speech Lang. Process. **16**(2), 467–476 (2008)

22. Weyand, T., Kostrikov, I., Philbin, J.: PlaNet-photo geolocation with convolutional neural networks. arXiv preprint (2016). arXiv:1602.05314
23. Xiao, J., Hays, J., Ehinger, K.A., Oliva, A., Torralba, A.: SUN database: large-scale scene recognition from abbey to zoo. In: Conference on Computer Vision and Pattern Recognition (CVPR), pp. 3485–3492. IEEE (2010)
24. Yang, J., Hauptmann, A.G.: (Un)reliability of video concept detection. In: Proceedings of the International Conference on Content-based Image and Video Retrieval (CIVR), pp. 85–94. ACM (2008)
25. Zahálka, J., Rudinac, S., Worring, M.: Analytic quality: evaluation of performance and insight in multimedia collection analysis. In: Proceedings of the International Conference on Multimedia (MM), pp. 231–240. ACM (2015)

Utterance Retrieval Based on Recurrent Surface Text Patterns

Guillaume Dubuisson Duplessis[1]([⊠]), Franck Charras[1], Vincent Letard[2],
Anne-Laure Ligozat[3], and Sophie Rosset[1]

[1] LIMSI, CNRS, Université Paris-Saclay, 91405 Orsay, France
{gdubuisson,charras,rosset}@limsi.fr
[2] LIMSI, CNRS, Univ. Paris-Sud, Université Paris-Saclay, 91405 Orsay, France
letard@limsi.fr
[3] LIMSI, CNRS, ENSIIE, Université Paris-Saclay, 91405 Orsay, France
annlor@limsi.fr

Abstract. This paper investigates the use of recurrent surface text patterns to represent and index open-domain dialogue utterances for a retrieval system that can be embedded in a conversational agent. This approach involves both the building of a database of such patterns by mining a corpus of written dialogic interactions, and the exploitation of this database in a generalised vector space model for utterance retrieval. It is a corpus-based, unsupervised, parameterless and language-independent process. Our study indicates that the proposed model performs objectively well comparatively to other retrieval models on a task of selection of dialogue examples derived from a large corpus of written dialogues.

Keywords: Dialogue utterance retrieval · Example-based dialogue modelling · Open-domain dialogue system · Evaluation

1 Introduction

Conversational systems are recently gaining a renewed attention in the research community, 50 years after the famous ELIZA system [18], as shown by the recent effort to generate and collect data from the (RE-)WOCHAT workshops[1]. This renewed attention is motivated by the opportunity of exploiting large amount of dialogue data to automatically author a dialogue strategy that can be used in conversational systems such as chatbots [2,3].

We consider the task of automatically authoring an open-domain conversational strategy from unlabelled dialogue data. The main goal is to provide a dialogue system with the ability to appropriately react to a large variety of unexpected out-of-domain human utterances by offering an engaging continuation to the dialogue. In this direction, approaches under study can be broken down into

---

This work was funded by the JOKER project (www.chistera.eu/projects/joker).
[1] See http://workshop.colips.org/re-wochat/ and http://workshop.colips.org/wochat/.

© Springer International Publishing AG 2017
J.M. Jose et al. (Eds.): ECIR 2017, LNCS 10193, pp. 199–211, 2017.
DOI: 10.1007/978-3-319-56608-5_16

generation-based approaches that aim at creating a response given a conversational context (e.g., [17]), and selection-based approaches that focus on selecting a response from a large set of utterances (e.g., [2,3,5]). This work focuses on the selection-based approach. More specifically, we view the problem as an instance of example-based dialogue modelling [8] where the goal is to rank dialogue examples from a large database in order to retrieve the best one. We are interested in the specific case where a dialogue example is an initiative (I)/response (R) pair (e.g., "(I) do you like paella? (R) yes, it's delicious."). The task aims at retrieving a dialogue pair given an input utterance in a large database of examples. The main idea is to rank initiative utterances from the database of examples against the input utterance to determine the dialogue example that fits best. In this paper, we propose to consider patterns of language use – occurring in a social, opportunistic and dynamic activity such as dialogue – to compare utterances. Our approach can be viewed as an instance of sequential pattern mining [11] applied to information retrieval in textual dialogues. The main contributions of this work are: (i) the extraction of recurrent surface text patterns (RSTP) from a corpus of written utterances; (ii) the representation of utterances as a bag-of-RSTPs; and (iii) the similarity measure between utterances that both takes into account the inverse frequency (IDF) of RSTPs and the relatedness between two RSTPs based on the Jaccard index. We assess this model on a task of utterance selection and show that it outperforms standard models.

Section 2 discusses related work. Section 3 describes the proposed model based on recurrent surface text patterns and outlines its main features. Next, Sect. 4 describes the adopted experimentation protocol along with the database of dialogue examples created in this work. Then, Sect. 5 presents and discusses the main results. Finally, Sect. 6 concludes this paper.

2   Related Work

Several retrieval models have been explored to select the most appropriate dialogue example from the database. The most common ones are vector-space models at the token level along with the cosine similarity [2] and classic Term Frequency-Inverse Document Frequency (TF-IDF) retrieval models [3,5]. This has also been framed as a multi-class classification problem, e.g., resolved with a perceptron model [5]. More recently, recurrent neural networks have also been proposed to predict if an utterance r is a response associated to a context c formed by a sequence of words [10]. Retrieving an appropriate utterance may also be considered as a short text retrieval problem, the query being the user initiative. From this point of view, the problem is close for example to a community question answering (cQA) problem [12], which aims at finding the existing questions that are semantically equivalent or relevant to the queried questions. Yet, contrary to the cQA problem, the surface form is at least as important as the semantic correspondence between the initiatives, and the objective is not necessarily to give relevant information, but to keep the user engaged in the conversation. Our approach aims at exploiting the recurrent surface text patterns of language use appearing across utterances to represent, index and efficiently retrieves similar utterances in a large database. Its main features are to implement a corpus-based, unsupervised, parameterless and language-independent process.

3 Recurrent Surface Text Pattern-Based Approach

We present a corpus-based process which aims at representing and indexing utterances for a retrieval system. This process is based on two main steps: (i) the building of a database of recurrent surface text patterns by mining a corpus of written dialogic interactions; and (ii) the exploitation of this database in a generalised vector-space model for utterance retrieval.

3.1 Mining of Recurrent Surface Text Patterns (RSTP)

An utterance is viewed as a sequence of tokens. For instance, the utterance "how do you usually introduce yourself ?" (u_1) involves 7 tokens. Similarly, the utterance "how do you know ?" (u_2) contains 5 tokens. We define a recurrent surface text pattern (RSTP) as being a contiguous sequence of tokens that appears in at least two utterances. For example, "how do you" is a RSTP appearing both in utterance u_1 and u_2. However, u_1 and utterance "hi !" do not share any RSTP. Intuitively, RSTPs are surface patterns of language use appearing across utterances in dialogue.

RSTPs are mined from a corpus to form a database further used to represent seen and unseen utterances. Our approach is an instance of sequential pattern mining [11]. The mining process consists in resolving the multiple common subsequence problem by using a generalised suffix tree [6] (resolution of this problem is usually performed to find common substrings in biological strings such as DNA, RNA or protein). Each utterance of the corpus is represented as a sequence of tokens. Let say we have K utterances which lengths sum to N (i.e., the corpus contains N tokens). Each utterance is inserted in the generalised suffix tree. Then, the tree is used to find the subsequences common to k utterances with k ranging from 2 to K. Each node in the tree keeps track of the number of utterances containing the subsequence in the corpus. Remarkably, this problem can be solved in linear time $O(N)$ where N is the total number of tokens in the corpus [6]. Before insertion, utterances are added special begin and end markers (noted, respectively, #B and #E). These markers allow to represent RSTPs starting or ending an utterance. For instance, the subsequence "#B how do you" is a RSTP of u_1 and u_2. However, a single marker is not considered as a RSTP (begin and end markers are excluded from 1-token RSTP).

RSTPs and the standard n-gram model both consider subsequences of tokens. However, they are not to be confused. Indeed, RSTPs belonging to a set of utterances are a subset of all the possible n-grams of this utterance set (with n varying from 1 to the maximum utterance length in the set). However, one important feature of a RSTP is to be recurrent. It means that it must appear in at least two utterances of a corpus (this is not necessary for a n-gram). Last but not least, a RSTP is not limited in size while a n-gram is by definition a contiguous sequence of n items. This work further empirically shows in Sect. 4.2 that the number of unique RSTPs in a corpus of around 3 million of utterances is comparable to the number of unique 3-grams.

3.2 RSTP-Based Model

From Vector Space Model to Generalised Vector Space Model. The vector space model (VSM) [15] has been widely adopted in information retrieval to determine the relevance of a document to a query. It relies on a set of terms t_i ($1 \leq i \leq n$) used to indexed a large amount of documents d_α ($1 \leq \alpha \leq p$). This model assumes that it exists a set of pairwise orthogonal term vectors t_i ($1 \leq i \leq n$) corresponding to the indexing terms. This set is assumed to be the generating set of the vector space. This vector space is then used to represent as linear combinations of the term vectors both the documents $d_\alpha = \sum_{i=1}^{n} a_{\alpha i} t_i$ and the query $q = \sum_{j=1}^{n} q_j t_j$. The similarity between a document and a query is based on their scalar product which is given in Eq. 1.

$$d_\alpha \cdot q = \sum_{i=1}^{n} a_{\alpha i} q_i \qquad (1) \qquad\qquad d_\alpha \cdot q = \sum_{j=1}^{n} \sum_{i=1}^{n} a_{\alpha i} q_j t_i \cdot t_j \qquad (2)$$

A standard retrieval strategy is to rank documents according to their similarity to the query (e.g., the cosine similarity). However, the orthogonality assumption of the VSM is often viewed as being too restrictive and unrealistic. Indeed, it does not take into account the relatedness between pair of terms whereas it might be argued that terms often relate to each other. The generalised vector space model (GVSM) has been proposed to incorporate a measure of similarity between terms into the retrieval process [19]. In doing so, it removes the pairwise orthogonality assumption. The similarity between a document and a query is based on the generalisation of the scalar product given in Eq. (2), which also is a measure of their similarity between two normalised vectors (the cosine similarity). Notably, if pairwise orthogonality is assumed, Eq. 2 becomes Eq. 1. To rank the documents, it is required to know (i) the components $a_{\alpha i}$ and q_j along the term vectors, and (ii) the similarity between every pair of term vectors expressed by $t_i \cdot t_j$ (the explicit representation of term vectors t_i is not required).

Representation of Utterances. We model utterances by a GVSM where terms are RSTPs. Utterances are represented by a bag of the most representative RSTPs they include. A RSTP r is representative of an utterance if it does not exist another RSTP r' included in the utterance such that r is a subsequence of r'. Formally, let R be the set of all RSTPs included in an utterance u. $r \in R$ is a representative RSTP of u iff $r \in R$ and $\forall r' \in R$, $r' \neq r$, $r \not\subseteq r'$. For example, let say we have a RSTP database D = { "how", "you know", "? #E", "#B how", "#B how do you", "#B Hi ! #E"}. The RSTPs included in u_2 = "how do you know ?" are: R = { "how", "you know", "? #E", "#B how", "#B how do you"}.

And the final representation keeping only the most representative RSTPs is: { "you know", "? #E", "#B how do you"}.

This representation ensures that there is not two RSTPs r and r' indexing an utterance such that $r \subset r'$. A particular case of this representation is a recurrent utterance (i.e. appearing several times in the corpus). In this case, the utterance is a RSTP and is thus represented by itself. In this work, we empirically show in Sect. 4.2 that this representation is sparse. One advantage of this representation is that it takes into account the word order to the extent of patterns (contrary, e.g., to a unigram model). Another one is that RSTPs are easily understandable from a human perspective.

In practical terms, finding RSTPs included in an utterance from a large database can be costly for a real-time interaction system if done naively. The first way is to search whether a RSTP is included in the utterance by taking each one of the RSTP in the database. This way can quickly become impractical if the database is very large. Another way consists in considering all the subsequences of the utterance and test whether this subsequence is a RSTP. This way is often more efficient because of the small size of utterances (some recent work reports that the maximum size of utterances is less than 30 tokens [3]).

Retrieval Strategy. The retrieval strategy takes into account relatedness between pairs of terms because RSTPs may be closely related (e.g., "#B how" and "#B how do you"). Similarity between two RSTPs is based on the following idea: the more the sequence of tokens of two RSTPs are similar, the more the RSTPs are similar. Conversely, two RSTPs are said to be orthogonal if they do not share a subsequence of tokens. Formally, we estimate $t_i \cdot t_j$ by a variant of the Jaccard index:

$$\frac{|lgcs(t_i, t_j)|}{|t_i| + |t_j| - |lgcs(t_i, t_j)|}$$

where $|lgcs(t_i, t_j)|$ is the size of the longest common subsequence between t_i and t_j. $t_i \cdot t_j$ is 0 when t_i and t_j do not share any token while it is 1 when $i = j$. Similarity between two utterances is given by Eq. 2. Let W_i be the weight assigned to RSTP t_i (the components of the vector). It is given by $W_i = TF(t_i) \times IDF(t_i)$ where $TF(t_i)$ is the raw frequency of t_i in the bag of RSTP representing the utterance (i.e., 0 or 1); and $IDF(t_i) = log(\frac{N}{n_i})$ where N is the total number of utterances mined to produce the RSTP database, and n_i is the number of mined utterances including t_i in their representation.

4   Experimentation

This experimentation aims at comparing selection methods on the task of retrieving a response utterance in a large corpus of open-domain textual dialogues from a given input utterance. The dialogue corpus consists of two main types of utterances: (i) *initiative* utterances that have at least one follow-up utterance; and (ii) *response* utterances that do not have a follow-up utterance. The retrieving process works as follows. Initiative utterances from the corpus are ranked against a given input utterance. Then, a random response is taken from the pool of response utterances of the highest ranked initiative utterance (in this work, note that 91% of the response pools are of size 1).

Evaluation aims at assessing (i) the ability of each selection method to find an initiative utterance that is close to the given input utterance, and (ii) the ability of each method to select an appropriate response to a given input utterance. This experiment compares a RSTP-based method with four other selection methods (described in Sect. 4.3) on a set of 1000 reference utterances. Reference utterances are the input utterances of the selection methods. Each reference utterance comes along with a (possibly large) predefined set of acceptable responses (detailed in Sect. 4.1). Notably, reference utterances do not appear in the selection corpus, that is, there is no initiative utterances that is strictly equal to any of the reference utterances.

For each method, assessment consists in comparing the selected response produced for a reference utterance against the list of acceptable responses associated with this reference utterance. The more similar the selected response is to *one of* the predefined acceptable responses, the better it is. To avoid a time-consuming, costly and possibly noisy human intervention at this step, we consider metrics coming from the machine translation domain such as BLEU [13] or TER [16]. The main idea behind these metrics is to measure the correspondence between a system output translation and a set of reference translations while maintaining an adequate correlation with human judgements of quality. The TER ("Translation Error Rate") metric is the most appropriate to the need of this experimentation since it targets cases where a large space of possible correct translations exists. In particular, it is not required for a selected response to be close to all the predefined acceptable responses but only to one of those. TER is defined as "the minimum number of edits needed to change a hypothesis so that it exactly matches one of the references, normalised by the average length of the references". Edits include insertion, deletion, substitution of single words and shifts of word sequences. For a given hypothesis utterance, it is given by the formula:

$$\text{TER} = \frac{\# \text{ of edits}}{\text{average } \# \text{ of reference words}}.$$

4.1   Selection Corpus and Reference Utterances

A subset of the English version of the OpenSubtitles2016 corpus [9] was used as the selection corpus. This corpus consists of a wide variety of subtitles of television dramas. It provides a large amount of pre-processed transcribed interactions that can be useful for dialogue modelling. Pre-processing includes subtitle encoding conversion, sentence segmentation, sentence tokenisation and corrections of spelling errors [9]. Subtitles are formatted as sequences of tokenised sentences with timing information and meta-data about the subtitle (e.g., identifiant of the TV episode). In this work, pre-processing was extended by applying a named entity (NE) recognition for each sentence. This was done with the Stanford NER [4]. NEs allows to generalise sentences by replacing person name, localisation and organisation (e.g., "My name is Alice." is turned into "My name is <person>."). Thus, NEs stay neutral for the similarity calculations undertaken

Table 1. Figures about the selection corpus (subset of OpenSubtitles2016 [9]) and about the dataset of reference utterances. T/U = Tokens per Utterance

	Selection corpus		Reference
	Initiative utterances	Response utterances	utterances
Unique utterance	3,174,606	2,481,369	1000
Tokens (unique)	23,148,094 (226,462)	19,557,246 (219,374)	5571 (348)
T/U: avg/median (std)	7.29/7.0 (6.58)	7.88/7.0 (5.67)	5.57/5.0 (0.78)
T/U: min/max	1/1431	1/1280	5/10

while ranking initiative utterances. However, the turn structure is missing from these subtitles which renders the OpenSubtitles2016 corpus noisy for dialogue modelling. To overcome this problem, a process similar to the one used to build the SubTle corpus [1] was carried out. It aims at extracting utterance pairs corresponding to an initiative and a response exchanged in a dyadic conversation. This heuristic helped to reduce the level of noise to approximately 25% of the conversational pairs on the SubTle corpus [1]. It is based on timing features about consecutive sentences, punctuation features (such as a sentence-initial dash) and the fact that sentences are shown on the same subtitle block (i.e., appearing on the same screen). This method allows to extract exchanges of utterances that are less noisy than the entire corpus. Table 1 presents some figures about the selection corpus. It includes more than 3 million of unique initiative utterances and around 2.4 million of unique response utterances.

The set of reference utterances along with their predefined set of acceptable responses has been automatically extracted from the subset of the OpenSubtitles2016 corpus. To this purpose, we extracted the 1000 most frequent utterances from the corpus which contains at least 5 tokens (inclusive). The high frequency of these utterances ensures that they are very likely to be used in a conversation by a human. The 5-token requirement follows recent observations showing that it is difficult for a human to reliably judge the validity of a conversational pair if the first part is too small [3]. Importantly, all the retained reference utterances have been discarded from the selection corpus. Table 1 presents some figures about the reference utterances. In average, a reference utterance has 191.11 acceptable responses (std = 426.94, median = 102, min = 62, max = 7677).

4.2   The RSTP-Based Method

The RSTP-based model was prepared by mining patterns on the set of initiative utterances of the selection corpus. Table 2 presents some figures about the RSTP database. First, the number of RSTP extracted from the corpus is less than 2 times the number of unique utterances. Indeed, the full database contains around 5.7 million unique patterns which amounts to 1.82 times the number of initiative utterances. If we only consider representative RSTPs that have been used to represent the initiative utterances of the selection corpus, it comes down to approx. 3.8 million of unique patterns (1.21 times the number of initiative utterances). In comparison, the number of unique trigrams extracted from the initiative utterances of the selection corpus is around 5.7 million items.

Besides, the representation of utterance with the RSTP-based model is sparse. Indeed, the number of patterns per utterance representation is in average 3.09 (std = 3.24, median = 3.0, min = 1, max = 582).

Figure 1 takes a closer look at the distribution of the size (in tokens) of RSTP used to represent initiative utterances from the selection corpus. It shows that the RSTP-based model effectively uses a wide variety of patterns in terms of size, contrary to a fixed n-gram model. Sizes of the patterns mostly range from 1 token to 8 tokens, with 50% of the patterns having a size between 3 and 5 tokens (median = 4 tokens).

Table 2. Figures about the RSTP database mined on the initiative utterances of the selection corpus and on the RSTP effectively used to represent the initiative utterances.

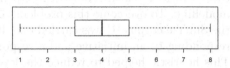

RSTP database	Full	Used
Size	5,776,901	3,846,956
Tokens per RSTP		
...avg/median	4.77/4.0	4.57/4.0
...std, min/max	2.23, 1/157	1.96, 1/157

Fig. 1. Distribution of the size of the RSTP effectively used to represent the initiative utterances (in tokens, including begin and end markers). For readability, outliers have been discarded.

4.3 Other Selection Methods

Four other selection methods are considered in this experimentation. These methods differ in their way to rank initiative utterances given a reference utterance. However, they follow the same process to pick the response utterance. First, the *random method* selects a random initiative utterances from the selection corpus following a uniform distribution. Thus, it does not take into account the reference utterance given as input. Secondly, the *TF-IDF method* implements a VSM at the token level (i.e. it considers unigram). It retrieves initiative utterances that are lexically close to the reference utterance but does not take into account word order. An utterance is represented by a TF-IDF weighted vector of the unigrams that occurred in it. Let W_i be the weight assigned to unigram u_i. It is given by $W_i = TF(u_i) \times IDF(u_i)$ where $TF(u_i)$ is the raw frequency of unigram u_i in the utterance; and $IDF(u_i) = log(\frac{N}{n_i})$ where N is the total number of initiative utterances, and n_i is the number of initiative utterances containing u_i. Similarity between two utterances is given by the cosine similarity of their vector representations. Then, the *trigram method* implements a VSM at the n-gram level with n = 3. It is equivalent to the previous model with the exception that it considers trigram instead of unigram and that begin and end markers are added to the utterance. This method takes into account lexical proximity between utterances and word order to the extent of trigrams. Finally, the last method relies on word and utterance embeddings using the *doc2vec model* [7]. Word and utterance embeddings are jointly learnt as the coefficients of a shallow neural network trained to predict a word given its context and the utterance it belongs to. We focused especially in harvesting the utterance embeddings as their cosine similarity can translate lexical and semantic similarity. The implementation provided by Gensim [14] is used with the length of the context window set to 2 and a vector dimension of 100. The model was trained on the entire selection corpus. Embeddings of the reference utterances are inferred and used to retrieve the closest initiative utterance with a nearest neighbour search.

5 Results

5.1 Ranking of Initiatives and Selection of Responses

We compare the results of the ranking process operated by each selection method. This process consists in finding an initiative utterance from the selection corpus that is close to a given reference utterance. For instance, for the reference utterance "what is this about?", the following initiatives were retrieved from the database: "– it looks like <person>!" (random), "– what about this?" (TF-IDF), "– <person>, what is this about?" (trigram), "– i don't. what is this about?" (doc2vec), and "– and what is this about?" (RSTP). For the reference utterance "good to see you.", the following results were retrieved: "– i'm not gonna do it this time." (random), "– good of you to see me." (TF-IDF), "– good to see you. thank you." (trigram), "– good good." (doc2vec), and "– good to see you. pleasure." (RSTP). It should be noted that in the vast majority of the cases, the ranking processes of the methods yielded a clear-cut initiative utterance matching the reference utterance. In some marginal cases, the TF-IDF, trigram and RSTP methods yielded more than one maximum result (at most 4 for the TF-IDF model, 2 for the others). In these cases, the result of the ranking process was a random choice between those maximum results. Table 3 (columns "I") takes a closer look at the common results between methods in the ranking

Table 3. Common results between methods in the ranking of the initiative utterance (I) and in the selection of the response utterance (R). Presented results are symmetric.

	Random		TF-IDF		Trigram		doc2vec		RSTP	
	I	R	I	R	I	R	I	R	I	R
Random	–	–	0%	0%	0%	0%	0%	0%	0%	0%
TF-IDF	0%	0%	–	–	17%	17.6%	5%	5%	8.3%	8.5%
Trigram	0%	0%	17%	17.6%	–	–	3%	3%	8%	8%
doc2vec	0%	0%	5%	5%	3%	3%	–	–	4%	4%
RSTP	0%	0%	8.3%	8.5%	8%	8%	4%	4%	–	–

process. Comparison consists in strict string equality. It turns out that the ranking step of the methods lead to different results. Methods share less than 10% of their ranking results, with the exception of the TF-IDF and trigram methods that share around 17% of their results. In particular, the random method does not share results with the other ones. It shows that each method has inherent characteristics making it more or less suited for utterance selection.

We now consider the impact of the methods on the quality of the selected response utterance. The selection process is the global procedure by which each method selects a response to a given reference utterance. For instance, the following responses were retrieved from the database for the reference utterance "you're not serious.": "– no." (random), "– i'm serious." (TF-IDF), "– listen to me. i am very serious." (trigram), "– i am." (doc2vec), and "– sorry to burst your bubble." (RSTP). However, results may be noisier. For example, the following results were retrieved for the reference utterance "can I help you?": "– a had accomplices." (random), "– we'll get her anyway." (TF-IDF), "– we'll get her anyway." (trigram), "– what are you doing?" (doc2vec), and "– yeah." (RSTP). Table 3 (columns "R") presents the common results between methods in terms of response selection. Methods select less than 10% of the same responses except for the TF-IDF and trigram methods that share 17.6% of their responses (consistently with their ranking results). Thus, methods mostly select different responses. Table 4 gives describing figures about the datasets of selected responses by each method. Sets of selected responses by the TF-IDF, trigram, doc2vec and RSTP methods are similar. They include between 85% (doc2vec method) and 89% (trigram, RSTP methods) of unique utterances. Responses contain around 5 tokens with a minimum of 1 token and a maximum between 35 (TF-IDF method) and 47 (RSTP method) tokens. Responses selected by the random method have a more variable size in terms of tokens per utterance as shown by a higher standard deviation and by a maximum size of 101 tokens.

Finally, we consider the quality of the response selected by each method. To avoid a time-consuming and labour-intensive human evaluation, we decided to assess the quality of a selected response by comparing it to the list of acceptable responses associated with each reference utterance. To this purpose, we chose to compute for each method the "Translation Error Rate" between a selected response to a reference utterance and the list of acceptable responses. This

Table 4. Figures about the datasets of selected responses for each method and their associated "Translation Error Rate" (TER). T/U = Tokens per Utterance

	Random	TF-IDF	Trigram	doc2vec	RSTP
Utterances (unique)	1000 (87%)	1000 (87%)	1000 (89%)	1000 (85%)	1000 (89%)
Tokens (unique)	5710 (1154)	5591 (1009)	5808 (1018)	5698 (1023)	5438 (1028)
T/U: avg/median	5.71/5.0	5.59/5.0	5.81/5.0	5.70/5.0	5.44/5.0
T/U: **std**, min/max	5.36, 1/101	3.51, 1/35	3.70, 1/42	3.77, 1/46	3.37, 1/47
TER	0.632	0.537	0.549	0.566	0.505

indicator computes the minimum number of edits needed to change a selected response so that it exactly matches one of the acceptable responses. Results are presented in Table 4. TER results range from 0.505 to 0.632. The worst TER is for the random method (0.632). The best rate is for the RSTP method (0.505). TF-IDF, trigram and doc2vec methods share comparable results (between 0.53 and 0.57). We performed a paired Wilcoxon test to check for statistically significant differences between methods. TER score for the RSTP method is significantly lower than the scores from the random ($p < 0.001$), trigram ($p < 0.05$) and doc2vec ($p < 0.01$) methods. However, it is not significantly lower than the score from the TF-IDF method. TER score for the random method is significantly higher than the scores from all the other methods. All other differences are not statistically significant at the 5% level.

5.2 Discussion

This experimentation has aimed at comparing four selection methods (a random one, two VSM based on unigram and trigram, a GVSM on RSTP and a word embeddings model) on a task of utterance selection in a large database of open-domain dialogue pairs. Results show that these methods are inherently different in the sense that they (i) mostly retrieve different initiative utterances given a reference one, and (ii) select different response utterances. Besides, we have measured the quality of utterances selected by each method in terms of the translation error rate (TER). This indicates that the RSTP-based method is a promising approach for utterance selection. However, these results should be taken with caution. First, the acceptability of an utterance is not entirely indicated by the TER score since it ignores the notion of semantic equivalence. Assessing the acceptability of each utterance would require a more costly evaluation based on human judges. Then, even though the TER score has allowed us to clearly distinguish the random model from the other ones, the error rates obtained by non-random methods are still high. We cannot exclude the possibility that the methods have selected valid responses that were not appearing in the list of acceptable ones (thus, increasing the error rate). Indeed, open-domain utterances may accept a huge space of possible responses that may be roughly estimated by our lists of acceptable responses. On the other hand, the database of dialogue example may still be noisy to a large extent despite our effort to reduce it. However, all selection methods are equally affected by this problem. Last but not least, this experimentation compares selection method on the basis of highly frequent reference utterances. An interesting extension of this work would consider the case of less frequent utterances. Nevertheless, this would require a database of those utterances along with their acceptable responses.

6 Conclusion and Future Work

This paper has presented a new corpus-based process that aims at finding and exploiting recurrent surface text patterns of language use to represent open-domain dialogue utterances for a retrieval task. Our approach provides the benefit of being corpus-based, unsupervised, parameterless, language-independent while exploiting patterns that are easily understandable from a human perspective. We have shown that this approach performs comparatively well to other retrieval models on a task of selection of dialogue examples derived from a large corpus of written dialogues. Future work includes the study of this approach on other corpora and other languages as well as the potential of our model to more generally model dialogue history involving several utterances.

References

1. Ameixa, D., Coheur, L., Redol, R.A.: From subtitles to human interactions: introducing the subtle corpus. Technical report, INESC-ID (2013)
2. Banchs, R.E., Li, H.: IRIS: a chat-oriented dialogue system based on the vector space model. In: Proceedings of the ACL 2012 Demonstrations, pp. 37–42 (2012)
3. Charras, F., Dubuisson Duplessis, G., Letard, V., Ligozat, A.L., Rosset, S.: Comparing system-response retrieval models for open-domain and casual conversational agent. In: Workshop on Chatbots and Conversational Agent Technologies (2016)
4. Finkel, J.R., Grenager, T., Manning, C.: Incorporating non-local information into information extraction systems by Gibbs sampling. In: Proceedings of the 43rd Annual Meeting on Association for Computational Linguistics, pp. 363–370 (2005)
5. Gandhe, S., Traum, D.R.: Surface text based dialogue models for virtual humans. In: Proceedings of the SIGDIAL (2013)
6. Gusfield, D.: Algorithms on Strings, Trees and Sequences. Cambridge University Press, Cambridge (1997)
7. Le, Q.V., Mikolov, T.: Distributed representations of sentences and documents. In: ICML, vol. 14, pp. 1188–1196 (2014)
8. Lee, C., Jung, S., Kim, S., Lee, G.G.: Example-based dialog modeling for practical multi-domain dialog system. Speech Commun. 51(5), 466–484 (2009)
9. Lison, P., Tiedemann, J.: OpenSubtitles2016: extracting large parallel corpora from movie and tv subtitles. In: 10th edition of the Language Resources and Evaluation Conference (LREC), Portorož, Slovenia, May 2016
10. Lowe, R., Pow, N., Serban, I.V., Pineau, J.: The ubuntu dialogue corpus: a large dataset for research in unstructured multi-turn dialogue systems. In: SIGDIAL, p. 285 (2015)
11. Mooney, C.H., Roddick, J.F.: Sequential pattern mining - approaches and algorithms. ACM Comput. Surv. 45(2), 19:1–19:39 (2013)
12. Nakov, P., Màrquez, L., Moschitti, A., Magdy, W., Mubarak, H., Freihat, A.A., Glass, J., Randeree, B.: Semeval-2016 task 3: community question answering. In: Proceedings of the 10th International Workshop on Semantic Evaluation (SemEval-2016), pp. 525–545 (2016)
13. Papineni, K., Roukos, S., Ward, T., Zhu, W.J.: BLEU: a method for automatic evaluation of machine translation. In: Proceedings of the 40th Annual Meeting on Association for Computational Linguistics, pp. 311–318 (2002)
14. Řehůřek, R., Sojka, P.: Software framework for topic modelling with large corpora. In: Proceedings of the LREC 2010 Workshop on New Challenges for NLP Frameworks, pp. 45–50. ELRA, Valletta, May 2010
15. Salton, G., McGill, M.J.: Introduction to Modern Information Retrieval. McGraw-Hill, Inc., New York (1986)

16. Snover, M., Dorr, B., Schwartz, R., Micciulla, L., Makhoul, J.: A study of translation edit rate with targeted human annotation. In: Proceedings of Association for Machine Translation in the Americas, vol. 200 (2006)
17. Sordoni, A., Galley, M., Auli, M., Brockett, C., Ji, Y., Mitchell, M., Nie, J.Y., Gao, J., Dolan, B.: A neural network approach to context-sensitive generation of conversational responses. CoRR abs/1506.06714 (2015)
18. Weizenbaum, J.: ELIZA - a computer program for the study of natural language communication between man and machine. Commun. ACM 9(1), 36–45 (1966)
19. Wong, S.K.M., Ziarko, W., Raghavan, V.V., Wong, P.: On modeling of information retrieval concepts in vector spaces. ACM Trans. Database Syst. 12(2), 299–321 (1987)

How Do Order and Proximity Impact the Readability of Event Summaries?

Arunav Mishra[1]([✉]) and Klaus Berberich[1,2]

[1] Max Planck Institute for Informatics, Saarbrücken, Germany
{amishra,kberberi}@mpi-inf.mpg.de
[2] htw saar, Saarbrücken, Germany

Abstract. Organizing the structure of fixed-length text summaries for events is important for their coherence and readability. However, typical measures used for evaluation in text summarization tasks often ignore the structure. In this paper, we conduct an empirical study on a crowdsourcing platform to get insights into regularities that make a text summary coherent and readable. For this, we generate four variants of human-written text summaries with 10 sentences for 100 seminal events, and conduct three experiments. Experiment 1 and 2 focus on analyzing the impact of sentence ordering and proximity between originally occurring adjacent sentences, respectively. Experiment 3 analyzes the feasibility of conducting such a study on a crowdsourcing platform. We release our data to facilitate future work like designing dedicated measures to evaluate summary structures.

1 Introduction

The World Wide Web has proven to be an effective platform for information dissemination for local and global news events. This has however resulted in an exponentially increasing amount of textual information made available over the Web. On one hand, the colossal amount of text data has aided various analytical tasks. On the other hand, it has contributed to *information overload* during retrospection of past events.

Automatic text summarization has been traditionally considered an effective tool to help users cope with large textual data [23]. The extractive text summarization task is often cast into a task of selecting sentences from a given set of documents and presenting them in a meaningful order [6,10,16,27]. In this realm, though a lot of focus has been given to the problem of selecting informative sentences to improve the content-quality of a summary [22], relatively less attention has been put into improving its structure [4,7,12,15] in terms of sentence ordering.

Traditionally, evaluation of automatically generated extractive text summaries has been considered a difficult task. This is primarily due to the absence of an "ideal" summary that can be used for comparison. On the one hand, there exist measures to estimate the content quality of text summaries, like ROUGE

© Springer International Publishing AG 2017
J.M. Jose et al. (Eds.): ECIR 2017, LNCS 10193, pp. 212–225, 2017.
DOI: 10.1007/978-3-319-56608-5_17

[17] and Pyramid [20], that are computed by comparing against multiple human-written *reference* summaries. On the other hand, text-quality measures [22], like readability, are often estimated by obtaining human preference judgments, which proves to be non-scalable.

What is missing is a corpus containing various orderings of fixed-length human-written focused text summaries on a large number of independent topics where the variants are annotated with human preference judgments based only on their readability.

With such a corpus made available, further insights can be obtained on regularities, like conventional proximity between sentences that make a summary more readable. Moreover, focused studies are needed to be conducted to analyze the impact of the sentence ordering on coherence and readability of text summaries in a systematic manner.

In this paper, we perform an empirical study through a crowdsourcing platform to get deeper insights into what makes a text summary of an event more readable and coherent. Crowdsourcing platforms, like Crowdflower [1], have made it possible to efficiently gather human judgments for various tasks. Typically, on these platforms any customer can design a job which is then performed by so called contributors. We use this platform to find answers to the following questions:

Q1 What is the impact of summary structure in terms of sentence order on the readability of summaries for past events?

Q2 Does changing the proximity between sentences in a coherent human-written summary affect its readability?

Q3 How feasible is it to evaluate the structure of a fixed-length summary for a past event through crowdsourcing?

To answer **Q1**, we design our first experiment where we present the contributors with two summaries that are differently ordered. Their task is to decide the one that is more readable and coherent. From this experiment, we intend to isolate the effect of sentence ordering on the overall quality of a summary. Insights gathered from this experiment can be used to design scalable methods to explicitly evaluate the structure of a summary. To answer **Q2**, we first generate a predetermined ordering for each summary that maximizes the *gap* (distance in terms of sentence positions) between original sentences. Then, in our second experiment, we present this variant alongside a randomly ordered summary and ask the contributors to choose the more coherent variant. Insights from this experiment can be used to infer inter-sentence relations that result in a better summary structure. To answer **Q3**, we design our third experiment to gather additional statistics to evaluate the feasibility of conducting such studies. For this, we analyze the difficulty, interestingness, rate of progress, quality of contributors, and overall contributor satisfaction level of the study. Insights gathered from this experiment can be used to design better studies on crowdsourcing platforms for other text summarization tasks.

Challenges include: (1) preparing the test data containing suitable variants of fixed-length summaries for a set of past events; (2) designing suitable user interfaces with appropriate quality control measures to ensure good judgments and filter out the contributors that try to cheat; and **(3)** finally, cleaning and performing suitable analysis on human assessments to answer **Q1**, **Q2**, and **Q3** that are described above.

Contributions made in this paper are as follows: **(1)** to the best of our knowledge, we are the first to present a crowdsourcing-based study to understand the effect of the summary structure on its readability; and **(2)** we release a corpus with four variants of 10-sentence summaries for 100 Wikipedia events along with pair-wise human preference judgments on their readability and coherence quality.

Organization of rest of the paper is as follows. In Sect. 2 we review related prior work from the literature. Section 3 describes the proposed methodology for conducting the user study. Section 4 gives details of the experimental setup and discusses the results obtained. Finally, we conclude and motivate future work in Sect. 5.

2   Related Work

Evaluation of automatically system-generated summaries is a hard task. This is primarily due to the absence of an *"ideal"* summary that can be leveraged as a ground truth. Commonly, the automatically generated summaries are compared with human-written or so-called *reference* summaries [18]. However, such an evaluation setting also poses several drawbacks as pointed out by Nenkova et al. [20]. The most prominent of them is low agreement in the reference summaries, that is, different sentences are selected by different humans while generating a summary. To deal with this, often multiple reference summaries are used for evaluating the content of a system-generated summary.

In intrinsic evaluations [25], there exist several metrics in the literature to measure the goodness of a generated summary. Largely, they can be categorized into text, co-selection, and content quality measures [22]. To evaluate content quality of a summary, measures like ROUGE [17], Pyramid [20], and longest common subsequence [22] are used. The co-selection and content quality measures can be automatically computed from gold-standard reference summaries. However, estimating the text quality of a summary, like coherence and readability often requires human judgements.

For event-related text or news article summarization tasks, the structure of a generated summary becomes important for its readability. However, lack of an ideal (correct) ordering for a given set of sentences in a summary makes the evaluation a challenge. In single document summarization, one possible ordering is provided by the source document itself. However, Jing [13] observed that extracted sentences may not retain the ordering of the documents. An alternative evaluation method can be to compare the prevalence of the discourse structure of the source documents represented by rhetorical [9] and coherence relations [11]. However, Ono et al. [21] discovered significant differences in accuracy when building a discourse representation from technical tutorial texts and from newspaper texts. Few prior works [8,20] have leveraged the small number of available reference summaries to evaluate a summary structure. Other approaches [3,4,15,25] have resorted to human preference judgements for system generated summaries which proves to be non-scalable.

Crowdsourcing services have been successfully used in various natural language processing (NLP) [24], information retrieval (IR) [2], and text summarization tasks [4,5,14,19]. Primarily, they have been leveraged to obtain human annotations for generating ground truths that are then used to evaluate automatic systems. In the context of evaluating short-fixed length summaries, Barzilay et al. [5] conduct a study to create a collection of multiple orderings generated by humans for about 9 sentences extracted from news articles. In another study, Kaisser et al. [14] study the effect of changing the summary length. In a more recent study, Lloret et al. [19] study the viability of using a crowdsourcing service to generate reference summaries.

In this work, we leverage a crowdsourcing service to gather human assessments for four possible orderings of 10 sentences for 100 events from Wikipedia. This is different from the work done by Barzilay et al. [5] as they generate a corpus of 10 orderings from sentences that are selected with their MultiGen system, from only 10 sets of news articles. We start with human-written summaries that are widely accepted as coherent, and use a crowdsourcing service to get insights into impact of altering the ordering and proximity between originally occurring adjacent sentences.

3   Setup

In general, it is important for any crowdsourcing-based evaluation to carefully design the experiments. For this work, we choose to use the Crowdflower platform for conducting experiments. Thus from here on, we use the terminology that is consistent with this platform. In this work we focus on summarization of news events, since in the past a lot of importance has been given to summarizing either blogs or news articles [18].

Event Selection. To generate a test set of fixed-length text summaries for our experiments, we begin by first selecting a set of seminal events in the past which have received considerable media coverage. This focuses the study on news events, and also enables to simulate query-focused summarization task by assuming the selected events as user queries. Moreover, it can be assumed that a Crowdflower contributor can better judge the structure of a summary if she has some prior knowledge on the event. Thus, to generate our test-event set, we leverage the Wikipedia page titled, *Timeline of modern history*[1] that lists a selected number of seminal events that occurred since the year 1901. We first split the textual description associated with each year into sentences each of which describes a single independent event. We then randomly sample a set of events and treat them as our test queries. An illustration of this process and a concrete example of an event from the test set is given in Fig. 1.

---

[1] https://en.wikipedia.org/wiki/Timeline_of_modern_history.

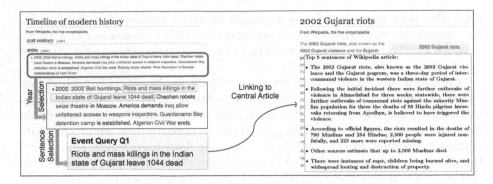

Fig. 1. Illustration of different test data preparation stages.

Summary Generation. We generate a fixed-length (in terms of sentence counts) summary for each of the events that are sampled from Wikipedia. For an unbiased study, it is crucial that each generated summary is: **(1)** human-written, i.e., it should not have any biases from any automatic method; **(2)** neutral, i.e., it should not reflect any point of view on the subject; **(3)** linguistically simple, i.e., it should be understandable by non-expert contributors. Thus, we leverage Wikipedia articles as a source to generate the test summaries. For each event selected for our study, we first manually identify the Wikipedia page that centrally describes the event in the query. We then select the first 10 sentences from the lead section and treat them as a fixed-length summary on the event, as illustrated in Fig. 1. It is easy to see that such a summary satisfies the three requirements that are described above.

Table 1. Instruction given to contributors

Overview
In this task you will compare two Summaries that are generated for a given Event. Both the summaries have exactly the same content but they differ in the ordering of the sentences. Your task is to read the two summaries and decide which of them presents a more readable and a coherent story. You have to also state the reasons behind your choice in the provided Text Area after the options. Finally, you have to rate the difficulty of each judgment on a scale of 5.

Help
- A good Readable and Coherent summary will order sentences in a way such that when read from top to bottom, it presents a smooth flow in the story.
- A bad Unreadable summary will order sentences in a way such that when read top to bottom, there are abrupt changes or jumps in the topics thus making the story confusing.
- In the rating, the scale goes from "very easy" to "very difficult". That is, 1=very easy; 2= easy; 3=not so easy; 4=difficult; 5=very difficult.

Process
1. Read the given event.
2. Read the two summaries of the event given as Summary A and Summary B from top to bottom.
3. Decide on which summary is more readable and has a coherent structure in terms of sentence ordering.
4. Write the reason behind your choice in the text box provided.
5. Finally, decide the difficulty level of judging the more readable summary.

Pro Tips
- The content of both Summary A and Summary B are exactly the same. They differ only in the Sentence Ordering.
- You have to read the sentences from top to bottom.
- You have to decide only the more readable or coherent summary.
- You do not judge the content of the summaries.
- Mark the difficultly level based on per summary and not the entire task

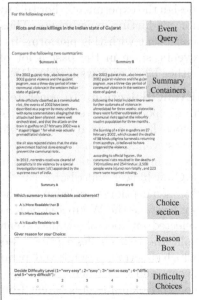

Fig. 2. Interface layout

Interface Design is regarded as the most important aspect of a crowdsourcing experiment and may not lead to any results if designed in an ad-hoc manner [2]. There are two main components of a user interface in Crowdflower. First, a questionnaire, and second, a set of instructions given to the contributor for completing the job.

There are several design decisions that have significant impact on the study: **(1)** number of summaries to compare in a single unit; **(2)** ease of access to the summary text; **(3)** the scale of graded judgments; **(4)** gather reasons behind a contributor's decision; and **(5)** finally, assess the difficulty of a judgment. To address these, our interface exhibits the layout illustrated in Fig. 2. *Event Query* section displays the query as to convey the general topic. The *Summary Container* compares two summaries as *Summary A* and *Summary B*, to gather preference judgments. The text is directly embedded into the interface to minimize the access time to the summaries. Graded judgments are gathered based on the option selected in the *Choice Section* where: *A is Equally Readable to B* denotes *0*; *A is More Readable than B* denotes *1*; and *B is More Readable than A* denotes *2*. We provide a text area where a contributor can specify the reason behind each decision and refer to it as *Reason Box*. Finally, we provide a set of radio buttons that are labeled with different difficulty levels to be chosen by a contributor for each unit they judge as illustrated in Fig. 2.

Instructions given to the contributors to guide them in accomplishing the task, should be clear, concise, and written in simple language without any jargon [2]. With this idea, our instruction set is divided into the following three sections: **(1)** *Help* defines various abstract concepts such as coherence, readability, and the different difficulty levels; *Process* lists step-by-step algorithm; and *Pro Tips* conveys the general dos and don'ts while making a judgment. The full instruction set is illustrated in Table 1.

Quality Control Mechanisms. Blindly trusting crowdsourcing services without proper quality control can result in unusable data with non-sense assessments. For this study, we take several measures to ensure result quality. As the first measure, we request contributors that are regarded as *highly performing* and account for 60% of monthly judgments across a variety of Crowdflower jobs. As a second measure, we design a set of test questions. Crowdflower offers a *Quiz Mode* where a small set of units are presented to the contributors before they go into the *Work Mode*. As test questions, we ourselves judge for about 10% of the total units by providing specific reasons for each which is displayed to the contributors in case they commit a mistake. In both modes, we set the minimum accuracy to be achieved as 70% otherwise the contributor is evicted. Though it is possible to set this higher [19], however excusing few mistakes also gives the contributors an opportunity to better educate themselves about the task through the reasons specified by us. As a third measure, we set the payment per unit relatively low as to attract contributors that are interested in the task as described by [19]. Finally, in order to detect contributors that passed the quiz mode and later tried to cheat, we introduce *traps* [26] with units containing exactly the same summaries.

4 Experiments

We design the following three experiments to answer **Q1**, **Q2**, and **Q3** in Sect. 1,

- *Experiment 1* analyzes the impact of sentence order on coherence.
- *Experiment 2* analyzes the impact of sentence proximity on coherence.
- *Experiment 3* analyzes the viability of using crowdsourcing for our study.

We make all our experiment data and results publicly available[2].

Test-Event Set. Following the method described in Sect. 3, we generate a test set containing 100 randomly sampled events that happened between 1987 and 2007. This time range was chosen considering the coverage of other public document corpora (like the New York Times annotated corpus) so as to make the data reusable.

Test-Summary Sets. To generate the test sets, we preprocess the raw summaries extracted from Wikipedia as described in Sect. 3. It is important to process the raw text summaries to make the sentences independent [5]. This is to prevent the contributors from basing their decisions on straightforward syntactic cues that originate from across-sentences dependencies. We perform two preprocessing steps: firstly, we resolve all co-references. Secondly, we transform all the sentences to lowercase. Figure 3 illustrates a concrete example. Finally, leveraging the preprocessed summaries, we generate the following four test sets:

1. **Original set** $\mathcal{O}$ summary, illustrated in Fig. 3, retains the original ordering.
2. **Reverse set** $\mathcal{R}$ summary, illustrated in Fig. 4, exhibits a reversed ordering.

[1] the 2002 gujarat riots, also known as the 2002 gujarat violence and the gujarat pogrom, was a three-day period of inter-communal violence in the western indian state of gujarat. [2] following the initial incident there were further outbreaks of violence in ahmedabad for three weeks ; statewide, there were further outbreaks of communal riots against the minority muslim population for three months. [3] the burning of a train in godhra on 27 february 2002, which caused the deaths of 58 hindu pilgrims karsevaks returning from ayodhya, is believed to have triggered the violence. [4] according to official figures, the **communal riots** resulted in the deaths of 790 muslims and 254 hindus ; 2,500 people were injured non-fatally, and 223 more were reported missing. [5] other sources estimate that up to 2,500 muslims died. [6] there were instances of rape, children being burned alive, and widespread looting and destruction of property. [7] the chief minister at that time, **narendra modi**, has been accused of initiating and condoning the violence, as have police and government officials who allegedly directed the rioters and gave lists of muslim-owned properties to them. [8] in 2012, **narendra modi** was cleared of complicity in the violence by a special investigation team (sit) appointed by the supreme court of india. [9] the sit also rejected claims that the state government had not done enough to prevent the **communal riots**. [10] while officially classified as a communalist riot, the events of 2002 have been described as a pogrom by many scholars, with some commentators alleging that the attacks had been planned, were well orchestrated, and that the attack on the **train in godhra on 27 february 2002** was a " staged trigger " for what was actually premeditated violence.

[1] the 2002 gujarat riots, also known as the 2002 gujarat violence and the gujarat pogrom, was a three-day period of inter-communal violence in the western indian state of gujarat. [2] while officially classified as a communalist riot, the events of 2002 have been described as a pogrom by many scholars, with some commentators alleging that the attacks had been planned, were well orchestrated, and that the attack on the train in godhra on 27 february 2002 was a " staged trigger " for what was actually premeditated violence. [3] the sit also rejected claims that the state government had not done enough to prevent the communal riots. [4] in 2012, narendra modi was cleared of complicity in the violence by a special investigation team (sit) appointed by the supreme court of india. [5] the chief minister at that time, narendra modi, has been accused of initiating and condoning the violence, as have police and government officials who allegedly directed the rioters and gave lists of muslim-owned properties to them. [6] there were instances of rape, children being burned alive, and widespread looting and destruction of property. [7] other sources estimate that up to 2,500 muslims died. [8] according to official figures, the communal riots resulted in the deaths of 790 muslims and 254 hindus ; 2,500 people were injured non-fatally, and 223 more were reported missing. [9] the burning of a train in godhra on 27 february 2002, which caused the deaths of 58 hindu pilgrims karsevaks returning from ayodhya, is believed to have triggered the violence. [10] following the initial incident there were further outbreaks of violence in ahmedabad for three weeks ; statewide, there were further outbreaks of communal riots against the minority muslim population for three months.

Fig. 3. Original set $\mathcal{O}$ summary

Fig. 4. Reverse set $\mathcal{R}$ summary

---

[2] http://resources.mpi-inf.mpg.de/d5/txtCoherence.

[1] the 2002 gujarat riots, also known as the 2002 gujarat violence and the gujarat pogrom, was a three-day period of inter-communal violence in the western indian state of gujarat. [2] there were instances of rape, children being burned alive, and widespread looting and destruction of property. [3] following the initial incident there were further outbreaks of violence in ahmedabad for three weeks ; statewide, there were further outbreaks of communal riots against the minority muslim population for three months. [4] while officially classified as a communalist riot, the events of 2002 have been described as a pogrom by many scholars, with some commentators alleging that the attacks had been planned, were well orchestrated, and that the attack on the train in godhra on 27 february 2002 was a " staged trigger " for what was actually premeditated violence. [5] the sit also rejected claims that the state government had not done enough to prevent the communal riots. [6] in 2012, narendra modi was cleared of complicity in the violence by a special investigation team (sit) appointed by the supreme court of india. [7] the burning of a train in godhra on 27 february 2002, which caused the deaths of 58 hindu pilgrims karsevaks returning from ayodhya, is believed to have triggered the violence. [8] the chief minister at that time, narendra modi, has been accused of initiating and condoning the violence, as have police and government officials who allegedly directed the rioters and gave lists of muslim-owned properties to them. [9] according to official figures, the communal riots resulted in the deaths of 790 muslims and 254 hindus ; 2,500 people were injured non-fatally, and 223 more were reported missing. [10] other sources estimate that up to 2,500 muslims died.

[1] the 2002 gujarat riots, also known as the 2002 gujarat violence and the gujarat pogrom, was a three-day period of inter-communal violence in the western indian state of gujarat. [2] there were instances of rape, children being burned alive, and widespread looting and destruction of property. [3] other sources estimate that up to 2,500 muslims died. [4] in 2012, narendra modi was cleared of complicity in the violence by a special investigation team (sit) appointed by the supreme court of india. [5] the burning of a train in godhra on 27 february 2002, which caused the deaths of 58 hindu pilgrims karsevaks returning from ayodhya, is believed to have triggered the violence. [6] while officially classified as a communalist riot, the events of 2002 have been described as a pogrom by many scholars, with some commentators alleging that the attacks had been planned, were well orchestrated, and that the attack on the train in godhra on 27 february 2002 was a " staged trigger " for what was actually premeditated violence. [7] following the initial incident there were further outbreaks of violence in ahmedabad for three weeks ; statewide, there were further outbreaks of communal riots against the minority muslim population for three months. [8] the sit also rejected claims that the state government had not done enough to prevent the communal riots. [9] according to official figures, the communal riots resulted in the deaths of 790 muslims and 254 hindus ; 2,500 people were injured non-fatally, and 223 more were reported missing. [10] the chief minister at that time, narendra modi, has been accused of initiating and condoning the violence, as have police and government officials who allegedly directed the rioters and gave lists of muslim-owned properties to them.

Fig. 5. Shuffled set S summary   **Fig. 6.** Proximity P summary

3. **Shuffled set S** summary, illustrated in Fig. 5, contains randomly shuffled ordering.
4. **Proximity maximizing set P** summary, illustrated in Fig. 6, exhibits an ordering that is generated by placing originally consecutive sentences as far as possible.

As an additional step, we keep the position of the first sentence across all the summaries unaltered. This is because we find that spotting the first sentence became a very easy cue for the contributors who desire to cheat by simply spotting the position of this sentence.

Crowdflower Settings. We created a single job with 700 units out of which 600 compared unique pairs of summaries for 100 queries from each set. The remaining 100 were introduced as *traps* that compared two identical summaries. In addition, our job had randomly selected 53 test units that were judged by us. For each unit we collected three judgments which summed up to a total of 2100. In a single *task* (page) we showed five units (rows) to the contributors, and paid $0.024 per unit per judgment, thus amounting to $0.012 per page. The total cost of the job was $83.47. The language requirement was set to *English* to ensure appropriate contributors. The performance setting was *high speed*, and the minimum accuracy in the test questions was set to 70%.

4.1 Experiment 1: Impact of Sentence Order

The main objective of this experiment is to analyze the impact of sentence order on the readability and coherence of fixed-length text summaries on past events.

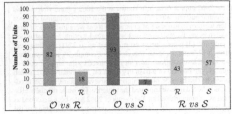
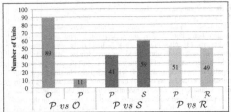

Fig. 7. Experiment 1 results   **Fig. 8.** Experiment 2 results

Summaries from the set $\mathcal{O}$, written by Wikipedians, should be most coherent. Reversing the order of the sentences as in set $\mathcal{R}$, should drastically affect the coherence of the text. We generate *300* units that pair-wise compare summaries from the sets $\mathcal{O}$, $\mathcal{R}$, and $\mathcal{S}$. A comparison with the set $\mathcal{S}$ summaries acts as a random test. We refer to the corresponding subsets, each containing 100 units as $\mathcal{O}$ *vs* $\mathcal{R}$, $\mathcal{O}$ *vs* $\mathcal{S}$, and $\mathcal{R}$ *vs* $\mathcal{S}$.

Results of our experiment are illustrated in Fig. 7. The final preference label of a unit is selected based on majority voting with three judgments. Across all the subsets under comparison, the contributors judge summaries from $\mathcal{O}$ to be the most coherent. Among the 100 units in $\mathcal{O}$ *vs* $\mathcal{R}$, 82 units set $\mathcal{O}$ summary are more coherent. For the $\mathcal{O}$ *vs* $\mathcal{S}$ subset, in 93 units the original summary was found to be better. We obtain an interesting result for the subset $\mathcal{R}$ *vs* $\mathcal{S}$ where in 57 units, the randomly shuffled set $\mathcal{S}$ summary was found to be more coherent. We found moderate agreement for the subsets $\mathcal{O}$ *vs* $\mathcal{R}$ and $\mathcal{O}$ *vs* $\mathcal{S}$, and fair agreement for $\mathcal{R}$ *vs* $\mathcal{S}$ with Fleiss' kappa scores as 0.42, 0.47, and 0.19 respectively. The longest text obtained from *Reason Box* across the 300 units under consideration consists of 54 words. However, the shortest description for a judgment is found to be just one word. The average length is 5.6 words.

Qualitative Analysis. It is concluded that sentence ordering has significant impact on the coherence quality of fixed-length summaries generated for past events. Reversing the original sentence ordering in fact proves to be the worst among the orders under comparison. Upon closer examination, we find that many of the summaries in the set $\mathcal{S}$ partially preserve the original ordering. This seems to make the summaries more coherent as compared to those in the $\mathcal{R}$ set where the ordering is completely reversed. This is also revealed by analyzing the comments of contributors from the *Reason Box*. It can be assumed that the original summaries follow an inherent structure that best conveys information on a specific event at hand. Reversing the original structure makes the summary more confusing. In Ex. 1 from Table 2, a contributor finds that topical jumps make a summary less coherent. However, sometimes a reverse chronological ordering seems to be better as in Ex. 3.

Table 2. Hand-picked examples of Reason Box text. Unit Id in the second column links to the released data for further reference of the readers. Summary A and B presented to contributor from the set indicated. The last column specifies the summary set that the contributor finds more coherent with the reason for the judgment.

Ex	Unit Id	Sum. A	Sum. B	Reason
1	1052032164	$\mathcal{O}$	$\mathcal{S}$	**Preference:** $\mathcal{O}$. **Reason:** The story makes sense. First, talk about the date, then about the consequences of the attacks and finish about the attack itself. B jumps from one issue to other, the link does not make sense sometimes
2	1052262816	$\mathcal{O}$	$\mathcal{R}$	**Preference:** $\mathcal{O}$. **Reason:** The summary A describes correctly the order in which the ministers of economy were named and replaced, while summary B is talking about what the third minister of economy made without referring to his predecessors
3	1051055147	$\mathcal{S}$	$\mathcal{R}$	**Preference:** $\mathcal{R}$. **Reason:** Again, difficult to choose but B starts with the cut of the power and finish with the restored, explaining the story in between
4	1052081752	$\mathcal{P}$	$\mathcal{O}$	**Preference:** $\mathcal{O}$. **Reason:** the order makes sense. It starts with the flight, the number of passengers and then talk about the flight. In text A, the author jump from the flight to the pilot to come back to the airplane to come back to the pilot
5	1051055286	$\mathcal{R}$	$\mathcal{P}$	**Preference:** $\mathcal{P}$. **Reason:** The two summaries have all paragraphs in wrong order. Both describes the causes of the accident at the end of the text when it should be at the beginning and both of them speaks about the doubts on the number of casualties before saying the official report of such amount
6	1051054921	$\mathcal{S}$	$\mathcal{P}$	**Preference:** $\mathcal{S}$. **Reason:** The "tower commission" is the main element around which everything revolves around in these summaries. In the Summary A. any sentence about "tower commission" is near the other about it so that's why I chose it

4.2   Experiment 2: Impact of Sentence Proximity

The main objective of this experiment is to analyze the impact of changing the proximity of the sentences that originally occur next to each other, on the readability of fixed-length text summaries on past events. The set $\mathcal{O}$ summaries are written by Wikipedians and can be assumed to exhibit a sentence grouping such that they present a coherent flow. For example, sentences on a single event aspect are placed next to each other. Thus, altering the ordering where such sentences are separated, and gap between them is maximized should deteriorate the coherence of the summary. We generate 200 units that pair-wise compare summaries from the sets $\mathcal{P}$, $\mathcal{O}$, and $\mathcal{S}$. We refer to the corresponding subsets with 100 units as $\mathcal{P}$ vs $\mathcal{O}$, and $\mathcal{P}$ vs $\mathcal{S}$. We generate an additional subset of 100 units as $\mathcal{P}$ vs $\mathcal{R}$ subset to isolate the impact of proximity by comparing to the worst ordering in set $\mathcal{R}$ summaries.

Results from our experiments are illustrated in Fig. 8. We select the final preference label for each unit based on a *majority vote* with three judgments. We find that the contributors judge the set $\mathcal{O}$ summaries to be the most coherent. Amongst the 100 $\mathcal{P}$ vs $\mathcal{O}$ subset units, in 89 the $\mathcal{O}$ summary is found to be more coherent. Only in 11 units, the set $\mathcal{P}$ summary is better. Across the $\mathcal{P}$ vs $\mathcal{S}$ subset, the randomly shuffled summary from $\mathcal{S}$ is found as more coherent in 59 units. An interesting result is obtained from the $\mathcal{P}$ vs $\mathcal{R}$ subset, where the summaries from the $\mathcal{P}$ and $\mathcal{R}$ sets are judged to be coherent in an almost equal number of units, i.e., 51 and 49 respectively. We found moderate agreement across the $\mathcal{P}$ vs $\mathcal{R}$, $\mathcal{P}$ vs $\mathcal{O}$, and $\mathcal{P}$ vs $\mathcal{S}$ subsets with Fleiss' kappa score of 0.27, 0.50, and 0.21 respectively. Analyzing the text obtained from the *Reason Box* across all the units, we find the longest to be of 50 words and the shortest is a single word. The average length is found to be 5.6 words.

Qualitative Analysis. We conclude that altering the proximity of the sentences in a summary reduced the coherence of the text. Results obtained from the $\mathcal{P}$ vs $\mathcal{R}$ show that contributors were divided between deciding the more coherent summary between these sets. Thus, we can conclude that altering the proximity deteriorates the coherence as much as reversing the order. The reason given by a contributor in Ex. 5 of Table 2 clearly indicates that both the orderings are equally bad. Another example that specifically highlights the effect of proximity is given in Example 6 in Table 2. In consistent with the first experiment, contributors find the more number of randomly shuffled summaries to be more coherent than the proximity altered ones. Upon closer examination, we find that in some randomly shuffled summaries few sentences retain their proximity by chance. These are marked as more coherent.

4.3   Experiment 3: Feasibility of Using Crowdflower

The main objective of this experiment is to evaluate the viability of conducting a Crowdflower study for evaluating a summarization task. To analyze the difficulty of the job, we ask the contributors to specify the difficulty level for each unit they judge. In the user interface this is given as a set of five radio buttons representing different difficulty levels. In addition, we analyze the interestingness of the job, rate of progress, and contributor satisfaction based on a survey provided by Crowdflower.

Results obtained for this experiment are illustrated in Fig. 9. Firstly, we look into the distribution of the difficulty for each unit specified by the contributors. Overall the judgments, we find that 13% of the units are marked as *very easy*, 33% are *easy*, 39% are *not so easy*, 10% are *difficult*, and only 3.9% are marked as *very difficult*. A closer examination of this distribution for each of the unit subsets are illustrated in Fig. 9a.

The entire job took approximately **77** h to complete. Figure 9b illustrates the rate of the judgments acquired over this time interval as obtained from Crowdflower. As shown in Fig. 9d, a single trusted judgment took about 1 min and 41 s on an average. An untrusted judgment takes comparable time of about 1 min and 11 s. The average time spent by a trusted contributor is 8 min and 27 s to judge 5 units whereas an untrusted contributor spends about 5 min and 58 s.

We consider the length of the textual description provided by the contributors as an indicator of the efforts put in by the contributors. The more interesting

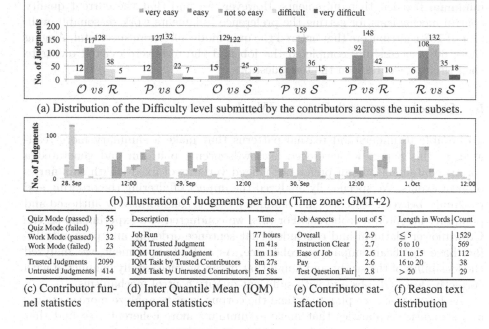

(a) Distribution of the Difficulty level submitted by the contributors across the unit subsets.

(b) Illustration of Judgments per hour (Time zone: GMT+2)

Quiz Mode (passed)	55
Quiz Mode (failed)	79
Work Mode (passed)	32
Work Mode (failed)	23
Trusted Judgments	2099
Untrusted Judgments	414

Description	Time
Job Run	77 hours
IQM Trusted Judgment	1m 41s
IQM Untrusted Judgment	1m 11s
IQM Task by Trusted Contributors	8m 27s
IQM Task by Untrusted Contributors	5m 58s

Job Aspects	out of 5
Overall	2.9
Instruction Clear	2.7
Ease of Job	2.6
Pay	2.6
Test Question Fair	2.8

Length in Words	Count
≤ 5	1529
6 to 10	569
11 to 15	112
16 to 20	38
> 20	29

(c) Contributor funnel statistics

(d) Inter Quantile Mean (IQM) temporal statistics

(e) Contributor satisfaction

(f) Reason text distribution

Fig. 9. Experiment results

a contributor finds the job, the more efforts she will put the task. This effort includes giving specific and descriptive reasons for her decision in the *Reason Box*. Table 2 shows the distributions of the textual description for all the units in terms of their word counts.

To assess the quality control mechanisms that we design for this task, we look into the number of contributors that were evicted from the job during the *Quiz* and *Work* modes of the job. These statistics are presented in Fig. 9c. We find no contributor falls into the *traps* set to judge identical summaries indicating the good quality of judgments.

Finally, we find reasonable contributor satisfaction for the job setting which was determined through a survey given to the contributor at the end of the job. All the features receive a score of more than 2.5 out of 5.

Qualitative Analysis. We conclude that a crowdsourcing service like Crowdflower, with correctly designed jobs, can be successfully leveraged to gather preference judgments for comparing summary structures for a text summarization task. Though the task was not found to be very easy by the contributors, they found the difficulty level within an acceptable range. The contributors spend reasonable time for judging per unit which indicates that they show good interest in the job. This is also reflected from the quality of the text input that we gather where they specify the reason for each decision even though, find a lot of they textual descriptions were very short (≤ 5 words). This suggests additional measures that need to be taken for future studies to improve its quality, like setting minimum text-length requirements. However, we find that the current quality control mechanisms for getting the preference judgments work reasonably well. Out of 133 contributors that entered the quiz mode 40% were successful. Among those 60% successfully completed the job and the rest were evicted owing to their drop in accuracy on the test questions.

5   Conclusion and Future Work

To be able to understand textual patterns that make a summary more readable, in this paper, we leveraged a crowdsourcing platform and generated a corpus containing human preference-judged variants of fixed-length summaries on past events. We specifically analyzed the impact of altering the ordering and proximity between sentences in a summary that is collaboratively authored and popularly accepted as coherent. For this, we conducted two experiments on the Crowdflower platform and concluded that sentence ordering and proximity infact have significant impact on coherence. As a third experiment, we analyzed the feasibility of the study, and found that it can be successfully conducted with correctly designed jobs and with sufficient quality control mechanisms.

As future work, we plan to extend the corpus so as to analyze more syntactic and semantic regularities that make a summary more coherent. We find that the current measures for evaluating text summarization task do not consider structural quality. In the future, we intend to look into the problem of designing measures that explicitly evaluate summary structures for the text summarization task.

References

1. Crowdflower platform. https://www.crowdflower.com
2. Alonso, O., Baeza-Yates, R.: Design and implementation of relevance assessments using crowdsourcing. In: Clough, P., Foley, C., Gurrin, C., Jones, G.J.F., Kraaij, W., Lee, H., Mudoch, V. (eds.) ECIR 2011. LNCS, vol. 6611, pp. 153–164. Springer, Heidelberg (2011). doi:10.1007/978-3-642-20161-5_16
3. Barzilay, R., et al.: Using lexical chains for text summarization. In: Advances in Automatic Text Summarization, pp. 111–121. MIT Press, Cambridge (1999)
4. Barzilay, R., et al.: Inferring strategies for sentence ordering in multidocument news summarization. J. Artif. Intell. Res. 17, 35–55 (2002)
5. Barzilay, R., et al.: Sentence ordering in multidocument summarization. In: HLT 2001
6. Baumel, T., et al.: Query-chain focused summarization. In: ACL 2014
7. Bollegala, D., et al.: A preference learning approach to sentence ordering for multi-document summarization. Inf. Sci. 217, 78–95 (2012)
8. Donghong, J., et al.: Sentence ordering based on cluster adjacency in multi-document summarization. In: IJCNLP 2008
9. Hobbs, J.R.: Literature and cognition. No. 21, Center for the Study of Language (1990)
10. Hu, P., Ji, D., et al.: Query-focused multi-document summarization using co-training based semi-supervised learning. In: PACLIC 2009

11. Hume, D.: Philosophical Essays Concerning Human Understanding. Georg Olms Verlag, Hildesheim (1750)
12. Ji, P.D., et al.: Sentence ordering with manifold-based classification in multi-document summarization. In: ECML 2006
13. Jing, H.: Summary generation through intelligent cutting and pasting of the input document. Technical report, Columbia University (1998)
14. Kaisser, M., et al.: Improving search results quality by customizing summary lengths. In: ACL 2008
15. Lapata, M.: Probabilistic text structuring: experiments with sentence ordering. In: ACL 2003
16. Li, Y., et al.: Query-focused multi-document summarization: combining a topic model with graph-based semi-supervised learning. In: COLING 2014
17. Lin, C.Y.: Rouge: a package for automatic evaluation of summaries. In: ACL-2004 Workshop (2004)
18. Litkowski, K.C.: Summarization experiments in DUC 2004. In: DUC 2004
19. Lloret, E., et al.: Analyzing the capabilities of crowdsourcing services for text summarization. Lang. Resour. Eval. 47(2), 337–369 (2013)
20. Nenkova, A., et al.: Evaluating content selection in summarization: the pyramid method. In: HLT-NAACL 2004
21. Ono, K., et al.: Abstract generation based on rhetorical structure extraction. In: NLP/CL 1994
22. Radev, D.R., et al.: Evaluation challenges in large-scale document summarization. In: ACL 2003
23. Salton, G., et al.: Automatic text structuring and summarization. Inf. Process. Manage. 33(2), 193–207 (1997)
24. Sorokin, A., et al.: Utility data annotation with Amazon mechanical Turk. Urbana 51(61), 820 (2008)
25. Steinberger, J., et al.: Evaluation measures for text summarization. Comput. Inform. 28(2), 251–275 (2012)
26. Tang, J., Sanderson, M.: Evaluation and user preference study on spatial diversity. In: Gurrin, C., He, Y., Kazai, G., Kruschwitz, U., Little, S., Roelleke, T., Rüger, S., Rijsbergen, K. (eds.) ECIR 2010. LNCS, vol. 5993, pp. 179–190. Springer, Heidelberg (2010). doi:10.1007/978-3-642-12275-0_18
27. Zhong, S., et al.: Query-oriented unsupervised multi-document summarization via deep learning model. Expert Syst. Appl. 42(21), 8146–8155 (2015)

Sentiment Propagation for Predicting Reputation Polarity

Anastasia Giachanou[1(✉)], Julio Gonzalo[2], Ida Mele[1], and Fabio Crestani[1]

[1] Faculty of Informatics, Università della Svizzera Italiana (USI), Lugano, Switzerland
{anastasia.giachanou,ida.mele,fabio.crestani}@usi.ch
[2] UNED NLP & IR Group, Madrid, Spain
julio.gonzalo@lsi.uned.es

Abstract. One of the core tasks of Online Reputation Monitoring is to determine whether a text mentioning the entity of interest has positive or negative implications for its reputation. A challenging aspect of the task is that many texts are polar facts, i.e. they do not convey sentiment but they do have reputational implications (e.g. *A Samsung smartphone exploded during flight* has negative implications for the reputation of Samsung). In this paper we explore the hypothesis that, in order to determine the reputation polarity of factual information, we can propagate sentiment from sentiment-bearing texts to factual texts that discuss the same issue. We test two approaches that implement such hypothesis: the first one is to directly propagate sentiment to similar texts, and the second one is to augment the polarity lexicon. Our results (i) confirm our propagation hypothesis, with improvements of up to 43% in weakly supervised settings and up to 59% with fully supervised methods; and (ii) indicate that building domain-specific polarity lexicons is a cost-effective strategy.

Keywords: Reputation polarity · Sentiment propagation

1 Introduction

One of the core tasks in online reputation management is to monitor what is posted online about an entity (a company, celebrity, etc.) and react in case there is an alert of a possible damage on the entity's reputation. Analysts have first to filter the stream of data and find the content that is relevant for the entity of interest. Then, they have to determine if a relevant post is likely to have positive, neutral or negative implications on the entity's reputation.

Reputation polarity is not a trivial task, and it is more challenging than sentiment analysis. A key problem is that there is a significant amount of tweets with positive or negative reputation polarity which do not explicitly express a sentiment. These tweets are known as *polar facts*. For example, the tweet *Chrysler recalls 919,000 Jeeps to fix air bags* does not convey any sentiment but it has negative impact on the reputation of *Chrysler*.

J.M. Jose et al. (Eds.): ECIR 2017, LNCS 10193, pp. 226–238, 2017.
DOI: 10.1007/978-3-319-56608-5_18

To address this challenge, we hypothesize that tweets that are about a specific topic should tend to have the same reputation polarity. In this way, if there are many tweets about a specific topic, then some of those tweets will explicitly express some sentiment towards the topic. Table 1 shows some example tweets relevant to the entity *HSBC* that are about the same topic (topic *accusations*). Table 1 also shows the actual (manually annotated) reputation polarity of each tweet, and the sentiment polarity as assigned by a state-of-the-art lexicon based approach. Note that there are some tweets (i.e. *t3*) that do not contain any sentiment word (*sentiment by lexicon* is *neutral*) but they have a negative impact on the entity's reputation, whereas other tweets in the same topic (i.e. *t1, t2*) have an explicit sentiment indicator. Propagation of sentiment across texts discussing the same issue might then be a way of annotating reputation polarity.

We consider two ways of propagating sentiment to sentiment-neutral texts: (i) direct propagation to texts with similar content; (ii) augmenting the lexicon with terms that indicate reputation polarity even if they do not convey sentiment polarity. Hence, we focus on two related research questions:

- *Can we use training material to detect terms with reputation polarity and use them to augment a general sentiment lexicon?* One of the state-of-art approaches in sentiment analysis is the lexicon based approach. However, the general lexicons are not effective for reputation polarity. Hence, we propose to augment general lexicons at different levels of granularity with terms extracted from training data to build reputation lexicons. An associated question is *what is the right level of generalization for a reputation lexicon.* We will explore three alternatives: (i) building a general purpose lexicon with all available training material; (ii) building domain-specific lexicons with training material for entities in a given domain (e.g. banking, automotive); (iii) building entity-specific lexicons with separated training material for each entity. In principle, the more specific a lexicon is, the most accurate results will give, but at a substantial cost, because we need more training examples. We want to investigate whether there is an optimal level of specificity that provides competitive results at a moderate cost.
- *Can we propagate sentiment to texts that are similar in terms of content to improve reputation polarity?* In order to answer this question we will consider two propagation alternatives: (i) first perform text clustering to detect topics, and then propagate sentiment within each topic; (ii) directly propagate sentiment from a sentiment-bearing text to other texts that are pairwise similar. In addition, we will also experiment with the use of a polar fact filter to avoid overpropagation to polarity-wise neutral texts.

2   Related Work

Although reputation polarity is substantial different to sentiment analysis, the two tasks have some similarities. To this end, past work on reputation polarity evolved from sentiment analysis. Previous work on opinion retrieval and sentiment analysis can be roughly divided into two categories: lexicon based and

Table 1. Examples of annotated tweets in the RepLab 2013 training dataset.

Oracle topic	Id	Tweet	Reputation polarity	Sentiment by lexicon
Accusations	t1	When I wake up I want to find these trending: Barclays, HSBC, executive arrests, fraud & Tory party. NOT Justin Bieber	Negative	Negative
Accusations	t2	THE CORPORATE POLITICIANS: 20 years of failure for Britain as they skimmed the system. #cnn, #times, #cnbc, #hsbc	Negative	Negative
Accusations	t3	@PoliticalPryers he's ceo of one of the banks involved. He high but not the top! By this time next week RBS, Llyods, HSBC will get same	Negative	Neutral

classification based approaches. The lexicon based approaches estimate the sentiment of a document using a list of opinion words [24,25] known as opinion lexicons. The presence of any opinionated word in a document is an indicator of sentiment. In its most typical scenario, lexicon based approach is unsupervised since it does not require any training data. More sophisticated approaches incorporate additional sentiment indicators such as proximity between query and opinion terms [7] or topic-based stylistic variations [9].

The classification based approaches use sets of features to build a classifier that can predict the sentiment polarity of a document [19]. The features range from simple n-grams to semantic features and from syntactic to medium's specific features. A number of researchers analyzed the impact of different features on Twitter sentiment analysis and established feature selection criteria [1,13,17]. The classification based approaches can be further divided into semi-supervised and supervised approaches. The major difference between the two categories is that the semi-supervised approaches combine labeled and unlabeled data. A comprehensive review on opinion retrieval and sentiment analysis can be found in a survey by Pang and Lee [18] whereas a comprehensive survey focused on Twitter sentiment analysis can be found by Giachanou and Crestani [8].

A number of proposed approaches for reputation polarity treated the task with methods similar to sentiment analysis' methods. Classifiers trained on sentiment and textual features showed to be very effective on RepLab evaluation campaign [2,3]. The best result on RepLab 2013 was achieved by Hangya and Farkas [10] who trained a Maximum Entropy classifier using sentiment lexicon, bigrams, number of negation words and character repetitions. Castellanos et al. [4] addressed the reputation polarity problem with an information retrieval based approach and found the most relevant class using the tweet's content as a query. Other approaches considered sentiment classifiers and lexicons [15,22].

Peetz et al. [20] assumed that understanding how the tweet is perceived is an important indicator for estimating the reputation polarity of a tweet. To this end, they proposed a supervised approach that also considered reception features such as tweet's replies and retweets. Their results showed that reception features were effective and their best result was obtained on entity dependent data.

Different form the previous work, we explore the hypothesis that texts that are about the same topic should share the same reputation polarity. To this end, we consider propagating sentiment using topically similar tweets. In addition, we are the first to consider a polar fact filter that is able to differentiate neutral tweets from polar facts.

3   Proposed Approach

Our starting point is a standard lexicon based approach for sentiment analysis. This approach detects the sentiment of a document by using a general list of words annotated with their sentiment polarity (*positive* or *negative*). The presence of any opinionated word in a document indicates the document's polarity. Hence, this approach generates a sentiment score for the document based on the number of opinionated terms it contains.

Let *polarity*(*d*) be the reputation polarity of a document *d*, where *polarity*(*d*) can take one of the values $\{-1, 0, 1\}$ referring to a positive, neutral and negative polarity respectively. Also, let S_d denote the sentiment score of a document *d* based on the sentiment scores of its terms, calculated as: $S_d = \sum_{t \in d} opinion(t)$, where $opinion(t)$ is the opinion score of the term based on an opinion lexicon. Then, according to the lexicon based approach the reputation polarity of a document is determined as follows:

$$polarity(d) = \begin{cases} 1, & \text{if } S_d > 0 \\ -1, & \text{if } S_d < 0 \\ 0, & otherwise \end{cases}$$

Here we should note that the sentiment score S_d depends on the number of opinionated words that appear in the document and for this reason the score is an integer value. One of the advantages of this method is that it does not require any training data. We use this method as our baseline.

In this paper we use the lexicon based approach as a starting point to find the sentiment of tweets and then we explore two different approaches to improve the reputation polarity. First, we extract terms that are closely related to positive or negative sentiment and use these words to augment a sentiment lexicon. Second, we propagate sentiment to factual tweets to determine their reputation polarity using the sentiment of tweets that are similar in terms of content.

3.1   Lexicon Expansion

One limitation of the lexicon based approaches is the word mismatch between the tweet and the general opinion lexicons. Tweets contain a lot of idiomatic words as with the case of the "elongated" words (e.g. *gooooood*). This problem is more evident for the reputation polarity task where there are a lot of tweets that do not contain any sentiment word but have an impact on the entity's reputation.

To address the problem of the word mismatch, we explore the effectiveness of lexicon augmentation. To learn new positive/negative words we use the training data provided in the collection. The positive/negative lexicons are expanded with the terms of the positive/negative tweets of the training set. We augment the lexicons on three different levels of granularity: *domain/entity independent*[1], *domain dependent* and *entity dependent*. After augmenting the lexicons, we use the lexicon based approach that uses the number of occurrences of opinionated terms to predict the reputation polarity of a document. This approach that we refer to it as *simple lexicon augmentation* considers only the presence of words as an indicator of reputation polarity.

In addition, we also investigate a fully supervised way to learn the words that indicate reputation polarity. This approach is based on the Pointwise Mutual Information (PMI) method originally proposed by Church and Hanks [6]. According to this approach, every term t is assigned a PMI score for each of the three reputation polarity classes: positive, neutral and negative. The sentiment score for a term t is calculated using the training data as follows:

$$PMI(d, positive) = \sum_{t \in d} PMI(t, positive)$$

$$PMI(t, positive) = \log_2 \frac{c(t, positive) * N}{c(t) * c(positive)}$$

where $c(t, positive)$ is the frequency of the term t in the positive tweets, N is the total number of words in the corpus, $c(t)$ is the frequency of the term in the corpus and $c(positive)$ is the number of terms in the positive tweets. The PMI of the terms for the negative and neutral classes is calculated in a similar way. Then these scores can be used to predict the polarity of the test documents. We assume that the polarity of a document is the one with the highest PMI score.

3.2   Polar Fact Filter

A limitation of propagation methods is that they may overestimate the number of tweets with reputation polarity (i.e. the sentiment polarity is potentially propagated to polar facts and to reputation-neutral tweets). A possible supervised solution is to first detect polar facts, building a classifier (*polar fact filter*) that takes a single tweet as an input and decides if the tweet is a polar fact or not. To this end, we address the task of identifying polar facts as a binary classification problem and do not differentiate between positive and negative tweets. We train a linear kernel Support Vector Machine (SVM) classifier to discriminate between

---

[1] In the rest of the paper we refer to this setting as *independent* for brevity.

polar facts and neutral tweets. SVM [5] is a state-of-art learning algorithm that has been effectively applied on text categorization tasks.

First, we separate the polar facts and the neutral tweets into two classes, $y_i \in \{-1, 1\}$, where N is the number of the labeled training data. The training examples are $(\mathbf{x}_1, y_1), \ldots, (\mathbf{x}_N, y_N), \mathbf{x} \in R^k$ where k is the number of features.

For the classification, we explored a number of different features that have proved to be effective for sentiment classification [12]. The features can be grouped in three classes as follows:

- n-grams: n-grams with $n \in [1, 4]$, character grams
- stylistic: number of capitalised words, number of elongated words, number of emoticons, number of exclamation and question marks
- lexicons: manual and automatic lexicons

We explore the effectiveness of the polar fact filter on three different training settings: *independent*, *domain dependent* and *entity dependent*.

3.3   Sentiment Propagation

As already mentioned, we assume that similar tweets in terms of content (topic) should tend to have the same polarity for reputation. Hence, we propose to propagate sentiment to tweets that are annotated as polar facts using the sentiment of similar tweets. We explore two different propagation approaches: *clustering* and *tweet to tweet similarity*. Also, we explore two different ways to propagate sentiment. The first method is based on the *maximum* sentiment of the similar tweets whereas the second is based on *tweet's similarity* to each of the reputation polarity classes.

To better describe our approach we introduce some notation. Let $D = \{d_1, \ldots, d_M\}$ be some tweets we want to predict their reputation polarity using a set of other tweets $D' = \{d'_1, \ldots, d'_N\}$ for which we already know their polarity. Also, let $D^+ = \{d_1^+, d_2^+, \ldots, d_K^+\}$, $D^\cdot = \{d_1^\cdot, d_2^\cdot, \ldots, d_V^\cdot\}$ and $D^- = \{d_1^-, d_2^-, \ldots, d_L^-\}$ be three different sets of tweets that are annotated as positive, neutral and negative respectively and $D' = D^+ \cup D^\cdot \cup D^-$.

To annotate a tweet d that belongs to D we count the number of tweets in D' that belong to each of the reputation polarity classes *positive, neutral* and *negative* denoted as $|D^+|, |D^\cdot|$ and $|D^-|$ respectively. The polarity of a document d is calculated as follows:

$$
polarity(d) = \begin{cases} 1, & \text{if } |D^+| = max\{freq(d)\} \\ -1, & \text{if } |D^-| = max\{freq(d)\} \\ 0, & otherwise \end{cases}
$$

where $max\{freq(d)\} = \max |D^+|, |D^\cdot|, |D^-|$. Here we should note that we propose to use the polar fact filter to differentiate between the tweets in D and in $D^\cdot$ and that $D \cap D' = \emptyset$.

The second approach to propagate sentiment is based on the tweet's similarity to each of the polarity classes. To annotate a tweet d that belongs to D, we first calculate the similarity to each of the three classes. For the positive class we calculate the similarity as follows:

$$sim^+(d) = \sum_{d_i \in D^+} sim(d, d_i^+)$$

The next step is to calculate the average similarity to the positive class as $avgSim^+(d) = sim^+(d)/|D^+|$ where $|D^+|$ is the number of positive tweets. We follow a similar way to calculate the similarities and the average similarity of the neutral and negative classes. Next, we calculate the maximum average among the three classes as

$$max\{avgSim(d)\} = \max avgSim^+(d), avgSim^{\cdot}(d), avgSim^-(d)$$

and finally we determine the polarity of the tweet d as:

$$polarity(d) = \begin{cases} 1, & \text{if } avgSim^+(d) = max\{avgSim(d)\} \\ -1, & \text{if } avgSim^-(d) = max\{avgSim(d)\} \\ 0, & \text{otherwise} \end{cases}$$

To determine D' (the set of tweets for which we already know the sentiment), we explore two different approaches: clustering and tweets' similarity. For clustering the tweets we used the approach that obtained the best result in Spina et al. [23]. This approach first trains a classifier to predict if two tweets belong to the same topic using term, semantic, metadata and temporal features and then uses a hierarchical agglomerative clustering algorithm to identify the clusters. The tweets' clusters are publicly available[2]. For the tweet to tweet similarity, we consider cosine similarity over a bag of terms representation.

4 Experimental Setup

Dataset. For this study, we use the RepLab 2013 [2] data set, which is the largest available test collection for the task of monitoring the reputation of entities (companies, organizations, celebrities, etc.) on Twitter. The RepLab 2013 collection contains 142,527 manually annotated tweets in English and Spanish. The tweets are about 61 different entities that belong to 4 domains: *automotive, banking, universities* and *music*.

Experimental Settings. We use publicly available word lexicons in English [16] and in Spanish [21] to identify the words that indicate positive or negative sentiment. We use information from tweets' metadata to identify the language of the tweet. We use the same tokenizer for English and Spanish tweets. For the results that are reported we considered the tweets that are relevant to an entity (tweets manually annotated as *related*) from the test set.

---

[2] https://github.com/damiano/learning-similarity-functions-ORM.

Polar Fact Filter. To build the polar fact filter we use a linear SVM classifier. As training data, we use the tweets in the training set which are annotated as neutral by the simple lexicon based approach. We explore a wide range of features such as n-grams, character grams, number of capitalised words, number of elongated words, number of emoticons, number of exclamation and question marks, automatic and manual lexicons. With respect to the lexicons explored for the polar fact filter, we consider Liu's lexicon [11], NRC emotion lexicon [14], MPQA lexicon [26] and Hashtag Sentiment Lexicon [12]. We explore three different levels of granularity for training the classifier: *independent, domain dependent* and *entity dependent*.

Evaluation. We present evaluation scores for our methods on all the three polarity classes, *positive, neutral* and *negative*, according to the instructions given at RepLab 2013. We report *F-score* for the proposed methods and the polar fact classification. We use the McNemar test to evaluate the statistical significance of differences, which is more appropriate for comparisons of nominal data.

5   Results and Discussion

In this section, we present the results of our proposed methodology on the reputation polarity task. First, we discuss the effectiveness of augmenting the lexicon at different levels of granularity, we continue with the performance of the polar fact filter and finally we present the results of sentiment propagation.

5.1   Lexicon Expansion

In order to address the first research question, we compare the results of augmenting the lexicon at different levels of granularity with the lexicon based approach (*baseline*). Results are displayed in Table 2. The main outcome is that augmenting the lexicon is effective at all levels of granularity, with improvements ranging from +17% in the general expansion to +25% if a specific lexicon is created for each individual entity. All improvements are statistically significant with respect to the baseline. Unsurprisingly, entity-specific lexicons give the best result, but note that the difference between domain and entity specific lexicons is thin (only 1%). This is an interesting observation, because it indicates that training data can be generalized for entities within a domain, and that is more cost-effective than having to annotate training data for every entity in a domain.

Alternatively, we also explore the effectiveness of PMI for predicting the reputation polarity. Similar to the simple lexicon augmentation approach, we use three different settings to learn the PMI scores: *independent* referring to all the training data, *domain dependent* referring to the setting where we learn PMI scores for each domain and *entity dependent* where we learn PMI scores for each entity. Table 3 displays the results. The conclusions are the same as for the previous method (the expansion substantially improves performance, entity-dependent expansion is the best but domain-dependent expansion is very close).

Table 2. Performance results of the lexicon based approach before and after augmenting the lexicon using independent, domain dependent and entity dependent data. A star(∗) indicates statistically significant improvement over the lexicon based approach.

Method	F-measure
Lexicon based	0.368
Lexicon augmentation - independent	0.431∗ (+17%)
Lexicon augmentation - domain dependent	0.455∗ (+24%)
Lexicon augmentation - entity dependent	**0.460∗ (+25%)**

Table 3. Performance results of the supervised method based on PMI, when trained on independent, domain dependent and entity dependent data. A star(∗) indicates statistically significant improvement over the lexicon based approach.

	F-score
Lexicon based	0.368
PMI - independent	0.547∗ (+49%)
PMI - domain dependent	0.572∗ (+55%)
PMI - entity dependent	**0.586∗ (+59%)**

The general performance of this method (which is fully supervised) is superior, and in fact entity-dependent PMI results are 5.6% better than the best results published to date on this dataset [20].

5.2 Polar Fact Filter

Table 4 presents the effectiveness of the polar fact filter when it is trained on different set of features and when it is trained on an *independent, domain dependent* or *entity dependent* setting. Similarly to the previous reported results, the best performance is obtained when the classifier is trained on the *entity dependent* setting. One interesting observation is that the best performance is obtained when the classifier is trained on *n-grams* and *character* grams using entity dependent data. This result was expected since this classifier aims to differentiate between polar fact tweets and neutral tweets and neither of them contain sentiment words.

However, the results indicate that sentiment lexicons are effective features for the polar fact filter when we use independent and domain dependent data. Note that for the polar fact filter we used 4 different lexicons that have been found to be effective for sentiment analysis [12] and which contain more information compared to the general lexicons. The results indicate that in case of independent and domain dependent data, sentiment lexicons can still provide useful information for reputation polarity. The model with the best performance (trained on *n-grams, character grams/entity-dependent*) is used in the rest of the experiments to detect the tweets that are polar facts and that have to be annotated with reputation polarity.

Table 4. Performance results (F-measure) of the polar fact filter classification when trained on independent, domain dependent and entity dependent data.

	Independent	Domain dependent	Entity dependent
n-grams	0.633	0.654	**0.692**
n-grams, stylistic	0.635	0.655	0.691
n-grams, stylistic, lexicons	**0.654**	**0.660**	0.668

5.3 Sentiment Propagation

For the second research question, we explore the effectiveness of propagating sentiment with the aim to improve reputation polarity. We compare the results of propagating sentiment using an automatic clustering and a cosine similarity approach. Table 5 presents the results of propagating sentiment to tweets that were annotated as polar facts. The results indicate that sentiment can be propagated topically to annotate tweets with reputation polarity: in all cases, the improvement is above 20% with respect to the no propagation baseline. For the best experimental setting (propagating to similar tweets using the max approach), the improvement is +43%. This confirms the hypothesis that tweets that share a similar (factual) content tend to share the same reputation polarity.

Table 5. Performance results (F-measure) of sentiment propagation approaches.

	Max	Similarity to class
No propagation	0.368	0.368
Cluster propagation	0.472 (+28%)	0.457 (+24%)
Similar tweets propagation	**0.526 (+43%)**	0.495 (+35%)

Finally, Table 6 compares the best results published until now for reputation polarity on the RepLab 2013 dataset (SVM trained on message and reception features and on an entity-dependent scenario) [20] with our best supervised and weakly-supervised approaches in terms of F-measure. The supervised approach based on PMI outperforms [20] with a 5.6% relative improvement in terms of F-measure (0.586 vs 0.553). This indicates that it is not necessary to use many features to get competitive results in reputation polarity. Unsurprisingly, we also see that fully supervised approaches outperform weakly supervised ones. Our best weakly supervised approach (propagation to similar tweets using max combination), however, is only 5% worse than [20] (0.526 vs 0.553). This small difference indicates that weakly supervised annotation of reputation polarity is feasible, which is a promising result as such methods are less dependent on the availability of training data.

Table 6. Comparison with state-of-the-art results.

System	F-measure
Peetz et al. 2016 (Best published result)	0.553
Supervised - PMI & Entity dependent	0.586
Weakly supervised - propagation (Tweets' similarity & max)	0.526

6   Conclusions and Future Work

The results of our experiments strongly support our initial hypothesis: sentiment signals can be used to annotate reputation polarity, starting with sentiment-bearing texts and propagating sentiment to sentiment-neutral similar texts. We have explored two approaches: augmenting the sentiment lexicon via propagation, and directly propagating sentiment to topically similar tweets.

Augmenting the sentiment lexicon in a weakly-supervised way improves results up to 25% if we generate a specific lexicon for each entity of interest. But, remarkably, generating domain-specific lexicons (which requires less training material) gives very similar results (24% improvement over the original sentiment lexicon). The conclusion is that sentiment lexicons can be augmented to create reputation polarity lexicons, and that the domain level is a cost-effective level of granularity for doing so. If we use a fully supervised approach to learn reputation polarity words (based on PMI scores), performance is 5.6% better than the best published result on the dataset so far [20]. This indicates that learning PMI values to predict reputation polarity is very effective.

Direct propagation of sentiment is also effective. In all conditions, the improvement is above 20% with respect to the no propagation baseline, and for the best setting (propagating to similar tweets using the max approach), the improvement is +43%. This is also a weakly supervised approach, because both the initial sentiment annotation and the propagation are unsupervised; the only supervised mechanism is the polar fact filter that prevents propagation to truly neutral tweets. Results, however, are only 5% worse than [20] (0.526 vs 0.553), which is a fully supervised approach. This small difference indicates that weakly supervised annotation of reputation polarity is feasible, which is a promising result as such methods are less dependent on the availability of training data.

Future work includes carefully analyzing the augmented vocabularies. We need to identify the percentage of erroneous additions, how frequently the new terms are sentiment-bearing terms that were absent from the initial vocabulary simply for lack of coverage, and non sentiment-bearing terms which specifically indicate factual polarity. We also plan to analyze different ways of propagating sentiment, and to explore the effectiveness of additional features (e.g. semantic, temporal) on finding the tweets that can be used for sentiment propagation.

Acknowledgments. This research was partially funded by the Swiss National Science Foundation (SNSF) under the project OpiTrack.

References

1. Agarwal, A., Xie, B., Vovsha, I., Rambow, O., Passonneau, R.: Sentiment analysis of Twitter data. In: LSM 2011, pp. 30–38. ACL (2011)
2. Amigó, E., et al.: Overview of RepLab 2013: evaluating online reputation monitoring systems. In: Forner, P., Müller, H., Paredes, R., Rosso, P., Stein, B. (eds.) CLEF 2013. LNCS, vol. 8138, pp. 333–352. Springer, Heidelberg (2013). doi:10.1007/978-3-642-40802-1_31
3. Amigó, E., Corujo, A., Gonzalo, J., Meij, E., de Rijke, M.: Overview of RepLab 2012: evaluating online reputation management systems. In: CLEF 2012 (2012)
4. Castellanos, A., Cigarrán, J., García-Serrano, A.: Modelling techniques for Twitter contents: a step beyond classification based approaches. In: CLEF 2013 (2013)
5. Chang, C.C., Lin, C.J.: LIBSVM: a library for support vector machines. Trans. Intell. Syst. Technol. 2(3), 27:1–27:27 (2011)
6. Church, K.W., Hanks, P.: Word association norms, mutual information, and lexicography. Comput. Linguist. 16(1), 22–29 (1990)
7. Gerani, S., Carman, M., Crestani, F.: Aggregation methods for proximity-based opinion retrieval. ACM Trans. Inf. Syst. (TOIS) 30(4), 1–36 (2012)
8. Giachanou, A., Crestani, F.: Like it or not: a survey of Twitter sentiment analysis methods. ACM Comput. Surv. (CSUR) 49(2), 28 (2016)
9. Giachanou, A., Harvey, M., Crestani, F.: Topic-specific stylistic variations for opinion retrieval on Twitter. In: Ferro, N., Crestani, F., Moens, M.-F., Mothe, J., Silvestri, F., Nunzio, G.M., Hauff, C., Silvello, G. (eds.) ECIR 2016. LNCS, vol. 9626, pp. 466–478. Springer, Cham (2016). doi:10.1007/978-3-319-30671-1_34
10. Hangya, V., Farkas, R.: Filtering and polarity detection for reputation management on tweets. In: CLEF 2013 (2013)
11. Hu, M., Liu, B.: Mining and summarizing customer reviews. In: KDD 2004, pp. 168–177. ACM (2004)
12. Kiritchenko, S., Zhu, X., Mohammad, S.M.: Sentiment analysis of short informal texts. J. Artif. Intell. Res. 50(1), 723–762 (2014)
13. Kouloumpis, E., Wilson, T., Moore, J.: Twitter sentiment analysis: the good the bad and the OMG! In: ICWSM 2011, pp. 538–541. AAAI Press (2011)
14. Mohammad, S.M., Turney, P.D.: Emotions evoked by common words and phrases: using mechanical turk to create an emotion lexicon. In: CAAGET 2010, pp. 26–34. ACL (2010)
15. Mosquera, A., Fernández, J., M. Gómez, J., Martínez-Barco, P., Moreda, P.: DLSI-Volvam at RepLab 2013: polarity classification on Twitter data. In: CLEF 2013 (2013)
16. Nielsen, F.: A new ANEW: evaluation of a word list for sentiment analysis of microblogs. In: ESWC 2011 Workshop on 'Making Sense of Microposts': Big Things Come in Small Packages, pp. 93–98 (2011)
17. Pak, A., Paroubek, P.: Twitter as a corpus for sentiment analysis and opinion mining. In: LREC 2010, pp. 1320–1326. ELRA (2010)
18. Pang, B., Lee, L.: Opinion mining and sentiment analysis. Found. Trends Inf. Retrieval 2(1–2), 1–135 (2008)
19. Pang, B., Lee, L., Vaithyanathan, S.: Thumbs up? Sentiment classification using machine learning techniques. In: EMNLP 2002, pp. 79–86. ACL (2002)
20. Peetz, M.H., de Rijke, M., Kaptein, R.: Estimating reputation polarity on microblog posts. Inf. Process. Manag. 52(2), 193–216 (2016)

21. Perez-Rosas, V., Banea, C., Mihalcea, R.: Learning sentiment lexicons in Spanish. In: LREC 2012. ELRA (2012)
22. Saias, J.: In search of reputation assessment: experiences with polarity classification in RepLab 2013. In: CLEF 2013 (2013)
23. Spina, D., Gonzalo, J., Amigó, E.: Learning similarity functions for topic detection in online reputation monitoring. In: SIGIR 2014, pp. 527–536. ACM (2014)
24. Taboada, M., Brooke, J., Tofiloski, M., Voll, K., Stede, M.: Lexicon-based methods for sentiment analysis. Comput. Linguist. 37(2), 267–307 (2011)
25. Turney, P.D.: Thumbs up or thumbs down? Semantic orientation applied to unsupervised classification of reviews. In: ACL 2002, pp. 417–424. ACL (2002)
26. Wilson, T., Wiebe, J., Hoffmann, P.: Recognizing contextual polarity in phrase-level sentiment analysis. In: HLT 2005, pp. 347–354. ACL (2005)

Transitivity, Time Consumption, and Quality of Preference Judgments in Crowdsourcing

Kai Hui[1,2(✉)] and Klaus Berberich[1,3]

[1] Max Planck Institute for Informatics, Saarbrücken, Germany
{khui,kberberi}@mpi-inf.mpg.de
[2] Saarbrücken Graduate School of Computer Science, Saarbrücken, Germany
[3] htw saar, Saarbrücken, Germany

Abstract. Preference judgments have been demonstrated as a better alternative to graded judgments to assess the relevance of documents relative to queries. Existing work has verified transitivity among preference judgments when collected from trained judges, which reduced the number of judgments dramatically. Moreover, strict preference judgments and weak preference judgments, where the latter additionally allow judges to state that two documents are equally relevant for a given query, are both widely used in literature. However, whether transitivity still holds when collected from crowdsourcing, i.e., whether the two kinds of preference judgments behave similarly remains unclear. In this work, we collect judgments from multiple judges using a crowdsourcing platform and aggregate them to compare the two kinds of preference judgments in terms of transitivity, time consumption, and quality. That is, we look into whether aggregated judgments are transitive, how long it takes judges to make them, and whether judges agree with each other and with judgments from TREC. Our key findings are that only strict preference judgments are transitive. Meanwhile, weak preference judgments behave differently in terms of transitivity, time consumption, as well as of the quality of judgment.

1 Introduction

Offline evaluation in information retrieval following the Cranfield [6] paradigm heavily relies on manual judgments to evaluate search results returned by competing systems. The traditional approach to judge the relevance of documents returned for a query, coined graded judgments, is to consider each document in isolation and assign a predefined grade (e.g., highly-relevant, relevant, or non-relevant) to it. More recently, preference judgments have been demonstrated [5,10,13] as a better alternative. Here, pairs of documents returned for a specific query are considered, and judges are asked to state their relative preference. Figure 1 illustrates these two approaches. Initiatives like TREC have typically relied on trained judges, who tend to provide high-quality judgments. Crowdsourcing platforms such as Amazon Mechanical Turk and CrowdFlower have emerged, providing a way to reach out to a large crowd of diverse workers for

© Springer International Publishing AG 2017
J.M. Jose et al. (Eds.): ECIR 2017, LNCS 10193, pp. 239–251, 2017.
DOI: 10.1007/978-3-319-56608-5_19

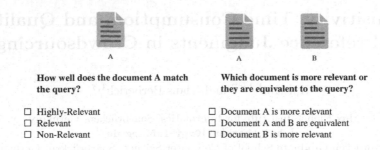

How well does the document A match the query?	**Which document is more relevant or they are equivalent to the query?**
☐ Highly-Relevant	☐ Document A is more relevant
☐ Relevant	☐ Document A and B are equivalent
☐ Non-Relevant	☐ Document B is more relevant

Fig. 1. Examples for graded (left) and preference judgments (right).

judgments. While inexpensive and scalable [1], judgments from those platforms are known to be of mixed quality [9,11,12].

Kazai et al. [10] demonstrated that preference judgments collected using crowdsourcing can be inexpensive yet high-quality. In their experiments preference judgments yielded better quality, getting close to the ones obtained from trained judges in terms of user satisfaction. Unfortunately, preference judgments are very expensive. To judge the relevance of n documents, $\mathcal{O}(n^2)$ preference judgments are needed, since pairs of documents have to be considered, whereas $\mathcal{O}(n)$ graded judgments suffice. Luckily, it has been shown that preference judgments are transitive [5,14] when collected from trained judges, which can be exploited to reduce their required number to $\mathcal{O}(n \log n)$. Whether transitivity still holds when preference judgments are collected using crowdsourcing is an open question as mentioned in [4]. In the aforementioned studies [5,14], trained judges stated their relative preference for all pairs of documents returned for a specific query. As a consequence, when considering a triple of documents, the same judge states relative preferences for all pairs of documents therein, making transitivity more of a matter of judges' self-consistency. When using crowdsourcing, in contrast, it is very unlikely that the same judge states relative preferences for all pairs of documents from a triple, given that workers typically only contribute a small fraction of work. Transitivity, if it exists, can thus only be a result of agreement among different judges. We examine whether transitivity holds when preference judgments are collected using crowdsourcing, when considering preference judgments aggregated from the stated preferences of multiple different judges.

Another difference between graded judgments and preference judgments, as reported by Carterette et al. [5], is that preference judgments tend to be less time consuming. Thus, in their experiments, trained judges took 40% less time to make individual preference judgments than to make individual graded judgments. We investigate whether this observation also holds when judgments are collected using crowdsourcing. If so, there is an opportunity to reduce cost by paying less for preference judgments.

Beyond that, previous works have considered different variants of preference judgments. When judges are asked to state strict preferences for two documents

d_1 and d_2, as done in [5,13,14], they can only indicate whether d_1 is preferred over d_2 ($d_1 \succ d_2$) or vice versa ($d_1 \prec d_2$). When asking for weak preferences, additional options are provided, allowing judges to state that the two documents are tied ($d_1 \sim d_2$) [10,15,16] or two documents are either equally relevant or equally non-relevant [4]. Allowing for ties is natural when judging search relevance, since it is unlikely that each of the possibly hundreds of returned documents has its own degree of relevance. We investigate whether weak preferences and strict preferences exhibit transitivity, and how they compare in terms of time consumption and quality.

Putting it together, we investigate the following research questions.

RQ1: *Do weak/strict preference judgments exhibit transitivity when collected using crowdsourcing?*

RQ2: *How do weak/strict preference judgments compare against graded judgments in terms of time consumption?*

RQ3: *Can weak/strict preference judgments collected using crowdsourcing replace judgments by trained judges?*

To answer these, we conduct an empirical study on CrowdFlower. Using topics and pooled documents from the TREC Web Track,[1] we collect graded judgments, weak preference judgments, and strict preference judgments. Akin to Carterette et al. [5], we examine transitivity by considering triples of documents. To analyze the time consumption for different kinds of judgments, our user interface is carefully instrumented to record the time that it takes judges to read documents and to make their judgment. We assess the inter-judge agreement for the different kinds of judgments and also examine to what extent they can replace judgments by trained judges from TREC.

We observe that transitivity holds over 90% for strict preference judgments collected using crowdsourcing; for weak preference judgments it only holds for about 75% of triples. In addition, we find that judges spend more time when asked for preference judgments than graded judgments in terms of total time consumption. Though time on making a single judgment is found to be lower for strict preference judgments. Finally, we see that preference judgments collected using crowdsourcing tend to show better agreement with TREC judges. Moreover, the agreement between strict preference judgments from crowdsourcing and judgments from TREC already match the agreement among trained judges reported from literature [5,10].

Organization. The rest of this paper is organized as follows. Section 2 recaps existing literature and puts our work in context. Following that, in Sect. 3, the setup of our empirical study is described. Section 4 describes its results and provides answers to the research questions stated above. Finally, in Sect. 5, we draw conclusions.

---

[1] http://trec.nist.gov/data/webmain.html.

2   Related Work

Preference judgments have been demonstrated as a better alternative to graded judgments, since there is no need to define graded levels [5], their higher inter-assessor agreement, and better quality [5,10,13]. Moreover, Carterette et al. [5] pointed out that preference judgments are less time-consuming than graded judgments.

Reduce the Number of Judgments in Preference Judgments. Assuming transitivity can dramatically bring down the number of judgments from $\mathcal{O}(n^2)$ to $\mathcal{O}(n \log n)$ [5]. To utilize transitivity, Rorvig [14] verified the transitivity among judgments from a group of undergraduates. Carterette et. al [5] tested transitivity among judgments from six trained judges, finding that the transitivity holds for 99% of document triples. Different from our settings, both works examined transitivity with trained judges, which is very different from the condition under crowdsourcing as indicated in Sect. 1. Moreover, both works applied strict preferences in their empirical studies. Meanwhile, follow-up works tend to extend this property to weak preferences [15]. Thus, in this work, we also examine the transitivity over weak preference judgments.

Weak Preferences Versus Strict Preferences. The choices between two kinds of preferences varied a lot among different works, even though some of them share similar motivations or research mythologies. Carterette et al. [5], Radinsky and Ailon [13] and Rorvig [14] employed strict preferences in their empirical studies for preference judgments. In the meantime, Kazai et al. [10] collected weak preference judgments from both trained judges and crowdsourcing workers to empirically explore the inter-assessors agreement and the agreement between the collected judgments and the real user satisfactions. Song et al. [15] introduced an option "same as" in the judging interface and assumed transitivity over the weak preferences in their QUICK-SORT-JUDGE method. Additionally, Zhu and Carterette [16] collected weak preferences through a "no preference" option in their research over the user preference for the layout of search results. It seems to us that the strict and weak preferences are regarded as interchangeable in existing works. However whether preference judgments with and without tie are the same in terms of judgment quality and judgment efforts remains unclear.

Crowdsourcing for Relevance Judgments. Existing works examined different ways to collect judgments from crowdsourcing [7] and provided a proper model to follow in collecting graded judgments from crowdsourcing [1]. Alonso and Mizzaro [2,3] demonstrated that it is possible to replace graded judgments from TREC using crowdsourcing. Additionally, Kazai et al. [10] compared graded and preference judgments from both trained judges and crowdsourcing, highlighting that preference judgments are especially recommended for crowdsourcing, where judgment quality can be close to the one from trained judges. Different from this work, Kaizai et al. [10] measured agreement based on individual judgments, instead of aggregated ones. As mentioned in [3], it is the aggregated judgments that can be used in practice. Moreover, the judgment quality is measured in terms of the agreement relative to user clicks, whereas in our work, the measurement is based on judgments from TREC Web Track. Thereby, in the regards of empirical analysis over judgment quality, our work can be regarded as an extension to both [3,10].

3   Empirical Study on CrowdFlower

User Interface. We display queries together with their description from the TREC Web Track 2013 & 2014. Judges are instructed to consider both the query and its corresponding description as in Fig. 1. To help them understanding the topic, we also display a link to run the query against a commercial web search engine. When collecting preference judgments, we show the query and description together with two documents (A and B) and ask judges "Which document is more relevant to the query?". When collecting strict preferences, judges can choose between the options "Document A is more relevant" and "Document B is more relevant". A third option "Document A and B are equivalent" is added, when collecting weak preferences. When collecting graded judgments, the query and description are shown together with a single document. Judges are asked "How well does the document match the query?" and can click on one of the grades "Non-Relevant", "Relevant", and "Highly Relevant". In our instructions we include the same definitions of grades from TREC.

Quality Control. Unique tasks, in our case judgments, are referred as rows in CrowdFlower. Multiple rows are grouped into a page, which is the basic unit for payment and quality control. The major means to control quality are test questions, that is, rows with a known expected input from workers. Test questions can be used to run a qualification quiz, which workers have to complete upfront. By thresholding on their accuracy in the qualification quiz, unreliable workers can be filtered out. Moreover, test questions can be interspersed with rows to continuously control the quality of work. Workers can thus be banned once their accuracy on interspersed test questions drops below a threshold. The accuracy threshold is set as 0.7, following the default on CrowdFlower.

Job Settings. When collecting graded judgments a page consists of eleven judgments and a test question, and workers are paid $0.10 on successful completion. When collecting preference judgments, we pack eight document pairs and a test question into each page, and pay workers $0.15 on successful completion. The rationale behind the different pays is that workers receive the same amount of $0.0083 per document read. Each row is shown to workers until three trusted judgments have been collected.

Selection of Queries and Documents. Queries and documents are sampled from the TREC Web Track 2013 & 2014. From the 100 available queries, we sample a subset of twelve queries.[2] Among the sampled queries, one query is marked as ambiguous by TREC, five queries are marked as unambiguous (single), and six queries are faceted. The original relevance judgments contain up to six relevance

---

[2] Queries are available in http://trec.nist.gov/data/webmain.html.

levels: junk pages (*Junk*), non-relevant (*NRel*), relevant (*Rel*), highly relevant (*HRel*), key pages (*Key*), and navigational pages (*Nav*), corresponding to six graded levels, i.e., $-2, 0, 1, 2, 3, 4$. Different from other grades, *Nav* indicates a document can satisfy a navigational user intend, making the comparison relative to other documents depend on the information intent from the crowdsourcing judges. Hence, in our work, documents labeled *Nav* together with documents labeled *Junk* are removed. Due to the limit occurrences, documents labeled *Key* and *HRel* are both regarded as highly relevant. For each query we determine two sets of documents. Each set consists of twelve documents selected uniformly across graded levels, resulting in four documents per graded level. The first set is used to collect judgments; the second set serves to create test questions. When collecting graded judgments, the selected documents are directly used. To collect preference judgments, we generate for each query all 66 pairs of documents and randomly permute each document pair. Test questions are generated treating the judgments from TREC as ground truth. To ensure that workers on Crowd-Flower see the same documents as trained judges from TREC, we host copies of ClueWeb12[3] documents on our own web server.

Time Consumption. To monitor the time consumed for reading documents and making judgments, we proceed as follows. We record the timestamp when judges start reading the shown document(s). To display available options for judging, workers have to click on a button "Click here to judge", and we record the instant when this happens. As a last timestamp, we record when the worker selects the submitted option. In recording timestamps, the order of clicks from judges are restricted by customized JavaScript, e.g., "Click here to judge" button is enabled only after document(s) is (are) read. We thus end up with three timestamps, allowing us to estimate the reading time, as the time passed between the first two timestamps, and the judgment time, as the time passed between the last two.

Judgment Aggregation. As mentioned, at least three trusted judgments are collected for each row. One straightforward option to aggregate them is to use majority voting as suggested by Alonso and Mizzaro [1]. However, in our setting, a simple majority vote may not break ties, given that there are more than two options to choose from. As a remedy we use workers' accuracies, as measured on test questions, in a weighted majority voting to break ties.

4   Results

We now report the results of our empirical study. After giving some general statistics about the collected judgments, we answer our three research questions, by comparing different groups of judgments over the same set of test queries employing statistical instruments like Student's t-test.

---

[3] http://lemurproject.org/clueweb12/index.php.

Table 1. General statistics about judgments collected using crowdsourcing.

	Graded Judgments	Strict Preferences	Weak Preferences
Total Cost	$9.36	$62.10	$76.80
#Judgments	919	2,760	2,931
#Judgments per Judge	28.80	55.00	20.10
Fleiss' κ	0.170	0.498	0.253
Distribution of Judgments			
"Highly-Relevant" 28%	$A \succ B$ 51%		$A \succ B$ 30%
"Relevant" 43%	$A \prec B$ 49%		$A \prec B$ 31%
"Non-Relevant" 29%	-		$A \sim B$ 39%

4.1 General Statistics

Table 1 summarizes general statistics about the collected judgments. The collected judgments are publicly available.[4]

Inter-Judge Agreement. Similar to [3], Fleiss' κ is computed over each query and average Fleiss' κ among all queries is reported in Table 1. To put our results in context, we merge "Highly-Relevant" with "Relevant" and convert graded to binary judgments, ending up with Fleiss' $\kappa = 0.269$, which is close to 0.195 reported in [3]. In addition, Kazai et al. [10] reported Fleiss' $\kappa = 0.24$ (cf. Table 2 PC (e) therein) among weak preference judgments from crowdsourcing, which approximates 0.253 in our work. We further conduct two-tailed Student's t-test in between the three kinds of judgments over different queries. The p-value between strict preferences and graded judgments is smaller than 0.001; between weak preferences and graded judgments is 0.314; whereas it is 0.005 between the two kinds of preference judgments. It can be seen that the judges achieve better inter-agreement for strict preferences than for the others, meanwhile there is no significant difference between weak preferences and graded judgments. This aligns with the observations from [5], that strict preferences exhibit higher inter-judges agreement. The introduction of "ties" reduces the inter-judges agreement, which might due to more options are available.

4.2 RQ1: Transitivity

In this section, transitivity is examined over both strict and weak preference judgments. Different from in [5] and in [14], we investigate transitivity based on aggregated judgments. This is because the aggregated judgments are the ultimate outcome from crowdsourcing, and also because, as mentioned in Sect. 1, triples from a single judge are too few over individual queries to lead to any conclusions. The results per query are summarized in Table 2. It can be seen that over strict preferences, transitivity holds for 96% triples on average, and the number is between 91% and 100% over individual query. This number is close to the transitivity reported in [5], where average transitivity is 99% and

---

[4] http://people.mpi-inf.mpg.de/~khui/data/ecir17empirical.

Table 2. Transitivity over aggregated judgments. The ratio of transitive triples out of triples in different types is reported. The numbers in the bracket are the number of transitive triples divides the total number of triples.

Query	Strict Preferences	Weak Preferences			
	asymTran	asymTran	s2aTran	s2sTran	Overall
216	100% ($^{220}/_{220}$)	96% ($^{78}/_{81}$)	89% ($^{90}/_{101}$)	8% ($^{3}/_{38}$)	78% ($^{171}/_{220}$)
222	99% ($^{218}/_{220}$)	100% ($^{40}/_{40}$)	98% ($^{117}/_{120}$)	50% ($^{30}/_{60}$)	85% ($^{187}/_{220}$)
226	96% ($^{210}/_{220}$)	98% ($^{39}/_{40}$)	87% ($^{86}/_{99}$)	24% ($^{19}/_{81}$)	66% ($^{144}/_{220}$)
231	98% ($^{216}/_{220}$)	100% ($^{17}/_{17}$)	95% ($^{107}/_{113}$)	30% ($^{27}/_{90}$)	69% ($^{151}/_{220}$)
241	99% ($^{217}/_{220}$)	100% ($^{52}/_{52}$)	99% ($^{112}/_{113}$)	31% ($^{17}/_{55}$)	82% ($^{181}/_{220}$)
253	91% ($^{199}/_{220}$)	100% ($^{24}/_{24}$)	86% ($^{66}/_{77}$)	38% ($^{45}/_{119}$)	61% ($^{135}/_{220}$)
254	99% ($^{218}/_{220}$)	100% ($^{39}/_{39}$)	97% ($^{105}/_{108}$)	36% ($^{26}/_{73}$)	77% ($^{170}/_{220}$)
257	95% ($^{208}/_{220}$)	97% ($^{88}/_{91}$)	86% ($^{87}/_{101}$)	11% ($^{3}/_{28}$)	81% ($^{178}/_{220}$)
266	94% ($^{207}/_{220}$)	100% ($^{69}/_{69}$)	98% ($^{123}/_{125}$)	50% ($^{13}/_{26}$)	93% ($^{205}/_{220}$)
277	91% ($^{200}/_{220}$)	100% ($^{37}/_{37}$)	82% ($^{109}/_{133}$)	54% ($^{27}/_{50}$)	79% ($^{173}/_{220}$)
280	99% ($^{218}/_{220}$)	100% ($^{37}/_{37}$)	85% ($^{85}/_{100}$)	29% ($^{24}/_{83}$)	66% ($^{146}/_{220}$)
296	96% ($^{212}/_{220}$)	90% ($^{35}/_{39}$)	77% ($^{82}/_{106}$)	19% ($^{14}/_{75}$)	60% ($^{131}/_{220}$)
Avg.	96% ($^{212}/_{220}$)	98% ($^{46}/_{47}$)	90% ($^{98}/_{108}$)	32% ($^{21}/_{65}$)	75% ($^{164}/_{220}$)

at least 98% triples from a single judge are transitive. Meanwhile, for weak preferences, this number is only 75% on average, and the minimum percentage is 60% from query 296, indicating that transitivity does not hold in general. To explore the reasons, we further decompose transitivity according to different types of preferences within unique document triples. In particular, the "better than" and "worse than" options are referred as asymmetric relationships and the "tie" option is referred as symmetric relationship [8]. The transitivity can be categorized as: *asymTran*, which lies among asymmetric relationships (no tie judgment in a triple); *s2aTran*, which lies in between symmetric and asymmetric relationships (only one tie judgment in a triple) and *s2sTran*, which lies among symmetric relationships (at least two tie judgments in a triple). Over each query, the 220 triples are thereby categorized according to the three types on which transitive percentage is computed. From Table 2, we can see that *asymTran* holds even better than in strict preferences, meanwhile, *s2aTran* holds for 90% on average. However, over *s2sTran*, the transitivity does not hold anymore: the transitive percentage drops to 32% on average.

Answer to RQ1: We conclude that transitivity holds for over 90% aggregated strict preference judgments. For weak preference judgments, though, transitivity only holds among non-tie judgments (*asymTran*) and in between tie and non-tie judgments (*s2aTran*). Thus, given judgments $d_1 \sim d_2$ and $d_2 \sim d_3$, we can not infer $d_1 \sim d_3$. We can see that, in terms of transitivity, weak and strict preference judgments exhibit differently, and extra caution must be taken when assuming transitivity when collecting weak preferences via crowdsourcing.

4.3   RQ2: Time Consumption

We compare time consumption for different kinds of judgments looking both at total time, which includes the time for reading document(s) and judgment

Table 3. Average time consumption (in seconds) and quartiles over twelve queries.

Time consumption		Average	25th percentile	Median	75th percentile
Graded judgments	Judgment	2.60	1.37	1.52	1.82
	Total	24.24	11.73	19.55	28.88
Strict preferences	Judgment	1.79	1.24	1.37	1.58
	Total	34.17	17.84	25.28	40.98
Weak preferences	Judgment	2.07	1.40	1.57	1.91
	Total	32.43	15.77	24.57	39.10

time. The results are summarized in Table 3, based on aggregated statistics from twelve queries. For judgment time, it can be seen that judges spend least time with strict preferences. The p-values from two-tailed Student's t-tests between the three kinds of judgments are as follows. P-value equals 0.055 between strict preferences and graded judgments, equals 0.196, between weak preferences and graded judgments, and equals 0.100 between the two kinds of preference judgments. We can conclude that judges are slightly but noticeably faster in making judgments with strict preferences than in making the other two kinds of judgments, meanwhile the difference between the time consumption with weak preferences and with graded judgments is insignificant. As for total time, Table 3 demonstrates that judges are significantly faster in finishing single graded judgments after considering reading time, with p-value from two-tailed Student's t-test is less than 0.001 relative to both preference judgments. However, there is no significant difference for judges with weak and strict preferences – the corresponding p-value equals 0.168.

Answer to RQ2: Judges are faster in making strict preference judgments. When considering total time, judges need to read two documents in preference judgments, making total time consumption higher. Moreover, when comparing the two kinds of preference judgments, judges take significantly less time with strict preferences, meanwhile there is no difference in terms of total time consumption. Compared with [5,14], time consumption is measured among judges from crowdsourcing, who are with more diverse reading and judging ability and might be less skillful than trained judges. Actually, the web pages being judged require more than 20 s on average to read, making reading time dominate the total time consumption.

4.4   RQ3: Quality

We compare the quality of three kinds of judgment collected via crowdsourcing in terms of their agreement with judgments from TREC (qrel). We employ both percentage agreement, which counts the agreed judgments and divides it by the number of total judgments, and Cohen's κ as in [3], and use the latter for two-tailed Student's t-tests. When evaluating preference judgments from crowdsourcing, judgments from TREC are first converted to preference judgments, by

Table 4. Agreement between graded judgments from crowdsourcing (columns) and TREC (rows).

TREC	Non-Relevant	Relevant	Highly-Relevant	#Total
Non-Relevant	56.3%	39.6%	4.1%	48
Relevant	14.6%	54.2%	31.2%	48
Highly-Relevant	14.6%	37.5%	47.9%	48

Table 5. Agreement between preference judgments from crowdsourcing (columns) and the one inferred from TREC judgments (rows).

(a) strict preferences

TREC	$A \prec B$	$A \succ B$	#Total
$A \prec B$	83.0%	17.0%	282
$A \sim B$	46.8%	53.2%	216
$A \succ B$	20.4%	79.6%	294

(b) weak preferences

TREC	$A \prec B$	$A \sim B$	$A \succ B$	#Total
$A \prec B$	62.8%	30.9%	6.3%	285
$A \sim B$	17.6%	59.7%	22.7%	216
$A \succ B$	7.6%	32.0%	60.5%	291

comparing labels over two documents, resulting in "better than", "worse than" or "tie". The percentage agreement over three kinds of judgment relative to judgments from TREC are summarized in Tables 4 and 5, where the percentage is normalized per row. To put our results in context, we first measure agreement based on binary judgments, by merging the grades *Relevant* and *High-Relevant* in both TREC judgments and graded judgments from crowdsourcing. In [3], percentage agreement equals 77% and Cohen's $\kappa = 0.478$, relative to judgments from TREC-7 and TREC-8. Meanwhile we obtain 75.7% and Cohen's $\kappa = 0.43$ – slightly lower values. We argue that is due to the document collections in use: ClueWeb12, used in our work, consists of web pages which are more diverse and noisy, making it harder to judge; whereas disk 4 & 5 used in TREC-7 and TREC-8 consist of cleaner articles.[5] When using three grades, graded judgments from crowdsourcing achieve 52.8% and Cohen's $\kappa = 0.292$ relative to judgments from TREC. And the percentage agreement is 59.1% and Cohen's $\kappa = 0.358$ for strict preferences, whereas for weak preferences the numbers are 61% and 0.419 respectively. Compared with graded judgments from crowdsourcing, the corresponding p-values from paired sample t-tests over Cohen's κ among queries are 0.259 and 0.052, indicating weak preference judgments agree with TREC judgments better.

Note that, however, for documents with the same grade in TREC a tie is inferred, whereas strict preferences do not permit tie judgments. From Table 5(a), it can be seen that 216 document pairs are inferred as tied, where agreement is zero for strict preferences currently. To mitigate this mismatch, in line with [5], tie judgments in inferred preference judgments are redistributed as "A is better" or "B is better". In this redistribution, an agreement is assumed, coined as *aar*. In other words, the 216 document pairs that are inferred as tied in Table 5(a) are redistributed so that 216 × *aar* random pairs are assigned with the same

---

[5] http://trec.nist.gov/data/docs_eng.html.

judgments as in collected strict preference judgments. The logic behind this is that the ground-truth strict preferences over these inferred ties are unknown and we need to assume an agreement over them to make strict preference judgments comparable. Thereby, two groups of agreement are reported for strict preference judgments at assumed agreement rates $aar = 50\%$ and 80%, respectively corresponding to random agreement and the average agreement under non-tie situations (average of 83% and 79.6% in Table 5(a)). Without influencing comparison results, graded judgments from crowdsourcing are also converted to preference judgments, making three kinds of judgments from crowdsourcing more comparable. In Table 6, it can be seen that Cohen's $\kappa = 0.530$ for strict preferences when assuming $aar = 50\%$, and the value for weak preferences is 0.419. Both preference judgments agree with TREC significantly better than graded judgments, with p-values from paired sample t-test equal 0.001 and 0.015 respectively. We further compare Cohen's κ from strict preferences ($aar = 50\%$) with the one from weak preferences, getting p-value from paired sample t-test equals 0.004, indicating strict preference judgments agree with judgments from TREC significantly better than weak preferences.

Answer to RQ3: From Table 6, it can be seen that agreement from strict preferences under $aar = 50\%$ and weak preferences are 88% and 49% higher than the collected graded judgments in terms of Cohen's κ. We further compare this agreement relative to TREC with the agreement among trained judges reported in literature, similar to [3]. Intuitively, if agreement between judgments from crowdsourcing and from TREC is comparable to the one among trained judges,

Table 6. Percentage agreement and Cohen's κ between inferred preference judgments from TREC and three kinds of judgments collected via crowdsourcing. For the column of strict preferences, tie judgments in the inferred judgments from TREC are redistributed by assuming different agreement rates. Results under $aar = 50\%$ and 80% are reported.

Query	Strict preferences				Weak preferences		Graded judgments	
	Break tie $aar = 50\%$		Break tie $aar = 80\%$					
	Percentage	Cohen's κ	Percentage	Cohen's κ	Percentage	Cohen's κ	Percentage	Cohen's κ
216	77%	0.594	85%	0.710	65%	0.466	53%	0.269
222	76%	0.569	83%	0.680	59%	0.391	65%	0.474
226	77%	0.589	79%	0.611	65%	0.473	62%	0.386
231	70%	0.494	83%	0.686	53%	0.310	65%	0.435
241	74%	0.557	83%	0.689	70%	0.543	59%	0.386
253	74%	0.533	77%	0.576	49%	0.248	36%	0.044
254	80%	0.649	91%	0.821	71%	0.573	65%	0.471
257	73%	0.529	83%	0.680	64%	0.445	61%	0.380
266	70%	0.459	73%	0.500	73%	0.588	38%	0.048
277	68%	0.397	70%	0.417	50%	0.261	38%	0.075
280	65%	0.389	74%	0.510	56%	0.345	44%	0.193
296	77%	0.601	85%	0.715	59%	0.386	50%	0.224
Avg	74%	0.530	81%	0.633	61%	0.419	53%	0.282

we can conclude that judgments from crowdsourcing are good enough to replace those from trained judges. Carterette et al. [5] reported agreement among six trained judges over preference judgments, and the percentage agreement is 74.5% (cf. Table 2(a) therein), whereas in our work agreement for strict preferences are 74% under $aar = 50\%$, and 81% under $aar = 80\%$. Kazai et al. [10] reported that Fleiss' κ among trained judges over preference judgments is 0.54 (cf. Table 2 PE (e) therein). Thus, we recompute the agreement between strict preference judgments and judgments from TREC in terms of Fleiss' κ, and get $\kappa = 0.504$ under $aar = 50\%$ and 0.637 under $aar = 80\%$. Note that strict preferences are collected in [5] and weak preferences are employed in [10]. Since the difference of these two kinds of preference judgments when collected from trained judges is unclear, we regard them the same. We can conclude that the agreement between strict preferences collected via crowdsourcing and TREC are comparable to the one among trained judges. Moreover, compared with strict preference judgments, we can conclude that judgment quality in crowdsourcing is significantly degraded when using weak preferences.

As reported in [2,3], we also observe judges from crowdsourcing can sometimes point out mistakes in TREC judgments. In total, we found around 20 such documents, especially via "test questions", by examining documents (or document pairs) that receive majority judgments opposing to the judgments from TREC. One example is clueweb12-0013wb-31-22050 and clueweb12-0806wb-32-26209 for query 280, "view my internet history". The former is labeled as "Highly-Relevant" and the latter is labeled as "Relevant" in qrel. However, none of them is relevant: the first page is a comprehensive list about history of internet & W3C, and the second page is a question on a forum about how to clean part of ones' browsing history.

4.5 Discussion

It has been demonstrated that weak and strict preferences are different in all three regards. To investigate the reasons, we reduce the number of options in weak preferences by merging "tie" with "A is better", merging "tie" with "B is better" or merging the two non-tie options, measuring the agreements among judges, getting Fleiss' $\kappa = 0.247, 0.266$, and 0.073 respectively. The corresponding p-values from two-tailed Student's t-tests relative to the one with three options are $0.913, 0.718$, and less than 0.001. It can be seen that judges tend to disagree more when making choices between ties and non-ties judgments. Put differently, the threshold to make a non-tie judgment is ambiguous and is varied among different judges. This implies that the tie option actually makes the judgments more complicated, namely, judges have to firstly determine whether the difference is large enough to be non-tied before judging the preferences.

5 Conclusion

In this work, we use crowdsourcing to collect graded judgments and two kinds of preference judgments. In terms of judgment quality, the three kinds of judgments can be sorted as follows, graded judgments < weak < strict preference judgments. Moreover, our position for tie judgments is: it can be used but must be with more cautions when collected via crowdsourcing, especially when attempting to assume transitivity.

References

1. Alonso, O., Baeza-Yates, R.: Design and implementation of relevance assessments using crowdsourcing. In: Clough, P., Foley, C., Gurrin, C., Jones, G.J.F., Kraaij, W., Lee, H., Mudoch, V. (eds.) ECIR 2011. LNCS, vol. 6611, pp. 153–164. Springer, Heidelberg (2011). doi:10.1007/978-3-642-20161-5_16
2. Alonso, O., Mizzaro, S.: Can we get rid of TREC assessors? Using mechanical turk for relevance assessment. In: SIGIR 2009 Workshop on the Future of IR Evaluation (2009)
3. Alonso, O., Mizzaro, S.: Using crowdsourcing for TREC relevance assessment. Inf. Process. Manag. 48(6), 1053–1066 (2012)
4. Bashir, M., Anderton, J., Wu, J., Golbus, P.B., Pavlu, V., Aslam, J.A.: A document rating system for preference judgements. In: SIGIR 2013 (2013)
5. Carterette, B., Bennett, P.N., Chickering, D.M., Dumais, S.T.: Here or there: preference judgments for relevance. In: Macdonald, C., Ounis, I., Plachouras, V., Ruthven, I., White, R.W. (eds.) ECIR 2008. LNCS, vol. 4956, pp. 16–27. Springer, Heidelberg (2008). doi:10.1007/978-3-540-78646-7_5
6. Cleverdon, C.: The cranfield tests on index language devices. In: Aslib Proceedings, vol. 19 (1967)
7. Grady, C., Lease, M.: Crowdsourcing document relevance assessment with mechanical turk. In: NAACL HLT 2010 Workshop on Creating Speech and Language Data with Amazon's Mechanical Turk (2010)
8. Hansson, S.O., Grne-Yanoff, T.: Preferences. In: Zalta, E.N. (ed.) The Stanford Encyclopedia of Philosophy (2012)
9. Kazai, G.: In search of quality in crowdsourcing for search engine evaluation. In: Clough, P., Foley, C., Gurrin, C., Jones, G.J.F., Kraaij, W., Lee, H., Mudoch, V. (eds.) ECIR 2011. LNCS, vol. 6611, pp. 165–176. Springer, Heidelberg (2011). doi:10.1007/978-3-642-20161-5_17
10. Kazai, G., Yilmaz, E., Craswell, N., Tahaghoghi, S.M.: User intent and assessor disagreement in web search evaluation. In: CIKM 2013 (2013)
11. Moshfeghi, Y., Huertas-Rosero, A.F., Jose, J.M.: Identifying careless workers in crowdsourcing platforms: a game theory approach. In: SIGIR 2016 (2016)
12. Moshfeghi, Y., Rosero, A.F.H., Jose, J.M.: A game-theory approach for effective crowdsource-based relevance assessment. ACM Trans. Intell. Syst. Technol. 7(4) (2016)
13. Radinsky, K., Ailon, N.: Ranking from pairs and triplets: information quality, evaluation methods and query complexity. In: WSDM 2011 (2011)
14. Rorvig, M.E.: The simple scalability of documents. J. Am. Soc. Inf. Sci. 41(8), 590–598 (1990)
15. Song, R., Guo, Q., Zhang, R., Xin, G., Wen, J.R., Yu, Y., Hon, H.W.: Select-the-best-ones: a new way to judge relative relevance. Inf. Process. Manag. 47(1), 37–52 (2011)
16. Zhu, D., Carterette, B.: An analysis of assessor behavior in crowdsourced preference judgments. In: SIGIR 2010 Workshop on Crowdsourcing for Search Evaluation (2010)

Exploring Time-Sensitive Variational Bayesian Inference LDA for Social Media Data

Anjie Fang[(✉)], Craig Macdonald, Iadh Ounis, Philip Habel, and Xiao Yang

University of Glasgow, Glasgow, UK
a.fang.1@research.gla.ac.uk,
{craig.macdonald,iadh.ounis,philip.habel,xiao.yang}@glasgow.ac.uk

Abstract. There is considerable interest among both researchers and the mass public in understanding the topics of discussion on social media as they occur over time. Scholars have thoroughly analysed sampling-based topic modelling approaches for various text corpora including social media; however, another LDA topic modelling implementation— Variational Bayesian (VB)—has not been well studied, despite its known efficiency and its adaptability to the volume and dynamics of social media data. In this paper, we examine the performance of the VB-based topic modelling approach for producing coherent topics, and further, we extend the VB approach by proposing a novel time-sensitive Variational Bayesian implementation, denoted as TVB. Our newly proposed TVB approach incorporates time so as to increase the quality of the generated topics. Using a Twitter dataset covering 8 events, our empirical results show that the coherence of the topics in our TVB model is improved by the integration of time. In particular, through a user study, we find that our TVB approach generates less mixed topics than state-of-the-art topic modelling approaches. Moreover, our proposed TVB approach can more accurately estimate topical trends, making it particularly suitable to assist end-users in tracking emerging topics on social media.

1 Introduction

Perhaps the greatest technological change over the past decade has been the advent and growth of social media. Yet despite social media's ubiquity, scholars still wrestle with the appropriate tools for best capturing the topics of discussion conveyed over these platforms [1–3]. To this end, researchers have employed various topic modelling approaches [1,4–8], e.g. Latent Dirichlet Allocation (LDA), but these efforts have proved challenging, as models applied to social media data can produce topics that are mixed and lack coherence, and are generally difficult to interpret [1]. To deal with the short nature of social media posts, LDA enhancement methods such as single topic assignment [1,9,10] and document pooling [2,11] have been proposed to improve the coherence of the generated topics within the sampling-based topic modelling approaches. However, another LDA implementation, the Variational Bayesian (VB)-based topic modelling approach, has not been well studied on social media posts. As the VB approach has been

J.M. Jose et al. (Eds.): ECIR 2017, LNCS 10193, pp. 252–265, 2017.
DOI: 10.1007/978-3-319-56608-5_20

shown to be more efficient for large datasets [6,12], it can be argued that VB can better handle the increasing volume and dynamicity of social media data.

It has been previously shown that the time dimension of documents (e.g. news articles) can help a topic modelling approach to provide more valuable information [7,8,13], for example, capturing the topic changes or topical trends over time. Apart from these additional benefits, we argue that distinguishing topical word usage over time can also help to generate more coherent and less mixed topics, thereby assisting the end-users in interpreting discussions on social media. We propose a time-sensitive VB (TVB) approach for social media data that embraces the time dimension of social media data. We extend the traditional VB approach by incorporating a Beta distribution, which is reported to fit various patterns [14]. The employed Beta continuous distribution is used to represent each topic's volume over time, i.e. the topical trend, similar to what has been used in [7]. However, we notice that time could have a negative bias on the topic inference when a Beta distribution does not fit the topics' trends. To solve this problem, we introduce a balance parameter to alleviate the bias of time.

To evaluate the performance of the proposed TVB approach, we create a ground truth Twitter dataset covering 8 large events. We evaluate our TVB approach together with several baselines from the literature (e.g. Twitter LDA (TLDA) [1], the Topic Over Time approach (TOT) [7]) in terms of topical coherence, the extent to which the generated topics are mixed, or the estimation errors of the topical trends. Our empirical results suggest that incorporating the time dimension does indeed help to enhance the coherence of the topics generated by the TVB approach compared with the traditional VB and sampling approaches. Moreover, we show that our TVB model can outperform the state-of-the-art LDA enhancement approaches (i.e. TLDA and TOT) in generating less mixed topics. This conclusion is further supported by conducting a user study to validate the results of the quantitative evaluation. Finally, we compare our TVB approach with the TOT approach when estimating the topical trends. We find that our proposed TVB model better estimates the topical trends.

The contributions of this paper are three-fold: (1) we study the VB approach and develop its enhancement for social media, (2) we propose a time-sensitive TVB approach by integrating the time dimension in the modelling process and (3) we show the advantages of the TVB approach in generating better quality topics and estimating more accurate topical trends.

The rest of this paper is organised as follow: Sect. 2 provides basic background on two LDA implementations, i.e. sampling & VB approaches, followed by a description of related work in Sect. 3. We describe our TVB approach in Sect. 4. Following that, we describe our dataset in Sect. 5 and the experimental setup in Sect. 6. The results are shown and discussed in Sect. 7. Finally, we provide concluding remarks in Sect. 8.

2   Two LDA Implementations: Sampling and VB Approaches

In topic modelling approaches, a topic k is represented by a distribution β_k (k is the topic index and K is the number of topics) over N terms drawn from a Dirichlet prior η, where N is the size of the vocabulary. A document in a corpus w is represented by $w_d = \{w_{d,1}, ..., w_{d,i}, ..., w_{d,N}\}$ (d is the document index and D is the number of documents in w) and has a topic belief distribution θ_d drawn from the Dirichlet prior α. A document w_d is associated with topic assignment $z_d = \{z_{d,1}, ..., z_{d,i}, ..., z_{d,N}\}$. The sampling approach [4,5], which is based on a Markov Chain Monte Carlo sampling, estimates the real posterior distributions (e.g. β_k & θ_d). In a typical sampling approach, such as the collapsed Gibbs sampling, each word is assigned a topic according to Eq. (1) in order to construct a Markov Chain on latent topics, where $n_{-(d,i),k}^{w_{d,i}}$ is the number of $w_{d,i}$ occurring in topic k and $n_{-(d,i),j}^d$ is the number of words from document w_d occurring in topic k not including the current one. After a number of iterations, β ($\{\beta_1, .., \beta_K\}$) and θ ($\{\theta_1, .., \theta_D\}$) can be estimated from the converged Markov Chain.

$$p(z_{d,i} = k|z_{-(d,i)}, w) = \frac{n_{-(d,i),k}^{w_{d,i}} + \eta}{n_{-(d,i),k} + N\eta} \times (n_{-(d,i),j}^d + \alpha) \tag{1}$$

The VB approach [6,12] approximates the variational distribution by minimising the distance from the true distribution. Specifically, an expectation maximization (EM) algorithm is used to maximise the lower bound of the log-likelihood of all documents, which equivalently minimises the distance between the variational distribution and the true posterior distribution. In the E step of EM, the variational Dirichlet prior γ_d of all documents are optimised together with $\phi_{D \times N \times K}$, which represents the words' topic belief within documents. In the M step of EM, $\phi_{D \times N \times K}$ is used to update the variational Dirichlet prior $\lambda_{K \times N}$ of β. The parameters' optimisation formulas in the EM algorithm are shown in Eq. (2). Finally, β and θ can be obtained when the lower bound converges. Importantly, since the VB approach does not have the topic assignment step, the single topic assignment strategy (mentioned in the introduction and discussed further in Sect. 3) cannot be applied. The main advantage of the VB approach is that the lower bound converges much more quickly than the sampling approach especially on large datasets [6,12]. Moreover, the VB approach can be intuitively implemented in parallel since the updates of γ_d & ϕ_d among documents do not impact each other, while the sampling approach cannot be easily parallelised as it is intrinsically sequential [15]. Because of the increasing volume of social media data and its dynamicity, it could be argued that the VB approach offers various advantages for those interested in interpreting discussions on social media as events transpire. In the next section, we review a number of existing methods, which aim to improve the quality of topic models and/or integrate the time dimension in the topic model.

$$\phi_{d,i,k} \propto exp\{E[log\beta_{k,i}] + E[log\theta_{d,k}]\}, \ \gamma_{d,k} = \alpha + \sum_{i,k} \phi_{d,i,k}, \ \lambda_{k,i} = \eta + \sum_{d,i,k} \phi_{d,i,k} \tag{2}$$

3   Related Work

Three methods are mainly used in the literature to adapt topic modelling approach for short social media posts: (1) A post is assigned to a single topic under the assumption that a post represents a single topic. This method was used in Twitter LDA as proposed by Zhao et al. [1] and later applied in [9,10]. Indeed, this method brings more coherent words for a topic. However, we argue that this method can generate multi-theme topics[1] since the underlying assumption cannot be upheld in all situations. For example, the same words can be used across multiple topics. Assigning all words in a tweet to a single topic could increase word overlaps and thus result in mixed topics. (2) Multiple posts are combined into a virtual document [2,11], also known as the pooling strategy (e.g. tweets from a single user are combined into a single document [2]). The pooling method can increase the number and occurrence of words, which makes it easier to apply a topic modelling approach. (3) Topical words are connected using word representations (e.g. word embedding). Sridhar [16] improved the topical coherence by applying soft clustering over word representations in a topic model. Nguyen et al. [17] introduced an additional word topic belief distribution calculated using word representations in the sampling approaches. Li et al. [18] assigned the semantically similar words under the same topic. All of these approaches improved the topical coherence by connecting similar words in order to overcome the shortness of posts on social media. We do not deploy the single topic assignment method in our approach since it cannot be applied in the VB approach, as mentioned in Sect. 2. Given that the central aim of this paper is to integrate time to the VB approach, we do not adopt the pooling method in our modelling process.

Early work on time-sensitive topic modelling by Blei and Lafferty [13] was based on a Markov assumption that the topic parameters are in a sequential structure over time. Later on, Blei et al. [19] used Brownian motion to estimate the topical evolutions over time. The proposed model was claimed to have a better predictive perplexity. However, these state-space models did not integrate the timestamps of documents in the generative process. Assuming that the topic proportion changes over time, Wang and McCallum [7] proposed a non-Markov topic model (TOT) using a Beta continuous distribution, which was reported to generate more interpretable topics and trends. Their work is based on a sampling approach, in which the timestamps of documents are incorporated in the generative process without considering time dependency for topics or words. Another recent work from [8] leveraged a time-dependent function to capture topical dynamics.

Although the sampling approach is still the preferred choice in analysing social media data, the advantages of the VB approach for a large corpus should not be ignored. For example, Hoffman et al. [20] and Braun and McAuliffe [12] recently proposed a VB-based solution to quickly inference a large number of documents. In this paper, we offer a solution to apply an enhanced VB approach (TVB) for

---

[1] A mixed topic contains keywords pertaining to multiple different topic themes.

social media data, which incorporates time in the topic modelling process. Our TVB approach is based on the same assumption as the TOT approach but is implemented using VB. In the next section, we introduce our TVB approach and elaborate further the differences between the TOT and TVB approaches.

4   Integrating the Time Dimension in the VB Approach

Our proposed TVB approach extends the traditional VB approach by integrating the time dimension of social media data. In this section, we explain how we implement the EM algorithm in our proposed TVB approach and compare it with the traditional VB and TOT approaches. To integrate the timestamps of social media posts, we deploy a continuous probability distribution τ for each topic. This time distribution τ_k represents the proportion of topic k over time. Theoretically, any continuous distribution can be used to simulate the topic proportion over time. However, to better estimate topical trends, the continuous distribution has to approximate the real topical trends. Indeed, recently, the Beta distribution has drawn a lot of attentions for accommodating a variety of shapes given an x-axis interval [14]. Therefore, we choose to use a Beta distribution since it can more accurately fit the various shapes of topical trends. Next, we describe the generative process and the EM implementation of our TVB approach.

Generative Process. Similar to the traditional LDA generative process, each word $w_{d,i}$ in a document d is assigned a topic assignment $z_{d,i}$ according to θ_d, where i is the word index. Since words (w) in social media posts are associated with timestamps (t), in the TVB approach, a pair $(w_{d,i}, t_{d,i})$ is drawn from $\beta_{z_{d,i}}$ and $\tau_{z_{d,i}}$, respectively, where a Beta distribution τ_k is parametrised by two shape parameters, ρ_k^1 and ρ_k^2. A similar strategy was previously applied in a time-sensitive sampling approach [7]. The process is defined as follows:

$$z_{d,i}|\theta_d \sim Dirichlet(\alpha), \quad w_{d,i}|z_{d,i}, \beta_{z_{d,i}} \sim Dirichlet(\eta), \quad t_{d,i}|z_{d,i}, \tau_{z_{d,i}} \sim Beta(\rho^1_{z_{d,i}}, \rho^2_{z_{d,i}})$$

EM Implementation of the TVB Approach. The core part of a variational inference is to minimise the distance between the variational distributions $q(\theta_d|\gamma_d)$ & $q(\beta_k|\lambda_k)$ and the two true posterior distributions $p(\theta_d|\alpha)$ & $q(\beta_k|\eta)$, i.e. maximising the lower bound of a document log-likelihood $p(w, t|\alpha, \eta)$ shown in Eq. (3). The right part of the equation is the lower bound of all documents, L. Commonly, the derivative of L is taken over parameters (γ, ϕ, λ) and thus the parameter optimisation formulas can be obtained by maximising the lower bound L. To achieve this, we first decompose all the items in L.

$$log\ p(w, t|\alpha, \eta) \geq L(w, t, \gamma, \lambda) = \sum_d E_q[log\ p(w_d, t_d, z_d, \theta_d, \beta, \tau|\alpha, \eta)]$$
$$= \sum_d (E_q[log\ p(w_d|z_d, \beta)] + E_q[log\ p(z_d|\theta_d)] + E_q[log\ p(\theta_d|\alpha)] + E_q[log\ p(\beta|\eta)] \quad (3)$$
$$+ E_q[log\ p(t_d|z_d, \tau)] - E_q[q(\theta_d, z_d, \beta)])$$

The sixth item of the lower bound L, the log-expectation of joint variational probability, is decomposed as shown in Eq. (4). These decomposed items together with the first five items in L can be expanded by leveraging the properties of the Dirichlet and Beta distributions. Finally, we have the expanded L shown in Eq. (5), where B is the Beta function and Γ is the Gamma function.

$$E_q[q(\theta_d, z_d, \beta)] = \sum_k E_q[log\ q(\theta_{d,k}|\gamma_{d,k})] + \sum_i E_q[log\ q(z_{d,i}|\phi_{i,k})] + \sum_{i,k} E_q[log\ q(\beta_{k,i}|\lambda_{k,i})] \quad (4)$$

$$
\begin{aligned}
L(w, t, \gamma, \lambda) = &\sum_d (\sum_{i,k} \phi_{d,i,k}\ E_q[log\ \beta_{k,i}] + \sum_{i,k} \phi_{d,i,k} E_q[log\ \theta_{d,k}] \\
&+ log\Gamma(K\alpha) - Klog\Gamma(\alpha) + \sum_k (\alpha - 1)E_q[log\theta_{d,k}] \\
&+ log\Gamma(\sum_{i,k} \eta) - \sum_{i,k} log\Gamma(\eta) + \sum_{i,k} (\eta - 1)E_q[log\ \beta_{k,i}] \\
&+ \sum_{i,k} \phi_{d,i,k}((\rho_k^1 - 1)log\ t_{d,i} + (\rho_k^2 - 1)log\ (1 - t_{d,i})) - \sum_k (\sum_i \phi_{d,i,k}\ log\ B(\rho_k^1, \rho_k^2)) \\
&- log\Gamma(\sum_k \gamma_k) + \sum_k log\Gamma(\gamma_k) - \sum_k (\gamma_k - 1)E_q[log\theta_{d,k}] \\
&- \sum_{i,k} \phi_{d,i,k}log\phi_{d,i,k} - log\Gamma(\sum_{i,k} \lambda_{k,i}) + \sum_{i,k} log\Gamma(\lambda_{k,i}) - \sum_{i,k} (\lambda_{k,i} - 1)E_q[log\ \beta_{k,i}])
\end{aligned}
\quad (5)
$$

To maximise L, we first optimise $\phi_{d,i,k}$ by setting $\frac{\partial L_{\phi_{d,i,k}}}{\partial \phi_{d,i,k}} = 0$ and obtain the $\phi_{d,i,k}$ optimisation formula shown in Eq. (6). Compared with the traditional VB approach, the third item in Eq. (6), the time statistics, is the additional feature we add to incorporate timestamps. Intuitively, the time statistics can have a direct impact on the term topic belief $\phi_{d,i,k}$. If a word $w_{d,i}$ is highly used in topic k at a time point t, $\phi_{d,i,k}$ is likely to be promoted if a post has the word $w_{d,i}$ with a timestamp t. However, the estimated time distribution may not always fit a topic's trend well. A drifted time distribution could give a negative bias on $\phi_{d,i,k}$. To solve this problem, we introduce a balance parameter δ, to control the impact of the time statistics on $\phi_{d,i,k}$ and alleviate such bias. Note that the influence of time in the TOT approach cannot be adjusted, e.g. through a δ parameter. Similar to $\phi_{d,i,k}$, we obtain the optimisation formula of γ and λ (shown in Eq. (7)) by setting their derivative of L to zero.

$$\phi_{d,i,k} \propto exp(E_q[log\beta_{k,i}] + E_q[log\theta_{d,k}] + \delta((\rho_k^1 - 1)log\ t_{d,i} + (\rho_k^2 - 1)log\ (1 - t_{d,i}) - log\ B(\rho_k^1, \rho_k^2))) \quad (6)$$

$$\gamma_{d,i} = \alpha + \sum_{i,k} \phi_{d,i,k}, \quad \lambda_{k,i} = \eta + \sum_{d,i,k} \phi_{d,i,k} \quad (7)$$

Meanwhile, to maximise L, we can also take the partial derivative with respect to the parameters of Beta distribution, ρ_k^1/ρ_k^2. Actually, this step is equivalent to maximising the likelihood of the timestamps in topics. By optimising ρ_k^1/ρ_k^2, we also obtain the estimated topical trends. Taking the derivative to zero, we obtain the optimisation formula of ρ_k^1/ρ_k^2 shown in Eq. (8). Since the Digamma function (ψ, log-derivative of Γ) is involved in the optimisation equation, it is impossible to calculate ρ_k^1/ρ_k^2 directly. In our TVB approach, we estimate ρ_k^1/ρ_k^2 using a parameter optimisation algorithm and we set their initial values following [21]. Note that, while we use EM to estimate ρ_k^1/ρ_k^2, the method of moment [7] is used in the TOT approach. In summary, in the iterative EM algorithm, we update ϕ and γ for each document (social media post) in the E step. In the M step, λ and ρ_k^1/ρ_k^2 are updated using the statistics information

---

Algorithm 1. Our TVB approach for Latent Dirichlet Allocation.

Initialize $\lambda_{N \times K}$, $\gamma_{D \times K}$
while L *not converges* **do**
$\quad$ E step:
$\quad$ **for** $d < D$ **do**
$\quad\quad$ **repeat**
$\quad\quad\quad$ **for** $i < N^d$ *&* $k < K$ **do**
$\quad\quad\quad\quad$ $\phi_{d,i,k} \propto exp(E_q[log\beta_{k,i}] + E_q[log\theta_{d,k}]$
$\quad\quad\quad\quad$ $+\delta((\rho_k^1 - 1)log\ t_{d,i} + (\rho_k^2 - 1)log\ (1 - t_{d,i}) - log\ B(\rho_k^1, \rho_k^2)))$
$\quad\quad\quad\quad$ $\gamma_{d,k} = \alpha + \Sigma_{i,k}\ \phi_{d,i,k}$
$\quad\quad$ **until** γ_d *converges*;
$\quad$ M step:
$\quad\quad$ $\psi(\rho_k^1) - \psi(\rho_k^1 - \rho_k^2) = \dfrac{\Sigma_{d,i,k}\ \phi_{d,i,k} log\ t_{d,i}}{\Sigma_{d,i,k}\ \phi_{d,i,k}}, \psi(\rho_k^2) - \psi(\rho_k^1 - \rho_k^2) = \dfrac{\Sigma_{d,i,k}\ \phi_{d,i,k} log\ (1 - t_{d,i})}{\Sigma_{d,i,k}\ \phi_{d,i,k}}$
$\quad\quad$ $\lambda_{k,i} = \eta + \Sigma_{d,i,k}\ \phi_{d,i,k}, \forall i \in N$

(ϕ) from all posts. At the same time, all the timestamps are taken into account to estimate ρ_k^1/ρ_k^2. Algorithm 1 shows the EM algorithm in our TVB approach.

$$\psi(\rho_k^1) - \psi(\rho_k^1 - \rho_k^2) = \frac{\Sigma_{d,i,k}\ \phi_{d,i,k} log\ t_{d,i}}{\Sigma_{d,i,k}\ \phi_{d,i,k}}, \ \psi(\rho_k^2) - \psi(\rho_k^1 - \rho_k^2) = \frac{\Sigma_{d,i,k}\ \phi_{d,i,k} log\ (1 - t_{d,i})}{\Sigma_{d,i,k}\ \phi_{d,i,k}} \quad (8)$$

5   Ground Truth Datasets

To evaluate our proposed TVB approach together with the existing topic modelling approaches, we create a Twitter dataset containing 8 selected popular hashtag-events that occurred in July and August 2016. This dataset was collected using Twitter API by searching for 8 hashtags: #gopconvention, #teamgb, #badminton, #gameofthrone, #juno, #nba, #pokemongo and #theresamay. For each hashtag-event, we randomly sample 2,000 tweets, hence we obtain a Twitter dataset containing 16,000 tweets. Such a balanced dataset has several advantages: (1) The reasonable size (16 K) of the Twitter corpus allows for the efficient conduct of our experiments, i.e. all approaches can quickly converge; (2) We avoid generating dominant and duplicated topics, thereby focusing the evaluation on the quality and coherence of the topics; (3) These predefined hashtags provide readily usable ground-truth labels, i.e. each hashtag-event is associated with the top 10 used words in its corresponding tweets. These labels of the 8 hashtag-events are used to match a generated topic with a hashtag-event. This enables us to evaluate how close the estimated topical trend to its real trend (further details are given in Sect. 6); (4) This ground truth dataset allows humans to more effectively examine the generated topics and to conduct a user study described in Sect. 7. Indeed, since this dataset contains a limited number of topics, it is more feasible for human interpreters to evaluate all the generated topics of a given topic model in the conducted user study. In the next section, we explain how we apply various topic modelling approaches on this dataset and the used metrics.

6   Experimental Setup

We compare our new proposed TVB approach to 4 baselines from the literature, namely TOT [7], TLDA [1], and the traditional sampling (Gibbs) [4] and VB [6] approaches. In particular, the TOT approach is included since it is the most closely related work that integrates the time dimension into the topic modelling process. We use 3 different metrics to evaluate the quality of the generated topic models: (1) the topical coherence, (2) the degree to which the topics are mixed and (3) the topical trends estimation error. In the following, we explain the experimental setup used for the topic modelling approaches and each of the metrics.

Topic Modelling Setup. For all approaches (Gibbs, TLDA, TOT, VB & TVB), η is set to 0.01 according to [4,5]. We do not follow the traditional setting for α ($\alpha = 50/K$), and set instead α to 0.4 for all approaches in our experiments, since in other separate preliminary experiments we noticed that a smaller α helps to generate topics with higher coherence for short texts. The number of topics is set to 10, which is slightly higher than the real number of topic (8 in our dataset corresponding to 8 hashtags) because a slightly higher number of topics assures that all hashtag-events can be extracted. As our Twitter dataset is not very big and contains distinguishable topics, all approaches can converge fast. Hence, for the sampling approaches (Gibbs, TLDA and TOT), we set the maximum number of iterations to 50. For the traditional VB and our proposed TVB approaches, we set the number of iterations to 10 as the VB approaches converge more quickly. Each experiment for each approach is repeated 10 times in order to conduct statistical significance. In TLDA, a document contains several tweets from a single Twitter user. However, most of users in our Twitter dataset have only one tweet. Hence, we create a virtual Twitter user by assigning 5 random tweets to this user. For all the other approaches, a document represents a single tweet.

Metric 1: Coherence Metric. A coherence metric is used to evaluate whether a generated topic is interpretable by humans. A higher score indicates that the topic is easier to understand. Following [22,23], we use a word embedding (WE) representations-based coherence metric to evaluate the coherence of the generated topics, which has been reported to have a high agreement with human judgments. In order to capture the semantic similarity of the latest hashtags and Twitter handle names, we train our WE model using 200 million English tweets posted from 08/2015 to 08/2016. The obtained WE model has 5 million tokens. We use the average coherence (**Aver**) to evaluate all topics in a topic model. Meanwhile, we also examine the top $2/7^2$ most coherent topics in a model for more effective coherence evaluation, i.e. **C@2** & **C@7** metrics, following to [24].

Metric 2: Topics Mixing Degree. A generated topic can be a mixture of several topic themes (multi-theme topics). The coherence score is calculated by

---

[2] Considering that the number of topics is 10, the top 2 and 7 most coherent topics are reasonable choices for a comprehensive coherence evaluation.

averaging the similarity of each two words in the top 10 words of a generated topic. Consider that if a topic contains two topic themes, as long as the coherence of words under a theme in this topic is high, the coherence metric can still yield a higher coherence. Although this multi-theme topic is interpretable by humans, a user expects to see the generated topics only containing a single topic theme. Therefore, it is necessary to identify the multi-theme topics. Since the generated multi-theme topics often contain the same topic theme, these multi-theme topics can be similar to each other. Thus, to quantify the extent to which a given topic in a model is mixed (MD), we use Eq. (9) to calculate the topic similarity in the entire topic model (containing K topics). The higher the similarity, the more likely that the model has more multi-theme topics. A similar methodology is used in [25] to identify the background topics.

$$MD(\boldsymbol{\beta}) = \sum_k \sum_{k'} cosine(\beta_k, \beta_{k'})/|\boldsymbol{\beta}|^2 \tag{9}$$

Metric 3: Topical Trends Estimate Error. Both the TOT and our TVB approaches estimate the topical trends. To evaluate the topical trends over time, we calculate the distance/error between the real topic trends and the estimated topical trends (using the Beta distributions in the TOT/TVB models). The error is calculated using the method shown in Eq. (10), where $PDF_k(t)$ is the probability density of the real timestamps of topics, which is obtained through the ground truth Twitter dataset. The ERR score ranges from 0 to 2. The generated topics are matched to the ground-truth topics if the top 10 words of a generated topic have at least 3^3 same words in the top 10 words of a hashtag event.

$$ERR(\tau) = \frac{\sum_k \int_0^1 |\tau_k(t) - PDF_k(t)| dt}{K} \tag{10}$$

7   Results

In this section, we analyse our experimental results shown in Table 1. The listed scores are the average scores of 10 models generated by each approach with respect to the 3 types of metrics (described in Sect. 6). For the coherence metrics (Aver, C@2 & C@7), a higher score means more coherent topics, whereas lower scores for the MD and ERR metrics indicate higher quality models. The subscript indicates whether a given approach is significantly[4] better than the other one. For example, the average coherence score (Aver) of TVB$_{\delta=0.8}$ (δ is the balance parameter) is significantly better than that of the VB approach, indicating that TVB generates topics with higher coherence than VB. To help understand the topical trends, we randomly choose one TOT and one TVB models and list their estimated topical trends together with the real trends in Fig. 1. Next, we will first analyse the results in terms of topical coherence and topical mixing degree. Then, we discuss the performances of the approaches in estimating topical trends.

---

[3] 3 mutual words in the top 10 words is a reasonable minimum number to indicate a similar topic.

[4] p-values (p < 0.05) are calculated by the t-test using 10 models of each two approaches.

Table 1. The topic coherence, mixing degree and topic trends estimation error.

Models	Coherence			MD	ERR
	Aver	C@2	C@7		
Gibbs (G)	0.154	0.204	0.168	$0.051_{W,T}$	×
TLDA (W)	$0.177_{G,V,T,T'}$	$0.248_{G,V,T,T'}$	$0.198_{G,V,T,T'}$	0.102_T	×
VB (V)	0.151	0.201	0.165	$0.049_{W,T}$	×
TOT (T)	$0.160_{G,V}$	0.205	0.175_V	0.149	1.358
TVB(T'), $\delta = 0.4$	0.152	0.202	0.165	$0.043_{W,T}$	1.211_T
TVB(T'), $\delta = 0.6$	0.153	0.204	0.166	$0.042_{W,T}$	1.256_T
TVB(T'), $\delta = 0.8$	0.158_V	$0.221_{G,V,T}$	0.174_V	$0.047_{W,T}$	1.206_T
TVB(T'), $\delta = 1.0$	0.156_V	0.209	0.170	$0.055_{W,T}$	1.168_T

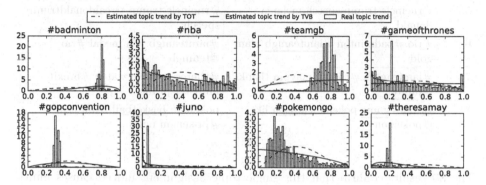

Fig. 1. The real and estimated topical trends, where x-axis and y-axis represent the timeline and density probability, respectively.

Topical Coherence and Topical Mixing Degree. Our experiments involve two types of topic modelling approaches: Sampling & VB approaches (shown as G, W, T & V, T' in Table 1) and two topic enhanced methods: single topic assignment (W) and incorporation of the time dimension (T, T'). First, for the topical coherence, it is clear that the single topic assignment, TLDA (W), significantly outperforms all of the other approaches. However, we can still see the positive impact of the time dimension in improving the coherence of models in both the TVB and TOT approaches. For example, the Aver coherence score of TOT is better than that of the Gibbs and VB approaches. In particular, for C@2, the TVB models outperform all of the other approaches, except the TLDA models with $\delta = 0.8$, while the TVB models with a lower/higher δ (T' with $\delta = 0.4, 0.6/1.0$) do not. This indicates that alleviating the bias of the time statistics (described in Sect. 4) helps to generate topics with a higher coherence. In terms of the MD metric, the TLDA models have higher mixing scores indicating they have more multi-theme topics. As argued in Sect. 3, aggressively assigning all words in a tweet under the same topic theme can result in multi-theme topics.

Table 2. Topic samples from TLDA and $TVB_{\delta=0.8}$ models, where the underlined words have a different topic theme from the others in a topic. Note that, we present a human with the top 10 words of a topic in our user study. We list the top 5 words for each topic in this table due to space limitations.

Topic	TLDA	$TVB_{\delta=0.8}$
1	#rio #badminton #olympics #iamteamgb wei	#badminton #rio #mas #olympics wei chong
2	#jupiter #juno @nasa orbit @nasajuno	#juno burn engine complete unlock #jupiter
3	#nbasummer nba #basketball @nba basketball	nba #basketball sign wire basketball
4	@gameofthrones #emmys season outstanding	thanks @gameofthrones #iamteamgb #emmys
5	#rncincle trump speech melania donald	#rncincle trump @realdonaldtrump speech
6	#rio #badminton #iamteamgb team gold	#iamteamgb win medal #rio @teamgb
7	#iamteamgb #theresamaypm thanks #jupiter	#theresamaypm watch #brexit minister prime
8	thrones game pokemon season like #pokemon	pokemon basketball team usa #pokemon news

The MD results confirm that this single topic assignment indeed causes multi-theme topics, which is the reason why the TLDA models exhibit a very high mixing degree. Besides, we notice that the TOT models have the highest topical mixing degree. This might be caused by the strong time bias in the sampling approach. Consider that if two topics have similar trends (topical proportions over time, e.g. #nba and #pokemongo in Fig. 1), it is likely that these two topics would mix, and thus it causes the generation of multi-theme topics in the TOT models. In this situation, reducing the importance of the time statistics by the balance parameter δ is equally increasing the importance of the words statistics (the first two items in Eq. 6), hence avoiding the negative bias of time statistics.

To verify that our generated TVB models have less multi-theme topics than TOT & TLDA, we also conduct a user study to compare the mixing degree of their generated topic models. Since the MD scores of the TOT models are significantly higher than those of the TLDA models, we choose to compare the mixing degree between the $TVB_{\delta=0.8}$ and TLDA models using human judgements. If the users confirm that the TVB approach generates less multi-theme topics than TLDA, it is reasonable to conclude that the TVB approach generates less multi-theme topics than TOT. In our user study, we ask 8 expert end-users whether a given topic contains multiple themes. Specifically, both the TVB and TLDA approaches generate 10 models. We pair these 20 models randomly and generate 10 pairs, where each pair has one model from TVB and another one from TLDA.

For each pair, we present a human with all the generated topics of the 2 models. The human is asked to identify all of the multi-theme topics from 2 given models (10 topics per model). A model in a pair is preferred (i.e. obtains a vote), if a human finds less multi-theme topics in this model pair. Each pair gets 3 judgements from 3 different humans. An approach gets a credit if its model in a pair obtains the majority votes out of 3. In the end, among the 10 pairs, our TVB approach gets 7 credits while the TLDA approach gets 2 credits, expect that 3 humans do not have agreement on one pair out of the 10. This user study confirms the results we obtained from MD that our TVB approach generates less multi-theme topics. We list two topic examples of our TLDA and $TVB_{\delta=0.8}$ models in Table 2. Both models generate human interpretable topics. However, we can see more multi-theme topics in the TLDA models, such as "badminton" (topic 1), "teamgb" (topic 6), "theresamaypm" (topic 7) and "pokemon" (topic 8), while the TVB model has less multi-theme topics: "gameofthrone" (topic 4) and "pokemon" (topic 8). In fact, it is easy to mix the topics "theresamaypm" and "teamgb" since they are all popular topics in the UK, and it is possible that the word usage in these two topics is similar. However, the topical trends of these two topics are not similar: "theresamaypm" was popular around 11/07/2016 when Theresa May became the new UK Prime Minister, while "teamgb" was highly discussed during the Olympic Games (from 05/08/2016 to 21/08/2016) (See the topical trends in Fig. 1). Our TVB approach can identify these different topical trends by integrating time.

Topical Trends Estimation Error. Both the TOT and TVB approaches estimate topical trends. The ERR metric indicates the distance between the real topical trends and the estimated ones (smaller distances are better). The ERR scores in Table 1 suggest that our TVB approach generates significantly more accurate topical trends than the TOT approach. The main reason is that the TOT approach has a very high mixing degree (see Table 1), which shows that it has more multi-theme topics similar to the TLDA approach. It could be difficult to match the real topics with the generated topics (explained in Sect. 6), and thus the multi-theme topics in the TOT model result in less accurate topical trends. Unlike the TOT/TLDA approaches, our TVB model has less multi-theme topics, which results in a more accurate estimation of the topical trends. In Fig. 1, both chosen models have duplicated topics, which are #badminton & #juno and #gameofthrone & #juno in the TOT and TVB models, respectively. Since the TOT models have more multi-theme topics, it is difficult to match the generated topics with the real ones. For example, the topic theme #nba is mixed with #pokemongo in the TOT model. As a result, the estimated trend of TOT for #nba is not accurate. Although both the TOT and TVB models do not exactly fit the real topical trends using Beta distributions, it is still clear that the estimated trends from the TVB model are closer to the real trends than those of the TOT model as illustrated in Fig. 1.

Apart from the used three metrics, it is worth recalling that all the VB and TVB models in our experiments are obtained by setting the maximum iteration to 10, while it is set to 50 for the TLDA, TOT & Gibbs models (see Sect. 6).

Using less iterations, our TVB approach can still provide very competitive results, which indicates its advantage in terms of convergence speed.

8 Conclusions

In this paper, we proposed a time-sensitive Variational Bayesian (TVB) topic modelling approach to improve the quality of generated topics and to estimate topical trends by leveraging the time dimension of social media posts. Our proposed TVB approach, extending the traditional Variational Bayesian approach, employed a Beta distribution to integrate time, where the time statistics were controlled by a balance parameter to alleviate bias. Through experimentation over a ground truth Twitter dataset covering 8 hashtag events, we showed that the time dimension helps to generate more coherent topics in our models with the set balance parameter. Backed by a user study, we find that our TVB approach generated less mixed topics compared with two state-of-the-art baselines. Moreover, our TVB approach can more accurately estimate the topical trends of social media posts.

References

1. Zhao, W.X., Jiang, J., Weng, J., He, J., Lim, E.-P., Yan, H., Li, X.: Comparing twitter and traditional media using topic models. In: Clough, P., Foley, C., Gurrin, C., Jones, G.J.F., Kraaij, W., Lee, H., Mudoch, V. (eds.) ECIR 2011. LNCS, vol. 6611, pp. 338–349. Springer, Heidelberg (2011). doi:10.1007/978-3-642-20161-5_34
2. Mehrotra, R., Sanner, S., Buntine, W., Xie, L.: Improving LDA topic models for microblogs via tweet pooling and automatic labeling. In: Proceedings of the SIGIR (2013)
3. Fang, A., Ounis, I., Habel, P., Macdonald, C., Limsopatham, N.: Topic-centric classification of Twitter user's political orientation. In: Proceedings of the SIGIR (2015)
4. Blei, D.M., Ng, A.Y., Jordan, M.I.: Latent Dirichlet allocation. J. Mach. Learn. Res. **3**, 993–1022 (2003)
5. Griffiths, T.L., Steyvers, M.: Finding scientific topics. Proc. Nat. Acad. Sci. **101**, 5228–5235 (2004)
6. Blei, D.M., Jordan, M.I.: Variational methods for the Dirichlet process. In: Proceedings of the ICML (2004)
7. Wang, X., McCallum, A.: Topics over time: a non-Markov continuous-time model of topical trends. In: Proceedings of SIGKDD (2006)
8. Hong, L., Dom, B., Gurumurthy, S., Tsioutsiouliklis, K.: A time-dependent topic model for multiple text streams. In: Proceedings of the SIGKDD (2011)
9. Cheng, X., Yan, X., Lan, Y., Guo, J.: BTM: topic modeling over short texts. In: Proceedings of the TKDE (2014)
10. Yan, X., Guo, J., Lan, Y., Xu, J., Cheng, X.: A probabilistic model for bursty topic discovery in microblogs. In: Proceedings of the AAAI (2015)
11. Weng, J., Lim, E.P., Jiang, J., He, Q.: TwitterRank: finding topic-sensitive influential twitterers. In: Proceedings of the ICWSM (2010)

12. Braun, M., McAuliffe, J.: Variational inference for large-scale models of discrete choice. J. Am. Stat. Assoc. **105**, 324–335 (2010)
13. Blei, D.M., Lafferty, J.D.: Dynamic topic models. In: Proceedings of the ICML (2006)
14. Guolo, A., Varin, C., et al.: Beta regression for time series analysis of bounded data. Ann. Appl. Stat. **8**, 74–88 (2014)
15. Asuncion, A., Welling, M., Smyth, P., Teh, Y.W.: On smoothing and inference for topic models. In: Proceedings of the CUAI, pp. 27–34 (2009)
16. Sridhar, V.K.R.: Unsupervised topic modeling for short texts using distributed representations of words. In: Proceedings of the NAACL-HLT (2015)
17. Nguyen, D.Q., Billingsley, R., Du, L., Johnson, M.: Improving topic models with latent feature word representations. In: Proceedings of the TACL (2015)
18. Li, C., Wang, H., Zhang, Z., Sun, A., Ma, Z.: Topic modeling for short texts with auxiliary word embeddings. In: Proceedings of the SIGIR (2016)
19. Wang, C., Blei, D., Heckerman, D.: Continuous time dynamic topic models. In: Proceeding of the CUAI (2008)
20. Hoffman, M., Bach, F.R., Blei, D.M.: Online learning for latent Dirichlet allocation. In: Proceedings of the NIPS (2010)
21. Johnson, N.L., Kotz, S., Balakrishnan, N.: Beta distributions. In: Continuous Univariate Distributions, vol. 2 (1995)
22. Fang, A., Macdonald, C., Ounis, I., Habel, P.: Topics in tweets: a user study of topic coherence metrics for Twitter data. In: Ferro, N., Crestani, F., Moens, M.-F., Mothe, J., Silvestri, F., Nunzio, G.M., Hauff, C., Silvello, G. (eds.) ECIR 2016. LNCS, vol. 9626, pp. 492–504. Springer, Cham (2016). doi:10.1007/978-3-319-30671-1_36
23. Fang, A., Macdonald, C., Ounis, I., Habel, P.: Using word embedding to evaluate the coherence of topics from twitter data. In: Proceedings of the SIGIR (2016)
24. Fang, A., Macdonald, C., Ounis, I., Habel, P.: Examining the coherence of the top ranked tweet topics. In: Proceedings of the SIGIR (2016)
25. AlSumait, L., Barbará, D., Gentle, J., Domeniconi, C.: Topic significance ranking of LDA generative models. In: Buntine, W., Grobelnik, M., Mladenić, D., Shawe-Taylor, J. (eds.) ECML PKDD 2009. LNCS (LNAI), vol. 5781, pp. 67–82. Springer, Heidelberg (2009). doi:10.1007/978-3-642-04180-8_22

E-Government and the Digital Divide: A Study of English-as-a-Second-Language Users' Information Behaviour

David Brazier and Morgan Harvey[✉]

Northumbria University, Newcastle upon Tyne, UK
{d.brazier,morgan.harvey}@northumbria.ac.uk

Abstract. Internet-based technologies are increasingly used by organisations and governments to offer services to consumers and the public in a quick and efficient manner, removing the need for face-to-face conversations and human advisors. Despite their obvious benefits for most users, these online systems may present barriers of access to certain groups in society which may lead to information poverty.

In this study we consider the information behaviour of ten ESL (English as a Second Language) participants as they conduct four search tasks designed to reflect actual information seeking situations. Our results suggest that, despite a perception that they have a good understanding of English, they often choose documents that are only partially or tangentially relevant. There were significant differences in the behaviour of participants given their perceived confidence in using English to perform search tasks. Those who were confident took riskier strategies and were less thorough, leading to them bookmarking a larger proportion of non-relevant documents. The results of this work have potentially profound repercussions for how e-government services are provided and how second-language speakers are assisted in their use of these.

1 Introduction

The rise of Internet-based technologies has transformed the ways in which society interacts with and utilises information. The proliferation of electronic services (e-services) in the wake of this has provided companies and users with 24 h access to a wide array of useful facilities. Although somewhat behind the consumer market, governments are slowly embracing this change, as seen by the UK government's 'digital by default' initiative, where a number of public services have been digitised and moved online [7].

The average user, who has access to the Internet via a plethora of technologies, and views it as an everyday tool, could and should see these changes as no great burden. For those in society, however, who are not aware of the existence of these changes; accepting or comfortable of the changes; or adept in the use of such technologies this raises concerns around the barriers that may be erected and the risk this poses of segregating service users, especially those in vulnerable groups [8].

© Springer International Publishing AG 2017
J.M. Jose et al. (Eds.): ECIR 2017, LNCS 10193, pp. 266–277, 2017.
DOI: 10.1007/978-3-319-56608-5_21

One such group, and the focus of this paper, are those who use English as a second language (ESL). Searching for information when you don't know the subject, the technical terminology or where to look are just some of the challenges that we can face in our day to day lives. However, consider the same challenges but when the user is in a unfamiliar culture or setting; has an incomplete grasp of the language, where even the slightest variations in meaning can significantly change search results; and may lack the awareness and experience of the reliability of sources being sought after and used.

There are many scenarios in which people may rely on face-to-face encounters with staff or the knowledge and experience of their friends, family or community members when their own was lacking. However, in instances where these social groups may not be attainable or their knowledge and experience is also deficient can have dire repercussions for those members of society who already face significant barriers [12]. In an effort to address these concerns the UK government are currently running both standard services (face-to-face, postal and over the telephone) and digital services simultaneously but it is not inconceivable that standard services will (eventually) be phased out.

Before such an eventuality, all attempts must be made to try and to facilitate those most at risk of being segregated and to understand any issues they may have in accessing and using these services. It is with this in mind that this paper seeks to identify the current information seeking behaviours of ESL users when performing e-government-related tasks, and to ascertain where and why issues arise during this process, in an effort to aid that facilitation.

2   Related Work

Recent research has considered the problems encountered by certain user groups when faced with the need to access and understand important sources of information. In a study of refugees trying to access e-government services, Lloyd et al. [12] found that the information poverty they experience was a product of the social exclusion of the participants as a result of barriers e-services can erect. Such information poverty can lead to "limited support networks, [an] inability to access the labour market, alienation from society and poorer educational outcomes" [15]. The study suggested that many issues stem from the fact that the community receiving the refugees has pre-existing assumptions about how information is best disseminated, assumptions which may not hold true for the refugees themselves.

A number of studies have looked at multilingual IR, with a rising number of studies investigating the information seeking behaviour of users in relation to language proficiency, notably when English is the foreign language [2,3,13]. Research by [2] focused on web content (or lack thereof) in the user's native language and the impact it had on user satisfaction through cognitive load. They also discussed the need for context (e.g. domain knowledge of the user group) when considering information quality and multilingualism. In related findings, Marlow et al. [13] showed that the perceived and actual difficulty of tasks increased as familiarity with the second language decreased.

The IT literacy and abilities of the user are important factors when it comes to searching in a foreign language [5]. When compared to searching in their native tongue, users required significantly more time, submitted more query reformulations and viewed/assessed a greater number of websites. Those with only an intermediate grasp of the foreign language struggled with query reformulation, although they did not find identification of relevant results quite so difficult.

In contrast to this, Bogers et al. [3] focused on the differences in behaviour between native and non-native English speakers when searching for books. Although the study found non-natives spent more time on task than native speakers, it revealed very little difference between the two groups in relation to the number of queries, query length, depth of results inspection or books added to the bookbag. They surmised this could be as a result of their users' experience in searching for books in English and having acceptable foreign language skills.

Jozsa et al. [10] considered the differences between native language and foreign language information seeking tasks. From the study they identified two different search strategies: superficial/cursory and in-depth, with little difference in performance when applying an in-depth strategy in both languages. Alternatively, it was found the superficial strategy in a foreign language performed much worse than in the native language. One explanation being that foreign language users, who may not be as familiar with nuances in the language, may miss signs of such subtle markers when not thoroughly analysing a document and thus may gather a lower quality result set.

Extant research in the field goes some way to disclosing the search behaviours of multilingual searchers but focuses predominately on "why" rather than "how" they search and how well they perform [14]. It is with this in mind that this paper looks to identify the ways in which second language users approach a number of important search tasks, the problems they face in doing so and which factors impact on these behaviours.

3   Methodology

To investigate the behaviour of (and ascertain the performance of) second-language speakers of English we required a number of contextually-relevant search tasks; the kinds that such users might need to conduct in a foreign country. Identification of the types of services ESL users would use was made by involving 7 international PhD students at a UK university, 6 of whom also took part in the study described in this work, in a pilot study. The students were recruited and tasked with identifying: what a government service entailed; which would be deemed most useful (to the group); and the information needs, information sources and skills that would be required to successfully utilise the e-service. From this information, four search tasks were designed to reflect realistic information seeking situations in an attempt to be relevant and a more interesting search experience for the participants [6]. The tasks in full are:

1. *Your friend from Peru and their family (2 members) are coming to visit you for 6 months while you are in the UK. Develop a list of instructions to help them apply for the necessary visas.*

2. *A family member is coming to the UK to live and wants information on housing. They have heard there are a number of options and have asked you for advice. Identify the options available to them and recommend which they should choose. Give reasons to support your recommendation.*

3. *Your friend just got back from a trip abroad and suddenly developed a high fever. A dry cough, chills, and breathing difficulties soon followed. What could they have? They have no insurance and have asked your advice on what to do. Provide them with recommended actions.*

4. *Your elderly neighbours have heard about the UK government's 'digital by default' initiative and are concerned about whether this will affect them and their friends at the local community centre. They have asked you to find out more about it. Use your best judgement to highlight what would impact them with reasons for your choices.*

All 4 tasks were assessed by the participants as being relevant or partially relevant to them with task 1 receiving the highest average relevance score and task 4 the lowest.

3.1 Procedure

Use of log data is common in IR studies but is limited when establishing context in the use of the search facility [1]. Therefore, in this study we take a mixed methods approach [5], utilising recorded observation to gather a rich data set of user searching strategies. Although perhaps viewed as being a poorer method than that of direct observation [9], it was preferred due to a desire to obtain both anecdotal and self-reported assessment of behaviour as well as query log information from the sessions. To further complement this data a semi-structured discussion was conducted post study with thematic analysis used to help explore participant experience and their search patterns.

Each session followed the same process of each participant filling in a demographic questionnaire which collected information on their area of study; age; gender; nationality; language(s) spoken and proficiency; IT use; search engine use in English and their native tongue; search engine competency and preference and their own awareness of existing UK governmental services.

Each task was allotted ten minutes for completion with up to five minutes for the participants to read the task and complete the pre- and post-questionnaires. This allowed for no more than one hour in total. Tasks were distributed to participants using a Latin square design to account for task fatigue and potential learning effects [11]. Prior to beginning each task, participants were asked to fill in a pre-task questionnaire [6] (see Table 1) to gauge their domain knowledge, interest in the topic and the perceived difficulty of the task using a five-point Likert scale where 1 is "Not at all" and 5 is "Very".

The participants were asked to read the description of each task and search for relevant documents/sources, bookmarking any website deemed relevant as they went. At the end of each task the participant was also required to complete a post-task questionnaire (again on a 5-point Likert scale), examples of which can be seen in Table 2.

Table 1. Pre-task questions.

Q1	I have searched about this topic before
Q2	I know about this topic
Q3	I am interested in this topic
Q4	It will be difficult to find information about this topic

Table 2. Selected post-task questions.

Q3	The task was relevant to me
Q6	I performed the task to the best of my ability
Q7	I found the task difficult
Q8	I'm confident the content I found satisfied the task
Q10	I'm confident I identified relevant websites
Q11	I'm confident in my ability to read the website content
Q12	I am confident in my ability to understand the content of the websites I visited

3.2 Participants

Participants for the study were sought via university mailing lists, paper adverts and face to face enquiry by the researcher, with the stipulation that contribution was voluntary. Face to face enquiry was the most successful with 70% recruited by this method.

The 10 study participants were all international PhD students from a large UK university who spoke ESL with 80% at a fluent level and 20% competent. All participants were from different countries across Europe (20%), Asia (70%) and Africa (10%) with a total of 11 languages spoken natively, and 15 languages in total up to a competent level. 40% of the participants were female with an average age of 31 ($SD = 3.56$) and 60% were male with an average age of 31.5 ($SD = 3.33$). Each was remunerated for their participation with a £10 Amazon voucher.

3.3 The Study

Morae Recorder was used to capture each participant's search session (as well as each post-study discussion) including audio and video, with four laptops available per session resulting in the maximum number of four participants per session. As a result there were a total of three sessions with two sessions of three students and one of four as dictated by participant and technical equipment availability. Using the Chrome browser, each participant was asked to use the Google Search Engine to start each task but were not limited to the search results page. Google Search was chosen over alternatives as it was the only search engine selected as being used by all participants, with the next best being Bing (20% of participants) and four other instances of an alternative search engine.

3.4 Measures and Metrics

Using Morae Manager each recorded session was manually tagged in order to establish several measures and metrics. Total task time was systematically logged when users clicked start task and end task; number of queries was the total number of times queries were submitted by participants or they clicked on a Google-related search link; length of query is the total number of terms per query; number of assisted terms are the number of query terms entered through the assistance functionality; length of time querying is the time from when they click on the search field up to the time they submit the query; time on the Search Engine Results Page (SERP) is calculated from when the SERP page is loaded to when the participant navigates away, either by SERP click or switching tab; link position is dependent on the listing number of the SERP link clicked assuming there are 10 links per SERP page; times bookmarked are the total number of documents bookmarked during that click-through session; The number of times in-site search and in-site link click are the total number per click-through session and the observational notes were key observations about participant search behaviour and are used to back up the quantitative nature of the log data.

To determine the relevance of the bookmarks logged by the participants, all bookmarks were assessed by two native English-speaking IR researchers [10] using a voting strategy and given scores between 1 and 4, where 1 is not relevant, 2 is tangentially relevant, 3 is partially relevant and 4 is totally relevant. Any bookmarks not assigned the same score by the two assessors in the first round were discussed and a single score was agreed, although this only occurred for a very small number of cases. To assess the classification of queries and reformulations, definitions after Chu et al. [5] were used and determined by the same researchers.

4 Results

In total participants bookmarked 267 pages, with an approximately equal split between governmental and non-governmental resources. Only 60.7% of the bookmarked pages were either partially or totally relevant, with 30.7% tangentially relevant and 8.5% non-relevant and there were no significant differences between the median number of bookmarks per task with each task receiving 8 or 9 per participant on average. Surprisingly, there was little difference in terms of relevance between governmental and non-governmental resources. This was mostly due to some participants bookmarking internal policy documents or documents discussing best practices for civil servant software engineers which were deemed to be only tangentially relevant and unlikely to be of help in the given contexts.

Performance. There was considerable variation in performance by different users with the bookmarks of five participants being only relevant in 50% or less of cases. There was also variation in the numbers of pages bookmarked; one participant only bookmarked 3 per task on average with the majority bookmarking

5 or more. Participant F acknowledged their limited bookmarks for the third task as in a real scenario they would not risk the health of another by self-diagnosing, and would instead only refer that person to a health professional in the first instance.

When viewing performance by task, the performance of participants was higher during task 1 as was the number of in-site links they clicked (Table 3). In post discussion it was noted that for those participants who found the visa section of the gov.uk website, which utilises a wizard to guide users, the process was simplified and informative and was the cause for the increased number of in-site clicks and performance. They also noted this facility had language selection, although no participant used an alternative language to English. This is found to confirm the notion that lower cognitive effort of the search option (in this case the wizard) can directly affect the preference of said search option [2]. It also highlights the point regarding the language of the in-site links' diminishing the multilinguality of the web. In this case when users were provided the option of other languages, they still preferred links in English.

Further insights into users predicted performance and their actual performance in this study were also documented in another paper [4].

Table 3. Table of performance by task and use of in-site link clicks.

Task	Average precision	In-site link clicks
1	0.91	2.37
2	0.70	0.46
3	0.54	0.24
4	0.38	0.37

Language Proficiency. The level of English proficiency was self-assessed [13] with 80% of the participants declaring themselves fluent and 20% competent with all participants using IT daily and formulating queries (on search engines) in English daily (90%) or a few times a week (10%). Half of the participants had used UK government e-services previously, 30% hadn't and 20% were unsure what was meant by the term. When judging their own abilities in formulating queries in English, identifying relevant search results and information on websites (all important skills for these tasks) five participants said they were "very confident" with the remaining five stating that they were less confident. Participants A and F were particularly lacking in confidence when it came to these abilities. It is worth noting that despite Participant F's low confidence, their self-assessed proficiency in the English Language was fluent. We will refer to the most confident group as *"confident"* and the other group as *"unconfident"* throughout the paper.

4.1   Reading Times

There was a considerable difference in reading times between participants, as shown in Table 4, which may be partially explained by the search strategies employed. Participant C in particular had a unique strategy for searching: in two tasks (2 and 4) they entered a URL directly (gov.uk in both instances), bypassing the search engine and using the in-site search functions and click-through to navigate the sites across only one tab. While in the other tasks (1 and 3) they only entered 1 query and again navigated through the use of in-site search and in-site click-through. This has direct influence on the amount of time spent on the SERP as well as the total time on documents, as seen in Table 4.

Table 4. Table of Time on documents vs. average precision of tasks.

UserID	Average precision	Time on documents
A	0.74	48.19
B	0.69	42.26
C	0.50	291.00
D	0.50	29.27
E	0.57	18.11
F	0.41	24.12
G	0.83	75.32
H	0.92	85.83
I	0.65	62.32
J	0.49	35.13

One might expect the amount of time needed to read documents to be inversely correlated with the reader's proficiency in finding relevant information in texts. Comparing the time spent reading documents by participants in the *confident* group with those in the *unconfident* one, we find that the former spent significantly less time (Wilcoxon signed rank test; $p = 0.005$; diff. between medians $= 24.5$ s). It is interesting that, once participant C is removed as an outlier, the time spent reading documents significantly predicts performance ($p = 0.001$, R-squared $= 0.754$) - for each additional second spent reading documents, the expected performance (in terms of precision) increases by 0.012. This suggests that when participants actually spent more time assessing the documents they were reading, they were able to more reliably assess relevance.

The strategy employed by participant F, who noted post study that they spent little time reading the documents in an attempt to try and get as many bookmarks, does little for success in the task. The findings of this superficial/cursory strategy would appear to support Jozsa et al. [10] but is contradicted by the findings of Rosza et al. [14] where users recommended skimming documents and employing the strategy of using the 'find' shortcut (ctr + F) to quickly find keywords on documents. From this perspective this study again supports Jozsa et al. as there was a distinct lack of use of the 'find' shortcut with only participant E utilising this function.

4.2 Confidence and Querying

It has already been noted that there is a difference in the reading time of documents amongst the participants - when grouped by confidence the *unconfident* spent significantly more time reading documents than the *confident*. This group also submitted significantly fewer queries ($p = 0.033$; diff. between medians = 2) which appears to contradict the study by Bogers et al. [3] which found non-native speakers to query much more. The lack of confidence also appears to effect query formulation time as well as the time spent reviewing SERPs with the *unconfident* taking significantly longer to submit a query ($p = 0.0025$; diff. = 4.5 s) and spending significantly longer on SERPs (p ll 0.01; diff. = 9.5 s), supporting the findings of Chu et al. [5].

Surprisingly the *unconfident* were found to use assistive functionality no more than the *confident*, but this was not significant. Although assistive functionality is discussed [14] and the participants recommended using Google suggestions to mitigate spelling mistakes, there is little in the literature on actual usage or lack thereof, and whether this is common among ESL communities. In our own study there were only nine instances of submitted terms with spelling mistakes by six users across tasks two, three and four. Such a small number was also noted by by Chu et al. [5] and may be explained by the fact that the *unconfident* submit shorter queries, a behaviour also noted in other studies [3,14].

The *confident* group had more failed queries (i.e. those with 0 clicks), perhaps suggesting they have the confidence to reject a query by assessing that results are poor. On a per-topic basis the *confident* users submitted an average of 1.6 failed queries, while the *unconfident* group only submitted 0.8. The *confident* group also tended to look deeper into the results lists than the other group - on average the two groups stopped clicking at rank positions 8 and 5 respectively.

4.3 Query Classification

We classified queries based on the definitions of Chu et al. [5] compared against the previously submitted query. "New Query" (1) = no terms in common. "Generalisation" (2) = same query, at least one term fewer. "Specialisation" (3) = same query, at least one term more. "Reformulation" (4) = at least one term in common and at least one term changed. "Synonym" (5) same as (4) but changed term is a synonym. "Content Change" (6) = same query but different content i.e. changing from "Web" to "News".

Although not significant, there were differences in the distribution of queries submitted by those in the *confident* and *unconfident* groups over the query classes (as shown in Fig. 1). *Confident* searchers used more "reformulations" (30%), "specialisations" (24%), "generalisations" (7%) and "spelling corrections" (2%). Whereas the *unconfident* searchers resorted to more frequently starting a "new

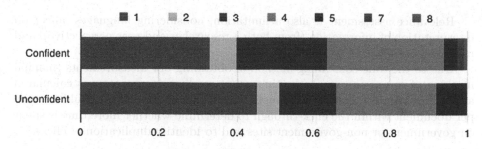

Fig. 1. Query classes by group.

query" (45%) and used more "synonyms" (8%). In contrast to the findings of Chu et al. [5], the distribution of classifications reveal that "new queries" and "reformulations" accounted for the majority of queries, approximately 66%.

Behaviours. Despite some participants knowing of their existence and acknowledging their usefulness, there were no instances of operator use in this study, although one participant did state that he "probably should use them more". Whether this is an effect of confidence is debatable, however it is interesting to note that Rozsa et al. [14] also found that participants encouraged the use of operators whilst not necessarily utilising the function themselves.

Most participants used multiple tabs. The extent ranged from intermittent (participants A, B, D, G, H, I) to extreme (participants E, F, J) with those at the lower end of the scale focussing mostly on just one tab with only occasional instances of switching between multiple tabs and the SERP. Those at the extreme end would alternate in short bursts between open documents on separate tabs (up to ten in one instance), SERPs and new search screens.

Four participants (A, B, C, J) used in-site search on websites with a total of six instances, two instances each for task 2, 3 and 4. They choose not to use in-site searches often because the general consensus was that Google was a reliable search facility and they could not say the same about individual websites. Participants E and G stated that they got better results from Google than any in-site search (in the past) and that it was just as quick to go back to the search and start again than use the website's in-built functionality.

5   Limitations

An obvious limitation of this study is the educational background and number of participants. Although no generalisable hypothesis can be drawn from this limited user representation, the results allow us some insights into the search behaviours of ESL users of E-Government services and, perhaps more interestingly, how perceived confidence relates to these behaviours. Self-assessment of language proficiency has clearly shown an impact with participants identifying concerns over ability, despite high proficiency in English.

Relevance assessment is also a limitation, considering languages' affect on interpretation of information (from both a researcher and user perspective), and must be considered in future studies.

Due to the time-consuming method of obtaining our measurements (manual marking of screen captures), the time spent reading documents was calculated as the the total time on click-through. This could and possibly should be time per document within the click-through to determine whether more time is spent on government or non-government sites and to identify duplication of clicks.

6   Conclusions and Future Work

This study expanded on previous work in multilingual IR from an information seeking behaviour perspective by examining the ways in which ESL users approach a number of important search tasks and the problems they face in doing so. We identified that even among our ESL participants, who had good overall proficiency in English, there were subgroups of participants who were confident and those who were less unconfident in their abilities to formulate queries in English, identify relevant search results and information on websites. We found that these levels of confidence had a number of key effects on the participants' behaviour when completing the tasks. The unconfident group spent more time assessing documents, more time formulating queries (yet submitted shorter queries) and queried less often. In spite of this, they had far fewer failed queries and actually performed better (in terms of precision). We also found differences in the kinds of queries submitted between the groups, with the confident users more likely to reformulate their queries than submit new ones.

The results point to many participants being overly-confident of their abilities and that this over-confidence may have resulted in them taking riskier strategies, being less thorough in their evaluations and, therefore, bookmarking a larger proportion of non-relevant documents. This echoes results from the literature on superficial searching strategies [10] and shows why such strategies might arise. Our results suggest that success in this context may be less dependent on second language proficiency, as one might expect, and may instead hinge on the search strategies employed and the fastidiousness of the user in assessing document relevance, elements which could be taught or where assistance could be given.

In future work we intend to run the same study with native speakers to determine whether their performance is indeed better, as one would expect and hope, and compare them with the non-native speakers. We would also expect that the behaviour of the native speakers would be more similar to the *confident* ESL participants, however they are likely to also display behaviours not demonstrated by the participants of this study.

References

1. Aula, A., Jhaveri, N., Käki, M.: Information search and re-access strategies of experienced web users. In: WWW 2005, pp. 583–592. ACM (2005)

2. Berendt, B., Kralisch, A.: A user-centric approach to identifying best deployment strategies for language tools: the impact of content and access language on web user behaviour and attitudes. Inf. Retr. **12**(3), 380–399 (2009)
3. Bogers, T., Gäde, M., Hall, M., Skov, M.: Analyzing the influence of language proficiency on interactive book search behavior. In: iSchools (2016)
4. Brazier, D., Harvey, M.: Strangers in a strange land: a study of second language speakers searching for e-services. In: CHIIR 2017 (2017)
5. Chu, P., Jozsa, E., Komlodi, A., Hercegfi, K.: An exploratory study on search behavior in different languages. In: IIiX 2012, pp. 318–321. ACM (2012)
6. Edwards, A., Kelly, D.: How does interest in a work task impact search behavior and engagement? In: ACM SIGCHI, pp. 249–252. ACM, March 2016
7. Freeguard, G., Andrews, E., Devine, D., Munro, R., Randall, J.: Whitehall monitor 2015 (2015). http://www.instituteforgovernment.org.uk/publications/whitehall-monitor-2015. Accessed 4 Oct 2016
8. Helbig, N., Gil-García, J.R., Ferro, E.: Understanding the complexity of electronic government: implications from the digital divide literature. Gov. Inf. Q. **26**(1), 89–97 (2009). From Implementation to Adoption: Challenges to Successful E-government Diffusion
9. Ingwersen, P., Järvelin, K.: The Turn: Integration of Information Seeking and Retrieval in Context, vol. 18. Springer Science & Business Media, Heidelberg (2006)
10. Józsa, E., Köles, M., Komlódi, A., Hercegfi, K., Chu, P.: Evaluation of search quality differences and the impact of personality styles in native and foreign language searching tasks. In: IIiX 2012, pp. 310–313. ACM (2012)
11. Kelly, D., Arguello, J., Edwards, A., Wu, W.: Development and evaluation of search tasks for IIR experiments using a cognitive complexity framework. In: ICTIR 2015, pp. 101–110. ACM, September 2015
12. Lloyd, A., Kennan, M.A., Thompson, K.M., Qayyum, A.: Connecting with new information landscapes: information literacy practices of refugees. J. Doc. **69**(1), 121–144 (2013)
13. Marlow, J., Clough, P., Recuero, J.C., Artiles, J.: Exploring the effects of language skills on multilingual web search. In: Macdonald, C., Ounis, I., Plachouras, V., Ruthven, I., White, R.W. (eds.) ECIR 2008. LNCS, vol. 4956, pp. 126–137. Springer, Heidelberg (2008). doi:10.1007/978-3-540-78646-7_14
14. Rózsa, G., Komlodi, A., Chu, P.: Online searching in English as a foreign language. In: WWW 2015 Companion, pp. 875–880 (2015)
15. Vinson, T.: The origins, meaning, definition and economic implications of the concept social inclusion/exclusion: incorporating the core indicators developed by the European union and other illustrative indicators that could identify and monitor social exclusion in Australia. Department of Education, Employment and Workplace Relations, Canberra (2009)

A Task Completion Engine to Enhance Search Session Support for Air Traffic Work Tasks

Yashar Moshfeghi[1]([⊠]), Raoul Rothfeld[1], Leif Azzopardi[2],
and Peter Triantafillou[1]

[1] University of Glasgow, Glasgow, UK
{Yashar.Moshfeghi,Raoul.Rothfeld,Peter.Triantafillou}@glasgow.ac.uk
[2] University of Strathclyde, Glasgow, UK
Leif.Azzopardi@strath.ac.uk

Abstract. Providing support for users during their search sessions has been hailed as a major challenge in interactive information retrieval (IIR). Providing such support requires considering the context of the search and facilitating the work task at hand. In this paper, we consider the work tasks associated with air traffic analysts, who perform numerous searches using a multifaceted search interface in order to acquire business intelligence regarding particular events and situations. In particular, we develop a novel task completion engine and seamlessly incorporated it within a current air traffic search system to facilitate the comparison of information objects found. In a study with 24 participants, we found that they completed the complex work task faster using the comparison feature, but for simple work tasks, participants were slower. However, participants reported (statistically) significantly higher satisfaction and had (statistically) significantly higher accuracy using the search system equipped with task completion engine. These findings help to steer systems to provide a better support to users in their search process.

1 Introduction

Searching is typically performed in the context of a task (usually a work task) [1,2], where the user desires to complete the task as efficiently and effectively as possible. While numerous search systems have been proposed to support the search process [3–10], providing effective task support is still a difficult and challenging problem in Interactive Information Retrieval (IIR) [11,12]. A promising direction to provide task support is the idea of a task completion engine [13], which explicitly goes beyond supporting the search task to facilitating it. This paper is one of the first attempts in this direction. A task completion engine builds on top of a search engine enabling the collection, collation and comparison of information found during the course of a search session. Essentially, the task completion engine aims to augment the user's cognitive capabilities in order to achieve a successful outcome: *reducing task completion times, improving decision making, decreasing the cognitive burden, and crucially reducing errors.*

© Springer International Publishing AG 2017
J.M. Jose et al. (Eds.): ECIR 2017, LNCS 10193, pp. 278–290, 2017.
DOI: 10.1007/978-3-319-56608-5_22

An important domain where task completion engines can potentially be of great use is within the Air Traffic industry, where finding relevant information efficiently is essential [14]. Flight analysts typically perform complex search tasks in order to find the relevant information (i.e. flight intelligence) to make informed decisions regarding flight performance management. In this context, common work tasks require the analysts to aggregate and compare the different information and data that is available to them. This typically involves posing many queries and examining a number of facets to acquire all the relevant information [14] (and thus is similar to most IIR search/work tasks [1,2,15]).

In this paper, we aim to study the effect of task completion engine in the effectiveness and efficiency of users in completing complex search tasks in this domain. To provide a use case for our investigation, we experiment with an air traffic search system [14], where analysts need to interact with information about aircraft, schedules, operators, airports, etc. through textual summaries (i.e. news, weather conditions, traffic conditions, airport notifications, etc.), structured data (i.e. flight times, temperatures, etc.) and visual representations (i.e. charts and graphs, etc.) in a timely manner. To do so, we seemingly incorporated a new feature to a real life air traffic search system to help task completion. Specifically, we proposed a contextualised comparison feature that first enables such systems to store the analysts' search state/results at different points of their search session. Second it allows analysts to compare their current search state/results to the stored one by automatically overlapping (superimposing) them across heterogeneous data visualisation.

This paper has three novel contributions: first we have investigated the effect of task completion engine in the context of a novel and specific domain [16], bringing Information Retrieval techniques to the problem of searching air traffic information [14]. Second we have provided evidence that the introduction of contextualised comparison feature has led to (statistically) significantly higher user satisfaction and accuracy in completing both simple and complex work tasks. Third, we have also found that incorporating task completion engines could introduce both benefits and limitations to search systems depending on the task difficulty faced by the users.

The remainder of the paper is organised as follows. Section 2 describes state-of-the-art works in task completion. Section 3 presents the approach of the paper. Section 4 describes our system. Sections 5 and 6 discuss the experimental methodology and results respectively. We then conclude and discuss future work.

2   Related Work

Search engines typically provide only limited support for users across their session(s) and often fail to help users complete satisfactorily more complex information search/work tasks [11,12]. However, there have been numerous attempts to improve the standard search interface to support searching e.g. [3–5,7,8,17,18]. For example, in [8], they augment search sessions by providing a viewable history of the pages that the user has interacted with during the course of a session. The history of pages are shown as thumbnails to provide users with a non-textual cue so that they can quickly re-access previously viewed pages and storing the pages in *WebBooks* [19]. They found that participants used the thumbnail history to view key hub pages, compare information on pages (i.e. hotel prices), and to obtain additional information related to the current page (e.g. to convert a currency). A similar augmentation was developed in [5] called *SearchBar*, which showed the list of pages visited but grouped by query, to enable easier navigation through the result history. In the context of a work task, to organise travel and trips, it was again found that the additional support helped in completing these complex search tasks. Following in this direction was the development of *Search-Pad* [17], which was devised to help searchers perform "research missions", i.e. complex search tasks, such as finding a good deal for a HDTV, the value of political parties, or collecting good recipes. *SearchPad* would enable users to take notes about various pages that they encountered through searching, so that they could make sense of the information that they had found, and invariably make a better decision (i.e. on what to buy, who to vote for, what to cook and eat, etc.).

Each of the examples above, highlights the need that people have to use the information that they have previously found in order to perform a work task, and try to augment the search engine/interface to provide cognitive support to help saving, collecting, and re-finding/re-accessing the information. On the other hand, other search interfaces have been devised to help support the exploration of results [3,4,9,10]. For example, Querium [4] provides users with numerous search features such as relevance feedback, query fusion, faceted search, and search histories, and facilitates collaborative search. The idea was to help users share, save, collaborate and revisit their information. *SearchPanel* [3], provides similar functionality through a web browser extension, to support people in their ongoing web information seeking tasks by mapping the space that has been explored. Rather than providing cognitive support in terms of histories and maps, its alternative approach is to help guide the users' querying process by providing facets and faceted search [9,10,20]. These interfaces, again, support the users across and through their session as they try to make sense of the information space and achieve a greater awareness of the topic of interest.

These developments have focused on helping users address their work tasks by augmenting the search engine. In [13], Balog sets out a vision for developing task completion engines that during the course of searching extracts out the salient entities and information from the pages, store this information, and facilitate decision making. This requires task modelling, understanding requests, resource representation and selection, and information retrieval, extraction and integration [13]. Key to this process is the information extraction of entities from the pages and the integration of information through semantic analysis with respect to the task at hand. For example, extracting different places to visit when on holidays, the different hotels and deals on offer, the different medicines and treatments available for a particular condition, etc. Thus, the development of such engines requires a significant amount of infrastructure. Here, since we focus on a specific domain, we are able to extract out the salient information based on semi-structured data, and thus can evaluate whether the addition of a contextualised comparison feature facilitates more efficient and effective completion of air traffic analyst work tasks.

3   Approach

As mentioned in Sect. 1, a task completion (TC) engine usually builds on top of a search engine enabling the collection, collation and comparison of information found during the course of a search session. The aim of such an engine is to augment the user's cognitive capabilities in order to achieve a successful outcome: reducing task completion times, improving decision making, decreasing the cognitive burden, and crucially reducing errors. While the concept of TC engines should by definition benefit users, developing an actual engine with such a functionality is not so easy. This is because this feature needs to be seamlessly merged with already existing functionalities of the search system. This is an important challenge that major search engine companies are facing when they are introducing new features, due to potential damage it can have on the revenue, etc.

With that in mind, we carefully identified an existing limitation in current search systems, i.e. users have to rely too much on their memories to accomplish the work task effectively and efficiently. This situation can become worse when the user has to memorise multiple data points or translate such data points across heterogeneous data representation. In order to tackle this challenge, we introduce the idea of storing search sessions and allowing users to retrieve the stored sessions at any time during the search process. While there exists a wealth of research on retrieving relevant information for a given query, to the best of our knowledge, there is no prior research on providing previous search session states to the user for cross comparison.

We also introduce the idea of highlighting the differences between the data stored and the one for the current search session. This is also a challenging task by itself, since it needs a deep understanding of the problem, various data representation and visualisation techniques that can facilitate users in their complex work task. To investigate our approach, as our use case, we focus on a novel search domain, i.e. an air traffic search system, where users have to perform complex task in a timely manner. In the rest of the section, we discuss how we implemented our approach in an operational air traffic search system.

4   Air Traffic System Task Completion Engine

The standard search engines used by an air traffic control analysis companies [14] are a multifaceted search system, consisting of the standard query input along with facets for selecting airports and airlines, in order to filter data. The results returned contain information objects of various modalities such as the number of flights and flight information including time, day, delays, weather, distances, etc.

Such numerical data is extracted out and associated with a particular entity (i.e. a flight, a carrier airline, an airport, etc.) and is used to make various air traffic decisions. While it appears quite different from a standard text based retrieval system, it is more on par with an interface for product search where users can compare prices and technical specifications about products, see ratings, etc. For the purposes of this study, we have seemingly incorporated a task completion engine into an air traffic search system which is representative of those used at a commercial air traffic analysis company. The rest of this section describes components of the system in detail.

Backend Component: During a search session, the backend receives several requests generated through either standard query input or from the interactions with the facets. The backend component then processes the queries, constructs filters, and applies them on the underlying data dimensions and thus, gradually reduces the presented amount of data to the desired subset.

User Interface Component: The search interface (as shown in Fig. 1) is composed of a querying interface that contains two drop-down menus, one for airports and the other for airlines (A), a selection reset button (B), and various interactive charts for conducting search queries (C–F). The user is presented with a line chart depicting the number of flight movements over time (C), a scatter plot illustrating all flights according to their time of day and delay in minutes (D), two row charts showing the number of flights per connected airports or weekday (E), and three bar charts showing the number of flights per delay in minutes, time of day, and flight distance in miles (F). Presented data is queried within a search session via mouse interaction on these charts.

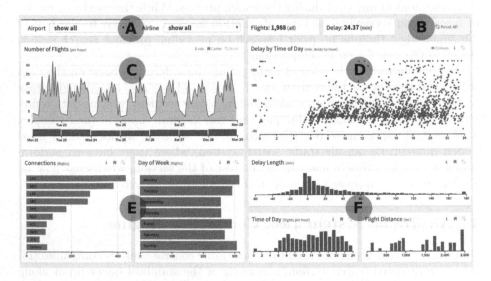

Fig. 1. The interface for air traffic control analysts. (A) querying component (B) new task/reset (C) flight movements chart (D) day/delay chart (E) flights per connect airport charts (F) flight delay charts.

User Tracking and Logging Component: User actions were monitored and logged by the system, including the number of interactions/clicks and time spent carrying out presented information retrieval tasks. Users were asked to indicate task completion by clicking a designated button.

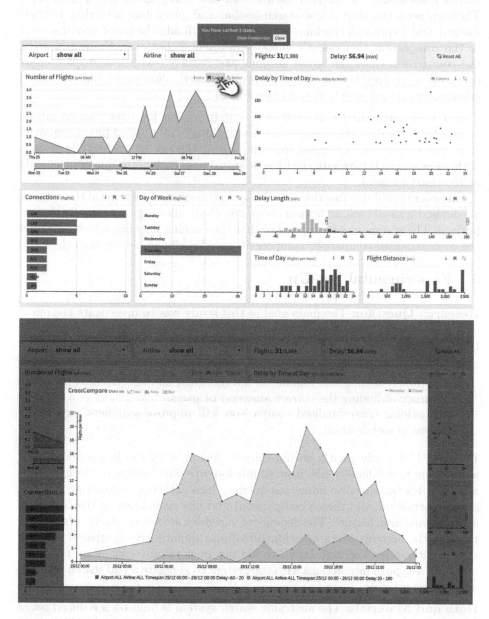

Fig. 2. Top: the interface after the use has saved one query in the session. Below: a comparison between the cached queries and the current query.

4.1 Contextualised Comparison Component

The contextualised comparison component can be activated through pressing designated caching buttons above each chart. Upon button press, the current search session and the queried data subset are being saved (see Fig. 1, top). Thereon, users can start a new search session and query data according to their interest and re-press the caching button, which will add the novel search session to the comparison component's memory (see Fig. 2, bottom). Upon activating the comparison chart, all saved sessions are rendered within a stacked chart overlay (e.g. grouped bar chart or multiple line chart), allowing for contextualised comparisons across search sessions (see Fig. 2).

Cross Comparison: Upon caching a search session for later contextualised comparison, the data dimensions and descriptions of all applied filters are saved in a queue. The user's request for comparison renders all cached dimensions within one stacked chart with the filter description of each of the queue's elements as the chart's legend.

In order to add this function seamlessly to the existing factions of the system, we devised a small caching button above any chart allowing contextualised comparison. The system was configured so that the caching buttons, which activated the contextualised comparison component, could easily be hidden from the view.

5 Experimental Set-Up

Research Question: The main goal of this study was to investigate the effect of a contextualised comparison feature added to the search interface to facilitate task completion, where we hypothesis that:

- $H1$: providing contextualised comparison will improve the efficacy (in terms of task completion time and number of interactions) and effectiveness (in terms of accuracy of finding the correct answers) of users.
- $H2$: providing contextualised comparison will improve searchers' experience (in terms of satisfaction).

Design: This study used a within-subject design, with the independent variables being task difficulty (i.e. from simple lookups/fact finding to more difficult and complex tasks involve numerous queries, data gathering, extracting relevant data/information, and then a comparison) and the availability of the contextualised comparison feature. The dependent variables are the qualitative (gathered through the accompanying questionnaires) and quantitative (gathered through system interaction logging) data. We did not perform any control on the time, number, or type of interactions with the system to simulate a real search scenario as much as possible.

Data and Materials: The air traffic search system is built on a reduced set of flight entries obtained from the American Statistical Association[1], which comprises heterogeneous flight information (such as origin and destination, date and

---

[1] http://www.amstat.org/publications/jse/jse_data_archive.htm.

time, airline code, or delays) for flights within the Unites States from 1987 to 2008. The various types of data are being made available to the user via the different charts of the user interface.

Tasks: Our commercial partner supplied a number of typical work tasks performed by their analysts. Using a similar approach to Brennan et al. [21], we selected ten work tasks, five of which we considered simple, and five of which we considered complex. The difficulty of these tasks was determined during a pilot study and measured by the number of interactions required to complete the task (i.e. apply query filter, view information object, note information/data, etc.). The complex tasks require participants to perform multiple queries and compare/contrast the information gathered from various filter states for each of these queries with each other to draw a final conclusion. Simple tasks can be answered via issuing a single query. Examples of the two types of tasks are:

Simple. *How many flights were operated per weekday?* This task required the participant to apply no filters.

Simple. *How many American (AA) airline flights were operated during the busiest hour at Chicago (ORD) airport, which had a delay of 0–60 and 60– 120 min?* This task required the participant to apply one airline and one airport filter, while switching between two delay filter states.

Complex. *For each day of a week, which airport has the most flights per weekday?* This task required the participant to cycle through all airports and contrast the weekday values.

Complex. *On Thursday the 25th, what is the difference in number of flights with a delay of less than 20 min compared to the number of flights with a delay of 20 or more minutes; at 8, 12, 16, and 20 o'clock?* This task required the participant to apply one time filter and compare the two different delay filter states at four points in time.

To counteract the order and fatigue effects we counter-balanced the task distribution using a Graeco Latin Square design.

Procedure: The ethics approval was obtained from the University of Glasgow. The formal meeting with the participants took place in an office setting. At the beginning of the session the participants were given an information sheet which explained the conditions of the experiment. The participants were notified that they have the right to withdraw from the experiment at any point during the study, without affecting their legal rights or benefits then asked to sign a Consent Form. Then, they were given an Entry Questionnaire to fill in.

The session proceeded with a brief tutorial on the use of the search interface with a short training task. After completion of the training task, each participant had to complete six search tasks (see Section Tasks), one for each level of task difficulties where the comparison feature is available or not (see Sect. 5). To negate the order and fatigue effects we counter-balanced the task distribution using a Graeco-Latin Square design.

The subjects were given 10 min to complete their task, during which they were left unattended to work. At the end of each task, the subjects were asked to complete a post-task questionnaire. Questions in the post-task questionnaire were randomised to avoid the effect of fatigue. Between each task, a cooling-off period was applied to avoid the carry-over effect. Finally, an exit questionnaire was administered at the end of the session.

Each study took approximately 120 min to complete; this is from the time they accepted the conditions until they finished answering the exit questionnaire. Users could only participate once in the study. The participants were all volunteers and did not received any compensation. The results of these studies are presented in Sect. 6.

Participants: 24 people were recruited to undertake the study, of which 8 were female and 16 were male. All participants were between 18 and 64 years old, with most between 18–24 (62.59%) and then 25–34 (25.00%). Most participants had at least bachelor degree (83.33%) at the time of the experiments. The majority of participants had knowledge about Computing and Information Technology, in particular search systems.

Baseline vs. Enriched System: For experimental purposes, we used two versions of the system, one without contextualised comparison component (i.e. Baseline) and one with (i.e. Enriched). The changes in the user interface (UI) between these two systems are minimal to avoid introducing any confound effect. In particular, for the Baseline system, the UI consists of all components, as shown in Fig. 1, without caching buttons above any chart – rendering the comparison functionality inaccessible. Whereas the UI of the Enriched system facilitates the use of the contextualised comparison component via caching buttons above each chart and, consequently, the comparison chart overlay.

Apparatus: For our experiment we used one desktop computer, equipped with a monitor, keyboard and mouse. The computer provided access to a custom-made air traffic search system which allowed the participants to perform their search tasks. The system was designed such that it logged participants' desktop actions, such as starting, finishing and elapsed times for interactions, mouse clicks using a common system time.

Questionnaires: At the beginning of the experiment, the participants completed an *entry questionnaire*, which gathered background and demographic information, and inquired about previous experience with online search systems and searching air traffic control data. At the end of each task, the participants completed a *post-task questionnaire*, where they were asked about their satisfaction with the system. Finally, an *exit questionnaire* was introduced at the end of the study gathering information about their general comments about the experiment.

6   Results

To compare the differences between the two systems we performed a paired t-test between the various measures taken for each system to check whether the

Enriched system (i.e. equipped with contextualised comparison component) was significantly different to the Baseline system. We use (*) and (**) to denote the level of significance where the confidence level is ($p < 0.05$) and ($p < 0.01$), respectively.

Log Analysis: Table 1 reports the mean (and standard deviation) of the time taken to complete the simple and complex tasks, along with the number of interactions (i.e. queries, clicks, facets, etc.) as well as the accuracy[2] at performing the tasks.

Table 1. Mean completion times, no. of interactions and accuracy per task. The value in parenthesis is the standard deviation. (*) and (**) denotes difference with the confidence levels ($p < 0.05$) and ($p < 0.01$) respectively.

Task	Task completion time		Interactions		Accuracy	
	Baseline	Enriched	Baseline	Enriched	Baseline	Enriched
Simple	**133.54** (80.34)	159.25 (96.55)	**15.92** (6.55)	26.87 (17.56)	83.33% (28.86)	**100%**** (0.0)
Complex	416.39 (212.30)	**156.38**** (78.35)	88.89 (38.12)	**23.55**** (14.13)	77.77% (25.45)	**88.88%*** (9.62)

The results indicate that for simple tasks (i.e. lookup based task), participants on the Baseline system completed the task with fewer interactions (15.9 vs 26.9) and did so in less time (133.5 vs 159.3 s). This could be due to participants' expertise with the Baseline system, although both results were not statistically different.

However, for the complex tasks, our results suggest that participants on the Enriched system performed significantly fewer interactions (23.5 vs 88.9) and completed their tasks in significantly less time (156.4 vs 416.4 s). In this case, both results were statistically different suggesting that for the more complex tasks that required participants to memorise several data points and cross compare them, the Enriched system provides a clear advantage. It appears that using the contextualised comparison component resulted in a slower performance in simple tasks but a quicker performance in complex tasks (addressing RQ1).

Interestingly, the participant's accuracy in performing their tasks significantly improved for both simple and complex tasks when the contextualised comparison component was used. These results were statistically different suggesting that participants made consistently fewer errors with the Enriched system. Our findings show that our task completion engine improved participants' effectiveness in performing their complex tasks which in such a domain could be extremely important (addressing RQ1). We now turn our attention to the questionnaire analysis to see if it reveals any further insights.

---

[2] The ratio of the number of correct answers to the total number of answers given.

Questionnaire Analysis: 23 out of 24 (95.8%) participants reported that they preferred using the Enriched system. Further all participants found it to be somewhat or very helpful. In addition, the majority of participants felt it was somewhat or very intuitive, except two participants (8.3%).

In terms of satisfaction, we asked participants to rate how easy it was to complete tasks with each system and how satisfied they were with the amount of time it took to complete tasks with each system. Table 2 shows the results for satisfaction, where the Enriched system was rated significantly higher on both counts (addressing RQ2). These suggests that even though participants took a little bit longer on average for simpler tasks they did not detract from their rating with respect to how satisfied with the time to complete tasks.

Table 2. Mean user satisfaction (SAT) per task on ease of completion and required amount of time, 1 (strongly disagree) to 5 (strongly agree). The value in parenthesis is the standard deviation. (**) denotes difference with the confidence levels ($p < 0.01$).

SAT with ease of completion		SAT with amount of time	
Baseline	Enriched	Baseline	Enriched
2.89 (1.22)	**4.66** (0.716)**	2.77 (1.14)	**4.72** (0.55)**

Comments from participants also confirmed this as they mentioned that it was *"easier"* and *"faster"* to complete tasks using Enriched system, while others mentioned that some of the complex tasks were *"laborious"* and *"infuriating"* to complete without the contextualised comparison component. *"[The] ability to store and then compare information significantly aided its interpretation"*, stated one participant with others agreeing that the contextualised comparison component was *"ideal for complex querying"* and for filtering out *"the factors that matter to you"*. However, multiple participants stated that the comparison feature did not benefit the completion of simple tasks. Others participants mentioned that even for simpler tasks the contextualised comparison component was useful as it enabled them to double check their results. This last comment was kind of unexpected, but suggests that the comparative component is useful to ensure accuracy.

7   Discussion and Conclusion

This paper investigated the effects of task completion engine on the efficiency, effectiveness and satisfaction of participants in completing complex work tasks. As a use case scenario, we considered the work tasks associated with air traffic analysts, who perform numerous searches in order to acquire business intelligence regarding particular events and situations. To support their work tasks, we seemingly incorporated such an engine, a contextualised comparison feature, into an air traffic search system.

Our findings reveal that participants on the search system equipped with contextualised comparison feature (Enriched system) completed complex tasks much more efficiently. However, on simpler tasks our participants took longer time to do so. This appeared to indicate that the Enriched system hindered their efficiency, but participants reported that they checked their answers, which took more time but ensured greater accuracy. Crucially participants had (statistically) significantly higher satisfaction and accuracy using the Enriched system. These findings show that introducing additional features such as contextualised comparison is generally positive, but it may increase the time to complete simpler tasks. This suggests that as we propose novel task completion engines, we need to be careful to determine when they help and when they hinder the user.

One of the main limitations of our study is that it is domain-specific and the use case of an air traffic search system is a rather industry-specific application. However, the notion of contextualised comparisons and search session caching can be applied to a multitude of information retrieval scenarios. Thus, it is expected to improve users' search sessions experiences in a wide range of information seeking tasks and lessen the user's cognitive load. However, this may come at the cost of reducing the efficiency at simpler tasks. Nonetheless, we have provided strong empirical evidence that the concept of contextualised comparison improves the search experience, efficacy and effectiveness lending weight to the progression from search engines towards task completion engines. Further work will be directed towards developing similar contextualised comparison component for other domains and tasks.

Acknowledgement. This work was supported by the Economic and Social Research Council [grant number ES/L011921/1].

References

1. Vakkari, P.: Task-based information searching. Annu. Rev. Info. Sci. Tech. **37**(1), 413–464 (2003)
2. Toms, E.G., Villa, R., McCay-Peet, L.: How is a search system used in work task completion? J. Inf. Sci. **39**(1), 15–25 (2013)
3. Qvarfordt, P., Tretter, S., Golovchinsky, G., Dunnigan, T.: SearchPanel: framing complex search needs. In: SIGIR 2014, pp. 495–504. ACM (2014)
4. Diriye, A., Golovchinsky, G.: Querium: a session-based collaborative search system. In: Baeza-Yates, R., Vries, A.P., Zaragoza, H., Cambazoglu, B.B., Murdock, V., Lempel, R., Silvestri, F. (eds.) ECIR 2012. LNCS, vol. 7224, pp. 583–584. Springer, Heidelberg (2012). doi:10.1007/978-3-642-28997-2_72
5. Morris, D., Morris, M.R., Venolia, G.: SearchBar: a search-centric web history for task resumption and information re-finding. In: CHI 2008, pp. 1207–1216. ACM (2008)
6. Sebrechts, M.M., Cugini, J.V., Laskowski, S.J., Vasilakis, J., Miller, M.S.: Visualization of search results: a comparative evaluation of text, 2D, and 3D interfaces. In: SIGIR 1999, pp. 3–10. ACM (1999)
7. Jones, W., Bruce, H., Dumais, S.: Keeping found things found on the Web. In: CIKM 2001, pp. 119–126. ACM (2001)

8. Jhaveri, N., Räihä, K.J.: The advantages of a cross-session web workspace. In: CHI EA 2005, pp. 1949–1952. ACM (2005)
9. Villa, R., Gildea, N., Jose, J.M.: FacetBrowser: a user interface for complex search tasks. In: MM 2008, pp. 489–498. ACM, New York (2008)
10. Kashyap, A., Hristidis, V., Petropoulos, M.: FACeTOR: cost-driven exploration of faceted query results. In: CIKM 2010, pp. 719–728. ACM (2010)
11. Belkin, N.J.: Some(what) grand challenges for information retrieval. SIGIR Forum **42**(1), 47–54 (2008)
12. Gäde, M., Hall, M.M., Huurdeman, H., Kamps, J., Koolen, M., Skove, M., Toms, E., Walsh, D.: Report on the first workshop on supporting complex search tasks. SIGIR Forum **49**(1), 50–56 (2015)
13. Balog, K.: Task-completion engines: a vision with a plan. In: Proceedings of the First International Workshop on Supporting Complex Search Tasks, vol. 1338 (2015)
14. Billings, C.: Aviation Automation: The Search for a Human-Centered Approach. Human Factors in Transportation. Lawrence Erlbaum Associates Publishers, Mahwah (1996)
15. Ruthven, I.: Interactive information retrieval. Annu. Rev. Info. Sci. Technol. **42**(1), 43–91 (2008)
16. Salampasis, M., Fuhr, N., Hanbury, A., Lupu, M., Larsen, B., Strindberg, H.: Integrating IR technologies for professional search. In: Serdyukov, P., Braslavski, P., Kuznetsov, S.O., Kamps, J., Rüger, S., Agichtein, E., Segalovich, I., Yilmaz, E. (eds.) ECIR 2013. LNCS, vol. 7814, pp. 882–885. Springer, Heidelberg (2013). doi:10.1007/978-3-642-36973-5_108
17. Donato, D., Bonchi, F., Chi, T., Maarek, Y.: Do you want to take notes? Identifying research missions in Yahoo! search pad. In: WWW 2010, pp. 321–330. ACM (2010)
18. Dziadosz, S., Chandrasekar, R.: Do thumbnail previews help users make better relevance decisions about web search results?. In: SIGIR 2002, pp. 365–366. ACM (2002)
19. Card, S.K., Robertson, P.G.G., York, W.: The WebBook and the Web Forager: an information workspace for the World-Wide Web. In: CHI 1996, pp. 111-ff. ACM (1996)
20. Tunkelang, D.: Faceted Search. Synthesis Lectures on Information Concepts, Retrieval, and Services. Morgan & Claypool Publishers, San Rafael (2009)
21. Brennan, K., Kelly, D., Arguello, J.: The effect of cognitive abilities on information search for tasks of varying levels of complexity. In: IIiX 2014, pp. 165–174. ACM (2014)

Personalized Keyword Boosting for Venue Suggestion Based on Multiple LBSNs

Mohammad Aliannejadi[1]([✉]), Dimitrios Rafailidis[2], and Fabio Crestani[1]

[1] Faculty of Informatics, Università della Svizzera Italiana, Lugano, Switzerland
{mohammad.alian.nejadi,fabio.crestani}@usi.ch
[2] Aristotle University of Thessaloniki, Thessaloniki, Greece
draf@csd.auth.gr

Abstract. Personalized venue suggestion plays a crucial role in satisfying the users needs on location-based social networks (LBSNs). In this study, we present a probabilistic generative model to map user tags to venue taste keywords. We study four approaches to address the data sparsity problem with the aid of such mapping: one model to boost venue taste keywords and three alternative models to predict user tags. Furthermore, we calculate different scores from multiple LBSNs and show how to incorporate new information from the mapping into a venue suggestion approach. The computed scores are then integrated adopting learning to rank techniques. The experimental results on two TREC collections demonstrate that our approach beats state-of-the-art strategies.

Keywords: Venue suggestion · User tags · Location-based social networks

1 Introduction

With the availability of location-based social networks (LBSNs), such as Yelp, TripAdvisor and Foursquare, users can share check-in data using their mobile devices. LBSNs collect valuable information about users' mobility records with check-in data including user feedback, such as ratings, tags and reviews. Being able to suggest personalized venues to a user plays a key role in satisfying the user needs on LBSNs, for example when exploring a new venue or visiting a city [4].

There is a number of different LBSNs that are widely used. However, a single LBSN does not have a comprehensive coverage over all venues and all types of information. For instance, Booking.com mainly focuses on hotels. Combining multimodal information e.g., ratings, tags, reviews of previously visited venues from different LBSNs improves the accuracy of venue suggestion [1]. For instance, the key idea of our best performing work in the TREC Contextual Suggestion Track 2015 [1] is to exploit multimodal information from multiple LBSNs, and combine them linearly to model the user preferences on venues, thus significantly beating the competitors that exploit information from a single LBSN.

© Springer International Publishing AG 2017
J.M. Jose et al. (Eds.): ECIR 2017, LNCS 10193, pp. 291–303, 2017.
DOI: 10.1007/978-3-319-56608-5_23

A main challenge for venue suggestion is how to model the user profile, built based on the user feedback on previously visited venues. For example, users may annotate a venue with predefined tags; even though the tag assignments are very sparse, tags contain valuable information which is worth exploiting when building user profiles. Relevant studies propose to model user profiles and generate recommendations based on the similarity between the user preferences and the venues' descriptions and categories [16]. Other studies leverage the opinions of users about a venue based on online reviews [2]. Some LBSNs, such as Foursquare, extract keywords from users' reviews. We refer to them as *venue taste keywords*. Another challenge for venue suggestion is how to calculate the correlation between user tags and information about a venue such as taste keywords. The correlated information could further be used to model the personalized user behavior for tagging venues and to solve the sparsity problem of user tags that often occurs in LBSNs.

In the effort to face these challenges, in this paper our main contribution can be summarized as follows:

1. We present a probabilistic generative approach to find the mapping between user tags and venue taste keywords, thus modeling more accurately the personalized opinion of users about venues.
2. We address the sparsity problem of user tags by performing personalized boosting of taste keywords of visited venues, so that our model is capable of predicting user tags for unexplored venues.
3. We examine several alternative machine learning models to perform tag prediction and evaluate the impact of our boosting approach on the venue suggestion task, comparing it with the alternative models.
4. We evaluate several learning to rank techniques to incorporate boosting and tag prediction into our venue suggestion model using information from multiple LBSNs.

The experiments on two benchmark datasets show that our proposed approach outperforms state-of-the-art strategies.

2   Related Work

Rikitianskii et al. [16] proposed a content-based approach to apply Part of Speech (POS) tagging to venues' descriptions, to get the most informative terms for a venue, which are then used to create positive and negative profiles when suggesting venues. Several rating-based collaborative filtering approaches have been proposed in the literature, which are based on finding common features among users' preferences and recommending venues to people with similar interests. These models are usually based on matrix factorization, exploiting check-in data for recommending venues, such as the studies reported in [5,9]. Factorization machines generalize matrix factorization techniques to leverage not only user feedback but also other types of information, such as contextual information in LBSNs [11]. In addition, some studies follow a review-based strategy, building enhanced user profiles based on their reviews. When a user writes a review about a venue there is a wealth of information which reveals the reasons why that particular user is interested in a venue or not. For example, Chen et al. [4] argued that reviews are helpful to deal with the sparsity problem in LBSNs.

There are also many studies that propose to annotate venues with taste keywords. For instance, He et al. [10] presented a latent-class probabilistic generative model to annotate venues with taste keywords. Ye et al. [17] trained a binary classifier for each venue taste keyword with a set of extracted features so that the trained classifiers can predict the taste keywords for a new venue. There are also other studies which use various types of information such as images and audio to annotate venues, such as the study in [6]. However, none of these studies exploit information from multiple LBSNs.

3 Proposed Approach

3.1 Personalized Keyword Boosting

Personalized Keyword-Tag Mapping. We propose a probabilistic framework to map user tags to venue taste keywords in a personalized manner. The goal of the mapping is to find the correlation between venue taste keywords and user tags. The fundamental assumption is that a user opts to assign a particular tag to a venue following a pattern that is related to the venue's content. We assume that a user chooses a particular tag for a particular venue based on its type and/or characteristics. For example, if a user tags a venue with *healthy-food* because the venue is a vegetarian restaurant, then we expect that the same user will tag other vegetarian restaurants with *healthy-food*. To model each user's personalized tag mapping, we propose a probabilistic generative model for mapping user tags and venue taste keywords. An example of such mapping is depicted in Fig. 1, with a sequence of two user tags and a set of four taste keywords for a venue. Our goal is to identify for each user the most probable mapping of venue taste keywords to user tags.

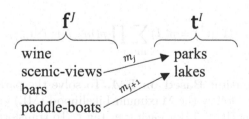

Fig. 1. An example of mapping of $J = 4$ venue taste keywords to $I = 2$ user tags.

Given a sequence without repetition of taste keywords $\mathbf{f}^J = \langle f_1 \ldots f_j \ldots f_J \rangle$, we have to compute the sequence of user tags $\mathbf{t}^I = \langle t_1 \ldots t_i \ldots t_I \rangle$ for each individual. Notice that $\mathbf{t}^I$ denotes a sequence named $\mathbf{t}$ whose length is I. Therefore, $\mathbf{t}^i$ would be of length i ($\mathbf{t}^i = \langle t_1 \ldots t_i \rangle$), while t_i denotes the i-th element of a sequence ($\mathbf{t}^I$). We aim at finding the sequence of tags, which maximizes $Pr(\mathbf{t}^I | \mathbf{f}^J)$:

$$\hat{\mathbf{t}}^I = \underset{\mathbf{t}^I}{\mathrm{argmax}}\{Pr(\mathbf{t}^I|\mathbf{f}^J)\} = \underset{\mathbf{t}^I}{\mathrm{argmax}}\{Pr(\mathbf{f}^J|\mathbf{t}^I)Pr(\mathbf{t}^I)\}, \tag{1}$$

where $Pr(\mathbf{t}^I)$ is the user tag model which assigns a probability to a given sequence of user tags. We consider user tags as independent of each other. Therefore, we rewrite the user tags model as follows:

$$Pr(\mathbf{t}^I) = p(I) \prod_{i=1}^{I} p(t_i | \mathbf{t}^{i-1}, I) = p(I) \prod_{i=1}^{I} p(t_i | I), \tag{2}$$

where $\mathbf{t}^{i-1} = \langle t_1 \ldots t_{i-1} \rangle$. We rewrite the probability $Pr(\mathbf{f}^J|\mathbf{t}^I)$ in Eq. 1 by marginalizing it over the latent variable which *maps* taste keywords to user tags: $\mathbf{m}^J = \langle m_1 \ldots m_j \ldots m_J \rangle$, with $m_j \in \{1, \ldots, I\}$:

$$Pr(\mathbf{f}^J|\mathbf{t}^I) = \sum_{\mathbf{m}^J} Pr(\mathbf{f}^J, \mathbf{m}^J|\mathbf{t}^I), \tag{3}$$

where

$$\begin{aligned} Pr(\mathbf{f}^J, \mathbf{m}^J|\mathbf{t}^I) &= p(\mathbf{m}^J|\mathbf{t}^I, I, J)p(\mathbf{f}^J|\mathbf{m}^J, \mathbf{t}^I, I, J) \\ &= p(J|\mathbf{t}^I) \prod_{j=1}^{J} [p(m_j|\mathbf{m}^{j-1}, J, \mathbf{t}^I, I)p(f_j|\mathbf{f}^{j-1}, \mathbf{m}^J, J, \mathbf{t}^I, I)], \end{aligned} \tag{4}$$

where $\mathbf{m}^{i-1} = \langle m_1 \ldots m_{i-1} \rangle$. In our model, we consider a zero-order dependence for both mappings m_j and taste keywords f_j, while $p(J|\mathbf{t}^I)$ depends only on J. Notice that m_j only depends on the length of the user tag sequence I and f_j depends only on its corresponding mapped user tag t_{m_j}. Therefore, our task is simplified to the estimation of the following probabilities:

$$Pr(\mathbf{f}^J|\mathbf{t}^I) = p(J) \sum_{\mathbf{m}^J} \prod_{j=1}^{J} p(m_j|I)p(f_j|t_{m_j}). \tag{5}$$

Parameter Estimation Based on EM. To solve the parameter estimation problem of Eq. 5, we follow the Maximum Likelihood (ML) criterion subject to the constraint $\sum_f p(f|t) = 1$, for each user tag t. To transform the constrained optimization problem to an unconstrained one, we use Lagrange multipliers. However, given that we have unobservable variables in our model (Eq. 3), there is no closed-form solution. According to the Expectation-Maximization (EM) algorithm, we adopt an iterative algorithm to estimate the parameters of our model. In the first step the model parameters are randomly initialized, while in the second step the initial model is used to optimize the model parameters. This process is repeated until the algorithm convergences.

Boosting Taste Keywords. Let $\mathbf{f}^I = \langle f_1 \dots f_I \rangle$ be the set of taste keywords of a venue and $\hat{\mathbf{f}} \in \mathbf{f}^I$ be the set of venue taste keywords which are mapped to user tags. According to our probabilistic approach, we consider that $\hat{\mathbf{f}}$ correlates to the user's interest more than the other venue taste keywords. Hence, we boost $\hat{\mathbf{f}}$ to model the user's interests. This helps us to address the data sparsity problem (see Sect. 1). The personalized boosted venue taste keywords are used for venue suggestion (see Sect. 3.2).

Predicting User Tags - Alternative Models. We also examine three alternative models for predicting user tags, where we use the statistical mapping ($\mathbf{m}$) between user tags and venue taste keywords to train a model and predict the user tags for a new venue. Given a list of taste keywords of a new venue, each alternative model predicts the most likely list of user tags. We examine the following models:

M1. In this model, we follow the ML criterion for Eq. 1 to find $\hat{\mathbf{t}}^I$ as described in Sect. 3.1.
M2. Assume that we have N sample pairs of user tags and venue taste keywords for each user: $S = \{(\mathbf{f}_{(1)}, \mathbf{t}_{(1)}), \dots, (\mathbf{f}_{(n)}, \mathbf{t}_{(n)}), \dots, (\mathbf{f}_{(N)}, \mathbf{t}_{(N)})\}$. We calculate N corresponding mappings: $M = \{\mathbf{m}_{(1)}, \dots, \mathbf{m}_{(n)}, \dots, \mathbf{m}_{(N)}\}$. M is then used to train a tagging model which is used to predict user tags for a new venue. In this model, we adopt Conditional Random Fields (CRF) [13] to predict user tags.
M3. This model is similar to M2. We adopt a SVM-based tagging model [12] instead of CRF to predict user tags.

3.2 Venue Suggestion Based on Multiple LBSNs

In this section, we briefly explain our method for performing venue suggestion, exploiting the scores from multiple LBSNs. We describe two sets of scores, the frequency and review-based scores, and show how to combine them to produce the final venue ranking using several learning to rank techniques.

Frequency-Based Scores. Frequency-based scores are based on the assumption that a user visits the venues that she likes more frequently than others and rates them positively[1]. We create positive and negative profiles based on contents of venues that a user has visited and calculate their corresponding normalized frequencies. A new venue is then compared with the user's profiles and we compute a similarity score. For simplicity, we only explain how to calculate the frequency-based score using venue categories. The method can be easily generalized to calculate the score for other types of information.

Given a user u and her history of rated venues $h_u = \{v_1, \dots, v_n\}$, each venue has a corresponding list of categories $C(v_i) = \{c_1, \dots, c_k\}$. We define the user category profile as follows:

---

[1] We consider reviews with rating [4, 5] as positive, 3 as neutral, and [1, 2] as negative.

Definition 1. *A **Positive-Category Profile** is the set of all unique categories belonging to venues that user u has previously rated positively. A **Negative-Category Profile** is defined analogously for venues that are rated negatively.*

We assign a user-level normalized frequency value to each category in the positive/negative category profile. The user-level normalized frequency for a positive/negative category profile is defined as follows:

Definition 2. *A **User-level Normalized Frequency** for an item (e.g., category) in a profile (e.g., positive-category profile) for user u is defined as:*

$$\mathrm{cf}_u^+(c_i) = \frac{\sum_{v_k \in h_u^+} \sum_{c_j \in C(v_k), c_j = c_i} 1}{\sum_{v_k \in h_u} \sum_{c_j \in C(v_k)} 1},$$

where h_u^+ is the set venues that users u has rated positively. A user-level normalized frequency for negative category profile, cf_u^-, is calculated analogously.

Based on Definitions 1 and 2 we create positive/negative category profiles for each user. Let u be a user and v be a candidate venue, then the category-based similarity score $S_{cat}(u, v)$ is calculated as follows:

$$S_{cat}(u, v) = \sum_{c_i \in C(v)} \mathrm{cf}_u^+(c_i) - \mathrm{cf}_u^-(c_i). \tag{6}$$

We use Eq. 6 to calculate a venue taste keyword score (S_{key}). As for boosting, for each user we generate positive and negative *boosted venue taste keyword profiles* following Definition 1 considering only the venue taste keywords that are mapped to user tags. Then, we calculate the frequency-based score for boosted keywords (S_{boost}) according to Definition 2 and Eq. 6.

We follow Definitions 1 and 2 to create positive and negative *user-tag profiles* for each user. The profiles contain user tags that each user has assigned to previously visited venues. However, since a new venue does not come with user-assigned tags, we predict user tags utilizing the alternative models of ML, CRF or SVM and consider them as user tags for the venue. We calculate a frequency-based score for the user tags predicted by ML, CRF or SVM similar to Eq. 6 and refer to them as S_{lm}, S_{crf} and S_{svm}, respectively.

Review-Based Score. To better understand the reasons a user gives a positive/negative rating to a venue, we need to dig into the reviews associated with the venue and examine what other users say about that same venue. We assume that if a user rated a venue positively, then she shares the same opinions with all other users who also gave a positive rating to that place. The same assumption also holds for negative ratings.

We adopt a binary classifier per user to learn why she likes/dislikes venues and to assign a score for a new venue[2]. The binary classifier is trained using the

---

[2] An alternative to binary classification would be a regression model, but in this case it is inappropriate due to the data sparsity, that degrades the accuracy of venue suggestion.

reviews from the venues a particular user has visited before. We use the positive training samples which are extracted from the positive reviews of liked example suggestions, and negative samples which are from the negative reviews of disliked example suggestions, analogously. Since the users' reviews may contain a lot of noise and off-topic terms, we calculate a TF-IDF score and use it as the feature vector for training the classifier. As classifier we use Support Vector Machine (SVM) [7] and consider the value of the SVM's decision function as the score (S_{rev}) since it approximates how relevant a venue is to a user profile.

Personalized Venue Ranking. Summarizing, our proposed model consists of the following scores $S_{cat}, S_{key}, S_{rev}$, and S_{boost} and is denoted as *PK-Boosting* (**P**ersonalized venue taste **K**eyword Boosting). After calculating the scores (Sect. 3.1) we combine them to produce a final ranking. We adopt several learning to rank[3] techniques for this purpose as they have proven to be effective before [14]. In particular, we examine the following learning to rank techniques: MART, RankNet, RankBoost, AdaRank, CoordinateAscent, LambdaMART, ListNet, and RandomForest. Regarding the three alternative models, we replace S_{boost} with the three other scores for each model as follows: S_{lm} for *UT-ML* (**U**ser **T**ag prediction using ML), S_{crf} for *UT-CRF*, and S_{svm} for *UT-SVM*.

4 Experimental Evaluation

4.1 Setup

Datasets. We evaluate our approach on two benchmark datasets, published by the Text REtrieval Conference (TREC). The datasets are those used in the *Batch Experiments/Phase 2* of the TREC Contextual Suggestion Track 2015 [8] and 2016[4]. We denote them as *TREC2015* and *TREC2016*, respectively. The task was to rerank a list of candidate venues in a new city for a user given her history of venue preferences in other cities. For both datasets each user has visited from 30 to 60 venues in one or two cities. Each user may have tagged venues to explain why she likes a venue. We crawled Yelp and Foursquare to get more information about venues, such as reviews and taste keywords. More in details, for each venue in the dataset, we created a query from the venue's name and location to find the corresponding Yelp and Foursquare profiles. To avoid noises in the dataset, we further verified the title and location of each result. The crawled data from both LBSNs have a big overlap. Yelp is crawled mainly for reviews, whereas Foursquare mainly for venue taste keywords. More details can be found in Table 1.

Evaluation Protocol. To perform a fair comparison, we use the official evaluation protocol and metrics of TREC for this task which are P@5 (precision at 5), nDCG@5 (Normalized Discounted Cumulative Gain at 5), and MRR (Mean

---

[3] We use the implementation of learning to rank named RankLib: https://sourceforge. net/p/lemur/wiki/RankLib/.

[4] https://sites.google.com/site/treccontext/.

Table 1. Statistical details of datasets

	TREC2015	TREC2016
Number of users	211	442
Number of venues	8,794	18,808
Number of venues crawled from Yelp	6,290	13,604
Number of venues crawled from Foursquare	5,534	13,704
Average reviews per venue	117.34	66.82
Average categories per venue	1.63	1.57
Average taste keywords per venue	8.73	7.89
Average user tags per user	1.46	3.61
Number of distinct user tags	186	150

Reciprocal Rank). As P@5 is considered the main metric for TREC2015, we also consider it as the main evaluation metric for our work. We conduct a 5-fold cross validation on the training data to tune our model.

Compared Methods. We consider our previous work [1] which is the best performing model of TREC 2015 as our baseline and denote it as *Baseline*. Baseline extracts information from different LBSNs and uses them to calculate two sets of scores based on: (1) user reviews and (2) venue content. The scores are then combined using linear interpolation. We choose this baseline for two reasons, firstly because it is the best performing run of TREC 2015, and secondly, because it also utilizes scores calculated from different LBSNs. We also compare our approach with the 3 alternative models of UT-ML, UT-CRF and UT-SVM, based on which the user tags can be predicted for a new venue (see Sect. 3.1).

4.2 Results

Performance Evaluation Against Compared Methods. Tables 2 and 3 demonstrate the performance of our approach against competitors for the TREC2015 and TREC2016 datasets, respectively. For each model we adopt the best performing learning to rank technique (see Tables 4 and 5), where the best learning to rank technique for PK-Boosting is ListNet [3]. Tables 2 and 3 show that PK-Boosting outperforms the competitors with respect to the three evaluation metrics. This shows that the proposed approach for boosting venue taste keywords improves the performance of venue suggestion. This happens because PK-Boosting solves the data sparsity problem, while at the same time it aids the ranking technique by capturing user preferences more accurately. The models UT-ML, UT-CRF and UT-SVM introduce a prediction error, when predicting user tags for a new venue, which is then propagated to venue ranking and subsequently degrades the models' performances.

Impact of Number of Visited Venues. Figure 2 reports P@5 of all models on TREC2015 and TREC2016. In this set of experiments, we vary the number

Table 2. Performance evaluation on TREC2015. Bold values denote the best scores, significant for $p < 0.05$ in paired t-test. Δ values (%) express the relative improvement, compared to the Baseline method. For each model we report the scores using the best learning to rank technique (Table 4).

	P@5	Δ(%)	nDCG@5	Δ(%)	MRR	Δ(%)
Baseline	.5858 ± .0073	-	.6055 ± .0061	-	.7404 ± .0055	-
UT-ML	.6114 ± .0048	4.37	.6241 ± .0029	3.07	.7380 ± .0023	−0.32
UT-CRF	.6161 ± .0041	5.17	.6212 ± .0043	2.59	.7302 ± .0016	−1.38
UT-SVM	.6152 ± .0058	5.02	.6256 ± .0033	3.31	.7419 ± .0014	0.20
PK-Boosting	**.6190 ± .0044**	**5.67**	**.6312 ± .0031**	**4.24**	**.7610 ± .0015**	**2.78**

Table 3. Performance evaluation on TREC2016. For each model we report the scores using the best learning to rank technique (Table 5).

	P@5	Δ(%)	nDCG@5	Δ(%)	MRR	Δ(%)
Baseline	.4656 ± .0064	-	.3055 ± .0059	-	.5975 ± .0050	-
UT-ML	.4852 ± .0036	4.21	.3239 ± .0046	6.02	.5824 ± .0010	−2.23
UT-CRF	.4820 ± .0056	3.52	.3153 ± .0038	3.21	.6214 ± .0013	4.31
UT-SVM	.4918 ± .0038	5.62	.3259 ± .0044	6.68	.6338 ± .0018	6.40
PK-Boosting	**.4951 ± .0046**	**6.36**	**.3259 ± .0032**	**6.68**	**.6480 ± .0015**	**8.78**

Table 4. Effect on P@5 for different learning to rank techniques in TREC2015. Bold values denote the best learning to rank technique per model.

	UT-ML	UT-CRF	UT-SVM	PK-Boosting
MART	.5829 ± .0026	.5915 ± .0030	.5886 ± .0028	.5943 ± .0024
RankNet	**.6114 ± .0048**	.6104 ± .0039	.6085 ± .0044	.6072 + .0027
RankBoost	.5934 ± .0029	.6019 ± .0036	.6019 ± .0039	.6038 ± .0031
AdaRank	.6028 ± .0054	.6009 + .0062	.6038 ± .0048	.5782 ± .0067
CoordinateAscent	.5924 ± .0044	.5858 ± .0048	.5848 ± .0036	.5896 ± .0063
LambdaMART	.5962 ± .0018	.5991 ± .0027	.5981 ± .0017	.6066 ± .0034
ListNet	.5991 ± .0036	**.6161 ± .0041**	**.6152 ± .0058**	**.6190 ± .0044**
RandomForests	.5736 ± .0043	.5877 ± .0062	.5810 ± .0061	.5870 ± .0043

of venues to map the taste keywords to the user tags. We calculate the scores of Sect. 3 with different number of venues and train the ranking model. Figure 2 shows that PK-Boosting achieves the highest accuracy, when compared with other models for all different number of venues. This result indicates that PK-Boosting is less prone to noise when the training set is smaller, whereas the prediction models ML and SVM are not very well trained using such a small training set. In fact, their performance is worse with smaller training sets.

Table 5. Effect on P@5 for different learning to rank techniques in TREC2016.

	UT-ML	UT-CRF	UT-SVM	PK-Boosting
MART	.4525 ± .0027	.3788 ± .0030	.3869 ± .0025	.4131 ± .0029
RankNet	.4820 ± .0042	.3672 ± .0034	**.4918 ± .0038**	.4754 ± .0039
RankBoost	.4000 ± .0025	.3574 ± .0020	.3574 ± .0027	.3574 ± .0022
AdaRank	.4557 ± .0064	.4557 ± .0050	.4557 ± .0062	.4557 ± .0059
CoordinateAscent	.4656 ± .0039	.4721 ± .0053	.4689 ± .0048	.4787 ± .0053
LambdaMART	**.4852 ± .0036**	.4557 ± .0018	.4492 ± .0028	.4820 ± .0025
ListNet	.4754 ± .0050	**.4820 ± .0056**	.4852 ± .0038	**.4951 ± .0046**
RandomForests	.4164 ± .0060	.4033 ± .0051	.4197 ± .0057	.3924 ± .0044

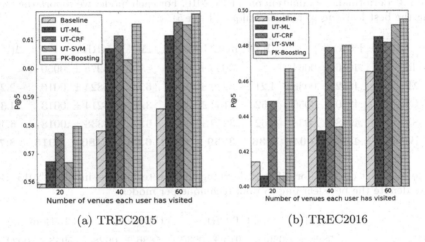

(a) TREC2015   (b) TREC2016

Fig. 2. Effect on P@5 by varying the number of venues that each user has visited for (a) TREC2015 and (b) TREC2016.

Using Information from Multiple LBSNs. Tables 6 and 7 evaluate the performance of the examined models after removing information from the different LBSNs. In this set of experiments, we report the relative drop in performance of different models when using information from the two different LBSNs. As we can see in almost all cases we observe a drop in the performance when a source of information is removed from the model. The average drop for TREC2015 is −4.96% and for TREC2016 is −10.59% which confirms the effectiveness of exploiting information from different LBSNs. This indicates that using multi-modal information from different LBSNs is a key to improve venue suggestion. For all different runs, the winning method is the proposed PK-Boosting approach, that uses a combination of information from both Foursquare and Yelp.

Table 6. Performance evaluation after removing information provided by Foursquare (F) and Yelp (Y) in the TREC2015 dataset. Δ values (%) express the relative drop, compared to the performance that each model has when using information from the two different LBSNs. (Average drop = -4.96%)

	F	Y	P@5	$\Delta(\%)$	nDCG@5	$\Delta(\%)$	MRR	$\Delta(\%)$
Baseline	✓	✓	.5858 ± .0073	-	.6055 ± .0061	-	.7404 ± .0055	-
	✓	✗	.5649 ± .0057	−3.57	.5860 ± .0062	−3.22	.7263 ± .0060	−1.90
	✗	✓	.5697 ± .0068	−2.75	.5917 ± .0056	−2.28	.7341 ± .0051	−0.85
UT-ML	✓	✓	.6114 ± .0048	-	.6241 ± .0029	-	.7380 ± .0023	-
	✓	✗	.5213 ± .0029	−14.73	.5220 ± .0043	−16.35	.6401 ± .0021	−13.26
	✗	✓	.5621 ± .0044	−8.06	.5653 ± .0035	−9.42	.6752 ± .0017	−8.51
UT-CRF	✓	✓	.6161 ± .0041	-	.6212 ± .0043	-	.7302 ± .0016	-
	✓	✗	.5621 ± .0023	−8.76	.5826 ± .0017	−6.21	.7226 ± .0027	-1.04
	✗	✓	.5991 ± .0029	−2.75	.6138 ± .0037	−1.19	.7388 ± .0029	1.17
UT-SVM	✓	✓	.6152 ± .0058	-	.6256 ± .0033	-	.7419 ± .0014	-
	✓	✗	.5640 ± .0040	−8.32	.5858 ± .0035	−6.36	.7277 ± .0039	−1.91
	✗	✓	.6047 ± .0037	−1.71	.6173 ± .0025	−1.33	.7413 ± .0026	−0.08
PK-Boosting	✓	✓	**.6190 ± .0044**	-	**.6312 ± .0031**	-	**.7610 ± .0015**	-
	✓	✗	.5630 ± .0039	−9.05	.5902 ± .0045	−6.50	.7458 ± .0026	−1.99
	✗	✓	.5934 ± .0013	−4.13	.6142 ± .0014	−2.69	.7518 ± .0013	−1.21

Table 7. Performance evaluation after removing information provided by Foursquare (F) and Yelp (Y) in the TREC2016 dataset. (Average drop = -10.59%)

	F	Y	P@5	$\Delta(\%)$	nDCG@5	$\Delta(\%)$	MRR	$\Delta(\%)$
Baseline	✓	✓	.4656 ± .0064	-	.3055 ± .0059	-	.5975 ± .0050	-
	✓	✗	.3967 ± .0053	−14.8	.2572 ± .0061	−15.81	.5916 ± .0052	−0.99
	✗	✓	.4525 ± .0051	−2.81	.2921 ± .0058	−4.39	.5736 ± .0060	−4.00
UT-ML	✓	✓	.4852 ± .0036	-	.3239 ± .0046	-	.5824 ± .0010	-
	✓	✗	.4557 ± .0017	-6.08	.3078 ± .0026	-4.97	.6458 ± .0037	10.89
	✗	✓	.4525 ± .0037	−6.74	.2971 ± .0033	−8.27	.6107 ± .0017	4.86
UT-CRF	✓	✓	.4820 ± .0056	-	.3153 ± .0038	-	.6214 ± .0013	-
	✓	✗	.3475 ± .0045	−27.91	.2213 ± .0030	−29.81	.4869 ± .0024	−21.65
	✗	✓	.4689 ± .0018	−2.72	.3162 ± .0051	0.29	.6254 ± .0021	0.64
UT-SVM	✓	✓	.4918 ± .0038	-	.3259 ± .0044	-	.6338 ± .0018	-
	✓	✗	.3508 ± .0042	−28.67	.2350 ± .0035	−27.89	.5358 ± .0036	−15.46
	✗	✓	.3836 ± .0043	−22.01	.2559 ± .0021	−21.48	.5315 ± .0038	−16.14
PK-Boosting	✓	✓	**.4951 ± .0046**	-	**.3259 ± .0032**	-	**.6480 ± .0015**	-
	✓	✗	.4459 ± .0044	−9.94	.3000 ± .0016	−7.95	.6025 ± .0017	−7.02
	✗	✓	.4557 ± .0020	−7.96	.2979 ± .0035	−8.59	.5960 ± .0019	−8.02

5   Conclusion

In this study, we presented a probabilistic generative model to map user tags to venue taste keywords. The resulted mapping allows to exploit several techniques to address the data sparsity problem for venue suggestion. We studied two directions: (1) the proposed PK-boosting model to boost venue taste keywords and (2) three alternative models to predict user tags for new venues. In addition, we explained how to incorporate the new information into a venue suggestion approach, calculating different scores from information from multiple LBSNs. Following learning to rank strategies, the final venue suggestion ranking is performed based on the computed scores. The experimental results on two benchmark datasets demonstrate that our approach beats state-of-the-art strategies. This confirms that the proposed approach, PK-Boosting, solves the data sparsity problem and captures user preferences more accurately. As future work, we plan to extend our model to capture the time dimension and perform time-aware venue suggestion, an important issue in LBSNs [15].

Acknowledgements. This work was partially supported by the Swiss National Science Foundation (SNSF) under the project "Relevance Criteria Combination for Mobile IR (RelMobIR)".

References

1. Aliannejadi, M., Bahrainian, S.A., Giachanou, A., Crestani, F.: University of Lugano at TREC 2015: contextual suggestion and temporal summarization tracks. In: Voorhees, E.M., Ellis, A. (eds.) TREC 2015. NIST (2015)
2. Aliannejadi, M., Mele, I., Crestani, F.: User model enrichment for venue recommendation. In: Ma, S., Wen, J.-R., Liu, Y., Dou, Z., Zhang, M., Chang, Y., Zhao, X. (eds.) AIRS 2016. LNCS, vol. 9994, pp. 212–223. Springer, Cham (2016). doi:10.1007/978-3-319-48051-0_16
3. Cao, Z., Qin, T., Liu, T.-Y., Tsai, M.-F., Li, H.: Learning to rank: from pairwise approach to listwise approach. In: Ghahramani, Z. (ed.) ICML 2007, vol. 227, pp. 129–136. ACM (2007)
4. Chen, L., Chen, G., Wang, F.: Recommender systems based on user reviews: the state of the art. User Model. User-Adap. Interact. **25**(2), 99–154 (2015)
5. Cheng, C., Yang, H., King, I., Lyu, M.R.: Fused matrix factorization with geographical and social influence in location-based social networks. In: Hoffmann, J., Selman, B. (eds.) AAAI 2012, pp. 17–23. AAAI Press (2012)
6. Chon, Y., Kim, Y., Cha, H.: Autonomous place naming system using opportunistic crowdsensing and knowledge from crowdsourcing. In: Abdelzaher, T.F., Römer, K., Rajkumar, R. (eds.) IPSN 2013, pp. 19–30. ACM (2013)
7. Cortes, C., Vapnik, V.: Support-vector networks. Mach. Learn. **20**(3), 273–297 (1995)
8. Dean-Hall, A., Clarke, C.L.A., Kamps, J., Kiseleva, J., Voorhees, E.M.: Overview of the TREC 2015 contextual suggestion track. In: Voorhees, E.M., Ellis, A. (eds.) TREC 2015. NIST (2015)

9. Griesner, J., Abdessalem, T., Naacke, H.: POI recommendation: towards fused matrix factorization with geographical and temporal influences. In: Werthner, H., Zanker, M., Golbeck, J., Semeraro, G. (eds.) RecSys 2015, pp. 301–304. ACM (2015)

10. He, T., Yin, H., Chen, Z., Zhou, X., Sadiq, S., Luo, B.: A spatial-temporal topic model for the semantic annotation of POIs in LBSNs. ACM TIST 8(1), 12:1–12:24 (2016)

11. Koren, Y., Bell, R.M., Volinsky, C.: Matrix factorization techniques for recommender systems. IEEE Comput. 42(8), 30–37 (2009)

12. Kudo, T., Matsumoto, Y.: Fast methods for kernel-based text analysis. In: Hinrichs, E.W., Roth, D. (eds.) ACL 2003, pp. 24–31. ACL (2003)

13. Lafferty, J.D., McCallum, A., Pereira, F.C.N.: Conditional random fields: probabilistic models for segmenting and labeling sequence data. In: Brodley, C.E., Danyluk, A.P. (eds.) ICML 2001, pp. 282–289. Morgan Kaufmann (2001)

14. Liu, T.: Learning to rank for information retrieval. Found. Trends Inf. Retr. 3(3), 225–331 (2009)

15. Liu, Y., Liu, C., Liu, B., Qu, M., Xiong, H.: Unified point-of-interest recommendation with temporal interval assessment. In: Krishnapuram, B., Shah, M., Smola, A.J., Aggarwal, C., Shen, D., Rastogi, R. (eds.) SIGKDD 2016, pp. 1015–1024. ACM (2016)

16. Rikitianskii, A., Harvey, M., Crestani, F.: A personalised recommendation system for context-aware suggestions. In: Rijke, M., Kenter, T., Vries, A.P., Zhai, C.X., Jong, F., Radinsky, K., Hofmann, K. (eds.) ECIR 2014. LNCS, vol. 8416, pp. 63–74. Springer, Cham (2014). doi:10.1007/978-3-319-06028-6_6

17. Ye, M., Shou, D., Lee, W., Yin, P., Janowicz, K.: On the semantic annotation of places in location-based social networks. In: Apté, C., Ghosh, J., Smyth, P. (eds.) SIGKDD 2011, pp. 520–528. ACM (2011)

Feature-Oriented Analysis of User Profile Completion Problem

Morteza Haghir Chehreghani[✉]

Xerox Research Centre Europe - XRCE, Meylan, France
morteza.chehreghani@xrce.xerox.com

Abstract. We study the user profile completion and enrichment problem, where the goal is to estimate the unknown values of user profiles. We investigate how the type of the features (categorical or continuous) suggests the use of a specific approach for this task. In particular, in this context, we validate the hypothesis that a classification method like K-nearest neighbor search fits better for categorical features and matrix factorization methods such as Non-negative Matrix Factorization perform superior on continuous features. We study different variants of K-nearest neighbor search (with different metrics) and demonstrate how they perform in different settings. Moreover, we investigate the impact of shifting the variables on the quality of (non-negative) factorization and the prediction error. We validate our methods via extensive experiments on real-world datasets and, finally, based on the results and observations, we discuss a hybrid approach to accomplish this task.

1 Introduction

In many applications, for example when dealing with user profiles, we encounter with incomplete datasets wherein a subset of fields/elements are unknown. For instance, in a transactional dataset, such as the transactions of account holders of a bank, some important demographic information might be unavailable. However, presence of such information might be critical to perform user behavior analysis and future prediction. Hence, an important task is to estimate correctly the unknown elements. Today, several techniques are available to accomplish this task, e.g., the methods proposed in [2,14,18,24]. A main category of such methods are developed in the context of matrix factorization, for example the Singular Value Decomposition (SVD) [12] technique. The assumption is that a low-rank representation can explain the data more sufficiently, thus, it can also be used to estimate the unknown values. The low-rank representation is robust with respect to the individual values or the entries of the dataset. Hence, the unknown values are filled by an initial value and then are estimated by the product of the respective factors. Several alternative methods have been proposed in this context, see for example [1,16,21,22,26,27,30,32].

On the other hand, one might formulate the problem as a classification task wherein the unknown element plays the role of the *target variable*. Then, the

© Springer International Publishing AG 2017
J.M. Jose et al. (Eds.): ECIR 2017, LNCS 10193, pp. 304–316, 2017.
DOI: 10.1007/978-3-319-56608-5_24

goal is to use the other fields as *training data* to estimate the value of this element. There exist a wide range of different classification methods. Examples are Support Vector Machines [5], Logistic Regression [10], and K-nearest neighbor classification. A main advantage of K-nearest neighbor method is that it does not require a training phase, thus it is more suitable for very large-scale applications, as well as for the cases wherein the training dataset is changing regularly (which is the case in our problem). K-nearest neighbor search has also applications in non-parametric density estimation and regression.

Thus, in this paper, we formulate this problem in two different ways and investigate in detail the suitability of each approach in different situations. In the first approach, we consider the problem as a classification task, where the unknown element (as the target variable) takes a value from C different possibilities. For this purpose, we investigate different variants of K-nearest neighbor (K-NN) classification and study in detail which distance measure is more appropriate for specific settings. In particular, we demonstrate that K-nearest neighbor search with Minimax distances performs better for very sparse user profiles, because of taking the transitive relations into account. Since C is preferred to be small, thus, we suppose that this approach is more appropriate for *categorical* features. In the second approach, we study the problem as a matrix factorization task and obtain a low-rank representation to estimate the unknown values. Thereby, due to non-negativity of the variables, we apply the Non-negative Matrix Factorization (NMF) method [21] and estimate the unknown values via multiplication of the respective factors. In order to improve the performance of this approach, we propose a regularized variant via shifting the elements of the user profile matrix. Our hypotheses is that matrix factorization methods are more suitable for *continuous* features, i.e., when C is large. We validate these methods via extensive and detailed experiments on real-world datasets. Finally, based on our experimental observations, we discuss a hybrid approach to accomplish this task.

The rest of the paper is organized as following. In Sect. 2, we introduce the methods and the evaluation criteria that we will use in this paper. In Sect. 3, we describe in detail our experiments and observations. Finally, we conclude the paper in Sect. 4, with a discussion on the choice of appropriate method(s) for user profile completion.

2   User Profile Completion Methods

In this section, we describe the methods and the evaluation criteria that will be used in our experiments. In particular, we discuss the different variants of K-nearest neighbor search and Non-negative Matrix Factorization.

2.1   Definitions and Problem Setup

We are given the $N \times D$ dataset **D** containing the information of N user profiles indexed by $i \in O = \{1, \ldots, N\}$. Each profile i is specified by D features (variables), where some of them might be unknown/missing. We might be given a matrix $\mathbf{X}$, where $\mathbf{X}_{ij}$ indicates the pairwise (e.g., the Euclidean) distance between the pair of user profiles i and j.[1] Then, the dataset can be represented by graph $\mathcal{G}(\mathbf{O}, \mathbf{X})$, whose nodes are the user profile indices $\mathbf{O}$ and the edge weights are the pairwise distances $\mathbf{X}$. $\mathcal{M}$ refers to the row and column indices of the unknown elements of $\mathbf{D}$, i.e.,

$$\mathcal{M} = \{(i, d) | \mathbf{D}_{id} \ is \ unknown\}. \tag{1}$$

2.2 Methods

As mentioned, in the first approach, we consider the problem as a classification task, where the unknown elements take a value from C different possibilities. Because the classification problem and as well as the training dataset changes from one unknown element to the other, thus, we aim to obviate the need for a training phase. For this purpose, we employ K-nearest neighbor (K-NN) classification method. In particular, we investigate three variants of K-nearest neighbor classification.

1. STND: Standard K-NN with the basic distance measure $\mathbf{X}$.
2. LINK: Link-based K-NN based on shortest distance algorithm [6].
3. Minimax: K-NN based on Minimax distances [19,20].

It has been shown that K-NN with the basic distance measure might fail to capture the correct underlying structure of the data (see, for example, [20]). Thus, instead, the use of Link-based methods [3,9] or Minimax distances [4,19,20] has been proposed. Link-based distance measures are usually obtained by inverting the Laplacian of the distance matrix treated as a kernel matrix [8,31]. However, this method is computationally expensive and its runtime is cubic with respect to the number of nodes of the graph [31]. Thereby, it is not applicable to large-scale datasets. Thus, in this paper we employ an approximate but computationally efficient Link-based method which is computed according to the shortest distance algorithm [6,29] and thus can be employed even on large-scale datasets.

The use of Minimax distances with K-nearest neighbor search was first investigated in [19]. Given graph $\mathcal{G}(\mathbf{O}, \mathbf{X})$, the goal is to compute the K nearest neighbors of the node (user profile) v based on Minimax distances. The Minimax distance between two nodes i and j is defined as the minimum of largest gap among all existing paths between them, i.e.,

$$\mathbf{X}_{i,j}^{MM} = \min_{r \in \mathcal{R}_{ij}(\mathcal{G})} \left\{ \max_{1 \le l \le |r|-1} \mathbf{X}_{r(l)r(l+1)} \right\}, \tag{2}$$

where $\mathcal{R}_{ij}(\mathcal{G})$ is the set of all paths between nodes i and j over graph $\mathcal{G}$. Each path r is identified by the sequence of the respective nodes belonging to

---

[1] The unknown elements might be filled by a default value before computing the pairwise distances.

that, i.e., r_l refers to the lth node index on the path. Note that there might exist exponentially many paths between i and j, which makes their explicit enumeration computationally infeasible.

The method in [19] proposes a computationally feasible algorithm based on message passing with forward and backward steps. The runtime of this method is $\mathcal{O}(KN)$, which is in principle identical to the runtime of the standard K-nearest neighbor search, but the algorithm performs several visits of the training dataset. In addition, this method requires to obtain a priori a minimum spanning tree (MST) over the graph which might take $\mathcal{O}(N^2)$ time. Later, a greedy algorithm has been developed that uses Fibonacci heaps to perform K-nearest neighbor search and its runtime is $\mathcal{O}(\log N + K \log K)$ [20]. Nevertheless, this method is limited to the sparse graphs built according to Euclidean distances. The recent method in [4] establishes a two-step procedure to compute the Minimax K nearest neighbors on general graphs with arbitrary distance measures, whose runtime is linear. This method can be interpreted according to the equivalence of pairwise Minimax distances over a graph and over any minimum spanning tree constructed on that [17]. Different minimum spanning tree algorithms usually follow a greedy procedure, which at each step, add a new node to the tree according to a greedy criterion [11]. Thereby, an efficient strategy would find only the first K (nearest) Minimax neighbors of the target node (user profile) v, instead of building a complete tree. Then, the question is which minimum spanning tree algorithm gives the first K Minimax neighbors of v, if it is started from v and is stopped after K steps? Theorem 1 shows that truncated Prim's algorithm (i.e., running the Prim's algorithm for only K steps), satisfies this condition.

Theorem 1. *When running the Prim's algorithm, the node visited earlier has a smaller (or equal) Minimax distance to/from the starting node v than a node visited later.*

Proof. Let assume node p is visited after node q, but its Minimax distance to v is smaller than the Minimax distance between v and q. Then, we show that this yields a contradiction. We consider two cases:

1. On the minimum spanning tree, p is connected to v via q. Then, the Minimax distance between p and v cannot be smaller than the Minimax distance between q and v, since there is only one path between p and v which passes through q. Then the largest weight on the path $p \to q \to v$ cannot be smaller than the largest weight on the path $q \to v$.
2. The path between p and v does not meet q. At each step, the algorithm adds a new node whose distance to the set $v \cap \mathcal{N}(v)$ is minimal. If the Minimax distance between p and v is smaller than the Minimax distance between q and v, then the weights of all the edges on the path $v \to p$ are smaller than the maximal weight on the path $v \to q$. This implies that the algorithm meets p before the edge with maximal weight on the path $v \to q$, i.e. it selects p before q, which is a contradiction. $\square$

The Prim's algorithm [28] starts from an initial node and expands the tree connected to that until it includes all other nodes, i.e., it grows and expands only a single tree instead of several of them. Therefore, to provide efficiency, we compute a truncated Prim minimum spanning tree starting from the target node v, which at the same time, gives its first K Minimax neighbors. In this approach, each new node constitutes the next Minimax nearest neighbor of the initial node, thus, performing exactly K growth steps is sufficient. The runtime of this method is $\mathcal{O}(KN)$ which is equal to the standard method.

In the second approach, we consider the task as a collaborative filtering problem and use the matrix factorization techniques to compute a low-rank representation for the original data. The data and the components are non-negative; thus, we employ the Non-negative Matrix Factorization (NMF) method [21,22]. NMF finds a decomposition of the data matrix $\mathbf{D}$ into two matrices $\mathbf{W}$ and $\mathbf{H}$ of non-negative elements, by optimizing for the squared Frobenius norm:

$$\arg\min_{\mathbf{W},\mathbf{H}} ||\mathbf{D} - \mathbf{WH}||^2 = \sum_{i,j}(\mathbf{D}_{ij} - \mathbf{W}_{i,.}\mathbf{H}_{.,j})^2. \qquad (3)$$

We, then, estimate an unknown element by multiplying the corresponding factors. However, there exist several variants of this factorization in the literature. In our experiments, we investigate some of them, i.e., the sparse variant [15], semi NMF [7], non-negative rank factorization [1], the total variation norm form [32], and with L_1 regularization [16]. However, the results are very similar, i.e., we do not observe a statistically significant difference among them on our datasets. On the other hand, in our datasets, the entries are non-negative, i.e., in some cases the variables are lower bounded by a positive number which is significantly larger than zero. Since we only require the non-negativity, thus, we propose to shift the elements of the dataset by a constant such that the minimum element in each column becomes zero. Then, after estimating an unknown element by the product of the respective factors, in order to obtain the unshifted value, we add the shift to the estimated value. This type of regularization, as our experiments will show, yields a significantly better prediction.

Notice that one must take two important considerations into account when comparing the K-NN variants and NMF: (i) K-NN is significantly faster than NMF, thereby we have more focus on this approach in this paper.[2] (ii) K-NN and NMF are not necessarily competitors. Rather, they might be orthogonal and complement of each other. NMF can be seen as a preprocessing step, which provides a possibly more suitable data representation. One can then apply K-NN on top of the NMF results.[3]

---

[2] Matrix factorization methods often require $\mathcal{O}(N^3)$ or $\mathcal{O}(N^2 \log N)$ runtime for training (and then they need to do matrix multiplication for estimation), whereas the runtime of different variants of K-NN is $\mathcal{O}(N)$ (more precisely $\mathcal{O}(KN|\mathcal{M}|)$ for $|\mathcal{M}|$ unknown elements).

[3] In our experiments, we did not observe a significant improvement when applying K-NN variants on the NMF results, instead of the original dataset.

2.3   Evaluation Criteria

To measure the quality of different methods, one can compare the estimated values of the unknown elements against the true values and report an error if they mismatch. Thereby, this *binary error* is defined as

$$error_{bin} = \frac{\sum_{(i,d)\in\mathcal{M}} \mathbb{I}_{\{\mathbf{D}_{id}\neq\widehat{\mathbf{D}_{id}}\}}}{|\mathcal{M}|}, \tag{4}$$

where $\mathbb{I}_{\{.\}}$ is the indicator function, and returns 1 is the condition is true and 0 otherwise. $\widehat{\mathbf{D}}_{id}$ shows the estimated value for the dth unknown element of the user profile i. $error_{bin}$ computes the error in a very strict way, i.e., it does not discriminate between a very good (but still not exact) estimation and a poor estimation. Thus, we take into account the quality of wrong estimates and compute a second type of prediction error as following:

$$error = \frac{\sum_{(i,d)\in\mathcal{M}} \frac{|\mathbf{D}_{id}-\widehat{\mathbf{D}_{id}}|}{r_d}}{|\mathcal{M}|}, \tag{5}$$

where, r_d shows the range of the dth feature, i.e., $r_d = \max_i \mathbf{D}_{id} - \min_i \mathbf{D}_{id}$.

3   Experiments

We perform our experiments on the following two datasets: (i) *CoIL*: This dataset includes the information of $5,823$ customers of an insurance company. The data consists of 86 variables and involves the product usage and socio-demographic data. This dataset was used in the CoIL 2000 Challenge [13]. (ii) *Wholesale*: This dataset contains the annual spending in monetary units on diverse product categories for 440 clients of a wholesale distributor. There are in total 8 categories about the product and geographical information. Among them, two variables (*Channel* and *Region*) are categorical and the rest are integer continuous variables. Additionally, we have performed similar experiments on a bank marketing dataset [25] as well as on the Statlog dataset (German credit data) [23]. However, since the observations and the results are similar and consistent. Thereby, due to space limit, we do not report them in this paper.

These datasets are complete, i.e., they do not include any unknown/missing element. Thus, we remove some elements according to a fixed parameter s which identifies the probability that an element is not given. We then employ the different methods and variants (K-NN or NMF) to estimate the unknown values and compute the prediction error. The matrix of pairwise distances $\mathbf{X}$ is obtained by computing the pairwise squared Euclidean distances between the user profiles. Before computing the pairwise distances, we initialize the unknown elements by a default value, e.g. zero, the mean of the column, the most frequent value of the column. These different initialization yield very similar results. We repeat the experiments for 50 different random realizations of the sparsity (w.r.t. a fixed s) and report the average results.

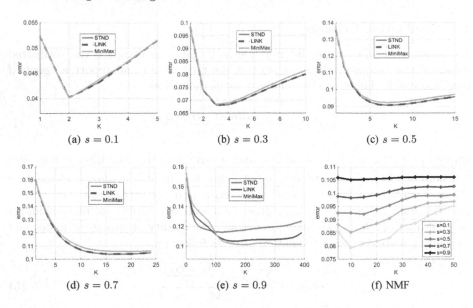

(a) $s = 0.1$ (b) $s = 0.3$ (c) $s = 0.5$

(d) $s = 0.7$ (e) $s = 0.9$ (f) NMF

Fig. 1. Prediction error of different variants of K-NN as well as NMF on complete the Coil data. In most cases, K-NN works better than NMF, since the variables are categorical.

3.1 CoIL Dataset

We first study the performance of different variants of K-NN for completing the unknown elements. Figure 1 shows the prediction error when we use different metrics with K-NN method, for different values of the sparsity parameter s and for different choices of K. We observe that: (i) For very small s, i.e. $s = 0.1, 0.3$, the methods perform very similarly, such that there is no significant distinction between them. In this case, the optimal K is 2 and 3. For small K the neighbors selected by different metrics are almost the same. This explains why the results are very close, since in this setting, the different variants compute similar neighbors. (ii) As we increase the sparsity, i.e., to $s = 0.5, 0.7$, then the standard K-NN with basic metric and with Link work slightly better. However, the improvement is not statistically distinguishable, i.e., we do not consistently observe this improvement among different realizations of sparsity. (iii) For the very sparse setting, i.e., when $s = 0.9$, K-NN with Minimax distances performs significantly better. The explanation is that when the data is very sparse, then the standard K-NN (or any other classification method) fails due to lack of availability of enough measurements. One effective way to improve data representation is to establish indirect or transitive relations. In other words, if user profile X is similar to Y, and Y is similar to Z, then we assume that X is similar to Z too, although they might not be similar based on their direst relation (perhaps due to sparsity). Minimax distances extend this kind of transitivity to an arbitrary number of intermediate nodes (user profiles). Thus, Minimax distances improve the measurements by establishing meaningful relations, and as a result, training and learning, for example for K-NN, can be performed in a more effective and proper way. (iv) In general, as we increase the sparsity s, the optimal number of nearest neighbors K increases too, in order to compensate the reduction in the number of appropriate measurements, which leads to increasing the prediction error as well.

Figure 1(f) shows the prediction error for the Non-negative Matrix Factorization approach. The results are shown for different sparsity s and for different number of components K. We observe: (i) Although the dimensionality of the original dataset is 86, there is a lower dimensional data representation which yields a smaller error, i.e., for $K = 10 \sim 20$ (for NMF, K indicates the rank of the new representation). (ii) K-NN and NMF behave in a consistent way. For small s, the error is smaller and as we increase the sparsity, the error increases too.

By comparing K-NN with NMF, we observe that K-NN performs better than NMF. The explanation is that the features of the CoIL dataset are *categorical*. Thus, considering the problem as a classification task where the different categories play the role of different class labels, seems a more reasonable strategy. For $s = 0.7$, NMF works slightly better, although it is slower. The reason could be that in this setting, the underlying structure changes such that a low dimensional embedding represents it more appropriately, thus NMF works better than K-NN. For $s = 0.9$, the data matrix is very sparse, thus, the classification method on the original data might fail, due to lack of enough measurements. Minimax distances provide a meaningful way to improve the pairwise distances through taking the transitive relations into account.

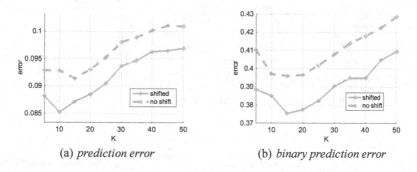

    (a) *prediction error*   (b) *binary prediction error*

Fig. 2. Shifting the features such that the lower bounds are is zero yields a better performance and reduces both types of the errors for the NMF method.

Shift of Features for NMF. In the original CoIL data, different variables have different lower bounds, e.g., some are bounded by zero and some others by one. However, since NMF requires the matrices to be only non-negative, therefore, we shift the entries of the original dataset such that the minimum of each feature (variable) becomes zero. Here, we study the impact of the shift and show why it is important and useful. Figure 2 shows that such a transformation reduces the error for $s = 0.3$. This observation in consistent for other values of s too. Hence, in all of our experiments, we first normalize the columns (features) such that their lower bounds equal 0.

3.2   Wholesale Dataset

In the CoIL dataset, the features are discrete (categorical) which explains why K-NN performs better than NMF. In the second experiment, we investigate the Wholesale dataset wherein some of the features are continuous and some others are categorical. This study is designed to validate the hypothesis that K-NN suits better for categorical variables and matrix factorization methods like NMF are more appropriate for continuous variables. Figure 3 shows the results on the whole Wholesale dataset. We observe: (i) K-NN works better than NMF. However, the difference in the prediction error of K-NN and NMF is less compared to the CoIL dataset. The reason is that the Wholesale dataset includes both categorical and continuous features, thus K-NN is not always the best option (i.e., not for all missing elements), as we will see later. (ii) Different variants of K-NN (STND, LINK or MiniMax) perform very similarly. (iii) For all methods and the different variants, as we increase the sparsity parameter s, the difficulty of the problem increases too, which in turn leads to increasing the prediction error, and as well as to a larger optimal K.

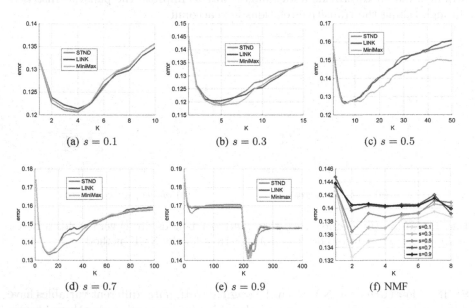

Fig. 3. Prediction error of different variants of K-NN and NMF for different sparsity s applied to the Wholesale dataset. The K-NN variants perform better than NMF.

In the following, we investigate the features of this dataset in more detail. We split the dataset into two subsets: (i) WholesaleCon: the subset that contains the continuous features, and (ii) WholesaleDis: the subset which includes the categorical features.

Figure 4 shows the results for the WholesaleCon subset. We observe: (i) NMF performs better than all different variants of K-NN. As mentioned before, the reason is that the features of this subset are continuous, thus the classification approach might not give the best solution. (ii) For all different variants of K-NN the optimal K is very large, almost close to the number of user profiles in the dataset. The explanation is that for such a setting, the number of categories (class labels), is very large (because the possible distinct values of the target variable is very large). Hence, K should be selected a very large number as well, to provide a sufficient distinction among different classes.

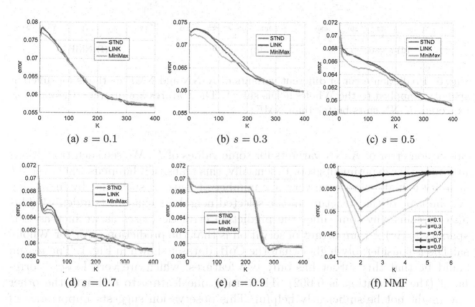

(a) $s = 0.1$ (b) $s = 0.3$ (c) $s = 0.5$

(d) $s = 0.7$ (e) $s = 0.9$ (f) NMF

Fig. 4. Prediction error of different variants of K-NN and NMF applied to the WholesaleCon subset. The features are continuous, so as expected, NMF performs better than K-NN.

Finally, we analyze the WholesaleDis subset in Fig. 5. This subset contains the categorical features. We observe: (i) The different K-NN variants perform better than NMF, since the features are categorical. (ii) The Link and Minimax variants of K-NN perform slightly better than the STND variant. The reason is that the WholesaleDis subset contains only two features. Thus, when we sparsify it, there do not exist many meaningful measurements left, as they are replaced by the initial value for the unknown values. The Link and the Minimax variants improve the direct relations, via investigating the transitive relations and exploring the indirect neighbors too. (iii) There are strong transitions in the

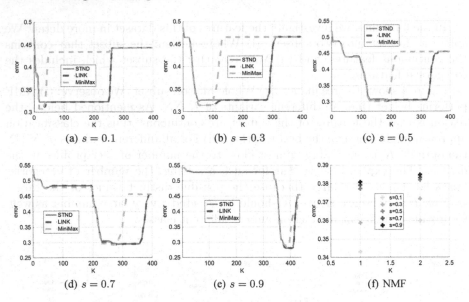

Fig. 5. Prediction error of different variants of K-NN and NMF methods for different sparsity s applied to the WholesaleDis subset. The features are discrete, therefore, as expected, K-NN suits better than NMF.

prediction error of K-NN variants for some values of K. We see such transitions in some other experiments too. Essentially, this transition happens whenever the sparsity is high or we have very few number of features such that by increasing K, the type of the structure in the selected neighbors changes (suddenly). This also explains why sometimes the prediction error decreases as we increase the sparsity s. (iv) An important observation is that the prediction error for WholesaleDis is considerably higher than the CoIL dataset shown in Fig. 1. The reason could be that this subset has only two features, which are very loosely correlated (the correlation is 0.062). Thus using one feature to estimate the other one might not be sufficiently helpful. This observation suggests importance of feature selection before performing K-NN.

4   Conclusion and Discussion

We studied the different aspects of the user profile completion problem, particularly the impact of the type of the features on the choice of appropriate methods. We showed that for categorical features a classification viewpoint (e.g., K-NN) works better, whereas for continuous variables matrix factorization methods (e.g., NMF) can yield a superior performance. We in particular analyzed the different aspects of these methods, e.g., the use of Minimax distances when the user profile matrix is very sparse, and shifting the features for NMF such that the lower bound becomes zero. Notice that shifting the features does not have any impact on the K-NN results. Moreover, the K-NN approach is significantly faster than the NMF approach, and it scales better to large datasets.

This study may suggest the following algorithmic procedure to accomplish the user profile completion task.

1. Check the features, shift them such that their lower bounds are zero.
2. Split the dataset into two subsets: dataCon (including continuous features) and dataDis (including categorical features).
3. For dataCon, use (the shift regularized) NMF to estimate the unknown values.
4. For dataDis,
   (a) if the subset is very sparse, then use K-NN with Minimax distances to complete the matrix.
   (b) otherwise, use the standard K-NN for this purpose.

However, we might add an initial step which checks the mutual dependency (e.g., the correlation coefficient) between the features in order to involve only those features in estimating an unknown element which have a strong impact.

References

1. Berman, A., Plemmons, R.J.: Nonnegative Matrices in the Mathematical Sciences. Academic Press, New York (1994)
2. Candes, E.J., Recht, B.: Exact matrix completion via convex optimization. Found. Computat. Math. **9**(6), 717–772 (2009)
3. Chebotarev, P.: A class of graph-geodetic distances generalizing the shortest-path and the resistance distances. Discrete Appl. Math. **159**(5), 295–302 (2011)
4. Chehreghani, M.H.: K-nearest neighbor search and outlier detection via minimax distances. In: Proceedings of the 2016 SIAM International Conference on Data Mining, Miami, Florida, USA, 5–7 May 2016, pp. 405–413 (2016)
5. Cortes, C., Vapnik, V.: Support-vector networks. Mach. Learn. **20**(3), 273–297 (1995)
6. Dijkstra, E.W.: A note on two problems in connexion with graphs. Numer. Math. **1**, 269–271 (1959)
7. Ding, C.H.Q., Li, T., Jordan, M.I.: Convex and semi-nonnegative matrix factorizations. IEEE Trans. Pattern Anal. Mach. Intell. **32**(1), 45–55 (2010)
8. Fouss, F., Francoisse, K., Yen, L., Pirotte, A., Saerens, M.: An experimental investigation of kernels on graphs for collaborative recommendation and semisupervised classification. Neural Networks **31**, 53–72 (2012)
9. Fouss, F., Pirotte, A., Renders, J.-M., Saerens, M.: Random-walk computation of similarities between nodes of a graph with application to collaborative recommendation. IEEE Trans. Knowl. Data Eng. **19**(3), 355–369 (2007)
10. Freedman, D.A.: Statistical Models: Theory and Practice. Cambridge University Press, Cambridge (2009)
11. Gabow, H.N., Galil, Z., Spencer, T., Tarjan, R.E.: Efficient algorithms for finding minimum spanning trees in undirected and directed graphs. Combinatorica **6**(2), 109–122 (1986)
12. Golub, G., Reinsch, C.E.R.: Singular value decomposition and least squares solutions. Numer. Math. **14**(5), 403–420 (1970)

13. Hettich, S., Bay, S.D.: The UCI KDD Archive, Irvine, CA. University of California, Department of Information and Computer Science (1999). http://kdd.ics.uci.edu

14. Hogben, L.: Graph theoretic methods for matrix completion problems. Linear Algebra Appl. **328**, 161–202 (2001)

15. Hoyer, P.O.: Non-negative matrix factorization with sparseness constraints. J. Mach. Learn. Res. **5**, 1457–1469 (2004)

16. Hsieh, C.-J., Dhillon, I.S.: Fast coordinate descent methods with variable selection for non-negative matrix factorization. In: 17th ACM SIGKDD International Conference on Knowledge Discovery and Data Mining, pp. 1064–1072 (2011)

17. Hu, T.C.: The maximum capacity route problem. Oper. Res. **9**, 898–900 (1961)

18. Johnson, C.R.: Matrix completion problems: a survey. Matrix Theory Appl. **40**, 171–176 (1990)

19. Kim, K.-H., Choi, S.: Neighbor search with global geometry: a minimax message passing algorithm. In: ICML, pp. 401–408 (2007)

20. Kim, K.-H., Choi, S.: Walking on minimax paths for K-NN search. In: AAAI (2013)

21. Lee, D.D., Seung, H.S.: Learning the parts of objects by nonnegative matrix factorization. Nature **401**, 788–791 (1999)

22. Lee, D.D., Seung, H.S.: Algorithms for non-negative matrix factorization. In: Advances in Neural Information Processing Systems 13, pp. 556–562. MIT Press (2001)

23. Lichman, M.: UCI Machine Learning Repository, Irvine, CA. University of California, School of Information and Computer Science (2013). http://archive.ics.uci.edu/ml

24. Meyer, C.D. (ed.): Matrix Analysis and Applied Linear Algebra. Society for Industrial and Applied Mathematics, Philadelphia (2000)

25. Moro, S., Cortez, P., Rita, P.: A data-driven approach to predict the success of bank telemarketing. Decis. Support Syst. **62**, 22–31 (2014)

26. Morse, A.S.: A gain matrix decomposition and some of its applications. Syst. Control Lett. **21**, 1–10 (1993)

27. Piziak, R., Odell, P.L.: Full rank factorization of matrices. Math. Mag. **72**, 193–201 (1999)

28. Prim, R.C.: Shortest connection networks and some generalizations. Bell Syst. Tech. J. **36**(6), 1389–1401 (1957)

29. Tenenbaum, J.B., de Silva, V., Langford, J.C.: A global geometric framework for nonlinear dimensionality reduction. Science **290**(5500), 2319 (2000)

30. Hongguo, X.: An SVD-like matrix decomposition and its applications. Linear Algebra Appl. **368**, 1–24 (2003)

31. Yen, L., Saerens, M., Mantrach, A., Shimbo, M.: A family of dissimilarity measures between nodes generalizing both the shortest-path and the commute-time distances. In: KDD, pp. 785–793 (2008)

32. Zhang, T., Fang, B., Liu, W., Tang, Y.Y., He, G., Wen, J.: Total variation norm-based nonnegative matrix factorization for identifying discriminant representation of image patterns. Neurocomputing **71**(10–12), 1824–1831 (2008)

Generating Descriptions of Entity Relationships

Nikos Voskarides[1](✉), Edgar Meij[2], and Maarten de Rijke[1]

[1] University of Amsterdam, Amsterdam, The Netherlands
{n.voskarides,derijke}@uva.nl
[2] Bloomberg L.P., London, UK
edgar.meij@acm.org

Abstract. Large-scale knowledge graphs (KGs) store relationships between entities that are increasingly being used to improve the user experience in search applications. The structured nature of the data in KGs is typically not suitable to show to an end user and applications that utilize KGs therefore benefit from human-readable textual descriptions of KG relationships. We present a method that automatically generates textual descriptions of entity relationships by combining textual and KG information. Our method creates sentence templates for a particular relationship and then generates a textual description of a relationship instance by selecting the best template and filling it with appropriate entities. Experimental results show that a supervised variation of our method outperforms other variations as it best captures the semantic similarity between a relationship instance and a template, whilst providing more contextual information.

1 Introduction

Results displayed on a modern search engine result page (SERP) are sourced from multiple, heterogeneous sources. For so-called organic results it has been known for a long time that result snippets, i.e., brief descriptions explaining the result item and its relation to the query, positively influence the user experience [20]. In this paper, we focus on generating descriptions for results sourced from another important ingredient of modern SERPs: knowledge graphs. Knowledge graphs (KGs) contain information about entities and their relationships. A large and diverse set of search applications utilize KGs to improve the user experience. For instance, web search engines try to identify KG entities in queries and augment their result pages with knowledge graph panels that provide contextual entity information [3, 12]. Such panels usually focus on a single entity and may include attributes of the entity and other, related entities.

Entities can be connected with more than one relationship in a KG, however. For example, two actors might have appeared in the same film, be born in the same country and also be partners. Recent work has focused on finding relationships between a pair of entities and ranking the relationships by a predefined relevance criterion [5]. When using relationships in real-world search applications, with SERPs being the prime example, a crucial problem is that they are

© Springer International Publishing AG 2017
J.M. Jose et al. (Eds.): ECIR 2017, LNCS 10193, pp. 317–330, 2017.
DOI: 10.1007/978-3-319-56608-5_25

typically represented in a formal manner that is not suitable to present to an end user. Instead, human-readable descriptions that verbalize and provide context about entity relationships are more natural to use [7]. They can be used, e.g., for entity recommendations [2] or for KG-based timeline generation [1].

Descriptions of KG relationships themselves are usually not included in large-scale knowledge graphs and previous work on automatically generating such descriptions has either relied on hand-crafted templates [1] or on external text corpora [22]. The main limitations of the former are that manually creating these templates is expensive, not generalizable, and thus it does not scale well. The latter approach is limited as the underlying text corpus may not contain descriptions for all certain relationship instances; it will not produce meaningful results for instances that do not appear in the text corpus.

We propose a method that overcomes these limitations by automatically generating descriptions of KG entity relationships. Since there exist textual descriptions of a certain relationship for some relationship instances, we aim to use these descriptions to learn how the relationship is generally expressed in text and use this information to generate descriptions for other instances of the same relationship. Existing relationship descriptions are usually complex and tailored to the entities they discuss. Also, it is likely that the KG does not contain all the information included in a description. For example, the KG might not contain any information about the second part of the following sentence: "*Catherine Zeta-Jones starred in the romantic comedy The Rebound, in which she played a 40-year-old mother of two ...*". Nevertheless, descriptions of the same relationship share patterns that are specific to that relationship. Therefore, we first create sentence templates for a certain relationship and then, for a new relationship instance, we select appropriate templates, which we formulate as a ranking problem, and fill them with the appropriate entities to generate a description.

We propose a method that generates descriptions of entity relationships for a relationship instance given a knowledge graph and a set of relationship instances coupled with their descriptions; we evaluate this method using an automatic and manual evaluation method, and release the datasets used to the community.[1] We show that we generate contextually rich relationship descriptions that are meant to be valid under the KG closed-world assumption. Moreover, our template-based method is naturally robust against KG incompleteness, since in the case of lack of contextual information about the relationship instance, it can still generate a basic description.

2   Related Work

Web search engine result pages (SERPs) can be augmented with information about the query and the documents from KGs in order to improve the user experience [12]. Also, SERPs can be augmented with textual descriptions and/or summaries with a prominent example being snippet generation for web search [20,21]. Closest to our setting, relationship descriptions have been studied in the context

---

[1] https://github.com/nickvosk/ecir2017-gder-dataset/.

of providing evidence for entity recommendation for web search [22] and timeline generation for knowledge base entities [1]. Our task, generating a description of a relationship instance given a KG, is similar to event headline generation, where the task is to generate a short sentence that summarizes a specific event. Similar to our templates, the headline patterns constructed in [17] consist of words and entity slots. Our method differs however, since relationships are more general than events and we thus have to deal with ambiguity at generation time when selecting which template matches a relationship instance.

Our task is also similar to concept-to-text generation, where the task is to generate a textual description given a set of database records [18]. In this context, our task is most closely related to [10, 19]. Saldanha et al. [19] use a template-based approach for generating company descriptions from Freebase. They construct sentence templates by replacing the entities in existing sentences by the Freebase relation of the entity to the company (e.g., ⟨company⟩ was founded by ⟨founder⟩). They add a preprocessing step where they remove phrases from the sentence that contain entities that are not connected to the company directly. At generation time, the authors replace the entity slots with the appropriate entities. Lebret et al. [10] propose a neural model to generate the first sentence of a person's biography in Wikipedia conditioned on Wikipedia infoboxes. Our setting is different from these papers since our generated descriptions are neither restricted to having entities that are directly connected to the subject entity in a KG nor need they be contained in a Wikipedia infobox.

3   Problem Definition

In this section we formally define the task of generating descriptions of entity relationships. Table 1 lists the main notation we use in the paper.

3.1   Prelimilaries

Let $\mathcal{E}$ be a set of entities and $\mathcal{P}$ a set of predicates. A *knowledge graph* $\mathcal{K}$ is a set of triples $\langle s, p, o \rangle$, where $s, o \in \mathcal{E}$ and $p \in \mathcal{P}$. We follow the closed-world assumption for $\mathcal{K}$ and use Freebase as our knowledge graph [4, 15]. A *sentence* a is a sequence of words $[v_1, \ldots, v_n]$, where each $v_i \in a$ is also in $\mathcal{V}$. Non-overlapping sub-sequences of a might refer to a single entity $e \in \mathcal{E}$.

A *relationship* r is a logical form in λ-calculus that consists of two lambda variables (x and y), at least one predicate, and zero or one existential variables [24]. Lambda variables can be substituted with Freebase entities, excluding compound value type (CVT) entities.[2] Existential variables, on the other hand, can be substituted with Freebase entities, including CVT entities. For example, the logical form of the relationship *starsInFilm* is $\lambda x. \lambda y. \exists z. actor\_film(x, z) \wedge film\_starring(z, y)$. Figure 1 shows the equivalent graphical representation of this relationship.

---

[2] CVT entities are special entities in Freebase that are used to model attributes of relationships (e.g., date of marriage).

Table 1. Glossary.

Symbol	Description
$\mathcal{K}$	Knowledge graph
$\mathcal{E}$	Set of entities
$\mathcal{P}$	Set of predicates
$\langle s, p, o \rangle$	Knowledge graph triple with $s, o \in \mathcal{E}$ and $p \in \mathcal{P}$
v	Word in vocabulary $\mathcal{V}$
a	Sentence
r_i	Relationship instance of relationship r
T_r	Set of templates $t \in T_r$ for relationship r
R_t	Set of relationship instances that support the template t
X	Set of pairs $\langle r_{i'}, y' \rangle$, where y' is a textual description (a single sentence)
C	Mapping from an entity to an entity cluster
K	Entity dependency graph of a sentence
G	Compression graph
P	Set of paths in G

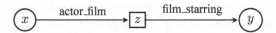

Fig. 1. Graphical representation of the logical form of the *starsInFilm* relationship. Lambda variables are shown in circles and existential variables in rectangles.

A pair $r_i = r\langle s, o \rangle$ is a *relationship instance* of r for entities $s, o \in \mathcal{E}$ if by substituting $x = s$ and $y = o$ in r and by executing the resulting logical form in the knowledge graph $\mathcal{K}$ we get at least one result. For example, *starsInFilm(BradPitt, Troy)* is a relationship instance of the *starsInFilm* relationship.

3.2 Task Definition

We assume that a relationship instance r_i can be expressed with a human-readable description (such as a single sentence) that contains mentions of both s and o and possibly other entities which may provide contextual information for the relationship r or the entities s and o. The task we address in this paper is to generate such a textual description y of the relationship instance r_i given the KG. For this we leverage a set of pairs X, where each $x \in X$ is a pair of $r_{i'}$ and y', and y' is the description of $r_{i'}$. We describe how we obtain this set in Sect. 5.

We aim to generate descriptions that are valid (expressing a relationship that can be found in the knowledge graph under the closed-world assumption), natural (grammatically correct), and informative, i.e., not just replicating the formal relationship but providing additional contextual information where possible.

We conclude our task definition with an example. Assume that we are given the relationship instance $starsInFilm(BradPitt, Troy)$. A possible description of this relationship instance is the following: "Brad Pitt appeared in the American epic adventure film Troy." This description not only contains mentions of the entities of the relationship instance and a verbalization of the relationship ("appeared in"), but also mentions of other entities that provide additional context. In particular, it contains mentions of Troy's type (Film), its genres (Epic, Adventure), and its country of origin.

4   Generating Textual Descriptions

In this section we detail our method which consists of three main steps. First, we enrich the description y' for each pair $\langle r_{i'}, y' \rangle \in X$ with additional entities from the KG (Sect. 4.1). Second, we use $\mathcal{K}$ and the set X to create a set of sentence templates T_r for the relationship r (Sect. 4.2). Third, given a new relationship instance, we use T_r and $\mathcal{K}$ to generate a description (Sect. 4.3).

4.1   Enriching the Textual Descriptions

In this step we perform entity linking to enrich the description y' for each pair $\langle r_{i'}, y' \rangle \in X$ with additional entities from the KG. This is done in order to facilitate the template creation step (Sect. 4.2). Each y' is a sentence that is about an entity $e \in \mathcal{E}$ and in the context of this paper we obtain these sentences from Wikipedia as our KG provides explicit links to Wikipedia articles. Although Wikipedia articles already contain explicit links to other articles and thus entities, these links are quite sparse. Therefore, we apply an algorithm for entity linking similar to [22].

Since y' originates from a Wikipedia article that is about a specific entity, we restrict the *candidate entities* (i.e., the entities that we consider adding to enrich y') to e itself, the in-links and out-links of the article of e in the Wikipedia structure, and the one-hop and two-hop neighbors of e in the KG. We infer the *surface forms* of each entity using the Wikipedia link structure, as is common in entity linking [14], and we also use the aliases of each entity provided by the KG.[3] In order to increase coverage for e, we enhance the set of surface forms of entity e using the rules in Table 2.

We iterate over the n-grams of the sentence that are not yet linked to an entity in decreasing order of length; if the n-gram matches a surface form of a candidate entity, we *link* the n-gram to the entity. If multiple entity candidates exist for a surface form, we rank the candidate entities by the number of entity neighbors they have in the sentence and select the top-ranked entity. Because of the very restricted set of candidate entities, the linking is usually unambiguous (with only one entity candidate per surface form).[4]

---

[3] We tag the sentences with POS tags and ignore unigram surface forms that are verbs.

[4] A manual evaluation of this algorithm on a held-out, random sample of 100 sentences in our dataset revealed an average of 93% precision and 85% recall per sentence.

Table 2. Additional surface forms per entity type.

Entity type	Surface form
Person	"he" or "she", person's surname
Film	"the film"
Music album	"the album"
Music composition	"the song", "the track"

Algorithm 1. Template creation

Input: A set X, the knowledge graph $\mathcal{K}$
Output: A set of templates T_r
1: $X' \leftarrow []$
2: **for** $\langle r_{i'}, y' \rangle \in X$ **do**
3:      $K \leftarrow$ BUILDENTITYDEPENDENCYGRAPH$(y', \mathcal{K})$
4:      $X'.append(\langle r_{i'}, y', K \rangle)$
5: $C \leftarrow$ CLUSTERENTITIES(X')
6: $G \leftarrow$ BUILDCOMPRESSIONGRAPH(X', C)
7: $P \leftarrow$ FINDVALIDPATHS(G)
8: $T_r \leftarrow \{\}$
9: **for** $p \in P$ **do**
10:      $t \leftarrow$ CONSTRUCTTEMPLATE(p, G, X')
11:      **if** $t \neq NULL$ **then**
12:          $T_r.add(t)$

4.2 Creating Sentence Templates

In this step, we create a set of templates T_r for a relationship r using the KG and the set of $\langle r_{i'}, y' \rangle$ pairs. The templates in T_r will be used in the next step to generate a novel description for the relationship instance r_i.

A *sentence template* t is a tuple (k, l, R_t), where (i) $k = [u_1 u_2 \ldots u_n]$ is a sequence, such that $\forall u_i \in l : u_i \in \mathcal{V} \cup \mathcal{E}_t$, (ii) l is a logical form in λ-calculus that consists of all the lambda variables in $\mathcal{E}_t$, at least one predicate and zero or more existential variables, and (iii) R_t is a set of relationship instances that support t.

The procedure we follow is outlined in Algorithm 1. First, we augment each $\langle r_{i'}, y' \rangle$ pair with an entity dependency graph K in order to capture dependencies between entities in a sentence (lines 1–4). Next, we build a mapping C that maps each entity in each sentence to a single cluster id (line 5). This is done in order to facilitate the detection of useful patterns in the sentences since each sentence describes a relationship for a particular entity pair. Then, we build a compression graph G (line 6) and use it to find valid paths P (line 7). Finally, for each path $p \in P$, we construct a template t and add it to the set of templates (lines 8–12). We now describe each procedure in Algorithm 1.

BUILDENTITYDEPENDENCYGRAPH(.) In order to build the graph K for a sentence y', we retrieve all paths between each pair of entities mentioned in y' from the KG and add them to K. We only consider 1-hop paths and 2-hop paths

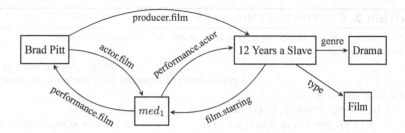

Fig. 2. Entity dependency graph for the sentence "Brad Pitt appeared in the drama film 12 Years a Slave". Nodes represent entities and edge labels represent predicates (med_1 is a CVT entity).

that pass through a CVT entity. Figure 2 shows the entity dependency graph for an example sentence.

CLUSTERENTITIES(.) In order to obtain C, we consider all $x' = \langle r_{i'}, y', K \rangle \in X'$ and map two entities in the same cluster if they share at least one incoming or outgoing edge label in their corresponding entity dependency graph K. For example, in the *starsInFilm* relationship, this procedure will create separate clusters for persons, films, dates and CVT entities.

BUILDCOMPRESSIONGRAPH(.) In this step, we build a compression graph $G = (V, E)$ using the sentence y' of each $\langle r_{i'}, y', K \rangle \in X'$. V is a set of nodes and E is a set of edges. We follow a similar procedure to [6], in which each node holds a list of $\langle sid, pid \rangle$ pairs, where sid is a sentence id and pid is the index of the word/entity in the sentence. In our case a node can be a word or an entity cluster. We map two words onto the same node if they have the same lowercase form and the same POS tag. We map two entities on the same node if they have the same cluster id.

FINDVALIDPATHS(.) In order to find valid paths in the graph G, we set all the entity cluster nodes as valid start/end nodes and traverse G to find a set of paths P from a start to an end node. In order to build templates that are natural we enforce the following constraints for the paths in P: (i) the path must contain a verb and (ii) the path must have been seen as a complete sentence at least once in the input sentences. For example, given the following sentences (the corresponding cluster id per entity are listed in brackets):

- y'_1: "Bruce_Willis[c_1] appeared in Moonrise_Kingdom[c_2]"
- y'_2: "Liam_Neeson[c_1] appeared in the action[c_3] film[c_4] Taken[c_2]"
- y'_3: "Brad_Pitt[c_1] appeared in the drama[c_3] film[c_4] 12_Years_a_Slave[c_2]"

we obtain the following valid paths by traversing the graph:

- p_1: "c_1 appeared in c_2"
- p_2: "c_1 appeared in the c_3 c_4 c_2"

CONSTRUCTTEMPLATE(.) Algorithm 2 outlines the procedure for constructing a template t from a path p. First, for each $\langle r_{i'}, y', K \rangle \in X'$, we check whether

---

Algorithm 2. CONSTRUCTTEMPLATE(.)

Input: A path p, the compression graph G, a set X', parameters α, β
Output: A template t
1: $D_g \leftarrow []$ ▷ entity dependency graphs
2: $R_t \leftarrow []$ ▷ relationship instances that support the template
3: **for** $\langle r_{i'}, y', K \rangle \in X'$ **do**
4:    **if** ISSUBSEQUENCE(p, y', G) **then**
5:       $h \leftarrow$ GETSUBSEQUENCE(p, y', G) ▷ get the actual subsequence
6:       $\langle s, o \rangle \leftarrow r_{i'}$ ▷ subject/object of the relationship instance
7:       **if** CONTAINSLINK(h, s) **and** CONTAINSLINK(h, o) **then**
8:          $D_g.append(K)$
9:          $R_t.append(r_{i'})$
10: **if** $|R_t| < \alpha$ **then**   ▷ too few relationship instances
11:    **return** $NULL$
12: $l \leftarrow$ BUILDLOGICALFORM(D_g, β) ▷ aggregate the entity dependency graphs
13: $k \leftarrow$ REPLACECLUSTERIDSWITHVARIABLES(p)
14: $t = (k, l, R_t)$

---

y' is a (possibly non-continuous) subsequence h of path p by using the positional information of each node in p from G.[5] If it is, we check whether h contains links to both the subject and the object of the relationship instance $r_{i'}$. If it does, we store the entity dependency graph and the relationship instance. Next, if the number of instances is less than a parameter α, we consider the template to be invalid. Subsequently, we build the logical form l by aggregating the entity dependency graphs D_g. Entity nodes that were part of the path p become lambda variables (nodes constructed from subject and object entities have special identifiers). Entity nodes that were not part of the path p (CVT entities) become existential variables. We ignore edges appearing in less than $|D_g| \cdot \beta$ entity dependency graphs. Lastly, we replace the cluster ids in p with the corresponding lambda variables to obtain a sequence k.

Figure 3 shows the logical form of a template constructed using the example sentences y'_1, y'_2 and y'_3 and their corresponding instances in graphical form ($\beta = 0.5$). Note that the edge "producer.film" has been eliminated since it only appears in one out of the three instances.

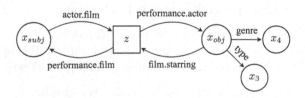

Fig. 3. Logical form of the template constructed using p_2 and y'_1, y'_2, y'_3 (with their corresponding relationship instances). $k =$"x_{subj} appeared in the x_3 x_4 x_{obj}". Lambda variables are shown in circles and existential variables in rectangles.

---

[5] For example, the path p_1 is a subsequence of y'_2.

4.3   Generating the Description

In this step we generate a novel description for a relationship instance r_i using the set of templates T_r and the knowledge graph $\mathcal{K}$. This comes down to selecting the template from T_r that best describes the relationship instance r_i and filling it with the appropriate entities.

The procedure is as follows. First, we rank the templates in T_r for the relationship instance using a scoring function $f(r_i, t)$. Subsequently, for each template $t = (k, l, R_t)$ we replace the subject and object lambda variables in l to obtain $l' = l[x_{subj} = s, x_{obj} = o]$. We then query the knowledge graph $\mathcal{K}$ using l' and if at least one instantiation of l' exists, we randomly pick one and replace all the entity variables in k with the entity names to generate the description y, otherwise we proceed to the next template. As an example, assume we are given the instance $r_i = starsInFilm(Ryan\_Reynolds, Deadpool)$ and we consider the template shown in Fig. 3. A possible instantiation of the template for this relationship instance will result in the description "Ryan Reynolds appeared in the comedy film Deadpool".[6]

The template scoring function $f(r_i, t)$ returns a score for a relationship instance r_i and template t. As we want to generate descriptions that are valid under the closed-world assumption of the KG, we promote templates that are semantically closest to the relationship instance. For a new relationship instance r_i we extract binary features for each entity in the r_i. Recall that r_i has two or more entities (subject s, object o and possibly a CVT entity z). For each entity e of r_i, we extract all triples $\langle e, p, e' \rangle$ from the KG $\mathcal{K}$. We restrict the feature space by discriminating between entity attributes and entity relations depending on the predicate p as in [13]. If the predicate p is an attribute (e.g., "gender"), we use the complete triple as a feature (e.g. $\langle s, gender, female \rangle$). If the predicate p is a relation (e.g., "date_of_death"), we only keep the subject and the predicate of the triple as a feature (e.g., $\langle e, person.date\_of\_death \rangle$). We also add a count feature for the relation predicates (e.g., $\langle s, person.children, 2 \rangle$, i.e., a person has two children). We denote the resulting binary vector for r_i as $vec(r_i)$. We obtain a vector $vec(t)$ for template t by summing the vectors of all the instances R_t of t. We also compute a vector $vec\_tfidf(t)$ that is a TF.IDF weighted vector of $vec(t)$, where IDF is calculated at the template level. Based on these ingredients, we define two scoring functions:

– **Cosine** Calculates the cosine similarity between vectors $vec(r_i)$ and $vec\_tfidf(t)$.
– **Supervised** Learns a scoring function using a supervised learning to rank algorithm. We treat r_i as a "query" and t as a "document."

We create training data for the supervised algorithm as follows. Recall that each r_i is coupled with a description y'. For each r_i, we assign a relevance label of 3 for templates that best match y (measured by the number of entities) and

---

[6] Note that there might be multiple instantiations (e.g., Deadpool is also a science fiction film) and selecting the optimal one depends on the application—we leave this for future work.

a relevance label of 2 for the rest of the templates that match y. In order to create "negative" training data, we sample templates that are dissimilar to the ones that match y in the following way. First, we calculate the average vector of all the templates that match y and build a distribution of templates based on the cosine distance from the average vector to each of the templates in T_r (excluding the ones that match y). Lastly, we sample at most the number of matching templates from the resulting distribution and assign them a relevance label of 1 (we ignore templates that have a cosine similarity to the average vector greater than 0.9). For the supervised model we use the following features: each element/value pair in $vec(r_i)$, the cosine similarity between vectors $vec(r_i)$ and $vec\_tfidf(t)$, the words in t, the number of entities in t and the size of R_t. We use LambdaMART [23] as the learning algorithm and optimize for NDCG@1.[7]

5   Experimental Setup

In this section we describe our experimental setup.

5.1   Datasets

We use an English Wikipedia dump dated 5 February 2015 as our document corpus. We perform sentence splitting and POS tagging using the Stanford CoreNLP toolkit. We use a subset of the last version of Freebase as our KG [4]: all the triples in the people, film and music domains, as these are well-represented in Freebase.

In order to create an evaluation dataset for our task, we first need a set of KG relationships. We rank the predicates in each domain by the number of instances and keep the 10 top-ranked predicates. We exclude trivial predicates such as "dateOfDeath". We then use the predicates to manually construct the logical forms of the relationships (see Fig. 1 for an example). Second, we need a set of $\langle r_{i'}, y' \rangle$ pairs for each relationship r, where $r_{i'} = r\langle s', o' \rangle$ is an instance of relationship r, s' and o' are entities and y' is a description of $r_{i'}$. To this end, for each relationship r, we randomly sample 12000 relationship instances from the KG. For each relationship instance $r_{i'}$, we pick the first sentence in the Wikipedia article of the subject entity s' that contains links to both s' and o'. If such a sentence does not exist, we proceed to the next instance. We manually inspected a subset of the sentences selected with this heuristic and the quality of the selected sentences was relatively good. Our final dataset contains 10 relationships and 90058 $\langle r_{i'}, y' \rangle$ instances in total and 8187 instances on average per relationship. We randomly select 80% of each relationship sub-dataset for training and 20% for testing.

5.2   Evaluation Metrics

We perform two types of evaluation: automatic and manual. For automatic evaluation we use METEOR [9], ROUGE-L [11] and BLEU-4 [16] as metrics.

---

[7] For this method we use 20% of the training data as validation data. The same test data is used for all methods.

METEOR was originally proposed in the context of machine translation but has also been used in a task similar to ours [19]. ROUGE is a standard metric in summarization and BLEU is widely used in machine translation and generation. As is common in text generation [8], we also employ manual evaluation. We ask human annotators to annotate each output sentence on three dimensions: validity under the KG closed-world assumption (0 or 1), informativeness (1–5) and grammaticality (1–5). One human annotator (not one of the authors) annotated 11 generated sentences per relationship per system (440 sentences in total).

5.3 Compared Approaches

We compare 4 variations of our method. The variations differ in the way they rank templates for a given relationship instance. The first variation (*Random*) ranks the templates randomly. The second (*Most-freq*) ranks templates by the number of relationship instances that support the template. The third (*Cosine*) ranks templates based on the cosine similarity between the vectors of the relationship instance and the template (Sect. 4.3). The fourth (*Supervised*) ranks templates using a learning to rank model (Sect. 4.3), for which we use LambdaMART with the default number of trees (1000). We set $\alpha = 20$ and $\beta = 0.5$ (Sect. 4.3). We depict a significant improvement in performance over *Random* with ▲ (paired two-tailed t-test, $p < 0.05$).

6 Results

In this section we describe our experimental results. We compare all methods discussed previously, using the automatic and manual setups, respectively.

6.1 Automatic Evaluation

Table 3 shows the automatic evaluation results. We observe that *Supervised* and *Cosine* outperform *Random* and *Most-freq* on all metrics. This is expected since the former two try to capture the semantic similarity between a relationship instance and a template. Although *Supervised* consistently outperforms *Cosine*, the differences between *Cosine* and *Supervised* are not significant.

We also observe that the scores for the automatic measures are relatively low. This is because of two reasons: (i) we generally generate much shorter sentences

Table 3. Automatic evaluation results, averaged per relationship.

Method	BLEU	METEOR	ROUGE
Random	1.14	16.56	24.13
Most-freq	0.13	13.99	21.96
Cosine	1.76▲	17.37	25.84▲
Supervised	**2.14▲**	**19.18▲**	**26.54▲**

Table 4. Manual evaluation results, averaged per relationship.

Method	Validity	Informativeness	Grammaticality
Random	0.4545	1.98	3.67
Most-freq	0.5000	1.60	3.62
Cosine	0.5636▲	2.05	**4.00**
Supervised	**0.5818**▲	**2.18**▲	3.90

than the reference sentence as not all information that appears in the reference sentence is represented in the KG, and (ii) since the reference sentences are extracted automatically, some of the reference sentences describe a minor aspect of the relationship or do not discuss the relationship at all.

6.2 Manual Evaluation

Table 4 shows the results for manual evaluation. The results follow a similar trend as in the automatic evaluation; *Supervised* and *Cosine* outperform *Random* and *Most-freq* on all metrics. *Supervised* significantly outperforms *Random* in terms of validity and informativeness. The differences between *Cosine* and *Supervised* are not significant.

6.3 Analysis

We have also examined specific examples and identify cases where the best performing approach (*Supervised*) succeeds or fails. In terms of validity, it succeeds in matching attributes of the relationship instance and the template. E.g., in the context of the relationship *parentOf*, it correctly figures out what the genders of the entities are and the semantically valid expression of the relationship between them, often better than *Cosine*, as illustrated by the following example:

(*Supervised*) "Emperor Francis I (1708 - 1765) was the father of Emperor Leopold II" (VALID)
(*Cosine*) "Emperor Francis I was the son of Emperor Leopold II" (INVALID)

Supervised benefits from training a model that combines multiple features such as the template words with attributes of the relationship instance to describe whether the relationship is still ongoing or not. One of the main cases where *Supervised* fails is in ranking a relationship instance in a temporal dimension with regards to other relationship instances, as illustrated by the following example for the *childOf* relationship:

"Thomas Howard was the second son of Henry Howard and Frances de Vere."
    (INVALID: Thomas Howard was the *first* son of Henry Howard)

The fact that our best performing approach (*Supervised*) has a relatively low validity score (0.5818) shows that there is room for improvement in capturing the semantic similarity between a relationship instance and a template.

In terms of informativeness, *Supervised* succeeds in offering contextual information about the relationship instance, such as dates, locations, occupations and film genres. The fact that informativeness scores are relatively low is because they are dependent on validity: when a generated sentence was assigned a validity of score 0, it was also assigned an informativeness score of just 1.

Grammaticality scores are high for all the systems with no significant differences. This is expected as the templates were generated using the same procedure for all the compared systems. Mainly, grammaticality is harmed when some entities in the generated sentence have the wrong surface form (e.g., 'Britain', 'British'), which is not surprising as we do simple surface realization (deciding which surface form of the entity best fits with the generated sentence) and only use the entity names as surface forms.

7   Conclusion

We have addressed the problem of generating descriptions of entity relationships from KGs. We have introduced a method that first creates sentence templates for a specific relationship, and then, for a new relationship instance, it generates a novel description by selecting the best template and filling the template slots with the appropriate entities from the KG. We have experimented with different scoring functions for ranking templates for a relationship instance and performed an automatic and a manual evaluation.

When using information about the relationship instance and the template taken from the KG, both automatic and manual evaluation outcomes are improved. A supervised method that uses both KG features and other template features (template words, number of entities) consistently outperforms an unsupervised method on all automatic evaluation metrics and also in terms of validity and informativeness.

As to future work, our error analysis showed that we need more sophisticated modeling for capturing the semantic similarity between a relationship instance and a template, especially for capturing temporal dimensions that also involve other relationship instances. We also want to explore more sophisticated methods for selecting the correct surface form for an entity to improve grammaticality. Finally, we aim to evaluate our method on generating descriptions for less popular KG relationships.

Acknowledgments. This research was supported by the Netherlands Institute for Sound and Vision and the Netherlands Organisation for Scientific Research (NWO) under project nr. CI-14-25. All content represents the opinion of the authors, which is not necessarily shared or endorsed by their respective employers and/or sponsors.

References

1. Althoff, T., Dong, X.L., Murphy, K., Alai, S., Dang, V., Zhang, W.: Timemachine: timeline generation for knowledge-base entities. In: KDD (2015)

2. Blanco, R., Cambazoglu, B.B., Mika, P., Torzec, N.: Entity recommendations in web search. In: Alani, H., et al. (eds.) ISWC 2013. LNCS, vol. 8219, pp. 33–48. Springer, Heidelberg (2013). doi:10.1007/978-3-642-41338-4_3
3. Blanco, R., Ottaviano, G., Meij, E.: Fast and space-efficient entity linking for queries. In: WSDM (2015)
4. Bollacker, K., Evans, C., Paritosh, P., Sturge, T., Taylor, J.: Freebase: a collaboratively created graph database for structuring human knowledge. In: SIGMOD (2008)
5. Fang, L., Sarma, A.D., Yu, C., Bohannon, P.: Rex: explaining relationships between entity pairs. In: VLDB (2011)
6. Ganesan, K., Zhai, C., Han, J.: Opinosis: a graph-based approach to abstractive summarization of highly redundant opinions. In: COLING (2010)
7. Gkatzia, D., Lemon, O., Rieser, V.: Natural language generation enhances human decision-making with uncertain information. In: ACL (2016)
8. Konstas, I., Lapata, M.: A global model for concept-to-text generation. JAIR **48**, 305–346 (2013)
9. Lavie, A., Agarwal, A.: METEOR: an automatic metric for MT evaluation with high levels of correlation with human judgments. In: WMT (2007)
10. Lebret, R., Grangier, D., Auli, M.: Neural text generation from structured data with application to the biography domain. In: EMNLP (2016)
11. Lin, C.-Y.: Rouge: a package for automatic evaluation of summaries. In: Text Summarization Branches Out: Proceedings of the ACL-04 Workshop (2004)
12. Lin, T., Pantel, P., Gamon, M., Kannan, A., Fuxman, A.: Active objects: actions for entity-centric search. In: WWW (2012)
13. Lin, Y., Liu, Z., Sun, M.: Knowledge representation learning with entities, attributes and relations. In: IJCAI (2016)
14. Meij, E., Weerkamp, W., de Rijke, M.: Adding semantics to microblog posts. In: WSDM 2012 (2012)
15. Nickel, M., Murphy, K., Tresp, V., Gabrilovich, E.: A review of relational machine learning for knowledge graphs: from multi-relational link prediction to automated knowledge graph construction. Proc. IEEE **104**(1), 11–33 (2016)
16. Papineni, K., Roukos, S., Ward, T., Zhu, W.-J.: BLEU: a method for automatic evaluation of machine translation. In: ACL (2002)
17. Pighin, D., Cornolti, M., Alfonseca, E., Filippova, K.: Modelling events through memory-based, Open-IE patterns for abstractive summarization. In: ACL (2014)
18. Reiter, E., Dale, R., Feng, Z.: Building Natural Language Generation Systems. MIT Press, Cambridge (2000)
19. Saldanha, G., Biran, O., McKeown, K., Gliozzo, A.: An entity-focused approach to generating company descriptions. In: ACL (2016)
20. Tombros, A., Sanderson, M.: Advantages of query biased summaries in information retrieval. In: SIGIR (1998)
21. Turpin, A., Tsegay, Y., Hawking, D., Williams, H.E.: Fast generation of result snippets in web search. In: SIGIR (2007)
22. Voskarides, N., Meij, E., Tsagkias, M., de Rijke, M., Weerkamp, W.: Learning to explain entity relationships in knowledge graphs. In: ACL-IJCNLP (2015)
23. Wu, Q., Burges, C.J., Svore, K.M., Gao, J.: Ranking, boosting, and model adaptation. Technical report, Microsoft Research (2008)
24. Yih, W.-T., Chang, M.-W., He, X., Gao, J.: Semantic parsing via staged query graph generation: question answering with knowledge base. In: ACL (2015)

Language Influences on Tweeter Geolocation

Ahmed Mourad$^{(\boxtimes)}$, Falk Scholer, and Mark Sanderson

RMIT University, Melbourne, Australia
{ahmed.mourad,falk.scholer,mark.sanderson}@rmit.edu.au

Abstract. We investigate the influence of language on the accuracy of geolocating Twitter users. Our analysis, using a large corpus of tweets written in thirteen languages, provides a new understanding of the reasons behind reported performance disparities between languages. The results show that data imbalance has a greater impact on accuracy than geographical coverage. A comparison between *micro* and *macro* averaging demonstrates that existing evaluation approaches are less appropriate than previously thought. Our results suggest both averaging approaches should be used to effectively evaluate geolocation.

Keywords: Geolocation · Language · Text-based · Tweeter

1 Introduction

Geolocating Twitter users (*tweeters*) is a service needed for many social media-based applications, such as finding an eyewitness to an event, managing natural crises, and personalizing regional ads. While tweeters can record their location on their Twitter profile, Hecht et al. [10] reported that >34% record fake or sarcastic locations. Twitter also allows tweeters to GPS locate their content, however, Han et al. [9] reported that <1% of tweets are geotagged. Inferring tweeter location based on features derived from tweet and profile content is therefore a field of investigation, which has included examination of social network analysis [2,11,18], event detection [19], geographic topic modeling [1,6], and language modeling [3,12,17,22]. Only a few researchers have considered the language in which a tweet is written as a feature to geolocate a tweeter [9,15].

Han et al. [9] observed that tweeters writing in some languages appeared to be easier to locate than those writing in others. They speculated that the geographical coverage of a language or the distribution of tweeters played an important role in determining location accuracy. So important was this role that accuracy might be largely predictable by considering language alone. However, in past work, correlations between such features and accuracy were not measured, and other features that might influence accuracy were not considered. The different evaluation measures that are typically employed to measure the output of a tweeter's geolocation system weren't considered either.

We conduct an evaluation of the features that impact the accuracy of a state-of-the-art geolocation technique, comparing different features across thirteen languages. Our results demonstrate the limitations of current evaluation

© Springer International Publishing AG 2017
J.M. Jose et al. (Eds.): ECIR 2017, LNCS 10193, pp. 331–342, 2017.
DOI: 10.1007/978-3-319-56608-5_26

approaches and lead us to propose an alternative perspective and framework for the evaluation of geolocation that is more closely aligned with the range of real-world problems for which geolocation is of interest.

2   Related Work

To the best of our knowledge, only two prior works have evaluated the impact of a language on geolocating tweeters [9] and tweets [15]. Both claimed that locating tweeters/tweets writing/written in languages with restricted regional coverage were easier to geolocate than those writing in widely used languages.

Priedhorsky et al. [15] examined the effect of a language as a feature in a multilingual model trained on a dataset of 13M geotagged tweets, showing that language is a valuable feature in geolocation prediction models. However, they did not evaluate their models on a per language basis.

Using a multilingual dataset of 23M geotagged tweets, Han et al. [9] showed that training separate per language models lead to higher accuracy. Han et al. noted that for some languages, geolocation accuracy was higher than for others. To explore tweeter distribution in the geographical region of that language, the authors measured the entropy of tweeters in cities on a per language basis. However, they did not correlate entropy with an evaluation measure, neither did they examine other features of languages that might impact on evaluation.

3   Methodology

To conduct our study, we required the following: a geolocation system, collections of tweeters on which to measure location accuracy, and evaluation measures.

From the existing geolocation approaches [3, 9, 12, 17, 22], we based our work on the research that addressed language influence, namely Han et al.'s system [9], which locates tweeters to one of 3,709 cities. We re-implemented the system, focusing on the part that uses Location Indicative Words (LIW) drawn from tweets, where mainstream noisy words were filtered out using their best reported feature selection method, Information Gain Ratio. Then we built a Multinomial Naïve Bayes (MNB) prediction model per language using scikit-learn [14].

We employed two global tweet collections: **WORLD**, spanning five months from late 2011 to early 2012 [9]; and **TwArchive** holding over four years of content[1] drawn from the 1% sample Twitter public API stream. Originally WORLD contained 23M geotagged tweets and 2.1M tweeters. In reconstructing it from the tweet IDs released by the authors, 27% and 30% of tweeters and tweets, were deleted. For TwArchive, we used a 2014 subset spanning nine months.

We separated languages in the collections using langid.py[2] [13]. We studied Arabic (ar), English (en), Spanish (es), French (fr), Indonesian (id), Italian (it),

---

[1] https://archive.org/details/twitterstream\&tab=collection.

[2] An open source language identification tool, trained over 97 languages, and tested over six European languages with an accuracy of 0.94. We accepted predictions with confidence ≥ 0.5 only.

Table 1. Number of tweeters, tweets, cities and countries after preprocessing.

		en	es	it	pt	id	nl	fr	ms	ko	ru	ar	th	tr
# Tweeters	WORLD	947k	242k	118k	111k	103k	94k	79k	64k	36k	29k	28k	27k	24k
	TwArchive	1.5M	541k	119k	284k	225k	59k	136k	136k	22k	73k	94k	49k	211k
# Tweets	WORLD	6.2M	1.2M	267k	670k	423k	381k	198k	222k	122k	196k	215k	156k	108k
	TwArchive	3.1M	1.1M	162k	836k	317k	74k	295k	179k	32k	147k	207k	127k	351k
# Cities	WORLD	2.9k	2.2k	2.1k	1.8k	1.9k	2k	2k	1.6k	1.1k	894	881	413	1.3k
	TwArchive	3.2k	2.3k	2.2k	1.9k	2k	2k	2.2k	1.7k	1.7k	1k	1.6k	727	1.6k
# Countries	WORLD	169	151	150	132	145	140	154	125	96	94	90	64	116
	TwArchive	173	159	156	139	147	148	164	142	129	107	139	80	147

Korean (ko), Malaysian (ms), Dutch (nl), Portuguese (pt), Russian (ru), Thai (th), and Turkish (tr). Text was tokenised using a Twitter specific tokeniser [8]. Arabic text was normalized using Tashaphyne[3] and an Arabic social media normalizer [4]. Normalization changed only the orthography of Arabic words. Use of the extra systems were necessary to reduce the sparsity of words. All non-alphabetical tokens and tokens with length <3 characters were removed.

We removed non-geotagged and duplicate tweets (using tweeter id and tweet text). Cities with fewer than fifty LIWs were removed to ensure a representative sample of words per city. Each tweeter was assigned a home city based on their geotagged tweets. We used a search library[4] released by Han et al. [9] that returns either the city corresponding to a GPS coordinate, or [none]. A tweeter's home city is the one associated with the simple majority of their tweets; in a tie, the first city is chosen. Tweeters with an unresolved home city (i.e. [none]) were removed from the corpus. Tweeters eligible for testing are required to have at least ten geotagged tweets. All previous processing steps were adopted from previous work [9] for a fair comparison, except for the Arabic normalization.

Table 1 shows that for all languages, tweeters are spread over thousands of cities and tens of countries. We found that around 25% of the tweeters in WORLD post in more than one language. The cumulative distribution (in WORLD) of tweeters over cities is shown, per language, in Fig. 1. Examining where the plot lines intersect the x-axis, we see that for *en*, *fr* and *it*, no single city contained more than 4% of all tweeters for that language. For languages, such as *tr*, *ko*, *th* and *ru*, one city contained more than 30% of tweeters. A similar pattern was found when examining cumulative distributions in the TwArchive.

To measure accuracy, we considered three evaluation metrics drawn from past work [3,6,9,12,17,22]: **(1) Acc**, city-level accuracy; **(2) Acc@161**, accuracy within 161 km (100 miles)[5]; **(3) MedErr**, median error distance between

---

[3] http://pythonhosted.org/Tashaphyne/.

[4] https://github.com/tq010or/acl2013.

[5] Although Cheng et al. [3] showed empirically that the percentage of tweeters within *x* miles increases as *x* increases, e.g., 30% of tweeters are placed within 16 km and 51% within 161 km, all subsequent research used an arbitrarily chosen 161 km. Note, Cheng et al. tested only on a US-based dataset, where the average distance between neighboring cities might be different from densely populated or small countries. Accuracy within 161 km might not be an effective evaluation measure from a language comparison perspective, however as it has been used in past work, we use it here.

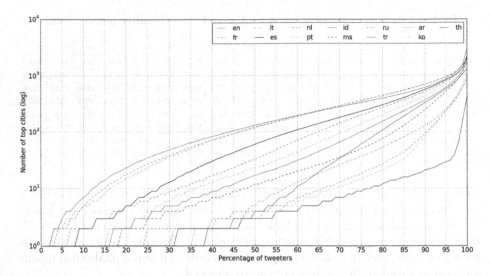

Fig. 1. Tweeters' cumulative distribution over cities in WORLD.

Table 2. Languages rank correlation τ_β between pairs of evaluation metrics.

	WORLD		TwArchive	
	Acc@161	Mederr	Acc@161	Mederr
Acc	0.00	−0.31	0.15	0.15
Acc@161	–	0.03	–	0.13

predicted and actual cities (km). We measured the agreement of the metrics on how they rank the accuracy of our geolocation system across the tweets of each language. Kendall's τ_β was used to measure the correlation between the ranks, see Table 2. There is no statistically significant rank correlation between any pair: the measures appear to be examining different aspects of geolocation. We therefore consider all three measures in our study.

4   Examination of Features

A range of features may influence geolocation accuracy. Although Han et al. speculated that distribution of tweeters was the reason for accuracy variation, many other differences were present in the language datasets they studied: the sets were of notably different sizes, written in different languages, and each contained different numbers of tweeters, tweets, and cities. Therefore, the features we explore are dataset size, a preliminary test of the impact of the language, and a range of individual features such as entropy and number of tweeters.

Table 3. Influence of dataset size, in terms of the slope of a linear regression model, on the evaluation measures for six languages in TwArchive.

	en	es	pt	fr	ar	tr
Acc	0.02	0.06	0.07	0.04	0.03	0.01
Acc@161	0.04	0.07	0.09	0.04	0.04	0.02
MedErr	−7.34	−1.17	−1.26	−0.31	−0.86	−0.10

4.1 Dataset Size

We focus on the six languages that have sufficient tweeters eligible for testing: two of which the geolocation system has low accuracy (*en* and *fr*), two with moderate accuracy (*es* and *pt*), and two with high accuracy (*ar* and *tr*). From each of the language sets, we randomly sample subsets of tweeters in decrements of 10%, from 100% down to 10%. Ten samples of each subset were created, and an average was taken. Table 3 shows that for *Acc*, there is a weak positive relationship between the number of tweeters and accuracy. We chose a slope, over a correlation measure, because it estimates the expected gain in accuracy with the increase in dataset size. While there is some variation across languages, the gradient of the slope is consistently small. The same pattern was found with *Acc@161*, while for the *MedErr*, the measure tends to decrease (improve) as the number of tweeters increases. The fact that the slope of the linear regression model is greater suggests that *MedErr* is more affected by the scale of the dataset than the accuracy measures. Hence, the *MedErr* is not an appropriate measure in the case of small datasets.

4.2 Preliminary Examination of Language

In past work, Han et al. noted that tweeters writing in some languages were easier to geolocate than those writing in others. We speculated that there may be something inherent in the way that tweets are written in each of the languages that causes the differences in geolocation accuracy. Because we had access to two collections covering the same 13 languages, we examined the relative geolocation accuracy per language across the two collections, shown in Table 4. Although the two collections vary in the number of tweeters, the previous result showed the impact of dataset scale was small. Therefore, if the language of tweets was impacting on accuracy, the relative accuracy across the two collections might be expected to be similar.

To determine the degree of agreement between the languages in the collections, we ranked the 13 languages by geolocation accuracy and calculated Kendall's τ_β between the two rankings. We found a statistically significant but moderate correlation of 0.46. The relative geolocation accuracy for a language changed notably across the two collections. The low correlation strongly suggests that differences in geolocation accuracy across languages are influenced by a property other than the actual language of the tweets.

Table 4. Accuracy of geolocation for the 13 languages in WORLD and TwArchive.

	en	es	pt	fr	ar	tr	id	it	nl	ru	ms	th	ko
WORLD	0.11	0.29	0.31	0.13	0.49	0.54	0.4	0.15	0.25	0.33	0.41	0.43	0.45
TwArchive	0.07	0.20	0.35	0.12	0.26	0.37	0.28	0.12	0.23	0.09	0.37	0.28	0.25

4.3 Correlation with Individual Features

In order to measure the impact of collection and tweeter/tweet features on geolocation accuracy per language, we measured the Pearson Correlation Coefficient between feature values and the relative accuracy of languages. The features used were entropy of tweeters distributed across all cities and a subset of cities, the total number of cities, the total number of tweeters, the number of LIWs per language, and the number of tweets. Both collections were used. In addition to Pearson, the coefficient of determination (R^2) was used to measure the explanatory power of the model. The results are shown in Table 5.

As can be seen, entropy has the strongest correlation with all three evaluation measures. Entropy over only the cities that had eligible test tweeters (entropy.test) was also calculated, and generally resulted in a higher correlation than entropy measured across all possible cities. For TwArchive, number of cities that had eligible test tweeters correlated strongest with *MedErr*.

Considering the average number of tweets per eligible test tweeter, if this number increases, accuracy should also increase, since tweeters reveal more information about their location [3]. The correlations with this feature appeared to contradict past work by being negative, however, they were not significant; note that the range of tweets per tweeter here was substantially smaller than the range Cheng et al. [3] examined. The number of LIW in a lexicon normalized by the number of tweets per language was also found not to correlate strongly with accuracy. The results shown earlier on the impact of dataset size (Table 5) can

Table 5. Pearson correlation between features and evaluation metrics; (* and † denote statistical significance with $p \leq 0.05$ and $p \leq 0.01$, respectively).

Feature	Acc				Acc@161				MedErr			
	WORLD		TwArchive		WORLD		TwArchive		WORLD		TwArchive	
	r	r^2	r	r^2	r	r^2	r	r^2	r	r^2	r	r^2
Entropy	**−0.87**†	**0.76**	−0.69†	0.47	−0.62*	0.38	−0.29	0.08	0.52	0.27	0.43	0.19
# Cities	−0.76†	0.57	−0.40	0.16	−0.57*	0.32	−0.26	0.07	0.54	0.30	0.57*	0.32
Entropy.test	−0.83†	0.69	−0.70†	0.49	**−0.85**†	**0.73**	−0.79†	**0.62**	0.82†	**0.68**	0.89†	0.79
# Cities.test	−0.55*	0.30	−0.51	0.26	−0.67*	0.45	−0.55*	0.30	0.81†	0.66	**0.93**†	**0.87**
Avg #tweets.test	−0.47	0.22	−0.51	0.26	−0.34	0.12	−0.10	0.01	0.34	0.12	0.12	0.01
# LIW words	0.40	0.16	0.37	0.14	–	–	–	–	–	–	–	–
# Tweeters	−0.57*	0.32	−0.39	0.15	−0.54	0.29	−0.46	0.21	0.76†	0.58	0.87†	0.76
# Tweets	−0.51	0.26	−0.38	0.15	−0.51	0.26	−0.47	0.22	0.76†	0.58	0.87†	0.75
Avg dist	–	–	–	–	0.12	0.01	0.51	0.26	−0.33	0.11	−0.30	0.09
Nbr avg dist	–	–	–	–	−0.46	0.21	−0.22	0.05	0.55*	0.31	0.53	0.28

also be seen here, as the number of tweeters and tweets per language correlate most strongly with *MedErr*, compared to the other evaluation measures.

Average distance measures were found to have a weak correlation with *Acc@161*. By measuring the average distance between neighboring cities, it was found to be in the range of 52–74 km (significantly less than the arbitrarily chosen 161 km as mentioned earlier in Sect. 3).

In summary, the correlation with different features showed that the distribution of tweeters has a greater impact on the accuracy of geolocation prediction than other features, especially geographical coverage. This is a different result described in previous research. It also shows that *Acc@161* is not an appropriate measure.

4.4   Considering Alternative Measures

The results in the previous section showed that the distribution of tweeters across cities (entropy) is a strong predictor of the accuracy of geolocation for different languages. However, the measures *Acc* and *Acc@161* are both heavily influenced by the accuracy of the geolocation system on a limited number of cities. As long as the system geolocates correctly on a few well populated cities, the accuracy will be high.

Evaluation measures are designed to estimate how well a system will do in a particular task. In the introduction, we stated that one example use of a geolocation system is finding eyewitnesses. It is perhaps worth asking if the distribution of eyewitnesses needed say by a news organization will match the distribution reflected in the accuracy measure. In this section, we explore alternative measures commonly used to evaluate classifiers when data is unbalanced [20]. We compare the way that different measures are affected by the different features of languages described above. First we describe the averaging methods, measures, and some default baselines to consider.

Averaging. When considering data imbalance, it is important to examine different averaging techniques: 1. *Micro* (μ) calculates the metric globally on absolute measures regardless of the city. This is the default averaging technique used to calculate the overall accuracy of previous geolocation prediction models. 2. *Weighted* (W) calculates the metric for each label and finds the average weighted by the frequency of each city in the training dataset. 3. *Macro* (M) calculates the metric for each city and finds their unweighted mean. It is the most appropriate for evaluating how classifiers behave on cities with a small number of tweeters, rather than *micro* averaging, which is influenced by big cities.

Measures. Although Precision (P) and recall (R), together with different averaging techniques, are the most common measures used in text categorization to evaluate the effectiveness of classifiers [20,23], they were never considered in prior tweeter geolocation work [2,3,6,9,17,22]. Sometimes *precision* is favored (e.g. when journalists are looking for eyewitnesses within a specific city [5]); at other times *recall* (e.g. when journalists are looking for eyewitnesses on the

Table 6. Comparison between Majority Class (MC) and Multinomial Naïve Bayes (MNB) models, in terms of *micro* precision (P_μ) and *macro* precision (P_M), for the top 13 languages in WORLD.

	en	es	pt	fr	ar	tr	id	it	nl	ru	ms	th	ko
MC P_μ	0.02	0.12	0.23	0.10	0.39	0.54	0.27	0.09	0.16	0.34	0.25	0.32	0.45
MNB P_μ	0.11	0.29	0.31	0.13	0.49	0.54	0.40	0.15	0.25	0.33	0.41	0.43	0.45
MC P_M	0.000	0.000	0.001	0.000	0.004	0.007	0.002	0.000	0.003	0.003	0.002	0.008	0.006
MNB P_M	0.047	0.027	0.036	0.033	0.059	0.027	0.079	0.018	0.077	0.006	0.086	0.267	0.046

ground and want to increase the search pool because eyewitnesses are rare in that case [21]). Both scenarios focus on a single location.

Baselines. Yang [23] pointed out that in the case of a very low average training instances per category (which applies here) the *majority class trivial classifier* tends to outperform all non-trivial classifiers. We therefore start by comparing our geolocation system against the Majority Class (MC) baseline.

Results. The first row of Table 6 shows that P_μ of MC for languages with the majority of tweeters originating from one city tend to match or outperform the MNB classifier, i.e. *tr*, *ru* and *ko*, in the WORLD data collection. For instance, a MC model for tweeters posting in Russian would fail to predict the location of any tweeter outside Moscow, although 70% of the tweeters are located in other cities (inside and outside Russia). The same pattern applies to TwArchive with one more biased language, than WORLD: Thai (th).

To evaluate classifiers at the level of each city, rather than overall performance, we compare precision based on *macro* averaging in the last two rows of Table 6. In contrast to P_μ, P_M shows that MNB classifiers outperform the MC for all languages.

While the result of the MC is obvious for languages like *tr*, *ru* and *ko* at the high end of the range of P_μ, given the data imbalance for such languages as shown in Fig. 1, it doesn't reflect the influence of imbalance on other languages like *en*, *fr* and *it* at the low end of the range, with other languages in between. To address this problem, we compare P_μ, to P_M, which shows an expected drop in performance in Table 6. In the case of *ru*, an MNB geolocation model would have a high accuracy of 33%, while having a poor average precision on the level of each city (0.6%). This contrast between *micro-macro* indicates the measures evaluate geolocation from different perspectives.

Correlation with Individual Features. Entropy was shown to have the highest correlation with *Acc* compared to other features. Here, we measure the correlation between the proposed alternative measures, using different averaging techniques, and the same set of features, excluding the poor ones. Correlations for the two data collections (WORLD and TwArchive) are displayed in Table 7. The *micro* columns are analogous to accuracy reported earlier in Table 6.

Table 7. Correlation between features and precision using different averages; (* and †) denote statistical significance with $p \leq 0.05$ and $p \leq 0.01$, respectively.

Feature	Micro				Weighted				Macro			
	WORLD		TwArchive		WORLD		TwArchive		WORLD		TwArchive	
	r	r^2	r	r^2	r	r^2	r	r^2	r	r^2	r	r^2
Entropy	$-0.87^\dagger$	**0.75**	$-0.69^\dagger$	0.47	$-0.79^\dagger$	0.62	$-0.78^\dagger$	0.61	-0.49	0.24	-0.63^*	0.40
# Cities	$-0.76^\dagger$	0.58	-0.40	0.16	-0.64^*	0.41	-0.42	0.18	-0.46	0.21	-0.43	0.18
Entropy.test	$-0.82^\dagger$	0.67	$-0.70^\dagger$	0.49	$-0.74^\dagger$	0.54	-0.52	0.27	-0.34	0.12	-0.49	0.24
# Cities.test	-0.54	0.29	-0.51	0.26	-0.44	0.19	-0.32	0.10	-0.24	0.06	-0.36	0.13
# Tweeters	-0.56^*	0.32	-0.39	0.15	-0.36	0.13	-0.21	0.05	-0.14	0.02	-0.27	0.07
# Tweets	-0.50	0.25	-0.38	0.15	-0.30	0.09	-0.20	0.04	-0.11	0.01	-0.29	0.09

In contrast to Acc and P_μ, entropy is not as strong an indicator of how well a geolocation model performs on the *macro* level. The moderate insignificant correlation between entropy and P_M aligns with the fact that *macro*-averaging should be independent of the distribution of tweeters across cities, i.e. all cities are treated uniformly. *Macro*-averaging generally has the lowest correlation with the different features. The same pattern applies to recall.

From a language perspective, we observed that the ranking of languages differs from one averaging technique to another and also from precision to recall. For instance, on the level of *micro-macro* precision, th remained among the top ranks while tr dropped to the bottom behind en. To measure the degree of agreement, we measured the τ_β correlations for all direct combinations of data collection, precision, recall, *micro*, *weighted* and *macro*, see Table 8.

For *precision*, the *micro* and *weighted* averages have a statistically significant, but *moderate* rank correlation in WORLD. In contrast, the *micro* and *weighted* averages for *recall* coincide, in both data collections. *Micro* and *macro* averages did not have a significant rank correlation. Finally, at the level of data collections, *micro* (precision and recall), and *weighted* recall have a statistically significant, albeit moderate, rank correlation.

The difference in precision between *micro* and *macro* averaging suggests that all languages are affected by the data imbalance. *Micro* averaging is biased towards big cities, while *macro* averaging assumes that all cities contribute

Table 8. Languages rank correlation τ_β for micro (μ), weighted (W), and macro (M) averaging; (* and †) denote statistical significance with $p \leq 0.05$ and $p \leq 0.01$, respectively.

(a) Across averaging techniques

	Precision				Recall			
	WORLD		TwArchive		WORLD		TwArchive	
	W	M	W	M	W	M	W	M
μ	$0.41^\dagger$	-0.08	0.38	0.08	$1.00^\dagger$	0.08	$1.00^\dagger$	0.15
M	0.00	–	0.08	–	0.05	–	0.15	–

(b) Across data collections

Precision			Recall		
μ	W	M	μ	W	M
0.46^*	0.13	0.00	0.46^*	0.49^*	0.03

equally to the metric. Some languages are still easier than others, but not because they are the only languages biased towards a small set of cities, and/or their usage is geographically limited to a specific region. All languages have a bias towards a small number of big cities; the difference between languages like *en* and *fr* compared to *ru* and *tr* is the number of big cities. For instance, the top 10 cities for *en* and *fr* in WORLD have a comparable number of tweeters (1–4%) of the total number, while the top city in *ru* and *tr* has more than 30% of tweeters and the second city drops down to less than 10% of tweeters.

In the end, the choice of which averaging technique to use in taking decisions depends on the application. However in the general case, we recommend using the *weighted* average instead of *micro* because it limits the dominance of big cities while maintaining their importance. At the same time, it reduces the potentially misleading evaluation when comparing languages.

5   Conclusion

We studied features that might influence the accuracy of a system that geolocates tweeters. Examining two large collections of tweets covering thirteen languages, we found substantial variation in accuracy across languages, a result that has been observed before but not studied or explained.

Our study is the first to show that the distribution of tweeters over cities is strongly correlated to accuracy. Past work suggested that the geographical coverage of a language may also be a factor, however, all the languages we studied were found to have a global coverage.

Our results can be used to influence future test set design. The scale of a test set was found to have little influence on accuracy. However, the distribution of tweeters was a strong influence. Although a geolocation system could potentially ground tweeters to one of few thousand cities, the skewed distribution present in the test sets meant that accuracy was influenced by only a few tens of cities. Current testing approaches are not as geographically broad ranging as one might imagine or expect. A consequence of the current testing regime is that a simplistic baseline, which grounds to one city per language, was measured to be as accurate as a state of the art system for more than one language.

To overcome such dataset limitations, we proposed using *macro* averaging. The contrast between it and *micro* averaging revealed that data imbalance affects all languages, even one that is extensively used, such as English. Our analysis demonstrated that reporting both *micro* and *macro* averaging, or using a *weighted* average, provides valuable additional insight.

For future work, we will consider evaluating other geolocation inference techniques from a language perspective, making use of a wide range of open source frameworks. For instance, Wing and Baldridge [22] demonstrated that probabilistic language models and hierarchical logistic regression outperform LIW and text-categorisation for English, but on a different representation of location (i.e. not cities). Jurgens et al. [11] released a framework for nine different network-based geolocation systems. Recently, Rahimi et al. [16] explored using a hybrid text and network based approach.

This work was originally motivated by studying the lexical variations of languages and their impact on geolocating tweeters. A simple feature represented by the number of LIW per language, due to the lack of enough resources, was found to have no impact. It was hard to assess the richness of the vocabulary associated with the different languages (English is the pivot), or dialects within the same language (no definitive list of dialects per language). Gonçalves and Sánchez [7] showed that Spanish varieties can be recognized in Twitter and categorized into regions covering urban cities versus rural areas and small towns. However, they acknowledged that English and Chinese are problematic. We consider focusing on Spanish as a starting point for such analysis.

Considering the data imbalance problem, we intend to explore building test sets that are more geographically balanced through geographically stratified sampling. We will also examine representing location using grids, which might lead to a more balanced distribution of tweeters. The evaluation, however, would be challenging because each representation would have a different set of classes (cities vs. grids).

A large number of parameters, including the error distance with a specific range (i.e. 161 km), and the threshold of the number of tweeters to represent a location, were found to be arbitrarily chosen in past work. We plan to estimate the optimal values for those parameters and develop more robust evaluation metrics for dynamic values as a step towards training language independent geo-inference models.

Acknowledgments. This work was made possible by NPRP grant# NPRP 6-1377-1-257 from the Qatar National Research Fund (a member of Qatar Foundation). The statements made herein are solely the responsibility of the authors.

We thank the anonymous reviewers for their careful reading of our manuscript and their many insightful comments and suggestions.

References

1. Ahmed, A., Hong, L., Smola, A.J.: Hierarchical geographical modeling of user locations from social media posts. In: Proceedings of WWW, pp. 25–36 (2013)
2. Backstrom, L., Sun, E., Marlow, C.: Find me if you can: improving geographical prediction with social and spatial proximity. In: Proceedings of WWW, pp. 61–70 (2010)
3. Cheng, Z., Caverlee, J., Lee, K.: You are where you tweet: a content-based approach to geo-locating Twitter users. In: Proceedings of CIKM, pp. 759–768 (2010)
4. Darwish, K., Magdy, W., Mourad, A.: Language processing for Arabic microblog retrieval. In: Proceedings of CIKM, pp. 2427–2430 (2012)
5. Diakopoulos, N., De Choudhury, M., Naaman, M.: Finding and assessing social media information sources in the context of journalism. In: Proceedings of SIGCHI, pp. 2451–2460 (2012)
6. Eisenstein, J., O'Connor, B., Smith, N.A., Xing, E.P.: A latent variable model for geographic lexical variation. In: Proceedings of EMNLP, pp. 1277–1287 (2010)
7. Gonçalves, B., Sánchez, D.: Crowdsourcing dialect characterization through Twitter. PloS One **9**(11), e112074 (2014)

8. Han, B., Baldwin, T.: Lexical normalisation of short text messages: Makn sens a# Twitter. In: Proceedings of ACL, pp. 368–378 (2011)
9. Han, B., Cook, P., Baldwin, T.: Text-based Twitter user geolocation prediction. J. Artif. Intell. Res. **49**, 451–500 (2014)
10. Hecht, B., Hong, L., Suh, B., Chi, E.H.: Tweets from Justin Bieber's heart: the dynamics of the location field in user profiles. In: Proceedings of SIGCHI, pp. 237–246 (2011)
11. Jurgens, D., Finethy, T., McCorriston, J., Xu, Y.T., Ruths, D.: Geolocation prediction in Twitter using social networks: a critical analysis and review of current practice. In: Proceedings of ICWSM (2015)
12. Kinsella, S., Murdock, V., O'Hare, N.: I'm eating a sandwich in Glasgow: modeling locations with tweets. In: Proceedings of the 3rd International Workshop on Search and Mining User-generated Contents, pp. 61–68 (2011)
13. Lui, M., Baldwin, T.: langid. py: an off-the-shelf language identification tool. In: Proceedings of ACL, pp. 25–30 (2012)
14. Pedregosa, F., Varoquaux, G., Gramfort, A., Michel, V., Thirion, B., Grisel, O., Blondel, M., Prettenhofer, P., Weiss, R., Dubourg, V., et al.: Scikit-learn: machine learning in Python. J. Mach. Learn. Res. **12**, 2825–2830 (2011)
15. Priedhorsky, R., Culotta, A., Del Valle, S.Y.: Inferring the origin locations of tweets with quantitative confidence. In: Proceedings of CSCW, pp. 1523–1536 (2014)
16. Rahimi, A., Cohn, T., Baldwin, T.: pigeo: a Python geotagging tool. In: Proceedings of ACL-2016 System Demonstrations, pp. 127–132 (2016)
17. Roller, S., Speriosu, M., Rallapalli, S., Wing, B., Baldridge, J.: Supervised text-based geolocation using language models on an adaptive grid. In: Proceedings of EMNLP, pp. 1500–1510 (2012)
18. Sadilek, A., Kautz, H., Bigham, J.P.: Finding your friends and following them to where you are. In: Proceedings of WSDM, pp. 723–732 (2012)
19. Sakaki, T., Okazaki, M., Matsuo, Y.: Earthquake shakes Twitter users: real-time event detection by social sensors. In: Proceedings of WWW, pp. 851–860 (2010)
20. Sebastiani, F.: Machine learning in automated text categorization. ACM Comput. Surv. (CSUR) **34**(1), 1–47 (2002)
21. Starbird, K., Muzny, G., Palen, L.: Learning from the crowd: collaborative filtering techniques for identifying on-the-ground Twitterers during mass disruptions. In: Proceedings of ISCRAM (2012)
22. Wing, B., Baldridge, J.: Hierarchical discriminative classification for text-based geolocation. In: Proceedings of EMNLP, pp. 336–348 (2014)
23. Yang, Y.: An evaluation of statistical approaches to text categorization. Inf. Retr. **1**(1–2), 69–90 (1999)

Human-Based Query Difficulty Prediction

Adrian-Gabriel Chifu[1], Sébastien Déjean[2], Stefano Mizzaro[3],
and Josiane Mothe[4(✉)]

[1] LSIS - UMR 7296 CNRS, Aix-Marseille Université, Marseille, France
adrian.chifu@lsis.org
[2] IMT UMR 5219 CNRS, Univ. de Toulouse, Univ. Paul Sabatier, Toulouse, France
sebastien.dejean@math.univ-toulouse.fr
[3] University of Udine, Udine, Italy
mizzaro@uniud.it
[4] IRIT UMR 5505 CNRS, ESPE, Univ. de Toulouse, UT2J, Toulouse, France
josiane.mothe@irit.fr

Abstract. The purpose of an automatic query difficulty predictor is to decide whether an information retrieval system is able to provide the most appropriate answer for a current query. Researchers have investigated many types of automatic query difficulty predictors. These are mostly related to how search engines process queries and documents: they are based on the inner workings of searching/ranking system functions, and therefore they do not provide any really insightful explanation as to the reasons for the difficulty, and they neglect user-oriented aspects. In this paper we study if humans can provide useful explanations, or reasons, of why they think a query will be easy or difficult for a search engine. We run two experiments with variations in the TREC reference collection, the amount of information available about the query, and the method of annotation generation. We examine the correlation between the human prediction, the reasons they provide, the automatic prediction, and the actual system effectiveness. The main findings of this study are twofold. First, we confirm the result of previous studies stating that human predictions correlate only weakly with system effectiveness. Second, and probably more important, after analyzing the reasons given by the annotators we find that: (i) overall, the reasons seem coherent, sensible, and informative; (ii) humans have an accurate picture of some query or term characteristics; and (iii) yet, they cannot reliably predict system/query difficulty.

1 Predicting Query Difficulty

The purpose of a query difficulty predictor is to decide whether an Information Retrieval (IR) system is able to properly answer a current query, that is to say, if it is capable of retrieving only the relevant documents that meet a user's information need as expressed through his or her query. Predicting query difficulty is a hot topic: if a search engine could predict its own chances of failure when processing a given query, it could adapt its processing strategies to increase the

© Springer International Publishing AG 2017
J.M. Jose et al. (Eds.): ECIR 2017, LNCS 10193, pp. 343–356, 2017.
DOI: 10.1007/978-3-319-56608-5_27

overall effectiveness, perhaps even by requesting more information directly from the user in order to better meet his or her needs. The example of an ambiguous term is a textbook case: the query "Orange" can be predicted as difficult because the term has various meanings attached to it. The system may decide to diversify its answers to encapsulate the various meanings of the word; or it may ask the user if he or she is interested in the telecom company, the color, the fruit, or something else; it may also derive the meaning from the user's past queries, if available. However, ambiguity is not the only reason for a query being difficult (the number of senses of query terms correlates only weakly with system effectiveness [12]).

Predicting query difficulty is challenging. Current automatic predictors are either computed before a search is carried out (pre-retrieval predictors, e.g., the inverse document frequency of the query terms [14]), or computed from a list of retrieved documents (post-retrieval predictors, e.g., the standard deviation between the top-retrieved document scores [13]). The literature reports slightly better correlations with actual system effectiveness when using post-retrieval predictors than pre-retrieval ones, although pre-retrieval predictors are the most interesting for real applications because they are cheaper to calculate. Still, these correlations are weak, even when the various predictors are combined [1,6,7,12]. Moreover, the current automatic predictors are founded on the way search engines process queries and documents, and the way they rank retrieved documents. They do not consider what causes the query to be difficult. Indeed, the features used to calculate automatic predictors are linked to inner functions of search engines, which do not necessarily reflect the human perception of difficulty.

In this paper our intention is to go one step further in query difficulty analysis and understanding, by taking into account the human perspective rather than the system perspective. Instead of considering how IR features could be used to predict difficulty, as it has usually been done so far, we focus on understanding what the *human* perception of query difficulty is and on *why* does a query sound as difficult. To do this, we conduct user studies where we ask human annotators[1] to predict query difficulty and explain the reasons for their prediction. We also aim to understand how different clues on the data and the amount of information provided to the human annotators affect the outcome. To this aim, in some cases the annotators receive only the query as submitted to the system (in our study we consider the *title* field of TREC topics as a query); in other cases the annotators receive, as a surrogate of the user's intent, a longer description of what the user requires (we consider the *descriptive* part of the TREC topic).

This paper is structured as follows. Section 2 introduces related work. Section 3 discusses the motivations of our approach and frames three research questions. Each of Sects. 4, 5, and 6 addresses each research question. Section 7 concludes the paper.

---

[1] We use the terms "predictors" and "annotators" or "participants" to distinguish between automatic and human prediction, respectively.

2   Related Work: Why Queries Are Difficult?

Evaluation in IR has a long history and programs such as TREC have brought many interesting clues on IR processes. One of them is the huge variability in terms of relevance of the retrieved documents, according to both topics and systems. The Reliable Information Access (RIA) workshop has been the first attempt to try to *understand* in a large scale the IR "black boxes". As stated by Harman and Buckley, "The goal of this workshop was to understand the contributions of both system variability factors and topic variability factors to overall retrieval variability" [4,5]. Harman and Buckley claim that understanding variability in results is difficult because it is due to three types of factors: topic statement, relationship between topics and documents and system features. The RIA workshop focused on the query expansion issue and analyzed both system and topic variability factors on TREC collections. By considering failure analysis, 10 classes of topics were identified manually, but no indications were given on how to automatically assign a topic to a category. One of the main conclusions of the failure analysis was that systems were missing an aspect of the query, generally the same aspect for all the systems. "The other major conclusion was that if a system can realize the problem associated with a given topic, then for well over half the topics studied, current technology should be able to improve results significantly."

Interactive IR studies are somehow related to our work as they involve users and analyze their behavior and difficulties while completing a search task. However, these studies are more oriented on analyzing the users' sessions, their successes and failures. The most related work is the one from Liu et al. [9,10], that aims at collecting and analyzing why users perceive a given task as difficult. In their study, the users were given complex tasks, such as collecting information to write a new entry in Wikipedia, which is quite different from TREC ad hoc task. Users were asked to provide reasons for pre-task difficulty perception and they mentioned time limitation, complexity and specific requirements. Other aspects were more related to users and interaction which is less related to our work. While Liu et al.'s research focuses on a few search topics and encapsulates the users' knowledge in their schema, we rather consider many search topics and focus on the reason why the system may fail given the query. Thus, we are more in line with RIA workshop, as a system failure analysis project.

Hauff *et al.* [8] analyzed the relationship between user ratings and system predictions using ClueWeb 2009. They study both the topic level and the query level. In the latter, the authors consider various queries for a single topic or information need and measure the users' ability to judge the query suggestion quality. The topic level is closer to ours: annotators who were provided with the topic title and description were asked to rate the quality of the queries on a five-level scale. The authors found that (i) the inter-annotator agreement is low (Cohen's Kappa between all possible pairs of annotators is between 0.12 and 0.54), (ii) the correlation of individual users and system performance is low (median correlation 0.31 for AP and 0.35 for P@30). Mizzaro and Mothe [11] confirmed, by a laboratory user study using TREC topics, that human prediction only weakly correlates with system performance. They also reports some results on why queries might be perceived as difficult by humans, but the analysis is very limited.

We go a step further: we try to study the reasons why a query is perceived as difficult by analyzing user comments. We also analyze the relationship between these reasons and human prediction of difficulty as well as with automatic predictors or query features and with actual system performance. We find interesting cues that can be reused either on query difficulty prediction or for improving users' information literacy. Finally, our results suggest that some reasons are good predictors of possible system failure.

3   Why Studying Human Query Difficulty Prediction?

In this paper, we go a step further in system failure and query difficulty analysis. Our main goal is to get cues on what users think a difficult query for a system is. These clues may differ from what the system actually finds as a difficult query. To this aim, we asked annotators to indicate both their prediction on query difficulty and their explanation for the reason they think the query is going to be easy or difficult for a search engine.

There are several motivations underlying our research and the user study approach that we have chosen. Current understanding of query difficulty and current query difficulty predictors are based on the way queries and documents are processed by the search engine. While we know that tf.idf of query terms has an effect on the system results, we do not know if humans are able to perceive other cues that the systems do not capture, nor if some of the human predictors are correlated to some automatic predictors, giving them more sense to humans. Reversely, it might be that some strategy used by human predictors could be a good automatic predictor, if calculated properly. Also, we do not know yet the theoretical possibilities of automatic query difficulty prediction. By studying how humans predict query difficulty we might be able to understand how difficult the task of automatic prediction is: for example, if predictions based on query terms only are much worse than full information need based predictions (or maybe even impossible at a satisfactory level), then we would have a more precise measure of how difficult (if not impossible) the task of the automatic prediction systems is.

Longer term objectives are: to define pre-retrieval predictors that are based on our findings about human perception and that, hopefully, will be at least as effective as automatic post-retrieval predictors, and better than current pre-retrieval ones; and derive some element for information literacy training. More explicit research questions are:

RQ1 Difficulty Reasons. *Why* is a query difficult? Can human annotators identify and express the reasons why a query is difficult? Are these reasons sound? Do these reasons correlate with automatic predictors, and/or with other query features?

Table 1. The two experiments. E1 uses TREC ad hoc collections while E2 use TREC Web track ClueWeb12 collection with TREC 2014 topics. (*) In E1 each participant chose which topics to annotate from the 150 available. (**) Free text annotations were recoded to derive re-coded reasons explaining difficulty, as explained in the text.

# of Particip.	Scale	Collection	# of topics	Metrics	Amount of info	Explan.	Topics
E1 38 (29 + 9)	3	TREC 6-8	91 (*)	AP	Q, Q+D	Free text (**)	321-350 in TREC 6, 351-381 in TREC 7, 421-450 in TREC 8 (*)
E2 22	5	TREC 2014	25	ERR@20 NDCG@20	Q, Q+D	Categories + Free text (**)	251 255 259 261 267 269 270 273 274 276 277 278 282 284 285 286 287 289 291 292 293 296 297 298 300

RQ2 Amount of Information. Automatic predictors use the query only since they cannot access the user's information need. Do human predictions depend on the amount and kind of information available? Do they evaluate queries in a different way when they know the query only and when they have a more complete description of the user's need?

RQ3 Links with Actual System Difficulty. Are these reasons accurate predictors of perceived or actual query difficulty? Do automatic predictors capture any difficulty reasons called upon by users?

Those research questions (of which RQ1 is probably the most interesting) are addressed by two experiments, named E1 and E2, performed in a laboratory settings and involving users. Overall, the main features of the two experiments are summarized in Table 1. We varied: the collection used (TREC 6-7-8 and TREC 2014, a.k.a. ClueWeb); the amount of information presented to the user (Query, Q, vs. Query and description, Q+D); and the collected annotations (level of difficulty on a 3 or 5 levels scale; explanation in free text, or explanation through five levels questions/answers). These two slightly different experimental designs allow us to study also two important issues: reason generation (E1) vs. identification (E2); and more longer and complete topic descriptions, but on quite old topics (E1) vs. shorter and less informative topic descriptions, but on more recent topics (E2). More details are presented as needed in the following sections.

4   RQ1: Difficulty Reasons

Our first objective is to know if users can explain why they think a query will be difficult or easy for a search engine.

4.1   Finding Reasons: First Experiment (E1)

The first experiment E1 aims to collect free text explanations that participants associate with query ease and difficulty. The participants to E1 were 38 Master's

Students (25 1st and 13 2nd year) in library and teaching studies; although they were trained to use specialized search engines, they had just an introduction class on how search systems work. Participants could choose as many topics as they wanted to annotate from the set of the 150 topics from TREC 6, 7, and 8 *adhoc* tracks, labeled as 301–450. Each participant was first shown the query only (Q in Table 1 and in the following; it corresponds to the Title part of the TREC topic) and asked to evaluate its difficulty on a three-level scale (Easy/Medium/Difficult), as well as to provide a mandatory explanation in free text. Since the query only might not reflect well the user's intent, the worker was then shown a more complete description of the query (Q+D, i.e., Descriptive and Narrative parts of TREC topic), and the worker again evaluated the query. This two-stages (Q followed by Q+D) prediction was repeated for the queries chosen by each participant.

Topics were displayed in different order to avoid any bias as the first topics may be treated differently because the task is new for annotators. Moreover, annotators could skip some topics if they wish; this was done to avoid them answering on a topic they did not understand or felt uncomfortable with. The drawback is that the number of annotations varies over topics, and that some topics are not assessed. However, our goal was to collect reasons that humans associate with ease and difficulty, and therefore an association with each topic was not needed. It was instead important to leave the participants free to generate any reason that they might come up with; this is why we used free text. Since the annotation process is difficult, we tried to provide to the students the most favorable conditions. Students could also choose between annotating the query only (and they were not shown the full topic description) or using both the Q part (before) and the full Q+D description (after). Of the 38 students, 29 annotated query difficulty considering Q only, whereas 9 students annotated using both Q and Q+D.

4.2   From Free Text to Re-coded Text

We analyzed difficulty reasons first using free text, then using re-coded free-text.

Manual Analysis of Free Text Comments. First, we analyzed the free text manually, with the objective of finding if there were some recurrent patterns. When we asked for free text comments we did not provide any comment writing guidance, except for using the keyword "Easy:" or "Difficult:" before any comment. We asked for free text explanations of their query difficulty predictions because we did not want to drive the results. Table 2(a) lists the most frequent words associated with ease and difficulty in the comments. In a few cases, the comments were difficult to understand or analyze because not explicit enough. This was for example the case when annotators wrote *vague* without detailing if it was a query term which they found vague or the topic itself. A typical example is the one of Query 417 from TREC 6-8 (Title: *creativity*) for which the 5 annotators considered the query as difficult using comments such as "too broad, not enough targeted", "far too vague", "far too vague topic", "keyword used very

Table 2. Most frequent: (a) words in free text comments; (b) comments after recoding.

(a)					(b)			
Easy because		Difficult because			Easy because		Difficult because	
Precise	113	Missing	64		Precise-Topic	66	Risk-Of-Noise	50
Clear	48	Broad	62		Many-Documents	45	Broad-Topic	43
Many	45	Risk	56		No-Polysemous-Word	31	Missing-Context	34
Polysemous	36	Context	34		Precise-Words	25	Polysemous-Words	22
Usual	16	Polysemous	33		Clear-Query	19	Several-Aspects	20
Specialist	15	Vague	26		Usual-Topic	16	Missing-Where	16
Simple	11	Many	21					

broad, risk of noise", and "a single search term, risk of getting too many results".
While some comments are quite explicit, others are difficult to interpret.

Re-coded Text. Automatic text analysis would have implied to apply advanced
natural language processing with no guarantee of success considering, for example, the specificity of the vocabulary, and the lack of data for training. For this
reason, we rather analyzed manually the free text and re-coded it; which is a
common practice in user studies. Table 3 shows some examples of the re-coding
we made.

Table 3. Examples of recoding.

Comment	Recoding
A single word in the query	One-Word
The term exploration is polysemous	Polysemous-Word
Far too vague topic	Too-Vague-Topic
Is it in US? Elsewhere?	Missing-Where
Few searches on this topic	Unusual-Topic
Risk of getting too many results	Too-Many-Documents
There are many documents on this	Many-Documents

Annotator Peculiarities. To check the correlation between annotators and
the annotations they provide (after re-coding), we used Correspondence Analysis (CA) [2] on the matrix that crosses annotators and re-coded comments (not
reported here because of space limits). Compared to more commonly used Principal Component Analysis, CA allows displaying on the same space the variables
and observations. We analyzed if some annotators used some specific comments
or have different ways of annotating difficulty reasons. We could not find very
strong peculiarities among the types of annotations the participants used that
would have justified a complementary experiment.

Comments Associated to Ease and Difficulty. Table 2(b) displays the most frequent re-coded reasons associated to ease (left part) and difficulty (right part). Remember that a given query can be annotated by some comments associated to both. Some phrases are associated both to ease and difficulty of a query (as, e.g., *Many-Documents*). Indeed, users may have in mind recall-oriented tasks and precision-oriented tasks.

While *Precise-Topic* is generally associated to ease (66 times), it is also associated to difficulty in 3 cases. In that cases it is associated to other comments, e.g. *The topic is very precise but it may be too specific.* In the same way, *Many-Documents* is mostly associated to ease and *Too-Many-Documents* to difficulty. When *Many-Documents* is used associated with difficulty, it is generally associated to *Risk-Of-Noise*.

This first analysis helped us in having a better idea on human perception of difficulty. However, E1 was not enough to study real effects because: (i) we had a different number of annotated topics per participants and a different number of annotators per topics; (ii) the free text expression was too hard to analyze; and (iii) the collection was not fully appropriate for humans to annotate query difficulty. We thus designed a second experiment addressing these issues, presented in the next section.

4.3   Reasons as Closed Questions: Second Experiment (E2)

We designed and performed a second experiment E2, with three main differences from E1 (presented in Sect. 4.1). First, we changed the collection from TREC 6-8 to ClueWeb12. TREC 6-8 collections are widely used and are appropriate for this kind of study, since they feature a large number of topics with a long and detailed description of the needs; these collections are still used for evaluation purposes [15]. But they are old: some participants had difficulties in annotating the queries just because of time reasons (although in the previous setting we made clear in the instructions that the collection contained documents from the 90s). For example in the 90s El Niño was a hot topic in News because it was one of the powerful oscillation events in history, but some of the 2015–16 young students did not hear about the phenomena and event from 1996–97. So in this second experiment we used a newer collection, the ClueWeb12 collection (Category A corpus) used in TREC 2014, which is a large and recent Web snapshot with more recent topics. As a consequence, topics were different too: we selected the 25 topics shown in the third row Table 1, that are the easiest 10, the most difficult 10, and the medium 5 according to the topic difficulty order presented in the TREC track overview paper [3]. One disadvantage of TREC 2014 (that is important to mention because it also justifies the previous experiment) is the rather short query Description. Second, we switched from three level difficulty to a five-level scale of difficulty ("Very Easy", "Easy", "Average", "Difficult", "Very Difficult") which is more standard.

Third, participants did not express difficulty reasons in free text as in E1, but using closed questions, that we designed on the basis of the free text comments gathered in E1 and of their re-coding. We were able to re-code the comments indicated by the users into reasons phrased as closed questions that could be answered following a scale of values. We used 32 reasons in total (denoted with Ri later on); they are listed in Table 4. We think we cover all the aspects we found in the E1 participants' annotations, i.e., any reason that was expressed in E1 can be expressed also in E2. These reasons were to be answered, both when annotating Q and Q+D, using a five-level scale, ranging from "-2 I strongly disagree" to "$+2$ I strongly agree".

Having the same number of predictions for each query makes the statistical analysis more smooth and sound; we thus consider the same number (8) of annotators for each topic. Participants were 22 volunteers recruited using generic emailing lists mainly from our research institutes, and they got a coupon for participating. Each of them was asked to annotate 10 queries (we took care of using the usual randomized experimental design). Annotators had to annotate the level of difficulty of the query using a five-level scale, but rather than asking to explain the reason of their grading in free text only, we asked them to answer the predefined closed questions. As in E1, Q was presented first, then Q+D. We collected 200 annotations of each type in total, with 8 annotators for each of the 25 topics (we removed annotations when we got more than 8). In the rest of the analysis we average the annotations over the participants for each annotated topic.

4.4 Closed-Reasons Analysis

Correlation Between Human Difficulty Perception and Closed Questions/Reasons. Table 4 shows the correlation between the values humans associate to a reason and the level of difficulty predicted, first when considering Q only, then when considering Q+D. These correlations are obtained after aggregating the results over the 8 annotators and the 25 topics. For example, "R19: None or very few relevant will be retrieved" is strongly correlated to human prediction of query difficulty, as R23 and R24 are, although negatively. Less correlated but still significantly, are R10 (unknown topic), R11 (too broad), R13 and R28 (various aspects), R27 (concrete query), R32 (vagueness). All these are interesting reasons that humans relates to difficulty. Other reasons are not correlated with their perception of difficulty such as R5, R6, R9, R12, and R15.

Correlation Between Closed Reasons. When we defined the closed questions/reasons from the free texts provided by E1 participants, we tried to avoid to use clearly correlated reasons, but we kept some that were not obviously correlated (e.g. R19, R23, and R24). Nonetheless, it is worth analyzing deeper the correlation between reasons. After having aggregated the data by topic over the annotators, we then calculate the Pearson correlation between reasons. We find that, for example, R1, R2 and R3 strongly correlate, the two first positively while the third negatively. We can also see other groups of correlated reasons: R12 and R15 (a mistake to have kept the two), R10 and R17, R19 and R24. Clearly, not all the reasons are independent, and in future experiments the highly correlated reasons can be removed.

Table 4. Closed reasons resulting from re-coding free text annotations and their correlations with human prediction on Q (column 2) and human prediction on Q+D (column 3).

Reason	Correlation	
	Q	Q+D
R1: The query contains vague word(s)	0.523	0.370
R2: The query contains polysemous/ambiguous word(s)	0.342	0.145
R3: The query contains word(s) that is (are) relevant to the topic/query	−0.410	−0.356
R4: The query contains generic word(s)	0.296	0.135
R5: The query contains proper nouns (persons, places, organizations, etc.)	−0.040	0.255
R6: The query contains uncommon word(s)	−0.005	0.024
R7: The query contains specialized word(s)	−0.238	−0.241
R8: The words in the query are inter-related or complementary	−0.028	0.187
R9: The query contains common word(s)	−0.089	0.006
R10: The topic is Unusual/uncommon/unknown	0.526	0.496
R11: The topic is too broad/general/large/vague	0.393	0.502
R12: The topic is specialized	−0.103	−0.136
R13: The topic has several/many aspects	0.614	0.708
R14: The topic is current/hot-topic	−0.118	−0.246
R15: The topic is Non-specialized	−0.017	0.037
R16: The topic is too precise/specific/focused/delimited/clear	−0.149	−0.237
R17: The topic is Usual/common/known	−0.627	−0.512
R18: The number of documents on the topic in the Web/collection is high	−0.693	−0.564
R19: None or very few relevant document will be retrieved	0.880	0.800
R20: Only relevant documents will be retrieved	−0.472	−0.604
R21: There will be different types of relevant documents in the Web/collection	−0.023	0.137
R22: Non-relevant sponsored links/documents will be shown	0.040	0.338
R23: Many of the relevant documents will be retrieved	−0.867	−0.763
R24: Many relevant documents will be retrieved	−0.873	−0.751
R25: Documents with various relevance levels can be retrieved	0.189	0.383
R26: The number of query words is too high	0.624	0.205
R27: The query is concrete/explicit	−0.390	−0.587
R28: The query concerns various aspects	0.458	0.681
R29: The number of query words is too low	0.185	0.353
R30: The query is clear	−0.532	−0.631
R31: The query is missing context	0.273	0.516
R32: The query is broad/vague	0.352	0.615

Can Human Reasons Be Explained by Query Features? Some of the closed-reasons are somehow associated with query features used in information retrieval studies. For example, "R2: The query contains polysemous/ambiguous word(s)" can be associated to the number of senses of query terms and thus to the *Synsets* query difficulty predictor, i.e., the number of senses in WordNet (http://wordnet.princeton.edu/) for the query terms proposed by Mothe and Tanguy [12]. "R4: The query contains generic word(s)" can also be associated

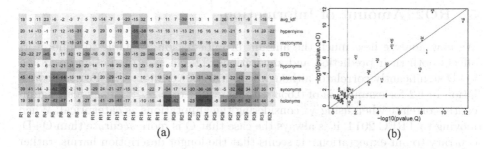

(a)   (b)

Fig. 1. (a) Correlations (x100) between reasons (X axis) and query features and automatic predictors (Y axis) when using Q. (b) Significance of the correlations between reasons and predicted difficulty (Q on X axis and Q+D on Y Axis). (Color figure online)

to linguistic characteristics that could be captured through WordNet. It is thus interesting to check if humans capture these features properly.

We thus analyze the correlation between the reasons and some query features. We consider the Synset linguistic feature which calculates the query term ambiguity based on WordNet [12]. We add other linguistics features extracted from WordNet. They correspond to the relations that exist between terms in WordNet resource: number of hyperonyms, meronyms, hyponyms, sister terms, synonyms, holonyms. These features were first calculated on each query term, then the median value is kept (we tried min, max, and avg also). We also consider two major statistical features used in the literature as query difficulty predictors: IDF, that measures the fact a term can discriminate relevant from non-relevant documents as a pre-retrieval predictor and STD, the standard deviation between the top-retrieved document scores [13], which is a post-retrieval predictor.

Results are presented in Fig. 1(a). The darker the color, the stronger the correlation. First, these results show that the human perception of ambiguity is not very strongly correlated to the ambiguity as capture by WordNet. This point will be worth analyzing deeper in future research that will imply an ad hoc user study. On the other hand, the number of query terms WordNet synonyms correlates with R28 (various aspects). That could be a way users express topic ambiguity as well. It also correlates with R6 (specialized word); this make sense since it is likely that specialized words have not many senses. The number of query terms holonyms (part-of relationships) in WordNet is strongly correlated with human perception of the number of possible relevant documents in the collection. This can be explained by the fact that if a term has a lot of holonyms, it is likely that a lot of documents will exist on the various parts of it.

When analyzing the correlation with IDF and STD automatic query difficulty predictors, we can observe from the figure that there are not very strong correlations with reasons. One of the strongest correlations is between R21 (different types of relevant documents) and IDF in one hand and between STD and R4 (generic words), R7 (specialized words), R17 (Usual/common/known) and STD on the other hand.

5   RQ2: Amount of Information

We now analyze how much the amount of information available to annotators affects both their prediction and annotations. The human prediction on Q and Q+D significantly correlate, although values are never high. The Pearson correlation in E2 for example is of 0.653 with a p-value of $2.2e^{-16}$. They also significantly correlate when using χ^2 considering the annotations as categorical. When moving to TREC 2014, it is always the case that Q is more accurate than Q+D, contrary to our expectations. It seems that the longer description harms, rather than helps, in TREC 2014. One possible explanation is that Q+D was much less detailed in TREC 2014 than in TREC 6-8. Also, a psychological effect might have happened: the participants were first shown the short description Q and then, when shown the Q+D, they might have assumed that "something has to be changed", thus worsening their prediction when it was good in first place.

Figure 1(b) reports the statistical significance of the correlation between the closed-reasons and the human prediction of the query difficulty. Each number represents the reason positioned according to its X and Y coordinates. X-axis corresponds to the p-values calculated on annotations collected using Q only while Y-axis corresponds to the p-values calculated when using Q+D, on log scale. The dotted lines (also in red) mark-up the 0.05 significance level. Reasons in the bottom-left rectangle defined by the dotted lines are not significantly correlated with the level of difficulty mentioned by the participants. For example R5 is in that corner; the value a human gives to it does not correlate with his perception of difficulty. On the other hand, reasons in the top-right rectangle are significantly correlated with it, both when considering Q and Q+D annotations (e.g., R19, R23). The value the user gives to the fact that there will be a lot of relevant documents (R23) correlates with his prediction of difficulty. On the bottom-right, the reasons given when considering Q only are significantly correlated with the level of difficulty humanly predicted on Q while the reasons provided on Q+D are not significantly correlated with the difficulty level predicted on Q+D (e.g. R2 and R4). For example "R2: Query contains polysemous words" significantly correlates with the predicted value of difficulty when considering Q but it is no more obvious when considering Q+D. The reverse phenomenon can be observed on the top-left rectangle (e.g., R9). Since we also observed (not reported in detail here due to lack of space) that the values given by the annotators on reasons when considering Q and Q+D highly correlate, what changed here is the perception of difficulty.

6   RQ3: Links with Actual System Difficulty

We also analyze the accuracy of human prediction: we calculate the correlation between human prediction and actual system effectiveness, considering the best system effectiveness for the corresponding TREC track, and using the official measures of the track. For space limits we cannot report detailed results, but all our attempts to detect correlation between human difficulty prediction and system effectiveness have failed. This result is consistent with the few related work that also focus on this topic [8,11].

7   Conclusion and Future Work

Compared to the RIA workshop [4,5], the annotators for this study are less-specialist in IR. Compared to Liu et al.'s studies [9,10], our study focuses on predicting query difficulty based on the query statement or on the intents of the user, but independently of the user's knowledge on the topic; even though it may have an influence on the annotation they provided. Compared to Hauff et al.'s work [8], we went a step further to understand the users' point of view on query difficulty.

When asking for free text reasons, we found that, overall, the reasons annotators provided seem coherent, sensible, and informative. Moreover, humans have an accurate picture of some query or term characteristics; for example regarding the ambiguity of terms, even if their perception of ambiguity is probably broader than what a linguistic resource can gather. But we also found that humans are bad to predict the difficulty a system will have to answer properly to a query. This result is consistent with the literature. Finally, we found that some reasons they answered through closed-questions are better correlated to actual system effectiveness than automatic predictors from the literature, opening new tracks for research on helping users to formulate their queries.

References

1. Bashir, S.: Combining pre-retrieval query quality predictors using genetic programming. Appl. Intell. **40**(3), 525–535 (2014)
2. Benzécri, J.-P., et al.: Correspondence Analysis Handbook. Marcel Dekker, New York (1992)
3. Collins-Thompson, K., Macdonald, C., Bennett, P., Diaz, F., Voorhees, E.: TREC 2014 Web track overview. In: Text REtrieval Conference. NIST (2015)
4. Harman, D., Buckley, C.: The NRRC reliable information access (RIA) workshop. In: SIGIR, pp. 528–529 (2004)
5. Harman, D., Buckley, C.: Overview of the reliable information access workshop. Inf. Retr. **12**(6), 615–641 (2009)
6. Hauff, C.: Predicting the effectiveness of queries and retrieval systems. Ph.D. thesis (2010)
7. Hauff, C., Hiemstra, D., de Jong, F.: A survey of pre-retrieval query performance predictors. In: CIKM, pp. 1419–1420 (2008)
8. Hauff, C., Kelly, D., Azzopardi, L.: A comparison of user and system query performance predictions. In: Conference on Information and Knowledge Management, CIKM, pp. 979–988 (2010)
9. Liu, J., Kim, C.S.: Why do users perceive search tasks as difficult? Exploring difficulty in different task types. In: Symposium on Human-Computer Interaction and Information Retrieval, USA, pp. 5:1–5:10 (2013)
10. Liu, J., Kim, C.S., Creel, C.: Exploring search task difficulty reasons in different task types and user knowledge groups. Inf. Process. Manag. **51**(3), 273–285 (2014)

11. Mizzaro, S., Mothe, J.: Why do you think this query is difficult? A user study on human query prediction. In: SIGIR, pp. 1073–1076. ACM (2016)
12. Mothe, J., Tanguy, L.: Linguistic features to predict query difficulty. In: Conference on Research and Development in IR, SIGIR, Predicting Query Difficulty Workshop, pp. 7–10 (2005)
13. Shtok, A., Kurland, O., Carmel, D.: Predicting query performance by query-drift estimation. In: Azzopardi, L., Kazai, G., Robertson, S., Rüger, S., Shokouhi, M., Song, D., Yilmaz, E. (eds.) ICTIR 2009. LNCS, vol. 5766, pp. 305–312. Springer, Heidelberg (2009). doi:10.1007/978-3-642-04417-5_30
14. Spärck Jones, K.: A statistical interpretation of term specificity and its application in retrieval. J. Doc. **28**(1), 11–21 (1972)
15. Tan, L., Clarke, C.L.: A family of rank similarity measures based on maximized effectiveness difference. arXiv preprint arXiv:1408.3587 (2014)

Fixed-Cost Pooling Strategies Based on IR Evaluation Measures

Aldo Lipani[1](✉), Joao Palotti[1], Mihai Lupu[1], Florina Piroi[1], Guido Zuccon[2], and Allan Hanbury[1]

[1] Institute of Software Technology and Interactive Systems,
TU Wien, Vienna, Austria
{aldo.lipani,joao.palotti,mihai.lupu,florina.piroi,
allan.hanbury}@tuwien.ac.at
[2] Faculty of Science and Engineering, Queensland University of Technology,
Brisbane, Australia
g.zuccon@qut.edu.au

Abstract. Recent studies have reconsidered the way we operationalise the pooling method, by considering the practical limitations often encountered by test collection builders. The biggest constraint is often the budget available for relevance assessments and the question is how best – in terms of the lowest pool bias – to select the documents to be assessed given a fixed budget. Here, we explore a series of 3 new pooling strategies introduced in this paper against 3 existing ones and a baseline. We show that there are significant differences depending on the evaluation measure ultimately used to assess the runs. We conclude that adaptive strategies are always best, but in their absence, for top-heavy evaluation measures we can continue to use the baseline, while for P@100 we should use any of the other non-adaptive strategies.

1 Introduction

Information Retrieval (IR) research relies heavily on well grounded empirical experiments that demonstrate the impact and merits of new techniques. The common framework of IR experimentation relies on the Cranfield paradigm [6, 22] of a test collection (a collection of documents, a set of topics, and a set of relevance assessments); this paradigm has predominantly driven the study and comparison of IR systems' effectiveness in the last decades of IR research.

In the first Cranfield experiment, relevance was modelled as a complete relation, i.e. a relevance judgement was expressed for each topic-document pair in the collection. However the large increase in size of document collections and the costs involved in obtaining relevance judgements soon rendered it impossible to source judgements for every topic-document pair in the collection. Even for a relatively small test collection with half a million documents (i.e. far from web-scale) and a few tens of topics, the effort to create a complete set of relevance

This research was partly funded by the Austrian Science Fund (FWF) project number P25905-N23 (ADmIRE).

© Springer International Publishing AG 2017
J.M. Jose et al. (Eds.): ECIR 2017, LNCS 10193, pp. 357–368, 2017.
DOI: 10.1007/978-3-319-56608-5_28

judgements would take more than a researcher's entire (hopefully long) lifetime. Until today, the most used method to avoid complete assessment is *pooling*.

The pooling method reduces the number of relevance judgements that are necessary in order to accurately assess the effectiveness of an IR system, or, more importantly, establishing the difference between the effectiveness of two systems. Pooling has been first introduced in the '70s [10], but has been used regularly only since the '90s with standardized IR benchmarking at the Text Retrieval Conference (TREC) [22]. Central to the use of pooling is that sufficiently many and sufficiently diverse systems have contributed to the creation of the *pool*, i.e. the set of documents that are collected for judgement. The most common and simplest pooling strategy is *Depth@k*, which prescribes the collection of relevance assessments only for the top k (referred to as the *pool depth*) documents from each of the document rankings of a number of IR systems.

Pooling, though used frequently to build test collections, was soon taken under scrutiny as it was observed that when the number of systems contributing to the pool was too low or the systems were not diverse enough, the identified set of relevant documents was not sufficient to reliably and accurately assess the effectiveness of an IR system that did not contribute to the pool [19]. This issue challenges the *re-usability of a test collection* as a tool for evaluating and comparing IR systems beyond those systems that contributed to the pool [21].

This test collection *bias* towards advantaging systems that participated in the pool creation over those that did not is ultimately due to an incomplete set of relevance judgements [11]. Zobel [26] first and Buckley et al. [2] later have shown that small test collections typically exhibit little bias, while large collections, such as modern web scale test collections, are affected by larger bias and thus such test collections may be rendered void when evaluating IR systems (especially those that did not contribute to the pool) if this bias is not controlled.

Research on controlling for pool bias follows two main approaches. On one hand there is work to reduce the bias at the test collection creation time. This has been done by devising alternative pooling strategies [4,14–16]. On the other hand, when the objective is the reuse of an existing test collection, research has explored the possibility of adjusting evaluation measures such that new systems can be fairly compared to the ones that contributed to the pool creation [12,23]. When the two approaches are combined, a new pooling strategy emerges, along with a matching evaluation measure [1,24], complying with the observation that performance measures are an intrinsic part of test collections [16].

This paper explores a family of strategies based on IR evaluation measures to identify documents to be placed in a pool of fixed size N, where the size is defined by a fixed budget, such that the test collection can be reliably used in later retrieval experiments. These strategies are: a baseline, $Take@N$; 3 pooling strategies as introduced by Moffat et al. [16], $RBPABased@N$, $RBPBBased@N$, and $RBPCBased@N$; and 3 newly proposed pooling strategies, $DCGBased@N$, $RRFBased@N$, and $PPBased@N$. These pooling strategies are empirically evaluated with respect to their impact on three common evaluation measures; the results are compared on a set of 11 TREC test collections.

2 Pooling Strategies

Our aim is to empirically study several strategies inspired by IR evaluation measures. In the following M denotes the function that associates a score to a given document d, retrieved by at least one run in the set of pooled runs R_p. The definition of M varies depending on the pooling strategy used and will be detailed in this section. The function $\rho(d, r)$ expresses the position (also called rank) of the document d in the run r.

The first fix-cost pooling strategy we present, $Take@N$, is also used as a baseline in the following experiments, similarly to previous study [15]. This strategy is based on the common $Depth@k$ pooling strategy, using the highest rank at which documents have been retrieved in the pooled runs to select the top N documents to assess. The strategies we present following $Take@N$ share the intuition behind it, replacing the choice by the mere document rank with the choice by a score, which is also function of the document rank. That is, the pooling strategies accumulate evidence of the importance of a document d for a given query based on both a) the rank $\rho(d, r)$ at which d has been retrieved in the pooled run $r \in R_p$, and b) on the particularities of a selection of evaluation measures. We describe now, in more detail, each of the pooling strategies with which we experiment in this paper.

Take@N (strategy T) creates, for each query, a global ranked list with the highest rank at which a retrieved document occurs in the R_p runs. The top N ranked documents for the query are selected into the pool. The $Take@N$ strategy is specified by the following definition for M:

$$M(d, R_p) = \max_{r \in R_p} \left(-\rho(d, r) \right) \tag{1}$$

This pooling strategy blindly takes into consideration the contribution of all pooled runs, whether they provide relevant documents or not. This behaviour is also the most fair among the pooling strategies, guaranteeing that every pooled run will have almost the same number of documents selected for assessment (the difference in the number of selected documents between runs is maximum 1).

DCGBased@N (strategy DCG) uses the discount function defined in the discounted cumulative gain to rank candidate documents to pool [9]. The discount is characterized by an inverse log_2 decay function and a gain value of 1. Formally documents for pooling are ranked in decreasing order by the values computed by M, where:

$$M(d, R_p) = \sum_{r \in R_p : d \in r} DCG(\rho(d, r)) = \sum_{r \in R_p : d \in r} \frac{1}{\log_2(\rho(d, r))} \tag{2}$$

RRFBased@N (strategy RRF) is rooted in the reciprocal rank (RR) evaluation measure, which is commonly used to assess system effectiveness in tasks such as known item search, question answering, or query auto completion [8]. A variant of RR, the reciprocal rank fusion (RRF), has been used in data fusion [7]. RRF makes use of an additional parameter, α, that controls the decay of the document contribution score as a function of rank. In this pooling strategy we employ the same idea, with $\alpha = 60$ as in Cormack et al. [7]; other values will be investigated in future work. Formally, candidate documents for the pool are ranked in decreasing order by the values computed with M where:

$$M(d, R_p) = \sum_{r \in R_p : d \in r} \text{RRF}(\rho(d, r)) = \sum_{r \in R_p : d \in r} \frac{1}{\rho(d, r) + \alpha} \tag{3}$$

PPBased@N (strategy PP, for *perfect precision*) is inspired by the family of measures that counts the number of relevant documents found at rank k divided by the number of documents up to rank k. Average Precision [3] and Sakai's Q-Measure [20] are examples of metrics belonging to this family. Since we model these measures as if all documents up to rank k were relevant, the rank score attributed to a document retrieved by runs in R_p is the number of runs that have retrieved that document:

$$M(d, R_p) = \sum_{r \in R_p : d \in r} \text{PP} = \sum_{r \in R_p : d \in r} 1 \tag{4}$$

This translates to a majority voting procedure to rank documents and select the top N.

RBPABased@N (strategy $RBPA$) computes pool document scores based on Rank Biased Precision (RBP) [17]. The RBP formula is characterized by a parameter p that models the user persistence, i.e. the likelihood that the user examines a document. The persistence parameter is effectively used to discount the contribution of a relevant document, similarly to other gain-discount based measures [5]. Pool candidate documents are ranked in decreasing order of the score computed by:

$$M(d, R_p) = \sum_{r \in R_p : d \in r} \text{RBPA}(\rho(d, r)) = \sum_{r \in R_p : d \in r} (1 - p) p^{\rho(d,r)-1} \tag{5}$$

In our experiments we use $p = 0.8$; this is akin to previous work that relied on RBP for evaluation [18,25] and for pooling [15,16]. The use of RBP as a document discount factor in weighting the contribution of documents to the pool creates a family of pooling strategies which, besides $RBPABased@N$, include $RBPBBased@N$ and $RBPCBased@N$ [16]. We next present the latter two.

RBPBBased@N (strategy $RBPB$) is an adaptive version of $RBPA$, which adds documents to the pool in an incremental way. By this strategy, for each run $r \in R_p$, we compute its residual $e(r)$, i.e. a value proportional to the number of not judged documents in the run. The residual is defined as:

$$e(r) = (1 - p) \sum_{d \in r : j(d)=?} p^{\rho(d,r)-1} \tag{6}$$

where $j(d)$ is 1 if the document d is judged relevant, 0 if judged as not relevant, and ? If the document is not judged.

With each new judgement the score $M(d, R_p)$ is recomputed as the runs' residuals have clearly changed (thus the adaptive nature of $RBPBBased@N$); this means recomputing the score:

$$M(d, R_p) = \sum_{r \in R_p : d \in r} RBPB(\rho(d, r)) = \sum_{r \in R_p : d \in r} (1 - p)p^{\rho(d,r)-1} \cdot e(r) \quad (7)$$

RBPCBased@N (strategy $RBPC$) is the second adaptive pooling strategy we present in this paper that uses both the RBP residuals, as $RBPBBased@N$, and the actual RBP score $b(r)$ of a run r, computed using a binary relevance:

$$b(r) = (1 - p) \sum_{d \in r : j(d)=1} p^{\rho(d,r)-1} \quad (8)$$

The candidate documents for pooling are decreasingly ranked by:

$$M(d, R_p) = \sum_{r \in R_p, d \in r} RBPC(\rho(d, r)) =$$

$$= \sum_{r \in R_p : d \in r} (1 - p)p^{\rho(d,r)-1} \cdot e(r) \cdot \left(b(r) + \frac{e(r)}{2} \right)^3 \quad (9)$$

Figure 1 shows the gain function variation with rank for the different pooling strategies, for one run r. The $RBPBBased@N$ and $RBPCBased@N$ strategies are not shown on this plot since, due to their adaptive nature, their shape changes with each judged document.

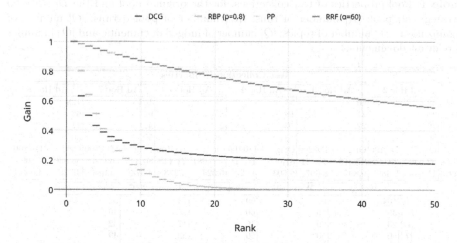

Fig. 1. Different pooling strategies score document ranks differently: This figure summarizes the gain functions in $DCGBased@N$, $RRFBased@N$, $PPBased@N$, and $RBPABased@N$ as functions of the rank position, for a run r.

3 Experiments and Results

The first part of this section describes the experimental set-up we have used. We list the test collections we made use of, the measures to assess the pool bias and the experimental methodology – similar to the one presented in previous studies [3, 12–15]. The second part presents the results of the experiments with a series of plots.

3.1 Experimental Setup

To test the pooling strategies we used a set of 11 test collections selected from different editions of the TREC evaluation campaigns. We used Ad Hoc 2–5, Ad Hoc 7–8, Web 9, Web 10, Web 11, Genomics 14 and Robust 14. These test collections have been built using a *Depth@k* strategy, but additional documents have later been judged when additional resources were available. Therefore to remove the influence of these spurious assessments we preprocessed the test collections to use a pure *Depth@k* pool. The pool details for each resulting test collection are shown in Table 1.

To evaluate[1] the selected pooling strategies we ran experiments that simulated the absence of a run from the pool. We did this for every run, in a *leave-one run-out* fashion, then we summarized the bias with the following pool bias measures: Mean Absolute Error (MAE), System Rank Error (SRE), and System Rank Error with Statistical Significance (SRE*). MAE measures the mean of the error between the run score when the run contributed to the pool and its

Table 1. Pool properties of test collections, for the original pool and the *Depth@100* (strategy D) pool; $|R|$ number of runs; $|R_p|$ number of pooled runs; $|O|$ number of organizations; $|T|$ number of topics; $|Q|$ number of judged documents; and $|Q_+|$ number of relevant documents.

Test Collection Properties										
	Ad Hoc 2		Ad Hoc 3		Ad Hoc 4		Ad Hoc 5		Ad Hoc 7	Ad Hoc 8
$|R|$	38		40		33		61		103	129
$|R_p|$	30		21		19		53		64	66
$|O|$	22		22		19		21		42	41
$|T|$	50		50		50		50		50	50
	Orig. → D@100		Orig. → D@200		Orig. → D@100		Orig. → D@100		Orig. → D@100	Orig. → D@100
$|Q|$	62,620	39,692	97,319	68,121	87,069	46,721	133,681	71,448	80,345 69,662	86,830 79,090
$|Q_+|$	11,645	9,489	9,805	8,607	6,503	4,622	5,524	4,333	4,674 3,986	4,728 4,090

	Web 9		Web 2001		Web 2002		Robust 2005		Genomics 2005	
$|R|$	104		97		69		74		62	
$|R_p|$	39		35		60		18		46	
$|O|$	23		29		16		17		32	
$|T|$	50		50		50		50		49	
	Orig. → D@100		Orig. → D@100		Orig. → D@50		Orig. → D@55		Orig. → D@60	
$|Q|$	70,070	49,161	70,400	46,135	56,650	53,318	37,798	22,173	39,958	32,013
$|Q_+|$	2,617	2,225	3,363	2,833	1,574	1,487	6,561	4,563	4,584	3,937

---

[1] The software used in this paper is available on the website of the first author.

score when left out. SRE is the sum of the rank error measured for each run, that is the difference in system ranking when the run contributed to the pool and when left out. SRE* is similar to SRE but counts the ranking difference only when statistical significance occurs (paired t-test $p < 0.05$).

To remove the influence that other contributing runs from the same organization may provide to the excluded run, we do instead a *leave-one organization-out*. We also remove the 25% of poorly performing runs, as done in previous studies [3,15]. To avoid also the discovery for each strategy of documents for which we do not know their relevance, that is they have not been judged in the original pool, we allow the selection of the documents to be pooled only from the top of the runs; we cut the runs at the depth k of the original *Depth@k* used to build the original pool.

To analyse the performance of each strategy at different fixed-cost budgets we test each strategy, varing the number of documents required to be judged from 5,000 to the size of the original pool in steps of 5,000. We selected three IR evaluation measures because: (1) they are common evaluation measures used in IR and (2) they present properties that are common across the majority of IR evaluation measures: top-heaviness (relevant documents at the top of the list are given more weight), utility based, and strongly correlated to the number of relevant documents retrieved. The IR measures are: MAP, NDCG, and P@100.

3.2   Results

Figure 2 shows the results we obtained for the TREC-8 Ad Hoc test collection, where we observe how the different pooling strategies behave for various numbers of total documents judged. Figure 3 shows the same data as Fig. 2 from a different view: it shows the performance of the different pooling strategies compared to the *Take@N* strategy. In Fig. 4 we show the performance for the NDCG measure on the other 10 test collections.

4   Discussion and Conclusion

In the paper we can distinguish two categories of strategies: (1) the non adaptive ones formed by *RRF*, *PP*, *RBPA*, and *DCG*, and (2) the adaptive ones formed by *RBPB* and *RBPC*. Note that *RBPC* not only uses information on whether a document is judged or not, but also concerning its relevance.

RBPC is the best performing strategy in all the test collections over MAP and NDCG as evaluation measures, and across all pool bias measures. Nevertheless it is the most difficult to operationalise as the pool needs to be built on the fly, a concern expressed before in the work of Lipani et al. [15].

Based on Fig. 3, we can clearly identify two different types of behaviour depending on the IR evaluation measure used. On one hand, both MAP and NDCG have similar behaviour. For these evaluation measures, *RBPB* and *RBPC* are the best strategies, followed by *RBPA*, *DCG*, *RRF* and *PP*. These last four pooling strategies have a similar behaviour characterized by a twist

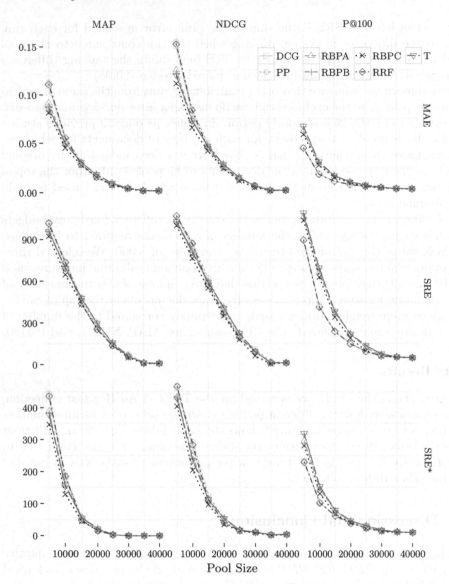

Fig. 2. Pool bias measured in terms of MAE, SRE, and SRE* for the pooling strategies on the Ad Hoc 8 test collection, for different pool sizes (i.e. number of documents that require relevance judgement).

between 10,000 and 15,000 judged documents in the case of the TREC-8 Ad Hoc collection. We also observe that a similar shape happens for the rest of the test collections, in Fig. 4 for NDCG.

The rank of the non-adaptive strategies is perfectly correlated with their speed of discount (change in reward for popularity) as observed in Fig. 1.

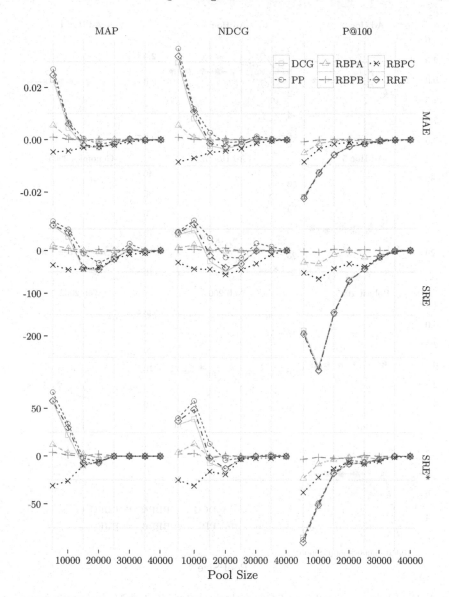

Fig. 3. Pool bias measured with respect to *Take@N* strategy in terms of MAE, SRE, and SRE* for the pooling strategies on the Ad Hoc 8 test collection, for different pool sizes (i.e. number of documents that require relevance judgement).

The linear and logarithmic discounts remove the rank information from the documents rewarding more popularity of a document among the runs. The relationship between the discount and the top-heaviness of the evaluation measures MAP and NDCG also explains the twist in preference, where *Take@N* is preferred for low N, then for higher N almost all non-adaptive methods outperform

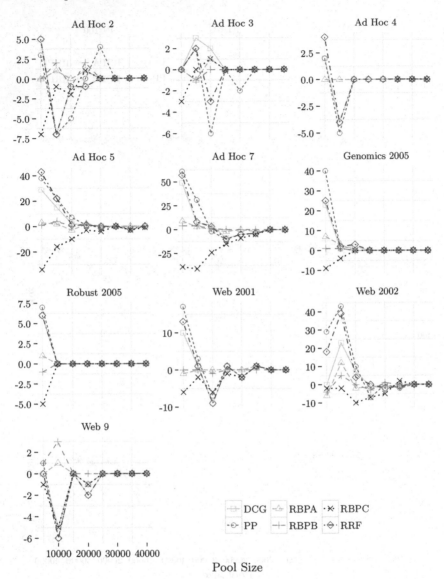

Fig. 4. Pool bias measured on NDCG with respect to *Take@N* strategy in terms of SRE* for the pooling strategies on the rest of the test collections, for different pool sizes (i.e. number of documents that require relevance judgement).

it, before they all converge to the same value. On the other hand, for P@100 we observe that *DCG*, *RRF*, and *PP* are the best, followed by *RBPC*, *RBPA*, and *RBPB*. The latter behaves very similarly to the baseline *Take@N*. P@100 correlates with the number of relevant documents discovered in this specific case because the size of the submitted runs equals the original depth of the pool, and we justify its different behaviour due to the absence of discount.

It is strange here that *RBPA* outperforms *RBPB*: a non-adaptive strategy outperforms an adaptive one. At this point we can only hypothesise that the exponential decay of *RBPA* fights popularity rewarding more the rank.

In the end, the conclusions we can draw at this point are as follows:

- for top-heavy metrics:
  - given a large budget, the *Take@N* strategy is guaranteed to be the least biased and therefore should be used;
  - given a small budget with which only very few documents can be assessed, then we should operationalise *RBPC*. It might take longer to create the assessments and it can only be done by one assessor per topic, but this would be likely in line with the budget constraints;
  - for a moderate budget and if we cannot operationalise *RBPC*, the non-adaptive strategies do not underperform *Take@N*, but neither do they consistently improve upon it.
- for *P@100* it appears that the non-adaptive methods always outperform the baseline *Take@N* and are therefore to be used. This is not only based on Ad Hoc 8 (Fig. 3), but is clearly visible for all test collections.

References

1. Aslam, J.A., Pavlu, V., Yilmaz, E.: A statistical method for system evaluation using incomplete judgments. In: Proceedings of SIGIR (2006)
2. Buckley, C., Dimmick, D., Soboroff, I., Voorhees, E.: Bias and the limits of pooling for large collections. Inf. Retr. **10**(6), 491–508 (2007)
3. Buckley, C., Voorhees, E.M.: Evaluating evaluation measure stability. In: Proceedings of SIGIR (2000)
4. Büttcher, S., Clarke, C.L.A., Yeung, P.C.K., Soboroff, I.: Reliable information retrieval evaluation with incomplete and biased judgments. In: Proceedings of SIGIR (2007)
5. Carterette, B.: System effectiveness, user models, and user utility: a conceptual framework for investigation. In: Proceedings of SIGIR (2011)
6. Cleverdon, C., Mills, J.: Factors determining the performance of indexing systems. In: Volume I - Design, Volume II - Test Results, ASLIB Cranfield Project (1966). (Reprinted in Sparck Jones, K., Willett, P. (eds.) Readings in Information Retrieval)
7. Cormack, G.V., Clarke, C.L.A., Buettcher, S.: Reciprocal rank fusion outperforms condorcet and individual rank learning methods. In: Proceedings of SIGIR (2009)
8. Dumais, S., Banko, M., Brill, E., Lin, J., Ng, A.: Web question answering: is more always better? In: Proceedings of SIGIR (2002)
9. Järvelin, K., Kekäläinen, J.: Cumulated gain-based evaluation of IR techniques. ACM Trans. Inf. Syst. **20**(4), 422–446 (2002)
10. Jones, K.S., van Rijsbergen, C.J.: Report on the need for and provision of an "ideal" information retrieval test collection. British Library Research and Development Report 5266, University of Cambridge (1975)
11. Lipani, A.: Fairness in information retrieval. In: Proceedings of SIGIR (2016)
12. Lipani, A., Lupu, M., Hanbury, A.: Splitting water: precision and anti-precision to reduce pool bias. In: Proceedings of SIGIR (2015)

13. Lipani, A., Lupu, M., Hanbury, A.: The curious incidence of bias corrections in the pool. In: Ferro, N., Crestani, F., Moens, M.-F., Mothe, J., Silvestri, F., Nunzio, G.M., Hauff, C., Silvello, G. (eds.) ECIR 2016. LNCS, vol. 9626, pp. 267–279. Springer, Cham (2016). doi:10.1007/978-3-319-30671-1_20

14. Lipani, A., Lupu, M., Palotti, J., Zuccon, G., Hanbury, A.: Fixed budget pooling strategies based on fusion methods. In: Proceedings of SAC (2017)

15. Lipani, A., Zuccon, G., Lupu, M., Koopman, B., Hanbury, A.: The impact of fixed-cost pooling strategies on test collection bias. In: Proceedings of ICTIR (2016)

16. Moffat, A., Webber, W., Zobel, J.: Strategic system comparisons via targeted relevance judgments. In: Proceedings of SIGIR (2007)

17. Moffat, A., Zobel, J.: Rank-biased precision for measurement of retrieval effectiveness. TOIS 27(1), 2 (2008)

18. Park, L.A., Zhang, Y.: On the distribution of user persistence for rank-biased precision. In: Proceedings of ADCS (2007)

19. Robertson, S.: On the history of evaluation in IR. J. Inf. Sci. 34(4), 439–456 (2008)

20. Sakai, T.: New performance metrics based on multigrade relevance: their application to question answering. In Proceedings of NTCIR (2004)

21. Soboroff, I.: A comparison of pooled and sampled relevance judgments. In: Proceedings of SIGIR (2007)

22. Voorhees, E.M., Harman, D.K.: TREC: Experiment and Evaluation in Information Retrieval. The MIT Press, Cambridge (2005)

23. Webber, W., Park, L.A.: Score adjustment for correction of pooling bias. In: Proceedings of SIGIR (2009)

24. Yilmaz, E., Aslam, J.A.: Estimating average precision with incomplete and imperfect judgments. In: Proceedings of CIKM (2006)

25. Zhang, Y., Park, L.A., Moffat, A.: Click-based evidence for decaying weight distributions in search effectiveness metrics. Inf. Retr. 13(1), 46–69 (2010)

26. Zobel, J.: How reliable are the results of large-scale information retrieval experiments? In: Proceedings of SIGIR (1998)

A Formal and Empirical Study of Unsupervised Signal Combination for Textual Similarity Tasks

Enrique Amigó[(✉)], Fernando Giner, Julio Gonzalo, and Felisa Verdejo

NLP & IR Group at UNED, Madrid, Spain
{enrique,julio,felisa}@lsi.uned.es
http://nlp.uned.es

Abstract. We present an in-depth formal and empirical comparison of unsupervised signal combination approaches in the context of tasks based on textual similarity. Our formal study introduces the concept of *Similarity Information Quantity*, and proves that the most salient combination methods are all estimations of *Similarity Information Quantity* under different statistical assumptions that simplify the computation. We also prove a *Minimal Voting Performance* theorem stating that, under certain plausible conditions, estimations of *Information Quantity* should at least match the performance of the best measure in the set. This explains, at least partially, why unsupervised combination methods perform robustly. Our empirical analysis compares a wide range of unsupervised combination methods in six different Information Access tasks based on textual similarity: Document Retrieval and Clustering, Textual Entailment, Semantic Textual Similarity, and the automatic evaluation of Machine Translation and Summarization systems. Empirical results on all datasets corroborate the result of the formal analysis and help establishing recommendations on which combining method to use depending on nature of the set of measures to be combined.

1 Introduction

Computing textual similarity is a core problem for many Information Access tasks: Document Retrieval compares queries and documents, Document Clustering groups together similar texts, Textual Entailment investigates whether the meaning of one text is implied in the meaning of another, etc. It is also crucial for the evaluation of systems that generate text, such as Machine Translation and Automatic Summarization systems, where similarity between system outputs and human references is measured.

It has often been reported that the combination of similarity signals (which can be full systems or similarity measures based on particular features of the textual content, such as n-grams, concepts, syntactic structures, text metadata, etc.) may provide better results than individual signals (or systems) in isolation. For instance, some results in the Information Retrieval field [23] report a substantial improvement of unsupervised system combinations with respect to the best single system in multiple TREC test beds. Some authors have also observed

© Springer International Publishing AG 2017
J.M. Jose et al. (Eds.): ECIR 2017, LNCS 10193, pp. 369–382, 2017.
DOI: 10.1007/978-3-319-56608-5_29

that human assessments can be replaced successfully by a voting combination of system outputs in an Information Retrieval evaluation campaign [7, 26]. In the context of Machine Translation evaluation, different authors have observed that mixing text similarity measures based on diverse linguistic levels (lexical, syntactic) may improve the correlation with human judgements [2, 3, 11, 22]. At a broader level, the benefits of unsupervised combination have been studied in the context of meta-classifiers [18, 23, 25].

Our primary goal is to establish the theoretical foundations for this effect; and our primary result is a *Minimal Voting Performance* theorem that states that, under certain (plausible) conditions, voting must give at least the same performance as the best individual system in the combination. Note that there are earlier formal efforts to explain why voting methods work, but they do not apply to our problem: (i) Condorcet's theorem (1785) assumes that measures are independent, which is often not the case in textual similarity problems (for instance, most systems agree on high-similarity items); and (ii) Kaniovski's theoretical explanation [17] assumes equal performance and homogenous correlation across signals, which, again, is unrealistic in text similarity problems.

The minimal voting performance theorem is complemented with an empirical validation in a wide range of textual similarity problems: Document Retrieval and Clustering, Text Entailment, Semantic Textual Similarity, and evaluation of Machine Translation and Automatic Summarization systems. Finally, we review the most salient unsupervised combination methods in the state of the art, both in terms of their formal properties and in terms of their comparative empirical performance.

2   Constraints for Text Similarity Combination Methods

As a first step towards formalizing the properties of an unsupervised combination of textual similarity signals, here we want to define a set of formal constraints to be satisfied by any similarity combining methods supported by observed phenomena in the state of the art.

For notation purposes, let Ω be the set of texts in a large collection, and $x \in \Omega^2$ a *similarity instance*, i.e., an ordered pair of texts, which it is the subject of study. A similarity measure, f, is a function which associates a similarity instance, $x \in \Omega^2$, with a real value, $f(x) \in \mathbb{R}$. We will denote the set of available measures to be combined as $\mathcal{F}$. And let us use $C_{\mathcal{F}}(x)$ to denote the similarity value assigned to instance x by the combination function C on the set of individual measures $\mathcal{F}$. The real similarity of an instance, $sim(x)$, is a random object which is conditioned by the values of the measures, $f \in \mathcal{F}$.

There is no clear consensus about the nature of similarity. There are, however, a number of common intuitions. One of them is that an increase in a certain similarity aspect is always a non-negative (≥ 0) signal of overall similarity. For instance, having a similar eye color, a similar skin type or a similar height are all evidence that make two people more similar to each other. This also applies to textual similarity: the fact that two texts share some feature (i.e. a word in common) is always a positive signal of similarity. According to this intuition we can postulate a *monotonicity axiom*:

Constraint 1 Monotonocity: *Given a finite set of similarity measures,* $\mathcal{F} = \{f^1, \ldots, f^m\}$, *an increase in one of them should produce an increase in the combination function:*

$$f^i(x) > f^i(y) \wedge \forall j \neq i(f^j(x) = f^j(y)) \Longrightarrow C_{\mathcal{F}}(x) > C_{\mathcal{F}}(y)$$

A second intuition is that the more a shared feature is unlikely, the higher its effect on similarity. Following the above example, sharing green eyes contributes to make two people similar more than sharing brown eyes, which are more common. This has been confirmed in psychological studies [27] and is the main reason to use tf.idf weighting schemes (uncommon words are more discriminative) and other similar strategies. Hence the following axiom:

Constraint 2 Unlikeliness: *The combination of a single measure should be inversely correlated with the likeliness of its values. Given a pair of measures,* $f, f' \in \mathcal{F}$, *such that,* $f(x) = v$ *and* $f'(x) = v'$:

$$\text{If } P_{y \in \Omega^2}\big(f(y) > v\big) < P_{y \in \Omega^2}\big(f'(y) > v'\big)$$
$$\text{then } C_{\{f\}}(x) > C_{\{f'\}}(x)$$

Another intuition is that an heterogenous (diverse) set of similarity aspects is a stronger evidence of overall similarity than an homogeneous (correlated) set. For instance, a coincidence of blue eyes and dark skin is stronger than a coincidence of blue eyes and pale skin, because the latter are more likely to occur together. There are examples for this in the context of meta-classifiers [19, 24] and in the context of text evaluation metrics [2, 3, 11, 22]. A boundary condition of this intuition is that evidence from redundant similarity measures do not increase the overall similarity: for instance, if two people share the skin color, then sharing also the skin color of the right hand does not add any similarity evidence. This is the basis for the following constraint:

Constraint 3 Redundancy: *Redundant measures should not add up when combined:*

$$\text{If } \forall y, z \in \Omega^2, f(y) > f(z) \leftrightarrow f'(y) > f'(z) \Longrightarrow$$
$$\text{then } C_{\{f^1..f^m, f\}}(x) = C_{\{f^1..f^m, f, f'\}}(x)$$

Any reliable procedure to combine textual similarity methods should be compliant with these three constraints, but, remarkably, in general they are not satisfied by combining functions in the state of the art. We will use them to analyze the formal properties of existing combining methods, and also to establish our central theorem in the following sections.

3   Minimal Voting Performance Theorem

We start by formulating a *Similarity Informativeness Hypothesis* that will be the starting point for our *Minimal Voting Performance* theorem, and is grounded on the concept of *joining area*:

Definition 1. *Let us consider a set of measures, $\mathcal{F} = \{f^1, \ldots, f^m\}$, and a set of real values, $\mathcal{U}$, which contains a value (a threshold), $u_i \in \mathcal{U}$, for each similarity measure, f^i, in the set. The* **joining area**, $\mathcal{A}_{\mathcal{U}}^{\mathcal{F}}$, *is the set of instances such that every measure, f^i, improves its threshold u_i:*

$$\mathcal{A}_{\mathcal{U}}^{\mathcal{F}} \equiv \{x \in \Omega^2 \mid f^i(x) \geq u_i, \ \forall i \in \{1, \ldots, m\}\}$$

For instance, if we set a high threshold for every measure, the joining area is the set of similarity instances which are very close according to all measures in the set. General speaking, as we make the thresholds higher, we make the joining area smaller, because less similarity instances are above all thresholds.

Then, a set of thresholds defines a joining area. Now, let us consider a similarity instance, x, and its similarity values, s_i, according to each measure, f^i. We can use the similarity values, s_i, as a set of thresholds, u_i. Then we obtain a joining area, which we will refer to as the joining area associated to x, $\mathcal{A}_{\{f^1(x), \ldots, f^m(x)\}}^{\mathcal{F}}$. Using this area, we now define the *Information Quantity* of a similarity instance:

Definition 2. *Given a set of similarity measures, $\mathcal{F} = \{f^1, \ldots, f^m\}$, the* **Information Quantity** *of a similarity instance x is the information quantity of its associated joining area:*

$$\mathcal{I}_{\mathcal{F}}(x) \equiv \mathcal{I}\left(\mathcal{A}_{\{f^1(x), \ldots, f^m(x)\}}^{\mathcal{F}}\right) = -log\left(P_{y \in \Omega^2}\left(f^j(y) \geq f^j(x). \forall f^j \in \mathcal{F}\right)\right)$$

Information quantity can be seen as a combining function for the set of similarity measures $\mathcal{F}$: it assigns a combined similarity score, $\mathcal{I}_{\mathcal{F}}(x)$, to every similarity instance, x, which depends on how unlikely is to find other instances, y, with higher similarity scores for all individual measures in $\mathcal{F}$. It can be proved that *information quantity* satisfies all three previous axioms: Increasing one f^i value necessarily decreases the joining area (axiom 1), joining areas in single measures corresponds with their unlikeliness (axiom 2) and redundant measures do not affect the joining area (axiom 3)[1]. We can now define the following hypothesis:

Similarity Informativeness Hypothesis: *Given a finite set of measures, $\mathcal{F} = \{f^1, \ldots, f^m\}$, and a threshold th of real similarity, the probability, for a given instance, x, of having a real similarity, $sim(x)$, larger than th is directly correlated with its information quantity:*

$$P\left(sim(x) > th \mid f^1(x), \ldots, f^m(x)\right) \simeq \mathcal{I}_{\mathcal{F}}(x)$$

In other words, we hypothesize that the probability of having a higher real similarity for a sample, x, is inversely correlated with the probability of finding other instances, y, with higher similarity values for all measures in $\mathcal{F}$. Under such

---

[1] Explicit proofs are avoided due to lack of space.

condition, it can be proved that the combination, $C_{\mathcal{F}}(x)$, is an estimation of the real similarity, $sim(x)$, at least as good as the best individual measure in $\mathcal{F}$:

Theorem 1. Minimal Voting Performance: *Assuming the Similarity Informativeness Hypothesis holds, the information quantity of similarity instances is a better predictor of real similarity than every single measure in the combination:*

$$Eff(\mathcal{I}_{\mathcal{F}}(x)) \geq Eff(f), \quad \forall f \in \mathcal{F}$$

where the effectiveness of a single signal, f, or a set of combined signals is defined as

$$Eff(f) \equiv P\left(f(x) > f(y) | sim(x) > sim(y)\right)$$

The proof is shown in Appendix A. What this theorem proves is that there is at least one way (using information quantity to combine individual measures) to ensure that a combination function performs, at least, as well as the best individual measure. But, Information Quantity requires many similarity samples to be estimated, given that the probability of unanimous measure improvements decrease exponentially with the number of measures. In the next section, we will see that many salient combination methods are, under certain statistical assumptions that simplify computation, equivalent to Information Quantity, and therefore the Minimal Voting Performance partially applies to them as well.

4 Formal Comparison of Unsupervised Combining Functions

We now examine the most salient combination methods in the literature, considering whether they satisfy our three axioms and how they relate to Information Quantity. The simplest way of unsupervised measure combining consists of applying traditional average schemes, such as the arithmetic, geometric, and harmonic means, and also the maximum and the minimum. Only the arithmetic mean is strictly monotonic[2] (first axiom). They can satisfy unlikeliness only if measures follow the same distribution of values. Finally, unlike the rest of averaging schemes, maximum and minimum trivially satisfy redundancy.

There is a correspondence between the average and the estimation of Similarity Information Quantity. We need to assume that the measures value distribution corresponds with its single information quantity $\left(f^{i}(x) \propto -log\left(P_{y}\right.\right.$ $\left.\left.(f^{i}(y) \geq f^{i}(x))\right)\right)$, and measures and instances are independent.

$$Avg(f^{i}(x)) \propto \sum_{i} f^{i}(x) \propto \sum -log\left(P_{y}(f^{i}(y) \geq f^{i}(x))\right) \simeq$$

$$-log\left(P_{y}\left(\forall i \left(f^{i}(y) \geq f^{i}(x)\right)\right)\right) = \mathcal{I}\left(\mathcal{A}^{\mathcal{F}}_{f^{1}(x),...f^{m}(x)}\right) = \mathcal{I}_{\mathcal{F}}(x)$$

---

[2] Note that a zero value avoids the effect of the rest of measures in geometric and harmonic means, and maximum and minimum only consider at the end one of the combined measures.

Voting and ranking fusion algorithms are also unsupervised combining methods. They include the Borda Count algorithm [14], the Reciprocal Rank Fusion [10], and the family of voting methods called *Condorcet* [15,23]. In general, voting algorithms satisfy unlikeliness, given that they consider the ranking position of measure values (voting preferences). The pure Condorcet voting method avoids the *Condorcet paradox* at the cost of sacrificing monotonicity, by ranking together instances belonging to the same transitivity cycle of majority voting (peer to peer). However, the Coppeland Condorcet variant satisfies monotonicity given that, instead of avoiding differences in cycles, it considers how many instances are improved by a majority of measures. Mean Reciprocal Rank does not comply with monotonicity, given that it only considers the maximum ordinal value across measures.

With respect to its correspondance with Similarity Information Quantity, Coppeland's method is equivalent when considering the amount of improved measures when estimating the probability of unanimous improvements in the case of contradiction between measures:

$$Copp_{\mathcal{F}}(x) = P_y(|\{f \in \mathcal{F}.f(x) \geq f(y)\}| \geq |\{f \in \mathcal{F}.f(x) \leq f(y)\}|)$$

Assuming that $P_y\left(\forall f \in \mathcal{F}\left(f(x) \geq f(y)\right)\right) \simeq \frac{|\{f \in \mathcal{F}.f(x) \geq f(y)\}|}{|\mathcal{F}|}$ then:

$$Copp_{\mathcal{F}}(x) \simeq P_y\left(\forall f \in \mathcal{F}\left(f(x) \geq f(y)\right)\right) \propto -log\left(P_y\left(\forall i\left(f^i(y) \geq f^i(x)\right)\right)\right) = \mathcal{I}_{\mathcal{F}}(x)$$

On the other hand, the most popular voting method, Borda count, can be expressed as the probability of outperforming other similarity instances according to single measures:

$$\text{Borda}^{\mathcal{F}}(x) = P_{\substack{y \in \Omega \\ f \in \mathcal{F}}}(f(x) \geq f(y)) \simeq \frac{1}{|\mathcal{F}|}\sum_{f \in \mathcal{F}}P_{y \in \Omega}(f(x) \geq f(y))$$

Borda count is therefore equivalent to Information Quantity if we assume independence across measures (symbol $\simeq$) and we consider the ranking position in a logarithm scale.

$$\text{Borda}^{\mathcal{F}}_{log}(x) = \frac{1}{|\mathcal{F}|}\sum_{f \in \mathcal{F}} -log(P_y(f(x) \geq f(y))) \propto \sum_{f \in \mathcal{F}}log(1 - P_y(f(x) \leq f(y))) =$$

$$-log\left(\prod_{f \in \mathcal{F}}P_{y \in \Omega}(f(x) \leq f(y))\right) \simeq -log(P(\forall f \in \mathcal{F}f(x) \leq f(y))) = \mathcal{I}_{\mathcal{F}}(x)$$

Given that Borda assumes independence, it is able to satisfy monotonicity and unlikeliness, but not redundancy.

The Unanimous Improvement Ratio (UIR) [4] is an unsupervised combining method for evaluation metrics such as precision and recall. It was presented as a complement to the F-measure (harmonic mean). UIR is based on how many instances are improved for all measures simultaneously.

$$\text{UIR}_{\mathcal{F}}(x) = P_y\left(\forall f\left(f(x) \geq f(y)\right)\right) - P_y\left(\forall f\left(f(y) \geq f(x)\right)\right).$$

Table 1. Formal comparison of unsupervised combining functions

Method	Maps into information quantity when	Satisfied axioms
UIR	Probability of unanimous improvement is 0.5 in the case of contradiction across measures	all axioms under enough available samples
Coppeland	Probability of unanimous improvement is proportional to the amount of improved measures in the case of contradiction across measures	all except redundancy
$Borda_{log}$	Individual measures are independent	all except redundancy
Average	Individual measures are independent and $\left(f^i(x) \propto -log\left(P_y(f^i(y) \geq f^i(x))\right)\right)$	all except redundancy
MRR, Harm., Geo., Max., Min, Pure Condorcet		all except strict monotonicity

UIR can be interpreted as an estimation of the joining area (Information Quantity) in such a way that the cases in which there is an increase of measures in both directions, the probability of unanimous increase is set to $\frac{1}{2}$:

$$\text{UIR}_{\mathcal{F}}(x) = P(\forall f \in \mathcal{F}\,(f(x) \geq f(y))) - P(\forall f \in \mathcal{F}\,(f(y) \geq f(y))) = P(\forall f\,(f(x) \geq f(y))) -$$
$$(1 - (P(\forall f\,(f(x) \leq f(y))) + P(\exists f, f'\,(f(x) < f(y) \wedge f'(x) > f'(y))))) =$$
$$= 2 \cdot P(\forall f \in \mathcal{F}\,(f(x) \geq f(y))) + P(\exists f, f' \in \mathcal{F}\,(f(x) < f(y) \wedge f'(x) > f'(y))) - 1 \propto$$
$$P(\forall f \in \mathcal{F}\,(f(x) \geq f(y))) + 0.5 \cdot P(\exists f, f' \in \mathcal{F}\,(f(x) < f(y) \wedge f'(x) > f'(y)))$$

In summary: the average, the Coppeland method, Borda and UIR can all be interpreted as estimations of Similarity Information Quantity under different statistical assumptions. In the light of the Minimal Voting Performance theorem, this is a possible explanation of why voting methods give robust results. Table 1 summarizes the formal analysis of measures in terms of their correspondence with Information Quantity and their compliance with the axioms.

5 Empirical Comparison of Unsupervised Combining Functions

We experiment with test collections corresponding to six tasks that involve computing text similarity:

Document Clustering (CL): We have used the WePS-1 data set [6], which contains around six thousands of manually grouped web pages. Here we have considered a set of 167 similarity measures, introduced in [5], that employ a wide range of features (from n-grams to different classes of named entities) and provide state-of-the-art results.

Semantic Textual Similarity (STS): We have employed the dataset provided by the pilot experience in SEMEVAL 2012 [1], which includes 3050 similarity instances distributed in four sets, and 88 runs (similarity measures). The similarity of pairs of sentences was rated on a 0–5 scale (low to high similarity) by human judges using Amazon Mechanical Turk.

Textual Entailment (TE): We have used the training set provided as part of the RTE-2 evaluation campaign [8], which consists of 800 text-hypothesis pairs. We have developed 102 similarity measures for this scenario (all of them are based on [21]). They all measure word overlap over different text components: levels in the parse tree, PoS tags, lemmas and relations. In order to preserve the formal properties of similarity measures, when sentences do not include a text component (e.g. a certain PoS tag), similarity is set to 0.5.

Document Retrieval (IR): We have used queries 701 to 750 in the GOV-2 collection used in the TREC 2004 Terabyte Track. Document/query similarity measures consist of 60 retrieval systems developed by the participants in the track. We consider the top 100 documents from a search engine output.

Machine Translation Evaluation (MT): We use data sets from the Arabic-to-English (AE) and Chinese-to-English (CE) NIST MT Evaluation campaigns in 2004 and 2005[3]. We take the sum of adequacy and fluency, both in a 1–5 scale, as a global manual assessment of quality [9]. These data sets include around 8000 similarity instances between MT outptus and human references. As similarity measures, we have used 64 automatic evaluation measures provided by the ASIYA Toolkit [16][4]. This set includes measures operating at different linguistic levels (lexical, syntactic, and semantic) and includes all popular measures (BLEU, NIST, GTM, METEOR, ROUGE, etc.)

Automatic Summarisation Evaluation (AS): We have used the DUC 2005/2006 test collections [5] [12,13]. At DUC, summaries were evaluated according to several criteria; here, we will focus on responsiveness judgements, for which the quality score was an integer between 1 and 5. We have employed standard variants of ROUGE [20] as similarity measures.

Using these datasets, we test the (comparative) ability of combining functions to predict the true similarity of instances. For this, we have conducted three experiments. In all of them, we consider pairs (x, y) of similarity instances in the data set with a difference in similarity according to humans $(sim(x) > sim(y))$. We test the ability of each method to combine measures and predict the closest similarity instance. The *effectiveness* $\text{Eff}(C_\mathcal{F})$ is computed as $P_{x,y}(C_\mathcal{F}(x) > C_\mathcal{F}(y)|sim(x) > sim(y))$. When the evaluated method returns the same value for both instances, we estimate effectiveness as 0.5. We have normalized single measures between 0 and 1 for averaging schemes.

---

[3] http://www.nist.gov/speech/tests/mt.
[4] http://www.lsi.upc.edu/~nlp/Asiya.
[5] http://duc.nist.gov/.

In the first experiment, we combine all available measures. We select randomly, for each data set, 10.000 pairs of similarity instances x and y from the same topic. In this experiment, we also compare with the best and the worst measure from the whole set. In this experiment we consider only two decimals values given that effectiveness is computed over 10.000 samples only.

In the second experiment we test the performance of combining functions over two measures only, in order to see how combining methods perform with a small number of individual measures. For each experimental case, we select randomly a measure pair. Then, for each pair of measures, we evaluate combining methods over every pair x, y of similarity instances in the data set with a difference in similarity according to humans (between 39.600 samples in the case of entailment and 101.251 instance pairs in the case of information retrieval) We compare the results of the combined measure with the best and the worst measure in the pair, according to their individual effectiveness over the rest of the data set. Note that when combining 2 measures, UIR and Coppeland are equivalent (two measures is majority and contradictory values for both measures is estimated as 0.5), and therefore their results are equivalent.

Finally, in the third experiment we replicate 5 times the less predictive measure, in order to test the ability of combining functions to accommodate redundant measures without introducing bias.

Table 2 shows the results. A salient aspect is that in all experiments, combining methods without correspondence with information quantity (MRR, Max, Min., Harm. and Geo.) achieve lower results than the rest of methods. This is a strong suggestion that the Minimal Voting Performance theorem can explain, at least partially, the behavior of unsupervised combining functions. In addition, in every dataset, there is at least one method that is able to achieve similar results than the best individual measure in the combination, in agreement with the Minimal Voting Performance theorem.

When combining all measures, UIR is less reliable than other methods, although it satisfies the three axioms. The reason is that the need for sample instances for its computation grows exponentially with the amount of measures when computing the unanimous improvements. If only two measures are combined, UIR performance improves drastically (achieving the best result in two datasets). And when redundant measures are added to the set, UIR is the best combination method for all datasets (because it is not affected by redundancy). Coppeland is equivalent to UIR for two measures, but when adding redundancy its effectiveness decreases.

In the absence of redundant measures, Coppeland's method is the best performer across all datasets, which indicates that its statistical assumptions correspond with actual data. On the other hand, the independence assumptions of Borda produce lower results in two data sets for both experiments. Borda with logarithmic transformation achieves slightly lower results than the original Borda, except in the case of redundant measures, where the logarithmic version is more robust (with a small but consistent difference across data sets).

Table 2. Empirical comparison of unsupervised combining functions

Combining all measures						
	Summarization	Retrieval	Entailment	Machine translation	Semantic Sim.	Clustering
Best measure	0.73	0.69	0.62	0.69	0.74	0.81
Worst measure	0.63	0.50	0.46	0.54	0.52	0.50
UIR	0,70	0,66	0,60	0,65	0,67	0,79
Coppeland	**0,72**	**0,67**	**0,67**	**0,69**	**0,75**	**0,83**
Borda	**0,72**	**0,67**	0,66	0,68	**0,75**	**0,83**
Borda$_{log}$	0,71	**0,67**	0,65	0,68	**0,75**	**0,83**
MRR	0,67	0,61	0,56	0,64	0,68	0,79
Avg	0,68	0,66	0,68	0,68	0,71	0,83
Geo	0,50	0,54	0,51	0,52	0,50	0,50
Harm	0,68	0,66	0,63	0,62	0,63	0,80
Combining two random measures						
Best measure	0.706	0.619	0.539	0.654	0.727	0.607
Alternative measure	0.679	0.559	0.505	0.613	0.671	0.529
UIR	**0.703**	**0.613**	**0.539**	**0.654**	**0.723**	**0.617**
Coppeland	**0.703**	**0.613**	**0.539**	**0.654**	**0.723**	**0.617**
Borda	**0.703**	0.612	**0.539**	**0.654**	**0.723**	0.616
Borda$_{log}$	0.700	0.611	0.538	**0.654**	0.720	0.616
MRR	0.693	0.604	0.532	0.645	0.710	0.615
Avg	0.696	0.606	0.539	0.653	0.723	0.616
Geo	0.693	0.607	0.539	0.647	0.716	0.616
Harm	0.692	0.587	0.539	0.644	0.712	0.616
Max	0.696	0.603	0.513	0.648	0.716	0.612
Min	0.688	0.581	0.537	0.624	0.698	0.554
Combining two random measures plus five redundant measures						
Best measure	0.706	0.619	0.539	0.654	0.727	0.607
Alternative measure	0.679	0.559	0.505	0.613	0.671	0.529
UIR	**0.703**	**0.613**	**0.539**	**0.654**	**0.723**	**0.617**
Coppeland	0.681	0.599	0.532	0.631	0.714	0.618
Borda	0.690	0.599	0.536	0.637	0.696	0.615
Borda$_{log}$	0.689	0.598	0.536	0.638	0.698	0.615
Avg	0.687	0.603	0.537	0.638	0.697	0.615
Geo	0.686	0.603	0.538	0.635	0.695	0.615
Harm	0.686	0.582	0.538	0.636	0.696	0.615
Max	0.696	0.603	0.513	0.648	0.716	0.612
Min	0.688	0.581	0.537	0.624	0.698	0.554

6   Conclusions

Empirical studies have corroborated repeatedly in different scenarios that unsupervised voting methods provide a performance equivalent to the best individual measure. Therefore, the unsupervised combination of measures is a powerful strategy, given that it avoids overfitting and guarantees robust results across different datasets (as compared to supervised strategies), even if the optimal measure is different in each dataset.

We have presented an in-depth formal and empirical comparison of unsupervised measure combination approaches in the context of tasks based on textual similarity. We have introduced the concept of *Similarity Information Quantity*, and our formal study has shown that the most salient combination methods are all estimations of *Similarity Information Quantity* under different statistical assumptions that simplify the computation. At the same time, we have proved a

Minimal Voting Performance theorem that states that, under certain plausible conditions, the *Similarity Information Quantity* measure combination matches at least the performance of the best measure in the set. Altogether, we have arrived at a theoretical result that helps understanding why and how voting methods perform robustly.

We have also performed a comprehensive comparison of the most salient combination methods in six different datasets, corresponding to six different Information Access tasks. In concordance with our theoretical study, methods which are estimations of *Information Quantity* perform better than other methods, achieving results close to the best individual measure across all datasets.

From a practical point of view, our experiments suggest that (i) if there are only a few measures, UIR should be the combination method of choice; (ii) otherwise, Coppeland seems to be the best method, unless the set of individual measures is significantly redundant. If there are many measures and potential redundancy, Borda is probably a good choice.

Our future work involves a careful study of the statistical assumptions underlying each of the combining functions and how the assumptions hold in typical datasets. We are also investigating how to generalize our results: what is the family of problems, beyond textual similarity, within the scope of our theoretical contribution to unsupervised combining functions.

Acknowledgments. This research was supported by the Spanish Ministry of Science and Innovation (VoxPopuli Project, TIN2013-47090-C3-1-P and Vemodalen, TIN2015-71785-R).

A   Appendix: Minimal Voting Performance Proof

Given two similarity instances, $x, y \in \Omega^2$, we will denote an increase in signal f, the information quantity $\mathcal{I}_{\mathcal{F}}(x)$ or the true similarity $sim(x)$ by $\Delta \mathcal{I}_{\mathcal{F}} \equiv \mathcal{I}_{\mathcal{F}}(x) > \mathcal{I}_{\mathcal{F}}(y)$ and $\Delta f \equiv f(x) > f(y)$ and $\Delta sim \equiv sim(x) > sim(y)$. Similarly, decreases will be denoted by ∇f. Therefore, the optimality theorem can be expressed as $P(\Delta \mathcal{I}_{\mathcal{F}}|\Delta sim) \geq P(\Delta f|\Delta sim), \forall f \in \mathcal{F}$. Assuming high granularity, we have $P(\Delta f) = P(\Delta \mathcal{I}_{\mathcal{F}}) = P(\Delta sim) = \frac{1}{2}$. Therefore $P(\Delta f|\Delta sim) = \frac{P(\Delta f|\Delta sim) \cdot P(\Delta sim)}{P(\Delta f)} = P(\Delta f|\Delta sim)$. This is valid for any other conditional probability. Therefore, the optimality theorem can be rewritten as:

$$P(\Delta sim|\Delta \mathcal{I}_{\mathcal{F}}) \geq P(\Delta sim|\Delta f) \equiv P(\Delta sim|\Delta \mathcal{I}_{\mathcal{F}}, \Delta f) \cdot P(\Delta \mathcal{I}_{\mathcal{F}}|\Delta f) + P(\Delta sim|\Delta \mathcal{I}_{\mathcal{F}}, \nabla f) \cdot P(\Delta \mathcal{I}_{\mathcal{F}}|\nabla f) \geq$$
$$P(\Delta sim|\Delta f, \Delta \mathcal{I}_{\mathcal{F}}) \cdot P(\Delta f|\Delta \mathcal{I}_{\mathcal{F}}) + P(\Delta sim|\Delta f, \nabla \mathcal{I}_{\mathcal{F}}) \cdot P(\nabla f|\Delta \mathcal{I}_{\mathcal{F}}) \equiv$$
$$P(\Delta sim|\Delta \mathcal{I}_{\mathcal{F}}, \Delta f) \cdot (P(\Delta \mathcal{I}_{\mathcal{F}}|\Delta f) - P(\Delta f|\Delta \mathcal{I}_{\mathcal{F}}))+$$
$$P(\Delta sim|\Delta \mathcal{I}_{\mathcal{F}}, \nabla f) \cdot P(\Delta \mathcal{I}_{\mathcal{F}}|\nabla f) - P(\Delta sim|\Delta f, \nabla \mathcal{I}_{\mathcal{F}}) \cdot P(\nabla f|\Delta \mathcal{I}_{\mathcal{F}}) \geq 0$$

Assuming high granularity, we have that $P(\Delta \mathcal{I}_{\mathcal{F}}|\Delta f) - P(\Delta f|\Delta \mathcal{I}_{\mathcal{F}}) = 0$ and the previous expression is equivalent to:

$$P(\Delta sim|\Delta \mathcal{I}_{\mathcal{F}}, \nabla f) \cdot P(\Delta \mathcal{I}_{\mathcal{F}}|\nabla f) - P(\Delta sim|\Delta f, \nabla \mathcal{I}_{\mathcal{F}}) \cdot P(\nabla f|\Delta \mathcal{I}_{\mathcal{F}}) \geq 0$$

On the other hand $P(\Delta sim|\Delta f, \nabla \mathcal{I}_{\mathcal{F}}) = 1 - P(\nabla sim|\Delta f \nabla \mathcal{I}_{\mathcal{F}}) = 1 - P(\Delta sim|\nabla f, \Delta \mathcal{I}_{\mathcal{F}})$. And assuming granularity $P(\nabla f|\Delta \mathcal{I}_{\mathcal{F}}) = P(\Delta \mathcal{I}_{\mathcal{F}}|\nabla f)$. Therefore, we need to prove that:

$$P(\Delta sim|\Delta \mathcal{I}_{\mathcal{F}} \nabla f) \cdot P(\Delta \mathcal{I}_{\mathcal{F}}|\nabla f) - (1 - P(\Delta sim|\nabla f, \Delta \mathcal{I}_{\mathcal{F}})) \cdot P(\Delta \mathcal{I}_{\mathcal{F}}|\nabla f) \geq 0 \equiv$$

$$P(\Delta sim|\Delta \mathcal{I}_{\mathcal{F}}, \nabla f) \cdot (P(\Delta \mathcal{I}_{\mathcal{F}}|\nabla f) + P(\Delta \mathcal{I}_{\mathcal{F}}|\nabla f)) - P(\Delta \mathcal{I}_{\mathcal{F}}|\nabla f) \geq 0 \equiv$$

$$P(\Delta sim|\Delta \mathcal{I}_{\mathcal{F}}, \nabla f) \cdot 2 \cdot P(\Delta \mathcal{I}_{\mathcal{F}}|\nabla f) - P(\Delta \mathcal{I}_{\mathcal{F}}|\nabla f) \geq 0 \equiv$$

$$(2 \cdot P(\Delta sim|\Delta \mathcal{I}_{\mathcal{F}} \nabla f) - 1) \cdot P(\Delta \mathcal{I}_{\mathcal{F}}|\nabla f) \geq 0 \equiv$$

$$2 \cdot P(\Delta sim|\Delta \mathcal{I}_{\mathcal{F}}, \nabla f) - 1 \geq 0 \equiv P(\Delta sim|\Delta \mathcal{I}_{\mathcal{F}}, \nabla f) \geq \frac{1}{2}$$

Then, we have to prove that $P(\Delta sim(x) \mid \Delta \mathcal{I}_{\mathcal{F}}, \nabla f) \geq \frac{1}{2}$. Assuming SIH, we have that:

$$P\left(sim(x) > th \mid f^1(x), \dots, f^n(x)\right) \simeq \mathcal{I}_{\mathcal{F}}(x) = \mathcal{I}\left(\mathcal{A}^{\mathcal{F}}_{\{f^1(x)..f^n(x)\}}\right).$$

Therefore, when $\mathcal{I}_{\mathcal{F}}(x) > \mathcal{I}_{\mathcal{F}}(y)$, we can infer that:

$$P\left(sim(x) > th \mid f^1(x), \dots, f^n(x)\right) > P\left(sim(y) > th \mid f^1(y), \dots, f^n(y)\right)$$

It is true for every th values, so we can infer that:

$$P\left(sim(x) > sim(y) \mid f^1(x), \dots, f^n(x), \ f^1(y), \dots, f^n(y), \mathcal{I}_{\mathcal{F}}(x) > \mathcal{I}_{\mathcal{F}}(y)\right) \geq \frac{1}{2}.$$

It is true even when a single measure decreases $f^i(x) < f^i(y)$, so we can derive that $P(\Delta sim|\Delta \mathcal{I}_{\mathcal{F}}, \nabla f) \geq \frac{1}{2}$.

References

1. Agirre, E., Cer, D., Diab, M., Gonzalez-Agirre, A.: Semeval-2012 task 6: A pilot on semantic textual similarity. In: *SEM 2012: The First Joint Conference on Lexical and Computational Semantics (SemEval 2012), pp. 385–393. Association for Computational Linguistics, Montréal, Canada, 7–8 June 2012
2. Akiba, Y., Imamura, K., Sumita, E.: Using multiple edit distances to automatically rank machine translation output. In: Proceedings of Machine Translation Summit VIII, pp. 15–20 (2001)
3. Albrecht, J., Hwa, R.: The role of pseudo references in MT evaluation. In: Proceedings of the Third Workshop on Statistical Machine Translation, pp. 187–190 (2008)
4. Amigó, E., Gonzalo, J., Artiles, J., Verdejo, F.: Combining evaluation metrics via the unanimous improvement ratio and its application to clustering tasks. J. Artif. Intell. Res. (JAIR) **42**, 689–718 (2011)
5. Artiles, J., Amigó, E., Gonzalo, J.: The role of named entities in web people search. In: Proceedings of the 2009 Conference on Empirical Methods in Natural Language Processing, vol. 2, EMNLP 2009, pp. 534–542. Association for Computational Linguistics (2009)

6. Artiles, J., Gonzalo, J., Sekine, S.: The SemEval-2007 WePS evaluation: establishing a benchmark for the web people search task. In: Proceedings of the 4th International Workshop on Semantic Evaluations, SemEval 2007, pp. 64–69. Association for Computational Linguistics, Stroudsburg (2007)
7. Aslam, J.A., Savell, R.: On the effectiveness of evaluating retrieval systems in the absence of relevance judgments. In: Proceedings of the 26th Annual International ACM SIGIR Conference on Research and Development in Information Retrieval, SIGIR 2003, pp. 361–362. ACM, New York (2003)
8. Bar-Haim, R., Dagan, I., Dolan, B., Ferro, L., Giampiccolo, D., Magnini, B., Szpektor, I.: The second PASCAL recognising textual entailment challenge. In: Proceedings of the Second PASCAL Challenges Workshop on Recognising Textual Entailment (2006)
9. Linguistic Data Consortium. Linguistic Data Annotation Specification: Assessment of Adequacy and Fluency in Translations. Revision 1.5. Technical report (2005)
10. Cormack, G.V., Clarke, C.L.A., Büttcher, S.: Reciprocal rank fusion outperforms condorcet and individual rank learning methods
11. Corston-Oliver, S., Gamon, M., Brockett, C.: A machine learning approach to the automatic evaluation of machine translation. In: Proceedings of the 39th Annual Meeting of the Association for Computational Linguistics (ACL), pp. 140–147 (2001)
12. Dang, H.T.: Overview of DUC 2005. In: Proceedings of the 2005 Document Understanding Workshop (2005)
13. Dang, H.T.: Overview of DUC 2006. In: Proceedings of the 2006 Document Understanding Workshop (2006)
14. de Borda, J.C.: Memoire sur les Elections au Scrutin. Histoire de l'Academie Royale des Sciences, Paris (1781)
15. de Condorcet, M.: Essai Sur l'Application de l'Analyse Á la Probabilite des Decisions Rendues e la Pluralite des Voix (1785)
16. Giménez, J., Màrquez, L.: Asiya: an open toolkit for automatic machine translation (meta-)evaluation. Prague Bull. Math. Linguist. **94**, 77–86 (2010)
17. Kaniovski, S., Zaigraev, A.: Optimal jury design for homogeneous juries with correlated votes. Theory Decis. **71**(4), 439–459 (2011)
18. Kuncheva, L.I., Whitaker, C.J., et al.: Is independence good for combining classifiers? pp. 168–171 (2000)
19. Kuncheva, L.I., Whitaker, C.J.: Measures of diversity in classifier ensembles and their relationship with the ensemble accuracy. Mach. Learn. **51**(2), 181–207 (2003)
20. Lin, C.-Y.: Rouge: a package for automatic evaluation of summaries. In: Moens, M.-F., Szpakowicz, S., (eds.), Text Summarization Branches Out: Proceedings of the ACL-2004 Workshop, pp. 74–81. Association for Computational Linguistics, Barcelona, Spain, July 2004
21. Lin, D.: Dependency-based evaluation of MINIPAR. In: Proceedings of Workshop on the Evaluation of Parsing Systems, Granada (1998)
22. Liu, D., Gildea, D.: Source-language features and maximum correlation training for machine translation evaluation. In: Proceedings of the 2007 Meeting of the North American Chapter of the Association for Computational Linguistics (NAACL), pp. 41–48 (2007)
23. Montague, M.H., Aslam, J.A.: Condorcet fusion for improved retrieval. In: Proceedings of the 2002 ACM CIKM International Conference on Information and Knowledge Management, McLean, VA, USA, 4–9 November 2002, pp. 538–548. ACM (2002)

24. Partridge, D., Krzanowski, W.: Software diversity: practical statistics for its measurement and exploitation. Inf. Softw. Technol. **39**(10), 707–717 (1997)
25. Sharkey, A.J. (ed.): Combining Artificial Neural Nets: Ensemble and Modular Multi-Net Systems, 1st edn. Springer-Verlag New York Inc., Secaucus (1999)
26. Soboroff, I., Nicholas, C., Cahan, P.: Ranking retrieval systems without relevance judgments. In: Proceedings of the 24th Annual International ACM SIGIR Conference on Research and Development in Information Retrieval, SIGIR 2001, pp. 66–73. ACM, New York (2001)
27. Tversky, A.: Features of similarity. Psychol. Rev. **84**, 327–352 (1977)

Paper2vec: Combining Graph and Text Information for Scientific Paper Representation

Soumyajit Ganguly[✉] and Vikram Pudi

International Institute of Information Technology Hyderabad, Hyderabad, India
soumyajit.ganguly@research.iiit.ac.in, vikram@iiit.ac.in

Abstract. We present Paper2vec, a novel neural network embedding based approach for creating scientific paper representations which make use of both textual and graph-based information. An academic citation network can be viewed as a graph where individual nodes contain rich textual information. With the current trend of open-access to most scientific literature, we presume that this full text of a scientific article contain vital source of information which aids in various recommendation and prediction tasks concerning this domain. To this end, we propose an approach, Paper2vec, which comprises of information from both the modalities and results in a rich representation for scientific papers. Over the recent past representation learning techniques have been studied extensively using neural networks. However, they are modeled independently for text and graph data. Paper2vec leverages recent research in the broader field of unsupervised feature learning from both graphs and text documents. We demonstrate the efficacy of our representations on three real world academic datasets in two tasks - node classification and link prediction where Paper2vec is able to outperform state-of-the-art by a considerable margin.

Keywords: Citation networks · Representation learning · Text and graph

1 Introduction and Related Work

Information mining from citation networks is a well studied problem but most research in this direction tackles it as a graph problem. This creates an outright loss of information as we drop the entire textual content from papers and consider them merely as nodes in a heterogeneous graph. Today, especially in the domain of Computer Science, almost all of the published full-length research articles are freely available from online websites like CiteSeerX and arXiv. This leaves us with ample opportunity to exploit the text information from these scientific papers. Being limited to merely the citation information in the graph can have drawbacks. While writing a paper, authors always have a space constraint and thus can only cite a limited number of prior literature. It is also not possible for a particular author to know or track each and every related research work to hers from this growing sea of knowledge. Often it happens that multiple leading-edge

© Springer International Publishing AG 2017
J.M. Jose et al. (Eds.): ECIR 2017, LNCS 10193, pp. 383–395, 2017.
DOI: 10.1007/978-3-319-56608-5_30

research being done on the same problem statement are unable to cite each other due to the close proximity of publication dates. All of these problems create more sparsity in the citation graph and we lose out on probable valuable edges.

In classical literature a text document is represented by its histogram based bag-of-words or N-gram model which are sparse and can suffer from high dimensionality. There have been later research which explore probabilistic generative models like LDA and pLSA which try to obtain document representations in the topic space instead. These typically result in richer and denser vectors of much fewer dimensions than bag-of-words. Throughout the last decade there have been some attempts at alleviating the network sparsity problem discussed above with the help of text information for all kinds of bibliographic, web and email networks. Some of those methods extend the probabilistic representation of text documents by exploiting their underlying network structure [7,10]. These algorithms show promise as they result in better performances than their content-only (text) or network-only (graph) counterparts on a range of classification tasks. However most of the approaches are semi-supervised and rely on the idea of label propagation throughout the graph and the representations thus created are specific to the task at hand. The notion of injecting textual information specific to an academic citation graph have been studied in [11]. Here the authors make use of potential citations by which they enrich the citation graph and reduce its sparsity. The algorithm proposed for finding these potential citations are based on collaborative filtering and matrix imputation based schemes. A recent approach called TADW [13] was proposed for learning network representations along with text data. To the best of our knowledge this has been the first attempt at tackling the problem of learning fully unsupervised representations of nodes in a graph where the nodes themselves are text data. The learning algorithm in TADW is based on matrix factorization techniques. We treat TADW as an important baseline in our experiments.

There has been a surge of unsupervised feature learning approaches of late which use deep neural networks to learn embeddings in a low dimensional latent vector space. These approaches originated in the field of computer vision and speech signal processing and are now being adopted extensively in other domains. For text, there came shallow neural network based approaches like word2vec [6] and paragraph vectors [5] which are dense word and document representations created using algorithms commonly known as Skip-gram [6]. These approaches are fully unsupervised and are based on the distributional hypothesis *"you shall know a word by the company it keeps"*. A flurry of research work in the last few years make use of the so called *word, document embeddings* and achieve state-of-the-art performances throughout the breadth of Natural Language Processing (NLP) tasks [2]. The Skip-gram algorithm introduced in [6] have been extended well beyond words and documents to create representations for nodes in a graph [3,8,12].

We harness the power of these neural networks in our quest of creating rich scientific paper embeddings and propose two novel ways by which we can combine the textual data from papers with the graph information from citation networks.

We evaluate Paper2vec against state-of-the-art representations for both graph and text in a multi-class node classification task and a binary link prediction task. Our main contributions in this paper are as follows:

- Introduce Paper2vec - a novel neural network based embedding for representing scientific papers. Propose two novel techniques to incorporate textual information in citation networks to create richer paper embeddings.
- Curate a large collection of almost half a million academic research papers with full text and citation information.
- Conduct experiments on three real world datasets of varied scales and discuss the performance achieved in two evaluation tasks.

The rest of this paper is organized as follows: in Sect. 2 we discuss the proposed methodology broken into 3 main steps, in Sect. 3 we give a thorough discussion of our datasets and the experiments conducted. We report our observations and analysis in Sect. 4 and finally in Sect. 5 we discuss possible future directions and conclude the paper.

2   The Paper2vec Approach

2.1   Problem Formulation

A citation network dataset can be represented as a graph $G = (V, E)$, where V represents vertices or in our case scientific papers and each edge $e \in E$ represents a citation which links a pair of vertices (v_i, v_j). Neither do we measure strength of citations nor do we differentiate between incoming and outgoing links and thus G is both undirected and unweighted. So we have $(v_i, v_j) \equiv (v_j, v_i)$ and $\omega_{v_i v_j} = 1$. At this point G need not be connected. While in theory, any scientific article is connected to other related works through citation links, in a more real world scenario we can have some papers whose citations are not openly available. In Sect. 3.1 we conduct our experiments on such a graph.

Let $f : V \rightarrow \mathbb{R}^D$ be the mapping function from the nodes to the representations which we wish to learn. D is the dimensionality of our latent space and $|V|$ is the total number of nodes in our graph (including the non-connected ones). f is a matrix of size $|V| \times D$ parameters. f is learned in two phases by similar optimization algorithms but with different objective functions. The first objective function f_1 aims at capturing the text information while the next f_2 aims at capturing the citation context information. Two ideas are discussed by us which combine information from these different modalities. Note that throughout this paper, we mention the terms vector, embedding or representation. All refer to the same idea of latent vector space for nodes or documents which we aim to learn. We describe each step in detail in the following subsections.

2.2   Phase 1: Learning from Text

As our first step we aim at finding good textual representations for all vertices $v_i \in G$. Since here we consider only the textual information of each paper, we can denote them by a set V $\{d_1, d_2 \ldots d_{|V|}\}$. We use w_i and d_k to denote word and document vectors in our corpus. The original idea for unsupervised learning of document representations [5] was an extension to [6] where we jointly learn document embeddings along with words. First we define a fixed context window $c_1 \in C_1$ over words in a sentence which is slided throughout our corpus. For all the possible contexts in $|C_1|$, we then train every word in a context to predict all words in the context given the document vector and the word vector itself. This results in the document vector contributing to the word prediction task. Effectively we want to maximize the average log probability given in Eq. 1.

$$\sum_{w_i, w_j \in c_1}^{|C_1|} log Pr(w_j | w_i, d_k) \tag{1}$$

where $Pr(w_j | w_i, d_k)$ is defined by the softmax function,

$$Pr(w_j | w_i, d_k) = \frac{exp(w_j^T w_i + w_j^T d_k)}{\sum_{t=1}^{C_1} exp(w_t^T w_i + w_t^T d_k)} \tag{2}$$

The above objective functions can be trained using the Skip-gram algorithm for learning our word and document embeddings. Skip-gram assumes that inside the context, all words are independent of each other and equally important. From Eq. 2 we can see that the update step of word vectors w_i require a summation over all the words in our vocabulary which can be huge. We can approximate this objective using positive word-context pairs and randomly sampled negative word-context pairs. An example for negative word-context (w_i, w_j) pair would be *(algorithm, ice-cream)* which is highly unlikely to appear in the same context window. This approach was coined SGNS (skip-gram negative sampling) and it can be used to train our objective function using a 1 hidden layer neural network. Training document embeddings along with words result in a rich representation which is generic in nature and can be further utilized for a variety of domain specific tasks [5]. Among the two architectures (DM, DBOW) proposed in [5], the DM model produced better performance for our evaluation tasks. After this training process, we have all our nodes in a latent space $\mathbb{R}^D$ where textually similar papers are situated closer to each other. We ignore the learned word vectors from here-on and proceed only with $d_{k, \forall k \in V}$.

2.3   Bridging Text Information with Graph

We propose two novel ways of combining the text information into the citation networks. The first method is based on creating artificial text-based links in the citation network. This notion of text-based links can be seen as somewhat analogous to the *potential-citations* introduced in [11]. All vertices in G (papers in our dataset) are already represented as vectors in $\mathbb{R}^D$ as mentioned in Sect. 2.2. For every node $v_i \in V$ we find the k nearest neighbours of v_i in $\mathbb{R}^D$ and connect them through edges. We call these artificial text-based edges E' and add them to $G' = (V, E \cup E')$. Thus we create links between similar papers in our dataset which are not already linked by citations. Our intuition here is that authors are not always fully aware of all current developments and might miss citing some important papers. In short-papers and poster papers, there is a strict space constraint and authors do not have much choice other than including a mere handful of citations. We claim that the textual similarity between two papers is important and thus aim to bring closer two papers which share similar text content but do not have a direct citation edge between them. In Sect. 4.1 we provide further details on selecting k with respect to dataset sizes.

Our second method is motivated from the works in the field of computer vision [9], where the authors used pre-trained neural network weights from a general task. They found that their network gave good invariant features when trained on a large dataset that were shown to be generic. Instead of random weight initialization for any domain specific task, they successfully used these pre-trained weights to obtain improved performance across several sub-domain tasks related to computer vision. We use our document vectors learnt from text in Phase 1 as initialization points and further refine them by a new objective function f_2 in Phase 2 which minimizes loss over edges as described in the following subsection.

Later in Fig. 1 we show through our empirical evaluations that the two aforementioned methods individually contribute to an increase in performance throughout our datasets. We combine both of them to get Paper2vec, a new state-of-the art technique for estimating scientific paper representations.

2.4   Learning from Graph

Henceforward we take G' as our input graph and first define the notion of context or neighbourhood inside G': a valid context $c_2 \in C_2$ for a node v_i is the collection of all nodes v_j that are at most h hops away from v_i. Value of h is determined by window size of c_2. Note that here we do not differentiate between (v_i, v_j) pairs on whether they are connected by citations or by text-based links. We obtain C_2 by sliding over random walk sequences starting from every node $v_{i,i \in V}$ in G'. Borrowing the same idea from Sect. 2.2, given a node vector v_i we try to predict its neighbouring nodes $v_{j,\forall j \in C_2}$ in the graph. This notion of converting a graph into a series of text documents has been motivated by the fact that word frequency in a document corpus and the visited node frequency during a random walk for a connected graph, both follow the power law distribution [8]. Using the same intuitions as before, we now try to maximize the likelihood function as shown in Eq. 3.

$$\sum_{v_i, v_j \in C_2}^{|C_2|} log Pr(v_j | v_i) \tag{3}$$

Once more for calculating $Pr(v_j | v_i)$ we can run into the computational problem of summing over all nodes in G' as shown in Eq. 4 which can be large. We approximate the objective function by taking sets of positve and negative (v_i, v_j) pairs. In this case, an example for a negative context for v_i would be some vertex

v_j which has a very low probability of being in h hop neighbourhood of v_i.

$$Pr(v_j|v_i) = \frac{exp(v_j{}^T v_i)}{\sum_{t=1}^{C_2} = exp(v_t{}^T v_i)} \tag{4}$$

Applying the CBOW algorithm [6] with negative sampling on our constructed vertex, context pairs would give us our desired representations. This is similar to the strategies discussed by Perozzi et al. in [8] with the difference being in weight initialization.

3   Experimental Study

In this section, we begin by discussing the dataset details and our two evaluation metrics. Next we provide a brief overview of all the algorithms we compared against Paper2vec before presenting the performance comparison. We provide discussion and analysis for our chosen methods wherever possible. Towards the end we briefly discuss about hyper-parameter tuning, practical issues, running-time and scalability.

3.1   Datasets

We chose to evaluate Paper2vec on three different academic citation datasets of increasing scale (small, medium, large) described as follows:

- **CORA ML subset**: This is a subset of only Machine Learning papers from the CORA dataset. There are 2,708 papers from 7 classes like *reinforcement learning, probabilistic methods, neural networks* etc. This dataset is connected with 5,429 citation links. For text information we had titles, abstracts and all sentences from a paper containing citations.
- **CORA full dataset**: This is the full CORA dataset containing 51,905 papers from 7 broad categories like *operating systems, databases, information retrieval* etc. We manually pruned out duplicate entities and papers which do not have any associated text information with it. This resulted in a dataset of 36,954 papers with 132,968 citation links within the dataset. We use the same text information as in CORA ML subset.
- **DBLP citation network** (version 2): This dataset is a large collection of computer science papers. DBLP only provides citation-link information and paper titles. For full text of these papers, we refer to a recent research by Zhou et al. [14] which has been crawled from CiteSeerX and is publicly available. This dataset is partly noisy with some duplicate paper information and there is a lack of unique one-to-one mapping from the DBLP paper ids to the actual text of that paper. During the creation of our final dataset, we either pruned out ambiguous papers or manually resolved the conflicts. We came up with a final set of 465,355 papers from the DBLP corpus for which we have full text available. In this set only 224,836 papers are connected by citations because most of the other cited links are outside DBLP (not from computer science

domain) and hence full text is not available. However our text-based linking strategy as discussed in Sect. 2.3 helps us in connecting the graph and getting a final vertex count of 412,806. With only the citations being considered, edge count comes to 2,301,292 (undirected). We gather the required class labels from the MAS dataset [1] by Chakraborty et al. This dataset contains 807,516 tagged research papers in computer science. Their tags are based on the sub-domains they belong to, in total there are 24 categories like *computer-graphics, operating-systems, databases, language and speech* etc. We took the intersection of these labels with our cleaned DBLP dataset and found 134,338 matches for our classification experiment.

3.2 Comparison to Previous Work

For comparison with Paper2vec first we chose both text-only and network-only algorithms. Along with this we compared against a recently proposed text and graph combined algorithm and a concatenated baseline method for combining text and graph representations. The algorithms are discussed below:

- **Deepwalk** [8] is a network-only algorithm which represents a graph as a series of text streams and learns the node representations by applying the SGNS algorithm.
- **TADW** [13] or Text Associated DeepWalk is a matrix factorization based approach to approximate Deepwalk. It also uses *tf-idf* features to fuse link and text data similar to ours.
- **LINE** [12] learns network-only graph representations in two phases - first order proximity and second order proximity. Their edge sampling algorithm is similar to our discussed negative sampling.
- **Paragraph Vector** [5] is the original algorithm proposed by Le et al. for learning latent representation for text documents. We use this algorithm in our text learning step to get pre-trained vectors. This serves as a text-only (content-only) baseline.
- **tf-idf** [4] is an improvement over the simple bag-of-words algorithm for representing documents in the vector space.
- **Concatenated baseline:** We concatenated Paragraph Vector with Deepwalk embeddings to serve as a baseline for our text-graph combination.

3.3 Evaluation Tasks

We chose two tasks to evaluate our learned embeddings. Across all our datasets, we thus conduct 6 sets of experiments to demonstrate the effectiveness of Paper2vec embeddings.

Node Classification: In this task we need to determine the class or label of a scientific paper given its representation. For the text-only methods, this problem can be treated as multi-class document classification. After the (unsupervised) feature learning phase from respective algorithms we evaluate the classification accuracy across our three datasets. We vary the training ratio from 10% to 50% (rest are treated as test-set) and report the scores for each trial averaged over 10 times. This is the exact experimental details found in [13].

Link Prediction: Here we are given a random pair of node representations (v_i, v_j) and we need to determine whether there should be a citation link between them. For every pair we have two representations - one for each node. We use the Hadamard operator to combine two node vectors into one edge representation. Our edge representations $f_2(E_{i,j}) = f_2(v_1) * f_2(v_2)$ remain in $\mathbb{R}^D$ as this operator performs element wise multiplication between the node vectors. Grover and Leskovec in [3] studied in detail this problem of edge (citation) representation by combining dense node embeddings. We remove 25% of citation links from each of our three datasets before starting the representation training phase. For creating the link prediction evaluation dataset, we have this 25% removed citation links (positive examples) and we add random links to each graph which were not originally present (negative examples). Now we have a binary link prediction problem where given two nodes, we need to predict the presence of a link between them. We took care not to mix our text-based links with this random negative samples. Our final link-prediction dataset contains 20,000, 60,000 and 440,000 examples for the small, medium and large datasets respectively. We report scores for 5-fold cross-validation on this binary classification task for all algorithms.

Table 1. Node classification performance on CORA (ML-subset) dataset.

Type of embedding	Algorithm (dimensions)	SVM training ratio				
		10%	20%	30%	40%	50%
Network-only	LINE (100)	68.32	71.68	74.92	7 8.06	79.95
	Deepwalk (100)	75.96	80.66	82.71	84.31	85.30
Content-only	Paragraph Vector (100)	76.83	81.12	82.82	83.54	84.41
	tf-idf (1433)	78.48	82.83	84.73	85.99	86.88
Combined	Concatenated baseline (200)	80.61	83.76	85.13	86.35	87.14
	TADW (160)	82.4	85.0	85.6	86.0	86.7
	Paper2vec (100)	83.41	86.49	87.46	88.26	88.85

3.4 Classifier Details

We use Support Vector Machines (SVM) similar to those in [13] for all our node classification tasks. In the link prediction tasks we present results with Logistic Regression as they performed better than SVM. We use the Python library *scikit-learn* for our classifier implementations. Since we aim at comparing the learned embeddings, we focus less on exact classifier settings. However we only report the best score achieved by every algorithm in each dataset (Tables 3 and 4).

Table 2. Node classification performance on CORA (full) dataset.

Type of embedding	Algorithms (dimensions)	SVM training ratio				
		10%	20%	30%	40%	50%
Network-only	LINE (200)	76.98	79.35	80.46	81.34	81.82
	Deepwalk (200)	77.97	80.24	81.29	82.17	82.65
Content-only	Paragraph Vector (200)	75.46	78.34	79.54	80.40	80.87
	tf-idf (5000)	76.29	77.21	78.24	79.14	80.34
Combined	Concatenated baseline (500)	80.16	81.72	82.68	83.37	83.88
	TADW (200)	77.51	79.69	80.68	81.17	81.36
	Paper2vec (200)	82.31	83.83	84.45	84.87	85.18

Table 3. Node classification performance on DBLP dataset. TADW scores are unavailable due to scalability issues.

Type of embedding	Algorithms (dimensions)	SVM training ratio				
		10%	20%	30%	40%	50%
Network-only	LINE (500)	61.36	62.38	62.74	63.11	63.61
	Deepwalk (500)	60.27	61.56	62.29	63.00	63.43
Content-only	Paragraph Vector (500)	57.26	58.03	59.35	60.12	60.45
	td-idf (50000)	57.57	58.59	58.87	59.5	60.09
Combined	Concatenated baseline (800)	64.56	65.78	66.35	66.98	67.77
	Paper2vec (500)	65.45	66.46	67.61	68.21	69.94

Table 4. Link prediction performance on all datasets. Presented scores are *micro-f1* with 5-fold cross-validation. All dimensions are kept same as node classification task with respect to dataset.

Type	Algorithm	CORA (ML)	CORA (full)	DBLP
Network-only	LINE	79.22	91.38	95.08
	Deepwalk	81.30	92.70	94.92
Content-only	Paragraph Vector	77.57	85.59	88.87
Combined	Concatenated baseline	83.13	93.36	95.32
	TADW	85.65	81.67	–
	Paper2vec	91.75	95.11	97.13

4 Results and Discussion

We set the dimension k in TADW at 160 as recommended by the authors in [13] for our CORA ML (small) dataset. In the CORA full (medium) dataset, we see an improvement in performance on increasing $k = 200$. For Deepwalk on all our datasets and evaluation tasks we found the number of walks started per vertex $\gamma = 10$ and window size $t = 5$ to perform best. Note that this is contradicting with the values set for Deepwalk ($\gamma = 80, t = 10$) in [13] and thus the results vary. We keep the dimensions of LINE, Deepwalk and Paragraph Vectors exactly the same as Paper2vec: 100 for small, 200 for medium and 500 for large. During our evaluation for *tf-idf* text based baselines we kept the maximum features (sorted by *df* value) as 1433 for our small [13] and 5000 for our medium dataset. We list below our major observations:

- Paper2vec is able to out-perform all baselines and competing algorithms throughout all datasets in both the tasks. It can be inferred from Fig. 1 that both our text infusion methods with graphs, individually increase the performance over baseline network-only methods.
- From our results in Tables 1 and 2, we can see that neural network based techniques are generally able to out-perform the matrix factorisation based methods. Similar trends have also been reported in [8,12].
- Our chosen concatenated baseline method of concatenating Deepwalk embeddings with Paragraph Vector is consistently able to outperform both individually and also TADW, though at the cost of (double) dimension size.
- Surprisingly Deepwalk embeddings are quite competitive with the content-only algorithms in our node classification task. This is quite astonishing given that unlike Paragraph Vector, Deepwalk was trained with no information about text. A behavior like this can be attributed to our domain of scientific data where authors tend to cite more papers from the same domain. LINE is able to perform better on the large DBLP dataset.
- Content-only algorithms resulted with better scores in our node classification task as expected. Their performance in link prediction fall short of the graph and combined embedding algorithms. It is interesting to observe that during node classification, *tf-idf* features perform quite well on the small and medium datasets. It's performance reaches the level of the combined method TADW. Results reported here for *tf-idf* are better than the binary feature based *bag-of-words* approach in [13].
- In the link prediction task, we can see on the CORA-ML (small) dataset, there is an improvement of above 11% for Paper2vec over Deepwalk. As we grow our dataset sizes, more citation links make the graph denser thus diminishing the effect of text information for this task.
- Our unsupervised representations are able to surpass prior work based on semi-supervised models like Collective Classification [10] in the first task.

Through these six experiments we show that in Paper2vec, we are able to successfully fuse text with graph without any trade-offs or loss of information. Our representations in $\mathbb{R}^D$ are able to perform better than concatenated features

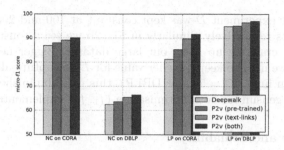

Fig. 1. Comparison of our text-infusion methods against baseline (Deepwalk). Here P2v refers to Paper2vec. NC and LP denotes node classification and link prediction respectively. The scores are based on 5-fold cross-validation.

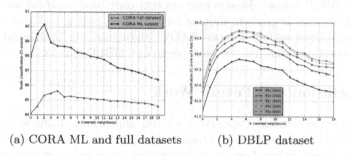

(a) CORA ML and full datasets   (b) DBLP dataset

Fig. 2. Node classification performance with varying k and D for Paper2vec.

in $\mathbb{R}^{2D}$ (baseline), previous state-of-the-art in link-text combination (TADW), all text-based methods for node classification and graph-based methods for link prediction. These results signify the vital role which textual data plays in creating rich embeddings for scientific papers.

4.1 Parameter Sensitivity

We used the DM model [5] and CBOW model [6] respectively for our text and graph learning stages. Varying window size of c_1 in text learning framework from 5 to 10 did not have notable impact on our document vectors, we fixed it at 10. However c_2, window size for the graph learning framework gave best results for the value of 5 on being varied from 2 to 10. This signifies that our paper similarity is best determined by its 5-hop neighbourhood distribution. During DM training for optimising f_1 we ran 10 epochs by setting a learning-rate decay of 0.02 after every epoch and keeping it constant throughout the epoch. The two remaining hyper-parameters to tune for Paper2vec are number of neighbors to connect k and our embedding dimensions D. From Figs. 2a and b we can see a steady increase of node classification performance as we connect $2, 4, 5$ neighbours for our small, medium and large datasets respectively. After this performance peak, there is a steady decline as we start connecting arbitrary (less similar) nodes together. In this experiment D was kept constant at 100 and 200 for small and medium datasets respectively. Similarly in Fig. 2b we see a constant increase in performance by cranking up D on our large dataset. Bigger networks contain more data and we need to keep a higher value for D to capture all of it. However beyond a certain limit ($D = 500$ for DBLP), this gain diminishes. We used the popular library gensim for both optimisation (f_1, f_2) implementations.

4.2 Runtime and Scalability

We conducted all our experiments on a single desktop PC with specifications: Intel Pentium G3220 processor and 8 GB memory. Every neural network based algorithm (Deepwalk, LINE, Paragraph Vector) including ours, were scalable to handle the DBLP dataset. However for our text-only baseline *tf-idf* we had to run mini-batch stochastic gradient descent for the classification task due to memory limitations. Unfortunately for the TADW algorithm, its matrix factorization based approach was not directly scalable to our DBLP dataset.

5   Conclusion and Future Work

In this paper, we present Paper2vec, a novel algorithmic framework for unsupervised learning of combined textual and graph features in academic citation networks. Our algorithm shows two ways by which we can incorporate the learned text representations with graph features and achieve a higher overall predictive performance in various tasks. For both our node classification and link prediction tasks, we find Paper2vec perform superior to state-of-the-art techniques across all datasets. A future research direction would be to devise an inference procedure for unseen papers not present within the training corpus. Through this we intend to explore the dynamics of our system in a more real world scenario where the incoming number of new papers is quite high and how we can incorporate them inside our model without loss of prior information. In addition to this, we hope to generalize our algorithm further and expand it to graphs beyond academic literature. Most of the web-data available today are connected by hyper-links and contain text information. We look forward to exploit information from different modalities and create rich representations across domains.

References

1. Chakraborty, T., Sikdar, S., Tammana, V., Ganguly, N.: Computer science fields as ground-truth communities: their impact, rise and fall. In: ASONAM (2013)
2. Collobert, R., Weston, J., Bottou, L., Karlen, M., Kavukcuoglu, K., Kuksa, P.P.: Natural language processing (almost) from scratch. CoRR abs/1103.0398 (2011)
3. Grover, A., Leskovec, J.: Scalable feature learning for networks. In: KDD (2016)
4. Joachims, T.: A probabilistic analysis of the Rocchio algorithm with TFIDF for text categorization. Technical report, DTIC Document (1996)
5. Le, Q.V., Mikolov, T.: Distributed representations of sentences and documents. In: ICML (2014)

6. Mikolov, T., Sutskever, I., Chen, K., Corrado, G.S., Dean, J.: Distributed representations of words and phrases and their compositionality. CoRR (2013)
7. Nallapati, R., Cohen, W.W.: Link-PLSA-LDA: a new unsupervised model for topics and influence of blogs. In: AAAI (2008)
8. Perozzi, B., Al-Rfou, R., Skiena, S.: DeepWalk: online learning of social representations. In: KDD (2014)
9. Razavian, A.S., Azizpour, H., Sullivan, J., Carlsson, S.: CNN features off-the-shelf: an astounding baseline for recognition. In: CVPR (2014)
10. Sen, P., Namata, G., Bilgic, M., Getoor, L., Gallagher, B., Eliassi-Rad, T.: Collective classification in network data. AI Mag. **29**, 1–24 (2008). http://eliassi.org/papers/ai-mag-tr08.pdf
11. Sugiyama, K., Kan, M.Y.: Exploiting potential citation papers in scholarly paper recommendation. In: JCDL (2013)
12. Tang, J., Qu, M., Wang, M., Zhang, M., Yan, J., Mei, Q.: LINE: large-scale information network embedding. In: WWW. ACM (2015)
13. Yang, C., Liu, Z., Zhao, D., Sun, M., Chang, E.Y.: Network representation learning with rich text information. In: IJCAI (2015)
14. Zhou, T., Zhang, Y., Lu, J.: Classifying computer science papers. In: IJCAI (2016)

Exploration of a Threshold for Similarity Based on Uncertainty in Word Embedding

Navid Rekabsaz[✉], Mihai Lupu, and Allan Hanbury

Institute of Software Technology and Interactive Systems,
Vienna University of Technology, 1040 Vienna, Austria
{rekabsaz_navid,lupu_mihai,hanbury_allan}@ifs.tuwien.ac.at

Abstract. Word embedding promises a quantification of the similarity between terms. However, it is not clear to what extent this similarity value can be of practical use for subsequent information access tasks. In particular, which range of similarity values is indicative of the actual term relatedness? We first observe and quantify the uncertainty of word embedding models with respect to the similarity values they generate. Based on this, we introduce a general threshold which effectively filters related terms. We explore the effect of dimensionality on this general threshold by conducting the experiments in different vector dimensions. Our evaluation on four test collections with four relevance scoring models supports the effectiveness of our approach, as the results of the proposed threshold are significantly better than the baseline while being equal to, or statistically indistinguishable from, the optimal results.

1 Introduction

Understanding the meaning of a word (semantics) and of its similarity to other words (relatedness) is the core of understanding text. An established method for quantifying this similarity is the use of *word embeddings*, where vectors are proxies of the meaning of words and distance functions are proxies of semantic and syntactic relatedness. Fundamentally, word embedding models exploit the contextual information of the target words to approximate their meaning, and hence their relations to other words.

Given the vectors representing words and a corresponding mathematical function, these models provide an approximation of the relatedness of any two terms, although this relatedness could be perceived as completely arbitrary in the language. This issue is pointed out by Karlgren et al. [9] in examples, showing that word embedding methods are too ready to provide answers to meaningless questions: *"What is more similar to a computer: a sparrow or a star?"*, or *"Is a cell more similar to a phone than a bird is to a compiler?"*. The emerging challenge here is: *how to identify whether the similarity score obtained from word embedding is really indicative of term relatedness?*

1.1 Related Work

The closest study to our work is Karlgren et al. [8], which explores the semantic topology of the vector space generated by Random Indexing. Based on their

© Springer International Publishing AG 2017
J.M. Jose et al. (Eds.): ECIR 2017, LNCS 10193, pp. 396–409, 2017.
DOI: 10.1007/978-3-319-56608-5_31

previous observations that the dimensionality of the semantic space appears different for different terms [9], Karlgren et al. now identify the different dimensionalities at different angles (i.e. distances) for a set of specific terms. It is however difficult to map these observations to specific criteria or guidelines for either future models or retrieval tasks. In fact, our observations provide a quantification on Karlgren's claim that "'close' *is interesting and* 'distant' *is not*" [9].

More recently, Cuba Gyllensten and Sahlgren [3] follow a data mining approach to represent the terms relatedness by a tree structure. While they suggest traversing the tree as a potential approach, they evaluate it only on the word sense induction tasks and its utility for retrieving similar words remains unanswered. They do point out however, that applying a nearest neighbour approach, where for every word we use the top k most similar words, is not theoretically justifiable. Rekabsaz et al. [17] recently showed this also experimentally in a retrieval task.

In general, different characteristics of term similarities have been explored in several studies: the concept of relatedness [10,12], the similarity measures [11], intrinsic/extrinsic evaluation of the models [1,4,19,21], or in sense induction task [3,5]. However, there is lack of understanding on the internal structure of word embedding, specifically how its similarity distribution reflects the relatedness of terms.

1.2 Motivation

Among the recent publications using word-embeddings for information retrieval, Rekabsaz et al. [17] do a brute-force search on similarity thresholds for the typical ad-hoc search task and evaluate their results against a set of TREC test collections. The parameter scan is obviously inefficient in general and we consider their work as the main motivation for the current study of a language-specific semantic similarity threshold.

In fact, we hypothesise that the "similar" words can be identified by a threshold on similarity values which separates the semantically related words from the non-related ones. We especially want to make this threshold independent of the terms and general on word embedding model. The reason for this choice is first the computational problem of term-specific thresholds as it puts burden on practical applications. Regardless of the efficiency issues, it is still reasonable to consider a general threshold. since it considers the centrality and neighbourhood of the terms by filtering different number of similar terms for each term.

Such a threshold has the potential to improve all studies that use similar/related words in different tasks i.e. query expansion [7], query auto-completion [14], document retrieval [16], learning to rank [20], language modelling in IR [6], or Cross-Lingual IR [22]. It should be noted though, that the meaning of "similar" also depends on the similarity function. We consider here the Cosine function as it is by far the most widely used word similarity function and leave the exploration of other functions for further studies. In fact, regardless of the similarity function, a threshold that separates the semantically related terms from the rest will always be an essential element to identify.

1.3 Approach

We explore the estimation of this potential threshold by first quantifying the uncertainty in the similarity values of embedding models. This uncertainty is an intrinsic characteristic of all the recent models, because they all start with some random initialization and eventually converge to a (local) solution. Therefore, even by training with the same parameters and on the same data, the created word embedding models result in slightly different word distributions and hence slightly different relatedness values. In the next step, using this observation, we provide a novel representation on the expected number of neighbours of an arbitrary term as a continuous function over similarity values, which is later used to estimate the general threshold.

In order to evaluate the effectiveness of the proposed threshold, we follow the approach previously introduced by Rekabsaz et al. [17] and test it in the context of a document retrieval task, on four different test collections, using the skip-gram with negative-sampling training word embeddings [13]. In the experiments, we apply the threshold to identify the set of terms to extend the query terms using both the Generalised Translation Model and the Extended Translation Model introduced by Rekabsaz et al. [17]. The results are compared with the optimal threshold, achieved as before by exhaustive search on the spectrum of threshold parameters. We show that in general using the proposed threshold performs either exactly the same as, or statistically indistinguishable from, the optimal threshold.

In summary, the main contributions of this paper are:

1. exploration of the uncertainty in word embedding models in different dimensions and similarity ranges.
2. introducing a general threshold for separating similar terms in different embedding dimensions.
3. extensive experiments on four test collections comparing different threshold values on different retrieval models.

The remainder of this work is structured as follows: We introduce the proposed threshold in Sect. 2. We present our experimental setup in Sect. 3, followed by discussing the results in Sect. 4. Section 5 summarises our observations and concludes the paper.

2 Global Term Similarity Threshold

We are looking for a threshold to separate the related terms from the rest. For this purpose, we start with an observation on the uncertainty of similarity in word embedding models, followed by defining a novel model of the expected number of neighbours for an arbitrary term, before we define our proposed threshold.

2.1 Uncertainty of Similarity

In this section we make a series of practical observations on word embeddings and the similarities computed based on them.

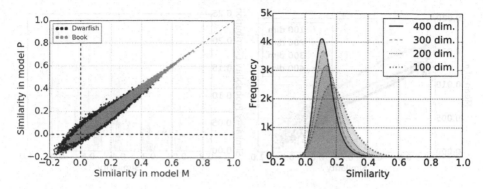

Fig. 1. (a) Comparison of similarity values of the terms *Book* and *Dwarfish* to 580k words between models M and P. (b) Histogram of similarity values of an arbitrary term to all the other words in the collection for 100, 200, 300, and 400 dimensions.

To observe the uncertainty, let us consider two models P and M. To create each instance, we trained the Skip-Gram with Negative-Sampling (SGNS) of the Word2Vec model with the sub-sampling parameter set to 10^{-5}, context windows of 5 words, epochs of 25, and word count threshold 20 on the Wikipedia dump file for August 2015, after applying the Porter stemmer. Each model has a vocabulary of approximately 580k terms. They are identical in all ways except their random starting point.

Figure 1a shows the distances between two terms and all other terms in the dictionary, for the two models, in this case of dimensionality 200. For each term we have approximately 580k points on the plot. As we can see, the difference between similarities calculated in the two models, appears (1) greater for low similarities, and (2) greater for a rare word (Dwarfish) than for a common word (Book). We can also observe that there are very few pairs of words with very high similarities.

Let us now explore the effect of dimensionality on similarity values and also uncertainty. Before that, in order to generalize the observations to an arbitrary term, we had to consider a set of "representative" terms. What exactly "representative" means is of course debatable. We took 100 terms recently introduced in the query inventory method by Schnabel et al. [19]. They claim that the selected terms are diverse in frequency and part of speech over the collection terms. In the remainder of the paper, we refer to *arbitrary* term as an aggregation over the representative terms i.e. each value related to the arbitrary term is the average of the values of the representative terms.

Figure 1b shows frequency histograms for the occurrence of similarity values for models of different dimensionalities. As we can see, similarities are in the $[-0.2, 1.0]$ range and have positive skewness (the right tail is longer). As the dimensionality of the model increases, the kurtosis also increases (the histogram has thinner tails).

Let us first suggest a concrete definition for uncertainty: We quantify the uncertainty of the similarity between two words as the standard deviation σ of

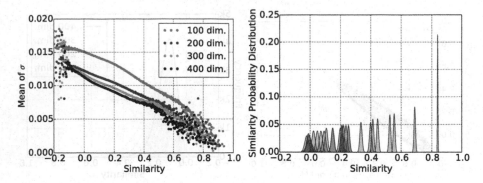

Fig. 2. (a) Standard deviation for similarity values. Points are the average over similarity intervals with equal lengths of 2.4×10^{-4} (b) Probability distribution of similarity values for the term *Book* to some other terms.

similarity values obtained from a set of identical models. We refer to identical models as the models created using the same method, parameters, and corpus. However as shown before, the similarity values of each word pair in each model are slightly different. The uncertainty of similarity between the words x and y is therefore formulated as follows:

$$\sigma_{x,y} = \sqrt{\frac{1}{|M|} \sum_{m \in M} (sim(\boldsymbol{x}_m - \boldsymbol{y}_m) - \mu)^2}, \quad \text{where} \quad \mu = \frac{\sum_{m \in M} sim(\boldsymbol{x}_m - \boldsymbol{y}_m)}{|M|}.$$

where M is the set of identical models and $\boldsymbol{x}_m$ is the vector representation of term x in model m and sim is a similarity function between two vectors.

To observe the changes in standard deviation, for every dimensionality, we create five identical SGNS models ($|M| = 5$).

Figure 2a plots the standard deviation, against the similarity values, for different model dimensionalities. For the sake of clarity in visualisation, we split the similarity values into 500 equal intervals (each 2.4×10^{-4}) and average the values in each interval.

The plots are smooth in the middle and scattered on the head and tail as the majority of similarity values are in the middle area of the plots and therefore the average values are consistent. However, we can observe that overall, as the similarity increases, the standard deviation, i.e. the uncertainty, decreases.

We also observe a decrease in standard deviation as the dimensionality of the model increases. On the other hand, the differences between models decrease as the dimension increases such that the models of dimension 300 and 400 seem very similar in comparison to 100 and 200. The observation shows a probable convergence in the uncertainty at higher dimensionalities.

These observations show that the similarity between terms is not an exact value but can be considered as an approximation whose variation is dependent on the dimensionality and similarity range. We use the outcome of these observations in the following.

2.2   Continuous Distribution of Neighbours

We have demonstrated that the similarity values of a pair of terms, obtained from identical embedding models are slightly different. In the absence of additional information, we assume that these similarity values follow a normal distribution.

To estimate this probability distribution, we use the mean and standard deviation values in Sect. 2.1. Figure 2b shows the probability distribution of similarities for term *Book* to 25 terms in different similarity ranges[1]. As observed before, by decreasing the similarity, the standard deviation of the probability distributions increases.

We use these probability distributions to provide a representation of the expected number of neighbours around an arbitrary term in the spectrum of similarity values: We first calculate the Cumulative Distribution Functions (CDF) of the probability distributions. We then subtract the CDF values from 1 which only reverses the direction of the distributions (from increasing left-to-right on X-axis to right-to-left). Finally, we accumulate all the cumulative distribution functions by summing all the values, shown in Fig. 3a. The values on this plot indicate the number of expected neighbours that have greater or equal similarity values to the term than the given similarity value. We can see the number of all the terms in the model (580k) in the lowest similarity value (−0.2) which then rapidly drops as the similarity increases. This representation of the expected number of neighbours in Fig. 3a has two benefits: (1) the estimation is continuous and monotonic, and (2) it considers the effect of uncertainty based on five models.

As noted before, the notion of *arbitrary* term is in fact an average over the 100 representative terms. Therefore, in calculating the representation of the expected number of neighbours, we also consider the confidence interval around the mean.

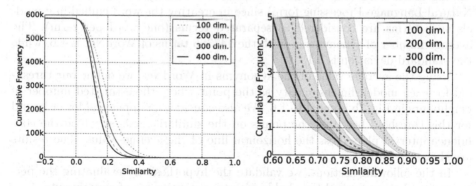

Fig. 3. (a) Mixture of cumulative probability distributions of similarities in different dimensions (b) Expected number of neighbours around an arbitrary term with confidence interval. The average number of synonyms in WordNet (1.6) is shown by the dash-line.

---

[1] we do not plot all the terms in the model to maintain the readability of the plot.

Table 1. Proposed thresholds for various dimensionalities

Dimensionality	Threshold boundaries		
	Lower	Main	Upper
100	0.802	**0.818**	0.829
200	0.737	**0.756**	0.767
300	0.692	**0.708**	0.726
400	0.655	**0.675**	0.693

This interval is shown in Fig. 3b. Here, the representation is zoomed on the lower right corner of Fig. 3a. The area around each plot shows the confidence interval of the estimation.

This continuous representation is used in the following for defining the threshold for the semantically related terms.

2.3 Similarity Threshold

Given the expected number of neighbours around the arbitrary term, represented in Fig. 3a and b, the question is *"what is the best threshold for filtering the related terms?"*. In order to address the question, we hypothesise that since this general threshold tries to separate related from unrelated terms, it can be estimated from the average number of synonyms over the terms. Therefore, we transform the above question into a new question: *"What is the expected number of synonyms for a word in English?"*

To answer this, we exploit WordNet. We consider the distinct terms in the related synsets to a term as its synonyms, while filtering the terms containing multi word (e.g. Natural Language Processing, shown in WordNet in Natural_Language_Processing form) since in creating the word embedding models such terms are considered as separated terms (one word per term). The average number of synonyms over all the 147306 terms of WordNet is 1.6, while the standard deviation is 3.1.

Using the average value of the synonyms in WordNet, we define our threshold for each model dimensionality as the point where the estimated number of neighbours in Fig. 3b is equal to 1.6. We also consider an upper and lower bound for this threshold based on the points on the similarity axis at which the confidence interval plots cross the horizontal line of the average value. The results are shown in Table 1.

In the following sections, we validate the hypothesis by evaluating the performance of the proposed thresholds with an extensive set of experiments.

3 Experimental Methodology

We test the effectiveness of our threshold in an Ad-hoc retrieval task on IR test collections by evaluating the results of applying various thresholds to retrieve the related terms.

Our relevance scoring approach is based on the query language model [15] and BM25 methods as two widely used and established methods in IR, which have shown competitive results in various domains. To use the additional information provided by word embeddings, we use the *Generalized Translation Model* and *Extended Translation Model* extensions introduced by Rekabsaz et al. [17], which build on top of the existing probabilistic models.

In the following, first we briefly explain the translation models when combined with word embedding similarity and then describe the details of our experimental setup.

3.1 Generalized and Extended Translation Model

In principle, a translation model introduces in the estimation of the relevance of the query term t a translation probability P_T, defined on the set of (related) terms $R(t)$, always used in its conditional form $P_T(t|t')$ and interpreted as the probability of observing term t, having observed term t'.

Translation models in IR were first introduced by Berger and Lafferty [2] as an extension to the language model. Recently, Rekabsaz et al. [17] extend the idea of translation model into four probabilistic relevance frameworks. Their approach is based on the observation that what one wants to compute in general in IR, and in particular in a probabilistic method, is the occurrence of concepts. Traditionally, these are represented by the words present in the text, quantified by term frequency (*tf*). Rekabsaz et al. posit that we can have a *tf* value lower than 1 when the term itself is not actually present, but another, similar term occurs in the text. They call this the Generalised Translation model (GT). However, in the probabilistic models, a series of other factors are computed based on *tf* (e.g. document length). Propagating the above changes to all the other statistics leads to even more changes in the scoring formulas. They refer to this as the Extended Translation model (ET).

In both translation models, they use word embedding to generate the $R(t)$ set by selecting the terms with the similarity value of greater than a given threshold to the query term t. In the following experiments we will show that the analytically obtained threshold described in the previous section is optimal for the ad-hoc retrieval task.

3.2 Experiment Setup

We evaluate our approach on four test collections: TREC-6, TREC-7, and TREC-8 of the AdHoc track, and TREC-2005 HARD track. Table 2 summarises the statistics of the test collections. For pre-processing, we apply the Porter stemmer and remove stop words using a small list of 127 common English terms.

In order to compare the performance of the thresholds, we test a variety of threshold values for each model. The thresholds cover a set of values on both sides of our introduced thresholds: for 100 dimension $\{0.67, 0.70, 0.74, 0.79, 0.81, 0.86, 0.91, 0.94, 0.96\}$, 200 dimension $\{0.63, 0.68, 0.71, 0.73, 0.74, 0.76, 0.78, 0.82\}$,

Table 2. Test collections used in this paper

Name	Collection	# documents
TREC 6	Disc4&5	551873
TREC 7 and 8	Disc4&5 without CR	523951
HARD 2005	AQUAINT	1033461

300 dimension $\{0.55, 0.60, 0.65, 0.68, 0.70, 0.71, 0.73, 0.75\}$, and 400 dimension $\{0.41, 0.54, 0.61, 0.64, 0.66, 0.68, 0.70, 0.71, 0.75\}$.

We set the basic models (language model or BM25) as baseline and test the statistical significance of the improvement of the translation models with respect to their basic models (indicated by the symbol †). Since the parameter μ for Dirichlet smoothing of the translation language model and also b, k_1, and k_3 for BM25 are shared between the methods, the choice of these parameters is not explored as part of this study and we use the same set of values as in Rekabsaz et al. [17]. The statistical significance test are done using the two sided paired t-test and statistical significance is reported for $p < 0.05$.

The evaluation of retrieval effectiveness is done with respect to Mean Average Precesion (MAP) and Normalized Discounted Cumulative Gain at cut-off 20 (NDCG@20), as standard measures in Adhoc information retrieval. Similar to Rekabsaz et al. [17] and in order to make the results comparable with this study, we consider MAP and NDCG over the condensed lists [18].

4   Results and Discussion

The evaluation results of the MAP and NDCG@20 measures of the BM25 Extended Translation (BM-ET) model on the four test collections, with vectors in 100, 200, 300, and 400 dimensions are shown in Fig. 4. Due to lack of space, we only show the detailed results of the BM-ET model as it has shown the best overall performance among the other translation models in Rekabsaz et al. [17]. For each dimension, our threshold and its boundaries (the interval between the lower and upper bound in Table 1) are shown with vertical lines. The baseline (basic BM25) is shown in the horizontal line. Significant differences of the results to the baseline are marked by the † symbol.

The plots show that the performance of the translation models are highly dependent on the choice of the threshold value. In general, we can see a trend in all the models: the results tend to improve until reaching a peak (optimal threshold) and then converges to the baseline. Based on this general behaviour, we can assume that including the terms whose similarity values are less than the optimal threshold introduces noise and deteriorates the results while using the cutting point greater than the optimal threshold filters the related terms too strictly. We test the statistical significance between the results of the optimal and proposed threshold in all the experiments (both evaluation measures, all

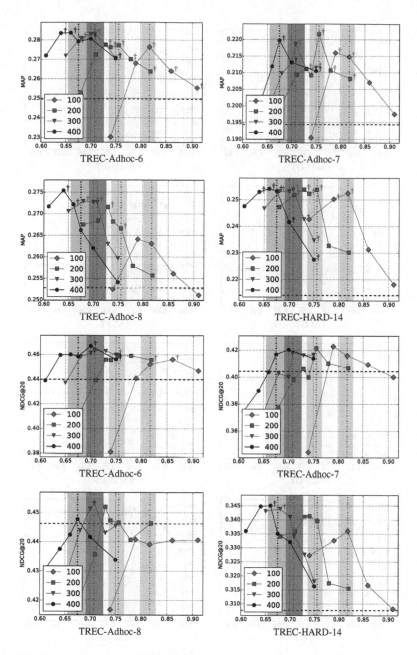

Fig. 4. MAP (above) and NDCG@20 (below) evaluation of the BM25 Extended Translation model on TREC-6, TREC-7, TREC-8 Adhoc, and TREC-2005 HARD for different thresholds (X-axes) and word embedding dimensions. Significance is shown by †. Vertical lines indicate our thresholds and their boundaries in different dimensions. The baseline is shown by the horizontal line. To maintain visibility, points with very low performance are not plotted.

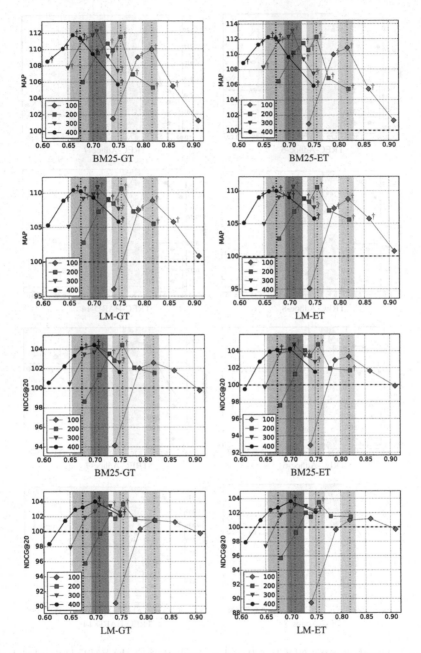

Fig. 5. Percentage of improvement of the relevance scoring models BM25 and Language Model (LM), combined with the Generalized Translation (GT) and Extended Translation (ET) models with respect to the baselines (standard LM and BM25) with the MAP (above) and NDCG@20 (below) evaluation measures for different thresholds, and word embedding dimensions, aggregated over all the collections.

Table 3. Examples of similar terms, selected with our threshold

book: publish, republish, foreword, reprint, essay
eagerness: hoping, anxious, eagerness, willing,wanting
novel: fiction, novelist, novellas, trilogy
microbiologist: biochemist, bacteriologist, virologist
shame: ashamed
guilt: remorse
Einstein: relativity
estimate, dwarfish, antagonize: no neighbours

relevance scoring models, collections, and dimensions), observing no significant difference in any of the cases.

In order to have an overview of all the models, we calculate the gain of each relevance scoring model for different thresholds and dimensionalities over its corresponding baseline and average the gains on the four collections. The scoring models are BM25 and Language Model (LM), combined with the Generalized Translation (GT) and Extended Translation (ET) models. The results for MAP and NDCG are depicted in Fig. 5. In all the translation models, our threshold is optimal for dimensions 100, 200, and 300. In dimension 400, the significance test between their results does not show any significant difference. These results justify the choice of the proposed threshold as a generally stable and effective cutting-point for identifying related terms.

To observe the effect of the proposed threshold, let us take a closer look at the terms, filtered as related terms. Table 3 shows some examples of the retrieved terms when using the word embedding model with 300 dimensions with our threshold (same as optimal in this dimension for all the translation models). As expected, the examples show the strong differences in the number of similar words for various terms. The mean and standard deviation of the number of similar terms for all the query terms of the tasks is 1.5 and 3.0 respectively. Almost half of the terms are not expanded at all. We can observe the similarity between this calculated mean and standard deviation and the aggregated number of synonyms we observed in WordNet in Sect. 2.3—mean of 1.6 and standard deviation of 3.1. It appears that although the two semantic resources (WordNet and Word2Vec) cast the notion of similarity in different ways and their provided sets of similar terms are different, they correspond to a similar distribution of the number of related terms.

5   Conclusion

We have analytically explored the thresholds on similarity values of word embedding to select related terms. Based on empirical observations on various models trained on the same data, we have introduced a method to identify the minimal cosine similarity value between two term vectors, allowing practical use of similarity values. The proposed threshold is estimated based on a novel representation of the neighbours around an arbitrary term, taking into account the variance of similarity values, captured from the values generated by different instances of identical models.

We extensively evaluate the application of the introduced threshold on four information retrieval collections using four state-of-the-art relevance scoring models. The results show that the proposed threshold is identical to the optimal threshold (obtained by parameter scan) in the sense that its results on ad-hoc retrieval tasks are either equal to or statistically indistinguishable from the optimal results.

Acknowledgement. This work is funded by: Self-Optimizer (FFG 852624) in the EUROSTARS programme, funded by EUREKA, the BMWFW and the European Union, and ADMIRE (P 25905-N23) by FWF. Thanks to Joni Sayeler and Linus Wretblad for their contributions in the SelfOptimizer project.

References

1. Baroni, M., Dinu, G., Kruszewski, G.: Don't count, predict! A systematic comparison of context-counting vs. context-predicting semantic vectors. In: Proceedings of ACL Conference (2014)
2. Berger, A., Lafferty, J.: Information retrieval as statistical translation. In: Proceedings of SIGIR (1999)
3. Cuba Gyllensten, A., Sahlgren, M.: Navigating the semantic horizon using relative neighborhood graphs. In: Proceedings of EMNLP, Lisbon, Portugal (2015)
4. De Vine, L., Zuccon, G., Koopman, B., Sitbon, L., Bruza, P.: Medical semantic similarity with a neural language model. In: Proceedings of CIKM
5. Erk, K., Padó, S.: Exemplar-based models for word meaning in context. In: Proceedings of ACL (2010)
6. Ganguly, D., Roy, D., Mitra, M., Jones, G.J.: Word embedding based generalized language model for information retrieval. In: Proceedings of SIGIR (2015)
7. Grbovic, M., Djuric, N., Radosavljevic, V., Silvestri, F., Bhamidipati, N.: Context- and content-aware embeddings for query rewriting in sponsored search. In: Proceedings of SIGIR (2015)
8. Karlgren, J., Bohman, M., Ekgren, A., Isheden, G., Kullmann, E., Nilsson, D.: Semantic topology. In: Proceedings of CIKM Conference (2014)
9. Karlgren, J., Holst, A., Sahlgren, M.: Filaments of meaning in word space. In: Macdonald, C., Ounis, I., Plachouras, V., Ruthven, I., White, R.W. (eds.) ECIR 2008. LNCS, vol. 4956, pp. 531–538. Springer, Heidelberg (2008). doi:10.1007/978-3-540-78646-7_52
10. Kiela, D., Hill, F., Clark, S.: Specializing word embeddings for similarity or relatedness. In: Proceedings of EMNLP (2015)
11. Koopman, B., Zuccon, G., Bruza, P., Sitbon, L., Lawley, M.: An evaluation of corpus-driven measures of medical concept similarity for information retrieval. In: Proceedings of CIKM (2012)
12. Kruszewski, G., Baroni, M.: So similar and yet incompatible: toward automated identification of semantically compatible words. In: Proceedings of NAACL (2015)
13. Mikolov, T., Chen, K., Corrado, G., Dean, J.: Efficient estimation of word representations in vector space. arXiv preprint arXiv:1301.3781 (2013)
14. Mitra, B.: Exploring session context using distributed representations of queries and reformulations. In: Proceedings of SIGIR (2015)
15. Ponte, J.M., Croft, W.B.: A language modeling approach to information retrieval. In: Proceedings of SIGIR (1998)

16. Rekabsaz, N., Bierig, R., Ionescu, B., Hanbury, A., Lupu, M.: On the use of statistical semantics for metadata-based social image retrieval. In: Proceedings of CBMI Conference (2015)
17. Rekabsaz, N., Lupu, M., Hanbury, A.: Generalizing translation models in the probabilistic relevance framework. In: Proceedings of CIKM (2016)
18. Sakai, T.: Alternatives to bpref. In: Proceedings of SIGIR (2007)
19. Schnabel, T., Labutov, I., Mimno, D., Joachims, T.: Evaluation methods for unsupervised word embeddings. In: Proceedings of EMNLP (2015)
20. Severyn, A., Moschitti, A.: Learning to rank short text pairs with convolutional deep neural networks. In: Proceedings of SIGIR (2015)
21. Tsvetkov, Y., Faruqui, M., Ling, W., Lample, G., Dyer, C.: Evaluation of word vector representations by subspace alignment. In: Proceedings of EMNLP (2015)
22. Vulić, I., Moens, M.-F.: Monolingual and cross-lingual information retrieval models based on (bilingual) word embeddings. In: Proceedings of SIGIR (2015)

Ancient Roman Coin Retrieval: A Systematic Examination of the Effects of Coin Grade

Callum Fare and Ognjen Arandjelović[(✉)]

University of St Andrews, St Andrews, UK
ognjen.arandjelovic@gmail.com

Abstract. Ancient coins are historical artefacts of great significance which attract the interest of scholars, and a large and growing number of amateur collectors. Computer vision based analysis and retrieval of ancient coins holds much promise in this realm, and has been the subject of an increasing amount of research. The present work is in great part motivated by the lack of systematic evaluation of the existing methods in the context of coin grade which is one of the key challenges both to humans and automatic methods. We describe a series of methods – some being adopted from previous work and others as extensions thereof – and perform the first thorough analysis to date.

1 Introduction

The present is an exciting time for computer vision: the field itself has matured, the hardware needed to support developed algorithms is affordable and pervasive, and the potential user base is greater than ever owing to the increasing recognition of the benefits that machine intelligence can offer. This technological and social climate has opened a vast field of potential new applications for computer vision, with many attractive and exciting problems emerging from its applications in arts and humanities. In this work we are interested in the application of computer vision to ancient numismatics.

1.1 Terminology

Considering the interdisciplinary nature of the present paper, it is important to explain the relevant numismatic terminology so that the specific task at hand and its challenges can be clearly understood. A succinct summary is presented next.

Firstly, when referring to a *coin*, the reference is made to a specific physical object i.e. a specimen. This is to be contrasted with a coin *type*. A coin type is a more abstract concept which is characterized by the semantic features shown on both sides of the coin (the obverse i.e. the "front", and the reverse i.e. the "back"). Multiple coins of the same type have the same visual elements e.g. the head or bust of a particular emperor with specific clothing (e.g. drapery or cuirass, crowned or laureate) and *legends* (textual inscriptions), a particular reverse motif etc. Notice that although the visual elements on coins of the same type are semantically the same, their depictions may differ somewhat. The reason lies in the fact that the same coin type was minted using dies created by different engravers. For example, observe in Fig. 1 which shows three specimens of the same type, that the spatial arrangements of the legend (by definition the same in all cases) is different between the very fine example in Fig. 1(b) and the extra fine example in Fig. 1(c). In the former case the break (space) in the legend is AEQVITA-SAVG, and in the latter AEQVI-TASAVG. Nevertheless the type is the same.

Condition Grades. As noted previously, at the focal point of the present work is the *condition* of a coin. Succinctly put the condition describes the degree of preservation of a coin, or equivalently the amount of damage it suffered since it was minted. The usual grading scale adopted in ancient numismatics includes the following main grades: (i) poor, (ii) fair, (iii) good, (iv) very good, (v) fine, (vi) very fine, and (vii) extremely fine. Virtually universally (i.e. save for extremely rare coin types) only the last three are considered of interest to collectors, that is fine (F), very fine (VF), and extremely fine (EF or XF). Note that less frequently used transitional grades can be derived from the main seven by qualifiers e.g. near or almost fine (nF, aF), better than fine (F+) etc.

An ancient coin in a fine condition displays all the main visual elements of the type, as illustrated with an example in Fig. 1(a). A very fine coin also has more subtle elements preserved such as clothing creases as exemplified in Fig. 1(b). An extremely fine condition coin is in approximately the same condition in which it was when it was minted, showing the entirety of the original detail, as can be seen in Fig. 1(c).

Miscellaneous. In order to appreciate the challenge of the task at hand, it is important to recognize a number of factors other than the condition which affect the appearance of a coin. These include die *centring*, surface metal changes (due to oxidation or other chemical reactions), and die wear.

Die centring refers to the degree to which the centre of the die coincides with the centre of the actual piece of metal against which it is struck to create the coin. A coin with poor centring may have salient design elements missing e.g. a part of the legend. An example of a somewhat poorly centred obverse can be seen in Fig. 1(c) and of a reverse in Fig. 2(a).

Depending on the presence of different substances in a coin's environment (soil, air etc.), the surface metal can change its colour and tone as it reacts with chemicals it is exposed to. Observe the difference in the tone of the coins in Fig. 1 as well as of those in Fig. 2.

Finally, it is worth noting that the appearance of a coin can be affected by die wear. Just as coins experience physical damage when handled and used, repeated use of a die in the minting process effects damage on the die. To a non-trained eye a coin minted with a worn die can seem identical to a worn coin minted with an intact die. However, a reasonably skilled (but not necessarily expert) numismatist can readily make a distinction, as subtler patterns of damage in the two cases are quite unlike one another. In addition, close inspection and the presence of oxidation or particles in ridges can be used for conclusive verification.

1.2   Previous Work

Most early and some more recent attempts at the use of computer vision for coin analysis have concentrated on modern coins [1–3]. This is understandable considering that modern coins are machine produced and as such pose less of a challenge than ancient coins. Modern coins do not exhibit variation due to centring issues, shape, different depictions of semantically identical elements, etc. From the point of view of computer vision, two modern coins at the time of production are identical. This far more restricted problem setting allows for visual analysis to be conducted using holistic representations such as raw appearance [4] or edges [5], and off-the-shelf learning methods such as principal component analysis [4] or conventional neural networks [6]. However such approaches offer little promise in the context of ancient numismatics.

The existing work on computer vision based ancient coin analysis can be categorized by the specific problem addressed as well as by the technical methodology. As regards the former categorization, some prior work focuses on coin instance recognition i.e. the recognition of a specific coin rather than a coin type. This problem is of limited practical interest, its use being limited to such tasks as the identification of stolen coins or the detection of repeated entries in digital collections. Other works focus on coin type recognition, which is a far more difficult problem [7–9]. Most of these methods are local feature based, employing local feature descriptors such as SIFT [10] or SURF [11]. The reported performance of these methods has been rather disappointing and a major factor appears to be the loss of spatial, geometric relationship in the aforementioned representations [12,13]. In an effort to overcome this limitation, a number of approaches which divide a coin into segments have been described [14]. These methods implicitly assume that coins have perfect centring, are registered accurately, and are nearly circular in shape. None of these assumptions are realistic. The sole method which does not make this set of assumptions builds meta-features which combine local appearance descriptors with their geometric relationships [9]. Though much more successful than the alternatives, the performance of this method is still insufficiently good for most practical applications.

All of the aforementioned work shares the same limitation of little use of domain knowledge. In particular, the general layout of the key elements of Roman imperial coins is generally fixed, save for few rare exceptions. Hence it makes sense to try to use this knowledge in analysis. The few attempts in the existing literature generally focus on the coin legend [15]. In broad terms this appears sensible as the legend carries a lot of information, much of which is shared with the coin's pictorial elements. For example, the obverse legend in almost all cases contains the name of the emperor depicted, and the reverse the name of the deity shown. The denarius of Antoninus Pius with Aequitas (goddess of justice and equality) in Fig. 1 illustrates this well, the obverse legend being **ANTONINVS**AVG**PIVS**PPTRPCOSIII, and the reverse **AEQVITAS**AVG. However, in spite of this, methods such as that described in [15] offer little promise for practical use. The key reason for this lies in the fact that the legend, with its fine detail, is one of the first elements of the coin to experience damage

(a) Fine   (b) Very fine   (c) Extra fine

Fig. 1. Specimens of Antoninus Pius's denarius (RIC 61) from our data set.

and wear. Coins with clearly legible legends are generally expensive and rare, and thus of little interest to most collectors. They are also the easiest to identify, by the very nature of their good preservation, and hence do not represent the target data well. Consequently, this class of algorithms is not of interest in the present paper.

The main purpose of this work is to provide a clear picture of the performance of existing methods on data representative of images likely to be used in practice. Moreover our aim is to provide the first systematic evaluation which looks specifically at the effects that coin grade has on coin type recognition accuracy. In particular, all work to date has been highly unstructured and *ad hoc* in its evaluation methodology. Some authors use data sets with coins in different conditions and unstated distributions thereof [9], and others very small data sets with coins in extremely rare, museum grade [8]. Hence the current understanding of different methods' behaviour is not very well understood at all.

2   Data

As noted in the previous section, one of the key motivating factors for the present work can be found in the lack of systematic evaluation of different algorithms described in the literature with respect to the condition of the coins present in the specific data sets used. Given that the condition of a coin by definition affects the visibility and even the very presence of elements depicted on the coin, it is unsurprising that it is a major factor which governs the ease (or lack thereof) that a human experiences when attempting to identify a coin. Understanding the behaviour of different methods when presented with this challenge, and in particular the effects of both the condition of the query coin as well as of the distribution of coin conditions in the so-called gallery corpus, should be a crucial consideration in directing future research efforts.

At this point in time there does not exist a data set structured in a manner which allows for the analysis outlined above to be conducted: none of the corpora used in previous work can be readily adopted for use to this end, nor are there any other readily available sources, to the best of our knowledge. Hence we collected a novel data set which we introduce for the first time in this paper – it will be made freely available after anonymity is no longer needed.

We collected our data by searching for images of coins sold by well known auction houses. In this manner we achieved two goals. Firstly, we could ensure

(a) Fine   (b) Very fine   (c) Extra fine

Fig. 2. Specimens of Caligula's denarius (RIC 2) from our data set.

(a) Fine   (b) Very fine   (c) Extra fine

Fig. 3. Specimens of Octavian's denarius (RIC 102) from our data set.

that the images are in the public domain and can thereafter be shared without restriction. Secondly, having been put up for sale by well known auction houses, the coins have been graded by professionals allowing us to associate reliable meta data with all images.

We collected 600 images in total. These represent 100 types of Roman imperial denarii, with six exemplars for each type: two in fine condition, two in very fine, and two in extremely fine. The period covered by the coins included in the data set starts with the beginning of the Empire and the rule of Octavian in 27 BC and ends with the end of the rule of Philip II (Philip the Arab) in 249 AD when the denarius ceases to be used due to economic and political crises. A few representative examples of different coin types in different grades from our corpus are shown in Figs. 1, 2 and 3.

3   Methods

This section describes the methods used in our evaluation.

3.1   Histogram Distance Measures

Given that all methods described in the sections that follow (and indeed most of the methods in the existing literature on computer vision based ancient coin analysis) employ histogram based representations, we start by detailing the histogram distance measures used in our experiments.

For the sake of continuity with previous work, the first distance measure we adopt is the standard Euclidean distance [16]. Given two L_2 normalized histograms $\mathbf{h}_1$ and $\mathbf{h}_2$, defined over some vocabulary of size n, the Euclidean distance $d_E(\mathbf{h}_1, \mathbf{h}_2)$ between them is given by the following expression:

$$d_E(\mathbf{h}_1, \mathbf{h}_2)^2 = \sum_{i=1}^{n} [h_1(i) - h_2(i)]^2, \tag{1}$$

where $h_j(i)$ denotes the i-th entry in histogram $\mathbf{h}_j$.

Notwithstanding the widespread use of the Euclidean distance metric, a recently proposed alternative which is based on Hellinger distance has universally been shown to yield superior performance [17]. The metric is just as simple and efficient to evaluate as the Euclidean one and is given by:

$$d_H(\mathbf{h}_1, \mathbf{h}_2)^2 = \sum_{i=1}^{n} \left[\sqrt{h_1(i)} - \sqrt{h_2(i)} \right]^2, \tag{2}$$

with an important difference that histograms should be L_1 normalized.

3.2 Baseline SIFT

The first algorithm we implemented and evaluated in this work is what we term 'Baseline SIFT' on the account of its widespread use in the existing literature [10,18]. As different elements of this algorithm are employed by the other approaches we also evaluated, we explain the key steps in some detail. In summary, Baseline SIFT first creates a visual dictionary by clustering SIFT descriptors from the coin gallery, uses the constructed dictionary to represent a single coin as a histogram of visual words, and performs matching using one of the distance metrics described previously.

Visual Dictionary Construction. Baseline SIFT starts the construction of a visual dictionary by detecting keypoints and extracting the corresponding SIFT descriptors from all coin images in the gallery of 'known' coins [10]. The extracted descriptors are then clustered using k-means clustering, with the parameter k set *a priori* (we will discuss the choice of k shortly). Given the stochastic nature of k-means, in order to obtain the best (most descriptive, for a set value of k) clustering, in this work we perform several clustering attempts and of those choose the one with the least average L_2 error measured between individual descriptors and their assigned cluster centres. The final k cluster centres are deemed the visual vocabulary which allows a single image of a coin to be represented using a fixed length representation. In particular, given the set of SIFT descriptors from a single coin, each descriptor is taken to be a representative of the visual word given by the closest cluster centre, and the entirety of the coin image represented by a histogram over the visual vocabulary.

On the Choice of Vocabulary Size. The choice of the visual vocabulary size k is an interesting and practically important one. Two different views on the approach taken can be put forward. A large value of k, commonly used in instance retrieval applications [19], can be considered as a way of hashing and matching noisy descriptors. Alternatively, smaller values of k, more often used for object class recognition, can be seen as a means of generalization. Although this choice has not been explicitly discussed in the existing literature of automatic ancient coin analysis, implicitly the latter view seems to be dominant [9]. Considering the lack of systematic analysis to date, we take no *a priori* stance and instead conduct experiments for a range of values of k.

3.3   Wedge SIFT

As discussed previously, Baseline SIFT and similar approaches suffer from a major limitation caused by the loss of spatial information. A specific SIFT descriptor affects the overall representation of the coin in the same manner regardless of its absolute location or indeed location relative to other descriptor loci. A number of methods in the literature attempt to address this problem by dividing a coin into segments. Given the approximately circular shape of coins, a natural way of segmenting a coin is into radial segments or wedges, constructing a histogram for each segment, and concatenating these into a single, higher dimensional vector used to represent the entirety of a coin [20]. We refer to this method as Wedge SIFT. An example is shown in Fig. 4(a) which displays keypoint loci colour coded for the corresponding segments, using four wedges.

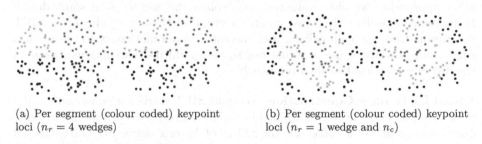

(a) Per segment (colour coded) keypoint loci ($n_r = 4$ wedges)

(b) Per segment (colour coded) keypoint loci ($n_r = 1$ wedge and n_c)

Fig. 4. Different coin segmentation approaches used to incorporate geometric information by grouping spatially related keypoints and the corresponding local appearance descriptors. (Color figure online)

3.4   Soft Wedge SIFT

All existing methods in the literature which attempt to combine sparse local feature based appearance information with geometric information by segmenting a coin do so using manually predefined segments and 'hard' segment membership i.e. a specific keypoint and the corresponding descriptor are strictly considered either to fall within a segment or not. Several problems emerge from this approach. Firstly, these methods implicitly assume that coins are perfectly registered both in terms of their translation and rotation. This is difficult to achieve by automatic means and indeed none of the existing work discusses this challenge. Yet, to perform this manually defeats the very premise of automatic coin analysis. What is more, the problem of exact rotational alignment is not even well posed as it is not objectively clear what the precise 'up' direction is in the first place. The hard membership of features compounds this problem – even a slight misalignment, translational or rotational, can greatly affect feature distributions in different segments. Hence in this work we also evaluate an extension of the original method by allowing soft feature membership within a wedge. In particular, we apply weighting using a triangular fuzzy number function which reaches its maximum value of 1 for a feature at the centre of a wedge (i.e. in the bisecting direction from the centre of the coin) and its minimum of 0 at the centres of the two neighbouring wedges [21, 22].

3.5 Wedge-Sector SIFT

The Wedge SIFT approach creates segments by diving the image of a coin (that is, more precisely, the images of its obverse and reverse) using radial boundaries. A complementary dividing methodology uses concentric circular boundaries using different radii, creating sectors. Therefore this algorithm has two free parameters, namely the number of radial boundaries n_r and the number of circular boundaries n_c, thereby dividing a coin into $n_r \times n_c$ segments. An example of keypoint assignments is shown in Fig. 4(b), for clarity using $n_r = 1$ wedges and $n_c = 4$ sectors.

3.6 Soft Wedge-Sector SIFT

The hard membership based Wedge-Sector SIFT segmentation suffers from the same limitations as those highlighted in the case Wedge SIFT. Hence we apply the same idea of soft membership of local features in Wedge-Sector SIFT by using weighting both in radial and angular directions.

3.7 Local Binary Patterns (LBPs)

Most methods in the existing literature on computer vision based ancient coin analysis rely on the use of sparse, local features. This is a reasonable choice given that the precise geometric layout between different elements of the same coin type can vary considerably across specimens minted with different dies. However, state of the art performance in problem domains where similar geometric flexibility is present, such as face recognition, has been achieved with the use of dense local features in the form of local binary patterns (LPBs) [23]. The LPB representation has proven to be very effective across a range of applications, including texture and face recognition [24, 25], and numerous others.

The elementary local LBP descriptor considers an image patch of size 3×3 pixels. By comparing the values of the 8 neighbouring pixels with the value of the central pixel, the neighbourhood is mapped to a series of binary digits (0 or 1) depending on whether a specific pixel has a smaller value than the central pixel or not, as illustrated. The 8 bit sequence corresponds to an integer in the range $[0, 127]$ and describes the local appearance. The description of an entire image (or a region of interest within it) is then obtained by creating histograms over local LBP descriptors within blocks into which the image is divided (the number of blocks is a free parameter, examined empirically in the next section).

3.8   Local Ternary Patterns (LTPs)

The thresholding of pixels at the heart of LBPs, similarly to the hard thresholding in terms of the spatial layout of local features discussed previously, is vulnerable to small perturbations when neighbouring pixels have values close to the central pixel. A generalization of the LPB descriptor in the form of a local ternary pattern (LTP) has demonstrated effectiveness in addressing this problem. In particular, instead of mapping neighbourhood pixels to binary digits, to produce a LTP the mapping is done to a ternary digit i.e. without loss of generality, to 0, 1, or 2. A pixel is mapped onto 0 or 2 respectively if its value is smaller or greater than that of the central pixel by at least a certain amount (this threshold is a free parameter), and to 1 otherwise. The latter, additional value can be seen as representing neighbourhood pixels sufficiently similar to the central one. The remainder of the method, that is the aggregation of local descriptors into histograms over blocks, and the concatenation of these to form a holistic representation, is performed just as in the original LBP based method.

4   Experiments

In this section we present our experiments. Specifically, we begin with a summary of automatic data preprocessing needed to prepare images for use in the algorithms described in the previous section, go on to explain our experimental methodology, and finally present and discuss our findings.

4.1   Automatic Data Preprocessing

The images in our database, being originally acquired for use at auctions, require additional processing before they can be used by the methods described in the previous section: their size is non-uniform, the photographs of a coin's obverse and reverse are shown within the same image and their locations are not in *a priori* known locations within the image. Here we describe a series of automatic pre-processing steps which normalize for these confounding sources of variation. In summary, we (i) detect and segment out image regions which correspond to the obverse and the reverse, (ii) estimate the size of the coin in the segmented images, and (iii) perform image rescaling to the canonical scale.

Obverse/Reverse Segmentation. The first step of our pre-processing pipeline concerns the separation of a coin's obverse and its reverse. This is achieved by keypoint localization using a Gaussian scale-space as described by Lowe [10], and then by clustering the loci using k-means for $k = 2$. This process readily leads to the identification of image areas which correspond to the two sides of a coin, as illustrated in Fig. 5.

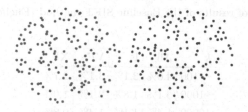

Fig. 5. Keypoint detection and clustering using k-means with $k = 2$ readily allows the areas of the image which correspond to the two coin sides to be separated. In this image keypoint loci are colour coded by their final cluster assignments. (Color figure online)

Scale Normalization. The second pre-processing step we conduct concerns image scale canonization i.e. rescaling to the uniform scale. This is needed because higher resolution images tend to produce higher numbers of keypoints which can clearly affect the representations described in the previous section. Following k-means clustering of loci of keypoints detected in raw images, coin scale can be determined in a simple manner. In particular we consider the median of the cluster to be the centre of the coin and estimate the average diameter of the coin by computing the mean distance of convex hull defining keypoints from this centre.

4.2 Experimental Methodology

In order to facilitate as thorough understanding of the effects of coin grade on the performance of different methods as possible, we conducted a series of experiments which vary the grade of coins used as query and as gallery. In particular, we evaluated all algorithms first using three experiments in which different coin grades (F, VF, or XF) were used to query a gallery with all conditions of coins present in it, and then three further experiments in which a query coin of a specific grade was matched against a gallery which includes coins of that grade only.

4.3 Results and Discussion

We began our analysis by looking at the simplest, Baseline SIFT method. The key results using different parameter values (in particular, the visual vocabulary size) for the two adopted histogram metrics are summarized in Tables 1 and 2. The first observation that emerges from the tables concerns the poor performance of the method which correctly recognized no more than 5% of query coins in the best case. This is consistent with previous reports in the literature [9], with the method showing any promise only in the context of the far simpler problem of coin instance recognition.

Table 1. Summary of results of the Baseline SIFT method (Euclidean histogram distance).

k	All	F	VF	XF
100	4.0%	3.1%	3.1%	2.7%
1000	3.1%	1.8%	2.1%	1.0%
10000	1.4%	1.1%	1.4%	0.8%

Table 2. Summary of results of the Baseline SIFT method (Hellinger histogram distance).

k	All	F	VF	XF
100	5.5%	3.4%	3.6%	3.6%
1000	3.4%	1.9%	2.6%	1.3%
10000	1.4%	1.0%	1.4%	0.6%

The second clear observation concerns the superiority of the Hellinger distance based histogram metric as compared with the more conventional Euclidean distance. This finding too is consistent with the reports in the literature on other recognition problems [17]. Therefore henceforth we adopt the use of this metric exclusively.

Next, notice that in both Tables 1 and 2, the best performances were achieved using the smallest value of k. This supports the idea of using coarse feature discretization as a means of providing generalization robustness, as discussed previously. Considering that the same trend was found in all experiments we conducted, henceforth all reported results are for $k = 100$.

As expected, results superior to those obtained with Baseline SIFT were attained through the use of geometric information and the two variants of Wedge SIFT, as illustrated in Tables 3 and 4. Nevertheless, on the absolute scale in the context of practical applicability, the recognition rates remain poor, not exceeding 10%.

Unsurprisingly, recognition was worst when poorest condition coins (F) were used. Interestingly though, in the case of the original Wedge SIFT, the use of the best condition coins (XF) did not effect an improvement over medium grade coins (VF) – rather, the performance worsened. A possible explanation for this may lie in the greater number of keypoints and the corresponding features detected in extremely fine coins. These features often correspond to idiosyncratic details specific to individual die engravers, rather than discriminative features in the context of coin type recognition. This is something that future research should bear in mind and which may be an interesting avenue to explore in the analysis of engraving style patterns.

Table 3. Summary of results of the Wedge SIFT method, using the Hellinger histogram distance metric and the visual vocabulary size $k = 100$ (see Table 2).

n_r	All	F	VF	XF
2	7.4%	6.6%	9.3%	6.2%
3	6.4%	7.7%	6.9%	4.8%
4	5.3%	5.6%	6.5%	3.8%

Table 4. Summary of results of the Soft Wedge SIFT method, using the Hellinger histogram distance metric and the visual vocabulary size $k = 100$ (see Table 2).

n_r	All	F	VF	XF
2	7.9%	6.6%	8.3%	8.7%
3	8.2%	6.4%	8.8%	9.1%
4	8.7%	6.1%	9.7%	10.1%

Table 5. Summary of results of the Sector SIFT method, using the Hellinger histogram distance metric and the visual vocabulary size $k = 100$ (see Table 2).

n_c	All	F	VF	XF
2	6.8%	5.6%	8.3%	6.2%
3	7.2%	6.1%	8.8%	6.7%
4	5.3%	7.1%	5.1%	3.8%

Table 6. Summary of results of the LBP based method.

n_b	All	F	VF	XF
3	3.6%	3.6%	4.6%	2.4%
4	4.8%	4.6%	5.1%	4.8%
5	4.4%	4.6%	4.6%	3.8%

Finally, the results obtained using the two Sector SIFT methods are summarized in Tables 5 and 6. While both methods performed better than Baseline SIFT, their recognition rates were lower than of Wedge SIFT and Soft Wedge SIFT. We expect that the key reason lies in the weaker geometric constraint imposed by this representation – in the context of the problem at hand, angular displacement is more informative than radial, as well as less sensitive to the precise localization of a coin's centre. Indeed, we highlighted the importance of the latter in our coverage of previous work, and the lack of consideration thereof in the previous work which proposed methods predicated on this information (Table 7).

Table **7.** Summary of results of the LTP based method.

t	All	F	VF	XF
5	5.6%	4.1%	6.9%	5.8%
6	5.8%	4.6%	6.0%	6.7%
7	4.5%	4.1%	4.2%	5.3%

5 Conclusions and Future Work

In this paper we focused on the problem of recognizing Roman imperial denarii
– a difficult computer vision problem which is of much interest to communities
interested in ancient numismatics. In particular our work was motivated by the
lack of systematic evaluation of the effects that coin grade has on the performance
of different algorithms. We described a series of different methods, some adopted
from previous work and others proposed as extensions thereof, and performed the
first thorough analysis in the existing literature. Our findings demonstrate the
difficulty of the problem and suggest that the existing methods still perform very
poorly on real world data. We analysed and discussed the behaviour of different
algorithms and their parameters, and highlighted a series of observations which
should guide future work. In particular our results suggest a focus on the use of
prior knowledge of the coin layout and line or edge based features [26,27] rather
than appearance.

References

1. Davidsson, P.: Coin classification using a novel technique for learning characteristic decision trees by controlling the degree of generalization. In: International
   Conference on Industrial and Engineering Applications of Artificial Intelligence
   and Expert Systems (1996)
2. Nölle, M., et al.: Dagobert - a new coin recognition and sorting system. In: International Conference on Digital Image Computing (2003)
3. Pan, X., Tougne, L.: Topology-based character recognition method for coin date
   detection. In: CAIP (2016)
4. Huber, R., et al.: Classification of coins using an eigenspace approach. PRL **26**(1),
   61–75 (2005)
5. van der Maaten, L., Boon, P.: COIN-O-MATIC: a fast system for reliable coin
   classification. In: MUSCLE CIS Coin Recognition Competition Workshop (2006)
6. Mitsukura, Y., et al.: Design and evaluation of neural networks for coin recognition
   by using GA and SA. In: IJCNN (2000)
7. Zaharieva, M., Kampel, M., Zambanini, S.: Image based recognition of ancient
   coins. In: Kropatsch, W.G., Kampel, M., Hanbury, A. (eds.) CAIP 2007. LNCS, vol.
   4673, pp. 547–554. Springer, Heidelberg (2007). doi:10.1007/978-3-540-74272-2_68
8. Kampel, M., Zaharieva, M.: Recognizing ancient coins based on local features. In:
   Bebis, G., et al. (eds.) ISVC 2008. LNCS, vol. 5358, pp. 11–22. Springer, Heidelberg
   (2008). doi:10.1007/978-3-540-89639-5_2

9. Arandjelović, O.: Automatic attribution of ancient Roman imperial coins. In: CVPR (2010)

10. Lowe, D.G.: Distinctive image features from scale-invariant keypoints. IJCV **60**, 91–110 (2003)

11. Bay, H., et al.: SURF: speeded up robust features. CVIU **110**, 346–359 (2008)

12. Rieutort-Louis, W., Arandjelović, O.: Bo(V)W models for object recognition from video. In: IWSSIP (2015)

13. Arandjelović, O.: Object matching using boundary descriptors. In: BMVC (2012)

14. Anwar, H., et al.: Coarse-grained ancient coin classification using image-based reverse side motif recognition. MVA **26**, 295–304 (2015)

15. Arandjelović, O.: Reading ancient coins: automatically identifying denarii using obverse legend seeded retrieval. In: Fitzgibbon, A., Lazebnik, S., Perona, P., Sato, Y., Schmid, C. (eds.) ECCV 2012. LNCS, vol. 7575, pp. 317–330. Springer, Heidelberg (2012). doi:10.1007/978-3-642-33765-9_23

16. Arandjelović, O.: Matching objects across the textured-smooth continuum. In: ACRA (2012)

17. Arandjelović, R., Zisserman, A.: Three things everyone should know to improve object retrieval. In: CVPR (2012)

18. Rieutort-Louis, W., Arandjelović, O.: Description transition tables for object retrieval using unconstrained cluttered video acquired using a consumer level hand-held mobile device. In: IJCNN (2016)

19. Sivic, J., Zisserman, A.: Video Google: a text retrieval approach to object matching in videos. In: ICCV (2003)

20. Anwar, H., Zambanini, S., Kampel, M.: Supporting ancient coin classification by image-based reverse side symbol recognition. In: Wilson, R., Hancock, E., Bors, A., Smith, W. (eds.) CAIP 2013. LNCS, vol. 8048, pp. 17–25. Springer, Heidelberg (2013). doi:10.1007/978-3-642-40246-3_3

21. Pedrycz, W.: Why triangular membership functions? Fuzzy Sets Syst. **64**, 21–30 (1994)

22. Bidder, O.R., et al.: A risky business or a safe BET? A fuzzy set event tree for estimating hazard in biotelemetry studies. Anim. Behav. **93**, 143–150 (2014)

23. Tang, H., et al.: 3D face recognition using local binary patterns. SP **93**, 2190–2198 (2013)

24. Heikkilä, M., et al.: Description of interest regions with local binary patterns. PR **42**, 425–436 (2009)

25. Ghiass, R.S., et al.: Infrared face recognition: a literature review. IJCNN (2013)

26. Arandjelović, O.: Gradient edge map features for frontal face recognition under extreme illumination changes. In: BMVC (2012)

27. Arandjelović, O.: Making the most of the self-quotient image in face recognition. In: FG (2013)

A Systematic Analysis of Sentence Update Detection for Temporal Summarization

Cristina Gârbacea[1] and Evangelos Kanoulas[2(✉)]

[1] University of Michigan, Ann Arbor, MI, USA
garbacea@umich.edu
[2] University of Amsterdam, Amsterdam, The Netherlands
e.kanoulas@uva.nl

Abstract. Temporal summarization algorithms filter large volumes of streaming documents and emit sentences that constitute salient event updates. Systems developed typically combine in an ad-hoc fashion traditional retrieval and document summarization algorithms to filter sentences inside documents. Retrieval and summarization algorithms however have been developed to operate on static document collections. Therefore, a deep understanding of the limitations of these approaches when applied to a temporal summarization task is necessary. In this work we present a systematic analysis of the methods used for retrieval of update sentences in temporal summarization, and demonstrate the limitations and potentials of these methods by examining the retrievability and the centrality of event updates, as well as the existence of intrinsic inherent characteristics in update versus non-update sentences.

Keywords: Temporal summarization · Content analysis · Event modeling

1 Introduction

Monitoring and analyzing the rich and continuously updated content in an online environment can yield valuable information that allows users and organizations gain useful knowledge about ongoing events and consequently, take immediate action. News streams, social media, weblogs, and forums constitute a dynamic source of information that allows individuals, corporations and government organizations not only to communicate information but also to stay informed on "what is happening right now". The dynamic nature of these sources calls for effective ways to accurately monitor and analyze the emergent information present in an online streaming setting.

TREC Temporal Summarization (TS) [5] facilitates research in monitoring and summarization of information associated with an event over time. Given an event query, the event type[1], and a high volume stream of input documents

---

[1] TREC TS focuses on large events with a wide impact, such as natural catastrophes (storms, earthquakes), conflicts (bombings, protests, riots, shootings) and accidents.

© Springer International Publishing AG 2017
J.M. Jose et al. (Eds.): ECIR 2017, LNCS 10193, pp. 424–436, 2017.
DOI: 10.1007/978-3-319-56608-5_33

discussing the event, a temporal summarization system is required to emit a series of event updates, in the form of sentences, over time, describing the named event. An optimal summary covers all the essential information about the event with no redundancy, and each new piece of information is added to the summary as soon as it becomes available.

Temporal summarization systems typically use a pipelined approach, (a) filtering documents to discard those that are not relevant to the event, (b) ranking and filtering sentences to identify those that contain significant updates around the event, and (c) deduplicating/removing redundant sentences that contain information that has already been emitted; some examples of the afore-described pipeline constitute systems submitted to TREC TS in past years [5]. In this work we are only focusing on identifying potential update sentences and their retrieval from a large corpus for summarization purposes; that is, we assume that all incoming documents are relevant to the event under consideration, and we deliberately choose to ignore the past history of what event updates have been emitted by the summarization system. These assumptions, which are ensured by the construction of our experiments and evaluations, provide a decomposition of the temporal summarization problem and allow a focus on fundamental theories behind understanding what constitutes a potential event update (from now on simply referred as update) and what not. We leave the study of the interplay of the three components as future work.

Event update identification algorithms fall under one of the categories below, or apply a combination of these methods [10]:

1. **Retrieval algorithms** that consider event updates as passages to be retrieved given a event query;
2. **Event update centrality** algorithms that assume update sentences are central in the documents that contain them, and hence algorithms that can aggregate sentences should be able to identify them;
3. **Event update modeling** methods that consider events bear inherit characteristics that are not encountered in non-update sentences, and hence algorithms that model event updates should be able to predict whether a sentence is an update or not.

In this work we present a systematic analysis of the limitations and potentials of the three approaches. We do **not** devise any new algorithm towards temporal summarization; our goal is to obtain a deeper **understanding of how and why** the aforementioned approaches fail, and what is required for a successful temporal summarization system. We believe that such an analysis is necessary and can shed light in developing more effective algorithms in the future.

The remainder of this paper is organized as follows: Sect. 2 describes prior initiatives and methods for temporal summarization of news events, Sect. 3 discusses the experimental design of our study, Sect. 4 describes the experimental results, and provides an analysis of these results around the limitations of the methods being tested, and last Sect. 5 outlines the conclusions of our work as well as future directions informed by these conclusions.

2   Related Work

Events play a central role in many online news summarization systems. Topic Detection and Tracking [2] has focused on monitoring broadcast news stories and issuing alerts about seminal events and related sub-events in a stream of news stories at document level. To retrieve text at different granularities, passage retrieval methods have been widely employed; see TREC HARD track [1] and INEX adhoc [11] initiatives for an overview. Passages are typically treated as documents, and existing language modeling techniques that take into account contextual information, the document structure or the hyperlinks contained inside the document are adapted for retrieval.

Single and multi-document summarization have been long studied by the natural language processing and information retrieval communities [4,10]. Such techniques take as input a set of documents on a certain topic, and output a fixed length summary of these documents. Clustering [20], topic modeling [3], and graph-based [7,14] approaches have been proposed to quantify the salience of a sentence within a document. McCreadie et al. [13] combine traditional document summarization methods with a supervised regression model trained on features related to the prevalence of the event, the novelty of the content, and the overall sentence quality. Kedzie et al. [12] also employ a supervised approach to predict the salience of sentences. Features combined include basic sentence quality features, query features, geotags and temporal features, but also features that represent the contrast between a general background corpus language model and a language model per event category. Gupta et al. [9] use background and foreground corpora to weight discriminative terms for topic-focused multi-document summarization. Finally, Chakrabarti et al. [6] and Gao et al. [8] combine evidence from event news and social media to model the different phases of an event using Hidden Markov Models and Topic Models.

3   Experimental Design

In this section we describe the experimental design used for our analysis. We consider three different approaches that have been adopted so far towards detecting event updates: (1) retrieval algorithms, (2) event update centrality algorithms, and (3) event update modeling methods.

1. Retrieval Algorithms: The primary goal of the experiments is to identify the limitations of retrieval algorithms towards temporal summarization of events. In the designed experiments we want to be as indifferent as possible to any particular retrieval algorithm; hence we focus on the fundamental component of any such algorithm which is the overlap between the language of an event query and the language of an event update in terms of shared vocabulary. If an event update does not contain any query term for instance, it is impossible to be retrieved by any lexical-based relevance model. This can give us a theoretical upper bound on the number of event updates that are at all retrievable. Clearly, even if an event update contains the query terms (we call that event update *covered*) it is still likely that it may not be retrieved, if for instance the query terms are not discriminative enough to separate the update from non-updates. Hence, we focus in our analysis on discriminative terms. To identify such terms, we compute word likelihood ratios. The log-likelihood ratio (LLR) [9,19] is an approach for identifying discriminative terms between corpora based on frequency profiling. To extract the most discriminative keywords that characterize events, we construct two corpora as follows. We consider all relevant annotated sentence updates from the gold standard as our foreground corpus, and a background corpus is assembled of all the non-update sentences from the relevant documents. Afterwards, for each term in the foreground corpus we compute its corresponding LLR score. In order to quantify which are the most discriminative terms in our collection, we rank the terms in descending order of their LLR scores and consider the top-N most discriminative in the rest of our experiments.

(**Query Expansion with Similar Terms**). We further want to understand the fundamendal reason behind any language mismatch between query and event updates. A first hypothesis is that such a mismatch is due to *different lexical representation* for the same semantics. Hence, in a second experiment we expand queries in two different ways: (a) we select a number of synonym terms using WordNet [16], and (b) we use a Word2Vec [15] model trained on the set of relevant gold standard updates from TREC TS 2013 and 2014; then similar to the previous experiment we test the limitations of such an approach examining whether the expanded query terms are also event update discriminative terms.

(**Query Expansion with Relevance Feedback**). A second hypothesis is that a vocabulary mismatch is due to a *topical drift* of the event updates. Imagine the case of the *"Boston Marathon Bombing"*. Early updates may contain all the query words, however when the topic drifts to the trial of the bombers or the treatment of the injured, it is expected that there will be a low overlap between the event query and the event updates due to the diverging vocabulary used. Such a vocabulary gap would be hard to fill by any synonym or related terms. However, if one were to consider how the vocabulary of the updates changes over time, one might be able to pick up new terms from past updates that could help in identifying new updates. This is a form of relevance feedback. To assess this hypothesis, given an update, we consider all the sentence updates that have appeared in documents prior to this update. Then we examine the vocabulary overlap between this current update and discriminative terms from past updates. A high overlap would designate that one can actually gradually track topical drift.

2. **Event Update Centrality:** Here we devise a set of experiments to test whether an event update is central in the documents that contain it. If this is the case, algorithms that can aggregate sentences should be able to identify relevant and informative updates. Graph-based ranking methods have been proposed for document summarization and keyword extraction tasks [7,14]. These methods construct a sentence network, assuming that important sentences are linked to

many other important sentences. The underlying model on which these methods are based is a random walk model on weighted graphs: an imaginary walker starts walking from a node chosen arbitrarily, and from that node continues moving towards one of its neighbouring nodes with a probability proportional to the weight of the edge connecting the two nodes. Eventually, probabilities of arriving at each node on the graph are produced; these denote the popularity, centrality, or importance of each node.

(Within Document Centrality). In this first experiment we are interested in testing whether an event update is central within the document that contains it. This scenario would be the ideal, since if this is the case, centrality algorithms running on incoming documents could emit event updates in a timely manner. To this end, we use LexRank [7], a state-of-the-art graph-based summarization algorithm, and examine the ranking of event updates within each document.

We pick LexRank to assess the salience of event updates as it is one of the best-known graph-based methods for multi-document summarization based on lexical centrality. Words and sentences can be modeled as nodes linked by their co-occurrence or content similarity. The complexity of mining the word network only depends on the scale of the vocabulary used inside the documents; it is often significantly reduced after applying term filtering. LexRank employs the idea of random walk within the graph to do prestige ranking as PageRank [17] does. We rely on the MEAD summarizer [18] implementation to extract the most central sentences in a multi-document cluster.

(Across Documents Centrality). Here we perform a maximal information experiment in which we are interested in assessing the ranking of sentence updates across documents. If this is the case, it signifies that even though sentence updates appear not to be central inside single documents, they become central as information is accumulated. We are aware that devising such an algorithm would not be providing users with timely updates, however in this experiment we want to identify the upper bound of centrality-based algorithms towards event summarization. Therefore we purposefully ignore the temporal aspect.

3. Event Update Modeling: We test the hypothesis that event updates bear inherent characteristics which are not encountered in non-update sentences. If this is indeed the case, then one might be able to devise a method that uses these inherent characteristics to predict whether a sentence is an update or not. We model the inherent characteristics of a general event update as the set of terms with high log-likelihood ratio, i.e. the set of the most discriminative event terms. Since extracting the most discriminative terms for an event at hand from the gold standard annotations would result in a form of overfitting – we learn from and predict on the same dataset – we devise two experiments.

(General Event Update Modeling). In the first experiment we test the hypothesis that an event update can be distiguished from a non-update independent of the event particulars or the event type. We define a general event as any event in our collection irrespective of the event type. We use the log-likelihood ratio test to identify the most discriminative terms in event updates vs. non-updates[2]. Afterwards we examine the degree of overlap between the extracted discriminative terms with the annotated updates for each test event.

(Event-Type Update Modeling). Given that different event types may be expressed using a different vocabulary, we repeat the experiment described above considering only events that have the same event type in common (as already mentioned, event types can be natural catastrophes, conflicts, accidents, etc.). Our goal is to learn discriminative LLR terms that are specific to a particular type of event. We use the annotated sentences from the gold standard for each event type in building our foreground corpus; the background corpus is made up of all non-update sentences from the relevant documents per event type.[3]

4 Results and Analysis

4.1 Datasets

In all our experiments we use the TREC KBA 2014 Stream Corpus[4] used by the TREC 2014 TS track. The corpus (4.5 TB) consists of timestamped documents from a variety of news and social media sources, and spans the time period October 2011–April 2013. Each document inside the corpus has a timestamp representing the moment when the respective document was crawled, and each sentence is uniquely identified by the combination document identifier and positional index of the sentence inside the document.

We run our experiments on two pre-filtered collections released by the TREC TS organizers based on the KBA corpus that are more likely to include relevant documents for our events of interest. The testsets provided contain 10 event queries for the TREC TS 2013 collection (event ids 1–10), and 15 event queries for the TREC TS 2014 collection (event ids 11–25). For each event, sentences in documents have been annotated as either updates or non-updates through an in depth-pooling experiment. Each event update contains one or more critical units of information, called information nuggets. The goal of a temporal summarization system is to emit event updates that cover all information nuggets. Information nuggets were extracted from update sentences by the TREC TS co-ordinators, and were used to further identify sentence updates not included in the original pool. In our evaluation we use this extended set of updates.

4.2 Retrieval Algorithms: Are Event Updates Retrievable?

The first question we want to answer is to what extent there is a language overlap between the event queries (and query expansions) with the event updates. To get a theoretical upper bound, we first examine how many event updates

---

[2] In total we extract 8,471 unigrams and 1,169,276 bigrams using the log-likelihood ratio weighting scheme.

[3] We discard event types for which there is not enough annotated data available.

[4] http://trec-kba.org/kba-stream-corpus-2014.shtml.

contain: *(i)* at least one query term, *(ii)* at least one query term after WordNet and Word2Vec query expansion, and *(iii)* at least one query term after query expansion with all the terms from event updates found in documents prior to the current update (relevance feedback). We observe that on average *24.4%* of event updates are guaranteed never to be retrieved by a traditional retrieval algorithm; this percentage remains unchanged when the query is expanded by WordNet synonyms, while it drops to *22.7%* of updates by a query expanded with Word2Vec[5]. Examples of expansion terms are shown in Table 1. Relevance feedback when using all query terms in past event updates lowers the amount of uncovered updates to *16%* on average across all event queries. Therefore, this also signifies that the upper bound performance for retrieval algorithms reaches approximately *84%* update coverage on average. Hence, retrieval algorithms with relevance feedback might be able to account for vocabulary gap and topic drift in the description of sub-events.

Table 1. Expansion terms and their rank on the basis of the log-likelihood ratio value (−1 designates that the term does not appear in the list of extracted LLR terms).

Event id	Query term rank	Any WordNet synonym rank	Any similar Word2Vec term rank
9	(guatemala, 2), (earthquak, 24)	(guatemala, 2), (earthquak, 24)	(guatemala, 2), (quak, 16), (earthquak, 24), (philippin, 36), (hit, 64), (7.4-magnitud, 112), (strong, 138), (struck, 201), (magnitud, 238), (strongest, 368), (caribbean, 586), (temblor, 5451), (tremor, 8303)
11	(concordia, 157), (costa, 183)	(concordia, 157), (costa, 183), (rib, −1)	(concordia, 157), (costa, 183), (shipwreck, 3636), (liner, 4793), (keel, 6252), (ill-fat, 6856), (vaus, 7721), (wreck, 8070), (genoa-bas, −1), (lean, −1), (7:13, −1), (raze, −1), (rica, −1)
22	(protest, 1), (bulgarian, 96)	(protest, 1), (bulgarian, 96)	(protest, 1), (resign, 74), (bulgarian, 96), (demonstr, 132), (bulgaria, 133),(dhaka, 167), (amid, 182), (ralli, 186), (shahbag, 235), (shahbagh, 478), (finmin, 547), (borissov, 630), (revok, 1183), (borisov, 1197), (boyko, 3284), (tender, 3469), (gerb, 4517), (activist, 8055)

In order to be realistic though, we compute likelihood ratios for words in our corpus. We first consider annotated updates as our foreground corpus, and non-updates as our background corpus. We rank terms on the basis of their

---

[5] Word2Vec was trained on the set of gold standard updates from the TREC TS 2013 and TREC TS 2014 collections.

discriminative power. In Table 1, in Column 1 we report on the query terms and their rankings among the most discriminative LLR terms extracted from the TREC TS 2013 and TREC TS 2014 collections. We observe that in general, query terms appear to be ranked high up in the list of discriminative terms. We repeat the same experiment after expanding query terms with WordNet and Word2Vec synonyms, and report on the ranks of the expanded query terms inside the list of LLR terms with high discriminative power in Table 1, in Columns 2 and 3. We observe that these query expansion terms are not very discriminative in general, although Word2Vec (trained on the test set) is able to pick up some discriminative terms.

Conclusion: Event query terms are central in event updates, however they cannot cover all updates, nor are they the most discriminative terms (e.g. see *"costa concordia"* in Table 1). A temporal analysis is necessary to identify whether the language gap is more evident as the event develops, however we leave this as future work. Further, based on the afore-described observations the language gap is not due to a lexical mismatch between the query and the updates, but rather due to topic drifting. Therefore, a dynamic algorithm that can adapt the lexical representation of a query – possibly by the means of relevance feedback – could bridge this gap.

4.3 Event Update Centrality: Do Event Updates Demonstrate Centrality?

Summarization methods applied at document level assume that event updates demonstrate centrality inside the documents they appear in. In the next set of experiments we test whether it is the case that event updates demonstrate centrality characteristics. Ideally, update sentences are central and salient inside the documents they are found in. This would allow a summarization algorithm to identify updates as soon as a document has streamed in.

To assess the within-document centrality of updates we run LexRank on each incoming document. We process the LexRank output to infer rankings inside documents for the set of relevant event updates. After ranking each sentence within a document, we compute three measures: precision at rank cut-off 1, precision at 10, and R-precision, where R is the number of update sentences within the document. The results of the experiment are shown as a heatmap in Fig. 1 – the first three columns, denoted as (A)[6]. The average precision values across the two collections can be found bellow the heatmap, while Table 2 shows the average values for each collections separately. For the TREC TS 2013 collection (events 1–10), we can see that it is rarely the case that event updates make it to the top of the ranking inside single documents. However, for the TREC TS 2014 dataset (events 11–25) we observe higher precision scores, especially in the top-10 positions.

---

[6] No documents were released for event 7, hence the white row in the heatmap.

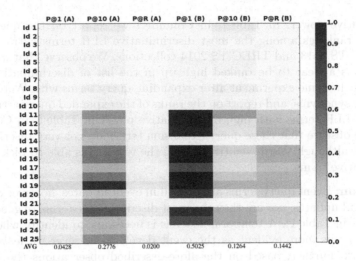

Fig. 1. Within – (A) – and across – (B) – document centrality scores based on LexRank.

To better understand the difference between the two collections we considered the case of a random algorithm that simply shuffles all sentences within a document, and ranks them by this permuted order. The intuition behind this experiment is that differences in document lengths betwen the two collections may affect the precision numbers observed - in short documents it is easier to achieve a higher precision. The *mean precision of the random algorithm at 10* for the 2013 collection is *0*, while for the 2014 collection is *0.028* - statistically significantly worse than the corresponding centrality scores. Hence, there is no clear reason for the observed differences between the two collections, and further investigation is required, that may also extend to missing judgement of sentences in the 2013 collection.

Table 2. Mean precision values for within – (A) – and across – (B) – document centrality for TREC TS datasets.

Average	P@1 (A)	P@10 (A)	P@R (A)	P@1 (B)	P@10 (B)	P@R (B)
2013	0.0045	0.0279	0.0003	0.0045	0.0279	0.0003
2014	0.0667	0.4366	0.0326	0.7151	0.1667	0.2028

We then take a retrospective look at the centrality of sentences by considering centrality scores across all relevant documents in each collection. The LexRank algorithm is now run over the entire corpus (multi-document sentence centrality), and sentences are then ranked with respect to the output scores. To make the two algorithms (within and across documents) comparable, we examine each document separately. First we rank the sentences within each document in accordance to the overall document ranking produced by LexRank, and then we compute the same three measures. The values can be seen in the form of a heatmap in Fig. 1 – the last three columns, denoted as (B).[7] First, we observe that the same pattern preserves for these two different collections. While computing centrality across documents does not change the precision values for the TREC TS 2013 dataset at all, for the TREC TS 2014 collection we can see a considerable increase. TREC TS 2014 annotated updates demonstrate centrality within and across documents, rendering them central in the development of the events under consideration. Furthermore, across-document centrality appears to bring some rather central updates at the very top of the ranked list, but within-document centrality appears to have a better effect on lower - up to 10 - ranks. Across-document centrality of sentences can also increase R-precision, demonstrating a robust behaviour.

Conclusion: Sentence centrality, when computed within a single document, does not appear to be a strong signal that can designate whether a sentence is an update or not. When computed across all documents, it consistently improves all measures. Such an algorithm, however, is not particularly useful since it has to wait for all documents to be streamed in before identifying any update sentences. One could, however, examine the minimum number of documents it takes for such a summarization algorithm before salient updates make it to the top of the ranking. We leave the construction of such an algorithm for future work.

4.4 Event Update Modeling: Do Event Updates Present Inherent Characteristics?

Given the results of the previous experiment, a hypothesis to test is whether knowing beforehand event discriminative terms can help in retrieving event updates. Clearly, different event types may have different inherent characteristics; for instance, it is likely that an event of type *accident* does not share the same characteristics as an event of type *protest*. Hence, we perform our analysis on different slices of the data.

First we create a general model of event updates by considering non-update sentences as a background corpus and update sentences as a foreground corpus. Then we compute the overlap between discriminative terms from this general model across all events and their types with the update sentences of the event under consideration. One can see in Fig. 2 – Column 1 that discriminative terms belonging to the general model appear on average in *95%* of the event updates. Note that this is not a theoretical upper bound, but rather an average case analysis, since terms with high LLR scores should in general be able to discriminate update from non-update sentences.

We repeat the same experiment, this time for each event type separately. We compute the overlap between the discriminative terms from the event type model

---

[7] For events 14, 21, 24 and 25 we cannot report on any centrality scores across relevant documents due to the size of the data and the inability of LexRank to handle it - hence the white rows in the heatmap in columns (B). The average values for the precision measures below the heatmap are computed excluding these events.

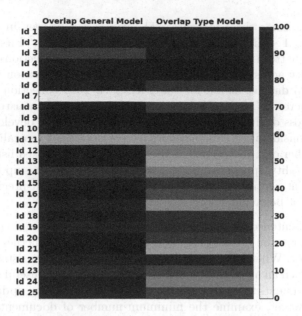

Fig. 2. Degree of overlap of discriminative terms with the TREC TS event updates.

and the annotated sentence updates, and present results for these experiments in Fig. 2 – Column 2. Interestingly, when mining event specific terms the degree of overlap drops to *72.84%* (it actually increases for the TREC TS 2013 collection to *94.28%* according to our intuition, but deteriorates for the TREC TS 2014 collection to *59.97 %*). This is against our hypothesis, as we were expecting that event specific discriminative terms will only increase the degree of overlap with the relevant sentence updates. We assume this happens due to the smaller size of the event type dataset used as a foreground corpus. The resulting event specific LLR terms are fewer but with a higher discriminative power, although we do not consider it when computing the overlap between the two models. In addition to this, we are using a fixed cut-off threshold (top 100) in our experiments for selecting terms from the discriminative list up until a specific rank. It could be that if we chose another threshold results would look different, however we leave the exploration of optimal cut-offs as future work towards devising effective algorithms.

Conclusion: Modeling event updates bears great promises towards devicing temporal summarization algorithms. It appears from our experiments that there is a number of discriminative keywords that can indicate the presence of an update sentence. The models built in this experiments somewhat overfit the data (all events were used to develop the models). A follow up experiment should perform a leave-one-out cross-validation to also test the predictive power of these terms. Nevertheless, it is clear from the results above this third approach in temporal summarization reserves more attention.

5 Conclusions

In conclusion, we have presented a systematic analysis of sentence retrieval for temporal summarization, and examined the retrievability, centrality, and inherent characteristics of event updates. We designed and ran a set of experiments on the theoretical upper bounds where possible, and on more realistic upper bounds with the use of discriminative terms obtained through likelihood ratio calculations. Our experimental design decisions are driven by abstraction whenever feasible, and state-of-the-art work where not possible.

Our results suggest that retrieval algorithms with query expansion have a theoretical upper bound that does not allow for the identification of all relevant event updates. A topical drift can be partially captured by (pseudo-)relevance feedback, however its performance is still bounded below 100% coverage. Further, we assessed sentence centrality with the use of graph-based methods and observed that update sentences are also salient sentences when enough documents are accumulated. The question that remains unanswered is what is the amount of information that needs to flow into the system before such salience can be reliably assessed. Last, modeling event updates through discriminative terms looks like a promising step towards improving the performance of a temporal summarization system. One thing that was not analyzed in this study is the interplay across these three categories of algorithms, and whether one could complement the other, or in which cases one is better than the other; we leave this as future work.

Finally, we believe that we provide evidence that can guide future research on the topic, and that our analysis is unique and original in the enormous space of temporal summarization research. We consider that certain directions have been outlined by our work, and we intend to explore these further in the future.

Acknowledgements. This research was supported by the Dutch national program COMMIT. All content represents the opinion of the authors, which is not necessarily shared or endorsed by their respective employers and/or sponsors.

References

1. Allan, J.: HARD track overview in TREC 2003 high accuracy retrieval from documents. Technical report, DTIC Document (2005)
2. Allan, J., Carbonell, J.G., Doddington, G., Yamron, J., Yang, Y.: Topic detection and tracking pilot study final report (1998)
3. Allan, J., Gupta, R., Khandelwal, V.: Topic models for summarizing novelty. In: ARDA Workshop on LMIR, Pennsylvania (2001)
4. Allan, J., Papka, R., Lavrenko, V.: On-line new event detection and tracking. In: Proceedings of the 21st ACM SIGIR Conference, pp. 37–45 (1998)
5. Aslam, J.A., Diaz, F., Ekstrand-Abueg, M., McCreadie, R., Pavlu, V., Sakai, T.: TREC 2015 temporal summarization. In: Proceedings of the 24th TREC Conference 2015, Gaithersburg, MD, USA (2015)
6. Chakrabarti, D., Punera, K.: Event summarization using Tweets. ICWSM **11**, 66–73 (2011)

7. Erkan, G., Radev, D.R.: Lexrank: graph-based lexical centrality as salience in text summarization. J. Artif. Intell. Res. **22**, 457–479 (2004)
8. Gao, W., Li, P., Darwish, K.: Joint topic modeling for event summarization across news and social media streams. In: Proceedings of the 21st ACM CIKM Conference, pp. 1173–1182. ACM (2012)
9. Gupta, S., Nenkova, A., Jurafsky, D.: Measuring importance and query relevance in topic-focused multi-document summarization. In: Proceedings of the 45th ACL Interactive Poster and Demonstration Sessions, pp. 193–196. ACL (2007)
10. Imran, M., Castillo, C., Diaz, F., Vieweg, S.: Processing social media messages in mass emergency: a survey. ACM Comput. Surv. (CSUR) **47**(4), 67 (2015)
11. Kamps, J., Pehcevski, J., Kazai, G., Lalmas, M., Robertson, S.: INEX 2007 evaluation measures. In: Fuhr, N., Kamps, J., Lalmas, M., Trotman, A. (eds.) INEX 2007. LNCS, vol. 4862, pp. 24–33. Springer, Heidelberg (2008). doi:10.1007/978-3-540-85902-4_2
12. Kedzie, C., McKeown, K., Diaz, F.: Predicting salient updates for disaster summarization. In: Proceedings of the 53rd Annual Meeting of the Association for Computational Linguistics, ACL, pp. 1608–1617 (2015)
13. McCreadie, R., Macdonald, C., Ounis, I.: Incremental update summarization: adaptive sentence selection based on prevalence and novelty. In: Proceedings of the 23rd ACM CIKM Conference, pp. 301–310. ACM (2014)
14. Mihalcea, R., Tarau, P.: Textrank: bringing order into texts. ACL (2004)
15. Mikolov, T., Sutskever, I., Chen, K., Corrado, G.S., Dean, J.: Distributed representations of words and phrases and their compositionality. In: Advances in NIPS, pp. 3111–3119 (2013)
16. Miller, G.A.: Wordnet: a lexical database for english. Commun. ACM **38**(11), 39–41 (1995)
17. Page, L., Brin, S., Motwani, R., Winograd, T.: The pagerank citation ranking: bringing order to the web (1999)
18. Radev, D.R., Allison, T., Blair-Goldensohn, S., Blitzer, J., Celebi, A., Dimitrov, S., Drabek, E., Hakim, A., Lam, W., Liu, D., et al.: Mead-a platform for multidocument multilingual text summarization. In: LREC (2004)
19. Rayson, P., Garside, R.: Comparing corpora using frequency profiling. In: Proceedings of the Workshop on Comparing Corpora, pp. 1–6. ACL (2000)
20. Vuurens, J.B.P., de Vries, A.P., Blanco, R., Mika, P.: Online news tracking for ad-hoc information needs. In: Proceedings of the 2015 lCTIR Conference, MA, USA, 27–30 September 2015, pp. 221–230 (2015)

A Multiple-Instance Learning Approach
to Sentence Selection for Question Ranking

Salvatore Romeo$^{(\boxtimes)}$, Giovanni Da San Martino, Alberto Barrón-Cedeño,
and Alessandro Moschitti

Qatar Computing Research Institute, HBKU, Doha, Qatar
{sromeo,gmartino,albarron,amoschitti}@hbku.edu.qa

Abstract. In example-based retrieval a system is queried with a document aiming to retrieve other similar or relevant documents. We address
an instance of this problem: question retrieval in community Question
Answering (cQA) forums. In this scenario, both the document collection
and the queries are relatively short multi-sentence documents subject to
noise and redundancy, which makes it harder for learning-to-rank algorithms to build upon the proper text representation.

In order to only exploit the relevant fragments of the query and collection documents, we treat them as a sequence of sentences, in a multiple-
instance learning fashion. By automatically pre-selecting the best sentences for our tree-kernel-based learning model, we improve over using
full text performance on the dataset of the 2016 SemEval cQA challenge
in terms of accuracy and speed, reaching the state of the art.

1 Introduction

The most common text-based search engines operate with relatively short
queries: a user inputs keywords or a short phrase into the engine expecting
to obtain a (small set of) document(s) satisfying her information need. In other
retrieval scenarios (e.g., in near-duplicate detection [29]), the query is yet another
document, similar in nature to those in the document collection. Unlike other
genres, in social media —such as cQA forums— the documents are short, informal, and noisy (e.g., ungrammatical, redundant, and off-topic). As a result, the
contents from both query and collection documents have to be carefully filtered
and selected in order to come out with proper representations for learning-to-
rank algorithms.

We experiment with the evaluation framework of the SemEval 2016 Task 3 on
cQA [26]. Task B of the challenge can be defined as follows. Let D be a collection
of questions, previously posted to the forum. Let q be a freshly-posted question.
Rank the documents in D according to their relevance against q. In general,
a document $d \in D$ has associated a thread of answers, previously posted by
other users. Therefore, retrieving a question $d \in D$ which is equivalent or similar
to q may fulfill the user's information need and may prevent the posting of a
near-duplicate question to the forum. We address this task as a learning-to-rank
problem. Our system relies on a paraphrase identification model based on tree

© Springer International Publishing AG 2017
J.M. Jose et al. (Eds.): ECIR 2017, LNCS 10193, pp. 437–449, 2017.
DOI: 10.1007/978-3-319-56608-5_34

kernels (TK) applied to relational syntactic structures [15]. Such approach was originally intended to deal with pairs of sentences, whereas questions in cQA are in general multi-sentence noisy paragraphs.

Our main contribution is the selection of the best sentences to learn the model upon, and we do it on the basis of a two-step multiple-instance learning strategy (MIL). Firstly, each question gathers together a number of instances (sentences) from which we learn a fast model for identifying the least-noisy, most-relevant ones only using vectorial representations. Secondly, we compute a more expensive syntactic and vectorial representations of the resulting text to learn a binary classifier at question level. We use the latter as a reranking function of our retrieval system. Sentence selection is performed with: (i) unsupervised methods based on scalar products with and without TF × IDF weights and (ii) supervised approaches based on an automatic selector of sentence pairs. Our experiments show that the MIL-based sentence selection model produces a better representation for the question re-ranking model based on TKs. Sentence selection allows our re-ranker to improve by up to 1.82 MAP points over using the full texts and potential improve the best system of the SemEval challenge.

The rest of the paper is distributed as follows. Section 2 puts the ground on tree kernels and multiple instance learning. Section 3 describes our multiple-instance learning approach to both sentence selection and question re-ranking. Section 4 discusses the experimental settings and the obtained results. Section 5 overviews related work. Finally, Sect. 6 includes conclusions and final remarks.

2   Background

In this section, we introduce the concepts that we use in the remainder of the paper: tree kernels in Sect. 2.1 and multiple-instance learning in Sect. 2.2.

2.1   Tree Kernel Models

Kernel methods do not require an explicit data representation in terms of feature vectors. The input of a kernel method is a function —called kernel function— representing the degree of similarity between two items. Kernel machines, e.g., SVM, can be expressed as a convex optimization problem, provided that the kernel function is positive semidefinite [8]. Tree kernels are functions that measure the similarity between tree structures. In this work, we apply the partial tree kernel [24], which computes the similarity between two trees in terms of the number of their shared subtrees, as follows:

$$K(T_1, T_2) = \sum_{n_1 \in N_{T_1}} \sum_{n_2 \in N_{T_2}} \Delta(n_1, n_2), \tag{1}$$

where N_{T_1} (N_{T_2}) is the set of nodes in tree T_1 (T_2). $\Delta(n_1, n_2)$ is computed as

$$\Delta(n_1, n_2) = \begin{cases} 0 & \text{if the labels in nodes } n_1 \text{ and } n_2 \text{ are different} \quad (2) \\ 1 + \sum_{\substack{B_1, B_2 \\ |B_1| = |B_2|}} \prod_{i=1}^{|B_1|} \Delta(c_{n_1}[B_{1i}], c_{n_2}[B_{2i}]) & \text{otherwise} \quad (3) \end{cases}$$

where $B_1 = \langle B_{11}, B_{12}, B_{13}, \dots \rangle$ and $B_2 = \langle B_{21}, B_{22}, B_{23}, \dots \rangle$ are index sequences associated with the ordered child sequences c_{n_1} of n_1 and c_{n_2} of n_2, respectively. B_{1i} and B_{2i} point to the i-th children in the two sequences, and $|B|$ represents the length of the sequence B.

2.2 Multiple Instance Learning

In multiple-instance learning examples are represented as sets (bags) of instances (feature vectors) [2]. In supervised learning, the bag has an associated target label, whereas the label of its members remains unknown. Indeed, some of the instances conforming a bag may be meaningless to discriminate the bag's target label. MIL can be formalized as follows. Let $\{X_1, \dots, X_L\} = \mathcal{X}$ be the set of examples (bags) and $\{x_1, \dots, x_l\}$ be the set of instances of an example $X \in \mathcal{X}$ (here l varies across examples). Given a training set $\{(X_1, Y_1), \dots, (X_L, Y_L)\}$, where $Y_i \in \mathcal{Y}$ is the label of X_i, the goal is to learn a function $F : \mathcal{X} \to \mathcal{Y}$.

MIL approaches can be roughly divided into instance- and bag-level. In the instance-level approaches, the decision $F(X)$ results from the aggregation of the decisions of local discriminative functions $f(x_i) \, \forall x_i \in X$ (cf. [6, 10] for examples). In the bag-level approaches X is mapped into a suitable representation and classified directly. Two bag-level classes have been proposed [2]:

(i) the *embedded space* paradigm, where all the instances are first mapped into a single feature vector and then a standard learning technique is applied. Typically, the representation is obtained by clustering the instances (e.g., k-means), and then forming a vectorial representation of the bag as a function of the clustering, e.g., a vector where the i-th element corresponds to the number of instances represented by the i-th cluster [28].

(ii) The *bag space* paradigm, which requires the definition of a distance or kernel function between bags for applying a learning algorithm, such as k-NN and SVMs. For example, [17] proposed the following kernel:

$$K(X, X') = \sum_{x \in X, x' \in X'} k(x, x')^p, \quad (4)$$

where $k(x, x')$ is a kernel function between instances and the kernel parameter p allows for combinations of features within the kernel $k()$.

We can cast question re-ranking as an instance of a MIL problem using kernels as similarity functions. The set of bags in our setting is composed of pairs of query and forum questions: $X = (q, d)$. Let $S_q = \{s_{q,1}, \ldots, s_{q,|S_q|}\}$ $(S_d = \{s_{d,1}, \ldots, s_{d,|S_d|}\})$ be the set of sentences in $q(d)$. Then the instances are all the pairs of sentences $x_{i,j} = (s_{q,i}, s_{d,j})$.

3   Question Re-ranking Model

3.1   Base Model

Our base learning model is a function $c : Q \times Q \to \mathbb{R}$. Since a document d in the collection is simply labeled as relevant or irrelevant with respect to q, we use a binary SVM [20] whose classification is the sign of the $c(\cdot)$ function. (We also explored with SVMrank [21], but the results were comparable.) The kernel function input to the SVM is a combination of two kernel functions on the parse-tree representations and vectors of similarities. We depart from the model proposed in [30], which combines the tree kernels K^T of Eq. (1):

$$K((q_I, d_I), (q_J, d_J)) = K^T(t(q_I, d_I), t(q_J, d_J)) + K^T(t(d_I, q_I), t(d_J, q_J)), \quad (5)$$

where d_I and d_J are the Ith and J^{th} retrieved questions and $t(x_1, x_2)$ extracts the syntactic tree from text x_1 and enriches it with REL tags. A REL tag is added to the words shared by x_1 and x_2. The REL tag is propagated up to the phrase level in the syntactic tree [15,30]. Figure 1 (bottom) shows an example. Equation (5) is the sum of two kernels applied to two $\{q, d\}$ pairs: one partial-tree kernel applied to the two query questions and one to the two forum questions.

To refine the outcome, we enhance the TK-based model on syntactic trees with 20 similarities $sim(q, d)$ at lexical level [27]. We use word n-grams $(n = [1, \ldots, 4])$, after stopword removal, to compute greedy string tiling [34], longest common subsequence [1], Jaccard coefficient [18], word containment [23], and cosine. We also include a similarity over the syntactic trees of the pair $\{q, d\}$ using the partial tree kernel, i.e., $K^T(t(q, d), t(d, q))$. Note that the operands of the kernel function are members of the same pair. The corpus includes the position of question d in the ranking obtained when the forum is queried with q with the Google search engine. We integrate this feature as the inverse of the position of d. All these similarities are used over an RBF kernel function [25].

3.2   A Multiple-Instance Approach to Question Re-Ranking

We integrate the model in Sect. 3.1 with a two-step MIL approach [17]. Firstly, we follow the instance-based paradigm, in which the instances are pairs of sentences $\{s_q, s_d\}$. Secondly, we follow the embedded space paradigm to build document-level classifiers, out of which the final ranking is computed.

Let $S \subseteq S_q \times S_d$ be a subset of size u of the Cartesian product between S_q and S_d; i.e., S is the set of selected sentences (we use S_X when we refer to a specific example X). Let $q^* = \prod(\{s_{q,i} | (s_{q,i}, s_{d,j}) \in S\})$ be the concatenation of the sentences in S_q appearing in S ($\prod$ denotes the concatenation operator). Similarly, let d^* be the concatenation of the sentences in S_d. We apply a kernel function to pairs (q^*, d^*) instead of pairs (q, d).

We now show the relationship between our approach and that of Eq. (4). The kernel in Eq. (5) is a combination of tree kernels, including the one in Eq. (1). For simplicity, we focus our discussion on Eq. (1), which can be decomposed as

$$\sum_{n_2 \in N_{T_2}} \Delta(r(T_1), n_2) + \sum_{n_1 \in N_{T_1} \setminus r(T_1)} \Delta(n_1, r(T_2)) + \sum_{n_1 \in N_{T_1} \setminus r(T_1)} \sum_{n_2 \in N_{T_2} \setminus r(T_2)} \Delta(n_1, n_2)$$

where $r(T)$ is the root of a tree T. The parse trees of all the sentences in the text hang from the root-labeled node, which is always the same and unique in every tree. As a consequence, considering the definition of $\Delta()$ in Eqs. (2) and (3), Eq. (1) can be further simplified as

$$\Delta(r(T_1), r(T_2)) + \sum_{n_1 \in N_{T_1} \setminus r(T_1), n_2 \in N_{T_2} \setminus r(T_2)} \Delta(n_1, n_2)$$

Thus, the kernel between two query questions, according to Eq. (1), would be

$$= \Delta(r(T(q_1^*)), r(T(q_2^*))) + \sum_{\{s_1 | (s_1, s_2) \in S_{q_1^*}\}} \sum_{\{s_1' | (s_1, s_2) \in S_{q_2^*}\}} \sum_{n_1 \in N_{T(s_1)}} \sum_{n_2 \in N_{T(s_2)}} \Delta(n_1, n_2)$$

$$= \Delta(r(T(q_1^*)), r(T(q_2^*))) + \sum_{\{s_1 | (s_1, s_2) \in S_{q_1^*}\}} \sum_{\{s_1' | (s_1, s_2) \in S_{q_2^*}\}} K^T(s_1, s_2)$$

where $T(x)$ is a function that creates a parse tree from a sentence x. As we are dealing with multiple-sentence documents, each T includes an additional root node that links together all the sentences' trees into a macro-tree. The second term of the summation resembles Eq. (4), but in this case the kernel is computed only on the top-u pairs.

The core function of the model is the TK and we select the texts representing q and d before feeding them into the model. We aim to identify those sentences which better represent each question towards the learning process to produce S. Our sentence-selection is based on a scoring function $c_s : S_q \times S_d \to \mathbb{R}$, which differs slightly from the $c(\cdot)$ function described in Sect. 3.1. The target label of the pair of sentences is the one of the corresponding bag [6]. We use the same similarities as in the question-level model —plus four new features: given the position of a sentence s in a question, we consider three Boolean features: whether s appears (i) in position 1, (ii) between positions 2 and 4 (inclusive), or (iii) after position 4. These features are duplicated for both s_q and s_d. An additional real-valued feature computes $1/position$. Hereinafter, we will call them *positional features*. We do not use TKs in the sentence-level classifier as in preliminary experiments (not reported), the outcome of the classifier deteriorated.

Finally, given a pair $\{q, d\}$, we compute $c(s_{q,i}, s_{d,j})$ and use the score to rank sentences: only the top-k sentence pairs are used to represent the question in the final re-ranking process. Figure 1 shows the automatically-selected sentences

q : **car taking to india.** I   d : Shiping CAR from Qatar to India. I am using Nis-
wish to take my Car(Toyota   san Altima for past two years. **I am planning to**
corolla 2003) to india; is it   **settle back India. Is it possible to ship my**
expensive?   **car to India?** Is it advisable. Any one did earlier.

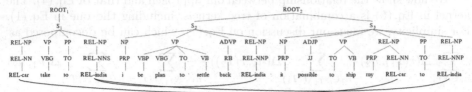

Fig. 1. Top: a pair of questions $\{q, d\}$ with automatically-selected sentences. One sentence is selected from q and two from d (highlighted). Bottom: representation of the
questions' selected sentences as syntactic macro-trees (including multiple sentences).
The representation is enriched with REL tags linking matches *car* and *india*.

from a pair $\{q, d\}$ and the resulting parse-tree representation for the ranking of
d. As observed, sentences which give context and are not essential to estimate
the relevance of d are discarded from the parse-tree representation. The scores
for the training set are computed by 5-fold cross validation. The scores for the
development and test sets are obtained by holdout, after learning on the training set. Our MIL approach lies between the two mentioned paradigms, since it
extracts a representation for the bag that depends not only on the instances
themselves, but also on the prediction scores of a classifier at instance level.

4   Experiments

In this section, we present and discuss the results obtained with our model. We
describe our evaluation framework in Sect. 4.1. The experiments both at sentence
and at question level are discussed in Sects. 4.2 and 4.3.

4.1   Evaluation Framework

We use the SemEval 2016 cQA corpus and evaluation settings to run our experiments [26]. This corpus contains a pool of 387 query questions, each of which
includes 10 potentially-related forum questions. The forum questions were originally gathered using the Google search engine, which represents the task baseline. The binary gold annotations —Relevant or not— were crowdsourced. The
class distribution is 40% relevant vs 60% irrelevant. We use the same train-
ing/dev/test partition as in the original dataset.[1] Following [26], we evaluate
with Mean Average Precision (MAP), and Mean Reciprocal Rank (MRR).

We employ binary SVMs using the KeLP toolkit [13] in all the experiments.
The TK over the parse trees is complemented with an RBF kernel over the
similarity features. In all the experiments we set the C parameter of the SVMs
to 1 and the parameters of tree and RBF kernels to the default values.

---

[1] This corpus is available at http://alt.qcri.org/semeval2016/task3/.

4.2   Selecting Sentences

First, we describe the experiments on sentence selection using the approaches from Sect. 3.2. We annotated sentence pairs from a subset of questions with CrowdFlower[2] to generate a gold standard to evaluate our sentence-level classifier. We selected only the 25 pairs of questions in the development set in which the forum question contained five or more sentences. The annotators were presented with one query-question sentence and five related-question sentences. The task consisted of determining which of the related sentences expressed the same information or idea as the query one. Each instance was annotated three times, with an inter-annotator agreement of 85.33.[3]

We selected sentences with SVMs, considering three different feature sets and kernel settings: (i) an RBF kernel on similarities (sim_{RBF}), (ii) a linear combination of similarities with a linear kernel on positional features ($sim_{RBF} + pos_{lin}$), and (iii) a linear combination of kernel (i) with an RBF kernel on positional features ($sim_{RBF} + pos_{lin}$). We attached the Google-provided position to the positional features. The score of the unsupervised model is computed as the cosine similarity between the TF × IDF vectors of each pair of sentences (TFIDF).

Table 1. Performance of the sentence-level classifier with various feature combinations.

Classifier	Acc	P	R	F1	MAP	MRR
TFIDF	-	-	-	-	60.83	63.43
sim_{RBF}	65.88	44.44	14.29	21.62	60.15	64.22
$sim_{RBF} + pos_{lin}$	68.24	53.85	25.00	34.15	61.13	64.22
$sim_{RBF} + pos_{RBF}$	71.76	59.09	46.43	52.00	62.84	66.67

Table 1 shows the performance of the different configurations. Comparing the models using the positional features or not, we observe that such features improve the performance w.r.t. all the evaluation metrics. The performance of TFIDF in terms of MAP is similar to the ones of classifiers sim_{RBF} and $sim_{RBF} + pos_{lin}$. Using the positional features in an RBF kernel produces a better performance than other models, obtaining an improvement in terms of MAP equal to 2.01, 2.62 and 1.71, w.r.t. TFIDF, sim_{RBF} and $sim_{RBF} + pos_{lin}$, respectively.

4.3   Ranking Questions

We focus the rest of the experiments on the impact of the sentence selection for generating smaller trees to be used in TKs. We ran one question re-ranker feeding the TKs with the outcome of each of the sentence classifiers at hand to find out

---

[2] http://www.crowdflower.com/.
[3] This dataset is available at http://alt.qcri.org/resources/iyas.

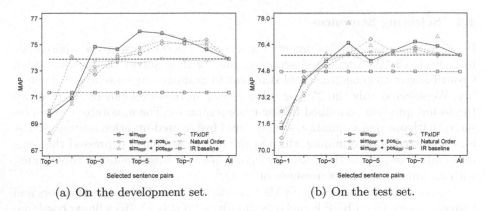

(a) On the development set.   (b) On the test set.

Fig. 2. MAP evolution for different sentence selection strategies. *All* stands for the system considering full texts (without sentence selection).

if MAP can be improved by selecting sentences. We kept the original input texts to compute the similarity features. Figures 2(a) and (b) show the results of the re-rankers obtained on the development and test sets with increasing number of selected sentences. For comparison, the MAP obtained when considering full texts —without any sentence selection— is 73.60 on the development set and 75.89 on the test set. They are represented in the converging points on the right-hand side of the plots. The natural order is our sentence selection baseline —k sentences are taken from left to right. Its best performance is achieved with 6 sentences: MAP of 73.92 and 76.02 on the development and test sets, respectively. On dev. set (Fig. 2(a)), the best model is sim_{RBF}, which performs best with 5 sentences, i.e., a MAP of 76.01. The second best system is $sim_{RBF} + pos_{RBF}$, reaching the best outcome with 6 sentences, for a MAP of 75.92. Models $sim_{RBF}+pos_{lin}$ and TFIDF show the best results only until 6 and 8 sentences are used, with MAP values of 75.27 and 75.36, respectively. In general, identifying the most similar sentence pairs in advance allows for the best results; and the least sentences considered, the faster the TK operates.

Regarding the results on the test set (Fig. 2(b)), the best performance is obtained by $sim_{RBF}+pos_{RBF}$ with only 4 sentences: MAP = 77.71. This shows that our approach can potentially highly improve the state of the art, i.e., 76.70 (see Table 2). However, the different model behavior observed in dev. and test sets suggest some challenges for estimating the optimal number of sentences.

The TFIDF, S_r and $sim_{RBF}+pos_{RBF}$ approaches have similar performance, i.e., 76.73, 76.57 and 76.26 of MAP, but after using 5 or more sentences. When our best sentence selector —$sim_{RBF}+pos_{RBF}$— is used, our model outperforms the best systems submitted to SemEval (cf. Table 2; Sect. 5) —being the only statistically different to the IR baseline (confidence = 90%).

Finally, it should be noted that selecting the sentences to represent q and d not only boosts the performance of our question ranker but, as a side effect, applying tree kernels to shorter text, makes training/testing up to 30% faster (e.g., when using our most accurate model).

5   Related Work

Different approaches have been proposed to overcome the lexical chasm when assessing the similarity between two questions. Early approaches used statistical machine translation (SMT) techniques to compute the semantic similarity between two questions. For instance, [19] used a language model based on word translation probabilities to compute the likelihood of generating a query question given a target (forum) question. [35] showed that models based on phrases are more effective than models based on words, as they are able to capture contextual information. However, approaches based on SMT typically require large amounts of data for parameter estimation.

Both [7,12] presented algorithms that try to go beyond simple text representation. [7] compute the similarity between two questions on Yahoo! Answers by using a smoothed language model that exploits the category structure of the forum. [12] searched for questions that are semantically similar to the user's question by identifying the question's topic and focus.

[33] presented an approach exploiting the questions' syntactic information. They proposed to find semantically-related questions by computing the similarity between their syntactic-tree representations. The tree similarity is computed as the number of sub-structures shared between two trees. The main difference with respect to our model is that we use more complex structural models, encoding relational structures and processing them by means of tree kernels. The latter captures effective structure relations, which boosts the performance of our re-ranker based on standard features.

Recent work has shown the effectiveness of neural models for question similarity [11] in cQA. For instance, [11] used CNN and bag-of-words (BOW) representations of query and forum questions to compute cosine similarity scores. Recently, [4] presented a neural attention model for machine translation and showed that the attention is helpful when dealing with long sentences.

The 2016 edition of the SemEval Task 3 on cQA [26] triggered a manifold of approaches to question retrieval. The top-three participants opted for SVMs as learning models. The top-ranked [16] used SVM^{rank} [22], the first [5] and second [14] runners up used KeLP [13] to combine various kernels. The amount of knowledge these models use is pretty different. [16] relies heavily on distributed representations and semantic information sources, such as Babelnet

Table 2. Performance of the best systems submitted to SemEval 2016 Task 3(B) on question ranking; i.e., on our test set (cf. Sect. 5 for models' details).

Classifier	MAP	MRR
UH-PRHLT-primary [16]	76.70	83.02
ConvKN-primary [5]	76.02	84.64
Kelp-primary [14]	75.83	82.71
IR Baseline [26]	74.75	83.79

and Framenet. The others do not. No statistically-significant differences were observed in the performance of these systems with respect to the baseline. Their performance is included in Table 2 for comparison with our results.

6 Conclusions

In this paper we described a learning-to-rank model based on tree kernels to rank a set of forum questions given a new question. Such a component allows Web forums to avoid posting near-duplicate questions and to answer to the user's information quest at no time. We proposed a model to pre-select a subset of the sentences composing each question in order to feed them into a tree-kernel-based question-ranking model. The reason is that tree-kernel models are affected by noisy text and redundant information, which is typically added by Web users when formulating or answering forum questions.

We expressed both the sentence selection and question ranking steps as a multiple-instance learning (MIL) instantiation. Our results on the SemEval 2016 cQA corpus showed that MIL models can improve the quality of the ranking by coming out with a better representation of the documents. As a result, our tree-kernel model learn better the parameters of the ranking function (as noise is filtered out from the texts), both boosting the performance of the ranker and speeding it up. Our proposed model outperforms the top systems submitted to the SemEval 2016 task on community Question Answering, however additional work is needed to reliably estimating the best number of sentences for each test set. In the future, we would like to explore more powerful kernels such as the smoothed partial tree kernel [9] as well as the most advanced tree kernel models applied in QA, e.g., [31,32].

References

1. Allison, L., Dix, T.: A bit-string longest-common-subsequence algorithm. Inf. Process. Lett. **23**(6), 305–310 (1986)
2. Amores, J.: Multiple instance classification: review, taxonomy and comparative study. Artif. Intell. **201**, 81–105 (2013)
3. Association for Computational Linguistics: Proceedings of the 10th International Workshop on Semantic Evaluation, SemEval 2016, June 2016
4. Bahdanau, D., Cho, K., Bengio, Y.: Neural machine translation by jointly learning to align and translate. arXiv preprint arXiv:1409.0473 (2014)
5. Barrón-Cedeño, A., Da San Martino, G., Joty, S., Moschitti, A., Al-Obaidli, F., Romeo, S., Tymoshenko, K., Uva, A.: ConvKN at SemEval-2016 Task 3: answer and question selection for question answering on Arabic and English fora. In: Proceedings of the 10th International Workshop on Semantic Evaluation [3], pp. 896–903
6. Bunescu, R.C., Mooney, R.J.: Multiple instance learning for sparse positive bags. In: Proceedings of the 24th International Conference on Machine Learning, pp. 105–112. ACM (2007)

7. Cao, X., Cong, G., Cui, B., Jensen, C.S., Zhang, C.: The use of categorization information in language models for question retrieval. In: Proceedings of the 18th ACM Conference on Information and Knowledge Management, pp. 265–274. ACM (2009)
8. Cristianini, N., Shawe-Taylor, J.: An Introduction to Support Vector Machines and Other Kernel-based Learning Methods, 1st edn. Cambridge University Press, Cambridge (2000)
9. Croce, D., Moschitti, A., Basili, R.: Structured lexical similarity via convolution kernels on dependency trees. In: Proceedings of the 2011 Conference on Empirical Methods in Natural Language Processing, pp. 1034–1046. Association for Computational Linguistics, Edinburgh, July 2011
10. Dietterich, T.G., Lathrop, R.H., Lozano-Pérez, T.: Solving the multiple instance problem with axis-parallel rectangles. Artif. Intell. **89**(1–2), 31–71 (1997)
11. dos Santos, C., Barbosa, L., Bogdanova, D., Zadrozny, B.: Learning hybrid representations to retrieve semantically equivalent questions. In: Zong and Strube [36], pp. 694–699
12. Duan, H., Cao, Y., Lin, C.Y., Yu, Y.: Searching questions by identifying question topic and question focus. In: Proceedings of the 46th Annual Meeting of the Association for Computational Linguistics and the Human Language Technology Conference, ACL-HLT 2008, pp. 156–164. Association for Computational Linguistics, Columbus, June 2008
13. Filice, S., Castellucci, G., Croce, D., Da San Martino, G., Moschitti, A., Basili, R.: KeLP: a kernel-based learning platform in Java. In: Proceedings of the Workshop on Machine Learning Open Source Software: Open Ecosystems. International Conference of Machine Learning, Lille (2015)
14. Filice, S., Croce, D., Moschitti, A., Basili, R.: KeLP at SemEval-2016 Task 3: learning semantic relations between questions and answers. In: Proceedings of the 10th International Workshop on Semantic Evaluation [3], pp. 1116–1123
15. Filice, S., Da San Martino, G., Moschitti, A.: Structural representations for learning relations between pairs of texts. In: Zong and Strube [36], pp. 1003–1013
16. Franco-Salvador, M., Kar, S., Solorio, T., Rosso, P.: UH-PRHLT at SemEval-2016 Task 3: combining lexical and semantic-based features for community question answering. In: Proceedings of the 10th International Workshop on Semantic Evaluation [3]
17. Gärtner, T., Flach, P.A., Kowalczyk, A., Smola, A.J.: Multi-instance kernels. In: Sammut, C., Hoffmann, A.G. (eds.) Machine Learning, Proceedings of the Nineteenth International Conference (ICML 2002), University of New South Wales, Sydney, Australia, 8–12 July 2002, pp. 179–186. Morgan Kaufmann, Burlington (2002)
18. Jaccard, P.: Étude comparative de la distribution florale dans une portion des Alpes et des Jura. Bulletin del la Société Vaudoise des Sciences Naturelles **37**, 547–579 (1901)
19. Jeon, J., Croft, W.B., Lee, J.H.: Finding similar questions in large question and answer archives. In: Herzog, O., Schek, H., Fuhr, N., Chowdhury, A., Teiken, W. (eds.) Proceedings of the 14th ACM International Conference on Information and Knowledge Management, Bremen, Germany, pp. 84–90 (2005)
20. Joachims, T.: Making large-scale support vector machine learning practical. In: Schölkopf, B., Burges, C.J.C., Smola, A.J. (eds.) Advances in Kernel Methods, pp. 169–184. MIT Press, Cambridge (1999)

448   S. Romeo et al.

21. Joachims, T.: Optimizing search engines using clickthrough data. In: Proceedings of the Eighth ACM SIGKDD International Conference on Knowledge Discovery and Data Mining, pp. 133–142. ACM, New York (2002)
22. Joachims, T.: Training linear SVMs in linear time. In: Proceedings of the 12th ACM SIGKDD International Conference on Knowledge Discovery and Data Mining, KDD 2006, pp. 217–226. ACM, New York (2006)
23. Lyon, C., Malcolm, J., Dickerson, B.: Detecting short passages of similar text in large document collections. In: Proceedings of the Conference on Empirical Methods in Natural Language Processing, EMNLP 2001, Pittsburgh, PA, pp. 118–125 (2001)
24. Moschitti, A.: Efficient convolution kernels for dependency and constituent syntactic trees. In: Fürnkranz, J., Scheffer, T., Spiliopoulou, M. (eds.) ECML 2006. LNCS (LNAI), vol. 4212, pp. 318–329. Springer, Heidelberg (2006). doi:10.1007/11871842_32
25. Müller, K.R., Mika, S., Rätsch, G., Tsuda, K., Schölkopf, B.: An introduction to kernel-based learning algorithms. IEEE Trans. Neural Netw./Publ. IEEE Neural Netw. Counc. 12(2), 181–201 (2001)
26. Nakov, P., Màrquez, L., Moschitti, A., Magdy, W., Mubarak, H., Freihat, A., Glass, J., Randeree, B.: SemEval-2016 Task 3: community question answering. In: Proceedings of the 10th International Workshop on Semantic Evaluation [3], pp. 525–545
27. Nicosia, M., Filice, S., Barrón-Cedeño, A., Saleh, I., Mubarak, H., Gao, W., Nakov, P., Da San Martino, G., Moschitti, A., Darwish, K., Màrquez, L., Joty, S., Magdy, W.: QCRI: answer selection for community question answering - experiments for Arabic and English. In: Proceedings of the 9th International Workshop on Semantic Evaluation, SemEval 2015. Association for Computational Linguistics, Denver (2015)
28. Nowak, E., Jurie, F., Triggs, B.: Sampling strategies for bag-of-features image classification. In: Leonardis, A., Bischof, H., Pinz, A. (eds.) ECCV 2006. LNCS, vol. 3954, pp. 490–503. Springer, Heidelberg (2006). doi:10.1007/11744085_38
29. Potthast, M., Stein, B.: New issues in near-duplicate detection. In: Preisach, C., Burkhardt, H., Schmidt-Thieme, L., Decker, R. (eds.) Data Analysis, Machine Learning and Applications. Selected Papers from the 31th Annual Conference of the German Classification Society (GFKL 2007). Studies in Classification, Data Analysis, and Knowledge Organization, pp. 601–609. Springer, Heidelberg (2008)
30. Severyn, A., Moschitti, A.: Structural relationships for large-scale learning of answer re-ranking. In: Proceedings of the 35th International ACM SIGIR Conference on Research and Development in Information Retrieval, SIGIR 2012, Portland, OR, pp. 741–750 (2012)
31. Tymoshenko, K., Bonadiman, D., Moschitti, A.: Convolutional neural networks vs. convolution kernels: feature engineering for answer sentence reranking. In: Proceedings of the 2016 Conference of the North American Chapter of the Association for Computational Linguistics: Human Language Technologies, pp. 1268–1278. Association for Computational Linguistics, San Diego, June 2016
32. Tymoshenko, K., Moschitti, A.: Assessing the impact of syntactic and semantic structures for answer passages reranking. In: Proceedings of the 24th ACM International Conference on Information and Knowledge Management, CIKM 2015, Melbourne, VIC, Australia, 19–23 October 2015, pp. 1451–1460 (2015)

33. Wang, K., Ming, Z., Chua, T.S.: A syntactic tree matching approach to finding similar questions in community-based QA services. In: Proceedings of the 32nd International ACM SIGIR Conference on Research and Development in Information Retrieval, pp. 187–194. ACM (2009)

34. Wise, M.: YAP3: improved detection of similarities in computer program and other texts. In: Proceedings of the Twenty-Seventh SIGCSE Technical Symposium on Computer Science Education, SIGCSE 1996, New York, NY, pp. 130–134 (1996)

35. Zhou, G., Cai, L., Zhao, J., Liu, K.: Phrase-based translation model for question retrieval in community question answer archives. In: Proceedings of the 49th Annual Meeting of the Association for Computational Linguistics: Human Language Technologies, vol. 1, pp. 653–662 (2011)

36. Zong, C., Strube, M. (eds.): Proceedings of the 53rd Annual Meeting of the Association for Computational Linguistics and the 7th International Joint Conference on Natural Language Processing, ACL-HLT 2015, Association for Computational Linguistics, Beijing, July 2015

Enhancing Sensitivity Classification with Semantic Features Using Word Embeddings

Graham McDonald[✉], Craig Macdonald, and Iadh Ounis

School of Computing Science, University of Glasgow, Glasgow G12 8QQ, UK
g.mcdonald.1@research.gla.ac.uk,
{craig.macdonald,iadh.ounis}@glasgow.ac.uk

Abstract. Government documents must be reviewed to identify any *sensitive* information they may contain, before they can be released to the public. However, traditional paper-based sensitivity review processes are not practical for reviewing born-digital documents. Therefore, there is a timely need for automatic sensitivity classification techniques, to assist the digital sensitivity review process. However, sensitivity is typically a product of the relations between combinations of terms, such as *who said what about whom*, therefore, automatic sensitivity classification is a difficult task. Vector representations of terms, such as word embeddings, have been shown to be effective at encoding latent term features that preserve semantic relations between terms, which can also be beneficial to sensitivity classification. In this work, we present a thorough evaluation of the effectiveness of semantic word embedding features, along with term and grammatical features, for sensitivity classification. On a test collection of government documents containing real sensitivities, we show that extending text classification with semantic features and additional term n-grams results in significant improvements in classification effectiveness, correctly classifying 9.99% more sensitive documents compared to the text classification baseline.

1 Introduction

Freedom of Information (FOI) laws[1,2] legislate that government documents should be opened to the public. However, many government documents contain *sensitive* information, such as *personal* or *confidential* information, that would be likely to cause harm to, or prejudice the interests of, an individual or organisation if the information were to be made public. Therefore, FOI laws provide exemptions that negate the obligation to release information that is of a sensitive nature.

To ensure that sensitive information is not made public, all government documents must be manually *sensitivity reviewed* prior to release. However, with the adoption of digital technologies, such as word processing and emails, the volume of government documents has increased and, moreover, documents are

---

[1] http://www.legislation.gov.uk/ukpga/2000/36/contents.
[2] http://www.foia.gov.

© Springer International Publishing AG 2017
J.M. Jose et al. (Eds.): ECIR 2017, LNCS 10193, pp. 450–463, 2017.
DOI: 10.1007/978-3-319-56608-5_35

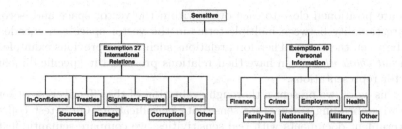

Fig. 1. The range of potential sensitivities relating to 2 of the 24 Freedom of Information Act 2000 (FOIA) exemptions, namely *International Relations* and *Personal Information.*

produced and stored in a more ad-hoc manner than the paper-based filing systems of previous decades. Therefore, the traditional sensitivity review process is not practical for the era of born-digital documents, and governments are facing an increasing backlog of digital documents awaiting review before they can be considered for release.

There is, therefore, a timely need for automatic sensitivity classification, to assist the digital sensitivity review process [1]. However, automatic sensitivity classification is a difficult task. For example, the UK Freedom of Information Act 2000 (FOIA) has 24 FOI exemptions[3], each with wide-ranging sub-categories of exemptions. Figure 1 illustrates the scope of potentially sensitive information from just 2 of these 24 exemptions, namely *International Relations* and *Personal Information.* As can be seen from Fig. 1, the scope of potentially sensitive information is broad. Moreover, a document can, potentially, contain many unrelated sensitivities. Therefore, in this work, we view sensitive information as a *composite* class of information that can be a result of one or more different types of sub-category sensitivities.

Text classification [2] is one approach that has been shown to be promising as a basis for automatic sensitivity identification algorithms [3,4]. Usually, a text classification model is learned by observing statistical patterns in the distributions of individual key terms from example documents. However, the potential effectiveness of sensitivity classification from single-term observations is limited, due to the fact that sensitivity classification is not a *topic-oriented* task [4] and, moreover, sensitivity tends to arise as a product of specific factors. For example, International Relations sensitivities are often a product of *who said what about whom*. It is, therefore, the relations between terms that can result in information being sensitive. One approach that has been shown to be effective at capturing the semantic relations between terms is word embeddings [5]. Word embeddings are vector space word representations, where each dimension maps to a latent feature of the word. We expect word embedding to be able to identify latent sensitivity in terms, due to two fundamental properties. Firstly, semantically *similar*

---

[3] 14 of the 24 FOIA exemptions apply to documents that are to be archived for public access.

terms are positioned close to each other within the vector space and, secondly, the directionality between multiple terms in the vector space can encode relations between the terms. Therefore, relations such as the previous example, *who said what about whom*, can have their relations preserved in specific dimensions of vector representations.

In this work, we present a thorough evaluation of the effectiveness of semantic word embedding features for sensitivity classification. On a test collection of government documents with real sensitivities, we compare semantic features with grammatical features derived from sequences of part-of-speech tags (POS) and term n-gram features. The contributions of this paper are two-fold. Firstly, we present the first in depth analysis of the effectiveness of word embeddings for sensitivity classification. Secondly, we show that semantic word embedding features can significantly improve the effectiveness of sensitivity classification. The combination of semantic word embeddings and term n-gram features correctly classified 9.99% more sensitive documents than the baseline text classification approach.

The remainder of this paper is structured as follows. In Sect. 2 we present work relating to sensitivity classification and word embeddings for text classification. In Sect. 3, we present the feature sets that we evaluate for sensitivity classification before, in Sect. 4, presenting our experimental setup. We present our results in Sect. 5, before providing some further analysis in Sect. 6, and conclusions in Sect. 7.

2   Related Work

Classifying sensitivities, such as FOI exemptions, to assist the sensitivity review of government documents, is a relatively new task. Moreover, it can be considered that the definition of sensitivity, in this context, is more broad than in most of the previous literature, e.g. preserving the privacy of personal data [6,7]. McDonald *et al.* [3] was the first work to address the automatic classification of FOI exemptions. In that work, the authors presented a proof-of-concept classifier for classifying specific FOI exemptions, and found that extending text classification with additional features, such as *the number of subjective sentences* and a *country risk score*, could improve the effectiveness of text classification for specific sensitivities. The work that we present in this paper differs from the work of [3] in a number of ways. Firstly, in [3], the authors deployed individual classifiers for each specific sensitivity, whereas our work addresses the more challenging task of classification of the composite class of sensitivity. Secondly, in [3], the authors extended text classification with *hand-crafted* features that were tailored for specific sensitivities. In this work, we present a fully automatic approach that could easily generalise to other collections or sensitivities.

Berardi *et al.* [4] built on the work of McDonald *et al.* [3] to optimise the cost-effectiveness of sensitivity reviewers. In that work, Berardi *et al.* deployed a *utility-theoretic* ranking approach for semi-automatic text classification [8]. Their approach ranks documents by the expected gain in accuracy that a classification system can achieve by having a reviewer correct mis-classified instances, i.e. if a reviewer validates a document that the classifier is least confident about, then the overall accuracy is increased. Berardi *et al.* found that their approach performed well at estimating the correctness of classification predictions from McDonald *et al.*'s approach, and achieved substantial improvements in overall classification ($+3\%$ to $+14\%$ F_2). However, these improvements were much smaller than their approach had achieved on other tasks and they concluded that the task of classifying by sensitivity is much harder than *topic-oriented* classification.

In other work, relating to FOI exemptions, McDonald *et al.* [9] investigated methods for identifying passages of text in documents that contained information that had been supplied *in confidence*. In that work, the authors identified confidential information by measuring the amount of sensitivity in specific part-of-speech (POS) n-grams. Inspired by the work of Lioma and Ounis [10], who showed that high frequency POS n-grams have a greater *content load*, McDonald *et al.* used POS n-grams with a high *sensitivity load* to train a Conditional Random Fields sequence tagger for predicting confidential sequences. Their work showed that POS n-grams could be effective for identifying a specific sensitivity. Therefore, we also use POS n-grams as classification features in this work. However, differently from the work of McDonald *et al.* [9], we test if POS n-grams are effective features for classifying the *composite* class of sensitivity and compare POS n-grams with the performance of word embeddings and term features.

As previously stated in Sect. 1, word embeddings are vector space representations of terms [5]. Word embeddings have low dimensionality, compared to the sparse vector representations more traditionally used in text classification. The dense vector formation of word embedding models allow them to capture semantic qualities of, and relations between, terms in a collection. This has resulted in word embeddings becoming very popular in natural language processing tasks, e.g. [11,12]. Moreover, there are a number of available word embedding frameworks, such as word2vec [13] and Glove [14], with models that are pre-trained on large corpora from different domains, such as Google News[4] or Wikipedia[5].

Recently, word embeddings have been shown to be effective in Information Retrieval and classification tasks, e.g. [15–17]. However, for classification, they have mostly been used for classifying short spans of text, such as tweets or sentences [17,18]. Typically, word embeddings have been used as an initialisation step for neural networks. However, recently, Balikas and Amini [19] presented a large scale study that integrated word embeddings as classification features for multi-class text classification. In that study, the authors obtained document vector representations by deploying simple composition functions (e.g. min, average, max) to construct vector representations of combinations of words, such as phrases or sentences, from term vector models [20]. They showed that these compositional document vectors could be effectively used as features to extend text classification and improve classification performance. In this work, we follow the methodology of [19,20] and compose document representations from word

---

[4] https://code.google.com/archive/p/word2vec/.
[5] http://nlp.stanford.edu/projects/glove/.

embeddings in the task of sensitivity classification. However, differently from Balikas and Amini [19], we show how these document representations combined with text features can be effective for discovering latent sensitivities.

3   Sensitivity Classification

In this section, we provide an overview of the feature sets that we test for sensitivity classification. Firstly, since term n-grams have not previously been studied for sensitivity classification, in Sect. 3.1, we briefly describe extending text classification with term n-gram features before, in Sect. 3.2, presenting the approach we deploy for generating grammatical features from POS sequences. Lastly, in Sect. 3.3, we present the approach that we deploy for generating semantic features using word embeddings.

The expected volumes of individual types of sensitivity vary between specific government departments. For example, in the UK, the Foreign and Commonwealth Office encounters many more *International Relations* sensitivities than the Department of Health. The approaches that we present in this section only depend on the terms in a collection and require no prior knowledge of specific sensitivities. Therefore, they could be deployed as part of a *first line of defense* across government departments.

3.1   Term Features

The first set of features that we evaluate are term features. Term features are a popular type of feature used for classifying textual documents. Indeed, using the frequencies of terms in documents to train classifiers, such as Support Vector Machines (SVM) [21], can be effective for many topic-oriented classification tasks [2].

Although sensitivity classification is not a topic-oriented task [4], text classification has been shown to be a strong baseline approach [3,4]. A popular, and effective, extension to text classification is to include additional n-gram term features [2]. N-gram features for text classification are, typically, a tuple of n contiguous terms from a larger ordered sequence of terms. Typically, text classification is extended with n-grams where $n \leq 4$. However, for sensitivity classification, we expect larger values of n to be more effective, since they have the potential to capture document structures that, in turn, can be an indicator of potential sensitivity. For example, table headings, such as *Name, Date of Birth, Residence*, can be a reliable indicator of Personal Information sensitivity. Therefore, in this work we test the effectiveness of larger term n-gram sequences, along with additional combinations of smaller values of n for completeness.

3.2   Grammatical Features

As previously mentioned in Sect. 2, part-of-speech (POS) n-grams have been shown to be effective for identifying text relating to *information supplied in confidence* [9]. However, as outlined in Sect. 1, sensitivity is a composite class containing many, more specific, types of sensitive information (such as confidential information) and the effectiveness of POS n-grams as features of sensitivity has not been fully studied for sensitivity classification. Therefore, in this work, we evaluate the effectiveness of POS n-grams as grammatical features for sensitivity classification.

POS n-gram features are derived similarly to the approach for term n-gram features. However, prior to selecting n-grams, a document is represented by the POS tags it contains. For example, the sentence "The informant provided the information" can be represented by the following POS tags "DT NN VB DT NN". When represented as POS 2-grams, the sentence becomes "DTNN NNVB VBDT DTNN". POS tags substantially reduce the vocabulary of a collection and provide a single representation of similar sentences. For example, sentences that are *about* different entities and actions but have the same grammatical structure have a single representation.

3.3   Semantic Features

In this section, we present the approach that we deploy for extending text classification with semantic features using word embeddings. As previously mentioned in Sect. 1, sensitivity is often a product of a combination of factors, such as *who said what about whom*. The common factors of these types of sensitivity are two-fold: Firstly, relations between terms are often preserved over multiple sensitivities. For example, in the sentences "the assailant denied offering the plans for the attack" and "The informant provided us the names of the suspect" the relation of Entity A giving something to Entity B is common to both sentences; The second common factor is that the entities or actions often have similar meaning, e.g. offering/provided or informant/assailant.

Word embedding models are trained by observing the contexts in which terms usually appear within large corpora, with the assumption that words occurring within similar contexts are semantically similar. The resulting word embedding models have two fundamental properties that can help us to identify relational sensitivities. Firstly, semantically similar terms tend to appear close to each other in the vector space (e.g. informant/assailant) and, secondly, the directionality between terms in the vector space can encode relations between terms (e.g. the direction of *assailant* to *offering* is close to parallel with *informant* to *provided*). This, in turn, means that semantically similar relations tend to have similar values in specific dimensions of their embedding representations.

To derive semantic features, we follow the approach of Balikas and Amini [19] to construct a document representation from word embeddings using a set of composition functions, *min*, *mean* and *max* [22,23]. For a given word embedding model, W, of term vectors, $V^{\text{term}} \in W$ and a document collection, C, a document vector representation, $V^{\text{doc}}, |v^{\text{doc}}| = |v^{\text{term}}|$, is composed by applying a composition function, $F \in \{min, mean, max\}$ to each document, $d \in C$. For example, using the composition function $F_{\max}$, the value of the nth dimension of the document representation, denoted as $V_{d,n}^{doc}$, is:

$$V_{d,n}^{doc} = max(V_{i,n}^{term}) \forall i \in C_d \tag{1}$$

Each dimension of V^{doc} can then be used as a single feature for the purposes of classification. Moreover, in addition to the composition functions *min*, *mean* and *max*, we also deploy the compound function *concat*, where the resulting document representation is:

$$Concat(d) = [min(d), mean(d), max(d)] \tag{2}$$

Word embedding models capture the semantic relations of terms *within a collection*. Therefore, it is possible that semantic relations which are important for identifying sensitivities within our test collection may not be present in our chosen model. To address this, we construct document representations using two word embedding models that have been trained on different domains, namely Google News[6] and Wikipedia[7]. To do this, we apply the selected composition function, F, to each model, w_i, separately, to obtain a document representation from each model. We concatenate the document representations and use each vector dimension as a separate classification feature, resulting in the document representation:

$$semantic\_representation(d) = [F(w_i, d), F(w_{(i+1)}, d), ...F(w_n, d)] \tag{3}$$

4 Experimental Setup

In this section we present our experimental setup for evaluating the effectiveness of *term*, *grammatical* and *semantic* features for sensitivity classification. The research questions that we address are two-fold. Firstly, **RQ1**: "Are semantic word embeddings features more effective for sensitivity classification than grammatical or term features?" and, secondly, **RQ2**: "Does using multiple word embedding models trained on different domains further improve the effectiveness of semantic features for sensitivity classification?". Table 1 presents the combinations of feature sets that we evaluate, and the abbreviations that we use to denote each combination in the remainder of this paper.

Collection: We use a test collection of 3801 government documents that contain real sensitivities. The documents were sensitivity reviewed by trained government sensitivity reviewers, who assessed the documents against 2 FOIA exemptions, namely *International Relations* and *Personal Information*. All documents that were judged as containing any Exemption 27 or Exemption 40 sensitivities were labeled as *sensitive*. Table 2 presents the resulting collection statistics, after stopword removal. We use a 5-fold Cross Validation to perform the binary classification *sensitive* vs. *not-sensitive*. To address the class imbalance in the collection (13.2% sensitive), we match the number of sensitive and not-sensitive training instances by randomly down-sampling the *not-sensitive* documents in each fold.

---

[6] https://code.google.com/archive/p/word2vec/.
[7] http://nlp.stanford.edu/projects/glove/.

Table 1. Experimental setup: feature set combinations and abbreviations.

Feature set	Stand alone	Extending baseline
Text classification (baseline)	Text	-
Term n-grams	TN	Text+TN
Grammatical	POS	Text+POS
Semantic	WE	Text+WE
Term & grammatical	TN+POS	Text+TN+POS
Term & semantic	TN+WE	Text+TN+WE
Grammatical & semantic	POS+WE	Text+POS+WE
Term & grammatical & semantic	TN+POS+WE	Text+TN+POS+WE

Table 2. Salient statistics of our test collection.

Total documents	Not sensitive	Sensitive				Unique terms	Avg. doc length
		International relations	Personal information	Both	Total		
3801	3299	231	156	115	502	122 348	710 terms

Baseline: We evaluate each of the feature sets against a baseline text classification system using bag-of-words uni-gram term features, denoted as Text. We remove stopwords and terms that appear in only 1, or more than half, of the training documents in a fold. Feature values are binary, i.e. term features are either present or not. When extending text classification, additional features are scaled in the range $[0, 1]$.

Term Features: For term features, presented in Sect. 3.1, we test for term n-grams where $n = \{2..10\}$. When testing for values of n, we include n-grams for all values $<n$, i.e. when $n = 3$ feature vectors are constructed from all bi-grams and tri-grams. In the remainder of this paper, we denote term features as TN_n (i.e. for the previous example, TN_3). Feature values are binary, i.e. either present or not.

Grammatical Features: For grammatical features, presented in Sect. 3.2, we use the TreeTagger[8] part-of-speech tagger to POS tag documents and use a reduced set of 15 POS tags following [9,10]. We test for POS n-grams where $n = \{1..10\}$. Following the experimental setup for term features, when testing for values of n, we include n-grams for all values $< n$. Grammatical features are denoted as POS_n.

Semantic Features: We use *pre-trained* word embedding models and test if using two word embeddings models trained on different domains improves the effectiveness of semantic features for sensitivity classification.

Table 3 presents the word embedding models that we test. For each model, we evaluate each of the composition functions presented in Sect. 3.3, *min, mean,*

---

[8] http://www.cis.uni-muenchen.de/~schmid/tools/TreeTagger/.

Table 3. Pre-trained word embedding models for deriving semantic features.

Model	Architecture	Vocabulary size	# Dimensions	Training	Context window	Ref
Google News	word2vec	3M	300	Negative sampling	BoW5	WE_{gn}
Wikipedia+Gigaword5	Glove	400,000	300	AdaGrad	10+10	WE_{wp}

max and *concat*. As can be seen from Table 3, the models have 300 dimensional vectors and, hence, the functions *min, mean* and *max* result in 300 document features (900 for *concat*).

Classification and Metrics: For pre-processing and classification, we use scikit-learn[9]. As our classifier, we use SVM with a linear kernel and $C = 1.0$, since this theoretically motivated, default, parameter setting has been shown to provide the best effectiveness for text classification [2,24]. We select F_2 as our main metric since sensitivity classification is a recall oriented task [3,4], where the consequences of miss-classifying a sensitive document are much greater than miss-classifying a not-sensitive document. We also report the standard F-Measure (F_1) and, to account for class imbalance, we report Balanced Accuracy (BAC), where 0.5 BAC is random. We also report Precision, True Positive Rate (TPR), True Negative Rate (TNR) and the area under the Receiver Operating Characteristic curve (auROC) which, when documents are ranked by the output of a classifier's decision function, denotes the probability that a randomly selected positive instance is ranked higher than a randomly selected negative instance.

We test statistical significance, $p < 0.05$, using McNemar's non-parametric test [25] which is calculated from the prediction contingency tables for a pair of classifiers. Significant improvements compared to the text classification baseline (Text) are denoted with †. Additionally, in Table 5, significant improvements compared to the text classification with additional term features (Text+TN) are denoted with ‡.

5 Results

In this section, to answer the two research questions elicited in Sect. 4, we present the results of our classification experiments, over two tables: Table 4 presents the classification performance for each combination of *textual, grammatical* and *semantic* feature sets as *stand-alone* features; Table 5 presents the performance of each combination of feature sets extending the text classification baseline.

The baseline text classification approach (Text) is shown at the top of Tables 4 and 5, followed by sections for single, paired and triple feature sets respectively. We present results for term features (TN), grammatical features

---

[9] http://scikit-learn.org/.

Table 4. Results for combinations of *textual*, *grammatical* and *semantic* feature sets, compared against the text classification baseline.

Configuration		Precision	TPR	TNR	F_1	F_2	BAC	auROC
Text		0.2410	0.6573	0.6841	0.3520	0.4874	0.6707	0.7419
TN_6	†	0.2607	0.6970	0.6975	0.3786	0.5207	0.6972	0.7626
POS_{10}		0.2149	0.6095	0.6611	0.3177	0.4456	0.6353	0.6861
WE_{wp}(concat)		0.2019	0.6055	0.6350	0.3025	0.4321	0.6203	0.6801
WE_{gn}(concat)		0.1959	0.6034	0.6226	0.2956	0.4258	0.6130	0.6434
$WE_{wp}+WE_{gn}$(concat)		0.2106	0.6235	0.6432	0.3146	0.4474	0.6334	0.6962
$TN_{10}+POS_{10}$		0.2647	0.5974	**0.7438**	0.3632	0.4724	0.6706	0.7407
$TN_{10}+WE_{wp}$(concat)	†	0.2634	0.7130	0.6948	0.3839	0.5302	0.7039	**0.7797**
TN_9+WE_{gn}(concat)	†	0.2552	0.7208	0.6778	0.3761	0.5267	0.6993	0.7638
$TN_8+WE_{wp}+WE_{gn}$(concat)	†	**0.2657**	**0.7309**	0.6911	**0.3890**	**0.5401**	**0.7110**	0.7772
$POS_{10}+WE_{wp}$(concat)		0.2174	0.6512	0.6405	0.3241	0.4619	0.6458	0.7120
$POS_{10}+WE_{gn}$(concat)		0.2081	0.6275	0.6356	0.3117	0.4455	0.6315	0.6956
$POS_{10}+WE_{wp}+WE_{gn}$(concat)		0.2199	0.6552	0.6462	0.3280	0.4670	0.6507	0.7202
$TN_{10}+POS_{10}+WE_{wp}$(concat)	†	0.2592	0.6931	0.6954	0.3760	0.5171	0.6942	0.7585
$TN_{10}+POS_{10}+WE_{gn}$(concat)	†	0.2474	0.6651	0.6863	0.3584	0.4937	0.6757	0.7472
$TN_9+POS_{10}+WE_{wp}+WE_{gn}$(concat)	†	0.2531	0.6850	0.6887	0.3679	0.5078	0.6868	0.7599

(POS) and semantic features (WE). For WE, we present the results of the single word embedding models, Wikipedia (WE_{wp}) and Google News (WE_{gn}), and when used together ($WE_{wp}+WE_{gn}$). Due to space constraints in Tables 4 and 5, we use F_2 as our preferred metric and present the best performing size of n-grams for TN and POS. For semantic features, we present the best performing composition function (*min*, *max*, *mean* or *concat*).

Firstly, we note that the text classification baseline (Text) achieves 0.4874 F_2 and 0.6707 BAC, markedly better than random (0.5 BAC). Addressing **RQ1**, from Table 4, we observe that semantic features (WE) on their own are competitive with, but do not out perform, the text classification baseline. Additionally, we can see that the *concat* composition function consistently performs best. These findings are in line with the findings of Balikas and Amini [19] on a different collection.

As single feature sets, only text n-gram features (TN) achieve significant improvements compared to the text classification baseline (0.5207 F_2 vs 0.4874 F_2), denoted as †. This shows that text features provide a strong foundation for sensitivity classification. Moreover, the best performing text n-gram size is $n = 6$, showing that larger sequences of text are indeed important for sensitivity classification. Adding semantic features to the text n-grams results in additional improvements, compared to the baseline, and $TN_8+WE_{wp}+WE_{gn}$(concat) achieves the best overall performance in Table 4.

From Table 5, we can see that extending text classification with semantic features significantly improves classification performance. The best performing configuration, Text+$WE_{wp}+WE_{gn}$(concat), achieves a 5.5% increase in F_2 score, compared with the baseline. However, extending text classification with term n-grams (Text+TN_9) achieves the best classification performance for single feature sets (+8.3% F_2).

Table 5. Results for combinations of *textual*, *grammatical* and *semantic* feature sets extending the text classification baseline.

Configuration			precision	TPR	TNR	F_1	F_2	BAC	auROC
Text			0.2410	0.6573	0.6841	0.3520	0.4874	0.6707	0.7419
Text+TN_9	†		0.2667	0.7010	0.7060	0.3858	0.5279	0.7035	0.7782
Text+POS_{10}	†		0.2596	0.6532	0.7160	0.3707	0.4999	0.6846	0.7498
Text+WE_{wp}(concat)	†		0.2474	0.6692	0.6905	0.3609	0.4984	0.6799	0.7584
Text+WE_{gn}(concat)	†		0.2435	0.6653	0.6850	0.3560	0.4933	0.6752	0.7459
Text+WE_{wp}+WE_{gn}(concat)	†		0.2557	0.6891	0.6947	0.3725	0.5138	0.6919	0.7594
Text+TN_6+POS_{10}	†		**0.2780**	0.6751	**0.7308**	0.3920	0.5224	0.7029	0.7725
Text+TN_9+WE_{wp}(concat)	†		0.2678	0.7090	0.7051	0.3881	0.5322	0.7070	**0.7874**
Text+TN_6+WE_{gn}(concat)	†		0.2699	0.7169	0.7044	0.3913	0.5371	0.7107	0.7784
Text+TN_7+WE_{wp}+WE_{gn}(concat)	†	‡	0.2730	**0.7229**	0.7069	**0.3956**	**0.5425**	**0.7149**	0.7859
Text+POS_{10}+WE_{wp}(concat)	†		0.2507	0.6493	0.7041	0.3609	0.4913	0.6767	0.7620
Text+POS_{10}+WE_{gn}(concat)	†		0.2515	0.6571	0.7020	0.3626	0.4950	0.6796	0.7546
Text+POS_{10}+WE_{wp}+WE_{gn}(concat)	†		0.2504	0.6532	0.7026	0.3612	0.4930	0.6779	0.7634
Text+TN_4+POS_{10}+WE_{wp}(concat)	†		0.2674	0.6811	0.7147	0.3827	0.5181	0.6979	0.7789
Text+TN_9+POS_{10}+WE_{gn}(concat)	†		0.2634	0.6830	0.7081	0.3786	0.5154	0.6955	0.7747
Text+TN_6+POS_{10}+WE_{wp}+WE_{gn}(concat)	†		0.2657	0.6910	0.7081	0.3825	0.5214	0.6995	0.7798

Overall, the best performance is achieved when text classification is extended with additional *term* and *semantic* features combined, Text+TN_7+WE_{wp}+WE_{gn}(concat). This combination achieves 0.5425 F_2 and 0.7229 TPR, correctly classifying 9.99% more sensitive documents than the text classification baseline. Notably, this combination also results in significant improvements compared to extending text classification with only term n-gram features (Text+TN_9), denoted as ‡ in Table 5.

In response to **RQ1**, firstly, we find that semantic word embedding features are, indeed, useful features for sensitivity classification. This is shown by the observation of significant improvements to classification effectiveness when they are added to the next best performing feature set, denoted by ‡ in Table 5. However, we conclude that the best overall classification performance is achieved when text classification is extended with additional *term n-gram* and *semantic* features. Moving to **RQ2**, Tables 4 and 5 show that using multiple embedding models, WE_{wp}+WE_{gn}, consistently out performs either of the single models, WE_{wp} or WE_{gn}, when they are used individually. Therefore, we conclude that using multiple word embedding models trained on different domains does, indeed, improve the effectiveness of semantic features for sensitivity classification.

6   Analysis

In this section, we provide analysis of the findings from our classification experiments. In Sect. 6.1, we discuss the classification predictions that are correct solely due to the word embedding features. In Sect. 6.2, we discuss the benefits for the sensitivity review process from extending text classification with semantic and term n-gram features.

6.1   Semantic Features

We now provide a short analysis of the documents we can correctly predict due to semantic features. We compare the best performing system, Text+TN+WE$_{wp}$+WE$_{gn}$, against text classification extended with term n-gram features, Text+TN.

Additional semantic features (from multiple domains) enable the classifier to convert 23 False Negative predictions to True Positive predictions, and 144 False Positive predictions to True Negative predictions. 13.77% of these converted predictions were sensitive documents. From the 23 converted sensitive documents, 15 are sensitive with respect to *International Relations*, 4 are sensitive with respect to *Personal Information* and 4 are sensitive with respect to both sensitivities.

Each of the documents with International Relations sensitivity contain multiple paragraphs that recount interactions and conversations between people and, moreover, the document's sensitivity is directly linked to these. This is in line with how we expect semantic features to enhance sensitivity classification, since these relations can be preserved in the dimensions of the vector representations. Interestingly, the sensitivities in documents relating to Personal Information also relate to actions, such as booking hotels, forced resignations and visa bans. Therefore, we intend to investigate such patterns of interaction relations further in future work, to develop classification rules for sensitivity and evaluate their cost/benefit trade-off for various sensitivity review user models.

6.2   Sensitivity Review

It is useful to provide sensitivity reviewers with a reliable way to predict how many sensitive documents remain in a partially reviewed collection. One way to approach this is to rank documents by a classifier's decision function output and review the ranking sequentially. We can then ask "how conservative does a classifier have to be, to correctly predict a certain percentage of sensitive documents?" In line with this user model, Fig. 2 presents the Receiver Operating Characteristic curve, and True Positive Rate vs classification threshold for our classifier with additional term and semantic features, compared against the baseline text classification.

As can be seen from Fig. 2(a), the additional features increase the True Positive Rate throughout the ranking. Therefore, a reviewer can have increased confidence in the system. Additionally, Fig. 2(b), shows that semantic and term features enable the classifier to be less conservative. For example, the gray dashed lines in Fig. 2(b) show that, with the additional features, we can correctly classify 95% of all sensitive documents by lowering the classification threshold to −0.46, whereas, the baseline would need to be set at −0.645. By using our approach, on this test collection, a reviewer would need to review 262 fewer documents to identify 95% of all sensitive documents.

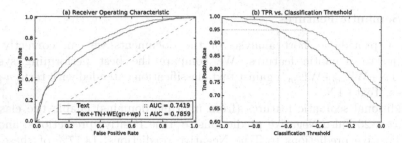

Fig. 2. (a) Receiver Operating Characteristic Curve. (b) True Positive Rate vs. Classification Threshold. The blue line shows the baseline text classification (Text) and the red line shows Text+TN$_7$+WE$_{wp}$+WE$_{gn}$(concat). The dashed line in (a) shows a random classifier. The dashed lines in (b) show the classification threshold required to achieve 0.95 TPR. (Color figure online)

7   Conclusions

In this work, we presented an effective approach for automatically classifying sensitive information in government documents, to assist the sensitivity review process. Our classifier deploys semantic features, derived from pre-trained word embedding models, to identify latent sensitive relations in documents. In a thorough evaluation, we compared the performance of the semantic features against grammatical and term features, as stand-alone features and extending text classification. We found that extending text classification with semantic features enabled our classifier to make significantly more accurate predictions, according to McNemar's test. Extending text classification with term n-gram and semantic features resulted in an 11.3% increase in F$_2$ score, correctly classifying 9.99% more sensitive documents than the baseline approach. Moreover, this approach markedly reduced the number of documents a reviewer would need to review to identify 95% of all sensitive documents in our collection (262 fewer documents).

Acknowledgements. The authors are thankful to the Foreign & Commonwealth Office and The National Archives of the UK for their support of this work.

References

1. DARPA: DARPA, new technologies to support declassification (2010). http://fas. org/sgp/news/2010/09/darpa-declass.pdf
2. Sebastiani, F.: Machine learning in automated text categorization. ACM Comput. Surv. **34**(1), 1–47 (2002)
3. McDonald, G., Macdonald, C., Ounis, I., Gollins, T.: Towards a classifier for digital sensitivity review. In: Rijke, M., Kenter, T., Vries, A.P., Zhai, C.X., Jong, F., Radinsky, K., Hofmann, K. (eds.) ECIR 2014. LNCS, vol. 8416, pp. 500–506. Springer, Cham (2014). doi:10.1007/978-3-319-06028-6_48
4. Berardi, G., Esuli, A., Macdonald, C., Ounis, I., Sebastiani, F.: Semi-automated text classification for sensitivity identification. In: Proceedings of CIKM (2015)

5. Harris, Z.S.: Distributional structure. Word **10**(2–3), 146–162 (1954)
6. Fung, B., Wang, K., Chen, R., Yu, P.S.: Privacy-preserving data publishing: a survey of recent developments. ACM Comput. Surv. (CSUR) **42**(4), 14 (2010)
7. Fang, Y., Godavarthy, A., Lu, H.: A utility maximization framework for privacy preservation of user generated content. In: Proceedings of ICTIR (2016)
8. Berardi, G., Esuli, A., Sebastiani, F.: A utility-theoretic ranking method for semi-automated text classification. In: Proceedings of SIGIR (2012)
9. McDonald, G., Macdonald, C., Ounis, I.: Using part-of-speech n-grams for sensitive-text classification. In: Proceedings of ICTIR (2015)
10. Lioma, C., Ounis, I.: Examining the content load of part-of-speech blocks for information retrieval. In: Proceedings of COLING/ACL (2006)
11. Pavlick, E., Rastogi, P., Ganitkevitch, J., Van Durme, B., Callison-Burch, C.: PPDB 2.0: better paraphrase ranking, fine-grained entailment relations, word embeddings, and style classification. In: Proceedings of ACL-IJCNLP (2015)
12. Ghosh, D., Guo, W., Muresan, S.: Sarcastic or not: word embeddings to predict the literal or sarcastic meaning of words. In: Proceedings of EMNLP (2015)
13. Mikolov, T., Sutskever, I., Chen, K., Corrado, G.S., Dean, J.: Distributed representations of words and phrases and their compositionality. In: Proceedings of NIPS (2013)
14. Pennington, J., Socher, R., Manning, C.D.: Glove: global vectors for word representation. In: Proceedings of EMNLP (2014)
15. Zheng, G., Callan, J.: Learning to reweight terms with distributed representations. In: Proceedings of SIGIR (2015)
16. Zuccon, G., Koopman, B., Bruza, P., Azzopardi, L.: Integrating and evaluating neural word embeddings in information retrieval. In: Proceedings of ADCS (2015)
17. Yang, X., Macdonald, C., Ounis, I.: Using word embeddings in Twitter election classification. CoRR abs/1606.07006 (2016)
18. Joulin, A., Grave, E., Bojanowski, P., Mikolov, T.: Bag of tricks for efficient text classification. CoRR abs/1607.01759 (2016)
19. Balikas, G., Amini, M.: An empirical study on large scale text classification with skip-gram embeddings. CoRR abs/1606.06623 (2016)
20. Mitchell, J., Lapata, M.: Composition in distributional models of semantics. Cogn. Sci. **34**(8), 1388–1429 (2010)
21. Cortes, C., Vapnik, V.: Support-vector networks. Mach. Learn. **20**(3), 273–297 (1995)
22. Collobert, R., Weston, J., Bottou, L., Karlen, M., Kavukcuoglu, K., Kuksa, P.P.: Natural language processing (almost) from scratch. JMLR **12**, 2493–2537 (2011)
23. Socher, R., Huang, E.H., Pennin, J., Manning, C.D., Ng, A.Y.: Dynamic pooling and unfolding recursive autoencoders for paraphrase detection. In: Proceedings of NIPS (2011)
24. Joachims, T.: Text categorization with support vector machines: learning with many relevant features. In: Nédellec, C., Rouveirol, C. (eds.) ECML 1998. LNCS, vol. 1398, pp. 137–142. Springer, Heidelberg (1998). doi:10.1007/BFb0026683
25. McNemar, Q.: Note on the sampling error of the difference between correlated proportions or percentages. Psychometrika **12**(2), 153–157 (1947)

Predicting Users' Future Interests on Twitter

Fattane Zarrinkalam[1,2]([✉]), Hossein Fani[1,3], Ebrahim Bagheri[1],
and Mohsen Kahani[2]

[1] Laboratory for Systems, Software and Semantics (LS3), Ryerson University,
Toronto, Canada
[2] Department of Computer Engineering, Ferdowsi University of Mashhad,
Mashhad, Iran
fattane.zarrinkalam@gamil.com
[3] Faculty of Computer Science, University of New Brunswick, Fredericton, Canada

Abstract. In this paper, we address the problem of predicting future
interests of users with regards to a set of *unobserved* topics in microblogging services which enables forward planning based on potential future
interests. Existing works in the literature that operate based on a known
interest space cannot be directly applied to solve this problem. Such
methods require at least a minimum user interaction with the topic to
perform prediction. To tackle this problem, we integrate the semantic
information derived from the Wikipedia category structure and the temporal evolution of user's interests into our prediction model. More specifically, to capture the temporal behaviour of the topics and user's interests, we consider discrete intervals and build user's topic profile in each
time interval separately. Then, we generalize users' interests that have
been observed over several time intervals by transferring them over the
Wikipedia category structure. Our approach not only allows us to generalize users' interests but also enables us to transfer users' interests across
different time intervals that do not necessarily have the same set of topics. Our experiments illustrate the superiority of our model compared to
the state of the art.

1 Introduction

Techniques for the identification and modeling of user interests based on
users' social presence have received much attention in the recent years [2,10].
Researchers have already explored ways in which user interests can be modeled
in social networks with special attention being given to Twitter. Existing works
often provide a view of users' interests with regards to a set of core themes. For
instance, some works have expressed users' interests in terms of bag of words,
Wikipedia entries or in relation to the current active topics on the social network.

While approaching the problem from different technical perspectives, most
of the existing works on social networks focus on modeling users' current interests and little work has been done on the prediction of users' potential future

© Springer International Publishing AG 2017
J.M. Jose et al. (Eds.): ECIR 2017, LNCS 10193, pp. 464–476, 2017.
DOI: 10.1007/978-3-319-56608-5_36

interests. In all these works, the interest space is assumed to be known *a priori*; therefore, various models of collaborative filtering and link prediction that require a known interest space can effectively be employed [3,23].

Our work in this paper aims to extend the state of the art by predicting users' interests with regards to future *unobserved* topics. In other words, our objective is to provide a solution for performing *what-if* analysis over potential future topics. For instance, we are interested in determining whether a given user would be interested in following the news about the release of a new mobile operating system that would compete with iOS. Our work will enable forward planning based on potential future interests. Given the focus of our work on unobserved topics, existing works in the literature that operate based on a known interest space cannot be directly applied to it. Those techniques would require at least some minimum user interactions [4].

To address the above problem statement, in this paper, we propose a prediction framework to integrate semantic information from knowledge bases such as Wikipedia and temporal evolution of each individual user's interests to predict user's future interests. Knowledge infused prediction algorithms have gained significant attention due to their competitive performance and ability to overcome the cold start problem [14,18]. However utilizing knowledge bases for improving user interest prediction methods in microblogging services is largely unexplored. Our prediction model is based on the intuition that, although it is possible that the topics of interest to the users dramatically change over time as influenced by real-world trends [1], users tend to incline towards topics and trends that are semantically or conceptually similar to a set of core interests. Therefore, in order to be able to achieve predictability, one would need to generalize each individual user's interests over several time intervals to gain a good insight into the user's overall mindset. To this end, we generalize users' interests that have been observed over several time intervals by transferring them onto the Wikipedia category structure. Generally, our approach utilizes the Wikipedia category structure to model high level user interests and takes the temporal evolution of user's interests into account in order to predict user's future interests. The key contributions of this paper are as follows:

- We propose a model that transfer user's interests from different time intervals onto the Wikipedia's category structure. In this process, we model high-level interests of users such that the evolution of user's interests over topics is captured.
- We illustrate how semantic information derived from the Wikipedia knowledge base as well as temporal information can be integrated in our model to predict user's interests with regards to unobserved topics of the future in Twitter.
- We perform experimentation to illustrate the impact of considering Wikipedia categories on the accuracy of predicting the future interests of users on Twitter. The experimental results demonstrate the superiority of our model compared to the state of the art methods which tackle cold item problem.

The rest of the paper is organized as follows: In Sect. 2 we describe the related work. Sections 3 and 4 are dedicated to the problem definition and the presentation of the details of our proposed approach. Section 5 presents the details of our experimental work. Finally, Sect. 6 concludes the paper.

2   Related Work

There is a rich line of research on user interest detection from social networks through the analysis of user generated textual content. To represent user interests, such works either use *Bag of Words*, *Topic Modeling* or *Bag of Concepts* approach. Since the *Bag of Words* [20] and *Topic Modeling* [19] approaches focus on terms without considering their semantics and the relationship between them, they do not necessarily utilize the underlying semantics of textual content. Furthermore, these approaches may not perform so well on short, noisy and informal texts like Twitter posts [6]. To address these issues, the *Bag of Concepts* approach utilizes external knowledge bases to enrich the representation of short textual content and model user interests through semantic entities (concepts) linked to external knowledge bases such as DBpedia. Since these knowledge bases represent entities and their relationships, they provide a way of inferring underlying semantics of content [13].

While existing work on microblogging services mainly focus on extracting users' current interests, little work has been done on predicting users' future interests. Bao et al. [3] have proposed a temporal and social probabilistic matrix factorization model that utilize users' sequential interest matrices at different time intervals and the users' friendships matrix to predict future users' interest in microblogging services. Their work is very similar to ours in a sense that we both try to predict future user interests in microblogging services by taking into account the temporal evolution of user interests. However, they are limited by the fact that they assume the topic set of the future to be known *a priori* and composed only of the set of topics that have been observed in the past. Therefore, they cannot predict user interests with regard to new topics since these topics have never received any feedbacks from users in the past.

Given users' interests change over time, temporal aspects have been widely used for the conventional recommendations and user modeling in online social networks [21]. Many researchers have focused on applying time decay functions over historical user generated content [8]. Based on time decay functions, the weight of each interest is calculated depending on its age. Recently, Piao and Breslin [15] have studied the effectiveness of different time decay functions for incorporating dynamics of user interests in the context of personalized link recommendations on Twitter. They have shown that using decay functions to build users' long-term profiles results in noticeable improvement in the quality of recommendations compared to user profiles without considering any decay of user interests. There is another line of related works that utilize knowledge base information to overcome the cold start problem in traditional algorithms in the context of recommender systems [11]. For example, Cheekula et al. [5] have proposed a content-based recommendation method that utilizes hierarchical user interests over Wikipedia category hierarchy to identify relevant entities. Their work is similar to ours in a sense that both model high-level interests of users over the Wikipedia category graph. However, they overlook the evolution of user's interests over time. Further, our work focuses on predicting user's interests over unobserved topics in the future as opposed to entity recommendation.

3   Preliminaries

3.1   User Interest Profile

In our work, we model users' interests in relation to the active topics of the social network. A topic z has traditionally been defined as a semantically coherent theme which has received substantial attention from the users.

Let t be a specified time interval, given $\mathbb{Z}^t = \{z_1^t, z_2^t, \ldots, z_K^t\}$ be K active topics in t, for each user $u \in \mathbb{U}$, we define her topic profile in time interval t, $TP^t(u)$, which is the distribution of u's interests over $\mathbb{Z}^t$, as follows:

Definition 1 *(Topic Profile). The topic profile of user $u \in \mathbb{U}$ in time interval t, with respect to $\mathbb{Z}^t$, denoted by $TP^t(u)$, is represented by a vector of weights over the K topics, i.e., $(f_u^t(z_1^t), \ldots, f_u^t(z_K^t))$, where $f_u^t(z_k^t)$ denotes the degree of u's interest in topic $z_k^t \in \mathbb{Z}^t$. A user topic profile is normalized so that the sum of all weights in a profile equals to 1.*

It should be noted that topic and user interest detection methods from microblogging services have already been well studied in the literature and therefore are not the focus of our work and we are able to work with any topic and interest detection method to extract $\mathbb{Z}^t$ and $TP^t(u)$.

3.2   Problem Definition

The objective of our work is to answer *what-if* questions by predicting user interests with regards to potentially trending topics of the future. To achieve this goal, we rely on temporal and historical user interest information in order to predict how users would react to future topics. Recent studies have already shown that trending topics on social networks can rapidly change in reaction to real world events and therefore, the set of topics might significantly change between different time intervals [1]. Therefore, to express the temporal dynamics of topics and user interests, we divide the users' historical data into L discrete time intervals $1 \leq t \leq L$ and extract L topic sets $\mathbb{Z}^1, \mathbb{Z}^2, \ldots, \mathbb{Z}^L$, in these time intervals using the microposts which are published in each time interval separately. More specifically, for each time interval $t : 1 \leq t \leq L$, we first extract active topics in that time interval $\mathbb{Z}^t$, and then for each user $u \in \mathbb{U}$, we build her topic profile in time interval t, $TP^t(u)$, as a result of which each user will have L user profiles, one for each of the time intervals. Informally speaking, our objective is to exploit the L historical topic profiles of a user u, to predict the user's inclination towards the topics of time interval $L + 1$.

Definition 2 (Future Topic Profile). *Given the topic profiles for each user u in each time interval of the historical data, $TP^1(u), \ldots, TP^L(u)$, and a set of topics in time interval $L+1$, $\mathbb{Z}^{L+1}$, which might not have been observed in the previous time intervals, we aim to predict $\widehat{TP}^{L+1}(u)$, the future topic profile of user u towards $\mathbb{Z}^{L+1}$.*

To address the challenge defined in Definition 2, we divide this problem into two subproblems: *historical user topic profile extraction* and *future interest prediction*, in which the output of the first subproblem becomes the input of the second one.

4   Proposed Approach

In this section, we first introduce our method to extract historical topic profile of users and then we describe our prediction model to predict future interests of users.

4.1   Historical User Topic Profile Extraction

As explained earlier, our work relies on each user's topic profiles within the past L intervals. Each user topic profile in a given time interval t is a distribution over the active topics in that time interval $\mathbb{Z}^t$, which is not necessarily the same as the topics in the previous or next time intervals. In order to extract $TP^t(u)$, the user topic profile for each user u in each time interval of the historical data, $1 \le t \le L$, we employ the LDA topic modeling approach.

Considering $\mathbb{M}^t$, the set of microposts as a text corpus published in time interval t, it is possible to extract topics $\mathbb{Z}^t$ using topic modeling methods. As proposed in [16], to obtain better topics from microblogging services without modifying the standard topic modeling methods, we enrich each micropost m from our corpus $\mathbb{M}^t$ by using an existing semantic annotator and employ the extracted entities, which can lead to the reduction of noisy content within the topic detection process. Therefore, in our work, each micropost is considered as a set of one or more semantic entities that collectively denote the underlying semantics of the microposts. Therefore, we view a topic, defined in Definition 3, as a distribution over Wikipedia entities.

Definition 3 (Topic). *Let $\mathbb{M}^t$ be a corpus of microposts published in time interval t and $\mathbb{E} = \{e_1, e_2, \ldots, e_{|\mathbb{E}|}\}$ be the vocabulary of Wikipedia entities, an active topic in time interval t, z^t, is defined to be a vector of weights, i.e., $(g_z^t(e_1), \ldots, g_z^t(e_{|\mathbb{E}|}))$, where $g_z^t(e_i)$ shows the participation score of term $e_i \in \mathbb{E}$ in forming topic z^t. Collectively, $\mathbb{Z}^t = \{z_1^t, z_2^t, \ldots, z_K^t\}$ denotes a set of K topics extracted from $\mathbb{M}^t$.*

To extract the topics from microposts using LDA, documents should naturally correspond to microposts. However, since our goal is to understand the topics that each user u is interested in rather than the topic that each single micropost is about, similar to previous works in the literature [17], we aggregate the published or retweeted microposts of a user u in time interval t, i.e., $\mathbb{M}_u^t$, into a single document. LDA has two parameters to be inferred from the corpus of documents: document-topic distributions θ, and the K topic-term distributions ϕ. Given that each document corresponds to a user u and Wikipedia entities $\mathbb{E}$ as the vocabulary of terms, by applying LDA over the microposts $\mathbb{M}^t$, the results produce the following two artifacts:

- K topic-entity distributions, where each topic entity distribution associated with a topic $z^t \in \mathbb{Z}^t$ represents active topics in $\mathbb{M}^t$, i.e., $(g_z^t(e_1), \ldots, g_z^t(e_{|\mathbb{E}|}))$
- $|\mathbb{U}|$ user-topic distributions, where each user-topic distribution associated with a user u, represents the topic profile of user u in time interval t, i.e., $TP^t(u) = (f_u^t(z_1^t), \ldots, f_u^t(z_K^t))$.

Now, given a corpus of microposts $\mathbb{M}$, we will break it down into L intervals and perform the above process separately on each of the intervals. This will produce $TP^1(u), \ldots, TP^L(u)$ for every user u in our user set, which is the required input for our future user interest prediction problem defined in Definition 2.

4.2 Future Interest Prediction

Given $TP^1(u), TP^2(u), \ldots, TP^L(u)$, our goal is to predict potential interests of each user u over $\mathbb{Z}^{L+1}$. It is important to point out that since $L+1$ is in the future, the topics $\mathbb{Z}^{L+1}$ have not yet been observed. Therefore, our work aims to answer important *what-if* questions in that it is able to predict how the users react to a given set of topics. This allows one to perform future planning by studying how users will react if certain topics emerge in the future. Our prediction model is based on the intuition that while user interests might change over time, they tend to revolve around some fundamental issues. More specifically, although user interests are driven by the shifts and changes in real world events and trends [1], they tend towards topics and trends that are semantically or conceptually similar. For this reason, we generalize users' interests that have been observed over several time intervals by transferring them over the Wikipedia category structure. This approach will not only allow us to generalize users' interests but also enables us to transfer users' interests across different time intervals that do not necessarily have the same set of topics.

Based on the above intuition, formally, for each user u, given the topic profiles of the user u in each time interval t, $TP^t(u)$, we utilize Wikipedia category structure to build a category profile for user u in each time interval t, denoted as $CP^t(u)$.

Definition 4 (Category Profile). *The category profile of user $u \in \mathbb{U}$ in time interval t toward Wikipedia categories $\mathbb{C} = \{c_1, c_2, \ldots, c_{|\mathbb{C}|}\}$, called $CP^t(u)$, is represented by a vector of weights, i.e., $(h_u^t(c_1), \ldots, h_u^t(c_{|\mathbb{C}|}))$, where $h_u^t(c)$ denotes the degree of u's interest in category $c \in \mathbb{C}$ at time interval t. A user category profile is normalized so that the sum of all weights in a profile equals to 1.*

Now, based on the Category Profiles of each user derived from the past L consecutive time intervals, $CP^1(u), \ldots, CP^L(u)$, we apply our model to predict $\widehat{TP}^{L+1}(u)$.

Category Profile Identification. In this section, we aim at utilizing the Wikipedia category structure to generalize the topic-based representation of user interests to category-based representation. To do so, there are two possible approaches through which we build the category profile of a user u at time interval t, $CP^t(u)$, given her topic profile $TP^t(u)$: *(1) attribution*, and *(2) hierarchical* approach.

In the *attribution* approach, for each user u, only those categories that are directly associated with the constituent entities of the user's topics of interest are considered as categories of interest. We essentially map $TP^t(u) = (f_u^t(z_1^t), \ldots, f_u^t(z_K^t))$ to $CP^t(u) = (h_u^t(c_1), \ldots, h_u^t(c_{|\mathbb{C}|}))$ as follows:

$$h_u^t(c) = \sum_{i=1}^{K} f_u^t(z_i^t) \times \Phi(z_i^t, c) \tag{1}$$

where $\Phi(z, c)$ denotes the degree of relatedness of topic $z^t = (g_z^t(e_1), \ldots, g_z^t(e_{|\mathbb{E}|}))$ to category $c \in \mathbb{C}$ and is calculated based on Eq. 2.

$$\Phi(z^t, c) = \sum_{i=1}^{|\mathbb{E}|} g_z^t(e_i) \times \delta_c(e_i) \tag{2}$$

Here, $\delta_c(e)$ is set to 1 if entity e is a Wikipedia page that belongs to the Wikipedia category c, otherwise it is zero and $g_z^t(e)$ is the distribution value of entity e in topic z^t, produced by applying LDA over $\mathbb{M}^t$ as described in Sect. 4.1. It is important to note that the reason why we can calculate the relatedness of each topic to each category is that we view each topic as a distribution over Wikipedia entities and in Wikipedia, each entry is already associated with one or more categories.

In the *hierarchical* approach, we assume that when a user is interested in a certain category, she might also be interested in broader related categories. Based on this, in the hierarchical approach, we first infer the broadly related categories of user interests by exploiting the hierarchy of the Wikipedia category structure. A major challenge in utilizing Wikipedia category structure as a hierarchy is that, it is a cyclic graph instead of a strict hierarchy [9]. Therefore, as a preprocess in the hierarchical approach, we transform the Wikipedia category structure into a hierarchy by adopting the approach proposed in [9]. The output of this process is a Wikipedia Category Hierarchy (WCH), a directed acyclic graph whose nodes are the Wikipedia categories $\mathbb{C}$ with an edge from $c_i \in \mathbb{C}$ to $c_j \in \mathbb{C}$ whenever c_i is a subcategory of c_j.

For a user u, given $TP^t(u) = (f_u^t(z_1^t), \ldots, f_u^t(z_K^t))$ and Wikipedia Category Hierarchy WCH as input, we infer the hierarchical interests of user u in time interval t, represented in the form of a category hierarchy. To do so, for each topic z_i^t, we first assign an initial score of $f_u^t(z_i^t) \times \Phi(z_i^t, c)$ to every category node $c \in \mathbb{C}$ similar to what is done in the attribution approach. Then, the score of each category node with a $score(c) > 0$ is propagated up the hierarchy as far as the root using a Spreading Activation function to calculate the new score of each node. We adopt the 'Bell Log' function as our spreading activation function as described in [9].

Now, given topic profiles of a user u in L consecutive time intervals of the historical data, i.e., $TP^1(u), \ldots, TP^L(u)$, we perform the above process separately on each of the intervals. This will produce $CP^1(u), \ldots, CP^L(u)$ for every user $u \in \mathbb{U}$, which is the input of our method described in the next section to predict $\widehat{TP}^{L+1}(u)$.

Interest Prediction. Given $CP^1(u), \ldots, CP^L(u)$, our first step to predict $\widehat{TP}^{L+1}(u)$ is calculating $CP^{L+1}(u)$. As already discussed in the literature, users' current interests are driven by their past interests, interactions and behavior where distant history has a lesser influence on the current interests compared to more recent events and activities [15]. Based on this observation, we employ a decay function in order to soften the impact of distant experiences on the users' future interests. We choose the exponential decay function which can describe this influence effectively [8]. More formally, we calculate the category profile of user u in time interval $L+1$, $CP^{T+1}(u) = (h_u^{L+1}(c_1), \ldots, h_u^{L+1}(c_{|\mathbb{C}|}))$, as follows:

$$h_u^{L+1}(c) = \sum_{t=1}^{L} exp(-\frac{L-t}{\alpha}) h_u^t(c) \tag{3}$$

where the value of $\alpha > 0$ presents the kernel parameter, and the value of L shows the number of time intervals that the historical data is divided to. In our experiments, we choose α as the length of each time interval t [12].

Given the high-level interests of user u in time interval $L+1$ represented over Wikipedia categories, $CP^{L+1}(u)$, and a set of unobserved topics (what-if subjects) for time interval $L+1$, $\mathbb{Z}^{L+1}$, we are interested in predicting a topic profile for user u, $\widehat{TP}^{L+1}(u) = (\hat{f}_u^{L+1}(z_1^{L+1}), \ldots, \hat{f}_u^{L+1}(z_K^{L+1}))$. We calculate $\hat{f}_u^{L+1}(z_i^{L+1})$ as follows:

$$\hat{f}_u^{L+1}(z_i^{L+1}) = \sum_{j=1}^{\mathbb{C}} \Phi(z_i^{L+1}, c_j) \times h_u^{L+1}(c_j) \tag{4}$$

where $\Phi(z, c)$ calculates the relatedness of topic z to category c based on Eq. 2.

5   Experiments

5.1   Dataset and Experimental Setup

In our experiments, we use an available Twitter dataset collected and published by Abel et al. [2]. It consists of approximately 3M tweets posted by 135,731 unique users. We annotated the text of each tweet with Wikipedia entities using the TAGME RESTful API[1], which resulted in 350,731 unique entities. We divide our dataset into $L+1$ fixed time intervals. The first L time intervals serve as our training data and the last is employed for testing. To prepare Wikipedia category graph, we downloaded the freely available English version of DBpedia, which is extracted from Wikipedia dumps dating from October 2015. This dataset consists of 968,350 categories with 2,225,459 subcategory relations between them. We preprocessed the Wikipedia category hierarchy as suggested in [9]. The outcome of this process is a hierarchy with a height of 20 and 824,033 categories with 1,506,292 links among them.

5.2   Evaluation Methodology and Metrics

Given the outputs of LDA over $L+1$ time intervals of our dataset, we consider the first L extracted topic profiles of each user u, $TP^1(u), TP^2(u), \ldots, TP^L(u)$, as her historical interests for training and $TP^{L+1}(u)$ as the golden truth of her interests for testing.

To evaluate $\widehat{TP}^{L+1}(u)$, we choose two popular metrics for evaluating the '*accuracy of predictions*': Mean Absolute Error (MAE) and Root Mean Squared Error (RMSE). A lower MAE or RMSE scores indicates more accurate prediction results. Further we calculate the Normalized Discounted Cumulative Gain (nDCG) as a well-known metric for evaluating the '*ranking quality*' of the results.

5.3   Comparison Methods

Our goal is to predict the degree of user interests over topics that emerge in the future, which have not been observed in the past. Among different recommendation strategies, collaborative filtering methods cannot recommend new items since these items have never received any user's feedbacks in the past. To tackle the cold item problem, content-based and hybrid approaches that incorporate item content are recommended [4]. Thus, we consider the following comparison methods:

SCRS (Semantic Content-based Recommender System) [14] extracts item features from Wikipedia to compute the semantic similarity of two items. The adoption of this approach in our context would need us to consider each topic of interest as an item and the constituent Wikipedia entities of a topic as its content. Then, we predict $\widehat{TP}^{L+1}(u) = (\hat{f}_u^{L+1}(z_1^{L+1}), \ldots, \hat{f}_u^{L+1}(z_K^{L+1}))$ as follows:

$$\hat{f}_u^{L+1}(z_i^{L+1}) = \frac{1}{K \times L} \sum_{t=1}^{L} \sum_{j=1}^{K} f_u^t(z_j^t) \times S(z_i^{L+1}, z_j^t) \tag{5}$$

where $S(z_1, z_2)$ denotes the similarity of two topics calculated by the cosine similarity of their respective entity weight distribution vectors defined in Definition 3.

---

[1] http://tagme.di.unipi.it/.

ACMF (Attribute Coupled Matrix Factorization) [22] is a hybrid approach that incorporates item-attribute information (item content) into the matrix factorization model to cope with the cold item problem. In our work, the items are the topics of all time intervals, i.e., $\mathbb{Z} = \bigcup_{1 \leq t \leq L+1} \mathbb{Z}^t$. Accordingly, the item relationship regularization term is adopted as follows:

$$\frac{\beta}{2} \sum_{i=1}^{|\mathbb{Z}|} \sum_{j=1}^{|\mathbb{Z}|} S(z_i, z_j) \|q_i - q_j\|_F^2 \tag{6}$$

where β is the regularization parameter to control the effect of the item (topic)-attribute information, $S(z_1, z_2)$ is the similarity between topics z_i and $z_j \in \mathbb{Z}$, as described for Eq. 5. Further, q is the topic latent feature vector, and $\|.\|_F^2$ is the Frobenius norm.

Attribution (Attribution-based future user interest prediction) is a variant of our proposed approach which uses the attribution method as described in Sect. 4.2 to build the category profile of a user.

Hierarchical (Hierarchical-based future user interest prediction) is a variant of our proposed approach which uses the Wikipedia Category Hierarchy as described in Sect. 4.2 to build category profile of a user.

5.4   Results and Discussion

In order to ensure that our experiments are generalizable and not impacted by the effect of parameter setting, we explore a range of values for the two possible variables that can affect the performance of our work, i.e., the length of the time intervals and the number of topics. We perform the evaluations for different lengths of time interval: 1, 3 and 7 days and for varying number of topics ranging from 20 to 50. We present the quality of the prediction results in Fig. 1 where we can observe that the two variants of our proposed approach, i.e., Attribution and Hierarchical methods, outperform SCRS and ACMF in terms of both MAE and RMSE. This observation confirms that utilizing Wikipedia category structure enables us to model user's high level interests more accurately and consequently can lead to improve the quality of user interest prediction with regards to new topics of the future. It is worth noting that this achievement is consistent in all different time interval sizes and the number of topics.

Figure 1 additionally shows that our method (Attribution) outperforms the other comparison methods in terms of the ranking metric (nDCG). This is an important observation when it is considered collectively with the results obtained from MAE and RMSE. It points to the fact that the Attribution method not only provides an accurate estimation of the degree of interest but is also able to accurately predict the ranking of user interests, which shows that we can estimate the *preference order* between user interests as well as the degree of difference between these interests for every given user. Now, when considering the other baseline methods, it is interesting to see that while SCRS performs the worst among the various methods in terms of MAE and RMSE, it produces accurate

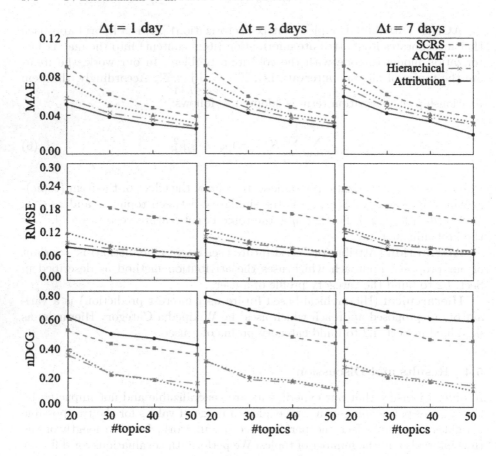

Fig. 1. Evaluation results in terms of MAE, RMSE and nDCG.

rankings. This can potentially be explained by the fact that while SCRS is not able to accurately predict the degree of user interests, it is able to estimate the preference order between the user interests. However, our proposed Attribution approach is still the best performing method in all three measures.

By comparing Attribution and Hierarchical variants of our proposed approach, one can observe that the Attribution method provides better results. Both methods model user high-level interests over Wikipedia categories. The difference is that, in the Attribution approach, only those categories that are directly associated with the constituent entities of the user's topics of interest are considered as categories of interest. However, in the Hierarchical approach, broadly related categories of user interests are also considered by applying a spreading activation function over the hierarchy of the Wikipedia category structure. Here we adopt the Bell-log activation function proposed in [9] for this purpose. We speculate the probable cause for the poor performance of the Hierarchical approach compared to Attribution approach is the Bell-log activation function. On the one hand, Bell-log activation function spreads all the scores from the leaves up to the root of the hierarchy in a way that broader categories receive higher scores. On the other hand, higher categories are usually common among majority of users' category profiles. In the prediction step, it may happen that a topic can belong to this very broad category. Hence, this topic will be predicted as a topic of interest for almost all users which leads to the above mentioned poor accuracy. We believe discrete time state space models [7] may alleviate the inappropriate score assignments by the Bell-log. Such models set category score based on a convex combination of its predecessors and successors. This will be another area for our future investigation.

Now, among the baselines and as shown in Fig. 1, ACMF, which is a hybrid recommender system that combines collaborative filtering and topic content, can achieve more accurate results in terms of MAE/RMSE in comparison with SCRS, which is solely based on topic content. This could indicate that incorporating user interests of other users might improve the accuracy of user interest predictions. Based on this observation, it seems promising to investigate collaborative extensions of our proposed approach as future work.

6 Conclusions

In this paper, we address the problem of predicting future interests of users with regards to a set of *unobserved* topics (*what-if* subjects) on Twitter. Our model is based on the intuition that while user interests might change over time in reaction to real world events, they tend to revolve around some fundamental issues that can be seen as the user's mindset. To capture the temporal behaviour of the topics and user's interests, we consider discrete time intervals and build user's topic profile in each time interval as the user's historical topic profiles. Then, we generalize each individual user's topic profile as we move through time from the oldest to the most recent interval to infer the user category profile using the Wikipedia category structure. Given a user category profile, we predict the degree of interest of a user to each unobserved topic based on the relatedness of each topic to the inferred category profile. Our experiments illustrate the superiority of our model compared to the state of the art.

References

1. Abel, F., Gao, Q., Houben, G., Tao, K.: Analyzing temporal dynamics in Twitter profiles for personalized recommendations in the social web. In: WebSci 2011, pp. 2:1–2:8 (2011)
2. Abel, F., Gao, Q., Houben, G.-J., Tao, K.: Analyzing user modeling on Twitter for personalized news recommendations. In: Konstan, J.A., Conejo, R., Marzo, J.L., Oliver, N. (eds.) UMAP 2011. LNCS, vol. 6787, pp. 1–12. Springer, Heidelberg (2011). doi:10.1007/978-3-642-22362-4_1
3. Bao, H., Li, Q., Liao, S.S., Song, S., Gao, H.: A new temporal and social pmf-based method to predict users' interests in micro-blogging. Decis. Support Syst. 55(3), 698–709 (2013)

4. Bobadilla, J., Ortega, F., Hernando, A., Gutiérrez, A.: Recommender systems survey. Knowl.-Based Syst. **46**, 109–132 (2013)
5. Cheekula, S.K., Kapanipathi, P., Doran, D., Jain, P., Sheth, A.P.: Entity recommendations using hierarchical knowledge bases. In: ESWC 2015 (2015)
6. Cheng, X., Yan, X., Lan, Y., Guo, J.: BTM: topic modeling over short texts. IEEE Trans. Knowl. Data Eng. **26**(12), 2928–2941 (2014)
7. Dahleh, M., Dahleh, M.A., Verghese, G.: Lectures on dynamic systems and control. Department of Electrical Engineering and Computer Science, Massachusetts Institute of Technology (2013)
8. Ding, Y., Li, X.: Time weight collaborative filtering. In: International Conference on Information and Knowledge Management, pp. 485–492 (2005)
9. Kapanipathi, P., Jain, P., Venkataramani, C., Sheth, A.: User interests identification on Twitter using a hierarchical knowledge base. In: Presutti, V., d'Amato, C., Gandon, F., d'Aquin, M., Staab, S., Tordai, A. (eds.) ESWC 2014. LNCS, vol. 8465, pp. 99–113. Springer, Cham (2014). doi:10.1007/978-3-319-07443-6_8
10. Kapanipathi, P., Orlandi, F., Sheth, A.P., Passant, A.: Personalized filtering of the Twitter stream. In: The Second Workshop on Semantic Personalized Information Management: Retrieval and Recommendation 2011, pp. 6–13 (2011)
11. Khrouf, H., Troncy, R.: Hybrid event recommendation using linked data and user diversity. In: RecSys 2013, pp. 185–192 (2013)
12. Li, L., Zheng, L., Yang, F., Li, T.: Modeling and broadening temporal user interest in personalized news recommendation. Expert Syst. Appl. **41**(7), 3168–3177 (2014)
13. Michelson, M., Macskassy,S.A.: Discovering users' topics of interest on Twitter: a first look. In: AND 2010, pp. 73–80 (2010)
14. Noia, T.D., Mirizzi, R., Ostuni, V.C., Romito, D., Zanker, M.: Linked open data to support content-based recommender systems. In: I-SEMANTICS 2012, pp. 1–8 (2012)
15. Piao, G., Breslin, J.G.: Exploring dynamics and semantics of user interests for user modeling on Twitter for link recommendations. In: SEMANTICS 2016, pp. 81–88 (2016)
16. Varga, A., Basave, A.E.C., Rowe, M., Ciravegna, F., He, Y.: Linked knowledge sources for topic classification of microposts: a semantic graph-based approach. J. Web Semant. **26**, 36–57 (2014)
17. Weng, J., Lim, E., Jiang, J., He, Q.: Twitterrank: finding topic-sensitive influential Twitterers. In: WSDM 2010, pp. 261–270 (2010)
18. Weng, L., Xu, Y., Li, Y., Nayak, R.: Exploiting item taxonomy for solving cold-start problem in recommendation making. In: ICTAI 2008, pp. 113–120 (2008)
19. Xu, Z., Lu, R., Xiang, L., Yang,Q.: Discovering user interest on Twitter with a modified author-topic model. In: WI 2011, pp. 422–429 (2011)
20. Yang, L., Sun, T., Zhang, M., Mei, Q.: WWW 2012, pp. 261–270 (2012)
21. Yin, H., Cui, B., Chen, L., Hu, Z., Zhou, X.: Dynamic user modeling in social media systems. ACM Trans. Inf. Syst. **33**(3), 10:1–10:44 (2015)
22. Yu, Y., Wang, C., Gao, Y.: Attributes coupling based item enhanced matrix factorization technique for recommender systems. CoRR, abs/1405.0770 (2014)
23. Zarrinkalam, F., Fani, H., Bagheri, E., Kahani, M.: Inferring implicit topical interests on Twitter. In: Ferro, N., Crestani, F., Moens, M.-F., Mothe, J., Silvestri, F., Nunzio, G.M., Hauff, C., Silvello, G. (eds.) ECIR 2016. LNCS, vol. 9626, pp. 479–491. Springer, Cham (2016). doi:10.1007/978-3-319-30671-1_35

A New Scheme for Scoring Phrases in Unsupervised Keyphrase Extraction

Corina Florescu[✉] and Cornelia Caragea

Computer Science and Engineering, University of North Texas, Denton, TX, USA
CorinaFlorescu@my.unt.edu, ccaragea@unt.edu

Abstract. Many unsupervised methods for keyphrase extraction typically compute a score for each word in a document based on various measures such as *tf-idf* or the PageRank score computed from the word graph built from the text document. The final score of a candidate phrase is then calculated by summing up the scores of its constituent words. A potential problem with the sum up scoring scheme is that the length of a phrase highly impacts its score. To reduce this impact and extract keyphrases of varied lengths, we propose a new scheme for scoring phrases which calculates the final score using the average of the scores of individual words weighted by the frequency of the phrase in the document. We show experimentally that the unsupervised approaches that use this new scheme outperform their counterparts that use the sum up scheme to score phrases.

1 Introduction

Keyphrase extraction is the task of automatically extracting descriptive phrases or concepts that represent the main topics of a document. Keyphrases provide a concise description of the topics of a document and are particularly useful in many applications ranging from information search and retrieval [1,2] to document summarization [3,4], classification [5], clustering [6], and recommendation [7] or simply to contextual advertisement [8]. In this paper, we aim at improving scoring of candidate phrases in unsupervised approaches to keyphrase extraction, using research papers as a case study.

Unsupervised approaches to keyphrase extraction have started to attract significant attention recently since, unlike supervised approaches, they do not require large human-annotated corpora, which are often expensive or impractical to acquire. Unsupervised keyphrase extraction is formulated as a ranking problem, where each candidate word of a target document receives a score based on various measures such as *tf-idf* [9] or PageRank [10]. Candidate words that have contiguous positions in a document are then concatenated into phrases. To compute the score of a phrase, many existing unsupervised approaches typically *sum up* the scores of its constituent words [10,11], and the top-ranked phrases are returned as keyphrases for the document. A potential problem with the *sum up* scoring scheme is that the length of a phrase highly impacts its score, with longer phrases receiving a higher score. For example, let us consider a research paper

© Springer International Publishing AG 2017
J.M. Jose et al. (Eds.): ECIR 2017, LNCS 10193, pp. 477–483, 2017.
DOI: 10.1007/978-3-319-56608-5_37

that contains the phrase "matrix factorization model" and has "matrix factorization" as one of its gold-standard author-annotated keyphrases. After running an unsupervised algorithm called SingleRank [11] on the paper, we obtain the scores for the individual words "matrix," "factorization," and "model" as follows: 0.047, 0.042, and 0.054, respectively. Since these words are adjacent in text, and hence, form a phrase, by summing up their scores, we obtain a score of 0.143 for the phrase "matrix factorization model," whereas the keyphrase "matrix factorization" receives a lower score of 0.089. We posit that the length of a phrase should not be the only factor that contributes to the *keyphraseness* of a phrase.

To reduce the impact of the length of a phrase on its score, we propose a new scheme for scoring phrases in unsupervised approaches. The new scheme uses *means*, e.g., the arithmetic or harmonic mean of the scores of its individual words, weighted by the frequency of the phrase in the document to quantify for the relevance of that phrase to the topics of the document. We incorporate this new scoring scheme into several representative unsupervised systems for keyphrase extraction and conduct experiments on three datasets of research papers. We show experimentally that the proposed scheme improves the performance of existing unsupervised approaches by as much as 76.28% (relative improvement in performance over current models).

2   Related Work

The unsupervised methods for keyphrase extraction have received a lot of attention and are becoming competitive with supervised approaches [12,13]. The PageRank algorithm is widely-used in keyphrase extraction models. Other centrality measures such as betweenness and degree centrality were also studied for keyphrase extraction [14]. However, based on recent experiments, the PageRank family of methods and *tf-idf* ranking are considered state-of-the-art for unsupervised keyphrase extraction [13,15].

Mihalcea and Tarau [10] proposed TextRank for scoring keyphrases using the PageRank values obtained on a word graph built from the adjacent words in a document. Wan and Xiao [11] extended TextRank to SingleRank by adding weighted edges between words co-occurring within a window size greater than 2. Unlike TextRank and SingleRank, where only the content of the target document is used for keyphrase extraction, textually-similar documents are included in the ranking process in ExpandRank [11]. Gollapalli and Caragea [16] extended ExpandRank to integrate information from the citation network where papers cite one another. Other approaches leverage clustering techniques on word graphs to improve keyphrase extraction [17,18]. Liu et al. [19] proposed TopicalPageRank, which decomposes a document into multiple topics, using topic models, and applies a separate PageRank for each topic. The PageRank scores of each topic are then combined into a single score, using as weights the topic proportions returned by topic models.

Several other approaches *directly* rank phrases, instead of first ranking individual words and then aggregating their scores to rank phrases. For example, the best performing keyphrase extraction system in SemEval 2010 [20] used statistical observations such as term frequencies to filter out phrases that are unlikely to be keyphrases. More precisely, thresholding on the frequency of phrases is applied, where the thresholds are estimated from the data. The candidate phrases are then ranked using the *tf-idf* model in conjunction with a boosting factor which aims at reducing the bias towards single word terms. Danesh et al. [21] computed an initial weight for each phrase based on a combination of heuristics such as the *tf-idf* score and the first position of a phrase in a document. Phrases and their initial weights are then incorporated into a graph-based algorithm which produces the final ranking of keyphrases. Word embeddings are employed as well to measure the relatedness between words in graph based models [22].

In this work, we propose a new scoring scheme for models that compute the score of a phrase by *summing up* the significance scores of its constituent words in order to rank phrases. The proposed scheme averages the significance scores of constituent words in order to limit the contribution of the length of a phrase to its score.

3   Proposed Scoring Scheme

We propose to compute the score of a phrase using *mean*tf*, which corresponds to the mean of the scores of the individual words weighted by the frequency of the phrase within a document. The *mean* reduces the score of a phrase and confers importance to shorter phrases as well. Both arithmetic and harmonic mean can be used to score phrases. The *tf* component in *mean*tf* aims at increasing the score of phrases that occur frequently in a document.

Consider again the example phrase provided in the introduction, "matrix factorization model" and its word scores 0.047, 0.042, and 0.054, respectively. Computing the harmonic mean of the score of the words within the two phrases, we obtain a score of 0.047 for "matrix factorization model" and a score of 0.044 for "matrix factorization," making the longer phrase still more likely to be retuned as a keyphrase. However, by incorporating the frequency of the two phrases, we obtain a score of 0.132 for "matrix factorization," whereas the score of "matrix factorization model" remains 0.047. In general, if a 3-word phrase would be a keyphrase for the document, its frequency is expected to be high (similar to that of the 2-word phrase), and hence, our proposed scoring scheme would return the longer phrase as a keyphrase.

Hence, we propose to score a multi-word phrase p as: $R(p) = mean(p)*tf(p)$, where $mean(p)$ is the mean of the scores of individual words within the phrase p and $tf(p)$ is the frequency of phrase p within the document. The *mean*tf* score is not a freestanding scoring scheme, but a step in unsupervised methods for keyphrase extraction. Therefore, we embed this scoring scheme into six well-known unsupervised algorithms, that first score words and then aggregate them to score phrases: Tf-Idf, TextRank, SingleRank, ExpandRank, CiteTextRank (CTR) and TopicalPageRank (TPR), which are briefly described below.

Tf-Idf [9]. In unsupervised methods for keyphrase extraction, *tf-idf* score is leveraged to rank candidate keyphrase. *TextRank* [10]. This method represents a document as a word graph according to adjacent words, then PageRank algorithm is used to measure the word importance within the document. *SingleRank* [11]. SingleRank extends TextRank adding weighted edges between words within a window of size greater than 2. *ExpandRank* [11]. In ExpandRank, textually-similar documents are included to enrich the knowledge in the word graph. *CTR* [16]. CTR extends ExpandRank by incorporating information from the citation network of a paper. *TPR* [19]. TPR runs multiple PageRanks on the word graph, one biased PageRank to each topic.

4   Experiments and Results

Datasets. We carried out experiments on three datasets of research papers. The first dataset was made available by Nguyen and Kan [23] and contains 211 research papers. The second and third datasets were made available by Gollapalli and Caragea [16] and consist of the proceedings of the ACM Conference on Knowledge Discovery and Data Mining (KDD) and the World Wide Web Conference (WWW), each with 834 and 1350 documents, respectively. In experiments, for all three datasets, we used the title and abstract of a research paper. The author-input keyphrases of a paper were used as gold-standard for evaluation. A summary of our datasets is provided in Table 1. For preprocessing, we used Porter Stemmer to reduce both extracted and gold-standard keyphrases to a base form. To train the topic model in

Table 1. A summary of our datasets.

Dataset	#Docs	#Kp	#AvgKp	1-grams	2-grams	3-grams	n-grams ($n \geq 4$)
Nguyen	211	882	4.18	260	457	132	33
KDD	834	3093	3.70	810	1770	471	42
WWW	1350	6405	4.74	2254	3139	931	81

TPR, we used $\approx 45,000$ papers extracted from the CiteSeer^x scholarly big dataset [24], compiled from the CiteSeer^x digital library. We evaluated the performance of the unsupervised models with *sum up*, *mean*, and *mean*tf* using the following metrics: Precision, Recall and F1-score, which are widely used in previous works [11,19]. We performed experiments using both harmonic and arithmetic mean, but no significant differences were found between the two means. Hence, we show results using the harmonic mean (*hmean*).

Results and Discussion. Table 2 compares Precision, Recall and F1-score at top 5 predicted keyphrases for the six unsupervised methods using all three scoring schemes: *sum up* (baseline), *hmean*, and *hmean*tf*, on all three datasets, Nguyen, WWW, and KDD. Note that CTR was run only on KDD and WWW since citation networks are not available for Nguyen.

As can be seen from the table, the models that use the aggregated score of a phrase based on *hmean*tf* substantially outperform their counterparts that use *sum up*. For example, on WWW, SingleRank with *hmean*tf* achieves an F1-score of 0.166 as compared with SingleRank with *sum up*, which achieves an F1-score of 0.097. Among all unsupervised models, TPR and CTR achieve the highest improvement in performance by replacing *sum up* with *hmean*tf*,

Table 2. Results of the comparison of various unsupervised models using *sum up*, *hmean* and *hmean*tf* to compute the compositional score of a phrase on three datasets, Nguyen, WWW, and KDD. The results are shown at top 5 predicted keyphrases. Best results are shown in **bold blue**.

Unsupervised method	Nguyen			WWW			KDD		
	Precision	Recall	F1-score	Precision	Recall	F1-score	Precision	Recall	F1-score
Tf-Idf - sum up	.099	.128	.108	.099	.115	.103	.093	.116	.100
Tf-Idf - hmean	.122	.154	.133	.141	.155	.142	.119	.151	.129
Tf-Idf - hmean*tf	**.147**	**.184**	**.159**	**.161**	**.180**	**.164**	**.147**	**.186**	**.159**
TextRank - sum up	.087	.115	.097	.094	.110	.097	.086	.108	.093
TextRank - hmean	.091	.116	.100	.104	.116	.106	.086	.111	.094
TextRank - hmean*tf	**.112**	**.144**	**.123**	**.126**	**.142**	**.129**	**.117**	**.149**	**.127**
SingleRank - sum up	.079	.103	.087	.094	.109	.097	.093	.116	.100
SingleRank - hmean	.112	.139	.121	.137	.151	.138	.11	.137	.118
SingleRank - hmean*tf	**.136**	**.171**	**.147**	**.163**	**.182**	**.166**	**.150**	**.187**	**.162**
ExpandRank - sum up	.095	.121	.103	.111	.126	.114	.100	.129	.109
ExpandRank - hmean	.107	.141	.119	.139	.151	.140	.109	.143	.120
ExpandRank - hmean*tf	**.141**	**.183**	**.155**	**.165**	**.184**	**.168**	**.147**	**.189**	**.161**
CTR - sum up	-	-	-	.114	.132	.118	.107	.138	.117
CTR - hmean	-	-	-	.151	.166	.152	.127	.167	.139
CTR - hmean*tf	-	-	-	**.186**	**.209**	**.189**	**.173**	**.223**	**.190**
TPR - sum up	.077	.100	.084	.089	.113	.097	.089	.113	.097
TPR - hmean	.111	.137	.12	.113	.140	.121	.113	.140	.121
TPR - hmean*tf	**.134**	**.168**	**.145**	**.158**	**.198**	**.171**	**.149**	**.186**	**.161**
Tf - phrase frequency	.104	.129	.112	.132	.142	.133	.098	.125	.106

whereas TextRank has the lowest improvement. For example, on the WWW collection, the relative improvement in performance for CTR, TPR, and TextRank models is 60.16%, 76.28%, and 32.98%, respectively.

The models that use the aggregated score of a phrase based on un-weighted *hmean* also outperform the *sum up* baselines, for all datasets. For example, on Nguyen, ExpandRank with *hmean* has an F1-score of 0.119 as compared with 0.103 F1-score of ExpandRank with *sum up*. However, the models that use only *hmean* perform worse compared with their counterparts that use the weighted version *hmean*tf*. For example, on the same dataset, ExpandRank with *hmean*tf* reaches an F1-score of 0.155 as compared with 0.110 F1-score of ExpandRank with *hmean*. Thus, the frequency of a phrase acts as an important component in computing the aggregated score of a phrase for unsupervised keyphrase extraction. Note that, in supervised models, the frequency of a phrase (or its *tf-idf*) is one of the top-ranked features by Information Gain [25]. To better understand the benefit of associating the *hmean* and *tf* scores, we also compare *hmean*tf* with *Tf-phrase frequency*. *Tf-phrase frequency* calculates the score of both single and multi-word phrases based on their number of occurrences in the target document. As can be seen in Table 2, leveraging only the term frequency of a phrase yields worse performance compared with the aggregated score based on *hmean*tf*.

With a paired T-test, our improvements in the evaluation metrics are statistically significant for *p*-values ≤ 0.05.

5   Conclusion and Future Work

In this paper, we proposed a new scheme for scoring phrases in unsupervised keyphrase extraction, showing the benefits of emphasizing both one-word and multi-word phrases. Instead of using the *sum* to compute the aggregated score of a phrase (as is commonly done in the literature), we proposed the use of weighted *means* to compute these scores. The results of our experiments using the harmonic mean weighted by the phrase frequency, $hmean\,^*tf$, show significant improvement in performance over the *sum* baseline on three datasets of research articles. Our findings can improve the performance of the keyphrase extraction task, which in turn, can improve indexing and retrieval of information in many application domains. In future, it would be interesting to explore the performance of $hmean\,^*tf$ on other types of datasets, e.g., news articles.

Acknowledgments. We very much thank our anonymous reviewers for their constructive comments and feedback. This research is supported by the NSF award #1423337.

References

1. Jones, S., Staveley, M.S.: Phrasier: a system for interactive document retrieval using keyphrases. In: Proceedings of the 22nd SIGIR, pp. 160–167 (1999)
2. Ritchie, A., Teufel, S., Robertson, S.: How to find better index terms through citations. In: Proceedings of the Workshop on How Can Computational Linguistics Improve Information Retrieval?, pp. 25–32. ACL (2006)
3. Zha, H.: Generic summarization and keyphrase extraction using mutual reinforcement principle and sentence clustering. In: Proceedings of the 25th SIGIR, pp. 113–120 (2002)
4. Qazvinian, V., Radev, D.R., Özgür, A.: Citation summarization through keyphrase extraction. In: Proceedings of the 23rd ACL, pp. 895–903 (2010)
5. Turney, P.D.: Coherent keyphrase extraction via web mining. In: Proceedings of the IJCAI, pp. 434–442 (2003)
6. Hammouda, K.M., Matute, D.N., Kamel, M.S.: CorePhrase: keyphrase extraction for document clustering. In: Perner, P., Imiya, A. (eds.) MLDM 2005. LNCS (LNAI), vol. 3587, pp. 265–274. Springer, Heidelberg (2005). doi:10.1007/11510888_26
7. Pudota, N., Dattolo, A., Baruzzo, A., Ferrara, F., Tasso, C.: Automatic keyphrase extraction and ontology mining for content-based tag recommendation. Int. J. Intell. Syst. **25**(12), 1158–1186 (2010)
8. Yih, W.t., Goodman, J., Carvalho, V.R.: Finding advertising keywords on web pages. In: Proceedings of the 15th WWW, pp. 213–222 (2006)
9. Zhang, Y., Milios, E., Zincir-Heywood, N.: A comparative study on key phrase extraction methods in automatic web site summarization. J. Digit. Inf. Manag. **5**(5), 323 (2007)
10. Mihalcea, R., Tarau, P.: Textrank: bringing order into text. In: Proceedings of the EMNLP, pp. 404–411 (2004)
11. Wan, X., Xiao, J.: Single document keyphrase extraction using neighborhood knowledge. In: Proceedings of the 23th AAAI, pp. 855–860 (2008)

12. Hasan, K.S., Ng, V.: Conundrums in unsupervised keyphrase extraction: making sense of the state-of-the-art. In: Proceedings of the 23rd ACL: Posters, pp. 365–373 (2010)
13. Hasan, K.S., Ng, V.: Automatic keyphrase extraction: a survey of the state of the art. In: Proceedings of the ACL, pp. 1262–1273 (2014)
14. Palshikar, G.K.: Keyword extraction from a single document using centrality measures. In: Ghosh, A., De, R.K., Pal, S.K. (eds.) PReMI 2007. LNCS, vol. 4815, pp. 503–510. Springer, Heidelberg (2007). doi:10.1007/978-3-540-77046-6_62
15. Kim, S.N., Medelyan, O., Kan, M.Y., Baldwin, T.: Automatic keyphrase extraction from scientific articles. Lang. Resour. Eval. **47**(3), 723–742 (2013)
16. Gollapalli, S.D., Caragea, C.: Extracting keyphrases from research papers using citation networks. In: Proceedings of the AAAI, pp. 1629–1635 (2014)
17. Grineva, M., Grinev, M., Lizorkin, D.: Extracting key terms from noisy and multitheme documents. In: Proceedings of WWW, pp. 661–670 (2009)
18. Liu, Z., Li, P., Zheng, Y., Sun, M.: Clustering to find exemplar terms for keyphrase extraction. In: Proceedings of the 2009 EMNLP, pp. 257–266 (2009)
19. Liu, Z., Huang, W., Zheng, Y., Sun, M.: Automatic keyphrase extraction via topic decomposition. In: Proceedings of the EMNLP, pp. 366–376 (2010)
20. El-Beltagy, S.R., Rafea, A.: Kp-miner: participation in semeval-2. In: Proceedings of the 5th International Workshop on Semantic Evaluation, Association for Computational Linguistics, pp. 190–193 (2010)
21. Danesh, S., Sumner, T., Martin, J.H.: Sgrank: combining statistical and graphical methods to improve the state of the art in unsupervised keyphrase extraction. In: Lexical and Computational Semantics, p. 117 (2015)
22. Wang, R., Liu, W., McDonald, C.: Corpus-independent generic keyphrase extraction using word embedding vectors. In: Software Engineering Research Conference, p. 39 (2014)
23. Nguyen, T.D., Kan, M.-Y.: Keyphrase extraction in scientific publications. In: Goh, D.H.-L., Cao, T.H., Sølvberg, I.T., Rasmussen, E. (eds.) ICADL 2007. LNCS, vol. 4822, pp. 317–326. Springer, Heidelberg (2007). doi:10.1007/978-3-540-77094-7_41
24. Caragea, C., Wu, J., Ciobanu, A., Williams, K., Fernandez-Ramirez, J., Chen, H.H., Wu, Z., Giles., C.L.: Citeseerx: a scholarly big dataset. In: ECIR (2014)
25. Caragea, C., Bulgarov, F.A., Godea, A., Gollapalli, S.D.: Citation-enhanced keyphrase extraction from research papers: a supervised approach. In: EMNLP, pp. 1435–1446 (2014)

JustEvents: A Crowdsourced Corpus for Event Validation with Strict Temporal Constraints

Andrea Ceroni[1(✉)], Ujwal Gadiraju[1], and Marco Fisichella[2]

[1] L3S Research Center, Appelstr. 9a, Hannover, Germany
{ceroni,gadiraju}@L3S.de
[2] Risk Ident GmbH, Zippelhaus 2, Hamburg, Germany
marco@riskident.com

Abstract. Inspecting text to affirm the occurrence of an event is a non-trivial task. Since events are tied to temporal attributes, this task is more complex than merely identifying evidence of entities acting together and thus defining the event in a document. Manual inspection is a typical solution, although it is an onerous task and becomes infeasible with an increasing scale of documents. Therefore, the task of automatically determining whether an event occurs in a document or corpus, named as *event validation*, has been recently investigated. In this paper, we present a dataset for benchmarking event validation methods. Events and documents are coupled in pairs, whose validity has been judged by human evaluators based on whether the document in the pair contains evidence of the given event. In contrast to the notion of relevance considered in available datasets for event detection, validity judgments in this work strictly consider whether a document reports an event within its timespan as well as the number of event participants reported in the document. These requirements make the generation of manual validity judgments an onerous procedure. The ground truth, made of multiple judgments for each pair, has been acquired through crowdsourcing.

Keywords: Event validation · Evaluation · Event detection · Crowdsourcing · Human computation

1 Introduction

Events are the crucial building blocks of all forms of news media. Since a large amount of online space is consumed in describing and discussing events, they are embedded within news articles, forums, blogs and different online social media. Inspecting documents to assess whether a set of events occur in them is a laborious task, which becomes unfeasible in scenarios where events are continuously and automatically detected on a large scale. The task of automatically determining whether a given event occurs in a given document or corpus, named as *event validation*, has recently been studied [2–4]. Automatic event validation can be applied (i) as a post processing step of event detection to improve the precision within the detected set of events, and (ii) to find documents to corroborate the

© Springer International Publishing AG 2017
J.M. Jose et al. (Eds.): ECIR 2017, LNCS 10193, pp. 484–492, 2017.
DOI: 10.1007/978-3-319-56608-5_38

occurrence of events and enrich available knowledge bases (e.g. [9]) with such event descriptions.

In line with both the definition of *event* given by the Topic Detection and Tracking (TDT) project [1] and more recent work on event detection, e.g. [6,8,12], we model events as a set of participants related within a given time period. For instance, the event {(*Novak Djokovic, Roger Federer, US Open*), *30/08/2015 to 15/09/2015*} represents the participation of two tennis players in the 2015 US Open. In this paper, we present a dataset for evaluating the performances of event validation methods, with the main goal of fostering advances and comparisons in this field. The dataset consists of 250 events and 6,457 candidate documents used as a base for assessing event occurrence. Events and documents are coupled in pairs and are associated with validity judgments, indicating the percentage of participants acting together within the event timespan in the document. These validity judgments were assigned by human evaluators via crowdsourcing.

Available datasets, for example those released within the context of TREC and TDT, are built mostly for ad-hoc retrieval, and therefore their high-level topics and corresponding relevance judgments are not suitable for event validation. Moreover, the validity judgments available in our dataset strictly take the number of event participants related together in the document and their conformation to the event timespan into account. Reading a document to assess (i) whether it contains a given event, (ii) how many event participants are mentioned and related together, and (iii) whether such relationships can be associated within the timespan of the event, is more complex than assessing the overall relevance of a document to an event/topic. The different judgment criteria and the effort to produce them make our ground truth more valuable. Event detection methods relying on the same event model as ours (e.g. [6,8]) can use our dataset as ground truth. Moreover, given the high focus on the temporal aspect and conformation during the event timespan, our dataset can be relevant to any event-related research, study, or analysis involving the temporal dimension in a strict manner. To the best of our knowledge, our dataset is the first publicly available corpus for event validation, where the occurrence of events in documents is assessed by taking both relations among event participants as well as their temporal validity into account.

2 Related Work

Available corpora for event detection like the TREC[1] and TDT[2] datasets are related to ours as they consist of annotated event-document pairs. The TREC 2014 Temporal Summarization Track used a set of 15 events along with related documents belonging to news streams. In the TREC Web and Ad-hoc Retrieval Tracks input queries (topics) can be regarded as high level events, and relevance judgments for documents are provided. The TREC Novelty Track dataset

---

[1] http://trec.nist.gov.
[2] http://www.itl.nist.gov/iad/mig/tests/tdt.

contains 25 event and 25 opinion topics, each one including 25 relevant documents and irrelevant ones in addition. The TDT5 corpus consists of 15 news "sources" in different languages, where documents are annotated with topic relevance judgments considering 250 topics (events), 126 for the English portion. This corpus is sparsely labeled: among over 250,000 English news articles only around 4,500 are annotated with topic relevance. The main difference with respect to our dataset resides in the meaning of topics and relevance judgments, since these corpora were designed for ad-hoc retrieval. Input topics can be regarded as high-level named events (e.g. "Costa Concordia disaster") and relevance judgments are given based on the classical query-document relevance in IR. Differently, the events in our dataset are characterized by a set of participants related together within a given time period, and the validity judgments for each event-document pair represent the degree of occurrence of the event within the document, measured with the number of participants that conformed together in the document, strictly within the event time span. Acquiring annotations by inspecting text and checking for mutual relationships and temporal constraints is a more onerous process than producing classical relevance judgments. Moreover, our dataset also contains false events (due to either unrelated participants or wrong time span) and partially true events, where one or more intruders are present along with the true event participants. This allows to test event validation methods not only in case of true and clean events, but also for false or ambiguous ones. The amount of manually annotated data in our dataset is comparable to the one in the above mentioned corpora. McMinn et al. [12] presented a corpus with more than 500 events and related tweets as a ground truth for event detection on Twitter, with an event model close to ours. Besides the different nature of documents (tweets instead of news articles), this work considers a more general notion of relevance, which does not count for either the number of event participants or the temporal validity.

Other works focus on collecting large sets of events, like YAGO2 [9], DBpedia, and Wikipedia Current Events [5], without particularly focusing on relations between events and supporting documents. Kuzey et al. [10] present methods for populating knowledge bases by extracting and organizing named events from news corpora. The ground truth used to evaluate the grouping of documents into events consists of around 100 named events and 1600 articles in Wikinews and news sources referenced in Wikipedia articles. Moreover, such ground truth is built without reporting mutual conformation of event participants with temporal constraints.

3   Dataset Description

We adopt the problem definition of event validation considered in [4]: given an event (its participants and timespan) and a document, the goal is assessing if and how many event participants conform together in the document within the event timespan. Therefore, in our publicly available dataset[3] *events* and *documents* are coupled into *pairs*, which are subject to validity assessment.

---

[3] http://github.com/xander7/JustEvents.

Table 1. Overall statistics of the event set.

Events	250
Distinct participants	456
Participants per event	2.94 ± 1.40

Table 2. Overall statistics of the document set.

Documents	6,457
Avg document length (char)	5,428
Documents per event	25.83 ± 7.40

Table 3. Distribution of events over different categories, with corresponding examples.

Category	%	Examples
Cinema	13%	{James Franco, Academy Award for Best Picture, 127 Hours, 83rd Academy Awards}, 25/01/2011–31/01/2011
Music	7%	{Jessie J, Price Tag, Who You Are}, 25/01/2011–31/01/2011
Nature and disasters	3%	{Rio de Janeiro, Floods, Mudslides}, 18/01/2011–24/01/2011
Sport	35%	{Kim Clijsters, Li Na, Caroline Wozniacki, Australian Open, Svetlana Kuznetsova}, 25/01/2011–31/01/2011
Politics	14%	{Gamal Nasser, Ahmed Shafik, Smartphone, Cairo, April 6 Youth Movement, Gamal Mubarak, National Democratic Party}, 18/01/2011–07/02/2011
Science and economics	4%	{World Economic Forum, Rosneft}, 25/01/2011–31/01/2011
TV and entertainment	16%	{John Cena, Booker T, Royal Rumble}, 25/01/2011–31/01/2011
Other	8%	{Andy Gray, Loose Women, Richard Keys}, 25/01/2011–31/01/2011

3.1 Events

Each event is made of (i) a set of participants, and (ii) a start and end date, indicating the timespan within which the event occurred. This follows event definitions used in previous works [6,8,12]. We applied the algorithm introduced by Tran et al. [13] to detect events, working on the Wikipedia Edit History of more than 1.8 million Wikipedia pages representing persons, locations, artifacts, and groups. Titles of Wikipedia pages are considered as event participants. The considered time period spans from 18th January 2011 to 7th February 2011. We chose this period because it covers newsworthy events, such as the Arab Spring, the Academy Awards Nominations, the Australian Open, the Super Bowl. The minimum granularity of event duration, a parameter of the applied algorithm, has been set to one week. In total, we detected 250 events, whose main characteristics are listed in Table 1. The distribution of the events over different categories, along with examples, is reported in Table 3. These categories were assigned manually based on the inspection of each of the 250 events. The considerable fraction of events related to sport (35%) is due to the actual occurrence of popular and newsworthy events within the considered time period, such as the Australian Open tennis tournament, the Super Bowl, and the Freestyle World Ski Championships. Moreover, complex events lasting many days or even weeks (such as the Australian Open) can trigger the detection of different sub-events within them.

Since events have been detected automatically, the event set also contains false events (due to either unrelated participants or wrong time span) as well as partially true events, where one or more intruders are present along with the true event participants. These events have been retained in the set and were also subject to manual evaluation since event validation has to deal with not only true and clean events, but also with false or ambiguous ones. Our comprehensive dataset thus supports evaluation of event validation methods for all potential cases, and contains a corresponding ground truth for them. This aspect will be further discussed in Sect. 3.3.

3.2 Documents

Documents in our dataset consist of Web pages that have been subject to scrutiny in order to assess the validity of events. We chose the Web as a source for documents due to its easy accessibility and wide event coverage. For each event, queries have been constructed by concatenating the name of event participants along with the months and year covered by the dataset (one distinct query for January 2011 and another for February 2011). We used the Bing Search API to perform queries and to retrieve the *top-20* Web pages for each query. Plain text has been extracted by using BoilerPipe[4], while Stanford CoreNLP[5] has been used for POS tagging, named entity recognition, and temporal expression extraction. After removing duplicates and discarding both non-crawlable Web pages and those with no content extractable by BoilerPipe, we have 6,457 documents corresponding to the 250 events. Titles and URLs of documents are provided along with plain texts. Some overall characteristics of the document set are summarized in Table 2.

Although the latest content of a Web page can be retrieved at any time via its URL, it might be different from the one considered at the time of the evaluation and available in our dataset. Therefore, validity judgments have to be related with the stored content of Web pages, not with the available content according to their latest versions. Moreover, due to the extraction of plain text via BoilerPipe, the stored content might slightly differ from the one of the Web pages.

3.3 Validity Judgments

To manually evaluate the validity of the 6,457 *(event, document)* pairs in the dataset, we decompose the task of assessing whether or not a document contains evidence of the occurrence of an event into atomic units and deploy them on CrowdFlower, a premier crowdsourcing platform. For each pair, workers were presented with the event (participants and timespan) and the document URL. The event timespan, specified by a start and end day, was strictly considered during the tasks. The workers were then asked to report the number of event

---

[4] http://code.google.com/p/boilerpipe/.
[5] http://nlp.stanford.edu/software/corenlp.shtml.

participants conforming to the same event in the document and within the event timespan. Workers also had the possibility to specify whether the temporal bearings of the document were unclear. The crowdsourced task can be previewed (see footnote 3) for the benefit of the reader. We followed task design guidelines to engage the workers [11]. And employed gold standard questions to detect untrustworthy workers [7]. We offered monetary rewards on successful task completion by paying 20 USD cents for each set of 10 pairs.

For each pair, we gathered at least 5 independent validity judgments resulting in over 32,285 responses in total. Based on these, we identified the most frequent judgment given by workers for the same pair as the aggregated validity judgment for each pair (in case of a tie, we considered the judgment that is closest to the average of all judgments for the same pair). These aggregated values give a more robust and intuitive indication of pair validity, coping with user disagreement and outliers. The independent judgments are made available in the dataset, for the remainder of the description we will refer to the aggregated judgments. Both the independent and the aggregated judgments can be utilized further depending on the application requirements. For instance, binary validity labels for pairs (i.e. *valid* or *invalid*) have been derived in [4] depending on whether the real-valued aggregated judgment exceeded a given validity threshold or not. This allowed to pose event validation as a classification problem. Among all the evaluated pairs, 6,336 (98.1%) have a proper aggregated judgment indicating how many event participants conform to the same event in the document and within the event timespan. For the other pairs, the Web page was either not available during the evaluation (110 pairs, 1.7%) or contained an unclear temporal setup (11 pairs, 0.2%). To show how pairs and events distribute over aggregated judgments, we present three cumulative frequency distributions (CFD) in Fig. 1. *Pairs (All Events)* represents the CFD of all the 6,336 pairs that received proper aggregated judgments. *Pairs (Positive Events)* is a CFD only considering pairs related to events that had at least one associated pair with the aggregated judgment greater than '0' within the entire dataset. *Events* is the CFD of events with respect to the maximum judgment over all their pairs.

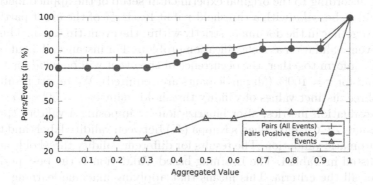

Fig. 1. Distribution of pairs and events over validity judgments.

Figure 1 shows a relatively low amount of pairs with validity greater than '0', despite the retrieved documents matching the event queries to an extent. As mentioned in Sect. 2, due to the fact that the events were generated by an automatic method, the event set also contains false events, which introduce only pairs with judgment equal to '0'. If false events are ignored (*Pairs (Positive Events) CFD*), the amount of pairs with judgment greater than '0' increases. Nevertheless, such increase is limited due to: (i) keyword matching considered to retrieve candidate documents is insufficient to ascertain document-level validity (as proved in [4]), (ii) the mutual conformation of participants has to satisfy (narrow) temporal constraints, and (iii) even true events might not occur in all the retrieved documents. Differently, when considering events and the maximum judgment that they received over their pairs (*Events CFD*), less than 30% of the events are completely false, i.e. those having all pairs with aggregated judgments equal to '0', while more than half of them have all the participants truly conforming together within the same time period (at least one associated pair with judgment equal to '1'). The remaining events are judged as having intermediate verity, i.e. only a subset of the participants conform together.

4   Experiments and Results

In this section we report the performances of three event validation methods applied to our dataset. *Keyword Matching* (KM) validates pairs by counting the percentage of event keywords present in documents. *Rule-based Validation* (RV) evaluates the occurrence of events in documents via pre-defined validation rules, estimating regions of text associated to dates within the event timestamps and returning the percentage of event participants present in these regions [3]. *Learning-based Validation* (LV) combines features extracted from events and documents and builds a model to predict event occurrence via supervised machine learning [4].

Binary validity labels for pairs (*valid* or *invalid*) have been derived by imposing a validity threshold on the real-valued aggregated judgments described in Sect. 3.3, according to the original experimental setup of the applied methods. A pair is said to be *valid* with a threshold τ if at least $\tau\%$ of the event participants conform together in the document strictly within the event timespan. This allows to pose event validation as a classification problem. For instance, if 3 out of 4 participants conform together, the occurrence of the event will be *valid* for $\tau = 50\%$ but *invalid* for $\tau = 100\%$ (all participants are required). We report results referring to three distinct values of validity threshold, namely $\tau = 50\%, 65\%, 100\%$.

The evaluation metrics, after binarization by imposing a validity threshold, are accuracy (ACC) and Cohen's Kappa (K) between validity labels and the output of automatic validation. The results for different validity thresholds and metrics are listed in Table 4. The Learning-based Validation is the best performing one under all the criteria. This means that applying machine learning, exploiting features from events and documents, is more effective than both considering mere keyword matching (*KM*) and manually designing validation rules (*RV*).

Table 4. Performances of different event validation methods.

	$\tau = 50\%$		$\tau = 65\%$		$\tau = 100\%$	
	ACC	K	ACC	K	ACC	K
KM	0.575	0.091	0.575	0.090	0.668	0.191
RV	0.867	0.401	0.868	0.400	0.870	0.286
LV	0.925	0.728	0.923	0.719	0.926	0.680

5   Scope and Limitations

We now elaborate on the scope and limitations of the presented dataset. Event detection methods relying on our event model (e.g. [6,8]) can also use our dataset as ground truth. Since the validity judgments provided by crowd workers were bound by strict temporal constraints that were laid down in accordance to the event definition, this dataset does not fit to scenarios where atemporal event validation is required. However, given the high focus on the temporal aspect and conformation during the event timespan, our dataset can be relevant to any event-related research involving time in a strict manner. The time period of events considered within this dataset is narrow, therefore it may be unsuitable for event validation purposes that contain events spanning a larger granularity of time. However, it is noteworthy that the complex and elaborate manner of task decomposition and consequent acquisition of human judgments makes the dataset a rich source of event validation for events with similar timespans. The quality and quantity of human judgments in our dataset makes it a valuable resource and follows the order of magnitude that is typical in works related to event detection and validation (Sect. 2).

6   Conclusion

We presented a crowdsourced corpus for evaluating the performances of event validation methods with respect to a given document corpus. The dataset comprises of 250 events and 6,457 corresponding documents. Each *(event, document)* pair is associated with at least 5 independent validity judgments, representing the number of event participants conforming together to the same event in the document and within the event timespan. The judgments are acquired through a crowdsourcing process where the manual task of event validation is decomposed to make it fit for the easy consumption of workers. To the best of our knowledge, this is the first dataset of its kind that is made publicly available: a corpus for event validation, where event occurrence in documents is a strict function of participants and their conformation within a specified timespan.

Acknowledgments. This work was partially funded by the European Commission in the context of the FP7 ICT project QualiMaster (grant number: 619525) and the H2020 ICT project AFEL (grant number: 687916).

References

1. Allan, J., Papka, R., Lavrenko, V.: On-line new event detection and tracking. In: Proceedings of the 21st Annual International ACM SIGIR Conference on Research and Development in Information Retrieval, SIGIR 1998 (1998)
2. Araki, J., Callan, J.: An annotation similarity model in passage ranking for historical fact validation. In: Proceedings of the 37th International SIGIR Conference on Research and Development in Information Retrieval, SIGIR 2014 (2014)
3. Ceroni, A., Fisichella, M.: Towards an entity–based automatic event validation. In: Rijke, M., Kenter, T., Vries, A.P., Zhai, C.X., Jong, F., Radinsky, K., Hofmann, K. (eds.) ECIR 2014. LNCS, vol. 8416, pp. 605–611. Springer, Heidelberg (2014). doi:10.1007/978-3-319-06028-6_64
4. Ceroni, A., Gadiraju, U., Fisichella, M.: Improving event detection by automatically assessing validity of event occurrence in text. In: Proceedings of the 24th ACM International Conference on Information and Knowledge Management, CIKM 2015 (2015)
5. Ceroni, A., Georgescu, M., Gadiraju, U., Naini, K.D., Fisichella, M.: Information evolution in Wikipedia. In: Proceedings of the International Symposium on Open Collaboration, OpenSym 2014 (2014)
6. Das Sarma, A., Jain, A., Yu, C.: Dynamic relationship and event discovery. In: Proceedings of the Fourth ACM International Conference on Web Search and Data Mining, WSDM 2011 (2011)
7. Eickhoff, C., de Vries, A.P.: Increasing cheat robustness of crowdsourcing tasks. Inf. Retrieval **16**, 121–137 (2013)
8. He, Q., Chang, K., Lim, E.-P.: Analyzing feature trajectories for event detection. In: Proceedings of the 30th Annual International ACM SIGIR Conference on Research and Development in Information Retrieval, SIGIR 2007 (2007)
9. Hoffart, J., Suchanek, F.M., Berberich, K., Weikum, G.: Yago2: a spatially and temporally enhanced knowledge base from Wikipedia. Artif. Intell. **194**, 28–61 (2012)
10. Kuzey, E., Vreeken, J., Weikum, G.: A fresh look on knowledge bases: distilling named events from news. In: Proceedings of the 23rd International Conference on Information and Knowledge Management, CIKM 2014 (2014)
11. Marshall, C.C., Shipman, F.M.: Experiences surveying the crowd: reflections on methods, participation, and reliability. In: Proceedings of the 5th Annual ACM Web Science Conference, WebSci 2013 (2013)
12. McMinn, A.J., Moshfeghi, Y., Jose, J.M.: Building a large-scale corpus for evaluating event detection on Twitter. In: Proceedings of the 22nd ACM International Conference on Information and Knowledge Management, CIKM 2013 (2013)
13. Tran, T., Ceroni, A., Georgescu, M., Djafari Naini, K., Fisichella, M.: WikipEvent: leveraging Wikipedia edit history for event detection. In: Benatallah, B., Bestavros, A., Manolopoulos, Y., Vakali, A., Zhang, Y. (eds.) WISE 2014. LNCS, vol. 8787, pp. 90–108. Springer, Heidelberg (2014). doi:10.1007/978-3-319-11746-1_7

Dimension Projection Among Languages Based on Pseudo-Relevant Documents for Query Translation

Javid Dadashkarimi, Mahsa S. Shahshahani, Amirhossein Tebbifakhr, Heshaam Faili, and Azadeh Shakery[✉]

School of ECE, College of Engineering, University of Tehran, Tehran, Iran
{dadashkarimi,ms.shahshahani,a.tebbifakhr,hfaili,shakery}@ut.ac.ir

Abstract. Using top-ranked documents retrieved in response to a query of a user has been shown to be an effective approach to improve the quality of query translation in dictionary-based cross-language information retrieval. In this paper, we propose a new method for dictionary-based query translation based on dimension projection of embedded vectors from the pseudo-relevant documents in the source language to their equivalents in the target language. To this end, first we learn low-dimensional vectors of the words in the pseudo-relevant collections separately and then aim to find a query-dependent transformation matrix between the vectors of translation pairs appeared in the collections. At the next step, the representation of each query term is projected to the target language and then, after using a softmax function, a translation model is built. Finally, the model is used for query translation. Our experiments on four CLEF collections in French, Spanish, German, and Italian demonstrate that the proposed method outperforms competitive baselines including a word embedding baseline based on bilingual shuffling. The proposed method reaches up to 87% performance of machine translation (MT) in short queries and considerable improvements in verbose queries.

Keywords: Cross-language information retrieval · Low-dimensional vectors · Dimension projection

1 Introduction

Pseudo-relevance feedback (PRF) has long been shown to be an effective approach for updating query language models in information retrieval (IR) [3–5,11]. In cross-language environments where there are a couple of document sets in different languages, it seems more interesting tailoring all the information written in multiple languages. To this end, cross-lingual relevance model (CLRM) and cross-lingual topical relevance model (CLTRLM) aim to find a way to transform knowledge of the sets to the query model [1,3,10]. Unlike CLRLM that depends on parallel corpora and bilingual lexicons, CLTRLM tailors comparable corpora.

© Springer International Publishing AG 2017
J.M. Jose et al. (Eds.): ECIR 2017, LNCS 10193, pp. 493–499, 2017.
DOI: 10.1007/978-3-319-56608-5_39

Ganguly et al. proposed to use this model for query translation and demonstrated that CLTRLM is an effective method particularly for resource-lean languages [1]. In CLTRLM, top-ranked documents $F_d^s = \{d_1^s, d_2^s, .., d_{|F_d^s|}^s\}$ retrieved in response to the source query ($\mathbf{q}^s$) and top-ranked documents $F_d^t = \{d_1^t, d_2^t, .., d_{|F_d^t|}^t\}$ retrieved in response to a translation of the query ($\mathbf{q}^t$) are assumed to be the relevant documents and then each word w^t in the target language is considered respect to be generated either from a target event or a source event as follows: $p(w^t|\mathbf{q}^s) = p(w^t|\mathbf{z}^t)p(\mathbf{z}^t|\mathbf{q}^s) + p(w^t|\mathbf{w}^s)p(\mathbf{w}^s|\mathbf{q}^s)$ in which $\mathbf{z}^t$ is a topical variable on F_w^t and $\mathbf{w}^s$ is a translation of w^t in the dictionary [1]. Recently, bilingual word embedding is tailored effectively where low-dimensional vectors are built after shuffling all the alignments [9]. However the effectiveness of this method has not been investigated in cross-language PRF yet.

In this paper we propose a new method for building translation models on pseudo-relevant collections using a neural network-based language model [2,6]. The proposed cross-lingual word embedding translation model (CLWETM) takes advantage of a query-dependent transformation matrix between low-dimensional vectors of the languages. Indeed, we aim to find a transformation matrix to bring the vector of each query term, built on the source collection, to dimensionality of the target language and then compute the translation probabilities based on a softmax function. To this aim, first we learn word representations of the pseudo-relevant collections separately and then focus on finding a transformation matrix minimizing a distance function between all translation pairs appeared at the collections. This method captures semantics of both the collections with a rotation and a scaling embedded in the matrix. Finally, a softmax function is used to build a query-dependent translation model based on the similarity of the transformed vector of each query term with the vectors of its translation candidates in the target language.

Unlike CLTRLM and the mixed word embedding translation model (MIXWETM) based on shuffling alignments in a comparable corpus ([9]), CLWETM considers sentence-level contexts of the words and therefore captures deeper levels of n-grams in both languages. The obtained model can be incorporated within a language modeling framework with collection dependent models.

Experimental results on four CLEF collections in French, German, Spanish, and Italian demonstrate that the proposed method outperforms all competitive baselines of dictionary-based cross-language information retrieval (CLIR) in language modeling when it is combined with a global translation model. The proposed method reaches up to 83% performance of the monolingual run and 87% performance of machine translation in short queries. CLWETM has better results in verbose queries and even improvements compared to MT in the Italian collection.

2   Linear Projection Between Languages Based on Pseudo-relevant Documents

In this section we introduce the proposed method in details. We employ an off-line approach for learning bilingual representations of the words by exploiting pseudo-relevant documents in both source and target languages. To this end, first low dimensional vectors of the words appeared in both collections are obtained separately and then the source vectors are transformed to the target language. A rotation alongside language specific scaling embedded in a matrix which is the focus of this section.

As shown in Eq. 1 our goal is to minimize f with respect to a transformation matrix $\mathbf{W} \in \mathbb{R}^{n \times n}$; f is defined as follow:

$$f(\mathbf{W}) = \sum_{(w^s, w^t)} \frac{1}{2} ||\mathbf{W}^T \mathbf{u}_{w^s} - \mathbf{v}_{w^t}||^2 \tag{1}$$

where, $w^t \in F_w^t$ is a translation of $w^s \in F_w^s$. $\mathbf{u}_{w^s} \in \mathbb{R}^{n \times 1}$ and $\mathbf{u}_{w^t} \in \mathbb{R}^{n \times 1}$ are the corresponding vectors respectively. To solve this problem we choose the stochastic gradient descent algorithm (i.e., $\frac{\partial f}{\partial \mathbf{W}} = 0$):

$$\mathbf{W}^{t+1} = \mathbf{W}^t - \eta (\mathbf{W}^T \mathbf{u}_{w^s} - \mathbf{v}_{w^t}) \mathbf{u}_{w^s}^T \tag{2}$$

where η is a constant learning rate.

2.1 Bilingual Representations and Translation Models

A transformation matrix rotates the source query and then scales it as follows: $\hat{\mathbf{u}}_{w^s} = \mathbf{W}^T \mathbf{u}_{w^s}$ The new translation model is built as follows:

$$p(w_t|w_s) = \frac{e^{\frac{\hat{\mathbf{u}}_{w^s} \cdot \mathbf{v}_{w^t}}{||\hat{\mathbf{u}}_{w^s}|| \; ||\mathbf{v}_{w^t}||}}}{\sum_{\bar{w}^t \in \mathcal{T}\{w^s\}} e^{\frac{\hat{\mathbf{u}}_{w^s} \cdot \mathbf{v}_{\bar{w}^t}}{||\hat{\mathbf{u}}_{w^s}|| \; ||\mathbf{v}_{\bar{w}^t}||}}} \tag{3}$$

where $\mathcal{T}\{w^s\}$ is the list of translations of w^s. Instead of topical information propagation taking place on CLTRLM and joint cross-lingual topical relevance model (JCLTRLM), CLWETM tailors semantic projection embedded in $\mathbf{W}$.

Combining Translation Models. Since the obtained model is a probabilistic translation model we can interpolate it with other models as follow: $p(w^t|w^s) = \alpha p_1(w^t|w^s) + (1 - \alpha)p_2(w^t|w^s)$ where α is a constant controling parameter.

3 Experiments

3.1 Experimental Setup

The overview of the collections used in our experiments is provided in Table 1. The source collection is a pool of Associated Press 1988-89, Los Angeles Times 1994, and Glasgow Herald 1995 collections that are used in previous TREC and CLEF evaluation campaigns.

In all experiments, we use the language modeling framework with the KL-divergence retrieval model and Dirichlet smoothing method to estimate the document language models, where we set the Dirichlet prior smoothing parameter

Table 1. Collection characteristics

Id	Lang.	Collection	Queries	#docs	#qrels
IT	Italy	La Stampa 94, AGZ 94	CLEF 2003–2003, Q:91–140	108,577	4,327
SP	Spanish	EFE 1994	CLEF 2002, Q:91–140	215,738	1,039
DE	German	Frankfurter Rundschau 94, SDA 94, Der Spiegel 94–95	CLEF 2002–03, Q:91–140	225,371	1,938
FR	French	Le Monde 94, SDA French 94-95	CLEF 2002–03, Q:251–350	129,806	3,524

μ to the typical value of 1000. To improve the retrieval performance, we use the mixture model for pseudo-relevance feedback with the feedback coefficient of 0.5 [12]. The number of feedback documents and feedback terms are set to the typical values of 10 and 50, respectively.

All European dictionaries, documents, and queries are normalized and stemmed using the Porter stemmer. Stopword removal is also performed.[1] The Lemur toolkit[2] is employed as the retrieval engine in our experiments.

We use the Google dictionaries in our experiments[3]. In the European languages, we do not transliterate out of vocabulary (OOV) terms of the source languages. The OOVs of the target language are used as their original forms in the source documents, since they are cognate languages. We assumed uniform distribution over the translation candidates of a source word as the initial translation model for retrieving top documents. It is worth mentioning that $p(w^t|\mathbf{w}^s)$ in CLTRLM is estimated by a bi-gram coherence translation model BiCTM introduced in [8]. Weights of the edges of the graph are estimated by $p(w_j|w_i)$ computed by SRILM toolkit [4]. BiCTM is also used as p_2 where α is set by 2-fold cross-validation (see Sect. 2).

As discussed in Sect. 2, we used stochastic gradient descent for learning $\mathbf{W}$ which is initialized with random values in $[-1, 1]$; η is set to a small value which also decreases after each iteration. $\mathbf{u}_{w^s}$ and $\mathbf{v}_{w^t}$ are computed based on negative sampling skip-gram introduced in [7]; the size of the window, the number of negative samples, and the size of the vectors are set to typical values of 10, 45, and 50 respectively.

It is shown that JCLTRLM outperforms CLTRLM [1] and therefore we opted JCLTRLM as a baseline. The parameters of LDA are set to the typical value of $\alpha_d = 0.5$ and $\beta_d = 0.01$. Number of topics in JCLTRLM is obtained by 2-fold cross-validation.

3.2   Performance Comparison and Discussion

In this section we compare effectiveness of a number of competitive methods in CLIR. We consider the following dictionary-based CLIR methods to evaluate the

---

[1] We use the stopword lists and the normalizing techniques available at http://memb ers.unine.ch/jacques.savoy/clef/.

[2] http://www.lemurproject.org/.

[3] http://translate.google.com.

[4] http://www.speech.sri.com/projects/srilm/.

Table 2. Comparison of different dictionary-based short query translation methods. Superscripts indicate that the improvements are statistically significant (2-tail t-test, $p \leq 0.05$). $n - m$ indicates all methods in range $[n, .., m]$.

ID	FR (short)			DE (short)			ES (short)			IT (short)		
	MAP	P@5	P@10	MAP	P@5	P@10	MAP	P@5	P@10	MAP	P@5	P@10
- MONO	0.3262	0.4121	0.3737	0.2675	0.4323	0.3688	0.3518	0.4962	0.4321	0.2949	0.3677	0.3115
1 MT	0.2858	0.3939	0.3394	0.2889	0.4375	0.3896	0.3339	0.4280	0.3800	0.2579	0.3510	0.3224
2 TOP-1	0.2211	0.3122	0.2735	0.2015	0.2531	0.2327	0.2749	0.3673	0.3265	0.1566	**0.2208**	0.1896
3 UNIF	0.1944	0.2694	0.2357	0.2148	0.2816	0.2367	0.2362	0.2939	0.2490	0.1526	0.2000	0.1562
4 STRUCT	0.1677	0.25	0.226	0.1492	0.2267	0.2044	0.2472	0.3348	0.3283	0.0994	0.1333	0.1178
5 BiCTM	0.2156	0.3143	0.2755	0.2126	0.2816	0.2612	0.2652*	0.3429	0.3163	0.1504	0.2167	0.1771
6 JCLTRLM	0.1735	0.2687	0.2417	0.1416	0.2178	0.1933	0.2358	0.3522	0.3283	0.1105	0.1733	0.1511
7 MIXWETM	0.2202	0.3143	0.2622	**0.2166**	**0.2166**	**0.2633**	0.2790	0.3755	0.3122	0.1587	0.2125	0.1833
8 CLWETM	**0.2312**$^{2-7}$	**0.3306**	**0.2806**	0.2158^{246}	0.2816	0.2551	**0.2915**$^{2-7}$	**0.3837**	**0.3367**	**0.163**$^{2-7}$	**0.2208**	**0.1937**

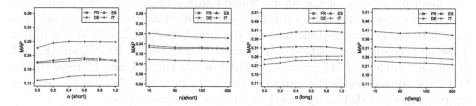

Fig. 1. MAP sensitivity of CLWETM to α and the number of feedback documents.

proposed method: (*1*) the top-1 translation of each term in the bilingual dictionaries (TOP-1), (*2*) all the possible translations of each term with equal weights (UNIFORM), (*3*) BiCTM proposed in [8], (*4*) JCLTRLM proposed in [1], and (*5*) MIXWETM [9]. As bases of comparisons we also provide results of the monolingual runs in each collection (MONO) and Google machine translator (MT). However, our main focus is to investigate superiority of the proposed method compared to the dictionary-based CLIR baselines which are available for many language pairs.

All the results on short queries obtained from the titles of the topics are summarized in Table 2 and Fig. 1 shows sensitivity of CLWETM to α and the number of feedback documents. As shown in the table, both MIXWETM and CLWETM outperform other methods in terms of MAP, P@5, and P@10 in all the collections. MIXWETM and CLWETM consistently achieved better results compared to others, but CLWETM is clearly more effective than MIXWETM in almost all the datasets. Although CLWETM lost the competition to MIXWETM in DE, but the differences are not statistically significant. One reason for this outcome is the lower performance of BiCTM compared to other collections (see Table 2, and Fig. 1). Another reason can be the lower sensitivity of the method to n in this collection. As shown in Fig. 1, top-ranked documents in DE are not as helpful as FR, ES, and IT and thus neither CLWETM nor MIXWETM has significant improvements.

Although the focus of this research is on dictionary-based CLIR, but it is clear that in the European collections with short queries the results of MT are higher than all the baselines. In the rest of the experiments, we shed light on

Table 3. Comparison of different dictionary-based long query translation methods.

	FR (long)			DE (long)			ES (long)			IT (long)		
ID	MAP	P@5	P@10	MAP	P@5	P@10	MAP	P@5	P@10	MAP	P@5	P@10
- MONO	0.4193	0.5354	0.4727	0.3938	0.5280	0.4780	0.5281	0.6720	0.5960	0.3947	0.5022	0.4356
1 MT	0.3395	0.4263	0.3747	0.3436	0.4400	0.4280	0.4208	0.5600	0.478	0.1376	0.1551	0.1306
2 TOP-1	0.3077	0.3960	0.3434	0.2242	0.3080	0.2500	0.3762	0.4800	0.4320	0.2195	0.2800	0.2622
3 UNIF	0.2709	0.3556	0.3091	0.2425	0.2840	0.2540	0.3243	0.3680	0.3340	0.2095	0.2311	0.2000
4 STRUCT	0.1800	0.2646	0.2394	0.2103	0.252	0.2500	0.2951	0.4000	0.3760	0.1942	0.2444	0.2244
5 BiCTM	0.3050	0.3899	0.3505	0.2442	0.3280	0.2780	0.3841	0.4640	0.4340	0.2172	0.2622	0.2422
6 JCLTRLM	0.2266	0.3414	0.2990	0.1520	0.2160	0.1880	0.2734	0.404	0.3500	0.1459	0.2133	0.1756
7 MIXWETM	0.2983	0.3919	0.3485	**0.2652**	0.3400	0.3040	0.3677	0.4280	0.4080	**0.2381**	**0.3022**	**0.2733**
8 CLWETM	**0.3167^{2-7}**	**0.4101**	**0.3657**	0.2622^{2-6}	**0.3480**	**0.3080**	**0.4029^{2-7}**	**0.500**	**0.4620**	0.2380^{2-6}	0.2978	0.2667

effectiveness of the methods on verbose queries obtained by concatenating title and description parts of the topics. Table 3 shows the results; the results also confirm the effectiveness of CLWETM over the dictionary-based methods. The most interesting point is decrements of the gaps between CLWETM and MT in quite all the collections. CLWETM reached 93.2%, 76.3%, 95.7%, and 172.9% of the performance of MT in terms of MAP in ES, DE, ES, and IT respectively. In IT we see noticeable decrement of the performance by MT on the verbose queries where the dictionary-based techniques are quite stable.

3.3   Parameter Sensitivity

We investigate the sensitivity of the proposed method to two parameters α and n in Fig. 1. We first fix one parameter to its optimal value and then try to get optimal value of the other one. The figure demonstrates that both parameters work stably across FR, ES and IT collections.

4   Conclusion and Future Works

In this paper we presented a translation model for cross-lingual information retrieval that uses feedback documents in source and target languages for creating word vectors in each language, and then learns a projection matrix to project word vectors in the source language to their translations in the target language and then build a translation model. Our method reaches considerable improvements in multiple CLEF collections. The proposed method reaches up to 87% performance of machine translation (MT) in short queries and considerable improvements in verbose queries.

References

1. Ganguly, D., Leveling, J., Jones, G.: Cross-lingual topical relevance models. In: COLING 2012 (2012)
2. Gouws, S., Bengio, Y., Corrado, G.: Bilbowa: fast bilingual distributed representations without word alignments (2014). arXiv preprint arXiv:1410.2455

3. Lavrenko, V., Choquette, M., Croft, W.B.: Cross-lingual relevance models. In: SIGIR 2002, pp. 175–182 (2002)
4. Lavrenko, V., Croft, W.B.: Relevance based language models. In: SIGIR 2001. ACM (2001)
5. Lv, Y., Zhai, C.: Revisiting the div. minimization feedback model. In: CIKM 2014, pp. 1863–1866 (2014)
6. Mikolov, T., Le, Q.V., Sutskever, I.: Exploiting similarities among languages for machine translation (2013). arXiv preprint arXiv: 1309.4168
7. Mikolov, T., Sutskever, I., Chen, K., Corrado, G.S., Dean, J.: Distributed representations of words and phrases and their compositionality. In: Advances in Neural Information Processing Systems, pp. 3111–3119 (2013)
8. Monz, C., Dorr, B.J.: Iterative translation disambiguation for cross-language information retrieval. In: SIGIR 2005, pp. 520–527 (2005)
9. Vulic, I., Moens, M.: Monolingual and cross-lingual information retrieval models based on (bilingual) word embeddings. In: SIGIR 2015, pp. 363–372 (2015)
10. Vulic, I., Smet, W.D., Tang, J., Moens, M.: Probabilistic topic modeling in multilingual settings: an overview of its methodology and applications. IP&M **51**(1), 111–147 (2015)
11. Zamani, H., Dadashkarimi, J., Shakery, A., Croft, W.B.: Pseudo-relevance feedback based on matrix factorization. In: CIKM 2016, pp. 1483–1492 (2016)
12. Zhai, C., Lafferty, J.: Model-based feedback in the language modeling approach to information retrieval. In: CIKM 2001, pp. 403–410. ACM (2001)

Labeling Topics with Images Using a Neural Network

Nikolaos Aletras[✉] and Arpit Mittal

Amazon.com, Cambridge, UK
{aletras,mitarpit}@amazon.com

Abstract. Topics generated by topic models are usually represented by lists of t terms or alternatively using short phrases or images. The current state-of-the-art work on labeling topics using images selects images by re-ranking a small set of candidates for a given topic. In this paper, we present a more generic method that can estimate the degree of association between any arbitrary pair of an unseen topic and image using a deep neural network. Our method achieves better runtime performance $O(n)$ compared to $O(n^2)$ for the current state-of-the-art method, and is also significantly more accurate.

Keywords: Topic models · Deep neural networks · Topic representation

1 Introduction

Topic models [5] are a popular method for organizing and interpreting large document collections by grouping documents into various thematic subjects (e.g. sports, politics or lifestyle) called topics. Topics are multinomial distributions over a predefined vocabulary whereas documents are represented as probability distributions over topics. Topic models have proven to be an elegant way to build exploratory interfaces (i.e. topic browsers) for visualizing document collections by presenting to the users lists of topics [6,14,15] where they select documents of a particular topic of interest.

A topic is traditionally represented by a list of t terms with the highest probability. In recent works, short phrases [4,11], images [3] or summaries [19] have been used as alternatives. Particularly, images offer a language independent representation of the topic which can also be complementary to textual labels. The visual representation of a topic has been shown to be as effective as the textual labels on retrieving information using a topic browser while it can be understood quickly by the users [1,2]. The task of labeling topics consists of two main components: (1) a candidate generation component where candidate labels are obtained for a given topic (usually using information retrieval techniques and knowledge bases [3,11]), and (2) a ranking (or label selection) component that scores the candidates according to their relevance to the topic. In the case of labeling topics with images the candidate labels consist of images.

© Springer International Publishing AG 2017
J.M. Jose et al. (Eds.): ECIR 2017, LNCS 10193, pp. 500–505, 2017.
DOI: 10.1007/978-3-319-56608-5_40

The method presented by [3] generates a graph where the candidate images are its nodes. The edges are weighted with a similarity score between the images that connect. Then, an image is selected by re-ranking the candidates using PageRank. The method is iterative and has a runtime complexity of $O(n^2)$ which makes it infeasible to run over large number of images. Hence, for efficiency the candidate images are selected a priori using an information retrieval engine. Thus the scope of this method gets limited to solving a *local problem* of re-ordering a small set of candidate images for a given topic. Furthermore, its accuracy is limited by the recall of the information retrieval engine. Finally, if new candidates appear, they should be added to the graph, the process of computing pairwise similarities and re-ranking of nodes is repeated.

In this work, we present a more generic method that directly estimates the appropriateness of any arbitrary pair of topic and image. We refer to this method as a *global method* to differentiate it from the localized approach described above. We utilize a Deep Neural Network (DNN) to estimate the suitability of an image for labeling a given topic. DNNs have proven to be effective in various IR and NLP tasks [7,16]. They combine multiple layers that perform non-linear transformations to the data allowing the automatic learning of high-level abstractions. At runtime our method computes dot products between various features and the model weights to obtain the relevance score, that gives it an order complexity of $O(n)$. Hence, it is suitable for using it over large image sources such as Flickr[1], Getty[2] or ImageNet [9]. The proposed model obtains state-of-the-art results for labeling completely unseen topics with images compared to previous methods and strong baselines.

2   Model

For a topic T and an image I, we want to compute a real value $s \in \mathbb{R}$ that denotes how good the image I is for representing the topic T. T consists of ten terms (t) with the highest probability for the topic. We denote the visual information of the image as V. The image is also associated with text in its caption, C.

For the topic $T = \{t_1, t_2, ..., t_{10}\}$ and the image caption $C = \{c_1, c_2, ..., c_n\}$, each term is transformed into a vector $\mathbf{x} \in \mathbb{R}^d$ where d is the dimensionality of the distributed semantic space. We use pre-computed dependency-based word embeddings [12] whose d is 300. The resulting representations of T and C are the mean vectors of their constituent words, $\mathbf{x_t}$ and $\mathbf{x_c}$ respectively.

The visual information from the image V is converted into a dense vectorized representation, $\mathbf{x_v}$. That is the output of the publicly available 16-layer VGG-net [13] trained over the ImageNet dataset [9]. VGG-net provides a 1000 dimensional vector which is the soft-max classification output of ImageNet classes.

The input to the network is the concatenation of topic, caption and visual vectors. i.e.,

$$X = [x_t || x_c || x_v] \tag{1}$$

This results in a 1600-dimensional input vector.

---

[1] http://www.flickr.com.
[2] http://www.gettyimages.co.uk.

Then, X is passed through a series of four hidden layers, $H_1, ..., H_4$. In this way the network learns a combined representation of topics and images and the non-linear relationships that they share.

$$h_i = g(W_i^T h_{i-1}) \tag{2}$$

where g is the rectified linear unit (ReLU) and $h_0 = X$. The output of each hidden layer is regularized using dropout [17]. The output size of H_1, H_2, H_3 and H_4 are set to 256, 128, 64 and 32 nodes respectively.

The output layer of the network maps the input to a real value $s \in \mathbb{R}$ that denotes how good the image I is for the topic T. The network is trained by minimizing the mean absolute error:

$$error = \frac{1}{n} \sum_{i=1}^{n} |W_o^T h_4 - s_g| \tag{3}$$

where s_g is the ground-truth relevance value. The network is optimized using a standard mini-batch gradient descent method with RMSProp adaptive learning rate algorithm [18].

3   Experimental Setup

We evaluate our model on the publicly available data set provided by [3]. It consists of 300 topics generated using Wikipedia articles and news articles taken from the New York Times. Each topic is represented by ten terms with the highest probability. They are also associated with 20 candidate image labels and their human ratings between 0 (lowest) and 3 (highest) denoting the appropriateness of these images for the topic. That results into a total of 6 K images and their associated textual metadata which are considered as captions. The task is to choose the image with the highest rating from the set of the 20 candidates for a given topic.

The 20 candidate image labels per topic are collected by [3] using an information retrieval engine (Google). Hence most of them are expected to be relevant to the topic. This jeopardizes the training of our supervised model due to the lack of sufficient negative examples. To address this issue we generate extra negative examples. For each topic we sample another 20 images from random topics in the training set and assign them a relevance score of 0. These extra images are added into the training data.

Our evaluation follows prior work [3,11] using two metrics. The **Top-1 average rating** is the average human rating assigned to the top-ranked label proposed by the topic labeling method. This metric provides an indication of the overall quality of the label selected and takes values from 0 (irrelevant) to 3 (relevant). The normalized discounted cumulative gain (**nDCG**) compares the label ranking proposed by the labeling method to the gold-standard ranking provided by the human annotators [8,10].

We set the dropout value to 0.2 which randomly sets 20% of the input units to 0 at each update during the training time. We train the model in a 5-fold cross-validation for 30 epochs and set the batch size for training data to 16. In each fold, data from 240 topics are used for training which results into 9,600 examples (20 original, 20 negative candidates per topic). The rest completely unseen 60 topics are used for testing which results into 1,200 test examples (note that we do not add negative examples in the test data).

4 Results and Discussion

We compare our approach to the state-of-the-art method that uses Personalized PageRank [3] to re-rank image candidates (**Local PPR**) and an adapted version that computes the PageRank scores of all the available images in the test set (**Global PPR**). We also test other baselines methods: (1) a relevant approach originally proposed for image annotation that learns a joint model of text and image features (**WSABIE**) [20], (2) linear regression and SVM models that use the concatenation of the topic, the caption and the image vectors as input, **LR (Topic+Caption+VGG)** and **SVM (Topic+Caption+VGG)** respectively. Finally, we test two versions of our own DNN using only either the caption (**DNN (Topic+Caption)**) or the visual information of the image (**DNN (Topic+VGG)**).

Table 1 shows the Top-1 average and nDCG scores obtained. First, we observe that the DNN methods perform better for both the evaluation metrics compared to the baseline methods. They achieve a Top-1 average rating between 1.94 and 2.12 better than the Global PPR, Local PPR, WSABIE, LR and SVM baselines. Specifically, the DNN (Topic+Caption+VGG) method significantly outperforms these models (paired t-test, $p < 0.01$). This demonstrates that our simple DNN model captures high-level associations between topics and images. We should

Table 1. Results obtained for the various topic labeling methods. †, ‡ and * denote statistically significant difference to Local PPR, Global PRR and WSABIE respectively (paired t-test, $p < 0.01$).

Model	Top-1 aver. rating	nDCG-1	nDCG-3	nDCG-5
Global PPR [3]	1.89	0.71	0.74	0.75
Local PPR [3]	2.00	0.74	0.75	0.76
WSABIE [20]	1.87	0.65	0.68	0.70
LR (Topic+Caption+VGG)	1.91	0.71	0.74	0.75
SVM (Topic+Caption+VGG)	1.94	0.72	0.75	0.76
DNN (Topic+Caption)	1.94	0.73	0.75	0.76
DNN (Topic+VGG)	$2.04^{‡*}$	0.76	0.79	0.80
DNN (Topic+Caption+VGG)	$\mathbf{2.12^{†‡*}}$	**0.79**	**0.80**	**0.81**
Human Perf. [3]	2.24	-	-	-

also highlight that the network has not seen either the topic or the image during training which is important for a generic model. In the WSABIE model, linear mappings are learned between the text and visual features. This restricts their effectiveness to capture non-linear similarities between the two modalities.

The DNN (Topic+Caption) model that uses only textual information, obtains a Top-1 Average performance of 1.94. Incorporating visual information (VGG) improves it to 2.12 (DNN (Topic+Caption+VGG)). An interesting finding is that using only the visual information (DNN (Topic+VGG)) achieves better results (2.04) compared to using only text. This demonstrates that images contain less noisy information compared to their captions for this particular task.

The DNN models also provide a better ranking for the image candidates. The nDCG scores for the majority of the DNN methods are higher than the other methods. DNN (Topic+Caption+VGG) consistently obtains the best nDCG scores, 0.79, 0.80 and 0.81 respectively. Figure 1 shows two topics and the top-3 images selected by the DNN (Topic+Caption+VGG) model from the candidate set. The labels selected for the topic #288 are all very relevant to a *Surgical operation*. On the other hand, the images selected for topic #99 are irrelevant to *Wedding photography*. For this topic the candidate set of labels do not contain any relevant images.

Topic #288: surgery, body, medical, medicine, surgical, blood, organ, transplant, health, patient

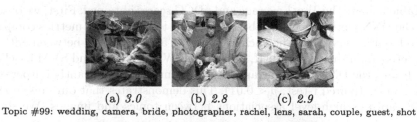

(a) *3.0*   (b) *2.8*   (c) *2.9*

Topic #99: wedding, camera, bride, photographer, rachel, lens, sarah, couple, guest, shot

(d) *0.4*   (e) *0.8*   (f) *0.8*

Fig. 1. A good and a bad example of topics and the top-3 images (left-to-right) selected by the DNN (Topic+Caption+VGG) model from the candidate set. Subcaptions denote average human ratings.

5   Conclusion

We presented a deep neural network that jointly models textual and visual information for the task of topic labeling with images. Our model is generic and works for any unseen pair of topic and image. Our evaluation results show that our proposed approach significantly outperforms the state-of-the-art method [3] and a relevant method originally utilized for image annotation [20].

References

1. Aletras, N., Baldwin, T., Lau, J.H., Stevenson, M.: Representing topics labels for exploring digital libraries. In: JCDL (2014)
2. Aletras, N., Baldwin, T., Lau, J.H., Stevenson, M.: Evaluating topic representations for exploring document collections. JASIST (2015)
3. Aletras, N., Stevenson, M.: Representing topics using images. In: NAACL-HLT, pp. 158–167 (2013)
4. Aletras, N., Stevenson, M.: Labelling topics using unsupervised graph-based methods. In: ACL (2014)
5. Blei, D.M., Ng, A.Y., Jordan, M.I.: Latent Dirichlet allocation. JMLR **3**, 993–1022 (2003)
6. Chaney, A.J.B., Blei, D.M.: Visualizing topic models. In: ICWSM, pp. 419–422 (2012)
7. Collobert, R., Weston, J.: A unified architecture for natural language processing: deep neural networks with multitask learning. In: ICML, pp. 160–167 (2008)
8. Croft, B.W., Metzler, D., Strohman, T.: Search Engines: Information Retrieval in Practice. Addison-Wesley, Boston (2009)
9. Deng, J., Dong, W., Socher, R., Li, L.J., Li, K., Fei-Fei, L.: ImageNet: a large-scale hierarchical image database. In: CVPR, pp. 248–255 (2009)
10. Järvelin, K., Kekäläinen, J.: Cumulated gain-based evaluation of IR techniques. ACM Trans. Inf. Syst. **20**(4), 422–446 (2002)
11. Lau, J.H., Grieser, K., Newman, D., Baldwin, T.: Automatic labelling of topic models. In: ACL-HLT, pp. 1536–1545 (2011)
12. Levy, O., Goldberg, Y.: Dependency-based word embeddings. In: ACL, pp. 302–308 (2014)
13. Simonyan, K., Zisserman, A.: Very deep convolutional networks for large-scale image recognition. arXiv preprint arXiv:1409.1556 (2014)
14. Smith, A., Hawes, T., Myers, M.: Hierarchy: visualization for hierarchical topic models. In: Workshop on Interactive Language Learning, Visualization, and Interfaces, pp. 71–78 (2014)
15. Snyder, J., Knowles, R., Dredze, M., Gormley, M., Wolfe, T.: Topic models and metadata for visualizing text corpora. In: NAACL-HLT, pp. 5–9 (2013)
16. Socher, R., Lin, C.C., Manning, C., Ng, A.Y.: Parsing natural scenes and natural language with recursive neural networks. In: ICML, pp. 129–136 (2011)
17. Srivastava, N., Hinton, G., Krizhevsky, A., Sutskever, I., Salakhutdinov, R.: Dropout: a simple way to prevent neural networks from overfitting. JMLR **15**(1), 1929–1958 (2014)
18. Tieleman, T., Hinton, G.: Lecture 6.5-RMSProp. COURSERA: Neural networks for machine learning (2012)
19. Wan, X., Wang, T.: Automatic labeling of topic models using text summaries. In: ACL, pp. 2297–2305 (2016)
20. Weston, J., Bengio, S., Usunier, N.: Large scale image annotation: learning to rank with joint word-image embeddings. Mach. Learn. **81**(1), 21–35 (2010)

Leveraging Site Search Logs to Identify Missing Content on Enterprise Webpages

Harsh Jhamtani[1]([✉]), Rishiraj Saha Roy[2], Niyati Chhaya[3], and Eric Nyberg[1]

[1] Language Technology Institute, Carnegie Mellon University, Pittsburgh, USA
{jharsh,ehn}@cs.cmu.edu
[2] Max Planck Institute for Informatics, Saarland Informatics Campus,
Saarbrücken, Germany
rishiraj@mpi-inf.mpg.de
[3] Big Data Experience Lab, Adobe Research, Bangalore, India
nchhaya@adobe.com

Abstract. Online visitors often do not find the content they were expecting on specific pages of a large enterprise website, and subsequently search for it in site's search box. In this paper, we propose methods to leverage website search logs to identify missing or expected content on webpages on the enterprise website, while showing how several scenarios make this a non-trivial problem. We further discuss how our methods can be easily extended to address concerns arising from the identified missing content.

1 Introduction

It is a common scenario when an online visitor navigates to a specific page on an enterprise website and does not find what she is looking for. In such cases, the visitor commonly issues a query indicative of the expected content to the *site search box*, which is provided by almost all large websites. In this research, we wish to leverage search logs from enterprise websites to determine such "missing content". However, site search can also be used for general site navigation without being driven by an absence of expected content on a webpage. For example, if a visitor's current search intent has been satisfied, she may issue a new site search query for her subsequent information need. This calls for an approach to automatically infer whether an issued query corresponds to missing content on a page or not. The contributions of this work are two-fold: (a) We propose an approach to address the novel problem of identifying missing content in enterprise sites at an individual page-level, using query logs and existing site content, and (b) we propose a decision tree for the classification of certain scenarios that would help the site administrator to take specific actions for dealing with user issues regarding the expected content.

Related Work. Existing work closest to ours' is by Yomtov et al. [6], who try to label those queries that have very low predicted MAP (mean average precision) in a standard evaluation setup as 'difficult queries'. In their work, such queries may be considered indicative of 'missing content' in the collection. However,

© Springer International Publishing AG 2017
J.M. Jose et al. (Eds.): ECIR 2017, LNCS 10193, pp. 506–512, 2017.
DOI: 10.1007/978-3-319-56608-5_41

they do not model the referral webpage from where the query was issued, which is only meaningful in an enterprise setting like ours' where queries correspond to specific referral pages (documents). Our system identifies content that is missing from particular webpages, and can suggest appropriate actions to the website administrators. Further, queries indicative of missing content in our case need not be "difficult" with respect to the collection and the associated search system. In other words, there are scenarios when content is indeed present on the website (and so the query is not 'difficult'), but not present where the user expected it.

Using our system may lead to placing new content or links on webpages to rectify missing content issues. From this perspective, previous research aimed at site navigation experience is relevant to our efforts. Cui and Hu [3] propose an approach for adding internal and external links in a website for improving search engine rankings. Lin and Liu [5] propose a method to optimize website link structure to improve site navigation. However, the goal of our work is very different from such works in the sense that our primary objective is not to improve search engine rankings of the site or minimize users' site navigation time. Rather, we aim to detect missing (but expected) content at a per-webpage granularity, which may however, indirectly result in the effects that past efforts tried to achieve.

2 Approach

If a visitor issues a query q (say, `photoshop student discount`) from a webpage w (say, www.adobe.com/products.html) belonging to an enterprise website W (say, www.adobe.com), then we call w the *referral webpage* for q. Our proposed method requires the following inputs: (a) *Query logs*, in the form of (w, q) pairs; and, (b) Textual content of all *webpages* in W. Our proposed method can be divided into two phases. In the first phase, we identify statistically significant (w, q) tuples These missing content tuples are then classified in the second phase for better interpretation.

2.1 Phase 1: Identifying Significant Tuples

Naïvely considering all (w, q) tuples with high frequencies as corresponding to missing content can be misleading. For example, in three days of adobe.com logs containing about $150,000$ queries, `flash download` accounts for more than 5% of queries issued from $\simeq 17\%$ of the referral webpages. It is unlikely that content corresponding to `flash download` is expected on all these pages. Website visitors, who land on the homepage or other pages on the site, try to reach the specific webpage to download `Flash`. Thus, the query `flash download` is not necessarily indicative of missing content on the corresponding referral webpages. We believe that queries corresponding to navigational efforts of visitors *without* experiencing absence of desired content will be spread out across pages

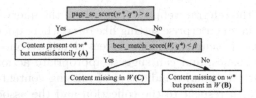

Fig. 1. Classification of significant tuples. Note that the cases in decision tree are exhaustive for all possible values of the two scores, because $best\_match\_score(q^*, W) \geq page\_se\_score(q^*, w^*)$ always holds.

uniformly at random. We thus try to identify (w, q) tuples that occur more often than expected by random chance. Hence, steps in Phase 1 are as follows:

1. Extract all (w, q) tuples from the raw query log.
2. Cluster queries with similar words but different orders together assuming them to bear the same search intent; for example, `adobe photoshop 7 download` and `photoshop download` can be grouped together.
3. Represent each cluster by its most frequent query, and replace all original queries in tuples by the representative queries of clusters to which the original queries belonged. Ties are broken arbitrarily.
4. Construct a matrix C of size $n \times m$ after clustering, such that $C[i][j]$ is the number of times the jth query (say, q) was issued from the ith webpage (say, w). Here, n and m are the numbers of distinct webpages in W and query clusters obtained from the log, respectively.
5. Compute the Pearson residual value [1] as a measure of statistical significance for each cell in C. If M is the total number of queries issued in W, the Pearson residual e_{ij} corresponding to (w, q) is given by:

$$e_{ij} = \frac{C[i][j] - \mu_{ij}}{\sqrt{\mu_{ij}(1 - p_{i+})(1 - p_{+j})}} \tag{1}$$

where, we obtain $p_{i+} = (\sum_{j=1}^{m} C[i][j])/M$, and $p_{+j} = (\sum_{i=1}^{n} C[i][j])/M$, and finally $\mu_{ij} = p_{i+} \times p_{+j} \times M$. Here, μ_{ij} refers to the expected number of occurrences of q for w. We observed that the number of times some $q \in Q$ (Q is the *set* of all queries issued to W) is issued from some $w \in W$ typically follows a Poisson distribution, which has the property that its standard deviation is equal to the square root of its mean. The denominator in Eq. 1 represents the standard deviation in the query frequency distribution, adjusted for the number of degrees of freedom.
6. Extract tuples (w^*, q^*) with residuals higher than a threshold δ as *significant tuples*. A large value of the Pearson residual for $C[i][j]$ means that q was issued from w much more than the expected number of times.

2.2   Phase 2: Classifying Significant Tuples

We now classify each (w^*, q^*) tuple into one of the three indicator classes:

1. **Unsatisfactorily present content (Class A):** Content relevant to q^* is present on w^* but does not satisfy user requirements properly.
2. **Missing content on page (Class B):** Content relevant to q^* is absent on w^* but present on other page(s) in W.
3. **Missing content on site (Class C):** Content good for q^* is absent in W.

These classes would be intuitive to the website administrator, as each class calls for a different action. For each (w^*, q^*), we calculate following scores:

1. **page_se_score** (w^*, q^*): Relevance **score** of **page** w^* for query q^*, as provided by the site search engine (a threshold on this score can capture whether content relevant to q^* is present on w^*).
2. **best_match_score** (q^*, W): The **score** of the **best match**ing page in W for q^*, again provided by the site search engine (a threshold on this score can detect if there is at least one page $\in W$ with content relevant to q^*).

Let α and β be two constants, so that classification is performed according to the decision tree shown in Fig. 1. Each of the three classes should be dealt with differently. Items in Class A can be referred back to the corresponding page authors for revision. For Class B, it is possible to leverage clickthrough data in order to identify which pages the visitor finally found the desired content. Content from these *relevant* pages can be used to update the original referral webpage w^*. Tuples in Class C require the addition of new content to W.

3   Experiments and Insights

Dataset. We obtained access to three days' enterprise search query logs from January 2015 for www.adobe.com, which uses Apache Lucene $v5.3$ as the site search engine. The total number of entries, i.e. (w, q) pairs in the provided log was $152,586$ ($25,936$ distinct queries; $2,081$ distinct pages; $26,727$ distinct tuples). We obtained $12,360$ query clusters by running a fast graph clustering algorithm Chinese Whispers [2] to cluster the queries. Input to the clustering algorithm is a graph where each distinct query was a node and there is an edge between two query nodes if their word-level overlap measured by the Jaccard index exceeded a threshold γ (chosen to be 0.7 by manual inspection).

Parameter Tuning. Both α and β can be interpreted as governing whether content corresponding to a query is present in a given text, so both take the same value. We obtained a set of 1000 binary-relevance judged (w, q) pairs (by humans, 500 relevant and non-relevant pairs each), such that $rel(w, q) = 1$ denotes "true" significance corresponding to missing content, and $rel(w, q) = 0$ otherwise. Optimal α^* (and equivalently, β^*) as per the MaxPCC criteria [4] can

Table 1. Proportions of tuples in the various classes.

Class	Counts	Percent tuples
Not statistically significant	18,639	69.74
Unsatisfactorily present content (A)	580	2.17
Missing content on page (B)	4,302	16.09
Missing content on site (C)	3,206	11.99

Table 2. Representative (webpage, query) tuples from each class.

Referral webpage	Query	Class
www.adobe.com/	photoshop	Insignificant
www.adobe.com/products/cs6/faq.html	education discount cs	A
www.adobe.com/support/downloads/help.html	removing acrobat 8.0	B
helpx.adobe.com/premiere-pro/topics.html	import not responding	C

be derived as $\alpha^* = \arg\max_\alpha (|A_0|+|A_1|)/|A|$, where, $A_0 = \{(w,q) \mid rel(w,q) = 0$ and $page\_se\_score(w,q) < \alpha\}$, $A_1 = \{(w,q) \mid rel(w,q) = 1$ and $page\_se\_score (w,q) \geq \alpha\}$, and $|A| = |A_0|+|A_1| = 1000$. α^* and β^* were thus both set to 0.21.

The choice of δ was guided by the distribution of Pearson residuals as follows. As mentioned earlier (Sect. 2.1), $e_{ij} > \delta$ are considered significant. Due to possible noise and randomness in data, we should be skeptical about small positive residuals that signify slightly higher-than-expected counts. Positive residuals were found to follow an exponential distribution with rate = 0.0139. The log likelihood of the fit, normalized by the number of values, was -5.28. We set δ as the mean of the distribution, which was 71.94, as we believe that positive values below the mean can be a result of noise.

Extraction and Classification of Tuples. First, we extracted significant tuples (w^*, q^*) using the Pearson residuals (Eq. 1), and then classified the tuples. Distribution of counts of tuples is shown in Table 1. We see that 30.26% of the tuples (sum of the last three rows) represent actionable items for the site administrator, thus showing the potential of our methods in highlighting scope for improvement in the website content placement and relevance. Since the Pearson residual value represents the degree of deviation from the *expected* behavior, so a tuple with a higher residual can be associated with a higher *surprise factor*, and can be prioritized over one with a lower value. Table 2 shows some examples of tuples belonging to different classes.

Comparison Between Raw and Normalized Counts: We conducted tests to check whether results provided by our approach could be obtained using raw counts of (w,q) tuples as well, instead of residuals. Some of our insights are:

1. Pearson's rank correlation coefficient (r) between the vectors of counts and residual values over all tuples was found to be very close to zero (-0.035).

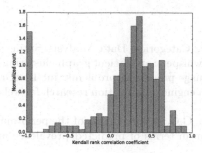

Fig. 2. Normalized τ-histogram for tuple rankings using residuals and counts.

Additionally, the Kendall rank correlation coefficient τ between the ranked lists when (w, q) tuples are ordered by frequency and residual value, was found to be -4.65×10^{-9}. This indicates almost no correlation between counts and residuals.

2. Subsequently, tuple counts were normalized as follows: if there are tuples (w, q_i), $i = 1 \ldots k$, with raw frequencies $f(w, q_i)$, $i - 1 \ldots k$, respectively, then the normalized frequencies are given by $nf(w, q_i) = \frac{f(w, q_i)}{\sum_{i=1}^{k} f(w, q_i)}$. Pearson's r and Kendall's τ for counts and residuals in this case were again found to be very low (0.03 and -1.02×10^{-6}, respectively).

3. For each referral webpage with at least five different queries, we calculated τ between the two ranked lists of queries for that page, obtained by sorting the queries with respect to counts and Pearson residuals in descending order. The distribution of τ values obtained over www.adobe.com ($\mu = 0.136, \sigma = 0.478$), is shown as normalized histogram in Fig. 2. τ coefficients are generally low, indicating little agreement between query rankings by counts and Pearson residuals for most webpages. From the above experiments, we conclude that raw or normalized (w, q) tuple counts do not produce results similar to those obtained using Pearson residuals.

4   Conclusion and Limitations

We have proposed a lightweight method for identifying page-specific missing content on large enterprise websites. We showed that using Pearson residuals are necessary alternatives to simple counts towards this goal. The proposed method ranks and classifies the significant query-webpage tuples into intuitive categories. Providing exact suggestions to address the missing-content issues, and a deployment-level evaluation, where, for example, we can observe whether query frequency on a specific page has reduced after making changes based on *reported* missing content issues, are the most promising future directions.

References

1. Agresti, A., Kateri, M.: Categorical Data Analysis. Springer, Heidelberg (2011)
2. Biemann, C.: Chinese whispers: an efficient graph clustering algorithm and its application to natural language processing problems. In: TextGraphs-1 (2006)
3. Cui, M., Hu, S.: Search engine optimization research for website promotion. In: ICM 2011 (2011)
4. Freeman, E.A., Moisen, G.G.: A comparison of the performance of threshold criteria for binary classification in terms of predicted prevalence and kappa. Ecol. Model. **217**(1), 48–58 (2008)
5. Lin, W., Liu, Y.: A novel website structure optimization model for more effective web navigation. In: WKDD 2008 (2008)
6. Yom-Tov, E., Fine, S., Carmel, D., Darlow, A.: Learning to estimate query difficulty including applications to missing content detection and distributed information retrieval. In: SIGIR 2005 (2005)

LTRo: Learning to Route Queries in Clustered P2P IR

Rami S. Alkhawaldeh[1(✉)], Deepak P.[2], Joemon M. Jose[3] (iD), and Fajie Yuan[3]

[1] Department of Computer Information Systems, University of Jordan-Aqaba,
Amman 77110, Jordan
r.alkhawaldeh85@gmail.com

[2] Centre for Data Sciences and Scalable Analytics, Queen's University Belfast,
University Road, Belfast BT7 1NN, UK
deepakp7@gmail.com

[3] School of Computing Science, University of Glasgow, University Avenue,
Glasgow G12 8QQ, UK
Joemon.Jose@glasgow.ac.uk, f.yuan.1@research.gla.ac.uk

Abstract. Query Routing is a critical step in P2P Information Retrieval. In this paper, we consider learning to rank approaches for query routing in the clustered P2P IR architecture. Our formulation, LTRo, scores resources based on the number of relevant documents for each training query, and uses that information to build a model that would then rank promising peers for a new query. Our empirical analysis over a variety of P2P IR testbeds illustrate the superiority of our method against the state-of-the-art methods for query routing.

1 Introduction

Query routing (*aka* resource selection) refers to the task of selecting a subset of resources to send each query to, in de-centralized search systems such as P2P IR and federated search systems. The considerations for P2P IR systems are typically different from those in federated search systems due to the asymmetry of document distribution across peers; for example, there could be peers with an order of magnitude more documents than others. Thus, methods which perform very well in federated search systems (e.g. CORI [1], logistic regression [2]) do not necessarily work that well for P2P IR. However, supervised approaches that make use of training data (i.e., past queries and information about peers deemed relevant for them) have not been explored much for the P2P IR query routing task.

In this paper, we consider the task of supervised query routing within the semi-structured cluster-based P2P IR architecture [3]. This architecture has been subject of recent interest [4,5], largely due to the presence of intra-peer content coherence at the query routing layer. For the first time, we consider learning-to-rank methods for supervised query routing within clustered P2P IR. Learning to Rank (LtR) techniques are supervised learning methods that can exploit training data in the form of a ranked list of objects [6]. Additionally, LtR approaches can

© Springer International Publishing AG 2017
J.M. Jose et al. (Eds.): ECIR 2017, LNCS 10193, pp. 513–519, 2017.
DOI: 10.1007/978-3-319-56608-5_42

also work with peer-specific [7], and peer-pairwise [8] relevance information. As an example, for our task of query routing, LtR approaches can be trained on a list of peers ordered according to their relevance to each query in the training set. In particular, we consider the following questions:

- Are LtR approaches applicable for the query routing problem in clustered P2P IR?
- How do LtR approaches compare against state-of-the-art models for query routing in clustered P2P IR?

2   Related Work

We now briefly survey related work on supervised resource selection. Among the first approaches for supervised resource selection was the method due to Arguello et al. [2] targeted towards the task of federated search; they propose usage of logistic regression to rank resources against queries. For every query-resource pair, the training feature vector is a concatenation of:

- *Query-dependent Corpus features:* A set of documents are sampled from each resource, and their relevance to the query is estimated using methods such as CORI [1] and ReDDE.top [9].
- *Query features:* These features encode query information such as the category of the query, and web documents that are deemed to be relevant to the query.

The relevance judgement is generated by firing training queries against the full dataset, i.e., the dataset across all resources. A resource is considered relevant if has more than a threshold (τ) number of documents among the top T documents from the full result. Hong et al. [10] extend this work for cases where a full dataset search is infeasible. Instead of the full dataset result, they build the 'full result' using just the top-T documents from each resource. In order to offset for inaccuracy in such approximation, they model and exploit similarities between resources in the query routing task. Thus, a resource which is not highly ranked against the queries using features may still be chosen by virtue of high resource-level similarity to other resources that are relevant to the query.

Cetintas et al. [11] propose a query routing approach that assesses resource relevance using the following formulation:

$$Rel(r_j|q) \propto \sum_{q' \in training} Rel(r_j|q') \times Sim(q', q)$$

Here, the relevance judgements for training queries are determined using the information as to whether the resource was selected for the query (using any resource selection method), whereas the similarity between queries are estimated using the correlation of their respective result sets.

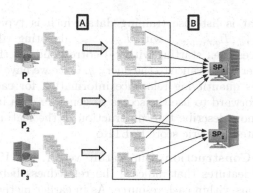

Fig. 1. Clustered P2P IR architecture

3   Clustered P2P IR Architecture

Figure 1 illustrates the construction of the clustered P2P IR architecture [3–5,12], our target architecture in this paper. Each of the peers maintain a subset of documents, as shown by the different P_is in the left side of the figure. The subset of documents within each peer are clustered independently (into k clusters, $k = 3$ in the figure), represented as Step **A**; we will call this as *intra-peer clustering*. Phase **B** clusters these intra-peer clusters, across peers, into a specified number (two, in the figure) of clusters. Each such cluster is managed by a super-peer (SP_i). Due to the clustering, not every super-peer necessarily would have representation from each peer; in our example, SP_2 does not have representation from P_1. Every query to the P2P IR system is sent to *each* of the super-peers, which would then employ the query routing approach to route the query to a subset of peers judged to be relevant to the query.

4   LTRo: Learning to Route

We now describe our LtR-based query routing approach, codenamed LTRo. General classification-based approaches such as those from [2,10] work with training data in the form of $[V_{q,r}, L_{q,r}]$ pairs. $V_{q,r}$ is a vector for the combination of query q and resource r, whereas $L_{q,r} \in \{-1, +1\}$ denotes whether the resource r is relevant for the query q or not. This is used to learn a mathematical model that can predict whether a resource is relevant to a query, thus enabling query routing:

$$\mathcal{F} : V_{q,r} \rightarrow \{+1, -1\}$$

LtR Training Data Formats: In addition to training data with binary relevance judgements as above, learning to rank approaches can exploit pairwise relevance judgements in the form of triplets like $[V_{q,r_1}, V_{q,r_2}, L_{q,r_1,r_2}]$ where L_{q,r_1,r_2} indicates whether resource r_1 is more relevant to q than r_2.

Yet another format is list-wise training data which is typically of the form $[V_{q,r_1}, V_{q,r_2}, \ldots, V_{q,r_m}, L_{q,r_1,\ldots,r_m}]$ with $L_{q,r_1,\ldots,r_m}$ denoting whether the chosen ordering of resources (i.e., starting with r_1) corresponds to the ordering in the non-increasing order of relevance to the query q. Once we have training data that has numeric values quantifying relevance information for each query-resource pair, it is straightforward to use the scores to generate data in any of the three forms above. We now describe the construction of the feature vector $V_{q,r}$ and that of the associated numeric score in LTRo.

Feature Vector Construction: Our feature vector, i.e., $V_{q,r}$, is constructed using a variety of features that indicate the relatedness between the training query and the corpus within each resource. As in earlier methods for supervised query routing, we sample documents from each resource, and use that to estimate the relatedness of the resource to each training query. The features we use are the concatenation of features from the following sources:

- Classical resource selection methods such as CORI [1] and CVV [13].
- Document Retrieval methods from various families, viz., (i) vector space models (TF and TF-IDF [14]), (ii) query relevance models (Language Modeling [15]), and (iii) divergence from randomness models (DFI0 [16], BB2 [17]). The usage of document retrieval methods is inspired by recent work [4] indicating their effectiveness for resource selection in the clustered P2P IR architecture.

Labelling: The labels associated with training data are critical to supervised learning. We now outline our method to associate numeric scores to each training vector $V_{q,r}$. Such numeric labels would then be converted, in a straightforward manner, to labels for appropriate choices of training data formats (pointwise, pairwise, or list-wise, as outlined earlier). We use the sampling-based approach for labeled data creation used in [10], whereby only a fixed sample of results (we set sample size to be 10) are obtained from each resource per training query. For every query-resource pair, we set the numeric score to the number of relevant retrieved documents in the sampled subset for the query.

LtR Models in LTRo: Having defined the construction of training vectors and associated scores, it is then simple to deploy any LtR algorithm for the task. We experimented with all the LtR models available in the RankLib[1] package, and did not find any perceivable difference in performance across them. Thus, we consistently employ the latest list-wise LtR technique from the RankLib library, i.e., co-ordinate ascent [18], in LTRo.

Testing Phase: For every new query (i.e., query from the test set), the LTRo model ranks the resources in the order of relevance to the query. We select the top-$k\%$ of all resources to route the query to. k is a parameter for the approach that may be varied; we experiment with values of k from $\{5\%, 10\%, 20\%, 30\%, 40\%, 50\%\}$ and report average of the evaluation measures across these values of k.

---

[1] https://sourceforge.net/p/lemur/wiki/RankLib/.

5   Experimental Study

We experimentally analyze LTRo against baseline approaches on several standard testbeds for P2P IR. We start by describing the setup and the baselines, and then go on to analyzing the experimental results.

Setup: We use several standard P2P IR testbeds from [19] in our evaluation. Each of these are based on the WT10g dataset[2], and model a variety of real-world data distributions with varying number of peers and varying skew of documents between peers. The characteristics of the various testbeds are summarized in Table 1. TREC 2000 and 2001 web track topics for the WT10g corpus are used as queries along with their ground truth relevance judgements. We selected 10,000 training query from 1.6 million known-item queries[3] leading to a choice of 18.82% single-term queries, 47% two-term queries, 19.7% three-term queries and the remaining 13.32% comprising four terms or more. The COMBMNZ [20] merging algorithm is used to combine the results from peers. We use the TREC 2001 query topics from 451–550 (these were excluded from training) as our test queries, thus replicating the setup from [4,5].

Baseline Methods: We have not come across supervised query routing methods that are specifically targeted to the clustered P2P IR architecture. Thus, we compare against the regression method from [2] (denoted as LR) as well as against a simple multi-layer perceptron based learner (MLP). In order to enable quantify the enhanced performance of the supervised approaches, we also report results from Taily [21], a recent unsupervised query routing method.

Experimental Results: Table 2 summarizes the comparative retrieval effectiveness of LTRo against the baseline approaches on each of the six testbeds, in terms of Precision (@top-1000), Recall (@1000), Precision@10 and MAP. The LTRo method is seen to outperform others in 75% of the metrics (18/24), and closely trails the leading method in the other cases (except for the DLWOR testbed, where the difference is more perceivable). The improvements achieved over the baseline approaches have also been indicated in the table. The results indicate that LTRo should be the method of choice for supervised query routing. This shows the effectiveness of going beyond binary relevance labeling and consequent usage of learning-to-rank approaches for the query routing problem in P2P IR.

Table 1. Test-beds general properties

Characteristics	ASISWOR	ASISWR	DLWOR	DLWR	UWOR	UWR
# Peers	11680	11680	1500	1500	11680	11680
# Docs	1692096	1788248	1692096	1740385	1692096	1788896
Avg. docs in peer	144.87	153.1	1128.54	1160.26	144.87	153.16

---

[2] http://ir.dcs.gla.ac.uk/test_collections/wt10g.html.
[3] http://boston.lti.cs.cmu.edu/callan/Data/P2P.

Table 2. LTRo retrieval effectiveness: two-paired statistically significant bootstrap t-test; $p \leq 0.01$ are denoted as • compared to Taily method.

DL*	DLWOR test-bed				DLWR test-bed			
Method	Precision	Recall	P@10	MAP	Precision	Recall	P@10	MAP
Taily	0.02815	0.52157	0.16050	0.08944	0.02519	0.48203	0.02367	0.02786
LR	0.04017	**0.57933**	0.22633	0.13322	**0.03687**	**0.54805**	0.05650	0.06389
MLP	**0.04028**	0.57889	0.22683	0.13135	0.03679	0.54706	0.05650	0.06389
LTRo	0.03972•	0.56756•	**0.23015•**	**0.13470•**	0.03668•	0.53493•	**0.05767•**	**0.06506•**
LTRo-LR	(−3.77%)	(−16.32%)	(+7.48%)	(+3.46%)	(−1.57%)	(−15.22%)	(+3.71%)	(+3.47%)
LTRo-MLP	(−4.8%)	(−18.77%)	(+6.04%)	(+7.55%)	(−0.94%)	(−14.23%)	(+3.71%)	(+3.48%)
ASIS*	ASISWOR test-bed				ASISWR test-bed			
Taily	0.02581	0.46064	0.15833	0.07046	0.01934	0.37833	0.01733	0.02042
LR	0.04356	0.54821	0.24500	0.12134	**0.03965**	0.52124	**0.06400**	0.06027
MLP	0.04354	0.54891	0.24400	0.12099	0.03954	0.52066	**0.06400**	0.06021
LTRo	**0.04412•**	**0.55917•**	**0.24600•**	**0.12454•**	0.03959•	**0.52462•**	0.06317•	**0.06194•**
LTRo-LR	(+2.76%)	(+6.73%)	(+1.07%)	(+5.34%)	(−0.28%)	(+2.46%)	(−1.72%)	(+4.18%)
LTRo-MLP	(+2.85%)	(+6.27%)	(+2.16%)	(+5.98%)	(+0.25%)	(+2.89%)	(−1.72%)	(+4.33%)
U*	UWOR test-bed				UWR test-bed			
Taily	0.02797	0.49474	0.18783	0.10229	0.02374	0.43882	0.01400	0.02451
LR	0.08842	0.73835	0.47133	0.32898	0.08488	0.71121	0.12317	0.14347
MLP	0.08842	0.73908	0.47233	0.32908	0.08485	0.71135	0.12317	0.14342
LTRo	**0.08856•**	**0.74407•**	**0.47400•**	**0.33283•**	**0.08498•**	**0.71152•**	**0.12783•**	**0.14663•**
LTRo-LR	(+0.24%)	(+2.33%)	(+0.98%)	(+1.7%)	(+0.17%)	(+0.12%)	(+4.31%)	(+2.71%)
LTRo-MLP	(+0.23%)	(+2.03%)	(+0.61%)	(+1.66%)	(+0.22%)	(+0.07%)	(+4.31%)	(+2.76%)

6   Conclusions and Future Work

In this paper, we considered the applicability of learning to rank methods for query routing within the clustered P2P IR architecture. Accordingly, we modeled the query routing problem within the learning to rank framework, and empirically evaluated it against state-of-the-art supervised and unsupervised algorithms for query routing. Our empirical analysis illustrates the superiority of our LtR approach, codenamed LTRo, in a large majority of scenarios, thus indicating that LTRo should be the method of choice for supervised query routing for clustered P2P IR.

References

1. Callan, J.P., Lu, Z., Croft, W.B.: Searching distributed collections with inference networks. In: Proceedings of SIGIR, pp. 21–28. ACM, New York (1995)
2. Arguello, J., Callan, J., Diaz, F.: Classification-based resource selection. In: Proceedings of CIKM 2009, pp. 1277–1286. ACM, New York (2009)
3. Klampanos, I.A., Jose, J.M.: An evaluation of a cluster-based architecture for peer-to-peer information retrieval. In: Wagner, R., Revell, N., Pernul, G. (eds.) DEXA 2007. LNCS, vol. 4653, pp. 380–391. Springer, Heidelberg (2007). doi:10.1007/978-3-540-74469-6_38
4. Alkhawaldeh, R., Jose, J., Padmanabhan, D.: Evaluating document retrieval methods for resource selection in clustered P2P IR. In: CIKM (2016)
5. Alkhawaldeh, R., Jose, J., Padmanabhan, D.: Clustering-based query routing in cooperative semi-structured peer to peer networks. In: ICTAI (2016)

6. Cao, Z., Qin, T., Liu, T.Y., Tsai, M.F., Li, H.: Learning to rank: from pairwise approach to listwise approach. Technical report MSR-TR-2007-40, Microsoft Research, April 2007
7. Friedman, J.H.: Greedy function approximation: a gradient boosting machine. Ann. Stat. **29**, 1189–1232 (2000)
8. Wu, Q., Burges, C.J., Svore, K., Gao, J.: Ranking, boosting, and model adaptation. Technical report MSR-TR-2008-109, Microsoft Research, October 2008
9. Arguello, J., Diaz, F., Callan, J., Crespo, J.F.: Sources of evidence for vertical selection. In: Proceedings of the 32nd International ACM SIGIR Conference on Research and Development in Information Retrieval. SIGIR 2009, pp. 315–322. ACM, New York (2009)
10. Hong, D., Si, L., Bracke, P., Witt, M., Juchcinski, T.: A joint probabilistic classification model for resource selection. In: Proceedings of SIGIR 2010, pp. 98–105. ACM, New York (2010)
11. Cetintas, S., Si, L., Yuan, H.: Learning from past queries for resource selection. In: Proceedings of CIKM 2009, pp. 1867–1870. ACM, New York (2009)
12. Alkhawaldeh, R.S., Jose, J.M.: Experimental study on semi-structured peer-to-peer information retrieval network. In: Mothe, J., Savoy, J., Kamps, J., Pinel-Sauvagnat, K., Jones, G.J.F., SanJuan, E., Cappellato, L., Ferro, N. (eds.) CLEF 2015. LNCS, vol. 9283, pp. 3–14. Springer, Cham (2015). doi:10.1007/978-3-319-24027-5_1
13. Yuwono, B., Lee, D.L.: Server ranking for distributed text retrieval systems on the Internet. In: DASFAA, pp. 41–50. World Scientific Press (1997)
14. Baeza-Yates, R.A., Ribeiro-Neto, B.: Modern Information Retrieval. Addison-Wesley Longman Publishing Co., Inc., Boston (1999)
15. Hiemstra, D.: Using language models for information retrieval. University of Twente (2001)
16. Dinçer, B.T., Kocabas, I., Karaoglan, B.: IRRA at TREC 2009: index term weighting based on divergence from independence model. In: TREC (2009)
17. Amati, G., Van Rijsbergen, C.J.: Probabilistic models of information retrieval based on measuring the divergence from randomness. ACM Trans. Inf. Syst. (TOIS) **20**(4), 357–389 (2002)
18. Metzler, D., Bruce Croft, W.: Linear feature-based models for information retrieval. Inf. Retr. **10**(3), 257–274 (2007)
19. Klampanos, I.A., Poznański, V., Jose, J.M., Dickman, P., Road, E.H.: A suite of testbeds for the realistic evaluation of peer-to-peer information retrieval systems. In: Losada, D.E., Fernández-Luna, J.M. (eds.) ECIR 2005. LNCS, vol. 3408, pp. 38–51. Springer, Heidelberg (2005). doi:10.1007/978-3-540-31865-1_4
20. Shaw, J.A., Fox, E.A.: Combination of multiple searches. In: Text REtrieval Conference, pp. 243–252 (1994)
21. Aly, R., Hiemstra, D., Demeester, T.: Taily: shard selection using the tail of score distributions. In: Proceedings of SIGIR 2013, pp. 673–682. ACM, New York (2013)

Faster K-Means Cluster Estimation

Siddhesh Khandelwal and Amit Awekar$^{(\boxtimes)}$

Indian Institute of Technology Guwahati, Guwahati, India
siddhesh166@gmail.com, awekar@iitg.ernet.in

Abstract. There has been considerable work on improving popular clustering algorithm 'K-means' in terms of mean squared error (MSE) and speed, both. However, most of the k-means variants tend to compute distance of each data point to each cluster centroid for every iteration. We propose a fast heuristic to overcome this bottleneck with only marginal increase in MSE. We observe that across all iterations of K-means, a data point changes its membership only among a small subset of clusters. Our heuristic predicts such clusters for each data point by looking at nearby clusters after the first iteration of k-means. We augment well known variants of k-means with our heuristic to demonstrate effectiveness of our heuristic. For various synthetic and real-world datasets, our heuristic achieves speed-up of up-to 3 times when compared to efficient variants of k-means.

Keywords: K-means · Clustering · Heuristic

1 Introduction

K-means is a popular clustering technique that is used in diverse fields such as humanities, bio-informatics, and astronomy. Given a dataset D with n data points in $\mathbb{R}^d$ space, K-means partitions D into k clusters with the objective to minimize the mean squared error (MSE). MSE is defined as the sum of the squared distance of each point from its corresponding centroid. The K-means problem is NP-hard. Polynomial time heuristics are commonly applied to obtain a local minimum.

One such popular heuristic is the Lloyd's algorithm [6] that selects certain initial centroids (also referred as seeds) at random from the dataset. Each data point is assigned to the cluster corresponding to the closest centroid. Each centroid is then recomputed as mean of the points assigned to that cluster. This procedure is repeated until convergence. Each iteration involves $n * k$ distance computations. Our contribution is to reduce this cost to $n * k'$, $(k' << k)$ by generating candidate cluster list (CCL) of size k' for each data point. The heuristic is based on the observation that across all iterations of K-means, a data point changes its membership only among a small subset of clusters. Our heuristic considers only a subset of nearby cluster as candidates for deciding membership for a data point. This heuristic has advantage of speeding up K-means clustering with marginal increase in MSE. We show effectiveness of our heuristic by extensive experimentation using various synthetic and real-world datasets.

© Springer International Publishing AG 2017
J.M. Jose et al. (Eds.): ECIR 2017, LNCS 10193, pp. 520–526, 2017.
DOI: 10.1007/978-3-319-56608-5_43

2 Our Work: Candidate Cluster List for Each Data Point

Our main contribution is in defining a heuristic that can be used as augmentation to current variants of k-means for faster cluster estimation. Let algorithm V be a variant of k-means and algorithm V' be the same variant augmented with our heuristic. Let T be the time required for V to converge to MSE value of E. Similarly, T' is the time required for V' to converge to MSE value of E'. We should satisfy following two conditions when we compare V with V':

- Condition 1: T' is lower than T, and
- Condition 2: E' is either lower or only marginally higher than E.

In short, these conditions state that a K-means variant augmented with our heuristic should converge faster without significant increase in final MSE.

Major bottleneck of K-means clustering is the computation of data point to cluster centroid distance in each iteration of K-means. For a dataset with n data points and k clusters, each iteration of K-means performs $n * k$ such distance computations. To overcome this bottleneck, we maintain a CCL of size k' for each data point. We assume that k' is significantly smaller than k. We discuss the effect of various choices for the size of CCL in Sect. 4. We build CCL based on top k' nearest clusters to the data point after first iteration of K-means. Now each iteration of K-means will perform only $n * k'$ distance computations.

Consider a data point p_1 and cluster centroids represented as $c_1, c_2..., c_k$. Initially all centroids are chosen randomly or using one of the seed selection algorithms mentioned in Sect. 3. Let us assume that $k' = 4$, and $k' << k$. After first iteration of K-means c_8, c_5, c_6, and c_1 are the top four closest centroids to p_1 in the increasing order of distance. This is the candidate cluster list for p_1. If we run K-means for second iteration, p_1 will compute distance to all k centroids. After second iteration, top four closest centroid list might change in two ways:

1. Members of the list do not change but only ranking changes among the members. For example, top four closest centroid list for p_1 might change to c_1, c_6, c_8, and c_5 in the increasing order of distance.
2. Some of the centroids in the previous list are replaced with other centroids which were not in the list. For example, top four closest list for p1 might change to c_5, c_2, c_9, and c_8 in the increasing order of distance.

For many synthetic and real world datasets we observe that the later case rarely happens. That is, the set of top few closest centroids for a data point remains almost unchanged even though order among them might change. Therefore, CCL is a good enough estimate for the closest cluster when K-means converges [1]. For each data point, our heuristic involves computation overhead of $O(k.log(k))$ for creating CCL and memory overhead of $O(k')$ to maintain CCL. For a sample dataset consisting 100,000 points in 54 dimensions and the value of $k = 100$ and $k' = 40$, this overhead is approximately 30 MB.

3   Related Work

In last three decades, there has been significant work on improving Lloyd's algorithm [6] both in terms of reducing MSE and running time. The follow up work
on Lloyd's algorithm can be broadly divided into three categories: Better seed
selection [2,5], Selecting ideal value for number of clusters [8], and Bounds on
data point to cluster centroid distance [3,4,7]. Arthur and Vassilvitskii [2] provided a better method for seed selection based on a probability distribution over
closest cluster centroid distances for each data point. Likas et al. [5] proposed
the Global k-means method for selecting one seed at a time to reduce final mean
squared error. Pham et al. [8] designed a novel function to evaluate goodness of
clustering for various potential values of number of clusters. Elkan [3] use triangle
inequality to avoid redundant computations of distance between data points and
cluster centroids. Pelleg and Moore [7] and Kanungo et al. [4] proposed similar
algorithms that use k-d trees. Both these algorithms construct a k-d tree over
the dataset to be clustered. Though these approaches have shown good results,
k-d trees perform poorly for datasets in higher dimensions.

Seed selection based K-means variants differ from Lloyd's algorithm only in
the method of seed selection. Our heuristic can be directly used in such algorithms. K-means variants that find appropriate number of clusters in data, evaluate the goodness of clustering for various potential values of number of clusters.
Such algorithms can use our heuristic while performing clustering for each potential value of k. K-means variants in third category compute exact distances only
to few centroids for each data point. However, they have to compute bounds on
distances to rest of the centroids for each data point. Our heuristic can help such
K-means variants to further reduce distance and bound calculations.

4   Experimental Results

Our heuristic can be augmented to multiple variants of K-means mentioned in
Sect. 3. When augmented to Lloyd's algorithm, our heuristic provides a speedup
of upto 9 times with the error within 0.2% of that of Lloyd's algorithm [1]. However to show the effectiveness of our heuristic, we present results of augmenting
it to faster variants of K-means such as K-means with triangle inequality (KMT)
[3]. Due to lack of space, we present results of augmenting our heuristic with only
this variant. Augmenting KMT with our heuristic is referred as algorithm HT.
Code and datasets used for our experiments are available for download [1].

During each iteration of KMT, a data point computes distance to the centroid of its current cluster. KMT uses triangle inequality to compute efficient
lower bounds on distances to all other centroids. A data point will compute
exact distance to any other centroid only when the lower bound on such distance is smaller than the distance to the centroid of its current cluster. During
each iteration of HT, a data point will also compute distance to the centroid
of its current cluster. However, HT will compute lower bounds on distances to
centroids only in its CCL. A data point will compute exact distance to any other

centroid in the candidate cluster list only when the lower bound on such distance is smaller than the distance to the centroid of its current cluster.

Experimental results are presented on five datasets, four of which were used by Elkan [3] to demonstrate the effectiveness of KMT and one is a synthetically generated dataset by us. These datasets vary in dimensionality from 2 to 784, indicating applicability of our heuristic for low as well as high dimensional data (please refer to Table 1). Our evaluation metrics are chosen based on two conditions mentioned in Sect. 2: Speedup to satisfy Condition 1 and Percentage Increase in MSE (PIM) to satisfy Condition 2. Speedup is calculated as T/T'. PIM is calculated as $(100 * (E' - E))/E$. We tried two different methods for initial seed selection: random [6] and K-means++ [2]. Both seed selection methods gave similar trends in results. To ensure fair comparison, the same initial seeds are used for both KMT and HT. For some experiments, HT achieves smaller MSE than KMT ($E' \leq E$). This happens because our heuristic jumps the local minima by not computing distance to every cluster centroid. Only in such cases, HT requires more iterations to converge and runs slower than KMT.

Effect of Varying k': Please refer to Table 2. The value of the total number of clusters k is set to 100 for all datasets. Running time and MSE of KMT is independent of value of k'. Speed up of HT over KMT increases with reduction in value of k'. This is expected as for small value of k', HT can avoid many redundant distance computations using small CCL. Speed up of HT over KMT is not same as the ratio k/k'. Reason for reduced speed up is that KMT also avoids some distance computations using its own filtering criteria of triangle inequality. Our heuristic achieves ideal speed of k/k' when compared against basic K-means algorithm [1]. E' increases with reduction in value of k'. However, E' is only marginally higher than E as PIM value never exceeds 1.5.

Effect of Varying k: Please refer to Table 3. Here, we report results for value of k' set to $0.4 * k$. With increasing value of k, HT achieves better speed up over KMT and difference between MSE of HT and MSE of KMT reduces. With increasing value of k, most of the centroid to data point distance calculations become redundant as data-point is assigned only to the closest centroid. In such scenario, our heuristic avoids distance computations with reduced PIM. This shows that our heuristic can be used for datasets having only few as well as large number of clusters.

Table 1. Datasets used in experiment

Name	Cardinality	Dimensionality	Description
Birch	100000	2	10 by 10 grid of Gaussian clusters
Covtype	150000	55	Remote soil cover measurements
Mnist	60000	784	Original NIST handwritten digit training data
KDDCup	95412	481	KDD Cup 1998 data
Synthetic	100000	100	Uniform random dataset

Effect of Seeding: Please refer to Tables 2 and 3. For each value of k' in Table 2 and k in Table 3, we used two different initial seedings - random (RND) and Kmeans++ [2]. If we compare the results, we observe that better seeding (KMeans++) generally gives better results in terms of PIM. Randomly selected seeds are not necessarily well distributed across the dataset. In such cases, successive iterations of K-means causes significant changes in cluster centroids. Improved seeding methods such as KMeans++ ensure that the initial centroids are spread out more uniformly. Thus centroids shift is less significant in successive iterations. In such scenario, CCL computed after first iteration is

Table 2. Effect of varying k' on HT performance. The value of $k = 100$. RND = Random initialization; KPP = Initialization using Kmeans++[2]

		$k' = 20$		$k' = 30$		$k' = 40$		$k' = 50$		$k' = 60$	
		RND	KPP	RND	KPP	RND	KPP	RND	KPP	RND	KPP
Birch	PIM (%)	-0.11	0	0.04	0	0	0	0	0	0	0
	Speedup	3.05	3.14	2.48	2.26	2.01	1.93	1.68	1.67	1.41	1.31
Covtype	PIM (%)	0.21	0.03	0.02	0	0	0	0	0	0	0
	Speedup	2.32	2.02	1.81	1.82	1.61	1.63	1.55	1.38	1.42	1.20
Mnist	PIM (%)	1.30	1.36	0.60	0.71	0.36	0.36	0.30	0.18	0.23	0.09
	Speedup	1.89	1.47	1.60	1.44	1.42	1.26	1.38	1.19	1.37	1.15
KDDCup	PIM (%)	0.81	0.70	0.11	0.15	0.08	0.02	-0.18	-0.01	0	0
	Speedup	1.44	1.60	1.33	1.15	1.42	1.02	0.88	0.99	1.18	1.02
Synthetic	PIM (%)	0.19	0.15	0.11	0.08	0.06	0.04	0.03	0.01	0.01	0.01
	Speedup	2.90	2.45	2.28	1.97	1.87	1.71	1.51	1.35	1.36	1.17

Table 3. Effect of varying k on HT performance. The value of $k' = 0.4 * k$. RND = Random initialization; KPP = Initialization using Kmeans++[2]

		$k = 50$		$k = 100$		$k = 500$		$k = 1000$	
		RND	KPP	RND	KPP	RND	KPP	RND	KPP
Birch	PIM (%)	0.31	0	0	0	0	0	0	0
	Speedup	1.65	1.71	1.98	1.97	2.14	2.10	2.12	2.15
Covtype	PIM (%)	0.01	0.02	0.26	0	0	0	0	0
	Speedup	1.35	1.31	1.65	1.50	1.94	1.87	1.97	1.90
Mnist	PIM (%)	0.94	0.87	0.38	0.52	0.09	0.23	0.13	0.07
	Speedup	1.10	1.20	1.23	1.45	1.28	1.24	1.29	1.19
KDDCup	PIM (%)	0.51	0.99	-0.06	0.15	0	0.03	0	0.02
	Speedup	1.02	1.38	0.85	1.18	1.13	1.33	1.19	1.37
Synthetic	PIM (%)	0.09	0.07	0.05	0.04	0.03	0.01	0.01	0.01
	Speedup	2.03	1.63	1.76	1.56	1.75	1.45	1.56	1.51

a better estimate for final cluster membership. Thus our heuristic is expected to perform better with newer variants of K-means that provide improved seeding.

Effect of Cluster Well-Separateness: We also performed experiments on synthetic datasets in two dimensions. These datasets were generated using a mixture of Gaussians. The Gaussian centers are placed at equal angles on a circle of radius r $(angle = \frac{2\pi}{k})$, and each center is assigned equal number of points $(\frac{n}{k})$. The experiment was done on synthetic datasets of 100000 points generated using the method described above with variance set to 0.25. The value of k is set to 100 and the value of k' is set to 40. We generated nine datasets by varying the radius from zero to forty in steps of five units. We ran KMT and HT over these nine datasets to check how our heuristic performs with change in well separateness of clusters. We observed that when clusters are close, both the algorithms converge quickly as initial seeds happen to be close to actual cluster centroids. With higher radius, initial seeds might be far off from the actual cluster centroids and KMT takes longer to converge. However, HT performs significantly better for higher values of radius as HT can quickly discard far away clusters. HT achieves a speedup of around 2.31 for higher radius values. For all experiments over these synthetic datasets, we observed that PIM value never exceeds 0.01 [1]. This indicates that our heuristic remains relevant even with variation in degree of separation among the clusters.

5   Conclusion

We presented a heuristic to attack the bottleneck of redundant distance computations in K-means. Our heuristic limits distance computations for each data point to CCL. Our heuristic can be augmented with diverse variants of K-means to converge faster without any significant increase in MSE. With extensive experiments on real world and synthetic datasets, we showed that our heuristic performs well with variations in dataset dimensionality, CCL size, number of clusters, and degree of separation among clusters. This work can be further improved by making the CCL dynamic to achieve better speed up while reducing the PIM value.

References

1. The code and dataset for the experiments can be found at: https://github.com/siddheshk/Faster-Kmeans
2. Arthur, D., Vassilvitskii, S.: k-means++: the advantages of careful seeding. In: ACM-SIAM Symposium on Discrete algorithms, pp. 1027–1035 (2007)
3. Elkan, C.: Using the triangle inequality to accelerate k-means. In: International Conference on Machine Learning, pp. 147–153 (2003)
4. Kanungo, T., Mount, D.M., Netanyahu, N.S., Piatko, C.D., Silverman, R., Wu, A.Y.: An efficient k-means clustering algorithm: analysis and implementation. IEEE Trans. Pattern Anal. Mach. Intell. **24**(7), 881–892 (2002)

5. Likas, A., Vlassis, N., Verbeek, J.J.: The global k-means clustering algorithm. Pattern Recogn. **36**(2), 451–461 (2003)
6. Lloyd, S.P.: Least squares quantization in PCM. IEEE Trans. Inf. Theory **28**(2), 129–137 (1982)
7. Pelleg, D., Moore, A.: Accelerating exact k-means algorithms with geometric reasoning. In: ACM SIGKDD, pp. 277–281. ACM (1999)
8. Pham, D.T., Dimov, S.S., Nguyen, C.: Selection of k in k-means clustering. J. Mech. Eng. Sci. **219**(1), 103–119 (2005)

Predicting Emotional Reaction in Social Networks

Jérémie Clos[1]([⊠]), Anil Bandhakavi[1]([⊠]), Nirmalie Wiratunga[1], and Guillaume Cabanac[2]

[1] Robert Gordon University, Aberdeen, UK
{j.Clos,a.Bandhakavi,n.Wiratunga}@rgu.ac.uk
[2] Université de Toulouse, Toulouse, France
guillaume.Cabanac@univ-tlse3.fr

Abstract. Online content has shifted from static and document-oriented to dynamic and discussion-oriented, leading users to spend an increasing amount of time navigating online discussions in order to participate in their social network. Recent work on emotional contagion in social networks has shown that information is not neutral and affects its receiver. In this work, we present an approach to detect the emotional impact of news, using a dataset extracted from the Facebook pages of a major news provider. The results of our approach significantly outperform our selected baselines.

1 Introduction

With the rise of the social web, a majority of online content has shifted from being static and document-oriented to being highly dynamic and discussion-oriented. With this shift, users have been spending more time navigating online discussions in order to stay informed with their social network. Recent work on emotion contagion in social networks [2] suggests that information is not neutral, and the way it is presented has an impact on the emotional state of its consumers. This demonstrates the importance of providing users with a way to control this content. In this work, we present a technique to predict the emotional impact of news on its consumers, using a dataset extracted from the Facebook pages of the New York Times, a major news network.

We highlight the novelty of our work with respect to existing research on textual emotion detection, before formalizing our problem and explaining our methodology. We evaluate our approach using two naive and two strong baselines. We conclude the paper by discussing our positive results and potential extensions of this work.

2 Related Work

Our work lies in the broader context of opinion mining. Most of the literature in this area aims to mine either the sentiment (*positive* and *negative*) or the basic

---

J. Clos and A. Bandhakavi—contributed equally to this work.

© Springer International Publishing AG 2017
J.M. Jose et al. (Eds.): ECIR 2017, LNCS 10193, pp. 527–533, 2017.
DOI: 10.1007/978-3-319-56608-5_44

emotions (*anger, joy, ...*) expressed in the content using computational models learned from labeled or distantly labeled sentiment or emotion corpora [1,4,7]. More recently work has also been done on the detection of emotion in a social network, but focusing on analyzing the emotion contained in text rather than its influence on others [5].

The originality of our work lies in predicting emotion reactions induced in readers by emotional text. Whilst harnessing emotion rated content (e. g., news stories) like in [6,8], to learn word-emotion lexicons, we also go a step further and propose methods to adopt such lexicons for predicting emotion reactions towards emotional text (e. g., news posts). The task described in this work is thus inherently harder because of the latent factors that are implied in the process, e. g., a joyful news might be received with anger by a certain population if they already have a negative predisposition towards the entity concerned by the news, and inversely. Analyzing this bias, however, is beyond the scope of this work and is reserved for future research.

3   Method

3.1   Problem Definition

We now give a formal outline to the problem of emotion reaction prediction. Given a set of posts P in a social network (e. g., Facebook) and their corresponding emotion rating vectors R, where R_i is the rating vector corresponding to the post P_i, we aim to predict the emotion rating vector r' for an unseen post p'. The emotion ratings for each post in P are normalized to form a probability distribution across the different emotions. For example, a post *friend met with an accident :(* and its emotion ratings vector ⟨*anger* : 0.35, *joy* : 0.0, *sadness* : 0.55, *surprise* : 0.15, *love* : 0.0⟩.

3.2   Methods

Our approach contains two different steps. First we learn an emotion lexicon from emotion rated Facebook posts, in order to model the emotion distribution of that particular post. Secondly we train a multi-linear regression (MLR) model using the emotion distribution as predictors. The regression model is used to predict the emotion reaction distribution on unseen posts, thus providing a mapping from the emotional state of the post to the emotional state of the users that are reacting to it.

3.3   Lexicon for Emotion Reaction Detection

In this section we describe our proposed unigram mixture model (UMM) applied to the task of emotion lexicon (EMOLEX) generation. We model real-world emotion data as a mixture of emotion bearing words and emotion-neutral (background) words. For example consider the tweet *going to Paris this Saturday*

#elated #joyous, which explicitly connotes emotion *joy*. However, the word *Saturday* is evidently not indicative of *joy*. Further *Paris* could be associated with emotions such as *love*. Therefore our generative model assumes a mixture of two unigram language models to account for such word mixtures in documents. More formally our generative model describes the generation of documents connoting emotion e_t as follows:

$$P(D_{e_t}, Z|\theta_{e_t}) = \prod_{i=1}^{|D_{e_t}|} \prod_{w \in d_i} [(1 - Z_w)\lambda_{e_t} P(w|\theta_{e_t})$$
$$+ (Z_w)(1 - \lambda_{e_t})P(w|N)]^{c(w,d_i)} \qquad (1)$$

where θ_{e_t} is the emotion language model and N is the background language model. λ_{e_t} the mixture parameter, $c(w, d_i)$ the number of times word w occurs in document d_i and Z_w a binary hidden variable which indicates the language model that generated the word w.

We can estimate parameters θ_{e_t} and Z using expectation maximization (EM), which iteratively maximizes the complete data (D_{e_t}, Z) by alternating between two steps: E-step and M-step. The E and M steps in our case are as follows:

E-step:

$$P(Z_w = 0|D_{e_t}, \theta_{e_t}^{(n)}) = \frac{\lambda_{e_t} P(w|\theta_{e_t}^{(n)})}{\lambda_{e_t} P(w|\theta_{e_t}^{(n)}) + (1 - \lambda_{e_t})P(w|N)} \qquad (2)$$

M-step:

$$P(w|\theta_{\theta_{e_t}}^{(n+1)}) = \frac{\sum_{i=1}^{|D_{e_t}|} P(Z_w = 0|D_{e_t}, \theta_{e_t}^{(n)})c(w, d_i)}{\sum_{w \in V} \sum_{i=1}^{|D_{e_t}|} P(Z_w = 0|D_{e_t}, \theta_{e_t}^{(n)})c(w, d_i)} \qquad (3)$$

where n indicates the EM iteration number. The EM iterations are terminated when an optimal estimate for the emotion language model θ_{e_t} is obtained. EM is used to estimate the parameters of the k mixture models corresponding to the emotions in E. The emotion lexicon *EmoLex* is learned by using the k emotion language models and the background model N as follows:

$$EmoLex(w_i, \theta_{e_j}) = \frac{P(w_i|\theta_{e_j}^{(n)})}{\sum_{t=1}^{k}[P(w_i|\theta_{e_t}^{(n)})] + P(w_i|N)} \qquad (4)$$

$$EmoLex(w_i, N) = \frac{P(w_i|N)}{\sum_{t=1}^{k}[P(w_i|\theta_{e_t}^{(n)})] + P(w_i|N)} \qquad (5)$$

where k is the number of emotions in the corpus, and *EmoLex* is a $|V| \times (k+1)$ matrix, where $|V|$ is the size of the vocabulary V.

3.4   Lexicon-Based Regression for Emotion Reaction Detection

In this section we describe the multilinear regression model built using feature vectors extracted using the EMOLEX emotion lexicon. The model is built in

two stages. In the first stage EMOLEX is used to extract features to represent
a post as a 5-dimensional emotion vector, using a simple average and aggregate
approach, meaning that each component of the feature vector is computed as
an average of the values of the corresponding component for each term in the
post. More formally the feature vector d_{vec} for a post d is extracted using the
formulation described in Eq. 6.

$$d_{vec} = \frac{\sum_{w \in d} EmoLex(w) \times count(w, d)}{|d|} \quad (6)$$

Here $EmoLex(w)$ represents the emotion vector corresponding to the word
w, $count(w, d)$ the frequency w in the post d and $|d|$ the length of the post. In the
second stage we build five separate MLR models, one for each target emotion.
We now describe the MLR model for an arbitrary emotion e_k.

Given a matrix of training vectors $D_{n \times 5} = d_{vec}^1, d_{vec}^2, \ldots, d_{vec}^n$, and their
corresponding user ratings vector $R_{n \times 1} = r_{e_k}^1, r_{e_k}^2, \ldots, r_{e_k}^n$, for emotion e_k, the
MLR model is defined in Eq. 7.

$$R = D \times W + \mathcal{E} \quad (7)$$

In this equation W represents the coefficient matrix, which when multiplied
with D becomes the fit of the regression model to the data. $\mathcal{E}$ is the vector that
captures the deviation of the model. The objective is to learn the coefficient
matrix W, which along with D, $\mathcal{E}$, best estimates (i. e., with a minimal training
error) the ratings vector R.

4   Evaluation

Given a set of emotionally charged Facebook posts, we investigate techniques to
estimate the emotional reactions towards them, captured in the form of numerical ratings: the number of times people clicked on an emotion emoticon. We
leverage a Facebook feature which allows users to react to any item published
on a user timeline using an emoticon as shown in Fig. 1.

Fig. 1. Emotional reactions in Facebook stories

We evaluated our method using a stratified k-fold cross validation with 5 folds
and the root mean square error (RMSE) as the performance metric. RMSE is a
standard performance metric used when estimating continuous quantities, and
is thus suited to our task. It is defined in Eq. 8 where Y is the vector of observed
values, $\hat{Y}$ the vector of predicted values and n the number of instances in the
dataset.

$$RMSE(Y, \hat{Y}) = \sqrt{\frac{\sum_{i=0}^{n} (\hat{Y}_i - Y_i)^2}{n}} \quad (8)$$

4.1 Baselines

We use two naive baseline methods based on general corpus statistics (UNIFORM and EMPIRICAL) which do not learn any computational model on the training posts in order to predict the emotion distribution of unobserved posts, as well as two stronger contenders: one based on a simple lexicon with a trivial mapping (EMOLEX) and one based on a linear regression trained on a WORD2VEC embedding (WORD2VEC+MLR).

1. UNIFORM assumes a completely uniform distribution over the target labels, so that no matter the input the output remains the following:

$$f(d) = \langle 0.2; 0.2; 0.2; 0.2; 0.2 \rangle$$

2. EMPIRICAL assumes that the distribution over the target labels is always the same as the empirical distribution observed in the training data, so that regardless of the input the output remains the following:

$$f(d) = \left\langle \frac{f(e_1)}{\sum_{i=0}^{|e|} f(e_i)}; \frac{f(e_2)}{\sum_{i=0}^{|e|} f(e_i)}; \frac{f(e_3)}{\sum_{i=0}^{|e|} f(e_i)}; \frac{f(e_4)}{\sum_{i=0}^{|e|} f(e_i)}; \frac{f(e_5)}{\sum_{i=0}^{|e|} f(e_i)} \right\rangle$$

where $f(e_i)$ is the frequency of emotion i in the training corpus.

3. EMOLEX simply uses the output of the emotion lexicon used to extract the feature vectors as a direct output.

$$f(d) = \langle \text{EMOLEX}_1(d); \text{EMOLEX}_2(d); \text{EMOLEX}_3(d); \text{EMOLEX}_4(d); \text{EMOLEX}_5(d) \rangle$$

where $\text{EMOLEX}_i(d)$ is the output of the lexicon for emotion i and document d.

4. WORD2VEC+MLR uses word vectors from a WORD2VEC embedding [3], computed on a 400-dimensional embedding with a skipgram-10 model on a Wikipedia corpus, and trains a MLR on it.

$$D\prime = \langle v(t_1); v(t_2); ...; v(t_n) \rangle$$

where $v(t_i)$ is the embedding vector for term i belonging to the document.

4.2 Dataset

We used a dataset crawled from the comments on the Facebook page of the New York Times. As detailed in Table 1 emotions are not uniformly distributed in the dataset itself, but the distribution of emotions in the Facebook posts is strongly correlated with the distribution of emotions in the reactions ($R = 0.8814$ on a Pearson test). We also note that the coverage of our emotion lexicon is close from the coverage of the WORD2VEC embedding despite the word embedding being computed on a general purpose resource.

Table 1. Descriptive statistics on the New York times dataset

Corpus statistics		Emotion probability distribution		
			Posts	Reaction
Number of posts	5367	Anger	0.192	0.220
Average terms/sentence	22.34	Joy	0.155	0.104
EMOLEX coverage	18792	Sadness	0.208	0.269
WORD2VEC coverage	16011	Surprise	0.178	0.100
		Love	0.264	0.304

4.3 Results

The results of our experiment, shown in Table 2 averaged over 5 folds show that our approach outperforms all the baselines. We note that while our approach outperforms all of the baselines by a significant margin ($p < 0.05$ on a pairwise two-tailed T-test computed on the 5 folds), the biggest margin remains between approaches that used an emotion mapping and approaches that did not. Hence, there is a correlation between the reactions of the users and the emotions displayed in the Facebook stories themselves, which leads more credence to preexisting works on online emotion contagion [2].

Table 2. Results (lower is better)

	Method	RMSE
Naive baselines	UNIFORM	0.578
	EMPIRICAL	0.532
Strong baselines	EMOLEX	0.510
	WORD2VEC+MLR	0.531
Approach	**EmoLex+MLR**	**0.492**

5 Conclusion

In this work we demonstrated the validity of our approach to predict the emotional reaction to a specific news item. We showed that the mapping from news item to an emotion space fed into a multilinear regression model outperformed both a direct mapping from the text (using WORD2VEC and a multilinear regression) and an estimation from the text (using the EMOLEX emotion lexicon). This work constitutes a first step towards building a generic model for estimating the emotional impact of news and providing users with a way to avoid being manipulated. Future extensions of this work will focus on diversifying the communication platforms used for spreading emotion-rich content, as well as studying the practical effect of such contagion on users.

References

1. Binali, H., Wu, C., Potdar, V.: Computational approaches for emotion detection in text. In: 4th IEEE International Conference on Digital Ecosystems and Technologies, pp. 172–177. IEEE (2010)
2. Kramer, A.D., Guillory, J.E., Hancock, J.T.: Experimental evidence of massive-scale emotional contagion through social networks. Proc. Natl. Acad. Sci. **111**(24), 8788–8790 (2014)
3. Mikolov, T., Sutskever, I., Chen, K., Corrado, G.S., Dean, J.: Distributed representations of words and phrases and their compositionality. In: Advances in NIPS, pp. 3111–3119 (2013)
4. Pang, B., Lee, L.: Opinion mining and sentiment analysis. Found. Trends Inf. Retrieval **2**(1–2), 1–135 (2008)
5. Qiyao, W., Zhengmin, L., Yuehui, J., Shiduan, C., Tan, Y.: ULM: a user-level model for emotion prediction in social networks. China Univ. Posts Telecommun. **23**, 63–88 (2016)
6. Rao, Y., Lei, J., Wenyin, L., Li, Q., Chen, M.: Building emotional dictionary for sentiment analysis of online news. World Wide Web **17**(4), 723–742 (2014)
7. Ribeiro, F.N., Araújo, M., Gonçalves, P., Gonçalves, M.A., Benevenuto, F.: Sentibench-a benchmark comparison of state-of-the-practice sentiment analysis methods. EPJ Data Sci. **5**(1), 1–29 (2016)
8. Staiano, J., Guerini, M.: DepecheMood: a lexicon for emotion analysis from crowd-annotated news. In: Proceedings of the 52nd Annual Meeting of the ACL, pp. 427–433 (2014)

Irony Detection with Attentive Recurrent Neural Networks

Yu-Hsiang Huang, Hen-Hsen Huang, and Hsin-Hsi Chen[(✉)]

Department of Computer Science and Information Engineering,
National Taiwan University, Taipei, Taiwan
{yhhuang,hhhuang}@nlg.csie.ntu.edu.tw, hhchen@ntu.edu.tw

Abstract. Automatic Irony Detection refers to making computer understand the real intentions of human behind the ironic language. Much work has been done using classic machine learning techniques applied on various features. In contrast to sophisticated feature engineering, this paper investigates how the deep learning can be applied to the intended task with the help of word embedding. Three different deep learning models, Convolutional Neural Network (CNN), Recurrent Neural Network (RNN), and Attentive RNN, are explored. It shows that the Attentive RNN achieves the state-of-the-art on Twitter datasets. Furthermore, with a closer look at the attention vectors generated by Attentive RNN, an insight into how the attention mechanism helps find out the linguistic clues of ironic utterances is provided.

Keywords: Irony detection · Neural networks · Sentiment analysis

1 Introduction

Authors/speakers often use ironic expressions to convey their strong feelings in some situations. An irony says something other than what it meant, or says the opposite of what it meant. For example, the utterance *"I love to be ignored"* means *"I hate to be ignored"* in general understanding. Automatic irony detection aims at realizing people's real intentions. It has many potential applications. In opinion mining and sentiment analysis, the polarities of opinionated expressions in reviews affect readers' decision-making on specific targets. Ironic expressions bring in much stronger comments and thus should have more effects.

Sarcasm and irony are very similar in surface form, but sarcasm ridicules on some victims [6]. In this paper, we focus on the phenomenon of using opposite literal meaning as a mean to strengthen one's point. In other words, we do not distinguish their strict differences. Irony detection is challenging, because, to understand the actual meaning, readers/listeners also need to consider context and background knowledge rather than just interpreting the expressions literally.

In this paper, we will explore Convolutional Neural Network (CNN), Recurrent Neural Network (RNN), and Attentive RNN in irony detection tasks, and compare them with the state-of-the-art feature engineering approaches. This

© Springer International Publishing AG 2017
J.M. Jose et al. (Eds.): ECIR 2017, LNCS 10193, pp. 534–540, 2017.
DOI: 10.1007/978-3-319-56608-5_45

paper is organized as follows. Section 2 surveys the related work. Section 3 presents how CNN, RNN and Attentive RNN model a sentence and classify it into one of the classes, i.e., Irony or Non-Irony. Section 4 introduces experimental datasets, shows and discusses the results. Section 5 further explains why Attentive RNN achieves the best performance with a case study.

2 Related Work

Reyes et al. [9] first collected an ironic tweet corpus by searching with hashtags to avoid labeling manually. Machine learning techniques with textual features were explored on the irony detection task [2,9]. The hashtag-based approaches are not always suitable for irony corpus construction for all languages. Tang and Chen [10] proposed a method to construct a Chinese irony corpus based on the use of emoticons, linguistic forms, and sentiment polarity. Recently, word embedding [8] is widely adopted in various NLP tasks for its power on semantic similarity between words. Ghosh et al. [3] proposed a maximum-valued matrix-element SVM kernel using word embedding to deal with word sense disambiguation task on an irony corpus.

3 Irony Detection Model

We regard the irony detection task as a binary classification problem. Words in a sentence are represented as a sequence of embedded word vectors with a table lookup in pre-trained vectors. We first encode each sentence into a vector with three different neural network models.

The first one is Convolutional Neural Network (CNN), which is introduced by LeCun et al. [7], and used as a sentence modeling method in NLP [5] with the use of word embedding [8]. Our CNN is applied with one-directional convolutions over the embedded word vectors with multiple filters in various sizes. After one-max-pooling applied over all the filters' outputs, the resulted scalars are concatenated together as the encoded vector.

The second model is Recurrent Neural Network (RNN), which is invented for the use of sequential data. In each instant of time, RNN generates an output vector which considers not only current input, but also the previous processed result (memory). The last output vector is taken as the encoded vector.

Attention mechanism is first used in NLP by Bahdanau et al. [1], and gets popular recently. Our third model is Attentive RNN, which makes a weighted combination of all the output vectors generated from the underlying RNN. In this way, our model takes a global view on all the past information. Given each output vector h_t resulting from time t and let Y be a matrix consisting of output vectors $[h_1;...;h_L]$ by concatenating them together, when the input sentence contains L words, we can get an attended vector h' via the following formula:

$$\alpha = softmax(w^T Y)$$
$$h' = Y\alpha^T$$

where w is a trained vector, and α is the attention weight vector. We get the encoded vector h' by summing up each output vector h_t weighted by the value in dimension t in α.

At the last, the sentence encoded vectors generated by the three neural networks are individually passed to a full-connected layer, along with a softmax layer, to project into the target space of the two classes, i.e., Irony and Non-Irony. The models are trained with the cross-entropy loss as objective function in an end-to-end way to make every part of the model optimized for this task.

4   Experiments

4.1   Datasets and Experiment Setup

The Twitter dataset collected by Ghosh et al. [3] is used in this paper. It includes 198,041 tweets in sarcastic (ironic) sense and 197,917 tweets in literal sentiment sense. Ghosh et al. first collected 37 target words (e.g. "genius") and search tweets containing these target words, i.e., each tweet belongs to one of the target words. Therefore this dataset can be viewed as 37 subdatasets. We take the best model of Ghosh et al., MVME kernel SVM with skip-gram word2vec, as our baseline. In the preprocessing step, the hashtags and username mentions are removed, and hyperlinks and out-of-vocabulary words are treated as special tokens. The skip-gram word2vec vectors pre-trained by Google are used[1], where the embedding would be fine tuned in the training process, and words not shown in the pre-trained vectors are initiated randomly. The CNN model is applied with filter sizes 2, 4, and 6, and each size has 500 filters. The RNN and Attentive RNN models are applied with the Long Short-Term Memory (LSTM) units [4]. All the hidden layers have a dimension size 400 in the three models. Dropout rate [12] is set to 0.2 in the last full-connected layer. RMSProp [11] is used for optimization with learning rate 0.001, rho 0.9, and epsilon 10^{-8}.

4.2   Results

Experimental results are shown in Table 1, including the micro average of Precision, Recall and F1 score among the target words. The maximum and the minimum F1 scores are also provided. We can find that CNN gets a really high average precision (91.5%) over all target words. RNN performs worse than CNN. When attention mechanism is introduced, Attentive RNN makes a progress in F1 score by 5.0% compared to the baseline and beats the other two models. It is interesting to see that the target word with minimum F1 score is "genius" in both models of Ghosh et al. and our Attentive RNN. This is because the target "genius" has less training instances in the dataset. However, the Attentive RNN still gets a better F1 score in this case.

---

[1] https://code.google.com/archive/p/word2vec/.

Table 1. Performance on Ghosh et al. dataset.

Model	Avg. P(%)	Avg. R(%)	Avg. F1(%)	Max. F1 (Target word)	Min. F1 (Target word)
Ghosh et al.	81.9 ± 3.8	88.1 ± 3.2	84.8 ± 3.0	88.8 (love)	74.2 (genius)
CNN	**91.5 ± 4.1**	86.2 ± 5.2	88.6 ± 3.3	93.9 (sweet)	81.6 (interested)
RNN	86.7 ± 4.2	89.9 ± 5.4	88.1 ± 3.5	93.9 (joy)	80.0 (shocked)
Attentive RNN	88.8 ± 4.3	**90.9 ± 2.9**	**89.8 ± 2.7**	**95.5 (beautiful)**	**83.6 (genius)**

We also conduct the experiment on Reyes et al. dataset [9]. The Attentive RNN increases the F1 scores to 92.3%, 89.5%, and 89.0% on Irony-{Education, Humour, Politic}, three subdatasets, respectively, which makes an improvement over the performance achieved by Barbieri and Saggion [2].

5 Discussion

In this section, we aim to get a clear insight into how attention mechanism improves irony detection performance.

5.1 Attention Weight Plots

We are interested in which word in a sentence the Attentive RNN will pay more attention to. Figure 1 presents the visualization of the attention weight vectors α generated under the classification process. Word having a higher attention weight is shown with a darker color, indicating that this model pays more attention to that word. We can observe that this model can easily capture the relatively negative terms, for instance, "ignored", "hurt", "bad", and "dumb" in 1i, 1g, 1f, and 1b. Besides, it also captures some swear words such as "fuck", "suck", and "shit" in 1a, 1c and 1e. Our model relies on the negative words in a potentially positive sentiment tweet (e.g., sentences with "glad", "like", "love", and "yeah") to determine whether it is ironic or not. It is interesting that the Attentive RNN

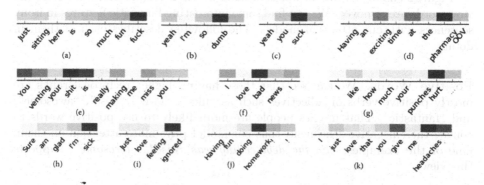

Fig. 1. Attention visualization

regards some words to have a negative sentiment, e.g., "headaches", "pharmacy", "homework" and "sick" in 1k, 1d, 1j, and 1h.

5.2   Attention Word Distribution

We also investigate on what kinds of words are more likely to be noticed by the Attentive RNN. Given a sentence s and an attention weight vector α, we make a ranking on words in s based on their corresponding attention weights in α. We calculate a reciprocal rank for each word w, which is the reciprocal of its ranking position i, and ignore those words with ranking position i higher than 3. We aim to find those words strongly emphasized by our model. We calculate each word's final reciprocal rank over all sentences in the dataset as follows.

$$ReciprocalRank_s^w = \begin{cases} \frac{1}{i}, & i \le 3 \\ 0, & otherwise \end{cases}$$

$$ReciprocalRank^w = \sum_{\forall s} ReciprocalRank_s^w$$

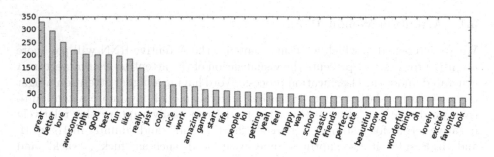

Fig. 2. Top 40 words ordered by the word reciprocal ranks.

Figure 2 shows the top 40 words ordered by their reciprocal ranks. Here the stop-words are removed. We can find some interesting linguistic features in irony sentences with our attention models. The following describes each of them in detail.

Positive Words. We can see that those having higher reciprocal ranks are mostly positive verbs or adjectives such as "like", "love", "best", "awesome", and "fantastic". That implies people are more likely to use positive words to convey their negative sentiment, thus our model pays more attention on them. One of the examples, *"So the debt ceiling was raised... awesome"*, supports this view.

Hyperbole. In addition, some interjections ("oh", "yeah", and "lol") and adverbs ("fucking", "really", and "just") get more weight than other words. The phenomenon suggests that people tend to use dramatic utterance to give a hint to their audience what they said may not follow the literal meaning. For instance, *"What a beautiful day to BE IN SCHOOL ALL DAY YEAH."*.

Fact. We can find that there are some nouns in our top 40 words. They are denoted as fact-related words. For fact-related words, we get "work", "game", "life", "school", "people", "friends", and "job". When people complain about those things hard to change or negotiate with, they describe their situations with an ironic utterance to accentuate their disappointment or other negative sentiment, such as, *"School until 6 pm today, got such a beautiful life."*

6   Conclusion

In this paper, the proposed Attentive RNN achieves the best performance with no further feature engineering. It increases the average F1 score to 89.8% in Ghosh et al. dataset. We further show that the attention vectors generated by our Attentive RNN captures specific words, which are useful to decide whether a tweet is ironic or not. With calculating the total reciprocal rank for each word, we find several linguistic styles popular in ironic utterances.

Acknowledgements. This research was partially supported by Ministry of Science and Technology, Taiwan, under grants MOST-104-2221-E-002-061-MY3 and MOST-105-2221-E-002-154-MY3.

References

1. Bahdanau, D., Cho, K., Bengio, Y.: Neural machine translation by jointly learning to align and translate. In: Proceedings of International Conference on Learning Representations (2015)
2. Barbieri, F., Saggion, H.: Modelling irony in Twitter. In: Proceedings of the Student Research Workshop at the 14th Conference of the European Chapter of the Association for Computational Linguistics, pp. 56–64 (2014)
3. Ghosh, D., Guo, W., Muresan, S.: Sarcastic or not: word embeddings to predict the literal or sarcastic meaning of words. In: Proceedings of the Conference on Empirical Methods in Natural Language Processing, pp. 1003–1012 (2015)
4. Hochreiter, S., Schmidhuber, J.: Long short-term memory. Neural Comput. **9**(8), 1735–1780 (1997)
5. Kim, Y.: Convolutional neural networks for sentence classification. In: Proceedings of the 2014 Conference on Empirical Methods in Natural Language Processing, pp. 1746–1751 (2014)
6. Kreuz, R., Glucksberg, S.: How to be sarcastic: the echoic reminder theory of verbal irony. J. Exp. Psychol. Gen. **118**(4), 374–386 (1989)

 7. LeCun, Y., Jackel, L., Bottou, L., Brunot, A., Cortes, C., Denker, J., Drucker, H., Guyon, I., Mller, U., Sckinger, E., Simard, P., Vapnik, V.: Comparison of learning algorithms for handwritten digit recognition. In: Proceedings of International Conference on Artificial Neural Networks, pp. 53–60 (1995)
 8. Mikolov, T., Sutskever, I., Chen, K., Corrado, G.S., Dean, J.: Distributed representations of words and phrases and their compositionality. In: Advances in Neural Information Processing Systems, pp. 3111–3119 (2013)
 9. Reyes, A., Rosso, P., Veale, T.: A multidimensional approach for detecting irony in Twitter. Lang. Resour. Eval. **47**(1), 239–268 (2013)
10. Tang, Y.-j., Chen, H.-H.: Chinese irony corpus construction and ironic structure analysis. In: Proceedings of the 25th International Conference on Computational Linguistics, pp. 1269–1278 (2014)
11. Tieleman, T., Hinton, G.: Lecture 6.5 - rmsprop. COURSERA: Neural Networks for Machine Learning, pp. 26–31 (2012)
12. Srivastava, N., Hinton, G., Krizhevsky, A., Sutskever, I., Salakhutdinov, R.: Dropout: a simple way to prevent neural networks from overfitting. J. Mach. Learn. Res. **15**(1), 1929–1958 (2014)

We Used Neural Networks to Detect Clickbaits: You Won't Believe What Happened Next!

Ankesh Anand[1]([✉]), Tanmoy Chakraborty[2], and Noseong Park[3]

[1] Indian Institute of Technology, Kharagpur, India
ankeshanand@iitkgp.ac.in
[2] University of Maryland, College Park, USA
tanchak@umiacs.umd.edu
[3] University of North Carolina, Charlotte, USA
npark2@uncc.edu

Abstract. Online content publishers often use catchy headlines for their articles in order to attract users to their websites. These headlines, popularly known as *clickbaits*, exploit a user's curiosity gap and lure them to click on links that often disappoint them. Existing methods for automatically detecting clickbaits rely on heavy feature engineering and domain knowledge. Here, we introduce a neural network architecture based on *Recurrent Neural Networks* for detecting clickbaits. Our model relies on distributed word representations learned from a large unannotated corpora, and character embeddings learned via Convolutional Neural Networks. Experimental results on a dataset of news headlines show that our model outperforms existing techniques for clickbait detection with an accuracy of 0.98 with F1-score of 0.98 and ROC-AUC of 0.99.

Keywords: Clickbait detection · Deep learning · Neural networks

1 Introduction

"Clickbait" is a term used to describe a news headline which will tempt a user to follow by using provocative and catchy content. They purposely withhold the information required to understand what the content of the article is, and often exaggerate the article to create misleading expectations for the reader. Some of the example of clickbaits are:

- "The Hot New Phone Everybody Is Talking About"
- "You'll Never Believe Who Tripped and Fell on the Red Carpet"

Clickbaits work by exploiting the insatiable appetite of humans to indulge their curiosity. According to the Loewenstein's information gap theory of curiosity [1], people feel a gap between what they know and what they want to know, and curiosity proceeds in two basic steps – first, a situation reveals a painful gap in our knowledge (that's the headline), and then we feel an urge to fill this gap and ease that pain (that's the click). Clickbaits clog up the social media news

© Springer International Publishing AG 2017
J.M. Jose et al. (Eds.): ECIR 2017, LNCS 10193, pp. 541–547, 2017.
DOI: 10.1007/978-3-319-56608-5_46

streams with low-quality content and violate general codes of ethics of journalism. Despite a huge amount of backlash and being a threat to journalism [2], their use has been rampant and thus it's important to develop techniques that automatically detect and combat clickbaits.

There is hardly any existing work on clickbait detection except Potthast et al. [3] (specific to the Twitter domain) and Chakraborty et al. [4]. The existing methods rely on a rich set of hand-crafted features by utilizing existing NLP toolkits and language specific lexicons. Consequently, it is often challenging to adapt them to multi-lingual or non-English settings since they require extensive linguistic knowledge for feature engineering and mature NLP toolkits/lexicons for extracting the features without severe error propagation. Extensive feature engineering is also time consuming and sometimes corpus dependent (for example features related to tweet meta-data are applicable only to Twitter corpora).

In contrast, recent research has shown that deep learning methods can minimize the reliance on feature engineering by automatically extracting meaningful features from raw text [5]. Thus, we propose to use distributed word embeddings (in order to capture lexical and semantic features) and character embeddings (in order to capture orthographic and morphological features) as features to our neural network models.

In order to capture contextual information outside individual or fixed sized window of words, we explore several Recurrent neural network (RNN) architectures such as Long Short Term Memory (LSTM), Gated Recurrent Units (GRU) and standard RNNs. Recurrent Neural Network models have been widely adopted for their ability to model sequential data such as speech and text well.

Finally, to evaluate the efficacy of our model, we conduct experiments on a dataset consisting of clickbait and non-clickbait headlines. We find that our proposed model achieves significant improvement over the state-of-the-art results in terms of accuracy, F1-score and ROC-AUC score. We plan to open-source the code used to build our model to enable reproducibility and also release the training weights of our model so that other developers can build tools on top of them.

2   Model

The network architecture of our model as illustrated in Fig. 1 has the following structure:

- **Embedding Layer:** This layer transforms each word into embedded features. The embedded features are a concatenation of the word's Distributed word embeddings and Character level word embeddings. The embedding layer acts as input to the hidden layer.
- **Hidden Layer:** The hidden layer consists of a Bi-Directional RNN. We study different types of RNN architectures (described briefly in Sect. 2.2). The output of the RNN is a fixed sized representation of its input.
- **Output Layer:** In the output layer, the representation learned from the RNN is passed through a fully connected neural network with a sigmoid output node that classifies the sentence as clickbait or non-clickbait.

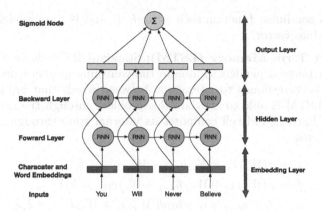

Fig. 1. BiDirectional RNN architecture for detecting clickbaits

2.1 Features

Two types of features are used in this experiment.

Distributed Word Embeddings: Distributed word embeddings map words in a language to high dimensional real-valued vectors in order to capture hidden semantic and syntactic properties of words. These embeddings are typically learned from large unlabeled text corpora. In our work, we use the pre-trained 300 dimensional word2vec embeddings which were trained on about 100B words from the Google News dataset using the Continuous Bag of Words architecture [6].

Character Level Word Embeddings: Character level word embeddings [7] have been used in several NLP tasks recently in order to incorporate character level inputs to build word embeddings. Apart from being able to capture orthographic and morphological features of a word, they also mitigate the problem of out-of-vocabulary-words as we can embed any word by its characters through character level embedding. In our work, we first initialize a vector for every character in the corpus. Then we learn the vector representation for any word by applying 3 layers of 1-dimensional CNN [8] with Rectified Linear Unites (ReLU) non-linearity on each vector of character sequence of that word and finally max-pooling across the sequence for each convolutional feature.

2.2 Recurrent Neural Network Models

Recurrent Neural Network (RNN) is a class of artificial neural networks which utilizes sequential information and maintains history through its intermediate layers. A standard RNN has an internal state whose output at each time-step is dependent on that of the previous time-steps. Expressed formally, given an input sequence x_t, a RNN computes it's internal state h_t by:

$$h_t = g(Uh_{t-1} + W_x x_t + b)$$

where g is a non-linear function such as $tanh$. U and W_x are model parameters and b is the bias vector.

Long Short Term Memory (LSTM): Standard RNNs have difficulty preserving long range dependencies due to the vanishing gradient problem [9]. In our case, this corresponds to interaction between words that are several steps apart. The LSTM is able to alleviate this problem through the use of a gating mechanism. Each LSTM cell computes its internal state through the following iterative process:

$$i_t = \sigma(W_{xi}x_t + W_{hi}h_{t-1} + W_{ci}c_{t-1} + b_i)$$
$$f_t = \sigma(W_{xf}x_t + W_{hf}h_{t-1} + W_{cf}c_{t-1} + b_f)$$
$$c_t = f_t \odot c_{t-1} + i_t \odot tanh(W_{xc}x_t + W_{hc}h_{t-1} + b_c)$$
$$o_t = \sigma(W_{xo}x_t + W_{ho}h_{t-1} + W_{co}c_t + b_o)$$
$$h_t = o_t \odot tanh(c_t)$$

where σ is the sigmoid function, and i_t, f_t, o_t and c_t are the input gate, forget gate, output gate, and memory cell activation vector at time step t respectively. $\odot$ denotes the element-wise vector product. W matrices with different subscripts are parameter matrices and b is the bias vector.

Gated Recurrent Unit (GRU): A gated recurrent unit (GRU) was proposed by Cho et al. [10] to make each recurrent unit adaptively capture dependencies of different time scales. Similarly to the LSTM unit, the GRU has gating units that modulate the flow of information inside the unit, however, without having a separate memory cells. A GRU cell computes it's internal state through the following iterative process:

$$z_t = \sigma(W_z x_t + U_z h_{t-1})$$
$$r_t = \sigma(W_r x_t + U_r h_{t-1})$$
$$\tilde{h}_t = tanh(W_h x_t + U(r_t \odot h_{t-1}))$$
$$h_t = (1 - z_t)\tilde{h}_{t-1} + z_t h_t$$

where z_t, r_t, $\tilde{h}_t$ and h_t are respectively, the update gate, reset gate, candidate activation, and memory cell activation vector at time step t. W_h, W_r, W_z, U_r and U_z are parameters of the GRU and $\odot$ denotes the element-wise vector product.

In our experiments, we use the Bi-directional variants of these architectures since they are able to capture contextual information in both forward and backward directions.

3   Evaluation

Dataset: We evaluate our method on a dataset of 15,000 news headlines released by Chakraborty et al. [4] which has an even distribution of 7,500 clickbait headlines and 7,500 non-clickbait headlines. The non-clickbait headlines in the dataset were sourced from Wikinews, and clickbait headlines were sourced from Buz-zFeed, Upworthy, ViralNova, Scoopwhoop and ViralStories. We perform all our experiments using 10-fold cross validation on this dataset to maintain consistency with the baseline methods.

Training Setup: For training our model, we use the mini-batch gradient descent technique with a batch size of 64, the ADAM optimizer for parameter updates and Binary Cross Entropy Loss as our loss function. To prevent overfitting, we use the dropout technique [11] with a rate of 0.3 for regularization. During training, the character embeddings are updated to learn effective representations for this specific task. Our implementation is based on the Keras [12] library using a TensorFlow backend.

Comparison of Different Architectures: We first evaluate the performance of different RNN architectures using Character Embeddings (CE), Word Embeddings (WE) and a combination of both (CE+WE). Table 1 shows the result obtained by various RNN models on different metrics (specifically Accuracy, Precision, Recall, F1, and ROC-AUC scores) after 10-fold cross validation.

Table 1. Performance of various RNN architectures after 10-fold cross validation. The 'Bi' prefix means that the architecture is Bi-directional.

Model	Accuracy	Precision	Recall	F1-Score	ROC-AUC
BiRNN (CE)	0.9629	0.9513	0.9757	0.9633	0.9929
BiRNN (WE)	0.9650	0.9722	0.9573	0.9647	0.9935
BiRNN (CE+WE)	0.9666	0.9530	0.9787	0.9655	0.9938
BiGRU (CE)	0.9661	0.9833	0.9482	0.9634	0.9945
BiGRU (WE)	0.9769	0.9761	0.9778	0.9770	0.9965
BiGRU (CE+WE)	0.9774	0.9662	**0.9893**	0.9776	0.9979
BiLSTM (CE)	0.9673	**0.9849**	0.9492	0.9667	0.9950
BiLSTM (WE)	0.9787	0.9759	0.9815	0.9787	0.9970
BiLSTM (CE+WE)	**0.9819**	0.9839	0.9799	**0.9819**	**0.9980**

We observe that BiLSTM(CE+WE) model slightly outperforms other models, and the BiLSTM architecture in general performs better than BiGRU and BiRNN. If we look at performance of an individual architecture using three different set of features, model using a combination of word embeddings and character embeddings consistently gives the best results, closely followed by model with only word embeddings.

Comparison with Existing Baselines: Finally, we compare our model with state-of-the-art results on this dataset as reported in Chakraborty et al. [4]. The models reported in [4] use a combination of structural, lexical and lexicon based features. In Table 2, we notice that our BiLSTM(CE+WE) model shows

Table 2. Comparison of our model with the baseline methods.

Model	Accuracy	Precision	Recall	F1-score	ROC-AUC
Chakraborty et al. (2016) (SVM)	0.93	0.95	0.90	0.93	0.97
Chakraborty et al. (2016) (Decision Tree)	0.90	0.91	0.89	0.90	0.90
Chakraborty et al. (2016) (Random Forest)	0.92	0.94	0.91	0.92	0.97
BiLSTM(CE+WE)	**0.98**	**0.98**	**0.98**	**0.98**	**0.99**

more than 5% improvement in terms of both accuracy and F1-score and more than 2% in terms of the ROC-AUC score over the best performing baseline (i.e. Chakraborty et al. [4] (SVM)).

4   Conclusion

In this paper, we introduced three different variants of Bidirectional Recurrent Neural Network model for detecting clickbaits using distributed word embeddings and character-level word embeddings. We showed that these models achieve significant improvement over the state-of-the-art in detecting clickbaits without relying on heavy feature engineering. In future, we would like to qualitatively visualize the internal states of our model and incorporate attention mechanism into our model.

References

1. Loewenstein, G.: The psychology of curiosity: a review and reinterpretation. Psychol. Bull. **116**(1), 75 (1994)
2. Dvorkin, J.: Why click-bait will be the death of journalism (2016). http://to.pbs.org/2gQ6mCN
3. Potthast, M., Köpsel, S., Stein, B., Hagen, M.: Clickbait detection. In: Ferro, N., Crestani, F., Moens, M.-F., Mothe, J., Silvestri, F., Nunzio, G.M., Hauff, C., Silvello, G. (eds.) ECIR 2016. LNCS, vol. 9626, pp. 810–817. Springer, Cham (2016). doi:10.1007/978-3-319-30671-1_72
4. Chakraborty, A., Paranjape, B., Kakarla, S., Ganguly, N.: Stop clickbait: Detecting and preventing clickbaits in online news media. In: ASONAM, pp. 9–16. San Fransisco, USA (2016)
5. Collobert, R., Weston, J., Bottou, L., Karlen, M., Kavukcuoglu, K., Kuksa, P.: Natural language processing (almost) from scratch. J. Mach. Learn. Res. **12**, 2493–2537 (2011)
6. Mikolov, T., Chen, K., Corrado, G., Dean, J.: Efficient estimation of word representations in vector space. arxiv preprint (2013). arXiv:1301.3781
7. dos Santos, C.N., Zadrozny, B.: Learning character-level representations for part-of-speech tagging. In: ICML, pp. 1818–1826 (2014)
8. Le Cun, B.B., Denker, J.S., Henderson, D., Howard, R.E., Hubbard, W., Jackel, L.D.: Handwritten digit recognition with a back-propagation network. In: Advances in neural information processing systems. Citeseer (1990)

9. Hochreiter, S., Schmidhuber, J.: Long short-term memory. Neural Comput. **9**(8), 1735–1780 (1997)
10. Cho, K., Van Merriënboer, B., Bahdanau, D., Bengio, Y.: On the properties of neural machine translation: encoder-decoder approaches. arxiv preprint (2014). arXiv:1409.1259
11. Srivastava, N., Hinton, G.E., Krizhevsky, A., Sutskever, I., Salakhutdinov, R.: Dropout: a simple way to prevent neural networks from overfitting. J. Mach. Learn. Res. **15**(1), 1929–1958 (2014)
12. Chollet, F.: Keras (2015). https://github.com/fchollet/keras

Learning to Classify Inappropriate Query-Completions

Parth Gupta[1] and Jose Santos[2]([⊠])

[1] Universitat Politecnica de Valencia, Valencia, Spain
pgupta@dsic.upv.es
[2] Microsoft, London, UK
jcsantos@microsoft.com

Abstract. Query auto-completion is a powerful feature anywhere users are querying and is nowadays omnipresent in many forms and entry points, e.g. search engines, social networks, web browsers, operating systems. Suggestions not only speed up the process of entering a query but also shape how users query and can make the difference between a successful search and a frustrated user. The main source of these query completions is past, aggregated, user queries. A non-negligible fraction of these queries contain offensive, adult, illegal or otherwise inappropriate content. Surfacing these completions can have legal implications, offend users and give the incorrect impression companies providing the query completion service condone these views. In this paper, we describe existing methods to identify inappropriate queries and present a novel machine learned approach that does not require expensive, human-curated, blocklists and is superior to these in recall and competitive in F1-score.

1 Introduction

Every day billions of queries are issued in commercial search engines in dozens of languages. These queries reflect users' needs, desires, behaviours, interests but also prejudices. These searches are also the main data source to build the query histogram models that power an auto-completion service [1]. Due to the organic nature of the query histogram model, we estimate 5–10% of the queries are inappropriate to surface to the end user as an auto-completion. The user is still able to type any query completely and get results.

We consider query suggestions inappropriate if they are offensive, condone violence or illegal actions or have a sexual intent. It should be noted a query may contain inappropriate terms but if the intent is clean the query should still be deemed OK e.g. "what is cocaine" vs "where to buy cocaine". A search engine deliberately wants to filter as many inappropriate suggestions as possible due to geopolitical and legal reasons, being preferable to incur in type I errors (false positives) than allowing a true inappropriate query go undetected (a type II error). Therefore, recall is preferred over precision.

© Springer International Publishing AG 2017
J.M. Jose et al. (Eds.): ECIR 2017, LNCS 10193, pp. 548–554, 2017.
DOI: 10.1007/978-3-319-56608-5_47

A typical method to deal with detecting inappropriate queries are substring and pattern match block lists [4–6]. Substring-match blocklists contain strings that can never appear in a query, e.g. swear words. Pattern-match blocklists assume various forms. One is the <entity> <qualifier> pattern where a list of entities e.g. person names, ethnic/religious/political groups with common associated derogatory expressions. When a query contains both a known entity and a derogatory qualifier associated with that entity type, the query is identified as offensive. For instance, a suggestion of the form "X are Y" is blocked by a pattern-match blocklist if $X \in$ {jews, christians, muslims, blacks} and $Y \in$ {stupid, idiots, retarded} while the individual X and Y terms may be acceptable on their own.

While the combination of both blocklist techniques performs acceptably, there are severe limitations: it is a semi-manual process requiring list curation and maintenance; all possible variations of an entity and derogatory terms must be provided, e.g. singular, plural, synonyms; no generalisation power.

The existing literature focuses on natural language or social media text and the techniques are of limited use when a very small context is available as in the case of a web query. In this paper, we propose a new model which learns to represent queries in different clusters of inappropriateness. Such representation is learnt through a supervised latent semantic projection algorithm based on deep neural networks. The proposed clustering method helps to uncover more inappropriate patterns as evidenced by high recall.

2   Approach

Our approach is to create an abstract offensive space where queries can be clustered. The abstract space is built using supervised latent projection methods. Supervised techniques such as deep structured semantic model (DSSM) [3] can incorporate the label information into projection learning. As we aim to learn an abstract space of offensiveness, we adapt DSSM model to incorporate offensive categories by injecting an objective function which clusters queries from same inappropriate categories as described in Sect. 2.1.

2.1   DSSM for Offensive Clusters

DSSM is structurally a deep neural network which models the queries to represent in a low-dimensional latent space.

Let $c_k \in C$ represent the kth inappropriate category where $|C| \geq 2$ and c_0 represents the appropriate (OK) category. Hence, $|C| = 2$ represents the binary setting with categories {OK, inappropriate}. Let $x_{q,c_k} \in \mathbb{R}^n$ be vector representation of query q labelled to belong inappropriate category c_k and n is input dimensionality. Queries are represented as word hashes because considering complete terms explodes the feature space while word-hashes have proven to be effective and efficient [3].

Query x_q is projected to $y_q = \phi(x_q)$ by DSSM (ϕ) where $y_q \in \mathbb{R}^m, m \ll n$ as shown in Eq. 1.

$$h_{q,l_1} = g(W_1 * x_q + b_1)$$
$$y_q = g(W_2 * h_q^{(l_1)} + b_2) \tag{1}$$

where, W_i and b_i represent i^{th} layer weights and bias parameters of the network, $h_q^{(l_1)}$ represent the hidden layer activities and g is hyperbolic tangent activation function. The DSSM is trained to maximise the objective function presented in Eq. 2 using backpropagation [3].

$$J(\theta) = \cos(y_q, y_q^+) - \cos(y_q, y_q^-) \tag{2}$$

where, y_q^+ represents same category query to that of y_q and y_q^- represents a different category query to that of y_q. The objective function $J(\theta)$ encourages those configurations θ which produce higher cosine similarity between queries belonging to the same category and lower cosine similarity between queries that belong to different categories.

Once the DSSM is trained, all the labelled queries are projected into the abstract space. Centroid for each category is calculated as shown in Eq. 3.

$$\mu_{c_k} = \frac{1}{m_k} \sum_i y_{q,c_k}^{(i)} \tag{3}$$

Now a new query y_q is classified to category c_k for which Euclidean distance $d(\mu_{c_k}, y_q)$ is minimum.

3   Experiments and Results

Here we present the experimental set-up to evaluate the effectiveness of the proposed models along with a couple of strong baselines.

3.1   Data

Our dataset consists of 79174 unique queries. These queries were derived from a prefix set biased towards inappropriate terms as follows. The prefix set was created by randomly sampling queries that contained offensive terms and keeping only the first half of the query. These prefix sets were then scraped against the auto-completion service of a commercial search engine, for the US market, and the resulting unique queries gathered.

The unique queries were then human judged for various inappropriate categories via a crowd-sourcing platform, with at least 5 judgements per query. Significant care was put to ensure the quality of judgments with real time audits and by limiting the number of queries a single judge could judge to a few hundred. Real time audits were done by randomly interspersing with the queries to judge a small percentage of non-contentious queries for which we know the

Table 1. Distribution of judgements, in thousands, over the 4 query categories in the dataset.

Cat.	Description	#	%
c_0	Okay	627.2	95.0%
c_1	Violence/illegal/self-harm	7.6	1.2%
c_2	Race/religion/sexual/gender	18.0	2.7%
c_3	Other offensive/profane	7.3	1.1%

ground truth. If a judge did not agree on at least 85% of these non-contentious queries, it would be disqualified a posteriori and all its judgments discarded.

In total, there were 660267 judgements (average 8.3 judgements per query), with the vast majority, 95.0%, being appropriate (OK). The statistics of the data is presented in Table 1.

From this labelled dataset, a query inappropriate score is computed as the ratio of inappropriate judgements over all the query judgements. This score is then converted into a binary label, Inappropriate if score ≥ 0.2, otherwise OK. There are 7284 inappropriate queries, 9.2% of the corpus. The dataset was randomly split into 70% for training and 30% for test.

3.2 Baselines

Blocklists. The blocklist-based approach uses a set of substring and pattern-matching techniques. Semi-manually built blocklists are generated by extracting common inappropriate patterns from user reported feedback and from crowd-sourcing tasks whose goal is to spot inappropriate query leakage in a commercial search engine auto completion service.

The aggregated size of the multiple substring blocklists is in the order of the tens of thousands of terms which block in the order of a few million queries. If a query matches any of the terms in any of the substring blocklists, it is deemed as inappropriate. The pattern-matching blocklists only block a query if it has terms in two complementary lists. There are multiple pattern-matching blocklists, one for each domain, such as {offensive, adult, illegal, violence}.

Table 2 contains a few entries from the English substring-match blocklist and the violence pattern-match blocklists. These lists are updated regularly, grow over time and require manual effort to maintain.

SVM with Word Hashes. We also trained a classifier with lexical features to test whether it can learn a classification boundary from the training data. First, we featurize the input queries by the word-hashing technique reported in [3] which codifies each input term into character 3-grams after marking the start and end of the word. This 3-gram featurization helps handling the sparseness of the features and keeps the feature space limited, especially to scale at web-level. A total of 9590 character 3-grams was obtained from the training data. Secondly,

Table 2. A few entries of the English: (1) offensive substring blocklist (1st column), (2) pattern-match violence block-lists (2nd and 3rd columns)

Offensive	Violence	Viol. modifiers
Beating newborn	Beheading	Video
Blacks should	Execution	Movie
Cannibal recipes	Hanged	Image

we trained a support vector machine (SVM) on the 3-gram featurized training data to obtain a classification boundary.

3.3 Results

We evaluate the baselines and proposed models on the test partition and measure the performance of predicting the **Inappropriate** class with the standard precision, recall and F1-score measures. The results are presented in Table 3.

Table 3. Precision, recall and F1-score of various models. † denotes statistical significance (p-value < 0.01) to corresponding blocklist metric

Model	Prec.	Rec.	F1
Blocklists	70.0%	49.0%	57.6%
SVM with 3-grams	70.4%	43.4%	53.8%†
DSSM with L2 norm	51.1%†	65.0%†	57.2%

The SVM baseline performs worse than blocklists in recall, because blocklists are highly curated while SVM tries to learn the discriminative patterns. Although blocklists and SVM achieve high precision, they obtain relatively low recall pointing to their poor generalization, mostly because they filter only on the lexical features. The DSSM-based clustering method introduced in Sect. 2.1 obtains higher recall. Higher recall points to its power to uncover more inappropriate patterns which was not possible with the lexical methods. An ensemble random forest classifier using 3-grams and distances to centroids with various norms, including cosine distance, as features was also trained but it only marginally improved F1-score over the much simpler DSSM with L2 norm.

It is worth noting the blocklists used are exhaustive, built over several years from larger corpus of queries than the one in this dataset, specifically to block inappropriate web search suggestions in a commercial auto-complete service. The presented blocklist is thus a very strong baseline. A fairer blocklist baseline would be humans crowdsource a blocklist based only on the training set queries and evaluate this blocklist performance on the test set. This was too expensive to do but would be required to filter inappropriate suggestions in a new language with the blocklist technology. In this regard the ML approaches are clearly superior as they can build the inappropriate model directly and inexpensively.

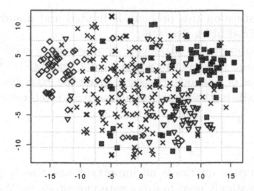

Fig. 1. t-sne 2D projection of the DSSM feature vectors, over-sampling the inappropriate classes for easier visualization. The $\times$ represents the OK category, c_0. $\Diamond$ represents c_2, ∇ represents c_1 and $\boxtimes$ represents c_3.

We show the discriminative power of DSSM technique by visualising the inappropriate clusters in Fig. 1. The 128-dimensional vectors are represented in 2-dimensional scatter plot using t-sne algorithm[1]. It can be noticed queries from particular inappropriate categories form dense well-separated clusters.

Table 4 presents a few correct and incorrect classifications made by the DSSM with L2 norm on the test set. There exists some noise in the judge scores which may be due to limited knowledge of the judges on specific topics. While analysing the false negatives, we noticed those queries are not as inappropriate as the correctly predicted ones. In fact, a false negative such as "batman jokes" is innocuous. In some cases, the query contains words which themselves are inappropriate but in the query context they are not, e.g. "UN office on crime and drug". The latent DSSM clustering features helped to identify such patterns.

Table 4. A few correct and incorrect predictions made by the DSSM.

Query	Judge score	Predicted class
Best guns for women	0.4	Inappropriate
Marijuana brownie recipe	1.0	Inappropriate
Batman jokes	0.2	OK
UN office on crime and drugs	0.2	OK

4   Related Work

Although offensive query classification has attained little research attention, offensive language identification in text is, in general, a well known problem.

---

[1] https://lvdmaaten.github.io/tsne/.

Nevertheless, literature on this problem is limited and mainly evolved around the use of a predefined block list [4–6].

Some work also exists on ML-based approaches to model lexical features for offensive language detection in social media text [2,7]. Our baseline SVM represents such class of methods. However, given the extremely short context in web queries, such ML methods are less attractive to web query classification.

5   Remarks

Detecting inappropriate queries is an important and timely problem with the internet being increasingly used to propagate violent views. We provided a principled solution for the inappropriate query classification problem based on deep neural networks. The experiments carried on a large labelled web query corpus suggest the DSSM approach significantly outperforms ML techniques based on lexical features. The DSSM approach is also superior in recall while being competitive in F1-score compared to expensive, human-curated blocklists.

References

1. Bar-Yossef, Z., Kraus, N.: Context-sensitive query auto-completion. In: Proceedings of WWW, pp. 107–116 (2011)
2. Gianfortoni, P., Adamson, D., Rosé, C.P.: Modeling of stylistic variation in social media with stretchy patterns. In: Proceedings of DIALECTS, pp. 49–59 (2011)
3. Huang, P.S., He, X., Gao, J., Deng, L., Acero, A., Heck, L.: Learning deep structured semantic models for web search using clickthrough data. In: Proceedings of CIKM, pp. 2333–2338 (2013)
4. Mahmud, A., Ahmed, K.Z., Khan, M.: Detecting flames and insults in text. In: Proceedings of ICON (2008)
5. Razavi, A.H., Inkpen, D., Uritsky, S., Matwin, S.: Offensive language detection using multi-level classification. In: Farzindar, A., Kešelj, V. (eds.) AI 2010. LNCS (LNAI), vol. 6085, pp. 16–27. Springer, Heidelberg (2010). doi:10.1007/978-3-642-13059-5_5
6. Spertus, E.: Smokey: automatic recognition of hostile messages. In: Proceedings of IAAI, pp. 1058–1065 (1997)
7. Xiang, G., Fan, B., Wang, L., Hong, J., Rose, C.: Detecting offensive tweets via topical feature discovery over a large scale twitter corpus. In: Proceedings of CIKM, pp. 1980–1984 (2012)

Counteracting Novelty Decay in First Story Detection

Yumeng Qin[1], Dominik Wurzer[2(✉)], Victor Lavrenko[2], and Cunchen Tang[1]

[1] Wuhan University - International School of Software, Wuhan, China
[2] Edinburgh University - School of Informatics, Edinburgh, UK
wurzer.dominik@gmail.com

Abstract. In this paper we explore the impact of processing unbounded data streams on First Story Detection (FSD) accuracy. In particular, we study three different types of FSD algorithms: comparison-based, LSH-based and k-term based FSD. Our experiments reveal for the first time that the novelty score of all three algorithms decay over time. We explain why the decay is linked to the increased space saturation and negatively affects detection accuracy. We provide a mathematical decay model, which allows compensating observed novelty scores by their expected decay. Our experiments show significantly increased performance when counteracting the novelty score decay.

1 Introduction

First Story Detection (FSD), also called New Event Detection, describes the task of identifying documents ("first-stories") that speak about an unknown event first. FSD systems process data streams and compute a novelty score for each encountered document, which indicates its novelty with respect to all previously encountered documents. If the novelty score falls above a fixed detection threshold, the document is considered to talk about a new event. FSD is part of the Topic Detection and Tracking initiative [1], and benefits financial institutes as well as reporters and homeland security agencies.

Previous research on FSD focused on increasing effectiveness or efficiency on public research data sets. To the best of our knowledge, no research up to this date considered the effect of processing more and more documents on detection accuracy. We show that novelty scores of FSD systems decay over time and explain why it is linked to increasing space saturation. Continuously decaying novelty scores have a direct negative effect on FSD accuracy, because detection is based on constant thresholds. We show how to counteract novelty score decay for three state-of-the-art FSD systems: the traditional comparison-based approach (UMass)[2], LSH-FSD [5] and a kterm-hashing based approach [6]. Our experiments show significantly improved accuracy when counteracting novelty decay.

© Springer International Publishing AG 2017
J.M. Jose et al. (Eds.): ECIR 2017, LNCS 10193, pp. 555–560, 2017.
DOI: 10.1007/978-3-319-56608-5_48

1.1 Related Work

Traditional FSD systems, like Umass [2], rely on exact vector proximity between each new document and all previously seen documents. This results in state-of-the-art accuracy at the cost of low efficiency. Recently, FSD was applied to unbounded social media streams [9–11]. To make FSD system applicable to high volume streams, research focused on scaling them by feature-reduction [3] or Locality Sensitive Hashing (LSH-FSD) [5]. LSH scales novelty computation by reducing the search space from the entire vector space to the size of a hash bin. K-term hashing [6], a memory-based novelty computation method, resulted in higher accuracy and effectiveness than [2,5]. Instead of relying on vector proxim-ity, k-term hashing builds a history, consisting of hashed kterms, that represent information about previously encountered documents. Novelty is computed by the proportion of unseen kterms with respect to the history. When FSD was first introduced, it was designed to operate on streaming data sets. However, official data sets are small (TDT: 15k–75k documents) and accuracy over time is still an overlooked area in TDT research. Our findings demonstrate that considering the impact of processing more and more documents on detection performance, allows increasing FSD accuracy significantly.

2 FSD on Millions of Documents

Figure 1 shows the cumulative average novelty score of UMass, LSH-FSD and k-term hashing, when processing 2 million documents. The curve of all three algo-rithms reveals a continuous decay of the average novelty score, as they process more and more documents. This decay has a direct impact on detection perfor-mance, which is based on constant thresholds. In particular during the first 1 mio document we observe a severe drop in average novelty scores. Consequently, FSD systems are more likely to recognize documents as "new events" during the first 1 mio documents, in comparison with the next 1 mio documents.

2.1 Exploring Causes for Novelty Score Decay Over Time

We explore the causes for the observed novelty score decay of 3 state-of-the-art FSD systems:

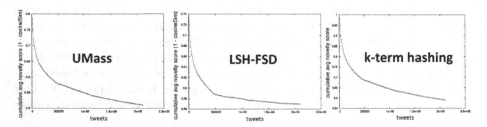

Fig. 1. Cumulative average novelty score of UMass, LSH-FSD and k-term hashing for 2 million tweets

Comparison Based FSD: UMass [2] compares each new arriving document with all previously seen documents. The novelty score depends on vector proximity to the closest previous document. As more documents arrive, the vector space fills up. The more saturated a space becomes, the more likely it becomes that additional objects are close to existing ones. The average novelty score decays with the increase in vector space saturation.

LSH Based FSD: LSH-FSD [5] shares the basic concept for computing novelty with UMass. The advantage of LSH-FSD over UMass resides in efficiency gains from limiting the search space from the entire vector space to $\frac{\#docs}{\#bins}$, the size of a hash bin. Although the search space is reduced, new documents added to it slowly increase its saturation. As a result, LSH suffers from the same novelty score decay as standard comparison based systems, as seen in Fig. 1.

K-term Hashing Based FSD: K-term hashing [6] forms for each document compounded terms (k-terms) and hashes them onto a bloom filter [8] to determine if they are new with respect to previously encountered documents. The fraction of unseen kterms determines the novelty score. To keep track of past information, every document adds its own k-terms to the bloom filter, which increases its space saturation. This resembles the principle of the saturated vector space, and causes the average novelty scores to decay over time.

3 Counteracting Novelty Score Decay Over Time

The novelty scores of FSD systems decays over time with the increase in space saturation. Unfortunately, one cannot simply remove data to avoid the space saturation, as this would cause a significantly reduction in detection accuracy [4]. Our approach to counteract novelty decay relies on compensating the score decay. We model the expected decay at a certain point in time (t) as a mathematical function and adapt the novelty score accordingly. In particular, we apply logarithmic, exponential and polynomial regression to the observed cumulative average novelty scores of the 52 mio random tweets, while optimizing the coefficient determinant (R^2). The lowest proportional variance and best generalisability is reached, when approximating the expected novelty score (EN) by an inverted natural logarithmic function, as seen in Eq. 1.

$$EN(t) = \gamma * (-)ln(t) + \delta \tag{1}$$

Parameter γ denotes the slope, and δ is the intercept on logarithmic scale. Both parameters are based on optimizing the coefficient determinant using 52 mio random tweets that act as training data. The parameter t describes a timestamp or a particular position within the stream. Figure 2 illustrates that the expected novelty decay based on the training data generalizes well, as it highly correlates with the observed novelty decay of the Cross-Twitter [4] data set. The coefficient determinant is $R^2 = 0.9987$. The high coefficient value indicates a low proportional variance between approximated and observed average novelty score.

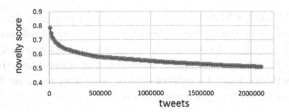

Fig. 2. The bold blue curve indicates the observed cumulative average novelty score when processing the Cross-Twitter data set; the red dotted curve resembles the expected cumulative average novelty score based on our mode, trained on 52 mio. random tweets. (Color figure online)

4   Experiments

In this section we explore the impact of counteracting novelty score decay on FSD accuracy.

Evaluation Metrics. We apply the standard TDT evaluation procedure [7] and the official TDT3 evaluation scripts with standard settings [1,2] for evaluating FSD accuracy. The Detection Error Trade-off (DET) curve shows the trade-off between miss and false alarm probability for the full range of novelty scores. Accuracy is measured by the minimum detection cost (C_{min}), which is the standard metric of TDT research publications. *Note:* lower values indicate higher accuracy.

Data Set. We use the official and publicly available Cross-Twitter[1] data set [4] that was also used by [4,6]. Cross-Twitter consists of 27 topics and 52 million tweets from the period of April till September 2011. We additionally use 52 million random tweets from the same time period as a training set for our decay model.

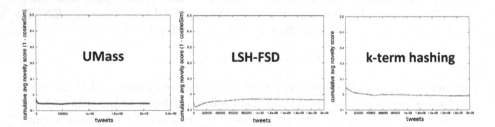

Fig. 3. Impact of adapting novelty scores on the cumulative average novelty score

---

[1] Available at: http://demeter.inf.ed.ac.uk/cross/.

4.1 Impact on Effectiveness

Figure 3 illustrates the average novelty score for UMass, LSH-FSD and k-term hashing, when compensating the observed novelty score according to the expected score, resulting from Eq. 1. The figure shows that score adjustment successfully counteracts novelty score decay, which results in constant average novelty scores for all three algorithms. Next we explore the impact of score adjustment on detection accuracy. All three systems are applied to Cross-Twitter and their scores are adjusted according to Eq. 1, whereas parameters are learned from 52 mio. random tweets. Table 1 shows the impact of counteracting the novelty decay on detection accuracy, measured by C_{min}. *Note*: lower values indicate higher accuracy. The table reveals that all three algorithms benefit from counteracting novelty decay and show accuracy gains of 4%. Additionally, we provide DET plots in Figs. 4, 5 and 6. The DET plots illustrate that the difference in accuracy is significant for the high precision area, where false alarm <15%. This is also the area, where all algorithms achieve their highest accuracy (C_{min}, illustrated by the red dot).

Table 1. Performance improvement through novelty score adjustment

Algorithm	Normal C_{min}	Score adjusted C_{min}	Difference
UMass	0.7981	0.7583	−5%
LSH-FSD	0.9061	0.8685	−4%
k-term hashing	0.7966	0.7645	−4%

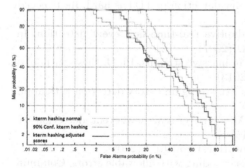 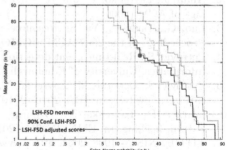

Fig. 4. DET plot for k-term hashing, showing significantly increased accuracy when counteracting novelty decay (Color figure online)

Fig. 5. DET plot for LSH-FSD, showing significantly increased accuracy when counteracting novelty decay (Color figure online)

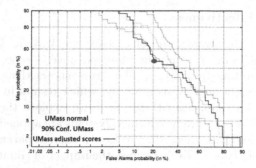

Fig. 6. DET plot for UMass, showing significantly increased accuracy when counteracting novelty decay (Color figure online)

5   Conclusion

We studied the behaviour of novelty scores from state-of-the-art FSD systems as they process more and more documents and revealed that they decay over time. We explained why the decay is connected to the increasing space saturation and provided a countermeasure based on mathematical decay model. Our experiments showed significantly increased detection accuracy when counteracting novelty decay using the proposed decay model.

References

1. Allan, J.: Topic Detection and Tracking: Event-Based Information Organization. Kluwer Academic Publishers, Norwell (2002)
2. Allan, J., Lavrenko, V., Jin, H.: First story detection in TDT is hard. In: Proceedings of ACM (2000)
3. Luo, G., Tang, C., Yu, S.: Resource-adaptive real-time new event detection. In: Proceedings of the 2007 ACM SIGMOD (2007)
4. Petrovic, S.: Real-time event detection in massive streams. Ph.D. thesis, School of Informatics, University of Edinburgh (2013)
5. Petrovic, S., Osborne, M., Lavrenko, V.: Streaming first story detection with application to Twitter. In: HLT 2010 (2010)
6. Wurzer, D., Lavrenko, V., Osborne, M.: Twitter-scal new event detection via K-term hashing. In: EMNLP 2015 (2015)
7. TDT by NIST - 1998–2004. http://www.itl.nist.gov/tdt/ (2008)
8. Bloom, B.H.: Space/time trade-offs in hash coding with allowable errors. Commun. ACM **13**(7), 422–426 (1970)
9. Cataldi, M., Caro, L.D., Schifanella, C.: Emerging topic detection on Twitter based on temporal and social terms evaluation. In: Proceedings of the 10th International Workshop on Multimedia Data Mining, pp. 4:1–4:10. ACM (2010)
10. Li, R., Lei, K.H., Khadiwala, R., Chang, K.C.: TEDAS: a Twitter-based event detection and analysis system. In: Proceedings of 28th International Conference on Data Engineering, pp. 1273–1276. IEEE Computer Society (2012)
11. Phuvipadawat, S., Murata, T.: Breaking news detection and tracking in Twitter. In: Proceedings of the 2010 IEEE/WIC/ACM International (2010)

Predicting Genre Preferences from Cultural and Socio-Economic Factors for Music Retrieval

Marcin Skowron[1(✉)], Florian Lemmerich[2], Bruce Ferwerda[3],
and Markus Schedl[3]

[1] Austrian Research Institute for Artificial Intelligence, Vienna, Austria
marcin.skowron@ofai.at
[2] GESIS - Leibniz Institute for the Social Sciences,
University of Koblenz-Landau, Mainz, Germany
florian.lemmerich@gesis.org
[3] Johannes Kepler University, Linz, Austria
{bruce.ferwerda,markus.schedl}@jku.at

Abstract. In absence of individual user information, knowledge about larger user groups (e.g., country characteristics) can be exploited for deriving user preferences in order to provide recommendations to users. In this short paper, we study how to mitigate the cold-start problem on a country level for music retrieval. Specifically, we investigate a large-scale dataset on user listening behavior and show that we can reduce the error for predicting the popularity of genres in a country by about 16.4% over a baseline model using cultural and socio-economics indicators.

1 Introduction

While research that considers individual, user-specific aspects to improve music retrieval and recommendation algorithms has received substantial attention in the past few years, cf. [1,8,10], studies on cultural differences between perception of music have not been conducted in the context of retrieval until quite recently [2,3,6,11]. The few existing works almost exclusively analyze the cultural differences in emotion or mood perceived when listening to music, with the aim to integrate such knowledge into music retrieval approaches [7,13].

Gaining a more fundamental knowledge about the differences in music taste in different countries and about how these differences relate to cultural and socio-economic dimensions can help building culture-aware and cross-cultural music retrieval systems, mitigating the cold-start problem, and improving search or recommendation results by considering the cultural background of users. In this short paper, we approach the cold-start problem in which we do *not* know anything about a new user or the overall music preferences in his country, but assume that country information can be easily inferred from basic user profile information. Given cultural and socio-economic factors that are publicly available, we aim at predicting the music taste for the user's country and by doing so infer an approximation of his music taste using his country's taste as a proxy. Therefore, the specific research question we address is to which extent we can predict the overall music taste in a country given cultural and socio-economic factors.

© Springer International Publishing AG 2017
J.M. Jose et al. (Eds.): ECIR 2017, LNCS 10193, pp. 561–567, 2017.
DOI: 10.1007/978-3-319-56608-5_49

2   Related Work

Cross-cultural research in the field of music retrieval is very limited. The studies that investigated cultural differences on users' music perception and consumption often limit themselves to a handful of cultures. For example, Hu and Lee [6] showed that there are differences between Americans and Chinese on mood perception in music, whereas Singhi and Brown [11] investigated the influence of lyrics between Canadians and Chinese. Although these findings confirm that cultural differences exist, they cannot easily be generalized to other cultures. More comprehensive studies were conducted by Ferwerda et al. [2,3] on cultural differences in the need for music diversity. By analyzing the music consumption of users in 97 countries, they identified distinct behavior that could be related to Hofstede's cultural dimensions. In this work, we explore to which extent listeners' music preferences can be predicted across countries using cultural and socio-economical aspects and state-of-the-art machine learning techniques.

3   Datasets

In the following sections we describe how we infer music preferences on the country level and how we model cultural and socio-economic aspects.

3.1   Modeling Music Preferences

We model music preference on the country level by utilizing the recently published LFM-1b dataset [9],[1] which offers demographic information and detailed listening histories for tens of thousands Last.fm users. We consider in our analysis only countries with at least 100 users in the LFM-1b dataset and for which *all* the Hofstede's cultural dimensions (cf. Sect. 3.2) are available. The 44 countries meeting both conditions are analyzed in the remaining of this work.

We define country-specific genre profiles that are used as a proxy for music taste as follows. First, the top tags assigned to each artist in the LFM-1b dataset are fetched via the respective Last.fm API endpoint.[2] These tags provide different pieces of information, including instruments ("guitar"), epochs ("80s"), places ("Chicago"), languages ("Swedish"), and personal opinions ("seen live" or "my favorite"). We then filter for tags that encode genre and style information. For this purpose, we use as index terms a dictionary of 20 genre names retrieved from Allmusic.[3] The genre profiles are eventually created as feature vectors describing the share of each genre among all listening events of the respective country's population, according to the LFM-1b dataset. More formally, the weight of genre g in country c is given as $w_{c,g} = \frac{\sum_{a \in A_g} le_{c,a}}{\sum_{a \in A} le_{c,a}}$, where A_g is the set of artists tagged with genre g, A is the entire set of artists, and $le_{c,a}$ is the number of listening events to artist a in country c.

---

[1] http://www.cp.jku.at/datasets/LFM-1b.
[2] http://www.last.fm/api/show/artist.getTopTags.
[3] http://www.allmusic.com.

3.2 Modeling Cultural Dimensions

For our study, we rely on Hofstede et al.'s cultural dimensions.[4] It is considered to be the most comprehensive and up to date framework for national cultures. They defined six dimensions to identify cultures [5]:

Power distance is defined as the extent to which power is distributed unequally by less powerful members of institutions (e.g., family). High power distance indicates that a hierarchy is clearly established and executed. Low power distance indicates that authority is questioned and attempted to distribute power equally.

Individualism measures the degree of integration of people into societal groups. High individualism is defined by loose social ties. The main emphasis is on the "I" instead of the "we", while opposite for low individualistic cultures.

Masculinity describes a society's preference for achievement, heroism, assertiveness and material rewards for success Low masculinity represents a preference for cooperation, modesty, caring for the weak and quality of life.

Uncertainty avoidance defines a society's tolerance for ambiguity. High scoring countries are more inclined to opt for stiff codes of behavior, guidelines, laws.

Long-term orientation is associated with the connection of the past with the current and future actions. Lower scoring countries tend to believe that traditions are honored and kept, and value steadfastness. High scoring countries believe more that adaptation and pragmatic problem-solving are necessary.

Indulgence denotes in general the happiness of a country. High indulgence is related to a society that allows relatively free gratification of basic and natural human desires related to enjoying life and having fun (e.g., be in control of their own life and emotions). Whereas low scoring countries show more controlled gratification of needs and regulate it by means of strict social norms.

3.3 Modeling Socio-Economic Dimensions

In addition, we investigate a range of socio-economic indicators to predict music taste. These indicators originate from the *Quality of Government* (QoG) dataset,[5] which collects approximately 2500 variables on country-level information from more than 100 data sources. From this dataset, we extract a subset of 181 variables for which all the scores are available for the set of analyzed countries (cf. Sect. 3.1). To give some examples, these attributes include information on *GDP, income inequality, agriculture's share of economy, unemployment rate* or *life expectancy*. Details on such variables are provided in [12].

---

[4] https://geert-hofstede.com/countries.html.
[5] http://qog.pol.gu.se/data/datadownloads/qogbasicdata.

4 Experiments and Results

4.1 Approach and Methods

We want to predict the popularity of each genre in a new country based on cultural and socio-economic data. For that purpose, we employ two ensemble regression methods: gradient boosting and random forests. Gradient boosting is an effective procedure applicable in regression problems offering a natural handling of heterogeneous features and robustness to outliers in output space. Random forests are known to show reasonable performance even with high amounts of noise visible in the features and can be used when the number of features is much larger than the number of observations. Additionally we tested Epsilon-Support Vector Regression with the linear and rbf kernels; their performance on the presented data-set was lower compared with the applied ensemble regression methods. For preprocessing, we tested a variety of techniques including univariate linear regression tests and kernel principle component analysis, but report only the best results here due to limited space. To train and evaluate the regressor, we use scores and features of the 44 countries. As comparative baseline, we consider also the average prevalence of that genre in the training set.

4.2 Results

Table 1 presents genre regressors performance over different sets of features: (i) Hofstede's dimensions, (ii) QoG dimensions, and (iii) the combined Hofstede's and QoG dimensions. We report the root–mean square error (RMSE) calculated over 5 independent, 10–fold cross–validation runs, one for each genre.

The regressors trained on the features inferred from Hofstede's and QoG dimensions outperformed the baseline approach in all the considered music genres. For 9 genres (alternative, pop, folk, rap, rnb, jazz, heavy metal, reggae, easy listening) the lowest RMSE was obtained using the QoG dimensions, in 3 cases (rock, punk, spoken word) the best performing regressors were trained only on Hofstede's dimensions, and for 6 genres (electronic, blues, country, classical, new age, world) the features were obtained from the both resources. The overall best performing regressor and resource type are the random forest regressor trained on the QoG dimensions. Here, the sum of RMSE for all the genres is at 0.1173, which constitutes a 12.2% improvement over the baseline approach. By selecting the best regressors for each genre we obtain an aggregated RMSE of 0.1117, a 16.4% reduction compared to the baseline. For variations that involve the QoG data, the improvements over the baseline are statistically significant according to Bonferroni-adjusted Wilcoxon signed rank tests with the null hypothesis that the error over each genre is equally distributed for the baseline and the respective classifier. This is not the case, if only Hofstede's dimensions are utilized.

In an additional analysis, we investigated which features influence the regressors most. While interpretation of socio-economical dimensions is often difficult, i.e., the best performing regressor uses a large number of relatively weak features of similar informative value, the features obtained from Hofstede's cultural

Table 1. Results: Music genre preferences regression accuracy from: Hofstede's dimensions, QoG dimensions and the combination of Hofstede's and QoG dimensions; cell values show information on the root mean squared error (RMSE) between the true and predicted genre popularities, for: (Baseline) - global average value for a genre, Gradient Boosting (G. Boost.) and Random Forest (R. Forest) regressors. Asterisk denotes the best performing method for each genre. The last line shows the Bonferroni-adjusted p-value of a Wilcoxon signed rank test compared to the baseline.

Genre	Baseline	Hofstede's dimensions		QoG dimensions		Hofstede's and QoG dimensions	
		G. Boost	R. Forest	G. Boost	R. Forest	G. Boost	R. Forest
Rock	0.02592	0.02131	*0.02042	0.02258	0.02255	0.02315	0.02288
Punk	0.00728	0.00695	*0.00648	0.00744	0.00724	0.00717	0.00725
Spoken word	0.00051	*0.00048	0.00049	0.00049	0.00049	0.00048	0.00048
Pop	0.01843	0.01783	0.01749	*0.01488	0.01551	0.01568	0.01626
Alternative	0.01823	0.01619	0.01674	0.01541	*0.01483	0.01562	0.01583
Folk	0.00930	0.00972	0.01029	0.00869	*0.00830	0.00833	0.00842
Rap	0.00573	0.00588	0.00585	*0.00510	0.00520	0.00527	0.00537
rnb	0.00325	0.00316	0.00310	0.00309	*0.00295	0.00305	0.00308
Jazz	0.00301	0.00274	0.00290	0.00279	*0.00254	0.00281	0.00272
Heavy metal	0.00256	0.00248	0.00246	0.00235	*0.00228	0.00237	0.00230
Reggae	0.00169	0.00186	0.00196	0.00150	*0.00139	0.00150	0.00150
Easy listening	0.00067	0.00062	0.00064	0.00056	*0.00053	0.00058	0.00058
Electronic	0.02557	0.02198	0.02162	0.02353	0.02321	0.02174	*0.02159
Blues	0.00508	0.00457	0.00469	0.00449	0.00452	*0.00431	0.00460
Country	0.00234	0.00249	0.00260	0.00219	0.00221	0.00222	*0.00218
Classical	0.00211	0.00228	0.00232	0.00193	0.00191	*0.00185	0.00188
New age	0.00111	0.00095	0.00094	0.00095	0.00093	0.00094	*0.00091
World	0.00085	0.00071	0.00069	0.00076	0.00072	0.00070	*0.00069
All genres	0.13364	0.12220	0.12168	0.11873	*0.11731	0.11777	0.11852
p-value		0.466	0.733	0.004	0.001	0.001	0.001

dimensions offer more consistent and interpretable results. Specifically, *Long Term Orientation* is the most informative feature for the largest number of genres, i.e.: rock, alternative, new age, rap, rnb, electronic and jazz; *Power Distance* is the most important feature for classical, blues and reggae genres; *Indulgence* for country, pop and folk; *Masculinity* for heavy metal; *Individualism* for punk and *Uncertainty Avoidance* for the spoken word genre.

5   Conclusions and Future Work

We presented an investigation of the predictive power of cultural and socio-economic dimensions to infer music genre preferences at the country level. We demonstrated that the application of cultural and socio-economics indicators lead to a significant reduction of the error for predicting the popularity of genres in a country by about 16.4% compared to the baseline approach, i.e., predicting the global, country-independent genre preferences. In this study we used a large-scale dataset on user listening behavior obtained from Last.fm user and analyzed how the cultural and socio-economical differences impact the users' music preferences. The study extends the scope of analysis compared to the previous works. In future work we will seek additional data sources, for instance, GPS-tagged microblogs [4], to obtain more fine grained results (e.g., at a regional level). Exploiting such precise data also enables the exploration of differences between rural and urban regions. Further, we will integrate the regressor proposed here into state-of-the-art recommendation algorithms and investigate its performance in comparison to other techniques to alleviate the cold-start problem.

Acknowledgments. This research is partially funded by the Austrian Science Fund (FWF) under grant no. P 27530.

References

1. Cheng, Z., Shen, J.: Just-for-me: an adaptive personalization system for location-aware social music recommendation. In: Proceedings of the 2014 ACM International Conference on Multimedia Retrieval, Glasgow, UK, April 2014
2. Ferwerda, B., Schedl, M.: Investigating the relationship between diversity in music consumption behavior and cultural dimensions: a cross-country analysis. In: Workshop on S, Halifax, Canada, July 2016
3. Ferwerda, B., Vall, A., Tkalčič, M., Schedl, M.: Exploring music diversity needs across countries. In: User Modeling, Adaptation and Personalization, Halifax, Canada (2016)
4. Hauger, D., Schedl, M., Košir, A., Tkalčič, M.: The million musical tweets dataset: what can we learn from microblogs. In: Proceedings of the 14th International Society for Music Information Retrieval Conference, Brazil, November 2013
5. Hofstede, G., Hofstede, G.J., Minkov, M.: Cultures and Organizations: Software of the Mind, 3rd edn. McGraw-Hill, New York (2010)
6. Hu, X., Lee, J.H.: A cross-cultural study of music mood perception between American and Chinese listeners. In: Proceedings of the 13th International Society for Music Information Retrieval Conference, Porto, Portugal, October 2012
7. Hu, X., Yang, Y.H.: Cross-dataset and cross-cultural music mood prediction: a case on Western and Chinese pop songs. IEEE Trans. Affect. Comput. (99) (2016)
8. Hu, Y., Ogihara, M.: NextOne player: A music recommendation system based on user behavior. In: Proceedings of the 12th International Society for Music Information Retrieval Conference, Miami, FL, USA, October 2011
9. Schedl, M.: The LFM-1b dataset for music retrieval and recommendation. In: Proceedings of the International Conference on Multimedia Retrieval, USA (2016)
10. Schedl, M., Stober, S., Gómez, E., Orio, N., Liem, C.C.: User-aware music retrieval. In: Müller, M., Goto, M., Schedl, M. (eds.) Multimodal Music Processing. Schloss Dagstuhl-Leibniz-Zentrum für Informatik, Germany (2012)
11. Singhi, A., Brown, D.G.: On cultural, textual and experiential aspects of music mood. In: Proceedings of the 15th International Society for Music Information Retrieval Conference (ISMIR), Taipei, Taiwan, October 2014

12. Teorell, J., Dahlberg, S., Holmberg, S., Rothstein, B., Khomenko, A., Svensson, R.: The Quality of Government Standard Dataset, version Jan16. University of Gothenburg, The Quality of Government Institute (2016)
13. Wang, J.-C., Yang, Y.-H., Wang, H.-M.: Affective music information retrieval. In: Tkalčič, M., De Carolis, B., de Gemmis, M., Odić, A., Košir, A. (eds.) Emotions and Personality in Personalized Services, pp. 227–261. Springer, Heidelberg (2016)

Batch Incremental Shared Nearest Neighbor Density Based Clustering Algorithm for Dynamic Datasets

Panthadeep Bhattacharjee and Amit Awekar$^{(\boxtimes)}$

Department of Computer Science and Engineering,
Indian Institute of Technology, Guwahati, India
{panthadeep,awekar}@iitg.ernet.in

Abstract. Incremental data mining algorithms process frequent updates to dynamic datasets efficiently by avoiding redundant computation. Existing incremental extension to shared nearest neighbor density based clustering (SNND) algorithm cannot handle deletions to dataset and handles insertions only one point at a time. We present an incremental algorithm to overcome both these bottlenecks by efficiently identifying affected parts of clusters while processing updates to dataset in batch mode. We show effectiveness of our algorithm by performing experiments on large synthetic as well as real world datasets. Our algorithm is up to four orders of magnitude faster than SNND and requires up to 60% extra memory than SNND while providing output identical to SNND.

Keywords: Clustering · Dynamic datasets

1 Introduction

Many popular clustering algorithms work on a static snapshot of dataset. However, many real-world applications such as search engines and recommender systems are expected to work over dynamic datasets. Such datasets undergo frequent changes where some points are added and some points are deleted. A naive method to get exact clustering over the changed dataset is to run the clustering algorithm again. Most of the computation in reclustering is going to be redudant. This problem becomes more severe with increase in frequency of updates to dynamic datasets. For large datasets, rerun of the algorithm might not finish before next batch of updates arrive. Incremental algorithms target this fundamental issue of redundant computation yet obtain output identical to their non-incremental counterpart. Incremental clustering problem can be defined more formally as follows. Consider a data set D alongwith its initial clustering $f : D \rightarrow \mathbb{N}$ and a sequence of n updates that convert D to D'. An incremental clustering is defined as a mapping $g : f, D' \rightarrow \mathbb{N}$ isomorphic to the one-time clustering $f(D')$ by the non-incremental algorithm.

SNND [1] is a widely used clustering algorithm for its robustness while finding clusters of varying densities and shapes even among high dimensional data.

© Springer International Publishing AG 2017
J.M. Jose et al. (Eds.): ECIR 2017, LNCS 10193, pp. 568–574, 2017.
DOI: 10.1007/978-3-319-56608-5_50

SNND converts a given data set to the SNN graph and clusters this graph. Efficiently computing changes to the SNN graph is the key challenge in designing an incremental extension to SNND. Existing incremental extension for SNND known as Incremental Shared nearest neighbor Density-based clustering (InSD) [4] has two main bottlenecks: First, it does not support deletion of points. Second, it supports addition of only one point at a time. If multiple updates are made to the dataset then, InSD treats each update independently and reclusters the dataset in response to each individual update. With increasing size of updates, InSD is only marginally faster than SNND and in some cases even slower.

Our goal is to design a Batch Incremental Shared Nearest Neighbor Density based Clustering (BISD) algorithm that overcomes both the bottlenecks of InSD and satisfies all three following conditions: 1. $T_{BISD} \leq T_{InSD}$, 2. $T_{BISD} \leq T_{SNND}$, and 3. $C_{BISD} = C_{SNND}$. Here T indicates running time of the algorithm and C indicates clustering provided by the algorithm. Unlike InSD, our algorithm integrates effect of multiple changes to the dataset by computing updates to the SNN graph in single execution. BISD achieves significant speedup at the cost of additional memory footprint while guaranteeing the same output. We show effectiveness of BISD by performing experiments over synthetic and real world datasets. Our datasets range in dimensionality from 2 to 70 and the number of points vary from 13000 to 100000. We observe that BISD is consistently multiple orders of magnitude faster than SNND while producing exact same output. BISD always maintains speed up over InSD.

2   Related Work

Density based clustering paradigm was introduced by Sanders et al. in their DBSCAN algorithm [2]. It computes density for each point in terms of number of points that are within certain distance. However, computing density using distance does not work well for high dimensional data and clusters with varying density. Shared nearest neighbor (SNN) based clustering paradigm was introduced by Jarvis and Patrick [3]. Their algorithm computes top k nearest neighbors (KNN) list for each point. Similarity between points is the extent of overlap between their corresponding KNN lists. Dataset is represented as SNN graph, which is a weighted, undirected, simple graph. Each vertex of the graph is a point. Edge exists between two vertices if they are in each other's KNN list. Weight of the edge is the similarity score computed as mentioned before. Edges with weight below the similarity threshold (δ_{sim}) are deleted from the SNN graph. Each component of the remaining graph is treated as a cluster.

SNND [1] algorithm combines two paradigms: density and SNN. It creates SNN graph similar to Jarvis & Patrick algorithm. However, it does not settle with simple idea of components as clusters. A vertex is labeled as core if its degree in SNN graph is above the strong link formation threshold (δ_{core}). A vertex is an outlier if it is neither core nor connected to any core vertex. Such vertices are not considered as part of any cluster. Rest of the vertices are labeled as non-core. Initially each core vertex is considered as a separate cluster. Two clusters are merged if at least one core vertex from one cluster is connected with a core vertex in other cluster. This process is repeated till no more clusters can be merged. A non-core vertex is assigned to a cluster having strongest link with it.

InSD [4] is an incremental extension of SNND. It does not handle deletion of points. InSD handles insertion of points only one point at a time. When a new point is added to the dataset, its KNN list is computed. This new point might occupy place in KNN lists of existing points, resulting in addition and deletion of edges from the SNN graph. InSD first goes through merge phase to identify clusters that will merge based on newly added edges to the SNN graph. Then InSD goes through split phase to identify clusters that will split into multiple fragments based on edges removed from the SNN graph. Output of InSD is identical to SNND. If multiple points are inserted in the dataset then, InSD processes each insertion independently and reclusters the whole dataset before handling next insertion.

3   Proposed Algorithm

Given a dataset D, initial clustering f, a sequence of n changes that converts D to D', extended nearest neighbors list for each point in D and SNN graph for D, our BISD algorithm provides clustering of D' that is identical to SNND. BISD partitions the n changes into set of points to be inserted (n_{add}) and deleted (n_{del}). First it handles all insertions, followed by all deletions to compute changes to the SNN graph and reclusters the data in single execution. BISD algorithm needs four parameters: δ_{sim}, δ_{core}, k, and w. First three parameters are identical to parameters used by SNND and InSD (Please refer Sect. 2). Fourth parameter w is the size of the extended KNN list ($w \geq k$). For each point BISD computes w nearest neighbors, but uses only top k out of them for clustering. Top k nearest neighbor list for a point P is denoted as $k(P)$ and its size as k. Extended nearest neighbor list for point P is denoted as $w(P)$ and its size as w. $k(P)$ is not stored separately. It is generated using top k elements of $w(P)$.

3.1   Insertion Phase

Insertion phase of BISD algorithm converts D to D'_{add} by adding all points in n_{add}. For each point P in n_{add}, $w(P)$ and $k(P)$ lists are created by computing distance with every point in D and every other point in n_{add}. Points in original dataset D are partitioned into three sets. A point in D is added to set $T1_{add}$, if any point in n_{add} becomes member of its KNN list. A point in D is added to set $T2_{add}$, if it is not already in $T1_{add}$ and at least one member of its KNN list is in $T1_{add}$. Rest of the points in D are unaffected by addition of new points. Such points are added to set U_{add}.

For example, please refer to Table 1. $P1$ is a point in D. Here size of k is two and size of w is four. Initially, $w(P1) = \{P2, P3, P4, P5\}$ and $k(P1) = \{P2, P3\}$. Let n_{add} consist of four points $N1$, $N2$, $N3$, and $N4$. After adding these four

Table 1. Example point

Point	P2	P3	P4	P5	N1	N2	N3	N4
Distance from P1	5	10	20	30	40	25	7	2

points, $w(P1) = \{N4, P2, P3, N2\}$ and $k(P1) = \{N4, P2\}$. Both $w(P1)$ and $k(P1)$ change. BISD will add $P1$ to $T1_{add}$ set. However, if n_{add} consisted of only points $N1$ and $N2$, then $w(P1) = \{P2, P3, P4, N2\}$ and $k(P1) = \{P2, P3\}$. Insertion of these two points affects $w(P1)$ but not $k(P1)$. In such case, $P1$ will not be added to $T1_{add}$ set.

3.2 Deletion Phase

During the deletion phase, points listed in n_{del} are removed from D'_{add} to get final dataset D'. For each point P in D', $w(P)$ and $k(P)$ are updated to remove any member that belongs to n_{del}. If size of $w(P)$ falls below the required size of KNN list (k), then $w(P)$ is recalculated for P. Otherwise, BIDS maintains truncated $w(P)$ as only top k entries in $w(P)$ are used to generate $k(P)$ and the SNN graph. The dataset D' is partitioned into three sets. Set $T1_{del}$ contains each point P in D' such that $k(P)$ was changed because of deletion of points. Set $T2_{del}$ contains each point P in D' such that P does not belong to set $T1_{del}$ and at least one member of $k(P)$ belongs to $T1_{del}$. Rest of the points in D' are added to the set U_{del}.

Let us continue the example in Sect. 3.1. At the end of insertion phase (after adding four points), $k(P1) = \{N4, P2\}$ and $w(P1) = \{N4, P2, P3, N2\}$. Consider first case where n_{del} consists of three points $N2$, $P2$, and $P3$. After removing these three points, size of $w(P1)$ is reduced to one. It is less than the required size of $k(P1)$. In such case, $w(P1)$ is recomputed and $P1$ is added to the set $T1_{del}$. Consider second case that n_{del} consists of only two points $N2$ and $P2$. In such scenario, $P1$ will still be added to set $T1_{del}$ but $w(P1)$ will not be recalculated as truncated $w(P1)$ is enough to generate $k(P1)$. Consider third case where n_{del} consists of only one point $P3$. In this case, $P1$ will not be added to set $T1_{del}$ as the deletion affects $w(P1)$ but $k(P1)$ remains unchanged.

3.3 Clustering Phase

Clustering phase aggregates changes from insertion and deletion phases to build updated SNN graph efficiently. First it builds set $T1 = (T1_{add} \cup T1_{del}) - n_{del}$. $T1$ set contains each point P that belongs to D' such that $k(P)$ was updated either in insertion or deletion phase. Then it builds set $T2 = ((T2_{add} \cup T2_{del}) - T1) - n_{del}$. This set contains each point R that belongs to D' such that $k(R)$ remained unchanged in both insertion and deletion phases, but at least one member of $k(R)$ belongs to $T1$. Points in D' that are not in $T1$ as well as $T2$, are unaffected by changes in the dataset.

We represent SNN graph using adjacency lists. For each point in n_{add}, new vertex is added to the SNN graph. Vertices corresponding to each point in n_{del} are removed from the SNN graph. For each point P in $T1$, its similarity score with every member of its $k(P)$ can potentially change. Therefore the adjacency lists of vertices in $T1$ are completely updated. However, for each point R in $T2$, its similarity score with a member of $k(R)$ can change only if that member is in $T1$. Hence, we only selectively update adjacency lists of vertices in $T2$. Points that are in $T1$ or $T2$ can possibly form new edges, update weights for some of the existing edges and lose some other existing edges. We can recluster this updated SNN graph using the clustering approach of $SNND$. After clustering, BISD stores the whole SNN graph, extended nearest neighbor list and cluster membership for all points to the secondary storage. This information is used during next invocation of BISD to handle new changes.

3.4 Differences from SNND and InSD

Consider the running example mentioned in Sects. 3.1 and 3.2. SNND will rerun whole computation over the complete dataset. SNND will ignore the fact that only few points are affected by these changes. SNND will compute top k nearest neighbors list for every point in D', build SNN graph from scratch and recluster this graph. As a result, SNND runs multiple orders of magnitude slower than BISD. However, BISD algorithm more efficiently rebuilds the SNN graph identical to SNND. Therefore clustering of BISD algorithm is identical to SNND.

InSD algorithm cannot handle deletion of three points. However to handle addition of four points, InSD will treat insertion of each point independently. InSD will recompute changes to SNN graph after insertion of each point and recluster the SNN graph before processing next point. InSD will not aggregate changes from multiple insertions and it will modify KNN lists more often than BISD algorithm. Therefore, InSD algorithm always runs slower than BISD.

4 Experiments

To demonstrate effectiveness of BISD algorithm, we conducted experiments on three real-world (Mopsi12, KDDCup'99, and KDDCup'04) and two synthetic (Birch3 and 5D) datasets. Please refer to Table 2. All algorithms were implemented in C++ on Linux platform. Experiments were conducted on a machine with 32 GB RAM and 1.56 GHz CPU. All code and datasets are available publicly for download[1].

Figure 1(a) and (b) show speed up of BISD algorithm over SNND and InSD respectively for all five datasets. X axis represents number of changes made to dataset as percentage of base dataset size. Y axis represents speed up of BISD. For Fig. 1(a), Y axis is log scaled to base ten. Speed up of BISD over SNND and InSD is calculated as T_{SNND}/T_{BISD} and T_{InSD}/T_{BISD} respectively. As number

---

[1] https://sourceforge.net/projects/bisd/.

Table 2. Datasets used

Dataset	Points	Dimensions	Description
Mopsi12	13000	2	Locations in Finland
5D	100000	5	Synthetic dataset
Birch3	100000	2	10 by 10 grid of gaussian clusters
KDDCup'04	60000	70	Identifying homologous proteins to native sequence
KDDCup'99	54000	41	Network intrusion detection

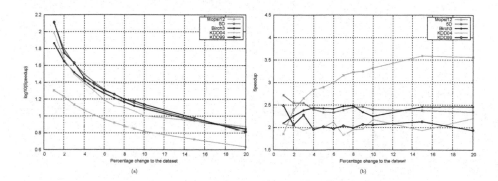

Fig. 1. Speedup of BISD algorithm over SNND and InSD

of changes to the dataset increase, insertion and deletion phases of BISD take more time. This happens because number of affected points either in sets $T1$ or $T2$ also increase with more number of changes. As a result, speed up of BISD over SNND reduces with increase in number of changes. BISD maintains extended KNN list in memory for all points. This memory overhead is up to 60% extra for BISD as compared to memory footprint of SNND. BISD achieves speed up of up to 4 over InSD as it aggregates changes over all insertions.

5   Conclusion and Future Work

We presented BISD algorithm that handles both addition and deletion to the dataset. It aggregates changes to the SNN graph across multiple changes to the dataset and reclusters dataset only once. BISD algorithm provides output identical to original SNND algorithm with bounded memory overheads. We plan to generalize our BISD algorithm for other density based clustering algorithms in future.

References

1. Ertöz, L., Steinbach, M., Kumar, V.: Finding clusters of different sizes, shapes, and densities in noisy, high dimensional data. In: SIAM ICDM, pp. 47–58 (2003)

2. Ester, M., Kriegel, H.-P., Sander, J., Xu, X.: A density-based algorithm for discovering clusters in large spatial databases with noise. In: ACM KDD, pp. 226–231 (1996)
3. Jarvis, R.A., Patrick, E.A.: Clustering using a similarity measure based on shared near neighbors. IEEE Trans. Comput. **C–22**(11), 1025–1034 (1973)
4. Singh, S., Awekar, A.: Incremental shared nearest neighbor density-based clustering. In: ACM CIKM, pp. 1533–1536 (2013)

Improving Tweet Representations Using Temporal and User Context

J. Ganesh[1(✉)], Manish Gupta[1,2], and Vasudeva Varma[1]

[1] IIIT, Hyderabad, India
ganesh.j@research.iiit.ac.in, vv@iiit.ac.in
[2] Microsoft, Hyderabad, India
gmanish@microsoft.com

Abstract. In this work we propose a novel representation learning model which computes semantic representations for tweets accurately. Our model systematically exploits the chronologically adjacent tweets ('context') from users' Twitter timelines for this task. Further, we make our model user-aware so that it can do well in modeling the target tweet by exploiting the rich knowledge about the user such as the way the user writes the post and also summarizing the topics on which the user writes. We empirically demonstrate that the proposed models outperform the state-of-the-art models in predicting the user profile attributes like spouse, education and job by 19.66%, 2.27% and 2.22% respectively.

1 Introduction

The short and noisy nature of tweets poses challenges in computing accurate latent tweet representations. We hypothesize that a principled usage of chronologically adjacent tweets from users' Twitter timelines can help in significantly improving the quality of the representation. We propose an attention based model that assigns a variable weight to each context tweet that captures the semantic correspondence between the target tweet and the context tweet. We further augment the attention model to be user-aware so that it can do well in modeling the target tweet by exploiting the rich knowledge about the user such as the way the user writes the post, and also summarizing the topics on which the user writes.

We summarize our main contributions below. In summary, our contributions are as follows. (1) Our work is the first to model the semantics of the tweet using the temporal context. (2) We introduce a novel attention based model that learns the weights for context tweets by back-propagating semantic loss. (3) We propose a novel way to learn user vector summarizing the content the user writes, which in turn helps in enriching the quality of the tweet embeddings. (4) We conduct quantitative analysis to showcase the application potential of the tweet representations learned from the model and also provide some interesting findings.

© Springer International Publishing AG 2017
J.M. Jose et al. (Eds.): ECIR 2017, LNCS 10193, pp. 575–581, 2017.
DOI: 10.1007/978-3-319-56608-5_51

2   Related Work

Le et al. [3] adapt Word2Vec to learn document representations which are good
in predicting the words present in the document. As seen in Sect. 5, for short
documents like tweets, the model tends to learn poor document representations
as the vector relies too much on the document content, resulting in overfitting.
Djuric et al. [1] learn document representations using word context (same as [3])
along with document stream context in a hierarchical fashion.

3   Problem Formulation

In this section we first introduce the notions of temporal context and attention,
and then provide a formal problem statement.

Temporal Context: Temporal context of a tweet $t(j)$ is the set of C_T tweets
posted before and after $t(j)$ by the same user. The value C_T is a user specified
parameter that defines the size of the temporal context to be considered to model
a given tweet. For example, in Fig. 1 we fix C_T as 2, the context tweets of $t(j)$
are $t(j-1)$, $t(j-2)$, $t(j+1)$ and $t(j+2)$.

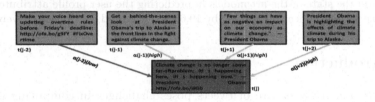

Fig. 1. $t(j-1)$, $t(j-2)$, $t(j+1)$ and $t(j+2)$ form the temporal context of $t(j)$. α's
denote the attention parameters of the proposed model.

Attention: An attention value is associated with a context tweet that defines
the degree of semantic similarity between the context tweet and the target tweet.
The more the latent semantic interactions between the tweets, the more is the
attention. We denote the attention of context tweet $t(j-1)$ as $\alpha(j-1)$. For
instance, in Fig. 1, the attention value of context tweet $t(j-2)$ should be lower
than that of context tweet $t(j-1)$ with respect to target tweet $t(j)$. In Fig. 1,
clearly $t(j-2)$ is not talking about the topic 'Climate Change' and so it makes
sense to have a lower attention value.

Problem Statement: Let the training tweets be given in the order in which
they are posted. In particular, we assume that we are given a user set U
of N_u tweet sequences, with each sequence $u(k) \in U$, containing N_t tweets,
$u(k) = \{t(1), .., t(j), .., t(N_t)\}$ posted by user $u(k)$. Moreover, each tweet $t(j)$
is a sequence of N_w words, $t(j) = \{w(j, 1), .., w(j, i), .., w(j, N_w)\}$. The problem
is to learn semantic low-dimensional representations for all the tweets in the
sequences in set U.

4 Proposed Models

Our model (Fig. 2) learns tweet representations in a hierarchical fashion: learning from the words present in the tweet using word context model (Fig. 2(a)) along with the temporal tweets present in the user stream using tweet context model (Fig. 2(b)). Both the models will be discussed in detail below. Let $\mathbf{w(j, i)}$, $\mathbf{t(j)}$ and $\mathbf{u(k)}$ denote the embedding for a word i from tweet j, tweet j and user $u(k)$ respectively, all of which have the size 'n'. We will discuss details about both of these models in this section.

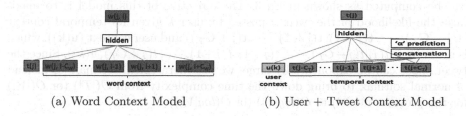

(a) Word Context Model   (b) User + Tweet Context Model

Fig. 2. Architecture diagram of our model.

4.1 Word Context Model

The goal of the word context model is to learn tweet representations which are good at predicting the words present in the tweet. The model has three layers. The first layer contains the word embeddings, $\mathbf{w(j, i - C_W)}, \cdots, \mathbf{w(j, i - 1)}$, $\mathbf{w(j, i + 1)}, \cdots, \mathbf{w(j, i + C_W)}$ near the ith target word in tweet j, which denote the word context for the word i (i.e., $w(j, i)$) along with the tweet embedding $\mathbf{t(j)}$. Secondly, there is a hidden layer with size equal to the number of words in the vocabulary ($|V|$). The final layer is a softmax layer which gives a well-defined probability distribution over words in the vocabulary. The input to the word context model is all pairs of word context of word i and tweet $t(j)$ in the corpus. The objective is to maximize the likelihood of the word $w(j, i)$ occurring given its context, i.e., $\mathbb{P}(w(j, i)|w(j, i - C_W), \cdots, w(j, i-1), w(j, i+1), \cdots, w(j, i+C_W), t(j))$. Equation 1 represents the forward propagation step in our 1-hidden layer feed forward model, where W_{WC} and T_{WC} denote the additional parameters of the model.

$$\hat{y}_{|V| \times 1}(j) = softmax(W_{WC} \times \sum_{l \in \{i - C_W, i + C_W\}\} \backslash i} \mathbf{w(j, l)} + T_{WC} \times \mathbf{t(j)}) \quad (1)$$

4.2 User + Tweet Context Model

The goal of this model is to enrich the tweet representation learned from the word context, by modeling the current tweet conditioned on its temporal context and the proposed user context. The user context makes our model user-aware by exploiting the user characteristics such as the way the user writes the post and also summarizing the topics on which the user writes. These user vectors are learned automatically from the set of tweets posted by the user through this model. As a naïve solution, we can directly adopt Djuric et al. [1]'s approach and apply on the Twitter stream. As discussed in Sect. 3, this assumption is too strong for social media streams. Can we assign attention levels to the context tweets with respect to the tweet being modeled? To learn the optimal values of attention $(\alpha(j))$, we introduce the attention parameters as shown in Eq. 2. The intuition is that semantic loss will be less if the weights of each of the temporal context tweets are learned accurately. The values of $\alpha(j)$'s can be computed as shown in Eq. 3. The objective of this model is to maximize the likelihood of the tweet j posted by user k given its temporal context $(\mathbf{t(j - C_T)}, \cdots, \mathbf{t(j-1)}, \mathbf{t(j+1)}, \cdots, \mathbf{t(j + C_T)})$ and user context $(\mathbf{u(k)})$, which is given by $\mathbb{P}(t(j)|t(j - C_T), \cdots, t(j-1), t(j+1), \cdots, t(j+C_T), u(k))$. Since the tweet space can be exponentially large, we use hierarchical softmax [5] instead of normal softmax to bring down the time complexity from $O(|T|)$ (or $O(|V|)$ for the previous model) to $O(log|T|)$ (or $O(log|V|)$).

$$\hat{y}_{|T| \times 1}(j) = softmax(U_{TC} \times \mathbf{u(k)} + T_{TC} \times \sum_{l \in \{j-C_T, j+C_T\} \backslash j} \alpha(l) \times \mathbf{t(l)}) \quad (2)$$

$$(\alpha(j - C_T) \cdots \alpha(j-1)\alpha(j+1) \cdots \alpha(j+C_T))$$
$$= softmax(A[\mathbf{t(j - C_T)}; \cdots; \mathbf{t(j-1)}; \mathbf{t(j+1)}; \cdots; \mathbf{t(j + C_T)};]) \quad (3)$$

where the parenthesis inside the softmax function represents concatenation of all context representations $((2 \times C_T \times n) \times 1$ in size). A is the additional weight matrix (of size $(2 \times C_T) \times (2 \times C_T \times n)$) added as parameters to the model. In practice, we observe that multiple passes ('epochs') on the training set are required to fine tune these attention values. The overall objective function intertwining both the models in a hierarchical fashion to be maximized can be summarized as shown in Eq. 4. We use the cross-entropy as the cost function between the predicted distribution $\hat{y}(j)$ and target distributions $t(j)$ and $w(j, i)$, for modeling using the temporal and word context respectively. We train the model using back-propagation [7] and Adam [2] optimizer.

$$\mathcal{L}(\theta) = \sum_{u(k) \in U} \left[\sum_{t(j) \in u(k)} \sum_{w(j,i) \in t(j)} \log \mathbb{P}(w(j, i)|w(j, i - C_W), \cdots, w(j, i-1), \right.$$
$$w(j, i+1), \cdots, w(j, i+C_W), t(j)) + \log \mathbb{P}(t(j)|w(j, 1), \cdots, w(j, N_w))$$
$$\left. + \log \mathbb{P}(t(j)|t(j - C_T), \cdots, t(j-1), t(j+1), \cdots, t(j+C_T), u(k)) \right]$$
$$+ \log \mathbb{P}(u(k)|t(1), \cdots, t(N_T)) \quad (4)$$

5   Experimental Evaluation

In this section we discuss details of our dataset, experiment, and then present quantitative analysis of the proposed models.

Table 1. User profile attribute classification - F1 score

Algorithm	Spouse	Education	Job
Paragraph2Vec [3]	0.3435	0.9259	0.5465
Simple Distance model (SD)	0.3704	0.9068	0.5872
HDV [1]	0.4526	0.8901	0.521
Ours (User = 0)	**0.5416**	0.9098	0.5935
Ours (User = 1)	0.4082	**0.9274**	**0.6067**

5.1   Dataset Description

We use the publicly available dataset described in Li et al. [4] for all the experiments. It contains tweets pertaining to three profile attributes (spouse, education and job) of a user. Specifically, it has a set of tweets from users' Twitter timelines, that talk about the attribute ('positive' tweets) and those that do not ('negative' tweets). We randomly sample 1600 users from the dataset and use 70-10-20 ratio to construct train, validation and test splits. Tweet embeddings are randomly initialized while the word embeddings are initialized with the pre-trained word vectors from Pennington et al. [6].

5.2   Experimental Protocol

We consider the binary task of predicting whether a given entity mention corresponds to particular users' profile attribute or not. We build our model to get the tweet vector and the entity vector by computing an average of all the tweet vectors for the entity. We tune the penalty parameter of a linear Support Vector Machine (SVM) on the validation set. Note that we use a linear classifier so as to minimize the effect of variance of non-linear methods on the classification performance and subsequently help in interpreting the results. We compare our model with three baselines: (1) Paragraph2Vec [3], (2) Simple Distance model (SD): A model that assigns attention weight to the context tweet which is inversely proportional to the distance of the tweet from the target tweet, (3) HDV [1], (4) Ours (User = 0): Our model when the user context is excluded from the temporal context, (5) Ours (User = 1): Our model when the user context is included in the temporal context. We empirically set n and C_W to 200 and 10 respectively for all the models. In case of SD, HDV and our models, we try values in {1, 2, 4, 6, 8, 10, 12, 14, 16} to fix the temporal context size parameter (i.e., C_T) which is crucial in improving the semantics of the tweet.

5.3   Comparative Analysis

From Table 1, we see that Paragraph2Vec overfits the validation set, resulting in poor accuracy during testing. HDV's assumption of giving equal attention value to the temporal context also results in lower accuracy compared with our

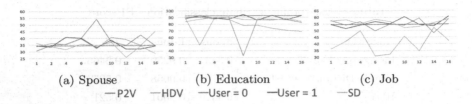

(a) Spouse   (b) Education   (c) Job

—P2V   —HDV   —User = 0   —User = 1   —SD

Fig. 3. Model performance w.r.t. temporal context size C_T.

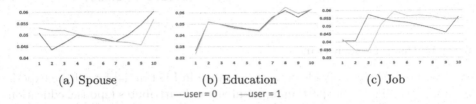

(a) Spouse   (b) Education   (c) Job

—user = 0   —user = 1

Fig. 4. Mean attention w.r.t. distance from the center C_T.

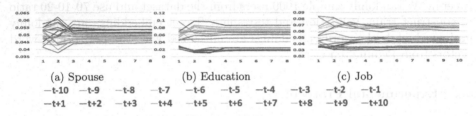

(a) Spouse   (b) Education   (c) Job

—t-10   —t-9   —t-8   —t-7   —t-6   —t-5   —t-4   —t-3   —t-2   —t-1
—t+1   —t+2   —t+3   —t+4   —t+5   —t+6   —t+7   —t+8   —t+9   —t+10

Fig. 5. Mean attention w.r.t. epoch for our model model when the user context is included in the temporal context.

models. SD model outperforms HDV in two tasks, which substantiates our claim against HDV's naïve assumption for social media. Our model with user vector outperforming the baselines for Education and Job attribute classification, shows the need to consider the user characteristics while modelling his/her tweets. The poor results for Spouse task suggest that this dataset has too many topic shifts and that the user vector turned out to be less accurate. Figure 3 displays the F1 results for different values of C_T, which is a vital parameter controlling the influence of temporal context. We observe that in some cases HDV outperforms the SD model, mainly due to the inability of the SD model to utilize the context information from farther tweets which are relevant with respect to the target tweet. Our models are 19.66%, 2.27% and 2.22% better compared to the baselines for the spouse, education and job attributes respectively.

5.4   Impact of Variable Attention

We plot the attention mean across each position of the context tweet with respect to the epoch number. From Fig. 5, we see that mean attention at each context position are approximately in the ballpark. Mean attention weights vary for each context position, exhibiting no relation with respect to the increase in distance (as seen in Fig. 4). These findings indicate the complexity of giving attention to tweets in the temporal context. Initially, we see that the mean attention weights are changing drastically indicating their sub-optimality. It is interesting to see the convergence of these weights to the optimal solution is fast (in terms of no. of epochs) in the model which uses user context when compared to the model that does not use it.

6 Conclusions

We proposed a model to learn generic tweet representations which have a wide range of applications in NLP and IR field. We discovered that the principled usage of the tweets in the temporal context is an important direction in enriching the representations. We also explored learning a novel user context vector to make our model user-aware while predicting the adjacent tweets. Through experimental analysis, we identified the cases when modeling the user characteristics help enhance the embedding quality. In future, we plan to understand the application potential of the user vector learned through our approach.

References

1. Djuric, N., Wu, H., Radosavljevic, V., Grbovic, M., Bhamidipati, N.: Hierarchical neural language models for joint representation of streaming documents and their content. In: WWW, pp. 248–255 (2015)
2. Kingma, D.P., Ba, J.: Adam: a method for stochastic optimization. CoRR abs/1412.6980 (2014)
3. Le, Q.V., Mikolov, T.: Distributed representations of sentences and documents. In: ICML, pp. 1188–1196 (2014)
4. Li, J., Ritter, A., Hovy, E.H.: Weakly supervised user profile extraction from twitter. In: ACL, pp. 165–174 (2014)
5. Morin, F., Bengio, Y.: Hierarchical probabilistic neural network language model. In: AISTATS (2005)
6. Pennington, J., Socher, R., Manning, C.D.: Glove: global vectors for word representation. In: ACL, pp. 1532–1543 (2014)
7. Rumelhart, D.E., Hinton, G.E., Williams, R.J.: Learning representations by back-propagating errors. Cogn. Model. 5(3), 1 (1988)

Personalized Parsimonious Language Models for User Modeling in Social Bookmaking Systems

Nawal Ould Amer[1,2], Philippe Mulhem[1(✉)], and Mathias Géry[2]

[1] Univ. Grenoble Alpes, CNRS, LIG, 38000 Grenoble, France
{Nawal.Ould-Amer,Philippe.Mulhem}@imag.fr
[2] Univ Lyon, UJM-Saint-Etienne, CNRS, Laboratoire Hubert Curien UMR 5516,
42023 Saint-Etienne, France
Mathias.Gery@univ-st-etienne.fr

Abstract. This paper focuses on building accurate profiles of users, based on bookmarking systems. To achieve this goal, we define personalized parsimonious language models that employ three main resources: the tags, the documents tagged by the user and word embeddings that handle general knowledge. Experiments completed on *Delicious* data show that our proposal outperforms state-of-the-art approaches and non-personalized parsimonious models.

Keywords: User profile · Parsimonious models · Words embeddings

1 Introduction and Related Works

Personalized search systems (PSS) define and manipulate users' representations, or *profiles*, to enhance query results quality. We focus here on the use of social bookmarking systems, as they are important textual sources of evidence about users' interests.

Two major sources of user information are investigated by PSS works on social bookmarks: the *tags* assigned by a user to a particular document, and the *content* of the tagged document. This information is then exploited to construct a user profile. For example, [13] models a user over his/her tags, where each tag is weighted using *tf-idf* values. The authors of [4] weight user tags using *tf-iuf*, where [3] proposed a variant of *tf-iuf* (cf. Sect. 3). Exploiting the content of the tagged documents (like web pages) is expected to broaden the profile vocabulary compared to tags. Indeed, previous studies on query log [11] have shown that document content is more useful. Most of works related to document content rely on *tf-idf* term weighting [5], or Latent Dirichlet Allocation (LDA) [6]. The main difficulty in modeling a user using his document content is to accurately filter the terms that come from the documents to keep only the important terms of the users' interests. This problem also occurs in relevance feedback models.

Parsimonious Language Models (PLM) [7] seek to build compact and precise term distributions by eliminating the stop words and nonessential terms. PLM

© Springer International Publishing AG 2017
J.M. Jose et al. (Eds.): ECIR 2017, LNCS 10193, pp. 582–588, 2017.
DOI: 10.1007/978-3-319-56608-5_52

was successfully applied for relevance feedback [8] to capture relevant terms from feedback document to expand a query. In this paper, we propose to adapt PLM to extract relevant terms from tagged document in order to model a user. To extract relevant terms, we use word embedding [1,9]. This paper introduces *Personalized Tagged Parsimonious Language Models* (PTPLM) that capture an accurate term distribution to model a user using his bookmarks. Our aim is to answer the following research questions: **RQ1:** Are user's tags effective to estimate important words of a user tagged document, then to model user's interests? **RQ2:** Are Personalized Tagged Parsimonious Language Models (PTPLM) able to improve state-of-the-art approaches? The paper is organized as follows. Section 2 details our PTPLM proposal. Section 3 presents the experiments conducted, and Sect. 4 is dedicated to the results and discussion. We conclude in Sect. 5.

2 Approach

2.1 PTPLM Estimation

In order to estimate personalized tagged parsimonious language models (PTPLM), we assume that each document d has a set of related tags which are assigned by a user u: $TG_u(d)$. Then, given d, its terms distribution θ_d and $TG_u(d)$, we re-estimate a new terms distribution for the document, noted θ_{d_u}. Let $d = \{t_1, t_2, \ldots, t_n\}$ the document tagged by a user u, and $TG_u(d) = \{tg_1, tg_2, \ldots, tg_P\}$ the set of tags given by u to d. We first estimate a document model as raw probabilistic estimation θ_d (i.e., first iteration in E-M algorithm) using maximum likelihood as follows: $P(t|\theta_d) = \frac{tf(t,d)}{|d|}$, where $tf(t,d)$ is the frequency of term t in d, of length $|d|$.

Now, taking inspiration from [7], we re-estimate the terms distribution by integrating the tags in the *E-Step*, where the terms related to the user tags should be important terms. In other words, if the term t (from the vocabulary V) in the document d is *related* to the tag tg used by a user u for the document d (using $P(t|\theta_{TG_{u_d}})$), then the term t is an important term.

$$E - Step : e_t = tf(t,d) \times P(t|\theta_{TG_{u_d}}) \times \frac{\lambda P(t|\theta_d)}{\lambda P(t|\theta_d) + (1-\lambda)P(t|\theta_C)} \quad (1)$$

$$M - Step : P(t|\theta_d) = \frac{e_t}{\sum_{t \in V} P(t|\theta_d)} \quad (2)$$

where $P(t|\theta_{TG_{d_u}})$ is estimated as follows:

$$P(t|\theta_{TG_{u_d}}) = \frac{1}{|TG_u(d)|} \sum_{tg \in TG_u(d)} P(t|tg) \quad (3)$$

where $P(t|tg)$ is the probability of term t given a user tag tg, and $P(t|tg)$ is the probability that a term t is related to the tag tg, estimated using the cosine similarity between the two embedded vectors corresponding to term t and tag tg as follow: $P(t|tg) = sim_{cos}(t, tg)$. The iteration is repeated until the estimates do not change significantly anymore. Then we obtain a new term distribution θ_d that we rename θ_{d_u}. This is a personalized document terms distribution.

2.2 Building Users' Profiles

Let $D_u = \{d_1, d_2, \ldots, d_N\}$ the set of documents tagged by a user u. After PTPLM estimation for each document d in D_u as described above, a document user profile θ_u is defined, as presented in the Algorithm 1. This algorithm builds the term user profile by averaging the term probabilities over the documents tagged by u (cf. line 6).

Algorithm 1. Estimation of User Model

Require:
$\quad D_u = \{d_1, d_2, .., d_N\}$: Set of document tagged by a user u.
$\quad TG_u(d) = \{tg_1, tg_2, .., tg_P\}$ Set of tags assigned to document d by a user u.
Ensure:
$\quad \theta_u$: User Model.
1: **for each** $d \in D_u$ **do**
2: $\quad \theta_{d_u} \leftarrow PTPLM(d, TG_u(d))$
3: **end for**
4: **for each** $t \in V$ **do**
5: $\quad$ **for each** $d \in D_u$ **do**
6: $\quad\quad P(t|\theta_u) = \frac{1}{|D_u|} \sum_{d \in D_u} P(t|\theta_{d_u})$
7: $\quad$ **end for**
8: **end for**

2.3 Ranking Model

To rank the documents, we use a query expansion model. We first select the terms related to the query (i.e. terms that are in the same context than the user query) from the user profile using the cosine similarities between q and t. Then, we expand the query using these terms with their weights $P(t|\theta_u)$. The ranking model is as follows: $RSV(q, d, u) = \alpha.RSV(q'_u, d) + (1 - \alpha).RSV(q'_u, TG(d))$, where q'_u is the expanded query of a user u, $TG(d)$ is the set of tags assigned to the document d by all users, and α is a parameter in $[0, 1]$.

3   Experiments

Dataset: We evaluate our proposal on the *Delicious* dataset [12]. We first perform a crawl of the English available web pages. For our experiment, we select only users with more than 100 unique tags for more than 100 unique bookmarks The resulting corpus contains 1,238,443 Web pages, 287,969 users and 204,505 unique tags.

Word Embeddings Train: We train a Continuous Bag-of-Word (CBOW) model [9] on Wikipedia corpus consisting of 20,151,102 documents and a vocabulary size of 2,451,307 words. The training parameters are set as follows: the output vectors size is set to 50, the width of the word-context window is set to 8, and the number of negative samples is set to 25.

Evaluation Methodology and Metrics: We use the evaluation framework for personalized search based on social annotation introduced by [2] and used in most of the state-of-the-art works [3,6,13]. This framework assumes that *"The users' bookmarking and tagging actions reflect their personal relevance judgment"*. Then, the tags are considered as queries. A document is assumed relevant for a tag t considered as a query q issued by a user u, if the document has been tagged by u with the tag t [2]. We split the dataset into training and testing subsets: the last 20% bookmarks (according to the timeline) for each user are for testing, where the first 80% bookmarks are used for learning the profiles. We generate 4,911 queries for 128 users and their relevance judgments. We use the Mean Average Precision (MAP), and P@5 as evaluation metrics.

Parameters Settings: We used the Terrier Information Retrieval framework to compute the matching. We choose to use BM25 [10] weighting model with its default parameters. For the PTPLM approach, we tested the different values of λ in Eq. (1). The retrieval performances are stable over its different values, we fix here $\lambda = 0.5$. In the M-step of PTPLM, the terms that receive a probability below a fixed threshold (i.e. 0.0001, as in [7]) are removed from the model. In Eq. 3, for the estimation of $P(t|tg)$ using $sim_{cos}(t, tg)$, we consider only positive values of similarity.

Baselines: We compare our proposal to three personalization state-of-the-art approaches: (**Xu**) where the weights of users' tags are based on TF-IDF values [13]; (**Cai**) where the weights of users' tags are based on user term frequency computed as follow: $w_t = \frac{TF(t)}{D_u}$, where D_u is the number of document tagged by a user [4]; (**Bouadjenek**) where the weights of users' tags are based on user terms frequency computed as follow: $w_t = TF(t) \times \log(\frac{|U|}{|U_t|})$, where U is number of users and $|U_t|$ is the number of users who used t [3]. We also consider classical (non-personalized) (**PLM**), as well as non-expanded queries (**Noexp**).

4   Results

4.1   Impact of User Tags on Parsimonious Language Models

To explore our first research question RQ1, Table 1 shows the estimation of top-5 terms distribution for the same document[1], assuming that the tags assigned by a user to the document are: *casino, games,* and *DangerouslyFun*. The distribution estimated using a standard language model is presented in column *Standard LM*. The *PLM* column displays the terms distribution using classical PLM [7], with the final probability for each term averaged over the user's document. The *PTPLM* column presents the distribution estimated as in Sect. 2.

As seen in Table 1, the PTPLM re-estimate the term probability according to the user tags: the model emphasizes the terms related to the user tags. For example, the probability of the term *casino* is boosted compared to other models. This example shows that PTPLM is able to capture more accurately the personalized view of documents.

---

[1] URL document: http://www.dangerouslyfun.com.

Table 1. Term distribution for one document

Standard LM		PLM		PTPLM	
Will	0.0689	Online	0.0777	Casino	0.1989
Online	0.0583	Casino	0.0670	Players	0.1277
Players	0.0477	Players	0.0555	Games	0.0978
Casino	0.0397	Games	0.0420	Casinos	0.0677
Can	0.0371	Casinos	0.0346	Gaming	0.0347

4.2 Evaluating User Profile Model: Comparison with Baselines

To answer our second research question RQ2, we compare the results of our proposed model PTPLM with those of the baseline and state-of-the-art user model approaches described in Sect. 3. We aim to assess the quality of the profiles, then we consider only *single-term expansions*, and we apply several cut-off points for the profiles (100, 200, 300, and 500 terms) according to the term weights. We tested all approaches over $\alpha \in [0,1]$, and we report the best configuration for each model in Table 2.

We see that PTPLM outperforms all state-of-the-art personalization models in term of MAP and P@5 for all user profile sizes. This shows that PTPLM is able to estimate a better terms distribution to describe user interests. The larger differences (in %) are obtained when keeping the top-100 terms from the profiles: this shows that our proposal is able to bring out important (relevant) terms more accurately than other approaches.

Table 2. One term expansion. Bold value: best query expansion system; $(x\%^{\nabla})$: significant MAP differences w.r.t. PTPLM, bilateral paired Student t-test, p< 0.05

Model	MAP				P@5			
Noexp	0.195				0.097			
Profile cutoff	100		200		300		500	
Models	MAP	P@5	MAP	P@5	MAP	P@5	MAP	P@5
PPLM	0.120 $(-34\%^{\nabla})$	0.058	0.157 $(-16\%^{\nabla})$	0.079	0.180 $(-6\%^{\nabla})$	0.090	0.184 $(-5\%^{\nabla})$	0.090
PTPLM	**0.181**	**0.092**	**0.188**	**0.094**	**0.192**	**0.096**	**0.194**	**0.097**
Bouadjenek	0.161 $(-10\%^{\nabla})$	0.081	0.177 $(-6\%^{\nabla})$	0.089	0.188 (-2%)	0.094	0.192 (-1%)	0.096
Cai	0.166 $(-8\%^{\nabla})$	0.084	0.178 $(-6\%^{\nabla})$	0.090	0.188 (-2%)	0.094	0.193 (-1%)	0.096
Xu	0.165 $(-8\%^{\nabla})$	0.083	0.178 $(-6\%^{\nabla})$	0.089	0.187 (-2%)	0.093	0.192 (-1%)	0.096

In Table 2 the *Noexp* runs are presented for the sake of completeness: we did not expect these runs to be outperformed by the naive and limited single-term expansions tested. Although, in a way to provide a fair framework when comparing Noexp and PTPLM results, we need to consider profiles that are potentially able to cover the many facets of the users' profiles. The Fig. 1 presents the 68

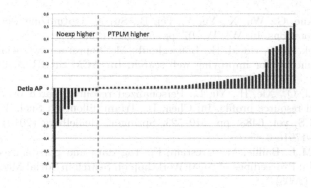

Fig. 1. Delta of AP values PTPLM top-500 profiles w.r.t. Noexp.

larger query-by-query AP differences, among the full set of queries, comparing the top-500 terms profiles from PTPLM and the Noexp results.

Our PTPLM proposal underperforms for 13 queries and outperforms for 55 queries the Noexp approach. So, even limited PTPLM-based expansions are able to play a positive role in many cases. We strongly believe that a more accurate usage of users' profiles will outperform the Noexp runs in the future.

5   Conclusion and Future Works

In this paper, we introduced the PTPLM approach, that exploits user tags to extract relevant terms from user tagged documents, expecting to obtain a better representation of user interests. According to our experiments conducted on *Delicious*, we found that PTPLM outperforms all state-of-the-art user modeling approaches. The PTPLM do not currently take benefit of user tags: we believe to gain effectiveness when using these tags. Our usage of profiles generated by PTPLM underperforms no-expansions runs. However, our analysis conducted query by query indicates that there is a great room for improving our usage of the generated profiles in the future. As future works, we are working on efficiently using our model to improve query expansion, and also on comparing our PTPLM with content based state-of-the-art approaches (e.g.: LDA).

Acknowledgements. This work is supported by the ReSPIr project of the région Auvergne Rhône-Alpes.

References

1. ALMasri, M., Berrut, C., Chevallet, J.-P.: A comparison of deep learning based query expansion with pseudo-relevance feedback and mutual information. In: Ferro, N., Crestani, F., Moens, M.-F., Mothe, J., Silvestri, F., Nunzio, G.M., Hauff, C., Silvello, G. (eds.) ECIR 2016. LNCS, vol. 9626, pp. 709–715. Springer, Heidelberg (2016). doi:10.1007/978-3-319-30671-1_57

2. Bao, S., Xue, G., Wu, X., Yu, Y., Fei, B., Su, Z.: Optimizing web search using social annotations. In: WWW 207, pp. 501–510 (2007)
3. Bouadjenek, M.R., Hacid, H., Bouzeghoub, M.: Sopra: a new social personalized ranking function for improving web search. In: ACM SIGIR 2013, pp. 861–864 (2013)
4. Cai, Y., Li, Q., Xie, H., Yu, L.: Personalized resource search by tag-based user profile and resource profile. In: Chen, L., Triantafillou, P., Suel, T. (eds.) WISE 2010. LNCS, vol. 6488, pp. 510–523. Springer, Heidelberg (2010). doi:10.1007/978-3-642-17616-6_45
5. Carman, M.J., Baillie, M., Crestani, F.: Tag data and personalized information retrieval. In: Proceedings of ACM Workshop on Search in Social Media, SSM 2008, pp. 27–34 (2008)
6. Harvey, M., Ruthven, I., Carman, M.J.: Improving social bookmark search using personalised latent variable language models. In: ACM WSDM 2011, pp. 485–494 (2011)
7. Hiemstra, D., Robertson, S., Zaragoza, H.: Parsimonious language models for information retrieval. In: SIGIR 2004, pp. 178–185 (2004)
8. Kaptein, R., Kamps, J., Hiemstra, D.D.: The impact of positive, negative and topical relevance feedback. In: 17th Text REtrieval Conference, TREC 2008 (2008)
9. Mikolov, T., Chen, K., Corrado, G., Dean, J.: Efficient estimation of word representations in vector space. CoRR abs/1301.3781 (2013)
10. Robertson, S.E., Walker, S., Jones, S., Hancock-Beaulieu, M.M., Gatford, M.: Okapi at TREC 3. In: Overview of 3rd Text REtrieval Conference, pp. 109–126 (1995)
11. Shen, X., Tan, B., Zhai, C.: Context-sensitive information retrieval using implicit feedback. In: ACM SIGIR 2005, pp. 43–50 (2005)
12. Wetzker, R., Zimmermann, C., Bauckhage, C.: Analyzing social bookmarking systems: a del.icio.us cookbook. In: Proceedings of ECAI Mining Social Data Workshop (2008)
13. Xu, S., Bao, S., Fei, B., Su, Z., Yu, Y.: Exploring folksonomy for personalized search. In: ACM SIGIR 2008, pp. 155–162 (2008)

A Novel Word Embedding Based Stemming Approach for Microblog Retrieval During Disasters

Moumita Basu[1,2], Anurag Roy[1], Kripabandhu Ghosh[3(✉)],
Somprakash Bandyopadhyay[2], and Saptarshi Ghosh[1,4]

[1] Indian Institute of Engineering Science and Technology,
Shibpur, Howrah, India
[2] Indian Institute of Management, Calcutta, Kolkata, India
[3] Indian Institute of Technology, Kanpur, Kanpur, India
kripa.ghosh@gmail.com
[4] Indian Institute of Technology, Kharagpur, Kharagpur, India
saptarshi@cse.iitkgp.ernet.in

Abstract. IR methods are increasingly being applied over microblogs to extract real-time information, such as during disaster events. In such sites, most of the user-generated content is written informally – the same word is often spelled differently by different users, and words are shortened arbitrarily due to the length limitations on microblogs. Stemming is a common step for improving retrieval performance by unifying different morphological variants of a word. In this study, we show that rule-based stemming meant for formal text often cannot capture the arbitrary variations of words in microblogs. We propose a context-specific stemming algorithm, based on word embeddings, which can capture many more variations of words than what can be detected by conventional stemmers. Experiments on a large set of English microblogs posted during a recent disaster event shows that, the proposed stemming gives considerably better retrieval performance compared to Porter stemming.

Keywords: Microblog retrieval · Stemming · Disasters · Word embedding · word2vec

1 Introduction

In recent years, microblogging sites (e.g., Twitter, Weibo), have become important sources of information on various topics and events, and Information Retrieval (IR) on microblogs (tweets) is now an important area of research. In such forums, the user-generated content is often written in informal, casual ways. Additionally, due to the strict limitation on the length of microblogs (140 characters at most), words are often abbreviated arbitrarily, i.e., without obeying any linguistic rules. Such arbitrary variations of words negatively affect the performance of IR methods. This factor is especially crucial in situations such

© Springer International Publishing AG 2017
J.M. Jose et al. (Eds.): ECIR 2017, LNCS 10193, pp. 589–597, 2017.
DOI: 10.1007/978-3-319-56608-5_53

as a disaster event (earthquake, flood, etc.), when it is important to retrieve all relevant information irrespective of the word variations.

In such scenarios, IR methods usually rely on stemming algorithms (stemmers), whose purpose is to improve retrieval performance by mapping the morphological variants (usually inflectional) of a word to a common *stem*. Some stemmers are language-specific (e.g., the popular Porter stemmer [7] for English), utilizing the rules of a natural language to identify variants of a word; while there has been language-independent algorithms also [4,5]. However, all the prior contributions in this area have been on text written in a formal way, and stemming on such informal, noisy text (like microblogs) has not been studied well in literature.

Motivation: We motivate the need for new stemming algorithms for microblogs through a case study. We collected a large set of microblogs posted during a particular disaster event – the earthquake in Nepal and India in April 2015 (see Sect. 3 for details of the dataset). We observed different variations of many words in this collection of microblogs, some of which are shown in Table 1. We found two broad types of variations: (1) *Different spellings of a word:* Several words are spelled differently by different users; such words include both English words (like 'epicentre' and 'epicenter') as well as non-English words (like 'gurudwara' and 'gurdwara'). Also, proper nouns like names of places are often spelled differently (e.g., 'Sindhupalchowk', 'Sindhupalchok', and 'Sindupalchowk'), as shown in Table 1. (2) *Arbitrarily shortened forms of a word:* Due to the strict limitation on the length of microblogs, words are often shortened arbitrarily, e.g., 'building' shortened to 'bldg', 'medical' shortened to 'med', and so on (see Table 1). Such variations do not conform to rules of the English language, and hence cannot be identified by standard stemmers like the Porter stemmer.

In this work, we propose a context-specific stemming algorithm that can identify arbitrary morphological variations of words in a given collection of microblogs. We view this as a stemming problem because we assume that the variations of a word will share some common initial characters (a *common prefix*). However, we consider this common prefix to be very short, preferably shorter than 3 characters (which was advocated by Paik *et al.* [5] for formally written text).

Note that stemming algorithms that use common prefix length with word association [5] has been found to out-perform the ones using only common prefix length [2]. The role of *context* becomes particularly important in the case of microblogs where non-standard word representations are ubiquitous (as evident from Table 1). Here, only a combination of common prefix and context can possibly group semantically related variants. In this work, we use the word-embedding tool word2vec [3] to harness the context of word variants, in conjunction with common prefix length and other string similarity measures, to identify inflectional variants of words. We compare retrieval performance over differently stemmed versions of a collection of English microblogs posted during the Nepal earthquake – using Porter stemming, and our proposed stemming

Table 1. Examples of morphological variations of words, and tweets containing the variations (from a collection of tweets related to the Nepal-India earthquake in April 2015).

Variations	Excerpts from tweets
Variations in spellings of a word	
Epicentre	6.7 magnitude #earthquake **epicentre** 49 km from Banepa in #Nepal says USGS
Epicenter	5.0 earthquake, 29 km SSW of Kodari, Nepal. Apr 26 13:11 at **epicenter**
Gurudwara	Delhi Sikh **Gurudwara** committee will send 25k food packet everyday to Nepal
Gurdwara	Delhi Sikh **Gurdwara** body to send Langar (food) for Nepal earthquake hit people
Sindhupalchowk	#**Sindhupalchowk** 1100+deaths and 99% Houses are Down
Sindhupalchok	Indian national Azhar 23, missing. Last location **Sindhupalchok**. Plz help
Sindupalchowk	Food distribution in **sindupalchowk**, sufficient for 7 days for 500 victims
Dharhara	Earthquake in Nepal: 180 bodies retrieved from **Dharhara** tower debris
Dharahara	Historic **Dharahara** Tower in #Kathmandu, has collapsed #earthquake
Dharara	**Dharara** Tower, built in 1832, collapses in #Kathmandu during earthquake
Arbitrarily shortened forms of a word	
Building	Earthquake destroyed hospital, road, **building** in Kavre district of Nepal
Bldg	Nepal quake stresses importance of earthquake resistant **bldg** designs in entire NCR
Secretary	Foreign **Secretary** statement on #Nepal earthquake available here [url]
Secy	Foreign **Secy** and Defence **Secy** giving latest updates on earthquake relief [url]
Medical	India: NDRF personnel, crack **medical** team with relief rushed to #Nepal
Med	4 planes to leave for #Nepal tmrw carry **meds**, **med** team, 30-Bed Hospital
IndianAirForce	#**IndianAirForce**/Army already helping with relief, food, medicines, and all calls to Nepal subsided
IAF	Drinking water plus emergency relief supplies headed to #Nepal by #**IAF** aircraft AP Photo [url]

– and demonstrate that our proposed stemming algorithm yields statistically significantly better retrieval performance for the same queries over microblogs.

2   Proposed Stemming Algorithm for Microblogs

This section describes our proposed stemming algorithm for microblogs. For a given collection of microblogs, let $\mathbb{L}$ be the lexicon (i.e., the set of all words case-folded to lower case) of the collection, excluding the stopwords, URLs, user-mentions, email-ids, and other non alpha-numeric words. We first describe how we identify 'similar' words, and then describe the stemming algorithm.

2.1   Measuring Word Similarity

To judge if two words w, $w^* \in \mathbb{L}$ are similar (i.e., likely variants of one another), we consider two basic types of similarity between the words, as follows.

(1) String similarity: This is checked in two steps. First, we check if the two words have a common prefix of length p (p is a positive integer). We consider p to be very short for informal, noisy microblogs, specifically $p \le 3$, since a common prefix of length 3 was advocated in [5] for *formally written* text. Second, we calculate the length of the Longest Common Subsequence [1] of the two words, denoted by $LCS_{length}(w, w^*)$.

(2) Contextual similarity: We trained word2vec [3] over the set of tweets.[1] The word2vec model gives a vector for each term in the corpus, which we refer to as the *term-vector*. Let $\overrightarrow{w}$ and $\overrightarrow{w^*}$ be the word2vec *term-vectors* of w and w^* respectively. The term-vector is expected to capture the context in which a word is used in the corpus [3]. Hence, contextual similarity of the two words is quantified as the cosine similarity of the corresponding word2vec term-vectors, denoted as $cos\_sim(\overrightarrow{w}, \overrightarrow{w^*})$.

Thus, we consider w^* to be a likely variant of w only if they have sufficient string similarity, as well as they have been used in a similar context in the corpus.

2.2   Proposed Stemming Algorithm

Our proposed stemming algorithm has the following two phases.

Phase 1: Identifying Possible Variants of Words: This phase is aimed at identifying the possible variants of words on the basis of *string similarity*, as described above. For each word $w \in \mathbb{L}$, we construct a set L_w that contains

---

[1] The Gensim implementation for word2vec was used – https://radimrehurek.com/gensim/models/word2vec.html. The continuous bag of words model is used for the training, along with Hierarchical softmax, with the following parameter values – Vector size: 2000, Context size: 5, Learning rate: 0.05.

all the words $w^* \in \mathbb{L}$ satisfying the following three conditions: (1) w^* has the same common prefix of length p as w, where $p \leq 3$. (2) $|w^*| \leq |w|$, i.e., w^* is of length less than or equal to length of w. We consider this condition because w^* is supposed to be a stem of w and conventionally a stem is smaller or equal in length as the original word. (3) The length of the Longest Common Subsequence of characters between w and w^*, $LCS_{length}(w, w^*) \geq \alpha|w^*|$, where $\alpha \in [0,1]$ is a parameter of the algorithm. This condition ensures that the variants of w have a common subsequence of at least a certain length with w. Thus, L_w contains the possible variants of w.

Phase 2: Identifying the Stem: In this phase, we look to filter out those variants of w from the set L_w which have high *contextual similarity* with w. We define the *Stemming Score* ($Stem_{score}$) between w and $w^* \in L_w$ as follows:

$$Stem_{score}(w, w^*) = \beta * cos\_sim(\overrightarrow{w}, \overrightarrow{w^*}) + (1 - \beta) * LCS_{length}(w, w^*) \qquad (1)$$

where, $\beta \in [0,1]$ is another parameter of the algorithm. Note that $Stem_{score}(w, w^*)$ is a measure of both the contextual similarity and string similarity of w and w^*. We choose only those $w^* \in L_w$ as the candidate stems for w for which $Stem_{score}(w, w^*) \geq \gamma$, where $\gamma \in [0,1]$ is another algorithmic parameter.

We construct a set L_w^s of candidate stems of w, comprising of only those words from L_w which satisfy the above condition ($L_w^s \subseteq L_w$). In case there are multiple words in L_w^s, the word in L_w^s with the *minimum length* is chosen as the stem for the set $\{w\} \cup L_w^s$ (in case of ties, we break ties arbitrarily).

Parameters of the Algorithm: The proposed stemming algorithm has four parameters – (i) p, the length of the common prefix ($p \leq 3$), (ii) α, a threshold on the string similarity, (iii) β, which decides the relative importance between string similarity and contextual similarity, and (iv) γ, the final threshold for considering a word as a candidate stem of another. We considered these parameters in order to make the algorithm generalizable for different types of text. The parameters can be decided based on factors such as, how noisy the text is, and how aggressively one wants to identify variants of a word.

Sample Output of the Algorithm: The algorithm identifies *groups of similar words which are stemmed to a common stem*, some examples of which are shown in Table 2. We see that the algorithm correctly identifies different types of word variations, including variations made following rules of English (e.g., 'donating', 'donated', 'donates' all stemmed to 'donate'), variations in spelling (e.g., 'gurudwara' and 'gurdwara', 'organisations' and 'organizations'), and arbitrarily shortened forms of words (e.g., 'organisations' and 'orgs', 'medicines' and 'meds', etc.). Evidently, standard stemmers will not be able to identify many of these variants.

Table 2. Examples of groups of words which were stemmed to a common stem, by the proposed stemming algorithm.

Group of words stemmed to a common stem	Stem
contribute, contributed, contribution, contributions	contribute
donating, donate, donated, donates, donation, donations	donate
collapse, collapsing, collapses, collapsed	collapse
gurudwaras, gurudwara, gurdwaras, gurdwara	gurdwara
organisations, organizations, organisation, organization, orgs, org	org
medical, medicine, medicines, medics, meds, med	med

3   Experiments and Results

We now apply the proposed stemming algorithm over a collection of microblogs, and report retrieval performance.

Microblog Collection: We consider a large collection of about 100K English tweets posted during a recent disaster event – the Nepal-India earthquake in April 2015.[2] After removing duplicate tweets, using a simplified version of the methodologies in [10], we obtained a set of 50,068 tweets. We pre-processed this set of tweets by removing a standard set of English stopwords, URLs, user-mentions, punctuation symbols, and case-folding. All experiments were performed over this pre-processed set of tweets.

Queries and Gold Standard Results: We consulted members of an NGO (Doctors For You) who participated in relief operations during the Nepal earthquake, to know the information requirements during the operations. They suggested some information needs – like resource needs (e.g., medicines, tents, medical teams), damage caused (e.g., damaged houses), the situation at specific locations, etc. – based on which 15 queries were formed as shown in Table 3.

To develop the gold standard, three human annotators were employed to find out the relevant tweets for each query. The annotators were asked to start with the query, observe the matching tweets to identify various variations of the terms in the query, and then retrieve tweets with the variations (if any) as well.

Parameter Setting for the Present Study: As stated earlier, we consider $p \leq 3$ for noisy microblogs. We experimented with $p = 1$ and $p = 2$. Evidently, more variants of words can be identified for $p = 1$; e.g., variants like 'Indian Air-Force' and 'IAF', 'building' and 'bldg' can only be identified for $p = 1$. However, such a low value of p also has the risk of identifying false positives. Hence, in this study, we use $p = 2$ for the experiments.

---

[2] https://en.wikipedia.org/wiki/April_2015_Nepal_earthquake.

Table 3. Comparison of the proposed method with Porter stemmer. Retrieval performance reported on three versions of the microblog collection – *unstemmed*, *Porter stemmed*, and stemmed using the proposed approach. The proposed approach significantly outperforms Porter stemmer ($p < 0.05$) in both the measures. Percentage improvements over *unstemmed* **and** *Porter stemmed* are also shown.

Query	Average precision			Recall@1000		
	Unstemmed	Porter	Proposed	Unstemmed	Porter	Proposed
Food send	0.1251	0.2356	**0.2542**	0.6214	**0.9660**	0.9563
Food packet distributed	0.1930	0.2283	**0.2645**	0.9515	**0.8835**	0.8350
House damage collapse	0.0065	0.0254	**0.0296**	0.2264	0.5283	**0.6226**
Medicine need	0.2029	**0.3528**	0.1390	0.4561	0.6140	**0.9298**
Tent need	0.1110	**0.5962**	0.5718	0.5195	0.9870	**1.0000**
Medicine medical send	0.1806	0.2851	**0.3775**	0.8333	**0.9808**	0.9744
Sindhupalchok	0.4457	0.4457	**0.9493**	0.4457	0.4457	**0.9620**
Medical treatment	**0.8003**	0.7998	0.7417	0.8471	0.8471	**1.0000**
Medical team send	0.5506	0.7358	**0.7548**	0.9290	0.9484	**0.9935**
NDRF operation	0.7337	0.9006	**0.9065**	0.9653	0.9653	**0.9722**
Rescue relief operation	0.5342	0.7205	**0.7440**	0.5846	0.8338	**0.9154**
Relief organization	0.2405	0.3015	**0.3293**	0.3448	**0.5460**	0.4598
Dharahara collapse	0.2659	0.6424	**0.9599**	0.7692	0.7692	**0.9780**
Epicentre	0.3613	0.3612	**0.9847**	0.3621	0.3642	**0.9853**
Gurudwara meal	0.2067	0.6116	**0.8429**	0.2671	0.7671	**0.9795**
All	0.3305	0.4828	**0.5900** (+78.5%, +22.2%)	0.6082	0.7631	**0.9042** (+48.7%, 18.5%)

The values of (α, β, γ) were determined using grid search over $0.5, 0.6, \ldots, 1.00 \times 0.5, 0.6, \ldots, 1.00 \times 0.5, 0.6, \ldots, 1.00$. The best performance values for our dataset were obtained for $\alpha = 0.6, \beta = 0.7, \gamma = 0.7$, which are being reported.

Baselines: Since we focus on English tweets, we compared our proposed approach with the Porter stemmer [7], perhaps the mostly widely used English language-specific stemming algorithm. We also compare with an unstemmed retrieval performance.

Evaluation Measures: We report the retrieval performance in terms of Average Precision (AP) and Recall@1000, for the queries as selected above.

Evaluation Results: We used the well-known Indri system [9] which uses a Language Modelling framework [6] for retrieval. Table 3 shows the query-wise retrieval performance over three versions of the microblog collection – unstemmed, Porter stemmed, and stemmed using the proposed approach.

Both Porter stemmer and the proposed stemmer enable better retrieval, than without stemming. Importantly, the proposed stemming approach gives statistically significantly better performance at 95% confidence level (p-value < 0.05) by Wilcoxon signed-rank test [8], than Porter stemming, for both the measures (22% better in terms of MAP, and 18.5% better in terms of Recall@1000, compared to Porter stemmer).

Evidently, the superior performance of the proposed methodology is because the proposed approach is able to identify arbitrary spelling variations which Porter stemmer (based on English language rules) could not identify.

4   Conclusion

We demonstrate that traditional rule-based stemming fails to identify many informal variations of words in microblogs. We also propose a novel context-specific stemming algorithm for microblogs, which takes into account both string similarity and contextual similarity among words. Through experiments on a collection of English microblogs posted during a disaster event, we demonstrate that the proposed stemming algorithm yields much better retrieval performance over the commonly used Porter stemmer.

Acknowledgement. This research was partially supported by a grant from the Information Technology Research Academy (ITRA), DeITY, Government of India (Ref. No.: ITRA/15 (58)/Mobile/DISARM/05).

References

1. Cormen, T., Leiserson, C., Rivest, R., Stein, C.: Introduction to Algorithms, 3rd edn. The MIT Press, Cambridge (2009)
2. Majumder, P., Mitra, M., Parui, S.K., Kole, G., Mitra, P., Datta, K.: Yass: yet another suffix stripper. ACM Trans. Inf. Syst. **25**(4), 18 (2007)
3. Mikolov, T., Yih, W., Zweig, G.: Linguistic regularities in continuous space word representations. In: NAACL HLT 2013 (2013)

4. Paik, J.H., Mitra, M., Parui, S.K., Järvelin, K.: Gras: an effective and efficient stemming algorithm for information retrieval. ACM Trans. Inf. Syst. **19:29**(4), 1–19:24 (2011)
5. Paik, J.H., Pal, D., Parui, S.K.: A novel corpus-based stemming algorithm using co-occurrence statistics. In: Proceedings of ACM SIGIR, pp. 863–872 (2011)
6. Ponte, J.M., Croft, W.B.: A language modeling approach to information retrieval. In: Proceedings of ACM SIGIR, pp. 275–281 (1998)
7. Porter, M.: An algorithm for suffix stripping. Program **14**(3), 130–137 (1980)
8. Siegel, S.: Nonparametric Statistics for the Behavioral Sciences. McGraw-Hill Series in Psychology. McGraw-Hill, New York (1956)
9. Strohman, T., Metzler, D., Turtle, H., Croft, W.B.: Indri: a language model-based search engine for complex queries. In: Proceedings of ICIA (2004). http://www.lemurproject.org/indri/
10. Tao, K., Abel, F., Hauff, C., Houben, G.J., Gadiraju, U.: Groundhog day: near-duplicate detection on Twitter. In: Proceedings of World Wide Web (WWW) (2013)

Search Personalization with Embeddings

Thanh Vu[1], Dat Quoc Nguyen[2(✉)], Mark Johnson[2], Dawei Song[1,3],
and Alistair Willis[1]

[1] The Open University, Milton Keynes, UK
{thanh.vu,dawei.song,alistair.willis}@open.ac.uk
[2] Department of Computing, Macquarie University, Sydney, Australia
dat.nguyen@students.mq.edu.au, mark.johnson@mq.edu.au
[3] Tianjin University, Tianjin, People's Republic of China

Abstract. Recent research has shown that the performance of search personalization depends on the richness of user profiles which normally represent the user's topical interests. In this paper, we propose a new embedding approach to learning user profiles, where users are embedded on a topical interest space. We then directly utilize the user profiles for search personalization. Experiments on query logs from a major commercial web search engine demonstrate that our embedding approach improves the performance of the search engine and also achieves better search performance than other strong baselines.

1 Introduction

Users' personal data, such as a user's historic interaction with the search engine (e.g., submitted queries, clicked documents), have been shown useful to personalize search results to the users' information need [1,15]. Crucial to effective search personalization is the construction of user profiles to represent individual users' interests [1,3,6,7,12]. A common approach is to use main topics discussed in the user's clicked documents [1,6,12,15], which can be obtained by using a human generated ontology [1,15] or using a topic modeling technique [6,12].

However, using the user profile to directly personalize a search has been not very successful with a *minor* improvement [6,12] or even *deteriorate* the search performance [5]. The reason is that each user profile is normally built using only the user's relevant documents (e.g., clicked documents), ignoring user interest-dependent information related to input queries. Alternatively, the user profile is utilized as a feature of a multi-feature learning-to-rank (L2R) framework [1, 13–15]. In this case, apart from the user profile, dozens of other features has been proposed as the input of an L2R algorithm [1]. Despite being successful in improving search quality, the contribution of the user profile is not very clear.

To handle these problems, in this paper, we propose a new *embedding* approach to constructing a user profile, using both the user's input queries and relevant documents. We represent each user profile using two projection matrices and a user embedding. The two projection matrices is to identify the user interest-dependent aspects of input queries and relevant documents while the

© Springer International Publishing AG 2017
J.M. Jose et al. (Eds.): ECIR 2017, LNCS 10193, pp. 598–604, 2017.
DOI: 10.1007/978-3-319-56608-5_54

user embedding is to capture the relationship between the queries and documents in this user interest-dependent subspace. We then *directly* utilize the user profile to re-rank the search results returned by a commercial search engine. Experiments on the query logs of a commercial web search engine demonstrate that modeling user profile with embeddings helps to significantly improve the performance of the search engine and also achieve better results than other comparative baselines [1,11,14] do.

2   Our Approach

We start with our new embedding approach to building user profiles in Sect. 2.1, using pre-learned document embeddings and query embeddings. We then detail the processes of using an unsupervised topic model (i.e., Latent Dirichlet Allocation (LDA) [2]) to learn document embeddings and query embeddings in Sects. 2.2 and 2.3, respectively. We finally use the user profiles to personalize the search results returned by a commercial search engine in Sect. 2.4.

2.1   Building User Profiles with Embeddings

Let $\mathcal{Q}$ denote the set of queries, $\mathcal{U}$ be the set of users, and $\mathcal{D}$ be the set of documents. Let (q, u, d) represent a triple (query, user, document). The query $q \in \mathcal{Q}$, user $u \in \mathcal{U}$ and document $d \in \mathcal{D}$ are represented by vector embeddings $\boldsymbol{v}_q$, $\boldsymbol{v}_u$ and $\boldsymbol{v}_d \in \mathbb{R}^k$, respectively.

Our goal is to select a *score function* f such that the implausibility value $f(q, u, d)$ of a correct triple (q, u, d) (i.e. d is a relevant document of u given q) is *smaller* than the implausibility value $f(q', u', d')$ of an incorrect triple (q', u', d') (i.e. d' is not a relevant document of u' given q'). Inspired by embedding models of entities and relationships in knowledge bases [9,10], the score function f is defined as follows:

$$f(q, u, d) = \|\mathbf{W}_{u,1}\boldsymbol{v}_q + \boldsymbol{v}_u - \mathbf{W}_{u,2}\boldsymbol{v}_d\|_{\ell_{1/2}} \tag{1}$$

here we represent the profile for the user u by two matrices $\mathbf{W}_{u,1}$ and $\mathbf{W}_{u,2} \in \mathbb{R}^{k \times k}$ and a vector embedding $\boldsymbol{v}_u$, which represents the user's topical interests. Specifically, we use the interest-specific matrices $\mathbf{W}_{u,1}$ and $\mathbf{W}_{u,2}$ to identify the interest-dependent aspects of both query q and document d, and use vector $\boldsymbol{v}_u$ to describe the relationship between q and d in this interest-dependent subspace.

In this paper, $\boldsymbol{v}_d$ and $\boldsymbol{v}_q$ are pre-determined by employing the LDA topic model [2], which are detailed in next Sects. 2.2 and 2.3. Our model parameters are only the user embeddings $\boldsymbol{v}_u$ and matrices $\mathbf{W}_{u,1}$ and $\mathbf{W}_{u,2}$. To learn these user embeddings and matrices, we minimize the margin-based objective function:

$$\mathcal{L} = \sum_{\substack{(q,u,d)\in\mathcal{G} \\ (q',u,d')\in\mathcal{G}'_{(q,u,d)}}} \max\left(0, \gamma + f(q, u, d) - f(q', u, d')\right) \tag{2}$$

where γ is the margin hyper-parameter, $\mathcal{G}$ is the training set that contains only correct triples, and $\mathcal{G}'_{(q,u,d)}$ is the set of incorrect triples generated by corrupting the correct triple (q, u, d) (i.e. replacing the relevant document/query d/q in (q, u, d) by irrelevant documents/queries d'/q'). We use Stochastic Gradient Descent (SGD) to minimize $\mathcal{L}$, and impose the following constraints during training: $\|v_u\|_2 \leqslant 1$, $\|\mathbf{W}_{u,1}v_q\|_2 \leqslant 1$ and $\|\mathbf{W}_{u,2}v_d\|_2 \leqslant 1$. First, we initialize user matrices as identity matrices and then fix them to only learn the randomly initialized user embeddings. Then in the next step, we fine-tune the user embeddings and user matrices together. In all experiments shown in Sect. 3, we train for 200 epochs during each two optimization step.

2.2 Using LDA to Learn Document Embeddings

In this paper, we model document embeddings by using topics extracted from relevant documents. We use LDA [2] to *automatically* learn k topics from the relevant document collection. After training an LDA model to calculate the probability distribution over topics for each document, we use the topic proportion vector of each document as its document embedding. Specifically, the zth element ($z = 1, 2, ..., k$) of the vector embedding for document d is: $v_{d,z} = \mathrm{P}(z \mid d)$ where $\mathrm{P}(z \mid d)$ is the probability of the topic z given the document d.

2.3 Modeling Search Queries with Embeddings

We also represent each query as a probability distribution v_q over topics, i.e. the zth element of the vector embedding for query q is defined as: $v_{q,z} = \mathrm{P}(z \mid q)$ where $\mathrm{P}(z \mid q)$ is the probability of the topic z given the query q. Following [1,14], we define $\mathrm{P}(z \mid q)$ as a mixture of LDA topic probabilities of z given documents related to q. Let $\mathcal{D}_q = \{d_1, d_2, ..., d_n\}$ be the set of top n ranked documents returned for a query q (in the experiments we select $n = 10$). We define $\mathrm{P}(z \mid q)$ as follows:

$$\mathrm{P}(z \mid q) = \sum_{i=1}^{n} \lambda_i \mathrm{P}(z \mid d_i) \tag{3}$$

where $\lambda_i = \dfrac{\delta^{i-1}}{\sum_{j=1}^{n} \delta^{j-1}}$ is the exponential decay function of i which is the rank of d_i in D_q. And δ is the decay hyper-parameter ($0 < \delta < 1$). The decay function is to specify the fact that a higher ranked document is more relevant to user in term of the lexical matching (i.e. we set the larger mixture weights to higher ranked documents).

2.4 Personalizing Search Results

We utilize the user profiles (i.e., the learned user embeddings and matrices) to re-rank the original list of documents produced by a commercial search engine as follows: (1) We download the top n ranked documents given the input query

q. We denote a downloaded document as d. (2) For each document d we apply the trained LDA model to infer the topic distribution v_d. We then model the query q as a topic distribution v_q as in Sect. 2.3. (3) For each triple (q, u, d), we calculate the implausibility value $f(q, u, d)$ as defined in Eq. 1. We then sort the values in the ascending order to achieve a new ranked list.

3   Experimental Methodology

Dataset: We evaluate our new approach using the search results returned by a commercial search engine. We use a dataset of query logs of 106 anonymous users in 15 days from 01 July 2012 to 15 July 2012. A log entity contains a user identifier, a query, top-10 URLs ranked by the search engine, and clicked URLs along with the user's dwell time. We also download the content documents of these URLs for training LDA [2] to learn document and query embeddings (Sects. 2.2 and 2.3).

Bennett et al. [1] indicate that short-term (i.e. session) profiles achieved better search performance than the longer-term profiles. Short-term profiles are usually constructed using the user's search interactions within a search session and used to personalize the search within the session [1]. To identify a search session, we use 30 min of user inactivity to demarcate the session boundary. In our experiments, we build short-term profiles and utilize the profiles to personalize the returned results. Specifically, we uniformly separate the last log entries within search sessions into a *test set* and a *validation set*. The remainder of log entities within search sessions are used for *training* (e.g. to learn user embeddings and matrices in our approach).

Evaluation Methodology: We use the SAT criteria detailed in [4] to identify whether a clicked URL is relevant from the query logs (i.e., a SAT click). That is either a click with a dwell time of at least 30 s or the last result click in a search session. We assign a positive (relevant) label to a returned URL if it is a SAT click. The remainder of the top-10 URLs is assigned negative (irrelevant) labels. We use the rank positions of the positive labeled URLs as the ground truth to evaluate the search performance before and after re-ranking. We also apply a simple pre-processing on these datasets as follows. At first, we remove the queries whose positive label set is empty from the dataset. After that, we discard the domain-related queries (e.g. Facebook, Youtube). To this end, the training set consists of 5,658 correct triples. The test and validation sets contain 1,210 and 1,184 correct triples, respectively. Table 1 presents the dataset statistics after pre-processing.

Table 1. Basic statistics of the dataset after pre-processing

#days	#users	#distinct queries	#SAT clicks	#sessions	#distinct documents
15	106	6,632	8,052	2,394	33,591

Evaluation Metrics: We use two standard evaluation metrics in document ranking [1,8]: *mean reciprocal rank* (**MRR**) and *precision* (**P@1**). For each metric, the higher value indicates the better ranking performance.

Baselines: We employ three comparative baselines with the same experimental setup: (1) **SE**: The original rank from the search engine (2) **CI**: We promote returned documents previously clicked by the user. This baseline is similar to the personalized navigation method in Teevan *et al.* [11]. (3) **SP**: The search personalization method using the short-term profile [1,14]. These are very comparative baselines given that they start with the ranking provided by the major search engine and add other signals (e.g., clicked documents) to get a better ranking performance [1,11].

Hyper-Parameter Tuning: We perform a grid search to select optimal hyper-parameters on the validation set. We train the LDA model[1] using only the relevant documents (i.e., SAT clicks) extracted from the query logs, with the number of topics (i.e. the number of vector dimensions) $k \in \{50, 100, 200\}$. We then apply the trained LDA model to infer document embeddings and query embeddings for all documents and queries. We then choose either the ℓ_1 or ℓ_2 norm in the score function f, and select SGD learning rate $\eta \in \{0.001, 0.005, 0.01\}$, the margin hyper-parameter $\gamma \in \{1, 3, 5\}$ and the decay hyper-parameter $\delta \in \{0.7, 0.8, 0.9\}$. The highest MRR on the validation set is obtained when using $k = 200$, ℓ_1 in f, $\eta = 0.005$, $\gamma = 5$, and $\delta = 0.8$.

4   Experimental Results

Table 2 shows the performances of the baselines and our proposed method. Using the previously clicked documents *CI* helps to significantly improve the search performance ($p < 0.05$ with the *paired t-test*) with the relative improvements of about $7^+\%$ in both MRR and P@1 metrics. With the use of short-term profiles as a feature of a learning-to-rank framework, *SP* [1,14] improves the MRR score over the original rank significantly ($p < 0.01$) and achieves a better performance than *CI*'s.

Table 2. Overall performances of the methods in the test set. **Our method**$_{-W}$ denotes the simplified version of our method. The subscripts denote the relative improvement over the baseline *SE*.

Metric	SE	CI [11]	SP [1,14]	Our method	Our method$_{-W}$
MRR	0.559	$0.597_{+6.9\%}$	$0.631_{+12.9\%}$	$\mathbf{0.656}_{+17.3\%}$	$0.645_{+15.4\%}$
P@1	0.385	$0.416_{+8.1\%}$	$0.452_{+17.4\%}$	$\mathbf{0.501}_{+30.3\%}$	$0.481_{+24.9\%}$

By directly learning user profiles and applying them to re-rank the search results, our embedding approach achieves the highest performance of search

---

[1] We use the LDA implementation in Mallet toolkit: http://mallet.cs.umass.edu/.

personalization. Specifically, our MRR score is significantly ($p < 0.05$) higher than that of *SP* (with the relative improvement of 4% over SP). Likewise, the P@1 score obtained by our approach is significantly higher than that of the baseline *SP* ($p < 0.01$) with the relative improvement of 11%.

In Table 2, we also present the performances of a simplified version of our embedding approach where we fix the user matrices as identity matrices and then only learn the user embeddings. Table 2 shows that our simplified version achieves second highest scores compared to all others.[2] Specifically, our simplified version obtains significantly higher P@1 score (with $p < 0.05$) than *SP*.

5   Conclusions

In this paper, we propose a new embedding approach to building user profiles. We model each user profile using a user embedding together with two user matrices. The user embedding and matrices are then learned using LDA-based vector embeddings of the user's relevant documents and submitted queries. Applying it to web search, we use the profile to re-rank search results returned by a commercial web search engine. Our experimental results show that the proposed method can stably and significantly improve the ranking quality.

Acknowledgments. The first two authors contributed equally to this work. Dat Quoc Nguyen is supported by an Australian Government Research Training Program Scholarship and a NICTA NRPA Top-Up Scholarship.

References

1. Bennett, P.N., White, R.W., Chu, W., Dumais, S.T., Bailey, P., Borisyuk, F., Cui, X.: Modeling the impact of short-and long-term behavior on search personalization. In: SIGIR (2012)
2. Blei, D.M., Ng, A.Y., Jordan, M.I.: Latent dirichlet allocation. J. Mach. Learn. Res. **3**, 993–1022 (2003)
3. Cheng, Z., Jialie, S., Hoi, S.C.: On effective personalized music retrieval by exploring online user behaviors. In: SIGIR (2016)
4. Fox, S., Karnawat, K., Mydland, M., Dumais, S., White, T.: Evaluating implicit measures to improve web search. ACM Trans. Inf. Syst. **23**(2), 147–168 (2005)
5. Harvey, M., Carman, M.J., Ruthven, I., Crestani, F.: Bayesian latent variable models for collaborative item rating prediction. In: CIKM (2011)
6. Harvey, M., Crestani, F., Carman, M.J.: Building user profiles from topic models for personalised search. In: CIKM (2013)
7. Liu, X.: Modeling users' dynamic preference for personalized recommendation. In: IJCAI (2015)
8. Manning, C.D., Raghavan, P., Schütze, H.: Introduction to Information Retrieval. Cambridge University Press, New York (2008)

---

[2] Our approach obtains significantly higher P@1 score ($p < 0.05$) than our simplified version with 4% relative improvement.

9. Nguyen, D.Q., Sirts, K., Qu, L., Johnson, M.: Neighborhood mixture model for knowledge base completion. In: CoNLL (2016)
10. Nguyen, D.Q., Sirts, K., Qu, L., Johnson, M.: STransE: a novel embedding model of entities and relationships in knowledge bases. In: NAACL-HLT (2016)
11. Teevan, J., Liebling, D.J., Ravichandran Geetha, G.: Understanding and predicting personal navigation. In: WSDM (2011)
12. Vu, T., Song, D., Willis, A., Tran, S.N., Li, J.: Improving search personalisation with dynamic group formation. In: SIGIR (2014)
13. Vu, T., Willis, A., Kruschwitz, U., Song, D.: Personalised query suggestion for intranet search with temporal user profiling. In: CHIIR (2017)
14. Vu, T., Willis, A., Tran, S.N., Song, D.: Temporal latent topic user profiles for search personalisation. In: Hanbury, A., Kazai, G., Rauber, A., Fuhr, N. (eds.) ECIR 2015. LNCS, vol. 9022, pp. 605–616. Springer, Heidelberg (2015). doi:10.1007/978-3-319-16354-3_67
15. White, R.W., Chu, W., Hassan, A., He, X., Song, Y., Wang, H.: Enhancing personalized search by mining and modeling task behavior. In: WWW (2013)

Do Easy Topics Predict Effectiveness Better Than Difficult Topics?

Kevin Roitero, Eddy Maddalena, and Stefano Mizzaro[✉]

Department of Mathematics, Computer Science, and Physics, University of Udine, Udine, Italy
roitero.kevin@spes.uniud.it, {eddy.maddalena,mizzaro}@uniud.it

Abstract. After a network-based analysis of TREC results, Mizzaro and Robertson [4] found the rather unpleasant result that topic ease (i.e., the average effectiveness of the participating systems, measured with average precision) correlates with the ability of topics to predict system effectiveness (defined as topic hubness). We address this issue by: (i) performing a more detailed analysis, and (ii) using three different datasets. Our results are threefold. First, we confirm that the original result is indeed correct and general across datasets. Second, we show that, however, that result is less worrying than what might seem at first glance, since it depends on considering the least effective systems in the analysis. In other terms, easy topics discriminate most and least effective systems, but when focussing on the most effective systems only this is no longer true. Third, we also clarify what happens when using the GMAP metric.

1 Introduction

Effectiveness evaluation of Information Retrieval (IR) systems is performed within many initiatives, such as TREC, NTCIR, INEX, CLEF, and others. Participants to these initiatives can test their own retrieval system over a set of *topics*, which are a representation of information needs. The effectiveness of each system is assessed considering various metrics such as Average Precision (AP) and Mean AP (MAP), that determine the final rank of systems. Interactions between the topics and the systems, and in particular between the difficulty of the topic and the final rank of the systems have been studied by Mizzaro and Robertson [4] considering link analysis techniques, and in particular the HITS algorithm [2]. More in detail, Mizzaro and Robertson investigate the correlation between topic ease and the ability to predict system effectiveness. They find that easier topics are better at estimating system effectiveness. In other terms, to be effective in TREC a system has to perform well on easy topics. This is undesirable since it is the difficult topics that are more interesting and it is by working on them that the state of the art of the discipline can advance most. In this paper we extend their analysis.

© Springer International Publishing AG 2017
J.M. Jose et al. (Eds.): ECIR 2017, LNCS 10193, pp. 605–611, 2017.
DOI: 10.1007/978-3-319-56608-5_55

2   Background

In an attempt to make this paper self contained, in this section we summarize the methodology and the relevant results of Mizzaro and Robertson [4] (for further details see the original paper). The output of the TREC competition can be represented as a table having as column the topics and as rows the systems (Fig. 1(b)). Each cell contains an effectiveness measure of each system on each topic. Effectiveness is measured according to some metric, and Average Precision (AP) is a common choice. In order to provide the final rank of systems, each row (i.e., in TREC terms, a *run*) is averaged to compute the MAP metric. Mizzaro and Robertson [4] introduce a dual metric, named Average AP (AAP), that is computed averaging over each column and represents the average value over the systems of the APs values for each topic: while MAP measures system effectiveness, AAP measures topic ease. The topic-system matrix of Fig. 1(b) is then normalized by transforming each AP value into $\overline{AP}_A(s_i, t_j)$ (Normalized AP according to AAP) and $\overline{AP}_M(s_i, t_j)$ (Normalized AP according to MAP):

$$\overline{AP}_A(s_i, t_j) = AP(s_i, t_j) - AAP(t_j) \text{ and}$$

$$\overline{AP}_M(s_i, t_j) = AP(s_i, t_j) - MAP(s_i).$$

The two matrices obtained from the normalization process (one obtained considering $\overline{AP}_A$ and one from $\overline{AP}_M$) are used to study interactions between systems and topics; this step is accomplished by building an adjacency matrix, and, consequently, the corresponding graph, made of the two normalized matrices; this process is summarized in Fig. 1(a). Each link between the system and the topic (see Fig. 1(c)) represents [4]:

- arc $s \rightarrow t$ with weight $\overline{AP}_M$: how much the system s "thinks" that the topic t is easy (or "un-easy" if the weight is negative);
- arc $s \leftarrow t$ with weight $\overline{AP}_A$: how much the topic t "thinks" that the system s is effective (or "un-effective" if the weight is negative).

The graph is used to compute *hubness* and *authority*, obtained using an extended version of the HITS algorithm [2] which allows to include the negative values for the arcs. The authority of a topic measures topic ease, while the authority of a system measures system effectiveness [4]. The hubness of a topic measures the topic capability to recognize effective systems, while the hubness of a system measures its ability to recognize easy topics [4]. When focussing on the values of AAP and topic hubness, as we do in our paper, Mizzaro and Robertson [4, pp. 483–484] state:

> "[...] easier topics are better at estimating system effectiveness. [the statement] is a bit worrying. It means that system effectiveness in TREC is affected more by easy topics than by difficult topics, which is rather undesirable for quite obvious reasons: a system capable of performing well on a difficult topic, i.e., on a topic on which the other systems perform badly,

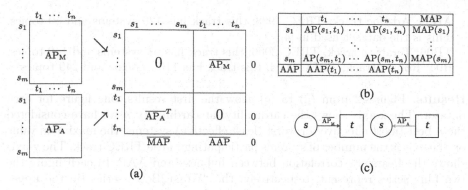

Fig. 1. (a) Construction of the adjacency matrix. $\overline{AP_A}^T$ is the transpose of $\overline{AP_A}$. (b) AP, MAP and AAP. (c) The relationships between systems and topics (from [4]).

would be an important result for IR effectiveness; conversely, a system capable of performing well on easy topics is just a confirmation of the state of the art."

This statement is obtained when commenting Fig. 5(d) in [4] (we present a slightly modified version of that figure in Fig. 2(g), analysed in more detail the following). It is also noted that the correlation between AAP and hubness disappears when using GMAP (Geometric MAP [5]) in place of MAP (and GAAP in place of AAP) [4, p. 484]:

"with GMAP[...] and GAAP [...] the correlation with hubness largely disappears"

3 Experiments

Aims and Settings. In this paper we further study the above results, analysing whether they hold if different subsets of systems are considered. We perform the same analyses, but we also repeat them using a subset of systems: we rank the systems according to their effectiveness (measured using MAP) and we select either the most or the least effective ones. More in detail, our procedure can be described as:

```
for cardinality n in range 1 to number of systems:
    order the systems according to MAP;
    select the first/last n systems;
    build the adjacency matrix;
    compute hubness (and authority) using HITS;
    compute Pearson's correlation between hubness and AAP;
```

Also, [4] used a single dataset; to study the generality of the results, in our experiments we consider the following three datasets:

– AH99: Ad-Hoc track, TREC 1998; this track has 129 systems and 50 topics. This is the dataset used in [4].
– TB06: Terabyte track, TREC 2006; this track has 61 systems and 149 topics.
– R04: Robust track, TREC 2004; this track has 110 systems and 249 topics.

Results. Figures 2 from (a) to (c) show the first results, one figure for each dataset. The x-axis shows the cardinality: at cardinality n we have considered the n best (most effective) or worst (least effective) systems; the maximum value on the axis is the number of systems participating in the TREC track. The y-axis shows the Pearson's correlation between hubness and AAP. In each figure the two blue series represent, respectively: the "Worst2Best" series (in the upper part of each figure) shows the correlation values when considering the systems ranked in ascending order of MAP (i.e., from least effective systems to most effective ones), and the "Best2Worst" series shows the correlation values when considering the systems ranked in descending order of MAP (i.e., from most effective systems to least effective ones). The red series are similar, with GMAP and GAAP in place of MAP and AAP.

Let us focus first on the top-right of the charts, i.e., on the maximum cardinality, where all the systems are considered (and the two series, of course, meet). That is the correlation value studied by Mizzaro and Robertson [4], and it is indeed high (a bit smaller for R04), substantially confirming previous results. The figures show something more, though, and in a consistent way over the three datasets. When considering the "Worst2Best" series (i.e., the systems ranked from the lowest to the highest MAP values), and moving towards left in the charts, the correlation values remain high when decreasing cardinality down to around 25% of the total systems, before decreasing and becoming noisy. This would still confirm the undesired effect, also taking into account that the noisy behaviour at low cardinalities can depend on the very low MAP of systems. However, when considering the "Best2Worst" series (i.e., the systems sorted from the highest to the lowest MAP values), the behaviour is different, and again similar for the three datasets: starting from low cardinalities, and moving towards the right in the charts, the correlations remain stable at near-zero values (let us say, in $[-0.5, 0.5]$) until about 90% of the maximum cardinality and then the correlation increases to the value obtained considering all the systems. Summarizing, the "undesired" feature that system effectiveness is affected mostly by easy topics manifests itself, with correlation values that are clearly different from zero, if and only if the very least effective systems (i.e., the bottom 10–25% or so) are considered, as shown by the "Best2Worst" series. The undesired feature is caused mostly by the least effective systems.

The scatterplots in Figs. 2(d)–(g) show the details of the AAP-hubness correlation for selected cardinalities (20, 100, 120, and 129, highlighted in Fig. 2(a) with the larger white dots) for the "Best2Worst" series of the AH99 dataset (other datasets are similar). The charts confirm that there is no correlation up to cardinality 100; some correlation exists at cardinality 120 and at full cardinality 129 we obtain the same result as [4, Fig. 5(d)]. Figure 2(h) shows the AAP-hubness correlation again at cardinality 20 and for the same dataset, but

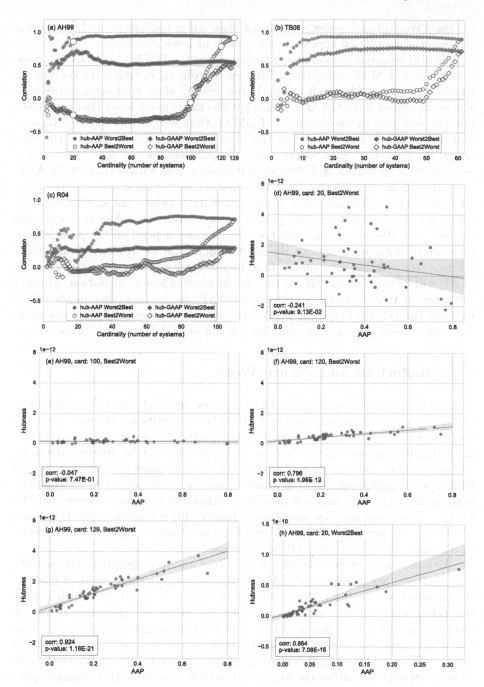

Fig. 2. Correlations for different systems subsets and metrics (a)–(c). Scatterplots for specific systems subsets at different cardinality values (d)–(h).

for the "Worst2Best" series, confirming that when considering only the worst systems the correlation appears much earlier, at lower cardinality.

We also repeated the same experiments using the GMAP (and GAAP) measure instead of MAP (and AAP). Differently from [4] in which the correlation largely disappears when using GMAP, our results show that some correlation still occurs (even if the GAAP-hubness correlation is much lower than the AAP-hubness one, due to the definition of GMAP that weighs less the easy topics), and the trend of the red series in the plots is comparable to that of the blue series, for both the "Best2Worst" and "Worst2Best" series. The effect of the least effective systems is still clear, even if smaller, also with GMAP and GAAP.

Although the general trends are the same across the three datasets, there are some differences. As mentioned above, correlation values are smaller for R04. Again for the same dataset, the growth of the correlation values is less sudden for the "Best2Worst" series. Both these results can depend on the peculiar features of R04: since it contains the most difficult topics from previous TREC editions, the systems have similar (and in general low) AP values, the variance of AP will be smaller, and this in turn might cause a lower correlation. Another difference is that correlation values for GAAP are much higher in the TB06 dataset: the benefical effect of GMAP reported in [4], besides being weaker in our experiments, almost disappears for this dataset. This requires further study.

4   Conclusions and Future Work

This paper presents an analysis that exploits some of the hidden relations between topics and systems used in TREC-like competitions. We have obtained three results: (i) we confirmed the original results that easier topics are better in distinguishing system effectiveness, and generalized it to different datasets; (ii) however, we also somehow disproved that result; more in detail, we showed that if we consider only the top ranked systems according to MAP (i.e., the systems for which evaluation is more interesting) there is no evidence that the ranking is affected only by easy topics; finally (iii) we proved that the above results are robust to the change of the metric used, even when GMAP, a metric that is more sensitive to low AP values, is used.

We leave plenty of space for future work. It would be interesting to investigate the effect of other metrics, like yaAP [6] which considers the number of relevant documents that can be related to topic difficulty; or NAP [3], which explicitly takes topic difficulty into consideration. It would be interesting to repeat the experiments considering a "dual" scenario, thus considering the topics and investigating the correlation between the systems and their ability to distinguish between easy and hard topics (according to some metric). This experiment might provide an explanation for the high AAP-hubness (and GAAP-hubness) correlation values in TB06. It is possible to think of a two-step evaluation of the systems (as it has already been suggested [3]): first we can use all the topics, or even the easy ones only, to evaluate the systems and provide a first rank; later, we can select the most effective systems and the most difficult topics, and use them to fine-tune the ranking of the most effective systems. This evaluation should not be affected from the undesired features that manifest with the classical one-step evaluation and has the potential of being more economic. The work on using fewer topics [1] is also to be taken into account when considering such an alternative two-step evaluation process. Finally, and more in detail, it would be interesting to investigate further about the features which make the correlation between topic ease and system effectiveness suddenly increase when considering about the 75% of the systems (sorted from more effective to less effective).

References

1. Berto, A., Mizzaro, S., Robertson, S.: On using fewer topics in information retrieval evaluations. In: Proceedings of the 2013 Conference on the Theory of Information Retrieval, ICTIR 2013 (2013)
2. Kleinberg, J.M.: Authoritative sources in a hyperlinked environment. J. ACM **46**(5), 604–632 (1999). http://dl.acm.org/citation.cfm?id=324140
3. Mizzaro, S.: The good, the bad, the difficult, and the easy: something wrong with information retrieval evaluation? In: Macdonald, C., Ounis, I., Plachouras, V., Ruthven, I., White, R.W. (eds.) ECIR 2008. LNCS, vol. 4956, pp. 642–646. Springer, Heidelberg (2008). doi:10.1007/978-3-540-78646-7_71
4. Mizzaro, S., Robertson, S.: HITS hits TREC: exploring IR evaluation results with network analysis. In: Proceedings of the 30th Annual International ACM SIGIR Conference on Research and Development in Information Retrieval (2007). http://dl.acm.org/citation.cfm?id=1277824&CFID=912758227&CFTOKEN=40185811
5. Robertson, S.: On GMAP: and other transformations. In: Proceedings of the 15th ACM International Conference on Information and Knowledge Management, CIKM 2006 (2006)
6. Robertson, S.: On smoothing average precision. In: Baeza-Yates, R., Vries, A.P., Zaragoza, H., Cambazoglu, B.B., Murdock, V., Lempel, R., Silvestri, F. (eds.) ECIR 2012. LNCS, vol. 7224, pp. 158–169. Springer, Heidelberg (2012). doi:10.1007/978-3-642-28997-2_14

Named Entity Classification Using Search Engine's Query Suggestions

Jayendra Barua[✉] and Dhaval Patel

Indian Institute of Technology, Roorkee, India
{jbarudec,patelfec}@iitr.ac.in
http://www.iitr.ac.in

Abstract. Existing Named Entity Recognition (NER) techniques uses external gazetteers lookup as a feature to improve classification accuracy of entity mentions in the text. However, gazetteers lookup suffer with low recall problem as creation and maintenance of gazetteers is a labor and cost intensive task. In this paper, we propose to use Search Engine's Query suggestion as external knowledge source instead of gazetteers for named entity classification in NER systems. Specifically, we build a Query Suggestion based Named Entity Classifier (QS-NEC), which learns entity types from Query Suggestions of Named Entities. We have used QS-NEC as an Entity Classification module in our NER framework. Our experiments on MSM Challenge dataset demonstrate that QS-NEC is efficient in classification of entity mentions and can be effectively used in NER systems.

1 Introduction

Named Entity Recognition (NER) is an important subtask of Information Extraction which aims to identify and classify named entity mentions in natural language text. Many existing machine learning NER techniques [10–12] incorporates external knowledge in the form of gazetteers or dictionaries lookup features to improve the classification accuracy of named entity mentions such as, Ratinov and Roth [11] uses 30 gazetteers containing over 1.5 million entities constructed from Wikipedia and web in Illinois NER system. Similarly, Ritter et al. [12] uses large entity dictionaries gathered from Freebase in their T-NER system. Manual construction and maintenance of high quality gazetteers is a laborious task. Therefore, many techniques have been proposed for automatic extraction of gazetteers from text [2,7,13]. However, main problem of using gazetteers or dictionary in NER systems is their "limited coverage" which leads to low recall problem [5,11]. The news and social media texts such as news headlines and Tweets often reports new entities. Therefore to perform NER on recent social media text, these techniques always needs to frequently update their gazetteers with newly emerging entities.

In this work, we propose to use Query Suggestions of Search Engines as external knowledge source in NER systems instead of gazetteers. Query suggestion is a common online feature in Search Engines which uses recent search logs [1,4] of

© Springer International Publishing AG 2017
J.M. Jose et al. (Eds.): ECIR 2017, LNCS 10193, pp. 612–618, 2017.
DOI: 10.1007/978-3-319-56608-5_56

users to provide query formulation assistance to users. The use of recent search logs in Search engines captures the user queries about the new emerging entities which can overcome the low recall and update problem of gazetteers.

Query Suggestions are short text and contains only highly relevant words and phrases to input query. We have observed that entities belongs to same entity type are often associated with same words in Query Suggestions, for e.g. Movie entities are often associated with words *movie review, trailor, movie download, watch online* and Location entities are associated with words *weather, map, airport, pin code, temperature* in Query Suggestions. This observation intuit that possible type of an entity mention can be determine by observing the entity into Query Suggestion search space. The key characteristics of Query Suggestions to infer entity types and to capture emerging entities motivated us to build a Query Suggestion based Named Entity Classifier (QS-NEC), which can learn the entity types from Query Suggestions of labeled entity mentions.

We have adopted the NER framework of existing NER techniques [6,9], where entity detection and entity classification are considered as separate modules. In our NER framework, we have used existing techniques [8,11,12] for entity detection task in text, whereas QS-NEC is used for entity classification task.

Our paper is organized as follows. In the next section, we describe Query Suggestion based Named Entity Classifier(QS-NEC). In Sect. 3, we describe our experiments followed by the conclusion in Sect. 4.

2   QS-NEC: Query Suggestion Based Named Entity Classifier

The classification of entities using *Query suggestions* can be seen as a document classification problem. Given an entity m, and predefined l entity types $C = \{c_1, c_2, c_3..c_l\}$, our problem is to classify m in one of the classes belongs to C by using a *Query suggestions* document $QS_m = \{s_1, s_2, s_3..s_n\}$ (with n suggestions) of m retrieved from a search engine. Given labeled examples of entities $M = \{m_1, m_2, m_3...m_k\}$ each having a Query suggestion document in $Sug\_set_M = \{QS_1, QS_2, QS_3...QS_k\}$, a variety of learning methods can be applied to learn entity types in C. We have used Support Vector Machine with linear kernel to learn entity types on the basis of word vector features derived from Query Suggestions of entities. To derive word vectors, we first pre-process each Query Suggestion document $QS_m \in Sug\_set_M$. The word vector wv_m for an entity m is created from QS_m using following procedure:

1. Initially, entity mention $= m$ and Word vector $wv_m = \phi$.
2. Add all possible n-grams of m to wv_m. For e.g. if $m = $ '*Reliance Ltd*', then terms '*Reliance*', '*Ltd*' and '*Reliance Ltd*' are added to wv_m.
3. **For** each Suggestion $s \in QS_m$, we detach mention m from s to result s'.
4. Derive all possible n-grams from s' and add them to wv_m.
5. **end-For.**
6. wv_m is the resultant word vector.

For e.g. Query Suggestions for movie 'Inferno' are : ⟨*inferno movie, inferno review, inferno trailer, inferno imdb, inferno movie review, inferno movie download, inferno film, inferno watch online*⟩.

Then, Word vector generated for '*Inferno*' are:

$$wv_{inferno} = \langle inferno, \; review, \; movie, \; trailer, \; imdb, \; movie \; review, \; movie,$$
$$review, \; movie \; download, \; movie, \; download, \; film, \; watch \; online, \; watch, \; online \rangle$$

In the aforementioned procedure, we extract the n-grams from Query Suggestions to capture the phrases in query suggestions as features. Because, many phrases repeatedly occur in Query Suggestions of entities of a particular type, such as, '*watch online*' and '*movie download*' are common phrases in Query Suggestions of movie entities. The n-grams of entity name can also be very useful as it may contain relevant terms about entity type, for e.g. in entity name '*Reliance Ltd*', the term '*Ltd*' is highly related to organization entity type. Such n-grams of entity names is crucial for predicting entity type in the cases, when no Query suggestion is retrieved for an entity.

For training purpose, we first extract the word vectors $WV_M = \{wv_1, wv_2, wv_3, ...wv_k\}$ by using aforementioned procedure from Query Suggestion of each labeled entity $m \in M$. Next, the word vectors WV_M are inputted to SVM (Linear Kernel), which learn entity types from WV_M to build Query suggestion based Named Entity Classifier (QS-NEC). In order to predict the type of an entity E using QS-NEC, we first retrieve Query Suggestions QS_E for entity E from search engine. Next, QS_E is preprocessed to generate word vector wv_E. Finally, the generated word vector wv_E is inputted to the learned QS-NEC, which outputs the predicted type of entity E.

NER with QS-NEC: To perform the NER task with QS-NEC, we have adopted the NER framework of Carreas et al. [6] and Habib et al. [9], where NER is performed in two separate task (1) Named Entity Detection, and (2) Named Entity Classification. Since, our focus is on Named Entity classification task therefore, for the named entity detection task, we are dependent on entity detection capability of existing techniques [8,11,12]. However, we have used QS-NEC for the Named Entity classification task. Given a text corpus, we first apply the named entity detection technique to extract the entity mentions from text. The extracted mentions further passed to QS-NEC, which retrieves the Query Suggestions for each Entity mentions from search engine and use it to classify the mention in a predefined entity type.

3   Experiments

We have performed experiment to compare classification accuracy of QS-NEC with state-of-the-art techniques and to demonstrate how QS-NEC can be used for NER task. We have performed named entity recognition on Twitter dataset taken from Making Sense of Microposts 2013 IE Challenge (MSM Challenge) [3]. MSM Challenge dataset[1] consisted of two parts: training part T_{train} with

---

[1] http://oak.dcs.shef.ac.uk/msm2013/ie_challenge/.

2815 manually annotated microposts and test part T_{test} with 1526 unannotated microposts. The entities in microposts are annotated with Person, Location, Organization and Miscellaneous entity types. Annotated microposts of T_{test} is also provided in the form of "*GoldStandard*" for evaluation purpose.

We compare classification ability of our QS-NEC with 4 existing state-of-the-art NER techniques: Stanford NER [8] (*4class classifier*), Illinois NER [11], Annie NER[2] and T-NER [12]. Illinois NER, Annie NER and T-NER uses external gazetteers look-ups as one of the key feature for named entity classification task, whereas Stanford NER only uses local and non-local features of text. The T-NER [12] actually classifies the twitter entities in 10 types. We merge[3] the similar entity types to result 4 basic types compatible to MSM dataset. Annie NER does not detect the 'Miscellaneous' entity type, therefore evaluation and comparison of Annie NER is done on the basis of 'Person','Location', and 'Organization' entity types only. We have used Query Suggestions of Google Search Engine available through Query Suggestion API [4]. The evaluation is done by calculating standard F1-Measure based on exact matching of detected mentions and class-labels with corresponding labeled mentions in *GoldStandard*. Along with the F1-measure of individual classes, we have also calculated weighted average and macro average (class average) of F1-Measure for evaluation.

Training QS-NEC: We have used annotated mentions in microposts of T_{train} to train our QS-NEC classifier. The total number of annotated mentions in T_{train} was 3195, which reduced to 1980 after removing duplicate mentions. Next, we retrieved the Google Query suggestions for each of the unique annotated mentions. Out of 1980 mentions, Google Query suggestion API returned suggestions for 1918 mentions (97% of all mentions), whereas no suggestions is retrieved for 62 mentions (3% of all mentions). Further, we use *Query suggestions* of each annotated mentions to generate the word vectors. We train SVM classifier using all word vectors of labeled mentions to build QS-NEC. To estimate the efficiency, we performed 10 fold cross-validation on trained QS-NEC classifier to evaluate its accuracy. Results are shown in Table 1. Overall high weighted and macro average is retrieved. However, lowest precision of 76% is recorded for ORG and lowest recall of 57% is recorded for MISC entities. The trained QS-NEC now can be used for classifying the entity mentions in T_{test}.

QS-NEC with Ideal Mention Detection: Application of an ideal mention detection technique on T_{test}, would have detected all the mentions present in *GoldStandard*. Considering an ideal mention detection technique in our NER framework, we applied the QS-NEC on entity mentions present in *GoldStandard* (ignoring the *GoldStandard* labels) to result QS-NEC labeled mentions. The QS-NEC labeled entity mentions are evaluated against the gold-annotated mentions

---

[2] http://services.gate.ac.uk/annie/.

[3] Person, Organization{band,company,sportsteam}, Location, Miscellaneous{facility, movie, product,tvshow,other}.

[4] https://www.google.com/support/enterprise/static/gsa/docs/admin/70/gsa_doc_set/xml_reference/query_suggestion.html.

Table 1. 10-fold cross validation of results QS-NEC on T_{train}

Entity type →	PER	LOC	ORG	MISC	Weighted average	Macro average
Precision	85.70%	82.30%	76.30%	79.40%	83.10%	**80.93%**
Recall	96.30%	76.20%	59.90%	57.50%	83.60%	**72.48%**
F-measure	**90.70%**	**79.20%**	**67.10%**	**66.70%**	**82.80%**	**75.93%**

present in *GoldStandard* to calculate the F1-score as shown in Table 2 under the heading 'QS-NEC(Ideal Mention Detection)'.

QS-NEC with State-of-the-Art NER Techniques: We first applied each Stanford NER, Illinois NER, Annie NER and T-NER individually on T_{test}. Each NER technique first detect the entity mention in tweet and then label it with one of the entity type. Thus detected and labeled mentions of each NER technique is evaluated against the gold-annotated mentions present in *GoldStandard* to calculate the F1-score as shown in Table 2 under headings Stanford NER, Illinois NER, Annie NER and T-NER. Now to compare QS-NEC on same set of mentions, we applied the QS-NEC separately on entity mentions detected by each of the state-of-the-art NER technique (we ignore the labels of the mentions tagged by NER techniques). The QS-NEC labeled entity mentions are evaluated against the gold-annotated mentions present in *GoldStandard* to calculate the standard F1-score as shown in Table 2 under the headings QS-NEC(Stanford NER), QS-NEC(Illinois NER), QS-NEC(Annie NER) and QS-NEC(T-NER).

Results and Discussion: Highest F1-score is obtained by 'QS-NEC(Ideal Mention Detection)' for all the entity types in Table 2 depicts that QS-NEC can perform entity classification with high accuracy if entities are correctly detected in the text. The difference between F1-scores of 'QS-NEC(Ideal Mention Detection)' and other QS-NEC configurations is due to 'Mention Detection' phase (see Table 3). Table 3 shows the count of correctly detected mentions by NER techniques and correctly classified mention by QS-NEC. The high classification accuracy on correctly detected mentions in all the NER techniques shows that QS-NEC can perform entity classification task with high accuracy in NER systems provided mentions are correctly detected.

Table 2 shows that, QS-NEC has better weighted and macro average F1 score in comparison with all the state-of-the-art NER techniques on same set of mentions. Considering the individually classes, Stanford NER, Illinois NER, and Annie NER have better accuracy than QS-NEC for 'PERSON' entity mentions, whereas QS-NEC excels in all other entity types. This also depicts that state-of-the-art NER methods are biased for 'Person' entity type. This is also shown by the macro average score, where QS-NEC dominate the state-of-the-art NER techniques by significant margin. Please note that 3 out of 4 state-of-the-art NER techniques significantly depends on gazetteers. Thus, overall we can conclude from results that Search engine's Query Suggestions can be used for entity classification instead of gazetteers.

Table 2. Comparison of F1-scores of QS-NEC with state-of-art NER techniques on MSM challenge dataset

Technique	PER	LOC	ORG	MISC	Weighted avg.	Macro avg.
QS-NEC(Ideal mention detection)	**95.90%**	**78.70%**	**83.90%**	**71.60%**	**91.50%**	**82.53%**
Stanford NER	72.97%	45.06%	19.16%	7.04%	55.55%	36.06%
QS-NEC(Stanford NER)	67.20%	48.57%	30.59%	21.18%	58.90%	41.89%
Illinois NER	**81.86%**	40.38%	22.81%	7.75%	62.30%	38.20%
QS-NEC(Illinois NER)	74.24%	50.85%	29.98%	**28.42%**	64.39%	45.87%
Annie NER	67.30%	49.62%	32.80%	-	60.52%	49.91%
QS-NEC(Annie NER)	66.91%	**56.50%**	**36.41%**	-	62.02%	53.27%
T-NER	75.71%	41.58%	21.80%	15.38%	61.03%	38.62%
QS-NEC(T-NER)	76.85%	49.50%	27.79%	24.52%	**66.11%**	44.67%

Table 3. Classification accuracy of QS-NEC on correctly detected entity mentions by NER techniques

NER technique	Count$_{GoldStandard}$	Count$_{CorrectlyDetected}$	Count$_{Correct-QS-NEC-labeled}$*
Ideal mention detection	1557	1557	1423 (91.39%)
Stanford NER	1557	1104	985 (89.22%)
Illinois NER	1557	1193	1089 (91.28%)
Annie NER	1462	960	929 (96.77%)
T-NER	1557	1155	1014 (87.79%)

*% shown is with respect to $Count_{CorrectlyDetected}$

4   Conclusion

In this paper, we proposed Query Suggestions based named entity classification which uses Search engine's Query Suggestions for classification of named entity mentions instead of static gazetteers in NER systems. Our approach leverages Search engine's Query Suggestions for the entity mention as external knowledge. Our experiment on MSM challenge 2013 dataset demonstrated that QS-NEC has better classification accuracy than gazetteer based NER techniques on same set of entity mentions in tweets. QS-NEC promises high classification accuracy in NER systems, provided entity mentions are correctly detected.

References

1. Baeza-Yates, R., Hurtado, C., Mendoza, M.: Query recommendation using query logs in search engines. In: Lindner, W., Mesiti, M., Türker, C., Tzitzikas, Y., Vakali, A.I. (eds.) EDBT 2004. LNCS, vol. 3268, pp. 588–596. Springer, Heidelberg (2004). doi:10.1007/978-3-540-30192-9_58
2. Boldyrev, A., Weikum, G., Theobald, M.: Dictionary-based named entity recognition. Ph.D. thesis, Masters Thesis in Computer Science (2013)
3. Cano Basave, A.E., Varga, A., Rowe, M., Stankovic, M., Dadzie, A.S.: Making sense of microposts (# msm2013) concept extraction challenge (2013)
4. Cao, H., Jiang, D., Pei, J., He, Q., Liao, Z., Chen, E., Li, H.: Context-aware query suggestion by mining click-through and session data. In: SIGKDD. ACM (2008)

5. Carlson, A., Gaffney, S., Vasile, F.: Learning a named entity tagger from gazetteers with the partial perceptron. In: AAAI Spring Symposium (2009)
6. Carreras, X., Màrquez, L., Padró, L.: A simple named entity extractor using adaboost. In: HLT-NAACL. ACL (2003)
7. Cohen, W.W., Sarawagi, S.: Exploiting dictionaries in named entity extraction: combining semi-Markov extraction processes and data integration methods. In: SIGKDD. ACM (2004)
8. Finkel, J.R., Grenager, T., Manning, C.: Incorporating non-local information into information extraction systems by gibbs sampling. In: ACL. ACL (2005)
9. Habib, M.B., Keulen, M., Zhu, Z.: Concept extraction challenge: University of Twente at #MSM2013 (2013)
10. He, Q., Spangler, W.: Semi-supervised data integration model for named entity classification, 22 March 2016. https://www.google.com/patents/US9292797. US Patent 9,292,797
11. Ratinov, L., Roth, D.: Design challenges and misconceptions in named entity recognition. In: CoNLL. ACL (2009)
12. Ritter, A., Clark, S., Etzioni, O.: Named entity recognition in tweets: an experimental study. In: EMNLP. ACL (2011)
13. Zamin, N., Oxley, A.: Building a corpus-derived gazetteer for named entity recognition. In: Zain, J.M., Wan Mohd, W.M., El-Qawasmeh, E. (eds.) ICSECS 2011. CCIS, vol. 180, pp. 73–80. Springer, Heidelberg (2011). doi:10.1007/978-3-642-22191-0_6

"When Was This Picture Taken?" – Image Date Estimation in the Wild

Eric Müller[1]([✉]), Matthias Springstein[1], and Ralph Ewerth[1,2]

[1] German National Library of Science and Technology (TIB), Hannover, Germany
{eric.mueller,matthias.springstein,ralph.ewerth}@tib.eu
[2] L3S Research Center, Hannover, Germany

Abstract. The problem of automatically estimating the creation date of photos has been addressed rarely in the past. In this paper, we introduce a novel dataset *Date Estimation in the Wild* for the task of predicting the acquisition year of images captured in the period from 1930 to 1999. In contrast to previous work, the dataset is neither restricted to color photography nor to specific visual concepts. The dataset consists of more than one million images crawled from Flickr and contains a large number of different motives. In addition, we propose two baseline approaches for regression and classification, respectively, relying on state-of-the-art deep convolutional neural networks. Experimental results demonstrate that these baselines are already superior to annotations of untrained humans.

1 Introduction

In recent years, huge datasets (e.g., *ImageNet* [8], *YFCC100M* [12]) were introduced fostering research for many computer vision tasks. In particular, such datasets are a prerequisite for the training of deep learning systems. However, estimating automatically the capturing time of (historical) photos has been rarely addressed yet and existing benchmark datasets do not contain enough images captured before 2000. But date estimation is an interesting and challenging task for historians, archivists, and even for sorting (digitized) personal photo collections chronologically. Existing approaches either rely on datasets solely containing historical color images [1,6,7] or focus on specific concepts like cities [10], cars [4], persons [2,9], or historical documents [3,5] and are therefore unable to learn the temporal differences of the broad variety of motives. For this reason, a huge dataset covering all kinds of concepts is necessary, which additionally enables the training of convolutional neural networks.

In this paper, we introduce a novel dataset *Date Estimation in the Wild* and make it publicly available to support further research. In contrast to existing datasets, it contains more than one million Flickr images captured in the period from 1930 to 1999. As shown in Fig. 1, the dataset covers a broad range of domains, e.g., city scenes, family photos, nature, and historical events. Two baseline approaches are proposed based on a deep convolutional neural network (GoogLeNet [11]) treating the task of dating images as a classification and

© The Author(s) 2017
J.M. Jose et al. (Eds.): ECIR 2017, LNCS 10193, pp. 619–625, 2017.
DOI: 10.1007/978-3-319-56608-5_57

Fig. 1. Some example images from the *Date Estimation in the Wild* dataset.

regression problem, respectively. Experimental results show the feasibility of the suggested approaches which are superior to annotations of untrained humans.

The remainder of the paper is organized as follows. Section 2 reviews related work on dating historical images. Section 3 introduces the *Date Estimation in the Wild* dataset as well as the baseline approaches in detail. The experimental setup and results are presented in Sect. 4 along with a comparison to human annotation performance. Section 5 concludes the paper.

2   Related Work

The first work that deals with dating historical images stemming from different decades has been introduced by Schindler et al. [10]. The authors present an approach to sort a collection of city-scape images temporally by reconstructing the 3D world, requiring many overlapping images of the same location. Jae et al. [4] identify style-sensitive groups of patches for cars and street view images in order to model stylistic differences across time and space. He et al. [3] and Li et al. [5] address the task of estimating the age of historical documents. While He et al. [3] explore contour and stroke fragments, Li et al. [5] apply convolutional neural networks in combination with optical character recognition. Ginosar et al. [2] and Salem et al. [9] model the differences of human appearance and clothing style in order to predict the date of photos in yearbooks.

More closely related to our work, Palermo et al. [7] suggest an approach to automatically estimate the age of historical color photos without restrictions to specific concepts. They combine different color descriptors to model the historical color film processes. The results on the proposed dataset, which contains 1375 images from 1930 to 1980, are further improved by Fernando et al. [1] by including color derivatives and angles. Martin et al. [6] treat date estimation as a binary task by deciding whether an image is older or newer than a reference image. However, the aforementioned approaches either rely on color photography, which was very uncommon before 1970, or focus on specific concepts.

3 Image Date Estimation in the Wild

In this section, the *Date Estimation in the Wild* dataset (Sect. 3.1) and the two proposed baseline approaches to predict the acquisition year of images (Sect. 3.2) are described in detail.

3.1 Image Date Estimation in the Wild Dataset

The Flickr API was utilized to download images for each year of the period from 1930 to 1999. We have observed that many historical images are supplemented with time information, either in the title or in the related tags and descriptions. Therefore, we used the current year as an additional query term to reduce the number of "spam" images. The only kind of filtering that we applied was restricting the search to photos. As a consequence, the dataset is noisy since it contains, for example, close-ups of plants or animals as well as historical documents. In order to avoid a bias towards more recent images, the maximum number of images per year was limited to 25000. Finally, the dataset consists of 1029710 images with a high diversity of concepts. Information about the granularity $g \in \{0, 4, 6, 8\}$ according to the Flickr annotation of the date entry is stored as well. The distribution of images per year and the related granularity of dates are depicted in Fig. 2.

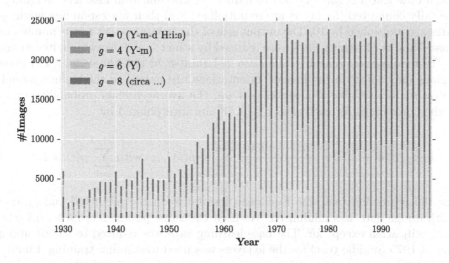

Fig. 2. Number of crawled images and the accuracy of the provided timestamps for each year in the *Date Estimation in the Wild* dataset.

In order to obtain reliable validation and test sets that match the dataset distribution, a maximum number of 75 *unique images* for 1930 to 1954 and 150 *unique images* for the remaining years were extracted. A *unique image* is defined as an image with a date granularity of $g = 0$ (Y-m-d H:i:s) or $g = 4$ (Y-m), for which no visual near-duplicates (detected by comparing the features from the last pooling layer of a GoogLeNet pre-trained on *ImageNet*) exist in the entire dataset. Subsequently, 8495 *unique images* were extracted for the validation set and another 16 per year were selected manually to obtain the test dataset containing 1120 images. The remaining 1020095 images constitute the training set. The dataset[1] is available at https://doi.org/10.22000/0001abcde.

3.2   Baseline Approaches

Two baseline approaches are realized by training a GoogLeNet [11] and treating image date estimation as a classification or regression problem, respectively.

Convolutional neural networks require many images per class c to learn appropriate models for the classification task. However, the dataset lacks images for the first three decades (Fig. 2). For this reason, we decided to use $|c| = 14$ classes by quantizing the image acquisition year into 5-year periods to reduce the classification complexity, while still maintaining a good temporal resolution. For the classification task, GoogLeNet was trained using Caffe on a pre-trained *ImageNet* model [8]. We randomly selected 128 images per batch for training, which were scaled by the ratio $256/\min(w, h)$ (w and h are image dimensions). To augment training data, the images were horizontally flipped and cropped randomly to fit in the reception field of $224 \times 224 \times 3$ pixel. The stochastic gradient descent algorithm was employed using 1M iterations with a momentum of 0.9 and a base learning rate of 0.001 to reduce the classification loss. The weights of the fully connected (fc) layers are re-initialized and their corresponding learning rates are multiplied by 10. The output size of the fc layers is set to the number of classes and the learning rates were reduced by a factor of 2 every 100k iterations.

Test images are scaled by the ratio $224/\min(w, h)$ and three 224×224 pixel regions depending on the images' orientations are passed to the trained model. To estimate a specific acquisition year y_E, the averaged class probabilities $p(c)$ of the three crops for each class $c \in [0, 13]$ are interpolated by:

$$y_E = 1930 + \left\lfloor 0.5 + \frac{1999 - 1930}{|c| - 1} \cdot \sum_{i=0}^{|c|-1} i \cdot p(i) \right\rfloor, \quad \text{with} \sum_{i=0}^{|c|-1} p(i) = 1. \quad (1)$$

For the regression task, the Euclidean loss between the predicted and ground truth image date was minimized. We used the same parameters for learning as in classification except for: The base learning rate was reduced to 0.0001 and a bias of 1975 (middle year) for the fc layers was used to stabilize training. Finally, the output size was set to 1 for regression to directly predict the year.

---

[1] Images or links (depending on the copyright status) and metadata are provided.

4   Experimental Results

In the experiments, the trained GoogLeNet models were applied to the test set. In contrast to Palermo et al. [7], we do not report the classification accuracy for predicting the correct 5-year period. For example, imagine that the ground truth date of an image is 1989 and the model predicts the class 1990–1994. Although the difference is possibly only one year the prediction would be false in this case. For this reason, we argue that the absolute mean error (ME) as well as the number of images with an absolute estimation error of at most n years (EE_n) are more meaningful for evaluation.

Table 1. Absolute mean error (ME) [y] and number of images estimated with an absolute estimation error of at most n years (EE_n) [%] for human annotators and for the baselines GoogLeNet classification (cls) and regression (reg) approaches on the *Date Estimation in the Wild* test set, with respect to each quantized 5-year period.

Year	Human performance				GoogLeNet cls				GoogLeNet reg			
	ME	EE_0	EE_5	EE_{10}	ME	EE_0	EE_5	EE_{10}	ME	EE_0	EE_5	EE_{10}
30–34	15.7	3.0	24.8	40.7	15.0	0.0	5.0	37.5	14.4	0.0	7.5	41.3
35–39	12.2	2.7	34.1	53.2	11.1	2.5	23.8	52.5	10.7	3.8	26.3	58.8
40–44	9.6	4.1	43.2	66.6	8.8	2.5	40.0	67.5	9.1	7.5	42.5	66.3
45–49	11.7	3.9	31.1	54.3	8.2	6.3	51.3	71.3	8.5	3.8	43.8	70.0
50–54	12.2	2.5	29.6	49.8	7.5	3.8	47.5	77.5	7.3	2.5	52.5	73.8
55–59	13.3	1.4	27.1	49.5	6.1	6.3	60.0	86.3	7.0	7.5	50.0	77.5
60–64	13.6	1.4	24.1	43.0	7.3	5.0	51.3	73.8	7.2	1.3	47.5	75.0
65–69	12.5	2.7	24.6	46.4	5.4	12.5	63.8	82.5	6.0	1.3	52.5	83.8
70–74	10.5	4.8	33.2	55.9	5.6	3.8	58.8	85.0	5.4	8.8	61.3	85.0
75–79	9.4	4.1	37.9	62.1	4.7	8.8	71.3	90.0	5.0	7.5	63.8	90.0
80–84	7.5	5.2	45.5	76.1	4.4	8.8	62.5	95.0	4.5	6.3	61.3	93.8
85–89	7.6	5.0	49.6	77.3	4.8	10.0	71.3	83.8	4.9	8.8	68.8	90.0
90–94	7.5	5.9	51.3	76.1	5.6	5.0	66.3	85.0	5.7	6.3	61.3	83.8
95–99	9.4	6.1	39.5	62.9	7.5	11.3	52.5	75.0	8.7	1.3	36.3	73.8
Overall	10.9	3.8	35.4	58.1	7.3	6.2	51.8	75.9	7.5	4.7	48.2	75.9

Human performance was investigated as well. Seven untrained annotators of different age (ranging from 26 to 58) were asked to label all 1120 images of the test set and to make a break after each batch of 100 images. The average human performance and the results of our baseline approaches are displayed in Table 1.

The results clearly show the feasibility of our baselines outperforming human annotations in nearly all periods and reducing the mean error by more than three years on the entire dataset. Another observation is that there is a correlation between the number of images and the results for each 5-year period. For this reason, an increased mean error for images between 1930 to 1964 is noticeable. Besides, the potential error can be higher for classes at the interval boundaries (1930 and 1999), which explains the slightly worse results for 1990 to 1999. A similar observation can be made for human annotations, since they are more familiar with images, TV material, and their own experiences starting from 1960. Interestingly, the human error is noticeably lower for images covering the period from 1940 and 1944, which frequently show scenes from World War II.

Despite the problem caused by the interval bounds of the entire time period which affects the interpolation step, the classification results are slightly better than for regression. This is attributed to the easier task of minimizing the classification loss of 14 classes compared to minimizing the Euclidean loss.

5   Conclusions

In this paper, we have introduced a novel dataset entitled *Date Estimation in the Wild* to foster research regarding the challenging task of image date estimation. In contrast to previous work, the dataset is neither restricted to color imagery nor to specific concepts, but includes images covering a broad range of motives for the period from 1930 to 1999. In a first attempt to tackle this challenging problem, we have proposed two approaches relying on deep convolutional neural networks to predict an image's acquisition year, considering the task as a classification as well as a regression problem. Both approaches achieved a mean error of less than 8 years and were superior to annotations of untrained humans. In the future, it is planned to exploit different specific classifiers for frequent concepts such as persons or cars to further enhance the performance of our systems.

References

1. Fernando, B., Muselet, D., Khan, R., Tuytelaars, T.: Color features for dating historical color images. In: IEEE International Conference on Image Processing, pp. 2589–2593 (2014)
2. Ginosar, S., Rakelly, K., Sachs, S., Yin, B., Efros, A.A.: A century of portraits: a visual historical record of American high school yearbooks. In: IEEE International Conference on Computer Vision Workshops, pp. 1–7 (2015)
3. He, S., Samara, P., Burgers, J., Schomaker, L.: Image-based historical manuscript dating using contour and stroke fragments. Pattern Recogn. **58**, 159–171 (2016)
4. Jae Lee, Y., Efros, A.A., Hebert, M.: Style-aware mid-level representation for discovering visual connections in space and time. In: IEEE International Conference on Computer Vision, pp. 1857–1864 (2013)
5. Li, Y., Genzel, D., Fujii, Y., Popat, A.C.: Publication date estimation for printed historical documents using convolutional neural networks. In: Third International Workshop on Historical Document Imaging and Processing, pp. 99–106 (2015)
6. Martin, P., Doucet, A., Jurie, F.: Dating color images with ordinal classification. In: International Conference on Multimedia Retrieval, pp. 447–450 (2014)
7. Palermo, F., Hays, J., Efros, A.A.: Dating historical color images. In: Fitzgibbon, A., Lazebnik, S., Perona, P., Sato, Y., Schmid, C. (eds.) ECCV 2012. LNCS, vol. 7577, pp. 499–512. Springer, Heidelberg (2012). doi:10.1007/978-3-642-33783-3_36

8. Russakovsky, O., Deng, J., Su, H., Krause, J., Satheesh, S., Ma, S., Huang, Z., Karpathy, A., et al.: Imagenet large scale visual recognition challenge. Int. J. Comput. Vis. **115**(3), 211–252 (2015)
9. Salem, T., Workman, S., Zhai, M., Jacobs, N.: Analyzing human appearance as a cue for dating images. In: 2016 IEEE Winter Conference on Applications of Computer Vision, pp. 1–8 (2016)
10. Schindler, G., Dellaert, F., Kang, S.B.: Inferring temporal order of images from 3D structure. In: IEEE Conference on Computer Vision and Pattern Recognition, pp. 1–7 (2007)
11. Szegedy, C., Liu, W., Jia, Y., Sermanet, P., Reed, S., et al.: Going deeper with convolutions. In: IEEE Conference on Computer Vision and Pattern Recognition, pp. 1–9 (2015)
12. Thomee, B., Shamma, D.A., Friedland, G., Elizalde, B., Ni, K., Poland, D., Borth, D., Li, L.J.: YFCC100M: the new data in multimedia research. Commun. ACM **59**(2), 64–73 (2016)

Low-Cost Preference Judgment via Ties

Kai Hui[1,2(✉)] and Klaus Berberich[1,3]

[1] Max Planck Institute for Informatics, Saarbrücken, Germany
{khui,kberberi}@mpi-inf.mpg.de
[2] Saarbrücken Graduate School of Computer Science, Saarbrücken, Germany
[3] htw saar, Saarbrücken, Germany

Abstract. Preference judgment, as an alternative to graded judgment, leads to more accurate labels and avoids the need to define relevance levels. However, it also requires a larger number of judgments. Prior research has successfully reduced that number to $\mathcal{O}(N_d \log N_d)$ for N_d documents by assuming transitivity, which is still too expensive in practice. In this work, by analytically deriving the number of judgments and by empirically simulating the ground-truth ranking of documents from TREC Web Track, we demonstrate that the number of judgments can be dramatically reduced when allowing for ties.

1 Introduction

Offline evaluation in information retrieval heavily relies on manual judgments to generate a ground-truth ranking of documents in response to a query. There exist two approaches to collect judgments, namely, graded judgments, where documents are labeled independently with predefined grades, and preference judgments, where judges provide a relative ranking for a pair of documents. For instance, given a test query, there are two rivaling systems s_1 and s_2, whose search results in response to the test query are (d_3, d_1, d_2) and (d_5, d_4, d_2) respectively. To compare these two search results, manual judgments are collected. When collecting graded judgments, the five documents are assessed by judges independently and are assigned predefined grades, say $d_1 : 0$, $d_2 : 1$, $d_3 : 1$, $d_4 : 2$, $d_5 : 2$; when collecting preference judgments, pairwise preferences over document pairs are collected, say $d_5 \sim d_4$, $d_5 \succ d_3$, $d_5 \succ d_2$, $d_5 \succ d_1$, $d_4 \succ d_3$, $d_4 \succ d_2$, $d_4 \succ d_1$, $d_3 \sim d_2$, $d_3 \succ d_1$, $d_2 \succ d_1$. We use $\succ$, $\prec$ and $\sim$ to denote the "better than", "worse than", and "tied with" relationships. Ultimately, with both approaches, a ground-truth ranking of documents can be determined. In this example, a same ground-truth ranking of documents is derived from graded and preference judgments: $d_5 \sim d_4 \succ d_3 \sim d_2 \succ d_1$.

Preference judgments have been demonstrated as a better alternative to the widely used graded judgments. Compared with graded judgments, preference judgments lead to better inter-assessors agreement, less time consumption per judgment [3] and better judgment quality in terms of agreement to user clicks [5]. Radinsky and Ailon [7] pointed out that these advantages come from the pairwise nature of preference judgments, i.e., the documents in the pair can mutually

© Springer International Publishing AG 2017
J.M. Jose et al. (Eds.): ECIR 2017, LNCS 10193, pp. 626–632, 2017.
DOI: 10.1007/978-3-319-56608-5_58

act as a "context", providing a reference for the judges. However, this pairwise nature also increases the number of judgments from $\mathcal{O}(N_d)$ to $\mathcal{O}(N_d^2)$ for N_d documents. Even after assuming transitivity, the number of judgments is still in $\mathcal{O}(N_d \log N_d)$ and hence much larger than the one from graded judgments, which is especially true for large N_d.

In this work, we highlight the ties in preference judgments, which have been introduced in existing works [5,8], but without noticing its potential in reducing the number of judgments. We assume transitivity among preference judgments as in [3,6,8], which might be over-optimistic in practice. We argue that, however, the collection of transitive judgments, and the design of judgment mechanisms that can tolerate intransitive judgments are orthogonal to this work. Moreover, the ultimate judgment cost should be the number of judgments times the cost per judgment, where a higher unit cost may lead to better transitivity. Instead we focus on demonstrating the potential of ties in reducing the number of judgments when transitivity is strictly observed. We investigate the number of judgments when allowing for ties analytically and empirically. In particular, we reexamine the number of preference judgments on N_d documents with established QUICK-SORT-JUDGE mechanism [8]. Moreover, we empirically investigate the number of judgments when simulating the ground truth from TREC Web Track 2011–2014. To this end, we argue that the tie is a compromise between the number of judgments and the judgment granularity. It clusters documents into tie partitions, and reduces the ranking of documents to the ranking of tie partitions. We demonstrate that the average number of judgments is reduced to $\mathcal{O}(N_t \log N_d)$, where N_t is the number of tie partitions. In addition, when simulating the ground truth from TREC, compared with graded judgments, only 43% more judgments are required when allowing for ties, whereas 773% more judgments are required in strict preference judgments. To the best of our knowledge, this is the first work to investigate and confirm the importance of ties in reducing the number of judgments.

Organization. Section 2 recaps existing literature and puts our work in context. Section 3 analyzes how the ties can help to reduce the number of judgments analytically and empirically. Finally, in Sect. 4 we draw conclusions.

2   Related Work

Reduce the Number of Judgments. Assuming transitivity among preference judgments, the complexity is reduced from $\mathcal{O}(N_d^2)$ to $\mathcal{O}(N_d \log N_d)$ [1,3,8], by avoiding a full comparison among all document pairs. Beyond transitivity, several attempts to further bring down the number of judgments were made. Carterette et al. [3] proposed to remove 20% "Bad" judgments by assigning them as worse than others. Niu et al. [6] addressed the expensiveness by only determining a full order for top-k search results, reducing the complexity to $\mathcal{O}(N_d \log k)$. Actually, the documents labeled as "Bad" in [3] and the documents out of top-k in [6] can be regarded as special cases of tie partitions–a single tie partition with low relevance documents. However, we argue that the reduction of the number of judgments is limited compared with a real tie option, which is especially true for the "Bad" judgments, given that the limited number of documents that are totally off-topic in practice. Moreover, the top-k ground-truth ranking from [6] is more suitable for learning to rank algorithms, and may lead to bias for evaluation purpose especially when smaller k is used. Other than that, no existing work has explicitly investigated the usage of ties in reducing the number of judgments.

QUICK-SORT-JUDGE. In our empirical analysis, we employ the labeling mechanism QUICK-SORT-JUDGE from [8], similar to a randomized *QuickSort* method. In QUICK-SORT-JUDGE, during each iteration, a document is randomly chosen as a pivot document, denoted as d_p. Thereafter, all remaining documents are grouped into worse than ($\prec d_p$), better than ($\succ d_p$) or tied with ($\sim d_p$) per manual judgments. The mechanism terminates when all documents have been recursively sorted. Note that, within each iteration, the documents on different sides of the pivot document are not manually judged, instead preferences between such document pairs are inferred with transitivity.

3 Number of Judgments

In this section, we investigate the average number of judgments required by preference judgments with ties analytically and empirically.

3.1 Theoretical Analysis

We reexamine the expected number of preference judgments when allowing for ties based on QUICK-SORT-JUDGE [8] as introduced in Sect. 2.

Notation. Given query q, we denote a set of documents as $\mathcal{D}$, and thus $N_d = |\mathcal{D}|$. Akin to the notation in [8], in the ground-truth ranking of documents on $\mathcal{D}$, documents that are mutually tied constitute N_t tie partitions, which are denoted as $t_1, t_2, \cdots, t_{N_t}$. Within individual tie partition t_i, documents are labeled with the same grade or are judged as mutually tied. For example, the ground-truth ranking of documents in the example from Sect. 1 can be represented as $t_1 \prec t_2 \prec t_3$, where $t_1 = \{d_1\}$, $t_2 = \{d_2, d_3\}$ and $t_3 = \{d_4, d_5\}$. Given tie partitions $t_i \prec t_j$, we use $\mathcal{D}_{ij}$ to denote documents which lie in between t_i and t_j in the ranking, namely, $\mathcal{D}_{ij} = \{d | t_i \prec d \prec t_j\}$. The set of tie partitions on $\mathcal{D}$ is denoted as $\mathcal{T}$. We introduce $\beta = \frac{N_d}{N_t}$, denoting the average number of documents per tie partition. Manual judgments can be categorized into two kinds: non-tie judgments, namely $\prec$and$\succ$, which sort tie partitions; and tie judgments, namely $\sim$, which cluster documents into tie partitions. Correspondingly, the total number of judgments, denoted as N_{jud}, can be split into the number of non-tie judgments, denoted as N_{ntie}, and the number of tie judgments, denoted as N_{tie}. And N_{ntie} can be further boiled down to judgments that determine relative order between a pair of tie partitions t_i and t_j, denoted as N_{ij}, namely, $N_{ntie} = \sum_{t_i, t_j \in \mathcal{T}} N_{ij}$.

Assumptions. As mentioned in Sect. 1, our analysis is based on transitivity assumption. The transitivity can be applied among tie partitions. For instance,

Table 1. The distribution and expectation of N_{ij}, namely, the number of judgments to determine the relative order of two tie partitions t_i and t_j.

Pivot document d_p	$t_i \prec d_p \prec t_j$	$d_p \in t_i$	$d_p \in t_j$
N_{ij}	0	$\|t_j\|$	$\|t_i\|$
$P(N_{ij})$	$\frac{\|\mathcal{D}_{ij}\|}{\|t_i\|+\|\mathcal{D}_{ij}\|+\|t_j\|}$	$\frac{\|t_i\|}{\|t_i\|+\|\mathcal{D}_{ij}\|+\|t_j\|}$	$\frac{\|t_j\|}{\|t_i\|+\|\mathcal{D}_{ij}\|+\|t_j\|}$
$E(N_{ij})$	$\frac{2\|t_i\|\|t_j\|}{\|t_i\|+\|\mathcal{D}_{ij}\|+\|t_j\|}$		

given t_i and t_j, by judging $d_k \in t_i$ and $d_l \in t_j$ as tied, one can get $t_i \sim t_j$ according to transitivity. In addition, we assume that $|t.| = \frac{N_d}{N_t} = \beta$, namely, tie partitions have the same size. Note that the size of different tie partitions is more skewed in practice, and this assumption is used to simplify Eq. 1.

Non-tie Judgments: Sort the Tie Partitions. For the non-tie judgments, the number of judgments is analyzed following the analysis for randomized *QuickSort* algorithm [4]. Akin to [4], conceptually, we index these tie partitions according to their ground-truth order, namely, $t_1 \prec t_2 \prec, \cdots, t_i \prec t_j, \cdots, t_{N_t}$. To approach this ground-truth order, one needs to determine relative order for each pair of tie partitions, say t_i and t_j. Therefore, one has to either select pivot document d_p from t_i or t_j, resulting in $|t_j|$ or $|t_i|$ judgments respectively, or select a pivot document d_p in between t_i and t_j, namely $d_p \in \mathcal{D}_{ij}$, leading to 0 judgments. In the former case, assuming $d_p \in t_i$, one needs to judge d_p relative to each document in t_j and make $|t_j|$ judgments. In the latter case, the relative order between t_i and t_j is inferred from the judgments between them and d_p, e.g., $t_i \prec d_p, t_j \succ d_p \implies t_i \prec t_j$. The distribution of the random variable N_{ij} is summarized in Table 1. And the expected total number of non-tie judgments $E(N_{ntie})$ can be computed as follows.

$$E(N_{ntie}) = E(\sum_{t_i, t_j \in \mathcal{T}} N_{ij}) = \sum_{i=1}^{N_t-1} \sum_{j=i+1}^{N_t} E(N_{ij})$$

$$= \sum_{i=1}^{N_t-1} \sum_{j=i+1}^{N_t} \frac{2|t_i||t_j|}{|t_i| + |\mathcal{D}_{ij}| + |t_j|} \tag{1}$$

Assuming that tie partitions have equal size, the complexity can be simplified as in Eq. 2, where $H_{N_t} = \sum_{k=1}^{N_t} \frac{1}{k}$ is the n_t-th harmonic number, which is in $\mathcal{O}(\log N_t)$ [4].

$$E(N_{ntie}) = \sum_{i=1}^{N_t-1} \sum_{j=i+1}^{N_t} \frac{2\beta^2}{\beta(j-i+1)}$$

$$= 2\beta \sum_{i=1}^{N_t-1} \sum_{k=2}^{N_t-i+1} \frac{1}{k} \tag{2}$$

$$< 2\beta \sum_{i=1}^{N_t} H_{N_t} = 2\beta N_t H_{N_t}$$

Tie Judgments: Generate Tie Partitions. When two documents are judged as tied, they are put into the same tie partition. For tie partition t_i, one needs to make $|t_i|$ tie judgments. Therefore, the total number of tie judgments is $E(N_{tie}) = \sum_{i=1}^{N_t} |t_i| = N_d$.

Total Number of Judgments. Henceforth, the expected total number of judgments equals the sum of the aforementioned two parts as in Eq. 3, which is in $\mathcal{O}(N_d \log N_t)$.

$$E(N_{jud}) = E(N_{ntie}) + E(N_{tie})$$
$$< 2\beta N_t H_{N_t} + N_d \tag{3}$$

3.2   Empirical Analysis

In this section, we empirically examine the number of judgments required in preference judgments to simulate the ground truth fromTREC.

Dataset. Our experiments are based on graded judgments from the 2011–2014 TREC Web Track[1] for adhoc task including 200 queries. The judgments contain at most six grades and one can sort them to establish a ground-truth ranking of documents.

Methods Under Comparison. We compare the number of judgments from three methods: graded judgments, preference judgments with ties and strict preference judgments. The number of judgments in graded judgments simply equals the number of documents. The preference judgments are simulated by randomly selecting document pairs with the established QUICK-SORT-JUDGE [8] as introduced in Sect. 2. Thereafter, in preference judgments with ties, the judgments are simulated by comparing the ground-truth labels of two documents from TREC. For strict preference judgments, given that ties are not allowed, the relative order between documents with the same labels from TREC are further determined by their string identifiers, which are unique and fixed among random experiments. We report the average number of judgments from 1000 repetitions of QUICK-SORT-JUDGE for both kinds of preference judgments.

Results. The results are summarized in Fig. 1. It can be seen that, the judgments from strict preferences are far more than the one when allowing for ties, namely, on average 500% more judgments are required. Compared with the number of judgments required by graded judgments, the numbers are 43% and 773% higher respectively when allowing and not allowing for ties.

3.3   Discussion

Results from Sects. 3.1 and 3.2 demonstrate that ties can dramatically reduce the number of judgments. Compared with strict preferences, ties actually produce coarser ground-truth rankings. This can be seen from the analytical results

---

[1] http://trec.nist.gov/tracks.html.

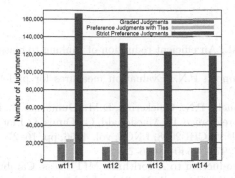

Fig. 1. The average number of judgments required by graded judgments and by preference judgments with/without ties on TREC Web Track. The x-axis is different years and y-axis represents the number of judgments. The averaged number of judgments from 1000 repetitions is reported as the actual number of judgments for both kinds of preference judgments.

$\mathcal{O}(N_d \log N_t)$ from Sect. 3.1: when $N_t = N_d$ ($\beta = 1$) it becomes strict preferences; and the number is reduced when $N_t < N_d$, where more documents are "squeezed" into a single tie partition. Meanwhile, the ground-truth ranking of documents is simplified to the ranking of tie partitions. In the example from Sect. 1, $d_2 \sim d_3$ and $d_4 \sim d_5$ are in the ground-truth ranking, meaning that the ground-truth relative rankings in between d_2 and d_3 and in between d_4 and d_5 are undetermined. In other words, the relative rankings between them are not considered in the evaluation as in [2]. Thus, the ties can be regarded as a compromise between the number of judgments and the judgment granularity.

Finally, we discuss whether there is potential to reduce the number of judgments with ties beyond QUICK-SORT-JUDGE. Similar to the strategy employed in [9], ideally, one can first make tie judgments to cluster documents, and thereafter make non-tie judgments to sort the tie partitions. By doing this, the number of tie judgments remains the same, namely N_d. Whereas for non-tie judgments, the number of judgments under $d_p \in t_i$ and $d_p \in t_j$ becomes 1 in Table 1, which means that one only needs to judge a pair of documents to determine the relative order of two established tie partitions. Accordingly, the number of judgments is reduced to $E(N_{jud}) = 2 \sum_{i=1}^{N_t-1} \sum_{k=2}^{N_t-i+1} \frac{1}{k} + N_d < 2N_t H_{N_t} + N_d$, which is in $\mathcal{O}(2N_t \log N_t + N_d)$ and is close to linear when $N_t \ll N_d$.

4   Conclusion

In this work, we analytically derive and empirically simulate the number of judgments required in preference judgments. We demonstrate that the number of judgments can be reduced by simply allowing for ties, from $N_d \log N_d$ to $N_d \log N_t$. For future works, as discussed in Sect. 3.3, novel judgment mechanisms are desired to better utilize ties.

References

1. Ailon, N., Mohri, M.: An efficient reduction of ranking to classification. arXiv 2007 (2007)
2. Carterette, B., Bennett, P.N.: Evaluation measures for preference judgments. In: SIGIR 2008 (2008)
3. Carterette, B., Bennett, P.N., Chickering, D.M., Dumais, S.T.: Here or there: preference judgments for relevance. In: Macdonald, C., Ounis, I., Plachouras, V., Ruthven, I., White, R.W. (eds.) ECIR 2008. LNCS, vol. 4956, pp. 16–27. Springer, Heidelberg (2008). doi:10.1007/978-3-540-78646-7_5
4. Cormen, T.H.: Introduction to Algorithms. MIT Press, Cambridge (2009)
5. Kazai, G., Yilmaz, E., Craswell, N., Tahaghoghi, S.M.: User intent and assessor disagreement in web search evaluation. In: CIKM 2013 (2013)
6. Niu, S., Guo, J., Lan, Y., Cheng, X.: Top-k learning to rank: labeling, ranking and evaluation. In: SIGIR 2012 (2012)
7. Radinsky, K., Ailon, N.: Ranking from pairs and triplets: information quality, evaluation methods and query complexity. In: WSDM 2011 (2011)
8. Song, R., Guo, Q., Zhang, R., Xin, G., Wen, J.-R., Yu, Y., Hon, H.-W.: Select-the-best-ones: a new way to judge relative relevance. Inf. Process. Manag. **47**(1), 37–52 (2011)
9. Wang, J., Li, G., Kraska, T., Franklin, M.J., Feng, J.: Leveraging transitive relations for crowdsourced joins. In: SIGMOD 2013 (2013)

Using Section Headings to Compute Cross-Lingual Similarity of Wikipedia Articles

Monica Lestari Paramita[1(✉)], Paul Clough[1], and Robert Gaizauskas[2]

[1] Information School, University of Sheffield, Sheffield, UK
{m.paramita,p.d.clough}@sheffield.ac.uk
[2] Computer Science Department, University of Sheffield, Sheffield, UK
r.gaizauskas@sheffield.ac.uk

Abstract. Measuring the similarity of interlanguage-linked Wikipedia articles often requires the use of suitable language resources (e.g., dictionaries and MT systems) which can be problematic for languages with limited or poor translation resources. The size of Wikipedia can also present computational demands when computing similarity. This paper presents a 'lightweight' approach to measure cross-lingual similarity in Wikipedia using section headings rather than the entire Wikipedia article, and language resources derived from Wikipedia and Wiktionary to perform translation. Using an existing dataset we evaluate the approach for 7 language pairs. Results show that the performance using section headings is comparable to using all article content, dictionaries derived from Wikipedia and Wiktionary are sufficient to compute cross-lingual similarity and combinations of features can further improve results.

Keywords: Wikipedia similarity · Cross-language similarity

1 Introduction

As the largest Web-based encyclopedia, Wikipedia contains millions of articles written in 295 languages and covering a large number of domains[1]. Many articles describe the same topic in different languages, connected via *interlanguage-links*. Measuring cross-lingual similarity within these articles is required for tasks, such as building comparable corpora [4]. However, this can be challenging due to the large number of Wikipedia language pairs and the limited availability of suitable language resources for some languages [7]. Language-independent methods for computing cross-lingual similarity have been proposed, for example based on character n-gram overlap, but the accuracy of such methods decreases significantly for dissimilar language pairs [2].

Based on previous work [6], we propose a method for computing similarity across languages using scalable, yet lightweight, approaches based on structural similarity (comparing section headings) and using translation resources built

---

[1] https://en.wikipedia.org/wiki/List_of_Wikipedias (20 Oct 2016).

© Springer International Publishing AG 2017
J.M. Jose et al. (Eds.): ECIR 2017, LNCS 10193, pp. 633–639, 2017.
DOI: 10.1007/978-3-319-56608-5_59

from Wikipedia and Wiktionary. This paper addresses the following research questions: (RQ1) How effective are section headings for computing article similarity compared to using the full content? and (RQ2) How effective is information derived from Wikipedia and Wiktionary for translating section headings compared to using high-quality translation resources?

2 Related Work

Since interlanguage-linked Wikipedia articles describe the same topic, they have often been assumed to contain similar content and have been utilised for various tasks, such as mining parallel sentences [1] and building bilingual dictionaries [3]. The similarity of these articles across languages, however, may vary widely and have not been thoroughly investigated in the past. One study that analysed Wikipedia similarity [6] identified characteristics contributing to cross-lingual similarity, including overlapping named entities and similar structure. Features, such as the overlap of links, character n-gram overlap and cognate overlap of the article contents have been investigated as ways to automatically identify cross-lingual similarity with promising results [2]. Previous work, however, have not explored structural similarity features to identify cross-lingual similarity of Wikipedia articles.

The approach we propose makes use of Wikipedia and Wiktionary to assist in translating section headings (previously identified as a possible indicator of an article's structural similarity [6]), prior to computing similarity. Both resources have been used to compute cross-lingual similarity [1,5] and semantic relatedness [8]. However, past work has often focused on highly-resourced language pairs. This study investigates the use of these resources for under-resourced language pairs.

3 Methodology

The contents of most Wikipedia articles are structured into sections and subsections, e.g. the Wikipedia article of "United Kingdom" includes the following section headings (titles): *Etymology, History, Geography,* etc. Our method aims to measure cross-lingual similarity between a document pair D_1 and D_2 in a non-English language (L_1) and English (L_2) by measuring the similarity between their section headings, which is computationally more efficient than comparing the entire content. We refer to these section headings as H_1 and H_2, respectively. The approach is described in Sect. 3.1 and evaluation setup in Sect. 3.2.

3.1 Proposed Approach to Compute Cross-Lingual Similarity

Dictionary Creation. Firstly, two dictionaries are built using Wikipedia and Wiktionary, a multilingual dictionary available in 152 languages[2]. An existing link-based bilingual lexicon method [1] was used to extract the titles of

---

[2] https://meta.wikimedia.org/wiki/Wiktionary#List_of_Wiktionaries (20 Oct 2016).

Wikipedia interlanguage-linked articles for each language pair, using them as dictionary entries. We supplemented this lexicon with entries from Wiktionary, as this contains more lexical knowledge compared to Wikipedia [5]. This was performed by collecting English Wiktionary entries and their translations in non-English language pairs.

Translation of Section Headings. Firstly, common headings that do not make useful contributions when computing article similarity, such as *References*, *External Links* and *See Also*, were filtered out. Stopwords were also removed using a list of frequent words gathered from Wikipedia (an average size of 871 words per language). Afterwards, the English section headings (H_2) are translated into L_1 (the non-English language), resulting in H'_2. For each section heading (h_1, h_2, ..., h_n) in H_2, the translation process is as follows:

1. If h_i exists in the dictionary, then extract all of its translations t_i.
2. If h_i does not exist as an entry in the dictionary:
   (a) If h_i includes > 1 word, split the heading h_i into each word (w_1, w_2, ..., w_n) and translate each word separately.
   (b) If no translation is found for a given word, trim 1 character from the end of the word and search for its translation. Perform this recursively until either a translation is found, or the original word has 4 characters left.
   (c) Perform step (a) for all words in h_i and concatenate the results.
3. Both h_i and t_i (if found) are then included in H'_2.
4. Steps 1-3 are repeated until all headings in H_2 have been translated.

Identification of Structural Similarity. In this stage, we aim to align similar section headings in both documents. Firstly, every source heading $s_i \in H_1$ is paired to every target heading $t_j \in H'_2$. For each s_i, we identify the most similar target heading t_n (allowing many-to-one alignments) using the following alignment and section similarity scoring ($secSimScore$) methods:

1. If s_i is contained in t_j, both headings are aligned; $secSimScore(s_i, t_j) = 1$.
2. If not, split heading s_i into each word ($w_1, w_2, ..., w_p$):
   (a) Find if w_m is included in t_j. If not, recursively trim w_m by 1 character until either it is included in t_j, or w_m has 4 characters left.
   (b) Perform step (a) for all words in s_i; $secSimScore(s_i, t_j)$ is calculated by measuring the proportion of words in s_i that are found in t_j.
3. Step 1-2 are performed between s_i and the remaining sections in H'_2. After which, the highest scoring pair is selected as the alignment for s_i.

After all the aligned sections in H_1 and H'_2 are identified, referred to as A_1 and A'_2, respectively ($A_1 \in H_1$ and $A'_2 \in H'_2$), the scores are aggregated to derive a structure similarity score for the document pair ($docSimScore$). Three different methods to measure the $docSimScore$ are investigated:

1. **align1**: This method does not take the *secSimScore* of the aligned sections into account, but instead relies on the number of aligned sections in both documents only:

$$docSimScore = \frac{(|A_1| + |A_2'|)}{(|H_1| + |H_2'|)} \tag{1}$$

where $|A_1|$ and $|A_2'|$ represent the number of aligned sections in H_1 and H_2', respectively, and $|H_1|$ and $|H_2'|$ are the number of sections in H_1 and H_2'.
2. **align2**: This method takes the *secSimScore* into account. In Eq. 1, $|A_1|$ is replaced with the sum of *secSimScore* for each aligned section in A_1.
3. **align3**: In this method, aligned sections with *secSimScore* < 1 are filtered out, prior to calculating *align3* using Eq. 1.

An additional feature, *the ratio of section length* (**sl**), is also extracted by measuring the ratio of number of section headings in both articles.

3.2 Evaluation Setup

To evaluate the approach we used an existing Wikipedia similarity corpus [6] containing 800 document pairs from 8 language pairs. Two annotators assessed the similarity of each document pair using a 5-point Likert Scale. Due to the unavailability of Wiktionary translation resource in Croatian-English, only 7 language pairs are used in this study: German (a *highly-resourced* language), and 6 *under-resourced* languages: Greek (EL), Estonian (ET), Lithuanian (LT), Latvian (LV), Romanian (RO) and Slovenian (SL); all paired to English (EN). Documents without section headings were removed for these experiments, resulting in 600 document pairs across the 7 language pairs. We compare the proposed methods to **c3g**, the tf-idf cosine similarity of the *char-3-gram overlap* between the article contents[3]. To investigate the effectiveness of Wikipedia-Wiktionary as translation resources, we use Google Translate as a state-of-the-art comparison.

4   Results and Discussion

(RQ1) How effective are section headings for computing article similarity compared to using the full content? We report the Spearman-rank correlations between similarity scores computed using methods from Sect. 3.1 and the average human-annotated similarity scores from the evaluation corpus in Table 1 ("Individual Features"). Results show that features based on section headings ($\rho = 0.36$ for *align1*) were able to achieve comparable overall correlations compared to using char-3-gram overlap (*c3g*) on the entire article contents ($\rho = 0.34$). Results using *align2* was similar ($\rho = 0.35$). The *align3* method, however, achieved significantly lower score ($\rho = 0.23$), suggesting that the strict alignment process may have lost valuable cross-lingual information. Section length

---

[3] This feature was previously identified as the best language-independent feature to identify cross-lingual similarity in Wikipedia [2].

Table 1. Correlation scores (Spearman's ρ) of individual and combined features

Lang	Individual Features					Combined Features		
	Section Headings (SH)				Article	SH	SH + Article	
	align1	align2	align3	sl	c3g	align1_sl	sl_c3g	align1_sl_c3g
DE	0.33*	0.28	−0.01	0.45*	**0.46***	0.42*	**0.67***	0.59*
EL	0.17	0.19	0.19	**0.42***	0.38*	0.36*	**0.56***	0.47*
ET	0.27*	0.29*	0.29*	0.37*	**0.57***	0.37*	**0.58***	0.54*
LT	0.43*	**0.44***	0.39*	0.40*	0.34*	0.54*	0.51*	**0.58***
LV	0.31*	0.33*	0.18	**0.34***	**0.34***	0.40*	0.46*	**0.49***
RO	**0.54***	**0.54***	0.51*	0.14	0.20	**0.40***	0.20	0.39*
SL	**0.41***	0.32*	0.00	0.33*	0.03	**0.44***	0.33*	0.42*
Avg	**0.36**	0.35	0.23	0.35	0.34	0.42	0.49	**0.50**

Note: *$p < 0.01$; the best results for the "Individual Features" and "Combined Features" are shown in bold; "Avg" score is calculated using Fisher transformation.

(sl) was shown to perform consistently across most language pairs ($\rho = 0.35$). The $c3g$ method, however, performed poorly for RO-EN and SL-EN ($\rho = 0.20$ and $\rho = 0.03$, not statistically significant), possibly due to dissimilar surface forms between languages. Section heading features were shown to achieve either the same or better correlation scores than $c3g$ in 5 of the 7 language pairs.

Our findings also suggest that a combination of features produces a more robust similarity measure. Table 1 ("Combined Features") reports the three best feature combinations. Firstly, a combination of only Section Headings (SH) features, $align1\_sl$, increases the correlation score to 0.42 ($\uparrow$16.67% compared to $align1$, the best individual feature). Correlation can further be increased by combining both SH and article features. We show that sl_c3g achieves $\rho = 0.49$ ($\uparrow$36.11%); considering that this feature can be computed without the need of a dictionary, this result is very promising. Lastly, the combination of three features, $align1\_sl\_c3g$, achieves the highest correlation score ($\rho = 0.50$; $\uparrow$38.89%).

(RQ2) How effective is information derived from Wikipedia and Wiktionary for translating section headings compared to using high-quality translation resources? Figure 1(a) shows the dictionary size derived from Wikipedia and Wiktionary used in this study, highlighting low numbers of entries for all under-resourced languages. To investigate the effect of different translation resources, we computed the $align1$ method using a high-quality translation resource: in this case Google Translate ($gAlign1$). The correlation scores of the original $align1$ method (using the Wiki resources) and $gAlign1$ are shown in Fig. 1b). Although a much higher $gAlign1$ correlation was achieved in EL-EN ($\rho = 0.46$, compared to $\rho = 0.17$ for $align1$), the correlation scores for the remaining language pairs are very similar. In some language pairs (DE-EN, ET-EN, and RO-EN), the use of Wikipedia-Wiktionary resources

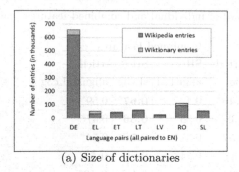
(a) Size of dictionaries

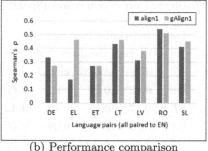
(b) Performance comparison

Fig. 1. Translation resources

achieved either the same or better correlation scores compared to using Google Translate. Our findings also show that the dictionary size does not significantly affect the performance of the section heading alignment methods. For example, LV-EN, which has the smallest dictionary (24.4 K entries) achieves similar *align1* correlation to DE-EN (the largest dictionary with 641 K entries). We also found that, although much smaller in size, an average of 66% of Wiktionary entries are not available in the Wikipedia lexicon; this shows the importance of Wiktionary in complementing the Wikipedia lexicon.

5   Conclusions and Future Work

This paper describes a 'lightweight' approach for identifying cross-lingual similarity of Wikipedia articles by measuring the structural similarity (i.e. similarity of section headings) of the articles. Results show that the section heading similarity feature (*align1*) and ratio of section length (*sl*) can be used to identify cross-lingual similarity with comparable performance to using the overlap of char-3-grams (*c3g*) on content from the entire article ($\rho = 0.36$, $\rho = 0.35$, and $\rho = 0.34$, respectively). A combination of these three features also further improves the results ($\rho = 0.50$). The use of Wikipedia-Wiktionary resource in this approach was shown to be as efficient to utilising Google Translate for many language pairs. These results are promising as these resources are freely available for a large number of languages. Future work will investigate more feature combinations and to measure similarity in Wikipedia in more language pairs.

References

1. Adafre, S.F., de Rijke, M.: Finding similar sentences across multiple languages in Wikipedia. In: Proceedings of EACL 2006, pp. 62–69, 4 April 2006
2. Barrón-Cedeño, A., Paramita, M.L., Clough, P., Rosso, P.: A comparison of approaches for measuring cross-lingual similarity of Wikipedia articles. In: Rijke, M., Kenter, T., Vries, A.P., Zhai, C.X., Jong, F., Radinsky, K., Hofmann, K. (eds.) ECIR 2014. LNCS, vol. 8416, pp. 424–429. Springer, Heidelberg (2014). doi:10.1007/978-3-319-06028-6_36

3. Erdmann, M., Nakayama, K., Hara, T., Nishio, S.: An approach for extracting bilingual terminology from Wikipedia. In: Haritsa, J.R., Kotagiri, R., Pudi, V. (eds.) DASFAA 2008. LNCS, vol. 4947, pp. 380–392. Springer, Heidelberg (2008). doi:10.1007/978-3-540-78568-2_28

4. Mohammadi, M., GhasemAghaee, N.: Building bilingual parallel corpora based on Wikipedia. In: ICCEA 2010, vol. 2, pp. 264–268. IEEE (2010)

5. Müller, C., Gurevych, I.: Using Wikipedia and Wiktionary in domain-specific information retrieval. In: Peters, C., Deselaers, T., Ferro, N., Gonzalo, J., Jones, G.J.F., Kurimo, M., Mandl, T., Peñas, A., Petras, V. (eds.) CLEF 2008. LNCS, vol. 5706, pp. 219–226. Springer, Heidelberg (2009). doi:10.1007/978-3-642-04447-2_28

6. Paramita, M.L., Clough, P., Aker, A., Gaizauskas, R.J.: Correlation between similarity measures for inter-language linked Wikipedia articles. In: LREC 2012, pp. 790–797 (2012)

7. Yasuda, K., Sumita, E.: Method for building sentence-aligned corpus from Wikipedia. In: Proceedings of WikiAI 2008, pp. 64–66, 13–14 July 2008

8. Zesch, T., Müller, C., Gurevych, I.: Using Wiktionary for computing semantic relatedness. In: AAAI Conference on Artificial Intelligence, pp. 861–866 (2008)

Learning to Rank for Consumer Health Search: A Semantic Approach

Luca Soldaini[✉] and Nazli Goharian

Information Retrieval Lab, Georgetown University, Washington, DC, USA
{luca,nazli}@ir.cs.georgetown.edu

Abstract. For many internet users, searching for health advice online is the first step in seeking treatment. We present a Learning to Rank system that uses a novel set of syntactic and semantic features to improve consumer health search. Our approach was evaluated on the 2016 CLEF eHealth dataset, outperforming the best method by 26.6% in NDCG@10.

Keywords: Learning to rank · Medical search · Consumer health search

1 Introduction

In recent years, the internet has become a primary resource for health information[1]. In searching medical information, lay people value access to trustworthy information [6], which has been shown to lead to a better understanding of health topics [9]. However, trustworthy health care resources—even those targeted at laypeople—use proper medical vocabulary, causing consumers to struggle [13].

In this paper, we propose a Learning to Rank (LtR) system that takes advantage of syntactic and semantic features to address the language gap between health seekers and medical resources. LtR algorithms have been successfully employed to improve retrieval of web pages [4]. In the health domain, they have been recently used to promote understandability in medical health queries [5] and retrieve medical literature [4]. The authors of this manuscript has previously experimented with the use of semantic relationships between terms in [7]. In this work we show how semantic features that capture the similarity between the query and retrieved documents can be effectively coupled with classic statistical features—such as those used in the LETOR dataset [4]—to promote relevant medical documents that answer consumer health queries.

Our approach is validated using the 2016 CLEF eHealth IR Task dataset [14], a collection of 300 medical queries designed to resemble laypeople health queries. Documents were retrieved from the category B subset of ClueWeb12[2]. We compared our approach to the best known baseline for this dataset, achieving a 26.6% improvement in terms of NDCG@10.

---

[1] http://www.pewinternet.org/2013/01/15/health-online-2013/.
[2] http://lemurproject.org/clueweb12/.

J.M. Jose et al. (Eds.): ECIR 2017, LNCS 10193, pp. 640–646, 2017.
DOI: 10.1007/978-3-319-56608-5_60

2   Methodology

2.1   Features

We proposed a combination of statistical and semantic features to train a LtR model. The feature set can be partitioned in five groups:

Statistical (STAT, 36 features): We considered a subset of features from the LETOR benchmark dataset, which have been shown to be useful in many LtR systems [4]. These features encode statistical information about the terms in the query and documents(e.g., term frequency (tf), inverse document frequency (idf)). We remand the reader to [4] for a complete list. We excluded some features because they are not available for our dataset (e.g., HITS scores). We also excluded all features that relied on the titles of webpages, as they showed poor correlation with relevance judgments in our tests.

Statistical Health (ST-HEALTH, 9 features): We expanded the set of statistical features by including health-specific features. We consider whether a document is certified by the *Health on Net Foundation*[3], an organization that publishes a code of good conduct for health websites. Such signal has been shown to be a good indicator of informative web sites [9]. We also extracted tf and idf of all terms in the document that can be found in the subset of health-related pages in Wikipedia, which were extracted following as in [9]. The average, variance, mode, and sum of tf and idf were used as features.

UMLS (UMLS, 26 features): The Unified Medical Language System[4] (UMLS) is a medical ontology maintained by the U.S. National Library of Medicine. Terms in this ontology are organized by concepts, each of which is associated with one or more semantic type. UMLS concepts are often present in queries issued by laypeople; thus, we explored their used as to identify relevant search results. To obtain the set of UMLS concepts in each document and in the query we used QuickUMLS [8], a medical concept extraction system. We match UMLS expressions belonging to 16 semantic types that are associated with symptoms, diagnostic tests, diagnoses, or treatments, as suggested in [8].

Latent Semantic Analysis (LSA, 2 features): To extract semantic relationships between terms, we built a 100-dimension Latent Semantic Analysis (LSA) model using a collection of 9,379 entries from the A.D.A.M. Medical Encyclopedia[5] (a consumer-oriented medical encyclopedia) and the MedScape[6] reference guide. The model was used to obtain vector representations of terms in the query and documents, which were summed using two strategies: simple sum and sum weighted by the probability of each term appearing in the health section of Wikipedia. This composition technique, while simple, has been shown to be very effective [1]. To extract LSA features, we computed the euclidean distance

---

[3] https://www.healthonnet.org/.
[4] https://www.nlm.nih.gov/research/umls/.
[5] https://medlineplus.gov/encyclopedia.html.
[6] http://reference.medscape.com/.

between the vector representing the query and the vector for the document. We used the similarity scores from the weighted and unweighted models as features.

Word Embeddings (w2v, 4 features): Similarly to [3], we used word embeddings trained on PubMed[7] and Google News[8] to obtain dense vector representations for terms in the document and in the query. Word embeddings from the medical domain provide a strong representation for medical terms, while general domain word embeddings should capture the terms lay people are be more familiar with. As in LSA, we used a sum and a weighted sum to compose the term vectors into the vector representation of the document or query. In total, 4 features were extracted: weighted and unweighted similarities between document and query using PubMed and Google News models.

2.2 Ranking Algorithms

LtR algorithms are typically partitioned in three groups: point-wise, pair-wise, and list-wise learners. We considered the following LtR algorithms: logistic regression, random forests, LambdaMART [11], AdaRank [12], and ListNet [2]. Logistic regression and random forests are point-wise algorithms; we trained them to predict, for each document, its likelihood of being relevant. LambdaMART, a pair-wise learner, is an ensemble method that aims at minimizing the number of inversions in ranking. ListNet and AdaRank are list-wise learners that are designed to find a permutation of the retrieved results such that the value of a loss function on the list of results is minimized. We used the implementation of LambdaMART, AdaRank, and ListNet available in RankLib[9] v.2.7.

3   Experimental Setup

Dataset: The proposed LtR approach to laypeople medical search was evaluated on the 2016 CLEF eHealth IR Task dataset [14]. The dataset consists of 300 queries modeled after 50 distinct topics. The topics were created by health professional from forum posts from the *AskDocs* section of Reddit; Results for the queries were retrieved from the ClueWeb12 category B dataset, a collection of 53 million web pages. In total, 25,000 documents were evaluated; to each one, a score between 0 and 2 was assigned. Because all queries created from the same forum post share the same information need, relevance judgments of queries on the same topic are identical. On average, 74.1 documents were deemed relevant for each query (min: 1; max: 335; median: 45; std.dev.: 74.7).

Experiments: Documents were indexed using the Terrier search engine, v. 4.0^{10}. As a baseline, we consider the BM25 scoring function defined by the CLEF

---

[7] https://github.com/cambridgeltl/BioNLP-2016/.
[8] https://code.google.com/archive/p/word2vec/.
[9] https://sourceforge.net/p/lemur/wiki/RankLib/.
[10] http://terrier.org/.

eHealth organizers in [14]. While simple, this baseline outperformed all 10 teams (29 runs) who participated in shared task. We use this baseline to retrieve up to 1,000 documents per query to train the LtR methods. All learners were trained under five fold cross validation and manually tuned using a separate validation set. Pair-wise and list-wise learners were configured to optimize NDCG@10 on the validation set. To avoid overfitting, we carefully generated the training, validation, and test set so that all queries from the same group are part of the same split. Finally, P@10 and NDCG@10 were used to evaluate all the approaches, as users of online search engines are more likely to pay attention to the first page of retrieved results than the subsequent ones.

4 Results

4.1 LtR Algorithms

We compare the LtR approaches from Sect. 2.2 with the baseline used in [14]. For all experiments, learners are trained on all the features described in Sect. 2.1; we will study the impact of individual features in Sect. 4.2.

Table 1. Performance of LtR algorithms on the dataset. Runs marked with * are significantly different from the baseline (Paired Student's t-test, $p < 0.05$).

Method	Type of approach	NDCG@10		P@10	
BM25 baseline [14]	n/a	0.241		0.291	
Random forests	Point-wise	0.249	(+3.3%)	0.293	(+0.6%)
Logistic regression	Point-wise	0.262*	(+8.7%)	0.317*	(+8.9%)
LambdaMART [11]	Pair-wise	**0.305***	**(+26.6%)**	**0.361 ***	**(+24.1%)**
AdaRank [12]	List-wise	0.239	(−0.8%)	0.292	(−0.7%)
ListNet [2]	List-wise	0.267*	(+10.8%)	0.333*	(+ 14.4%)

Of all learners reported in Table 1, LambdaMART achieves the best performance (+26.6% NDCG@10, +24.1% P@10 over the baseline). This demonstrates that LtR can be successfully exploited to improve the access to relevant medical resources that satisfy the need of online health seekers. As expected, LambdaMART outperforms point-wise LtR approaches, as it is often the case [4]. LambdaMART also achieves better performance than the two list-wise methods, AdaRank and ListNet (difference is statistically significant for both). This is to be expected, as previous work found LambdaMART to be very competitive in LtR tasks on web results when optimizing for NDCG@10 [10].

4.2   Feature Analysis

The performance of the model trained on each set of features is presented in Table 2. We observe that the model trained only on the statistical features (STAT) obtains better performances than models trained on other sets of features. This is to be expected, as statistical features were modeled after the LETOR feature set, which has been shown to be very effective for LtR tasks [4]. The model trained solely on statistical health features (ST-HEALTH) ranks second, suggesting that the presence and frequency of health terms plays an important role in identifying relevant results. This intuition is reinforced by the findings shown in Table 3, where ST-HEALTH features are among the highest ranked in terms of importance.

Table 2. Performance of LambdaMART trained on each set of features separately. All runs are significantly different from the best method (Paired Student's t-test, $p < 0.05$).

Features group	NDCG@10	P@10
BM25 baseline	0.241	0.291
STAT	0.274	0.322
ST-HEALTH	0.260	0.311
UMLS	0.253	0.307
w2v	0.160	0.210
LSA	0.121	0.188
All features	**0.305**	**0.361**

Table 3. Top 10 features ranked by weight (normalized). The weight of each feature was computed by averaging their information gain.

Feature	Group	Weight
Avg. *idf* in health Wikipedia	ST-HEALTH	0.0995
# of matching UMLS concepts in document	UMLS	0.0776
Avg. *tf* in health Wikipedia	ST-HEALTH	0.0616
BM25 similarity score	STAT	0.0605
# concepts in *"Sign or Symptom"* UMLS semantic type	UMLS	0.0579
Similarity weighted word embeddings PubMed	w2v	0.0521
# concepts in *"Injury or Poisoning"* UMLS semantic type	UMLS	0.0418
LM similarity score	STAT	0.0408
Similarity weighted word embeddings Google News	w2v	0.0393
Spam scores	STAT	0.0335

The UMLS features set shows limited improvements over the BM25 baseline. However, based on their ranking in Table 3, we argue that they have an important role in model built using all features, as they capture information about symptoms and diseases mentioned in the queries.

Lastly, we note that neither word embedding similarity features (w2v) nor latent semantic analysis similarity features (LSA) features are enough to train an effective LtR model by themselves. This outcome could be due to the fact that these features sets, which contain just 4 and 2 features, do not encode enough information to train a comprehensive model. However, while w2v features improve the effectiveness of the model when combined with other features (Table 3), LSA features have less of an impact on the model built by LambdaMART. This might be due to the fact that the set of 9,379 pages the LSA model was trained on is too small to capture the semantic similarity between queries and the retrieved documents. Conversely, similarity features derived by dense word representations are effective for this task as long as the model used to derive them is accurate.

4.3 Query Performance

In this section, we compare the per-query performance of the baseline with the best ranker from Table 1. Results are shown in Fig. 1. Rather than reporting the individual NDCG@10 for each query, we average the results of all queries that belong to the same query group. This approach is motivated by the fact that all queries in the same group share the same information need (and document relevance judgments). Therefore, by averaging the performance of all queries in the same group, we can study whether the performance of the best ranker relative to the baseline is due to the information need associated with each query. To convince the reader that this representation is justified, the variance for each query group is shown in Fig. 1. As the variance for each topic is moderate, we conclude that our approach is appropriate.

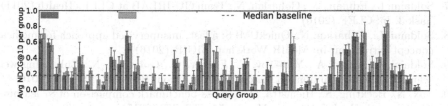

Fig. 1. NDCG@10 of the baseline and the best performing method of Table 1. To increase the clarity of the figure, we averaged the value of NDCG@10 of all queries from the same query group (i.e., all queries sharing the same information need.)

The proposed ranker outperforms the baseline on 36 out of 50 topics. Interestingly, LambdaMART outperforms the baseline in all but one query whose NDCG@10 is below median. In other words, there exists a statistically significant correlation between the performance of the baseline on each query and the difference between the NDCG@10 of the baseline and LambdaMART (Spearman's rank correlation, $r_s = -0.38$, $p < 0.05$). This suggests that LtR is a viable strategy for addressing difficult queries; however, its performance are still bounded by the quality of results retrieved by the baseline.

5 Conclusions

In this paper we proposed a novel set of syntactic and semantic features for LtR for consumer health queries. The proposed approach led to a 26.2% increase in NDCG@10 over existing methods. The impact of several Learning to Rank algorithms was studied; furthermore, we discussed the effectiveness of our proposed features. This work demonstrates that semantic features can be effectively exploited for LtR in laypeople health search.

References

1. Blacoe, W., Lapata, M.: A comparison of vector-based representations for semantic composition. In: ACL. Association for Computational Linguistics (2012)
2. Cao, Z., Qin, T., Liu, T.-Y., Tsai, M.-F., Li, H.: Learning to rank: from pairwise approach to listwise approach. In: ICML (2007)
3. Cohan, A., Fong, A., Goharian, N., Ratwani, R.: A neural attention model for categorizing patient safety events. In: ECIR (2017)
4. Liu, T.-Y., Xu, J., Qin, T., Xiong, W., Li, H.: LETOR: benchmark dataset for research on learning to rank for information retrieval. In: LTR Workshop, SIGIR (2007)
5. Palotti, J., Goeuriot, L., Zuccon, G., Hanbury, A.: Ranking health web pages with relevance and understandability. In: SIGIR (2016)
6. Palotti, J., Hanbury, A., Müller, H., Kahn, C.E.: How users search and what they search for in the medical domain. IR J. **19**, 189–224 (2016)
7. Soldaini, L., Edman, W., Goharian, N.: Team GU-IRLAB at CLEF eHealth (2011): Task 3. In: CLEF (2016)
8. Soldaini, L., Goharian, N.: QuickUMLS: a fast, unsupervised approach for medical concept extraction. In: MedIR Workshop, SIGIR (2016)
9. Soldaini, L., Yates, A., Yom-Tov, E., Frieder, O., Goharian, N.: Enhancing web search in the medical domain via query clarification. IR J. **19**, 149–173 (2016)
10. Tax, N., Bockting, S., Hiemstra, D.: A cross-benchmark comparison of 87 learning to rank methods. Inf. Process. Manag. **51**, 757–772 (2015)
11. Wu, Q., Burges, C.J., Svore, K.M., Gao, J.: Adapting boosting for information retrieval measures. IR J. **13**, 254–270 (2010)
12. Xu, J., Li, H.: AdaRank: a boosting algorithm for information retrieval. In: SIGIR (2007)
13. Zeng, Q.T., Kogan, S., Plovnick, R.M., Crowell, J., Lacroix, E.-M., Greenes, R.A.: Positive attitudes, failed queries: an exploration of the conundrums of consumer health information retrieval. J. Med. Inf. **73**, 45–55 (2004)
14. Zuccon, G., Palotti, J., Goeuriot, L., Kelly, L., Lupu, M., Pecina, P., Mueller, H., Budaher, J., Deacon, A.: The IR task at the CLEF eHealth evaluation lab (2011): user-centred health information retrieval. In: CLEF (2016)

Matrix Factorisation with Word Embeddings for Rating Prediction on Location-Based Social Networks

Jarana Manotumruksa[✉], Craig Macdonald, and Iadh Ounis

University of Glasgow, Glasgow, UK
j.manotumruksa.1@research.gla.ac.uk,
{craig.macdonald,iadh.ounis}@glasgow.ac.uk

Abstract. With vast amounts of data being created on location-based social networks (LBSNs) such as Yelp and Foursquare, making effective personalised suggestions to users is an essential functionality. Matrix Factorisation (MF) is a collaborative filtering-based approach that is widely used to generate suggestions relevant to user's preferences. In this paper, we address the problem of predicting the rating that users give to venues they visit. Previous works have proposed MF-based approaches that consider auxiliary information (e.g. social information and users' comments on venues) to improve the accuracy of rating predictions. Such approaches leverage the users' friends' preferences, extracted from either ratings or comments, to *regularise* the complexity of MF-based models and to avoid over-fitting. However, social information may not be available, e.g. due to privacy concerns. To overcome this limitation, in this paper, we propose a novel MF-based approach that exploits word embeddings to effectively model users' preferences and the characteristics of venues from the textual content of comments left by users, regardless of their relationship. Experiments conducted on a large dataset of LBSN ratings demonstrate the effectiveness of our proposed approach compared to various state-of-the-art rating prediction approaches.

1 Introduction

In recent years, location-based social networks (LBSNs) such as Yelp and Foursquare have emerged as popular platforms that allow users to search for Point-of-Interest and post ratings as well as their opinions/comments about venues they have visited. This makes LBSN data very suitable for making recommendations of venues for users to visit. Matrix Factorisation (MF) – based on collaborative filtering – is a popular technique used to effectively recommend items to users by assuming that users who share similar preferences (rating positively or negatively the same items) are likely to prefer similar items [1]. Previous works [2,3] proposed MF-based approaches that leverage such users' explicit feedback (e.g. ratings) to model their preferences, and thereby effectively suggest new venues for users to visit. However, rating data is sparse in nature,

© Springer International Publishing AG 2017
J.M. Jose et al. (Eds.): ECIR 2017, LNCS 10193, pp. 647–654, 2017.
DOI: 10.1007/978-3-319-56608-5_61

i.e. users/venues have very few ratings, hindering the quality of venue suggestions. To alleviate the sparsity problem, various MF-based approaches [3,4] have
been proposed to consider auxiliary information such as the ratings of each user's
friends to effectively predict the user's ratings. In particular, Ma *et al.* [4] proposed a social-based regularisation technique that *regularises* the complexity of
a MF model, by assuming that users are likely to be influenced by their friends
who rate similar venues with similar scores.

Apart from the social information, the comments associated with ratings on
venues left by users can provide insights about why they rated a given venue
positively or negatively, while also reflecting characteristics of each venue. Previous works [2,3,5,6] have shown that the textual content of comments can be
leveraged to effectively model user's preferences and characteristics of venues.
Recently, word embeddings are being increasingly applied in many applications
due to their effectiveness in capturing semantic properties of textual content,
such as text classification [7] as well as recommendation system [3,8]. In particular, Musto *et al.* [8] apply several word embedding techniques to enhance the
effectiveness of content-based collaborative approaches for tasks such as book
and movie recommendation. Moreover, Manotumruksa *et al.* [3] extended the
regularisation technique proposed by Ma *et al.* [4], by exploiting word embeddings to estimate the similarity between friends from their comments of venues
they both visited. Their assumption is that users are not only influenced by
friends who like/dislike similar venues but are also influenced by friends who
share similar tastes, which can be extracted from the explicit textual feedback
in the form of comments they have left on venues. Unlike previous works mentioned above [2–4], this paper contributes a novel MF-based approach that jointly
models user's preferences and characteristic of venues from the textual content of
comments to effectively predict user-venue ratings. Experiments on a large real-
word dataset demonstrate the effectiveness of our proposed approach in comparison with various state-of-the-art user-venue rating prediction approaches. Our
proposed approach is as effective as state-of-the-art rating prediction approaches
that consider social information and textual content of comments.

2   Joint Linear Combinations of Matrix Factorisations

In this section, we first describe the problem of predicting user-venue rating and
detail traditional Matrix Factorisation (MF) techniques. Next, we explain how
we exploit word embeddings to enable MF to consider textual content of users'
comments.

2.1   Traditional Matrix Factorisation

We formally describe the notations and the problem of predicting user's rating
on venues. First, user-venue ratings are represented as a matrix $R \in \mathbb{R}^{m \times n}$ where
m and n are the number of users and venues. Let $r_{i,j}$ and $c_{i,j}$ denote the 1–5
scale rating and textual content of comment of user i on venue j, respectively.

Using the historical ratings R_i of user i, the task is then to predict the rating this user would give to venue j.

Matrix Factorisation (MF) is a collaborative filtering technique that assumes that users who share similar preferences (e.g. visiting similar venues and rating these venues similarly) are likely to influence each other [1]. In particular, the goal of MF is to reconstruct the rating matrix R by calculating the dot product of latent factors of users $U \in \mathbb{R}^{m \times d}$ and venues $V \in \mathbb{R}^{n \times d}$ where d is the number of latent dimensions:

$$R \approx \hat{R} = U^T V \qquad (1)$$

The MF model is trained by minimising a loss function L, which consists of sum-of-squared-error terms between the actual ratings and predicted ratings, as follows:

$$L(U,V) = \min_{U,V} \frac{1}{2} \sum_{i=1}^{m} \sum_{j=1}^{n} I_{i,j} \cdot (r_{i,j} - \hat{r}_{i,j})^2 + \frac{\lambda}{2}(\|U\|_F^2 + \|V\|_F^2) \qquad (2)$$

where λ is a regularisation parameter and $\|.\|_F^2$ denotes the Frobenius norm, used to avoid overfitting. $I_{i,j}$ is an indicator, which gives 1 if user i rated venue j, otherwise 0.

2.2   Combining Matrix Factorisations with Word Embeddings

As mentioned in Sect. 1, explicit feedback by users in comments can provide insights into why users rate a venue positively or negatively and also reflect the characteristics of the venue. Unlike previous works [2,3,5], we propose a MF approach that leverages the textual content of comments by other users, regardless of their relationship, to alleviate the sparsity problem. A straightforward approach would be to represent comments using a bag-of-words approach, however this would not consider the context in which terms occur, and hence not model the semantic properties of comments. Instead, we follow [6,8], by exploiting word embeddings to semantically model the user's preferences $S_u \in \mathbb{R}^{m \times k}$ and characteristics of venue $S_v \in \mathbb{R}^{n \times k}$ from the comments in a low-dimensional space, where k is the number of word embedding dimensions, as follows:

$$S_{u_i} = \sum_{c_{i,j} \in C_{u_i}} \sum_{t \in c_{i,j}} w2v(t) \times r_{i,j} \qquad S_{v_j} = \sum_{c_{i,j} \in C_{v_j}} \sum_{t \in c_{i,j}} w2v(t) \times r_{i,j} \qquad (3)$$

where C_{u_i} and C_{v_j} are the sets of user i's and venue j's comments and $w2v(t) \in \mathbb{R}^k$ is a function that returns a word embedding representation of term t. Note that the $w2v()$ function in Eq. (3) can be replaced with a word representation generated by more complex convolutional or recurrent neural networks (e.g. [7]). However, we consider other formulations beyond the scope of this paper and leave these as future work.

We note that the MF-based approach proposed by Hu *et al.* [2] leverages the comments to decompose latent factors of venues V into a combination of latent factors of comment terms, which can alleviate the sparsity problem for venues that have few ratings. However, their proposed approach lacks flexibility, because it requires the latent factors of comment's terms to be in the same space as the latent factors of venues V (i.e. the dimension d of these latent factors need to be equal). Instead, we argue that those two latent factors do not necessarily share the same space due to different nature of venues and comments. Intuitively, similar venues can be recognised by services provided by the venues, while similar comments can be recognised by terms appearing in the comments and their *semantics*. Therefore, the latent factors of venues and comments should not share the same dimensions, indeed the latent factors of comments should be larger due to the complexity of comments. Our work also differs from [3] in that comments are considered inherent to the matrix factorisations rather than the regularisation.

Hence – and unlike previous works [2,3] – to incorporate the representations of users' preferences S_u and characteristics of venues S_v within a joint MF model, we modify Eq. (1) to linearly combine matrix factorisations:

$$R \approx \hat{R} = \alpha U^T V + (1 - \alpha)(U_s^T S_u + V_s^T S_v) \tag{4}$$

where $U_s \in \mathbb{R}^{m \times k}$ and $V_s \in \mathbb{R}^{n \times k}$ are semantic latent factors of users and venues respectively and α is a parameter that controls the influence between the latent factors (U and V) and the semantic latent factors (U_s and V_s). To avoid overfitting, we regularise the model based on the complexity of the semantic latent factors, as follows:

$$L(U, V, U_s, V_s) = L(U, V) + \frac{\lambda}{2}(\|U_s\|_F^2 + \|V_s\|_F^2) \tag{5}$$

Finally, we apply Stochastic Gradient Descent (SGD) to find a local minimum of the loss function (Eq. (5)), by optimising each of the latent factor matrices U, V, U_s, V_s, while fixing the other, until convergence as follows:

$$\frac{\partial L(U,V)}{\partial U_i} \overset{W2V}{=} \sum_{j=1}^{n} I_{i,j}(r_{i,j} - \hat{r}_{i,j})\alpha V_j + \lambda U_i \qquad \frac{\partial L(U,V)}{\partial V_j} \overset{W2V}{=} \sum_{i=1}^{m} I_{i,j}(r_{i,j} - \hat{r}_{i,j})\alpha U_i + \lambda V_j$$

$$\frac{\partial L(U,V)}{\partial U_{s_i}} \overset{W2V}{=} \sum_{j=1}^{n} I_{i,j}(r_{i,j} - \hat{r}_{i,j})(1 - \alpha)S_{v_j} + \lambda U_{s_i}$$

$$\frac{\partial L(U,V)}{\partial V_{s_j}} \overset{W2V}{=} \sum_{i=1}^{m} I_{i,j}(r_{i,j} - \hat{r}_{i,j})(1 - \alpha)S_{u_i} + \lambda V_{s_j}$$

Table 1. Overview of user-rating prediction approaches.

Models	Social	Comments	Params	Intuitions
MF [1]	×	×	λ	Users are likely to prefer venues rated that other similar users rate highly
VMF [2], JMF [5]	×	✓	λ	+ Users are likely to prefer venues that share similar characteristics (according to textually similar comments)
SoReg [4]	✓	×	λ, α	+ Users are likely to prefer venues that their friends rate highly
BoWReg, DeepReg [3]	✓	✓	λ, α	+ Users are likely to prefer venues visited by their friends who have similar tastes
MFw2v	×	✓	λ, α	+ Users' preferences can be extracted from their comments on venues and users are likely to prefer venues that share similar characteristics

3 Evaluation

In this section, we evaluate the effectiveness of our proposed model $MFw2v$ by comparing with state-of-the-art rating prediction approaches. In particular, we aim to address the following research question **RQ**: Can we exploit word embeddings to effectively model user's preferences and characteristics of venues and improve the prediction accuracy of traditional MF-based approach?

3.1 Experimental Setup

We first describe the experimental setup used to evaluate the effectiveness of our proposed approach (MFw2v) and summarise baselines in details. We conducted experiments using the publicly available Yelp dataset[1], which consists of 2,225,214 ratings by 552,339 users for 77,079 venues. We conduct 5-fold cross-validation experiment where each fold has 60% training, 20% validation and 20% testing. We implement all experiments using LibRec [9], a Java library for recommendation systems. For each fold, the α in Eq. (4) is determined using the validation set. Following [2,5], we set the dimension of latent factors d to 10 and $\lambda = 0.001$. For word embeddings, we use the Word2Vec tool[2], to train a skip-gram model [10] using the default settings (window size 5 and word embedding dimensions $k = 100$) on the Yelp dataset. Previous work by Mikolov et al. [11]

---

[1] www.yelp.com/dataset_challenge.
[2] code.google.com/archive/p/word2vec.

showed that the skip-gram model performs better than or equally to the CBOW model. The baselines used in this our experiments are summarised below, while their parameters and sources of evidence are highlighted in Table 1.

MF [1] is the traditional matrix factorisation approach, which only considers the user-venue matrix to predict the ratings (described in Sect. 2.1).

VMF [2] is a state-of-the-art bag-of-words based MF approach that considers geographical and textual information (i.e. comments about a venue). To permit a fair evaluation, we re-implement their approach to consider only textual information, and ignore the geographical location of venues, in common with our own proposed approach.

JMF [5] is a state-of-the-art rating prediction approach that jointly models comments and user's ratings by exploiting skip-thought vectors [12][3] to represent the content of comments. Instead of skip-thought vectors, we re-implement their approach to exploit word embeddings to permit a fair comparison with our proposed approach.

SoReg [4] is a social-based regularisation approach that enhances the rating prediction of MF-based approaches by assuming that friends who rate similar venues with similar scores can be influenced by each other.

BoWReg & **DeepReg** [3] are textual-social based regularisation approaches that leverage both social information and comments to reduce the complexity of MF-based approaches. In particular, DeepReg exploits word embeddings to estimate similarity between friends; BowReg is an orthogonal bag-of-words based regularisation approach.

Finally, Mean Absolute Error (MAE) and Root Mean Square Error (RMSE) are used to measure rating prediction accuracy (for both metrics, lower is better).

3.2   Experimental Results

Table 2 reports the rating prediction accuracy, in terms of MAE and RMSE, of our proposed MFw2v and other baselines. Firstly, we observe that our proposed approach, MFw2v, outperforms all MF baselines in terms of MAE and is comparable with DeepReg in terms of RMSE. In particular, comparing with the traditional MF (MF), the prediction accuracy of MFw2v is ~12% more effective than MF for both MAE and RMSE. This implies that the users' preferences and the characteristics of venues extracted from the textual content of comments using word embeddings can enhance the rating prediction accuracy of traditional MF. Indeed, textual content of comments of venues are publicly available in LBSNs, while social information maybe not available for privacy reasons. Our proposed approach (MFw2v), which exploits comments of venues to model characteristics of venues, outperforms SoReg for both metrics (9.52% and 2.97% for MAE and RMSE respectively). Moreover, by comparing MFw2v and DeepReg, the experimental results demonstrate that MFw2v is as effective

---

[3] A state-of-the-art deep learning approach.

Table 2. Prediction accuracy in terms of MAE and RMSE of various approaches. Percentage differences of prediction accuracy are calculated with respect to the best performance achieved for that metric, which are highlighted in bold.

Metrics	MF	VMF	JMF	SoReg	BoWReg	DeepReg	MFw2v
MAE	1.1640	1.2198	1.1795	1.1260	1.1004	1.0781	**1.0188**
Δ	12.47%	16.48%	15.77%	9.52%	7.42%	5.50%	
RMSE	1.5243	1.5006	1.5073	1.3870	1.4354	**1.3456**	1.3458
Δ	11.72%	10.33%	12.02%	2.99%	6.26%		0.01%

as the state-of-the-art rating prediction approach (DeepReg). Note that MFw2v only takes user's comments into account, while DeepReg considers both social information and users' comments. Although the improvements in Table 2 are relatively small, Koren [1] pointed that small improvements in MAE and RMSE can lead to marked improvements in the quality of recommendations in practice.

4   Conclusion

This paper proposed a MF-based approach that leverages the textual content of comments to effectively model users' preferences and characteristic of venues, and exploits word embeddings to captures the semantic properties of comments. Our comprehensive experiments conducted using a large existing dataset (2.2M ratings from 500k users) demonstrate the effectiveness of our proposed approach in comparison with state-of-the-art rating prediction approaches. Indeed, our proposed approach (MFw2v) is shown to be as effective as state-of-the-art approaches, while only requires venues' comments as auxiliary information. For future work, we plan to apply various deep learning techniques to our proposed approach and explore the impact of word embedding parameters.

References

1. Koren, Y.: Factor in the neighbors: scalable and accurate collaborative filtering. Trans. Knowl. Discov. Data (TKDD) **4**(1), 1:1–1:24 (2010)
2. Hu, L., Sun, A., Liu, Y.: Your neighbors affect your ratings: on geographical neighborhood influence to rating prediction. In: Proceedings of SIGIR (2014)
3. Manotumruksa, J., Macdonald, C., Ounis, I.: Regularising factorised models for venue recommendation using friends and their comments. In: Proceedings of CIKM (2016)
4. Ma, H., Zhou, D., Liu, C., Lyu, M.R., King, I.: Recommender systems with social regularization. In: Proceedings of WSDM (2011)
5. Jin, Z., Li, Q., Zeng, D.D., Zhan, Y., Liu, R., Wang, L., Ma, H.: Jointly modeling review content and aspect ratings for review rating prediction. In: Proceedings of SIGIR (2016)

6. Manotumruksa, J., Macdonald, C., Ounis, I.: Modelling user preferences using word embeddings for context-aware venue recommendation. arXiv preprint arXiv:1606.07828 (2016)
7. Kim, Y.: Convolutional neural networks for sentence classification. arXiv preprint arXiv:1408.5882 (2014)
8. Musto, C., Semeraro, G., Gemmis, M., Lops, P.: Learning word embeddings from wikipedia for content-based recommender systems. In: Ferro, N., Crestani, F., Moens, M.-F., Mothe, J., Silvestri, F., Nunzio, G.M., Hauff, C., Silvello, G. (eds.) ECIR 2016. LNCS, vol. 9626, pp. 729–734. Springer, Heidelberg (2016). doi:10.1007/978-3-319-30671-1_60
9. Guo, G., Zhang, J., Sun, Z., Yorke-Smith, N.: LibRec: a Java library for recommender systems. In: Proceedings of UMAP (2015)
10. Mikolov, T., Chen, K., Corrado, G., Dean, J.: Efficient estimation of word representations in vector space. arXiv:1301.3781 (2013)
11. Mikolov, T., Sutskever, I., Chen, K., Corrado, G.S., Dean, J.: Distributed representations of words and phrases and their compositionality. In: Proceedings of NIPS (2013)
12. Kiros, R., Zhu, Y., Salakhutdinov, R.R., Zemel, R., Urtasun, R., Torralba, A., Fidler, S.: Skip-thought vectors. In: Proceedings of NIPS (2015)

Word Similarity Based Model for Tweet Stream Prospective Notification

Abdelhamid Chellal[✉], Mohand Boughanem, and Bernard Dousset

IRIT University of Toulouse UPS, Toulouse, France
{abdelhamid.chellal,mohand.boughanem,bernard.dousset}@irit.fr

Abstract. The prospective notification on tweet streams is a challenge task in which the user wishes to receive timely, relevant, and non-redundant update notification to remain up-to-date. To be effective the system attempts to optimize the aforementioned properties (timeliness, relevance, novelty and redundancy) and find a trade-off between pushing too many and pushing too few tweets. We propose an adaptation of the extended Boolean model based on word similarity to estimate the relevance score of tweets. We take advantage of the word2vec model to capture the similarity between query terms and tweet terms. Experiments on the TREC MB RTF 2015 dataset show that our approach outperforms all considered baselines.

Keywords: Prospective notification · Tweet summarization · word2vec

1 Introduction

User generated content (UGC) in social media streams provides valuable information about what is happening in the world and covers both scheduled and unscheduled events. In many cases, Twitter provides the latest news before traditional media, especially for unscheduled events. Hence, social media streams seem to be the appropriate source of information to fulfill the information need of a user who is looking for updates to be timely pushed for topics of interest. In this context, two different tasks can be defined: retrospective summarization and prospective notification [4]. In the first, documents (tweets) are known in advance while, in the second, the stream is filtered in real time and relevant and non-redundant updates are pushed immediately to the user.

However, unlike traditional data sources, UGC in social media streams is characterized by the volume, velocity and variety of the published information. Indeed, the published posts can vary significantly in terms of quality. These features make prospective notification of social media streams a challenging task.

TREC 2015 Microblog Real-Time Filtering (MB-RTF) [2] and TREC Real-Time Summarization 2016[1] are two evaluation campaigns. The objective of the task is to identify relevant tweets from the stream and send those updates directly

---

[1] http://trecrts.github.io/TREC2016-RTS-guidelines.html.

© Springer International Publishing AG 2017
J.M. Jose et al. (Eds.): ECIR 2017, LNCS 10193, pp. 655–661, 2017.
DOI: 10.1007/978-3-319-56608-5_62

to the user's mobile phone. The main issue in this task is to find a trade-off between pushing too many and pushing too few tweets. In the later case, the user may miss important updates and in the former case, the user may be over-whelmed by irrelevant and/or redundant information.

This paper explores a novel approach for prospective notification in tweet streams that pushes in a real time fashion the most salient (relevant and non-redundant) information related to an ongoing event as soon as it occurs in the stream. Knowing that an effective system needs to optimize three properties: The relevance with respect to the topic of interest, the novelty/redundancy and the latency between the publication time and the notification time of selected tweets. To fulfill these requirements, the proposed approach consists of three filters that are adjusted sequentially and in which the decision to select/ignore a tweet is made immediately. The first filter is a simple tweet quality and topicality filter, the second filter is related to the relevance and the third one is for novelty control in order to avoid pushing redundant information to the user.

The main contribution of this paper is the proposition of an adaptation of the Extended Boolean Model (EBM) [5] based on word similarity to estimate the relevance of the incoming tweet with respect to the topic of interest. In addition, instead of using the TF-IDF weighting technique, the query term weight is esti-mated by taking advantage of the word2vec model [3]. It is estimated through its similarity with the tweet's terms, computed by cosine similarity between word vector generated by the word2vec model. Indeed, the novelty score of the incom-ing tweet is measured using word overlap with respect to words of tweets already pushed. The defined novelty function avoids a pairwise comparison allowing to reduce the computational complexity. The experiments conducted on the TREC MB RTF 2015 dataset show that our approach outperforms all the baselines.

2   Related Work

Prospective notification in social media streams is the task in which user wishes to receive timely, relevant and non-redundant updates [4]. The TREC MB RTF-2015 official results reveal that runs PKUICSTRunA2 [1] and UWaterlooATDK [6] are the two best performing ones among 37 runs from 14 groups [2]. In the former, the relevance score of tweets is evaluated by using the normalized KL-divergence distance and the decision to select a tweet is based on a predefined threshold set using human intervention. They manually scan the ranked list of top-10 selected tweets of previous day from top to bottom, and the relevance score of the first irrelevant tweet is chosen as a threshold in the next day for the related topic. In UWaterlooATDK run, the relevance score is based on the query term occurrence in the tweet. The threshold is fixed for each day according to the score of the top-10 tweets returned in the previous day. In [7] authors improve the effectiveness of their approach (UWaterlooATDK) by using a daily feedback strategy to estimate the relevance threshold for the next day. However, one can argue that a daily interaction for ongoing feedback judgment might be too onerous in practice. We show in this work that the result reported in [7] is outperformed by the proposed approach, in which the threshold is set adaptively without the use of feedback.

3   Real-Time Tweet Filtering

Our approach acts like a filter with three levels related to the topicality of tweet, its relevance and its novelty respectively. The decision of pushing/ignoring the incoming tweet is made immediately as the tweet occurs. Tweets that pass all these filters are selected as a summary which is denoted by S.

3.1   Tweet Quality and Topicality Filter

The first filter eliminates non-English tweets and those containing less than three tokens. It also drops all incoming tweets that do not contain a predefined number of query words. The incoming tweet T is considered as a candidate tweet if its number of overlapping words with the query title is higher than the minimum of either a predefined constant (K) or the number of words in the query title $min(K, |Q^t|)$. Pilot experiments on TREC MB RTF dataset revealed that the filter $k = 2$ captures about 40% of relevant tweets while the filter $k = 1$ returns 74% of relevant tweets but it also brings up a lot of noise. These results motivated our choice to set $k = 2$.

3.2   Relevance Filter

Tweets have a limit length of 140 characters, are noisy and ungrammatical, which implies that the statistical features such as term frequency may be less useful. We believe that the similarity between the tweet words and query words is the key feature. Hence, to evaluate the relevance score of the incoming tweet with respect to the query, we propose (i) to use the extended Boolean model [5] to evaluate the relevance score of a tweet; (ii) to use the similarity score between query words and tweet words to evaluate the weight of query words.

Assume that the query Q (user interest) consists of a title Q^t and description Q^d of the information need. The query title Q^t represents "**AND**ed terms" while Q^d represents "**OR**ed terms". In the Extended Boolean Model, the relevance scores of tweet $T = \{t_1, ..., t_n\}$ to "AND query" Q^t and "OR query" Q^d are estimated respectively as follows:

$$RSV(T, Q^t_{and}) = 1 - \sqrt{\frac{\sum_{q^t_i \in Q^t}(1 - W_T(q^t_i))^2}{|Q^t|}} \tag{1}$$

$$RSV(T, Q^d_{or}) = \sqrt{\frac{\sum_{q^d_i \in Q^d}(W_T(q^d_i))^2}{|Q^d|}} \tag{2}$$

where $W_T(q)$ is the weight of the query term q in the tweet T. $|Q^t|$ and $|Q^d|$ are the length of the title and the description of the query respectively. q stands for the term q^t_i in the query title Q^t or the term q^d_i in the query description Q^d.

Instead of using the TF-IDF like weighting schema, we propose to estimate the weight $W_T(q)$ by evaluating the similarity between the query term q and all the terms of tweet T as follows:

$$W_T(q) = \max_{t_i \in T}[w2vsim(t_i, q)] \tag{3}$$

where $w2vsim(t_i, q)$ is the similarity between tweet word t_i and query word q. We propose to represent terms using their word2vec [3] representation and the similarity between two terms is measured by cosine similarity between their word2vec vectors. The intuition behind this proposition is that tweets that have words sharing many contexts with the query words will be more relevant. The main advantage of using word2vec model to estimate the similarity between two words is that a query word which does not appear in a tweet but shares many contexts with the tweet words will get a weight different from 0. Indeed, the relevance score of an incoming tweet is evaluated at the time the new tweet arrives, independently of tweets previously seen in the stream and without the need for indexing the tweet stream. The word vectors used in our experiments to estimate the word similarity were generated using tweets crawled during 9 days before evaluation period.

With $RSV(T, Q_{and}^t)$ and $RSV(T, Q_{or}^d)$, we got two relevance scores for tweet T regarding the title and the description of the query, respectively. The final relevance score of tweet T is measured by combining the aforementioned scores linearly with title terms having greater weight than description terms as follows:

$$RSV(T, Q) = \lambda \times RSV(T, QT_{and}) + (1 - \lambda) \times RSV(T, QD_{or}) \tag{4}$$

where $\lambda \in [0, 1]$ is a parameter determining the trade-off between the query title's words and the description's words. Based on pilot experiments where λ was varied from 0 to 1 in increments of 0.1, the weight λ was set to 0.8.

A tweet passes the relevance filter if its score is above a certain threshold. This threshold is estimated at the decision time based on the previous values. Our thresholding strategy is to consider the average of the previously seen values of the relevance score. However, we do not lower the threshold under a global minimum threshold GT. Hence the relevance threshold is defined by $max(GT, avg(RSV(T, Q)))$.

3.3   Novelty Filter

The intuitive way to estimate the novelty of an incoming tweet is to conduct a pairwise comparison with previously seen tweets in the stream using a standard similarity function such as cosine similarity or KL-divergence. Due to the limited length of tweets, meaningful words rarely occur more than once which implies that aforementioned similarity functions are less useful for evaluating the distance between two tweets. Indeed, a pairwise comparison does not fit a real-time scenario. For these reasons, we propose to merge all tweets already selected in the summary into a "summary word set" and evaluate the novelty score using the number of overlapping words between the incoming tweet and summary word set. Assume that SW is the set of words that occur in current summary, then the novelty score of the incoming tweet T is evaluated as follows:

$$NS(T, SW) = 1 - \frac{|SW \cap T|}{|T|} \qquad (5)$$

Tweets with novelty score less than 0.6 were discarded. We set the novelty threshold experimentally using TREC MB RTF 2015 dataset.

4   Experimental Evaluation and Results

Experiments were conducted on the TREC 2015 Microblog Real Time Filtering (MB RTF) track dataset by using replay mechanism over tweets captured during the evaluation period. This collection was generated by each participant independently using Twitter's streaming API during the 10 days of the evaluation period (20 July to 29 July 2015). In our experiments, we focus on the scenario "Push notifications" which corresponds to a real-time task and where a maximum of 10 tweets per day per topic are returned. The organizers defined two evaluation measures that consider both the relevance and the time at which they were pushed [2]. The primary metric is the expected latency-discounted gain (ELG) in which a latency penalty is applied. The second metric is the normalized cumulative gain (nCG). These two metrics are defined as follows:

$$ELG(S) = \frac{1}{N} \times \sum_{T \in S} G(T) \times max(0, (100 - delay)/100) \qquad (6)$$

$$nCG(S) = \frac{1}{Z} \times \sum_{T \in S} G(T) \qquad (7)$$

where S is the generated summary, N is the number of returned tweets and Z is the maximum possible gain (given the 10 tweet per day limit). The delay is the latency (in minutes) between the tweet creation time and the time the system decides to push it. G(T) is the gain of each tweet which is set as follows: irrelevant tweets receive a gain of 0, relevant tweets receive a gain of 0.5 and highly relevant tweets receive a gain of 1.0.

Thresholding Impact: To better understand the impact of the threshold used in the relevance filter, we plot in Fig. 1 the effectiveness of our system in terms of ELG and nCG across a range of global threshold values. The baseline in this experiment is the empty run and the oracle run which represents the run where only relevant tweets that pass the first filter are selected. Figure 1 shows that the best results in terms of both metrics are achieved with global threshold GT = 0.5. The threshold controls the number of pushed tweets. Note that the empty run is a challenging baseline that many systems in TREC 2015 failed to beat. For some days, no relevant tweets occur and the system should push nothing in this case. Also, the comparison between the best results and the oracle run

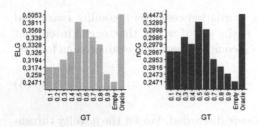

Fig. 1. ELG and nCG for different thresholds, the oracle run and the empty run.

Table 1. Comparative evaluation with state-of-the-art.

Method	ELG	nCG	%ELG
WSEBM	**0.3811**	**0.3289**	
EBM	0.2583	0.2544	+32.22%
Tan et al. [7]	0.3678	-	+3.48%
TREC MB RTF 2015 official Results			
PKUICSTRunA2	0.3175	0.3127	+16.68%
UWaterlooATDK	0.3150	0.2679	+17.34%

Note. % indicates improvements in terms of ELG.

reveals that more improvements can be achieved through better filtering and threshold setting.

Comparative Evaluation with State-of-the-Art Approaches: In this section, we compare our approach (denoted by WSEBM for Word Similarity EBM) against the two high-performing official results from the TREC MB-RTF 2015 PKUICSTRunA2 [1] and UWaterlooATDK [6] and against the approach described in [7] in which they improve their results obtained in TREC 2015. In addition, in order to evaluate the impact of using word similarity as weighting technique, we compare our method with standard EBM. This baseline is based on the proposed functions (4) in which we consider the query term number of occurrences in a tweet as the term's weight. Table 1 reports the results in terms of ELG and nCG. As shown in this table, the WSEBM outperforms all baselines overall metrics. We found performance improvements up to ELG values of about 16% for the best run in TREC MB 2015 task and of about 3.4% for the approach based on feedback strategy to set the relevance threshold.

5   Conclusion

In this paper, we introduced a new approach for prospective notification in which we show that word similarity matching and simple thresholding strategy achieve good results in terms of expected and cumulative gain. The use of the semantic word relationships improves the efficiency of the relevance filter. The proposed relevance function enables the use of simple threshold across all topics. The results showed that better results can be achieved if the threshold is appropriately set. In future work, we plan to leverage social signals to filter incoming tweets.

References

1. Fan, F., Fei, Y., Lv, C., Yao, L., Yang, J., Zhao, D.: Pkuicst at TREC 2015 microblog track: query-biased adaptive filtering in real-time microblog stream. In: Text Retrieval Conference, TREC, Gaithersburg, USA, 17–20 November (2015)
2. Lin, J., Efron, M., Wang, Y., Sherman, G., McCreadie, R., Sakai, T.: Overview of the TREC 2015 microblog track. In: Text Retrieval Conference, TREC, Gaithersburg, USA, 17–20 November (2015)

3. Mikolov, T., Chen, K., Corrado, G., Dean, J.: Efficient estimation of word representations in vector space. CoRR abs/1301.3781 (2013)
4. Qian, X., Lin, J., Roegiest, A.: Interleaved evaluation for retrospective summarization and prospective notification on document streams. In: Proceedings of the 39th International ACM SIGIR Conference on Research and Development in Information Retrieval, SIGIR 2016, pp. 175–184 (2016)
5. Salton, G., Fox, E.A., Wu, H.: Extended Boolean information retrieval. Commun. ACM **26**(11), 1022–1036 (1983). http://doi.acm.org/10.1145/182.358466
6. Tan, L., Roegiest, A., Clarke, C.L.: University of waterloo at TREC 2015 microblog track. In: Text Retrieval Conference, TREC, Gaithersburg, USA, 17–20 November (2015)
7. Tan, L., Roegiest, A., Clarke, C.L., Lin, J.: Simple dynamic emission strategies for microblog filtering. In: Proceedings of the 39th International ACM SIGIR Conference on Research and Development in Information Retrieval, SIGIR 2016, pp. 1009–1012 (2016)

Negative Feedback in the Language Modeling Framework for Text Recommendation

Hossein Rahmatizadeh Zagheli[✉], Mozhdeh Ariannezhad,
and Azadeh Shakery

School of Electrical and Computer Engineering, College of Engineering,
University of Tehran, Tehran, Iran
{rahmatizadeh,m.ariannezhad,shakery}@ut.ac.ir

Abstract. Text recommendation is the task of delivering sets of documents to users with respect to their profiles. One of the most important components of these systems is the filtering component. The filtering component decides about the relevancy of a document to a profile, which specifies the user interests, by comparing the similarity score between them with a predetermined threshold. In this paper, we propose a filtering approach which exploits the negative feedback from the user in a language modeling framework to compute the relevancy score of new documents. In other words, the negative feedback from the user is considered as the representative of the documents that he dislikes and leads the system to avoid suggesting such documents in the future. Our experiments on CLEF 2008–09 INFILE Track collection demonstrate the effectiveness of our proposed method and indicate that using negative feedback results in significant improvements over baselines.

Keywords: Content-based recommender system · Text recommendation · Language modeling · Negative feedback · Text filtering

1 Introduction

Nowadays with the rapid growth of information on the Web, satisfying users with specific long-term information needs is an important challenge. As a solution to this problem, textual recommender systems can be used to manage large information flows and expose users to only the information that they actually need. These systems, which are generally based on statistical models (e.g., probabilistic, inference network and vector space), compute a numeric score to show how well a document matches the user profile, and recommend a document if its relevancy score is greater than a threshold. Textual recommender systems can learn from user feedback and become more accurate over time. Collaborative filtering is one of the most successful techniques that utilizes ratings and usage data from a community of users in order to generate recommendations. However, this information is not always available and/or it is not obvious how it should be combined with item content in order to make a recommendation. In many domains where

© Springer International Publishing AG 2017
J.M. Jose et al. (Eds.): ECIR 2017, LNCS 10193, pp. 662–668, 2017.
DOI: 10.1007/978-3-319-56608-5_63

textual information is abundant (e.g., books, news) and content plays a key role in recommendation, content-based filtering systems can be more useful.

In a stream of incoming documents, a text recommendation system selects the documents that are estimated to be relevant to users with respect to their profiles, which describe their information needs. To decide about the relevancy of a new document, the filtering component of the system relies on the profile-document similarity and users' past preferences implied through feedback. While user profiles lead the recommendation system to find what the users are looking for, their past preferences can drive the system to avoid recommending the documents that they dislike. Considering negative feedback, i.e., the previously recommended documents that user marked as non-relevant, as well as user profile, can be advantageous in computing the relevancy of a new document.

In this paper, we propose a filtering approach based on language modeling (LM) that employs user's negative feedback in a formal framework. In this framework, the score of a newly arrived document with respect to a given profile depends on the difference between two Kullback-Leibler divergence (KL-divergence) values: KL-divergence between the language model of the profile and the language model of the document, and the KL-divergence between the language model of the profile and the language model of non-relevant documents. The language model of non-relevant documents is estimated based on the negative feedback from the user. Our experiments demonstrate the effectiveness of using negative feedback through this model and our proposed filtering approach outperforms state-of-the-art filtering methods.

2   Related Works and Background

Previous work shows the usefulness of language models in content-based recommender systems and information filtering. Bogers and van den Bosch [4] surveyed the efficiency of using language modeling in news recommendation task and demonstrated its advantage over a tf-idf weighting method. Relevance-based language modeling is considered in news recommendation in [6,10]. In collaborative filtering systems, Bansal et al. present a method leveraging recurrent neural networks to represent text items [2] and Valcarce explored how to apply language models for finding user or item neighbourhoods [9]. Parapar et al. proposed a LM based recommendation approach where recommendation is modelled as a profile expansion process [8].

Even though the concept of exploiting LM techniques for recommendation systems is in common between the mentioned approaches and our work, our focus in this paper is on using negative feedback in the language modeling framework to examine the relevancy of arriving documents to users. Different studies in information retrieval have shown the effectiveness of using negative feedback in computing query-document relevancy [11,12]. Moreover, the impact of using non-relevant documents in filtering systems has been studied in [5], where a negative profile is used to express the features of non-relevant documents in a filtering system based on vector space model. In addition, considering feedback from users for profile updating and threshold optimization has led to improvement of filtering performance [14,15].

3   Negative Feedback in the Language Modeling Framework for Recommendation

User feedback is usually used in the process of profile updating and threshold setting in textual recommender systems [14,15]. Exploiting a language modeling framework in the filtering component enables a system to consider users' feedback directly, without changing the profile. In other words, when computing the similarity between a document and a given profile, it is desirable to have a look back at the previous documents recommended to user. In an incoming stream of documents, number of non-relevant documents, i.e., the documents that a given user dislikes, is usually more than relevant ones [1]. The idea is to use them as a signal to avoid similar non-relevant documents in the future recommendations.

In the language modeling framework for the filtering component, the divergence between the language model of profile P, i.e., θ_P, and the language model of document D, i.e., θ_D, can be used as the similarity measure between P and D [13]. Since each language model is a probabilistic distribution, we can use KL-divergence to compute the score of D with respect to P as follows:

$$score(P, D) = -D(\theta_P || \theta_D) = -\sum_{w \in P} p(w|\theta_P) \log \frac{p(w|\theta_P)}{p(w|\theta_D)}, \tag{1}$$

where $p(w|\theta_P)$ and $p(w|\theta_D)$ are the probability of word w in the language models of P and D, respectively. These probabilities are calculated using maximum likelihood estimation and smoothed using Dirichlet prior smoothing technique.

In order to compute the similarity score between a new document and the given user profile considering the negative feedback from the user, we propose the idea of using the language model of non-relevant documents directly in the scoring function to avoid recommending the documents that have a rather high similarity score but are potentially non-relevant to user's information need. We assume for each document D, there exist a complement document $\bar{D}$ that a user would like if he does not like D. We believe by proper estimation of this complement document, by considering the non-relevant documents, the filtering system can push better suggestions to users; since the closer the language model of D is to the language model of P and the farther away the language model of $\bar{D}(\theta_{\bar{D}})$ is from the language model of P, the higher similarity between D and P will be. In that direction, the difference between two KL-divergence values is used to score D with respect to P: the KL-divergence between θ_P and θ_D, and the KL-divergence between θ_P and $\theta_{\bar{D}}$. Hence, the score of D with respect to P can be computed as follows:

$$score(P, D) = D(\theta_P || \theta_{\bar{D}}) - D(\theta_P || \theta_D)$$

$$= \sum_{w \in V} p(w|\theta_P) \log \frac{p(w|\theta_P)}{p(w|\theta_{\bar{D}})} - \sum_{w \in V} p(w|\theta_P) \log \frac{p(w|\theta_P)}{p(w|\theta_D)}$$

$$= \sum_{w \in V} p(w|\theta_P) \log \frac{p(w|\theta_D)}{p(w|\theta_{\bar{D}})}. \tag{2}$$

Now the challenge lies in the estimation of $\theta_{\bar{D}}$, i.e., the negative document language model of D. Recently, Lv and Zhai proposed the concept of negative document language models to take into account negative query generation that was ignored in basic retrieval functions [7]. In the case of information retrieval, documents that a user dislikes can be thought as the ones in the search results that he didn't click on. On the other hand, the score of all documents have to be computed before showing them to the user, which makes it impossible to use the retrieval results to approximate $\theta_{\bar{D}}$, that is needed to compute the score of D. Based on the intuition that almost all other documents in the collection are complementary to D, the authors in [7] assume that $\bar{D}$ is a document containing all of the terms in the collection, excluding the ones which have occurred in D. We refer to this negative document as $\bar{D}_{coll}$. Each word in $\bar{D}_{coll}$ has the same frequency $\delta > 0$ which maximizes the information entropy under the only prior data D. Formally,

$$c(w, \bar{D}_{coll}) = \begin{cases} 0 & c(w, D) > 0 \\ \delta & \text{otherwise} \end{cases} \tag{3}$$

and the document length of $\bar{D}_{coll}$ is thus approximated as $|\bar{D}_{coll}| = \sum_{w \in V} c(w, \bar{D}_{coll}) \approx \delta |V|$, where V is the set of unique words in the whole document collection C.

However, the filtering component of a text recommendation system does not rank the documents. Instead, the score of each document is compared with a threshold and if higher, the document is recommended to the user. Moreover, the system maintains a set of non-relevant documents that has been shown to the user previously and can be used as a representative for the documents that the user dislikes. We refer to this set as $NonRel$. Initially, $NonRel$ is empty and it grows as the user marks a recommended document as non-relevant. Intuitively, using $NonRel$ should result in a more accurate estimation of $\theta_{\bar{D}}$. In that direction, following the approach proposed in [7], we can assume the same frequency $\delta > 0$ for all the words that have occurred in $NonRel$. We refer to this negative document as $\bar{D}_{NonRel}$ and $c(w, \bar{D}_{NonRel})$ is calculated as follows:

$$c(w, \bar{D}_{NonRel}) = \begin{cases} 0 & c(w, D) > 0 \quad \text{or} \quad c(w, NonRel) = 0 \\ \delta & \text{otherwise} \end{cases} \tag{4}$$

where $c(w, NonRel)$ is defined as the sum of counts of the word w in all documents of $NonRel$. The document length of $\bar{D}_{NonRel}$ is approximated as $|\bar{D}_{NonRel}| = \sum_{w \in V_{NonRel}} c(w, \bar{D}_{NonRel}) \approx \delta |V_{NonRel}|$, where V_{NonRel} is the set of unique words in the document set $NonRel$. Regardless of the method we choose to model $\bar{D}$, the language model $\theta_{\bar{D}}$ can be calculated according to the maximum likelihood estimation. After computing the language model of $\theta_{\bar{D}}$ using any of the two proposed methods, we can now calculate the score of a document D with respect to a given profile P using Eq. 2 and perform recommendation.

4   Experiments

We use INFILE dataset [3] in our evaluations. This collection has about 1.5 million news article, provided by Agence France Presse (AFP) in three languages Arabic, English, and French. The collection contains a set of 50 profiles for each language, which were used to evaluate monolingual and cross-lingual filtering systems in CLEF 2008 and 2009 INFILE Track. In our experiments, only the English documents and topics are used and the title and keywords of each topic are considered. Porter stemming is done on documents and topics. All the experiments are carried out using the Lemur toolkit[1].

Following CLEF INFILE Track, precision, recall, and F-measure are used for evaluating the methods. Statistical significance of the differences between the corresponding means are calculated using two-tailed paired t-test at a 95% confidence level. We evaluate the performance of the following filtering methods:

- **LM:** in this method, similarity between the profile and the document is calculated using Eq. 1 and the negative feedback from user is not considered in computing the language model of documents.
- **Coll:** similarity between the profile and the document is calculated using Eq. 2 and the language model of non-relevant documents is computed according to Eq. 3, i.e., the collection is considered as the complement of all documents.
- **NonRel:** similarity between the profile and the document is calculated similar to Coll, but the language model of non-relevant documents is computed according to Eq. 4, i.e., for each profile its negative feedback documents are considered as the complement of each document.

In the mentioned methods, user profiles are not updated during filtering and for threshold updating, we perform LAUTO threshold optimization algorithm, proposed in [13]. The algorithm starts with an initial threshold and checks the status of the system. If the system is recommending non-relevant documents, the threshold is probably smaller than its optimal value and if the system is rejecting a number of continuous received documents, the threshold is likely to be higher than its optimal value. According to these situations, the current value of the threshold is updated. The parameters of threshold updating method are set using 5-fold cross validation over the collection. Dirichlet smoothing parameter μ is set to the average document length in the collection. Parameter δ of the proposed methods is set to 0.001 empirically.

Table 1 summarizes the results achieved by the proposed methods and the standard language modeling approach. As shown in the table, the proposed

---

[1] http://www.lemurproject.org/.

Table 1. Performance of proposed methods compared to language modeling approach. */• shows significance over LM/Coll methods, respectively.

Method	F-measure	Precision	Recall
LM	0.3384	0.3446	0.4525
Coll	0.4134*	0.4236*	0.5446*
NonRel	0.4166*•	0.4264*•	0.5482*•

Table 2. Comparison of the best performing proposed method with the baselines. */• shows significance over MLE/NFB methods, respectively.

Method	F-measure	Precision	Recall
MLE	0.3042	0.3232	0.5275
BasicNFB+QTE	0.3588	0.3903	0.6227
NonRel	0.4166*•	0.4264*	0.5482

methods perform significantly better than standard language modeling in terms of precision, recall and F-measure, and estimating the language model of complement documents using negative feedback further improves the recommendation performance.

Moreover, we compare our best performing method with MLE and Basic-NFB+QTE methods as baselines. The MLE method uses maximum likelihood estimation for threshold updating [15]. In order to set dissemination thresholds, the MLE approach assumes that relevant document scores are distributed normally and non-relevant document scores are distributed exponentially and jointly estimates the parameters of the two density distributions. Basic-NFB+QTE method penalizes the documents that are similar to the known non-relevant documents in the language modeling framework, using an approach which is in spirit similar to Rocchio for the vector space model [11]. The results are reported in Table 2. As shown in the table, NonRel performs significantly better than baselines in terms of F-measure. However, we observe no significant improvements over precision and recall. A possible reason for this result is that F-measure is considered as the main evaluation measure in our experiments, and the parameters are tuned in order to optimize F-measure.

5 Conclusions and Future Work

In this paper, we examine the effect of using the user negative feedback in a text recommender system. We assume for each document D, there exist a complement document $\bar{D}$ that a user would like if he does not like D and its language model is estimated using the non-relevant documents from user feedback. Our experiments demonstrate the effectiveness of this approach compared to baselines. There are many possible directions to extend this work. The proposed language modeling approach for filtering can be further improved by considering positive feedback. Moreover, proposing a more accurate estimation for complement documents, i.e., other than uniform, may improve the filtering performance. Additionally, since negative feedback examples might be quite diverse, a single negative language model may not be optimal to represent all of the non-relevant documents. Exploiting a topic modeling approach to estimate multiple negative model documents may be able to enhance the performance of the filtering system.

References

1. Algarni, A., Li, Y., Xu, Y., Lau, R.Y.: An effective model of using negative relevance feedback for information filtering. In: CIKM, pp. 1605–1608 (2009)
2. Bansal, T., Belanger, D., McCallum, A.: Ask the GRU: multi-task learning for deep text recommendations. In: RecSys, pp. 107–114 (2016)
3. Besançon, R., Chaudiron, S., Mostefa, D., Timimi, I., Choukri, K., Laïb, M.: Information filtering evaluation: overview of CLEF 2009 INFILE track. In: Peters, C., Nunzio, G.M., Kurimo, M., Mandl, T., Mostefa, D., Peñas, A., Roda, G. (eds.) CLEF 2009. LNCS, vol. 6241, pp. 342–353. Springer, Heidelberg (2010). doi:10.1007/978-3-642-15754-7_41
4. Bogers, T., van den Bosch, A.: Comparing and evaluating information retrieval algorithms for news recommendation. In: RecSys, pp. 141–144 (2007)
5. Hoashi, K., Matsumoto, K., Inoue, N., Hashimoto, K.: Document filtering method using non-relevant information profile. In: SIGIR, pp. 176–183 (2000)
6. Lavrenko, V., Schmill, M., Lawrie, D., Ogilvie, P., Jensen, D., Allan, J.: Language models for financial news recommendation. In: CIKM, pp. 389–396 (2000)
7. Lv, Y., Zhai, C.: Negative query generation: bridging the gap between query likelihood retrieval models and relevance. Inf. Retr. J. $18(4)$, 359–378 (2015)
8. Parapar, J., Bellogín, A., Castells, P., Barreiro, Á.: Relevance-based language modelling for recommender systems. Inf. Process. Manag. $49(4)$, 966–980 (2013)
9. Valcarce, D., Parapar, J., Barreiro, Á.: Language models for collaborative filtering neighbourhoods. In: Ferro, N., Crestani, F., Moens, M.-F., Mothe, J., Silvestri, F., Nunzio, G.M., Hauff, C., Silvello, G. (eds.) ECIR 2016. LNCS, vol. 9626, pp. 614–625. Springer, Heidelberg (2016). doi:10.1007/978-3-319-30671-1_45
10. Wang, J., Li, Q., Chen, Y.P., Liu, J., Zhang, C., Lin, Z.: News recommendation in forum-based social media. In: AAAI, pp. 1449–1454 (2010)
11. Wang, X., Fang, H., Zhai, C.: Improve retrieval accuracy for difficult queries using negative feedback. In: CIKM, pp. 991–994 (2007)
12. Wang, X., Fang, H., Zhai, C.: A study of methods for negative relevance feedback. In: SIGIR, pp. 219–226 (2008)
13. Zamani, H.: Recommender systems for multi-publisher environments. Master's thesis, University of Tehran (2015)
14. Zhai, C., Jansen, P., Stoica, E., Grot, N., Evans, D.A.: Threshold calibration in CLARIT adaptive filtering. In: TREC, pp. 96–103 (1998)
15. Zhang, Y., Callan, J.: Maximum likelihood estimation for filtering thresholds. In: SIGIR, pp. 294–302 (2001)

Plagiarism Detection in Texts Obfuscated with Homoglyphs

Faisal Alvi[1,2]([✉]), Mark Stevenson[1], and Paul Clough[1]

[1] University of Sheffield, Sheffield S10 2TN, UK
{mark.stevenson,p.d.clough}@sheffield.ac.uk
[2] King Fahd University of Petroleum and Minerals, Dhahran, Saudi Arabia
falvi1@sheffield.ac.uk

Abstract. Homoglyphs can be used for disguising plagiarized text by replacing letters in source texts with visually identical letters from other scripts. Most current plagiarism detection systems are not able to detect plagiarism when text has been obfuscated using homoglyphs. In this work, we present two alternative approaches for detecting plagiarism in homoglyph obfuscated texts. The first approach utilizes the Unicode list of confusables to replace homoglyphs with visually identical letters, while the second approach uses a similarity score computed using normalized hamming distance to match homoglyph obfuscated words with source words. Empirical testing on datasets from PAN-2015 shows that both approaches perform equally well for plagiarism detection in homoglyph obfuscated texts.

1 Introduction

The notion of 'Disguised Plagiarism' refers to a class of methods used for intentionally hiding text that has been copied [8]. Furthermore, 'Technical Disguise' is a particular form of disguised plagiarism, wherein obfuscation techniques are used in order to evade the detection of plagiarized text by changing the computational representation of text. An important method for technically disguising text is to substitute characters visually identical to other characters in some other script (i.e., homoglyphs) [5]. For example, the Latin character 'p' (Unicode U+160) and the Cyrillic 'р' (Unicode U+0440) have identical glyphs but distinct Unicode values, making the words 'paypal' and 'paypal' appear identical to a human evaluator, but undetectable to an automated plagiarism detection system that has not been designed to deal with such changes. In tests of several leading plagiarism detection systems most were unable to detect similarities between source and plagiarized texts obfuscated using homoglyphs [7,12].

In this work we present two alternate approaches for plagiarism detection in homoglyph obfuscated texts: (1) by using the Unicode list of 'confusables' to find and replace homoglyphs with visually identical ASCII letters; and (2) by using a measure of similarity based on normalized hamming distance to match homoglyph obfuscated words with source words. Our work shows both approaches perform equally well for detecting plagiarism in homoglyph obfuscated texts. Both approaches have their particular advantages and limitations and may therefore be applicable in specific application scenarios in homoglyph obfuscated texts.

© Springer International Publishing AG 2017
J.M. Jose et al. (Eds.): ECIR 2017, LNCS 10193, pp. 669–675, 2017.
DOI: 10.1007/978-3-319-56608-5_64

2   Related Work

Homoglyph substitution has been used as part of standard tests for plagiarism detection systems. For example, Gillam et al. [5] used homoglyph substitution as an obfuscation strategy for testing plagiarism detection systems. Their results demonstrated that six out of seven plagiarism detection systems were unable to detect any similarity between source and substituted text. Figure 1 gives a list of characters used in their work, with the number of instances of each character visually detected by human evaluators stated as well. In the annual 'Plagiarism Detection Software Test' by Weber Wulff et al. [12], 13 out of 15 plagiarism detection systems failed to report any similarity between a given text source and its homoglyph substituted version.

Replacement letters	e - e	h - h	v - v	l - l	u - u	i - i	p - ρ	k - κ
Found	0/20	0/20	3/20	4/20	6/20	9/20	12/20	14/20
Risk of detection	0%	0%	15%	20%	30%	45%	60%	70%

Fig. 1. Replacement letters for visually similar characters from [5]

Heather [6] describe a variety of techniques for technically disguising plagiarized text, which include: modifying the character map, rearranging the glyphs in fonts, replacing text with graphical symbols, and inserting characters in background (white) font between words. Kakoneen and Mozgovoy [7] also discuss a number of 'technical tricks' that can be used to obfuscate texts, including: (1) the insertion of similar looking characters from foreign alphabets (homoglyph substitution); (2) the insertion of background colored characters in between spaces; and (3) the use of scanned images in place of text. According to their results, *"None of the evaluated systems were able to detect any instances of plagiarism from the documents."*

In addition to text obfuscation during plagiarism, homoglyphs have also been used in IDN (Internationalized Domain Name) homograph attacks[1] used to direct users towards alternative websites. With such an attack, users could be directed towards the website 'paypal.com' which is a Cyrillic substituted version of the Latin 'paypal.com'. Existing approaches to deal with IDN homoglyph attacks include: (1) Punycode [3] that converts non-ASCII characters into ASCII characters irreversibly (e.g., Góoglé is converted to the ASCII 'xn–oole-ksbc'); (2) coloring-based strategies that distinguish homoglyphs by assigning various colors to foreign script characters [13]; and (3) a Unicode character similarity list (UC-SimList) [4] to detect homoglyphs in URLs. Some of these approaches might not be useful for plagiarism detection e.g. Punycode results in loss of information, and coloring requires visual inspection. However, the idea of using a list of Unicode equivalents for detecting IDN homograph attacks can be utilized for plagiarism detection in homoglyph obfuscated texts.

---

[1] https://en.wikipedia.org/wiki/IDN_homograph_attack.

3   Methodology

3.1   Resources

The Unicode List of Confusable Characters. Several lists of homoglyph-alphabet pairs are freely available (e.g., homoglyphs.net). The Unicode consortium has released a list of *confusables*, which is a list of visually similar character pairs that includes homoglyphs and their corresponding Latin letters [1]. We use Version 9.0.0 of the list of confusables containing 6167 pairs of confusable characters. Figure 2 shows a partial list of letters similar to the letter 'p' taken from this list.

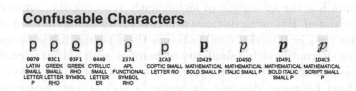

Fig. 2. Visually confusable characters for 'p' from the Unicode list of confusables

Evaluation Dataset. We use PAN-2015 evaluation lab [11] dataset submission by Palkovskii and Belov [10] which is based on the PAN-2013 training dataset with characters in the suspicious documents replaced with homoglyphs. This dataset consists of 5185 document pairs divided into five categories of 'no plagiarism', 'no obfuscation', 'translation', 'random' and 'summary' obfuscation.

3.2   Approaches

Approach 1: Unicode Confusables. In our first approach (shown in Fig. 3), we find and replace every non-ASCII character in the suspicious documents with the corresponding visually matching character from the list of confusable characters. This process replaces homoglyphs in the text of the suspicious documents with visually similar ASCII characters. The resulting suspicious documents can then be compared with the source documents for similarity. In our approach we use word trigram similarity as the seeding strategy, with merging and filtering to discard small matches as false positives [2].

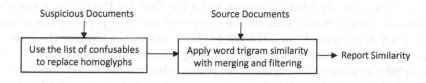

Fig. 3. Block diagram for plagiarism detection using the list of confusables

Approach 2: Normalized Hamming Distance. Our second approach uses normalized hamming distance as an approximate string matching technique. Such techniques are well-suited for this task since homoglyph substitution may partially change the structure of a word. *Hamming Distance* (when applied to strings of characters) detects the number of substitutions (replacements) from one string into another by finding the number of positions where the two strings differ [9]. We use a similarity score (sim_h) computed using normalized hamming distance, defined between two words w_1, w_2 of equal length as:

$$sim_h(w_1, w_2) = 1 - \texttt{Number of substitutions}(w_1, w_2)/\texttt{length}(w_1).$$

Compared to other approximate string similarity measures, normalized hamming distance has the advantage of significantly reducing the number of false positives generated. Hamming distance is undefined for strings of unequal length, (we consider $sim_h = 0$ in this case), whereas these strings might be marked as similar using alternative string similarity techniques, such as character skip gram matching. For example, sim_h(play, plays) = 0, while sim_h(play, play) = 0.75.

Normalized hamming distance similarity (sim_h) is used to compare each word in the suspicious document with the words in the source document. If a pair of words have a value of sim_h greater than or equal to a particular threshold, we consider them as similar. The threshold value depends on the extent of homoglyph substitution in the dataset. For example, if most of the letters in each word have been replaced by homoglyphs, then a lower threshold value will be required to match these words.

Using this procedure for approximate matching of words instead of exact matching, we apply word trigram similarity with merging and filtering (as used in the list-based approach) to find the plagdet score between the source and suspicious documents. We conduct our experiments on the PAN-2015 dataset used in the list-based approach. Regarding the threshold value of sim_h for matching words in our experiments, we do not pre-select a value for this threshold. Instead we calculate plagdet scores for the entire dataset for a range of values of sim_h as shown in Fig. 4.

4   Results and Discussion

4.1   Approach 1: Unicode Confusables

Table 1 shows the results of plagiarism detection in terms of Precision, Recall and Plagdet [11] scores. It can be seen that except for summary obfuscation, Plagdet scores for all other categories including that for the entire dataset are moderately high ($\geq$0.60). This can be compared with the performance of most of the PAN approaches from 2012–2014 [11] on this dataset where the reported Plagdet scores were mostly 0, suggesting a significant improvement.

During the homoglyph replacement phase using the list-based approach, we observed that a number of replacements were also made for non-Latin characters in suspicious documents which were not intended as homoglyphs in source

Table 1. Plagdet-scores using the homoglyph replacement approach

	Dataset	No obf.	Random obf.	Transl. obf.	Summ. obf.
Precision	0.772	0.663	0.953	0.781	0.826
Recall	0.727	0.988	0.667	0.643	0.107
Plagdet	**0.670**	**0.717**	**0.707**	**0.632**	**0.150**

documents. For example, the currency symbol '¢' was replaced by a 'c'. This observation suggests that the proposed approach of using a list of homoglyph-alphabet pairs to replace characters may not work well when the source text contains a large number of foreign characters, since these might be converted to ASCII characters in the substitution phase. However, the approach can be improved by searching through the source documents to distinguish homoglyphs from true source non-Latin characters, at the cost of increased computation time.

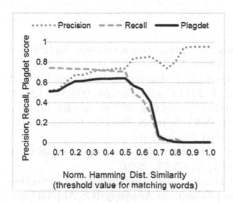

Fig. 4. Plagdet scores for plagiarism detection using normalized hamming distance similarity

Table 2. Plagdet scores for the case when $sim_h = 0.450$

Category	Plagdet
Entire dataset	0.644
No obfuscation	0.662
Random Obf.	0.688
Translate. Obf.	0.626
Summary Obf.	0.142

4.2   Approach 2: Normalized Hamming Distance

Figure 4 shows Precision, Recall and Plagdet scores for various values of sim_h threshold. It can be seen that a threshold value for $sim_h \approx 0.45$ is giving the highest Plagdet score of 0.644. Table 2 gives Plagdet scores for each category of plagiarism in the dataset for a threshold value of 0.45. Similar to Table 1, we observe that except for summary obfuscation, most of these values are moderately high (≥ 0.6). Although the scores in Table 1 are somewhat higher than those in Table 2, the differences are small enough to consider performance of the approaches to be similar for detecting plagiarism in homoglyph obfuscated texts. From Fig. 4 we observe that a careful selection of threshold value for sim_h is important. The Plagdet score rapidly decreases after $sim_h = 0.5$ since higher threshold values increase the number of true matches being rejected. For large datasets, this problem can alleviated by first applying the approach on a smaller collection of training documents to obtain a suitable initial estimate for the threshold value.

5 Conclusions and Future Work

The development of techniques for automated plagiarism detection continues to be an active area of research. In this work we presented two approaches for plagiarism detection in homoglyph obfuscated texts which perform equally well for plagiarism detection. One approach utilizes the Unicode list of confusables to replace homoglyphs with visually identical letters; the other approach uses a similarity score computed using normalized hamming distance. For future work, improvised versions of these approaches can be incorporated into a set of approaches for detecting multiple forms of technical disguise.

References

1. Unicode List of Visually Confusable Characters. http://www.unicode.org/Public/security/9.0.0/confusables.txt. Accessed 19 Oct 2016
2. Alvi, F., Stevenson, M., Clough, P.D.: Hashing and merging heuristics for text reuse detection. In: Working Notes for CLEF 2014 Conference, pp. 939–946 (2014)
3. Costello, A.: RFC3492-Punycode: a bootstring encoding of Unicode for internationalized domain names in applications (IDNA). Network Working Group (2003). http://www.ietf.org/rfc/rfc3492.txt. Accessed 19 Oct 2016
4. Fu, A.Y., Deng, X., Wenyin, L.: REGAP: a tool for Unicode-based web identity fraud detection. J. Digital Forensic Pract. 1(2), 83–97 (2006)
5. Gillam, L., Marinuzzi, J., Ioannou, P.: Turnitoff-defeating plagiarism detection systems. In: Proceedings of the 11th Higher Education Academy-ICS Annual Conference. Higher Education Academy (2010)
6. Heather, J.: Turnitoff: identifying and fixing a hole in current plagiarism detection software. Assess. Eval. High. Educ. 35(6), 647–660 (2010)
7. Kakkonen, T., Mozgovoy, M.: Hermetic and web plagiarism detection systems for student essays an evaluation of the state-of-the-art. J. Educ. Comput. Res. 42(2), 135–159 (2010)
8. Meuschke, N., Gipp, B.: State-of-the-art in detecting academic plagiarism. Int. J. Educ. Integrity 9(1), 50–71 (2013)
9. Navarro, G.: A guided tour to approximate string matching. ACM Comput. Surv. (CSUR) 33(1), 31–88 (2001)
10. Palkovskii, Y., Belov, A.: Submission to the 7th International Competition on Plagiarism Detection (2015). http://www.uni-weimar.de/medien/webis/events/pan-15. Accessed 15 Oct 2016
11. Potthast, M., Göring, S., Rosso, P., Stein, B.: Towards data submissions for shared tasks: first experiences for the task of text alignment. In: Working Notes Papers of the CLEF 2015 Evaluation Labs, CEUR Workshop Proceedings, September 2015

12. Weber-Wulff, D., Möer, C., Touras, J., Zincke, E.: Plagiarism Detection Software Test 2013 (2013). http://plagiat.htw-berlin.de/software-en/test2013/report-2013/. Accessed 15 Oct 2016
13. Wenyin, L., Fu, A.Y., Deng, X.: Exposing homograph obfuscation intentions by coloring unicode strings. In: Zhang, Y., Yu, G., Bertino, E., Xu, G. (eds.) APWeb 2008. LNCS, vol. 4976, pp. 275–286. Springer, Heidelberg (2008). doi:10.1007/978-3-540-78849-2_29

Iterative Estimation of Document Relevance Score for Pseudo-Relevance Feedback

Mozhdeh Ariannezhad[1]([⊠]), Ali Montazeralghaem[1], Hamed Zamani[2], and Azadeh Shakery[1]

[1] School of Electrical and Computer Engineering, College of Engineering, University of Tehran, Tehran, Iran
{m.ariannezhad,ali.montazer,shakery}@ut.ac.ir
[2] Center for Intelligent Information Retrieval, College of Information and Computer Sciences, University of Massachusetts Amherst, Amherst, MA, USA
zamani@cs.umass.edu

Abstract. Pseudo-relevance feedback (PRF) is an effective technique for improving the retrieval performance through updating the query model using the top retrieved documents. Previous work shows that estimating the effectiveness of feedback documents can substantially affect the PRF performance. Following the recent studies on theoretical analysis of PRF models, in this paper, we introduce a new constraint which states that the documents containing more informative terms for PRF should have higher relevance scores. Furthermore, we provide a general iterative algorithm that can be applied to any PRF model to ensure the satisfaction of the proposed constraint. In this regard, the algorithm computes the feedback weight of terms and the relevance score of feedback documents, simultaneously. To study the effectiveness of the proposed algorithm, we modify the log-logistic feedback model, a state-of-the-art PRF model, as a case study. Our experiments on three TREC collections demonstrate that the modified log-logistic significantly outperforms competitive baselines, with up to 12% MAP improvement over the original log-logistic model.

Keywords: Pseudo-relevance feedback · Document effectiveness · Axiomatic analysis · Query expansion

1 Introduction

Search queries are usually too short to precisely express the underlying information need, which leads to poor retrieval performance. To address this problem, pseudo-relevance feedback (PRF) technique updates the query model using the top retrieved documents that are assumed to be relevant to the initial query. PRF has been shown to be highly effective in improving the retrieval performance [2,7,8,11,12]. In order to theoretically analyze PRF models, previous work [2,8,9] has proposed various constraints (axioms) that they should satisfy.

© Springer International Publishing AG 2017
J.M. Jose et al. (Eds.): ECIR 2017, LNCS 10193, pp. 676–683, 2017.
DOI: 10.1007/978-3-319-56608-5_65

To satisfy the PRF constraints and thus to improve the accuracy of PRF models, different modifications have been suggested for well-established PRF models, such as mixture model [12] and geometric relevance model [10].

Pal et al. [9] proposed the "relevance effect" constraint as follows: the terms in the feedback documents with higher relevance scores should get higher weights in the feedback model. To satisfy this constraint, they used the initial relevance score of documents as their weight in the feedback model, similar to relevance models [7]. On the other hand, Keikha et al. [5] showed that the initial retrieval score of a document is not a good indicator for its effectiveness in the feedback model. They proposed a supervised algorithm to predict the document effectiveness for this task. In this paper, we argue that the relevance score of feedback documents can be better estimated using the feedback weights of the terms they contain. The intuition is that a document is more useful for PRF if it contains more informative terms for PRF. To this end, we propose the "feedback weight effect" constraint that implies the documents containing terms with higher weights in the feedback model should have higher relevance scores. State-of-the-art PRF models, such as relevance model [7], mixture model [12], matrix factorization-based model [11], and log-logistic feedback model [1], do not satisfy this constraint. In order to satisfy the introduced constraint, we propose a *general* iterative unsupervised algorithm that can be applied to any PRF model. In each iteration, the algorithm alternates between two steps: (1) computing the relevance scores of documents based on the feedback weights of their terms, and (2) computing the feedback weights of the terms with regard to the relevance scores of the documents they appear in.

To study the effectiveness of the proposed algorithm, we modify the log-logistic feedback model [1] to satisfy the feedback weight effect constraint using our iterative algorithm. Log-logistic model is a state-of-the-art PRF model that was previously shown to satisfy many PRF constraints and outperform competitive baselines, including geometric relevance model [10] and mixture model [2]. The experiments on three TREC collections demonstrate that our modification significantly outperforms the baselines.

2   Methodology

In this section, we introduce the "feedback weight effect" constraint and propose an iterative reinforcement algorithm to simultaneously compute the feedback weights of terms and the relevance scores of feedback documents. We use the notation previously used in [2,8]. $FW(w, F)$ and $RS(d, q)$ denote the feedback weight of term w in the feedback set F and the relevance score of document d for a given query q, respectively. $TF(w, d)$ denotes the frequency of term w in document d and $IDF(w)$ represents the inverse document frequency of term w.

2.1   PRF Constraints for Relevance Score of Feedback Documents

Pal et al. [9] introduced the relevance effect constraint for PRF models as follows:

Relevance effect: If a term w occurs in two documents $d_1, d_2 \in F$, and $RS(d_1, q) > RS(d_2, q)$, then: $FW(w, F \setminus \{d_1\}) < FW(w, F \setminus \{d_2\})$.

The relevance effect constraint indicates that the terms in the feedback documents with higher relevance scores should have higher weights in the feedback model compared to those in the documents with lower relevance scores. An important issue here is how to compute the relevance score? Pal et al. [9] followed the idea behind the relevance models [7] and used the initial retrieval score of feedback documents (e.g., the query likelihood score) as their relevance score. On the other hand, Keikha et al. [5] showed that the initial retrieval score is not an optimal indicator of document effectiveness for query expansion. Based on their observations, we provide a theoretical axiom for estimating the effectiveness of documents for feedback. We argue that the relevance score of feedback documents should depend on the feedback weights of the terms they contain. Since the feedback weight of a term demonstrates the usefulness of the term for PRF, a document that contains more informative terms is more useful for PRF. As a result, such a document should have a higher relevance score. In this regard, we define the feedback weight effect constraint as follows:

Feedback weight effect: If $d \in F$ and w_1 and w_2 are two feedback terms where $TF(w_1, d) = TF(w_2, d) \geq 1$, $IDF(w_1) = IDF(w_2)$ and $FW(w_1, F) > FW(w_2, F)$, then: $RS(d \setminus \{w_1\}, q) < RS(d \setminus \{w_2\}, q)$.

Note that the feedback weight effect is a constraint for the relevance score and can be satisfied regardless of whether the PRF model enforces the relevance effect or not.

2.2 Relevance Score Estimation via an Iterative Reinforcement Model

To satisfy the aforementioned constraints, we provide an iterative approach that simultaneously computes the feedback weight of terms and the relevance score of feedback documents. The relevance effect states that a term should have a high feedback weight if it appears in many feedback documents with high relevance scores, and the feedback weight effect implies that a feedback document should have a high relevance score if it contains many terms with high feedback weights. In other words, the feedback weight of a term is determined by the relevance score of the feedback documents it appears in, and the relevance score of a feedback document is determined by the feedback weights of the terms it contains. For simplicity, we respectively use $FW(w)$ and $RS(d)$ instead of $FW(w, F)$ and $RS(d, q)$, in the equations. The following steps are alternated until convergence, with a uniform initialization for the document and term scores:

1. Computing feedback term weights:

$$\forall w \in V_F : FW(w)^{(n)} = Com(w) \sum_{d \in F} TW(w, d, q) RS(d)^{(n-1)}, \qquad (1)$$

2. Computing document relevance scores:

$$\forall d \in F : RS(d)^{(n)} = \frac{1}{|d|} \sum_{w \in d} TW(w, d, q) FW(w)^{(n-1)}. \tag{2}$$

In the above equations, $TW(w, d, q)$ is a term weighting function that demonstrates the importance of term w in document d with respect to the query q, $|d|$ denotes the length of document d, and V_F represents the set of feedback terms. $FW(w)^{(n)}$ and $RS(d)^{(n)}$ respectively denote the feedback term weight and the document relevance score computed in the nth iteration. In the first equation, $Com(w) = \frac{|F_w|}{|F|}$ ($|F_w|$ denotes the number of feedback documents that contain w) shows how common w is in the feedback documents. $Com(w)$ was previously used in [5] and leads to satisfying the DF effect constraint [2]. Note that in each iteration, the feedback weights and the relevance scores should be normalized subject to $\sum_{w \in V_F} FW(w)^{(n)} = 1$ and $\sum_{d \in F} RS(d)^{(n)} = 1$. The proposed algorithm differs from the one introduced in [4], in that it does not calculate the relevance scores for the feedback documents.

Similar ideas regarding iterative computation of related variables have been used in different tasks, such as in the HITS algorithm [6]. The convergence of our algorithm can be proven, similar to the proof presented in [6].

2.3 Case Study: Log-Logistic Feedback Model

As mentioned above, the proposed constraint and algorithm are general and independent of the feedback model. In this paper, we consider the log-logistic feedback model [1], a state-of-the-art PRF model. Clinchant and Gaussier [2] showed that the log-logistic model satisfies their PRF constraints and outperforms many feedback models, including the geometric relevance model [10] and the mixture model [12]. The log-logistic model calculates the feedback weight of a term w in a document d, as follows:

$$TW_{LL}(w, d, q) = \log \left(\frac{t(w, d) + \lambda_w}{\lambda_w} \right), \tag{3}$$

where $\lambda_w = \frac{N_w}{N}$ (N_w is the number of documents in the collection that contain w and N is the total number of documents in the collection), and $t(w, d)$ is the normalized term frequency component defined as: $t(w, d) = TF(t, d) \log(1 + c\frac{avg_l}{|d|})$, where avg_l denotes the average document length and c is a free hyperparameter. It is shown that log-logistic satisfies TF, IDF, DF, concavity, and document length constraints [2]. Recently, Montazeralghaem et al. [8] modified the log-logistic model as follows, in order to satisfy the relevance effect constraint:

$$TW_{LLR}(w, d, q) = RS_{init}(q, d) \times TW_{LL}(w, d, q), \tag{4}$$

where $RS_{init}(q, d)$ denotes the initial retrieval score of document d. Both the original log-logistic feedback model and the modified version in [8] use the mean of $TW(w, d, q)$ over all the feedback documents as $FW(w, F)$.

We use the term weight definition provided in Eq. (4) in our iterative algorithm proposed in Sect. 2.2. This method is referred to as *LLIR*. In order to prove that LLIR satisfies the proposed axiom, we consider two terms w_1 and w_2 and a feedback document d. When $TF(w_1, d) = TF(w_2, d) \geq 1$ and $IDF(w_1) = IDF(w_2)$, it can be shown that $TW_{LLR}(w_1, d, q) = TW_{LLR}(w_2, d, q)$. Considering the case where $FW(w_1, F) > FW(w_2, F)$, it is obvious that $TW_{LLR}(w_1, d, q)FW(w_1, F) > TW_{LLR}(w_2, d, q)FW(w_2, F)$, which implies RS $(d \setminus \{w_1\}, q) < RS(d \setminus \{w_2\}, q)$, if we use Eq. 2 to compute the score of feedback documents.

3 Experiments

Collections. In our experiments, we used three standard TREC collections whose statistics are provided in Table 1. AP and Robust are newswire collections, whereas WT10g is a Web collection containing more noisy documents.

Experimental Setup. We used the titles of TREC topics as queries. All indexes and topics were stopped using the standard INQUERY stopword list and stemmed using the Porter stemmer. All experiments were carried out using the Lemur toolkit[1]. Initial retrieval results were obtained using the query likelihood model with Dirichlet prior smoothing ($\mu = 1000$).

Parameter Setting. The number of feedback documents, the number of feedback terms, the feedback coefficient, and the parameter c are set using 2-fold cross validation over the queries of each collection, for all methods. We swept the number of feedback documents between $\{10, 25, 50, 75, 100\}$ and the number of feedback terms between $\{10, 50, 100, 150, 200\}$. We changed the feedback coefficient from 0 to 1 in the increment of 0.1, and the parameter c from 1 to 10 in the increment of 1.

Evaluation Metrics. We use three metrics to measure the retrieval quality: (1) mean average precision (MAP) of the top-ranked 1000 documents, (2) the precision of the top 10 retrieved documents (P@10), and (3) the robustness index (RI) [3]. Statistically significant differences of performance are determined using the two-tailed paired t-test at a 95% confidence level.

3.1 Results and Discussion

We consider three baselines: (1) the document retrieval model without pseudo-relevance feedback (NoPRF), (2) the original log-logistic feedback model (LL) [1], and (3) the enhanced log-logistic model (LLR) which satisfies the relevance effect, proposed in [8]. Furthermore, we also report the results achieved by the proposed method both after one iteration (LLIR-1-iteration) and after convergence (LLIR-converged).

---

[1] http://lemurproject.org/.

Table 1. Summary of TREC collections and topics.

ID	Collection	Queries	#docs
AP	TREC 1-3 Ad-hoc track, Associated press 88–89	Topics 51–200	165k
Robust	TREC 2004 Robust track collection	Topics 301–450 & 601–700	528k
WT10g	TREC 9-10 Web track collection	Topics 451–550	1692k

Table 2. Retrieval effectiveness of the iterative model compared to the baselines. Superscripts 0/1/2/3 indicate that the improvements over NoPRF/LL/LLR/LLIR-1-iteration are significant.

Method	AP			Robust			WT10g		
	MAP	P@10	RI	MAP	P@10	RI	MAP	P@10	RI
NoPRF	0.2663	0.4309	–	0.2490	0.4237	–	0.2080	0.3030	–
LL	0.3300^0	0.4691	0.44	0.2798^0	0.4394	0.29	0.2089	0.3071	0.08
LLR	0.3381^{01}	0.4624	**0.47**	0.2822^0	**0.4450**	0.29	0.2230^{01}	0.3101	0.17
LLIR-1-iteration	0.3406^{01}	0.4698	0.42	0.2876^{012}	0.4365	0.28	0.2219^{01}	0.3101	0.17
LLIR-converged	$\mathbf{0.3507^{0123}}$	**0.4765**	0.45	$\mathbf{0.2926^{0123}}$	0.4442	**0.31**	$\mathbf{0.2344^{0123}}$	**0.3121**	**0.21**

Table 2 summarizes the results achieved by the proposed method and the baselines. As shown in the table, LL performs significantly better than NoPRF on all the collections, which shows the effectiveness of the log-logistic model. LLIR-converged outperforms NoPRF and LL on all collections, indicating the importance of the proposed constraint for PRF. The significant improvements achieved by LLIR (after convergence) over LLR show that the document scores estimated by our model are more effective than the initial retrieval scores which are used by LLR. According to Table 2, the LLIR results after convergence are significantly higher than those obtained after the first iteration. It is worth mentioning that LLIR converges after 8 to 10 iterations, indicating its efficiency and low computational cost. The performance of LLIR-converged in terms of P@10 and RI is also superior to the baselines, except in two cases (RI on AP and P@10 on Robust) where the results are comparable to the highest values. In general, the results show that our method have impressive overall ranking performance, e.g., the MAP value achieved by LLIR-converged is up to 12% higher than those obtained by the original log-logistic model (LL).

Figure 1 plots the sensitivity of the proposed method with respect to the number of feedback terms added to the query. According to this figure, after 50 terms, the performance becomes stable in the newswire collections (AP and Robust), while by increasing the number of terms, we lose the performance in the WT10g collection. To have an insight into the term weights computed by the proposed method, Table 3 reports the top 10 terms for the query "gulf war syndrom" in the Robust collection, computed by the LLR and the proposed method. As shown in the table, the order of the terms have changed and also

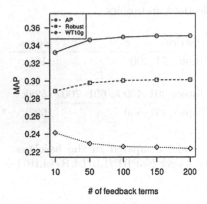

Fig. 1. Sensitivity of the LLIR method to the number of feedback terms.

Table 3. The top terms added to the query "gulf war syndrom" (topic 630) by LLR and LLIR methods.

LLR		LLIR	
Syndrom	0.1862	Syndrom	0.2452
Gulf	0.1507	Gulf	0.2015
War	0.1126	War	0.1317
Veteran	0.0932	Veteran	0.0754
Vietnam	0.0867	Defenc	0.0730
Desert	0.0804	Militari	0.0694
Defenc	0.0757	Desert	0.0570
Soldier	0.0735	Serv	0.0501
Militari	0.0734	Time	0.0494
Diarrhoea	0.0677	American	0.0473

the term "vietnam" which is irrelevant to the initial query does not appear in the list of top terms estimated by LLIR, while a relevant term ("american") is added to the list. For this query, the average precision achieved by LLIR is 0.6034 which is much higher than the one obtained by LLR (i.e., 0.2867).

4   Conclusions

In this paper, we introduced a new constraint concerning the relevance score of the feedback documents. The constraint states that the documents containing more informative terms for PRF should have higher relevance scores. We further proposed a general iterative algorithm that can be applied to any PRF model in order to guarantee the satisfaction of the proposed constraint. We applied our algorithm to the log-logistic feedback model as a case study. Our experiments on three TREC collections showed that the proposed modification significantly improves the results.

Acknowledgements. This work was supported in part by the Center for Intelligent Information Retrieval. Any opinions, findings and conclusions or recommendations expressed in this material are those of the authors and do not necessarily reflect those of the sponsor.

References

1. Clinchant, S., Gaussier, E.: Information-based models for ad hoc IR. In: SIGIR (2010)
2. Clinchant, S., Gaussier, E.: A theoretical analysis of pseudo-relevance feedback models. In: ICTIR (2013)

3. Collins-Thompson, K.: Reducing the risk of query expansion via robust constrained optimization. In: CIKM (2009)
4. Dehghani, M., Azarbonyad, H., Kamps, J., Hiemstra, D., Marx, M.: Luhn revisited: significant words language models. In: CIKM (2016)
5. Keikha, M., Seo, J., Croft, W.B., Crestani, F.: Predicting document effectiveness in pseudo relevance feedback. In: CIKM (2011)
6. Kleinberg, J.M.: Authoritative sources in a hyperlinked environment. J. ACM 46(5), 604–632 (1999)
7. Lavrenko, V., Croft, W.B.: Relevance based language models. In: SIGIR (2001)
8. Montazeralghaem, A., Zamani, H., Shakery, A.: Axiomatic analysis for improving the log-logistic feedback model. In: SIGIR (2016)
9. Pal, D., Mitra, M., Bhattacharya, S.: Improving pseudo relevance feedback in the divergence from randomness model. In: ICTIR (2015)
10. Seo, J., Croft, W.B.: Geometric representations for multiple documents. In: SIGIR (2010)
11. Zamani, H., Dadashkarimi, J., Shakery, A., Croft, W.B.: Pseudo-relevance feedback based on matrix factorization. In: CIKM (2016)
12. Zhai, C., Lafferty, J.: Model-based feedback in the language modeling approach to information retrieval. In: CIKM (2001)

Design Patterns for Fusion-Based Object Retrieval

Shuo Zhang and Krisztian Balog[(✉)]

University of Stavanger, Stavanger, Norway
{shuo.zhang,krisztian.balog}@uis.no

Abstract. We address the task of ranking objects (such as people, blogs, or verticals) that, unlike documents, do not have direct term-based representations. To be able to match them against keyword queries, evidence needs to be amassed from documents that are associated with the given object. We present two design patterns, i.e., general reusable retrieval strategies, which are able to encompass most existing approaches from the past. One strategy combines evidence on the term level (early fusion), while the other does it on the document level (late fusion). We demonstrate the generality of these patterns by applying them to three different object retrieval tasks: expert finding, blog distillation, and vertical ranking.

1 Introduction

Viewed broadly, information retrieval is about matching information objects against information needs. In the classical ad hoc document retrieval task, information objects are documents and information needs are expressed as keyword queries. This task has been a main focal point since the inception of the field. The past decade, however, has seen a move beyond documents as units of retrieval to other types of objects. Examples of object retrieval tasks studied at the Text REtrieval Conference (TREC) include ranking people (experts) [1,4], blogs [10, 11], and verticals [5,6]. Common to these tasks is that objects do not have direct representations that could be matched against the search query. Instead, they are associated with documents, which are used as a proxy to connect objects and queries. See Fig. 1 for an illustration. The main question, then, is how to combine evidence from documents that are associated with a given object.

Most approaches that have been proposed for object retrieval can be categorized into two main groups of retrieval strategies: (1) *object-centric* methods build a term-based representation of objects by aggregating term counts across the set of documents associated with the objects; (2) *document-centric* methods first retrieve documents relevant to the query, then consider the objects associated with these documents. Viewed abstractly, the object retrieval task is about fusing or blending information about a given object. This fusion may happen early on in the retrieval process, on the term level (i.e., object-centric methods), or later, on the document level (i.e., document-centric methods). Using either of the two strategies, two main shared components can be distilled:

© Springer International Publishing AG 2017
J.M. Jose et al. (Eds.): ECIR 2017, LNCS 10193, pp. 684–690, 2017.
DOI: 10.1007/978-3-319-56608-5_66

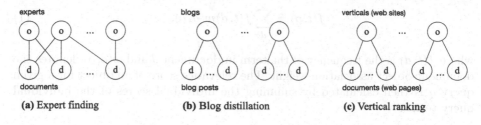

experts   blogs   verticals (web sites)

documents   blog posts   documents (web pages)

(a) Expert finding   (b) Blog distillation   (c) Vertical ranking

Fig. 1. Illustration of various object retrieval tasks.

the underlying term-based retrieval model (e.g., language models, BM25, DFR, etc.) and the document-object association method. Various instantiations (i.e., choice of retrieval strategy, retrieval model, and document-object associations) have been studied, but always in the context of a particular object retrieval task, see, e.g., [2,7,9,13].

We show in this paper, as our main contribution, that further generalizations are possible. We present two *design patterns* for object retrieval, that is, general repeatable solutions that can easily emulate most previously proposed approaches. We call these design patterns to emphasize that they can be used in many different situations. The second contribution of this work is an experimental evaluation performed for three different object retrieval tasks: expert finding, blog distillation, and vertical ranking. Using standard TREC collections, we demonstrate that the early and late fusion patterns are indeed widely applicable and deliver competitive performance without resorting to any task-specific tailoring. The implementation of our models is available at http://bit. ly/ecir2017-fusion.

2 Fusion-Based Object Retrieval Methods

Object retrieval is the task of returning a ranked list of objects in response to a keyword query. We assume a scenario where objects do not have direct term-based representations, but each object is associated with one or more documents. These documents are used as a bridge between queries and objects. We present two design patterns, i.e., general retrieval strategies, in the following two subsections. Both strategies consider the relationship between a document and an object; we detail this element in Sect. 2.3

2.1 Early Fusion

According to the early fusion (or object-centric) strategy a term-based representation is created for each object. That is, the fusion happens on the term level. One can think of this approach as creating a pseudo document for each object; once those object description documents are created, they can be ranked using standard document retrieval models. We define the (pseudo) frequency of a term t for an object o as follows:

$$\tilde{f}(t,o) = \sum_d f(t,d)w(d,o), \tag{1}$$

where $f(t,d)$ is the frequency of the term in document d and $w(d,o)$ denotes the document-object association weight. The relevance score of an object for a given query q is then calculated by summing the individual scores of the individual query terms:

$$score(o,q) = \sum_{i=1}^{|q|} score(q_i,o) = \sum_{i=1}^{|q|} score(q_i, \tilde{f}, \varphi),$$

where φ holds all parameters of the underlying retrieval model (e.g., k_1 and b for BM25). For computing $score(t,o)$, any existing retrieval model can be used. Specifically, using language models with Jelinek-Mercer smoothing it is:

$$score_{LM}(t,o) = \log\left((1-\lambda)\frac{\tilde{f}(t,o)}{|o|} + \lambda P(t)\right),$$

where $|o|$ is the length of the object ($|o| = \sum_t \tilde{f}(t,o)$), $P(t)$ is the background language model, and λ is the smoothing parameter. Using BM25, the term score is computed as:

$$score_{BM25}(t,o) = \frac{\tilde{f}(t,o)(k_1+1)}{\tilde{f}(t,o) + k_1(1 - b + b\frac{|o|}{avg(o)})} IDF(t),$$

where $IDF(t)$ is computed as $\log\frac{N}{|\{o:\tilde{f}(t,o)>0\}|}$ and $avg(o)$ is the average object length.

Table 1 lists exiting approaches for different search tasks, which can be classified as early fusion. Due to space constraints, we only highlight one specific method for each of the object ranking tasks we consider.

Table 1. Examples of early fusion approaches. Notice that the aggregation happens on the term level. (Computing the log probabilities turns the product into a summation over query terms.)

Task	Model	Equation
Expert finding	Profile-based [8]	$P(q|\theta_c, R=1) = \prod_{t \in q} p(t|\theta_c, R=1)^{n(t,q)}$
Blog distillation	Blogger model [3]	$P(q|\theta_{blog}) = \prod_{t \in q} P(t|\theta_{blog})^{n(t,q)}$
Vertical ranking	CVV [12]	$Goodness(c,q) = \sum_{i=1}^{|q|} CVV_i \times df_{i,c}$

Table 2. Examples of late fusion approaches. Notice that aggregation happens on the document level; each formula contains a term that expresses the document's relevance.

Task	Model	Equation
Expert finding	Voting model [9]	$score\_cand\_RR(e,q) = \sum_{d \in R(q) \cap profile(e)} \frac{1}{rank(d,q)}$
Blog distillation	Posting model [3]	$P(q|blog) = \sum_{post \in blog} P(q|\theta_{post})P(post|blog)$
Vertical ranking	ReDDE [12]	$R(c,q) = \sum_{d \in c} P(R|d)P(d|c)|c|$

2.2 Late Fusion

Instead of creating a direct term-based representation for objects, the late fusion (or document-centric) strategy models and queries individual documents, then aggregates their relevance estimates. Formally:

$$score(o,q) = \sum_d score(d,q)w(d,o), \qquad (2)$$

where $score(d,q)$ expresses the document's relevance to the query and can be computed using any existing document retrieval method, such as language models or BM25. As before, $w(d,o)$ is the weight of document d for the given object. The efficiency of this approach can be further improved by restricting the summation to the top-K relevant documents. Table 2 shows three exiting models for different search tasks, which can be catalogued as late fusion strategies.

2.3 Document-Object Associations

Using either the early or the late fusion strategy, they share the component $w(d,o)$, cf. Eqs. (1) and (2). This document-object association score determines the weight with which a particular document contributes to the relevance score of a given object. In this paper, we consider two simple ways for setting this weight. We introduce the shorthand notation $d \in o$ to indicate that document d is associated with object o (i.e., there is an edge between d and o in Fig. 1). According to the *binary* method, $w(d,o)$ can take only two values: it is 1 if $d \in o$ and 0 otherwise. Alternatively, the *uniform* method assigns the value $\frac{1}{len(o)}$ if $d \in o$, where $len(o)$ is the total number of documents associated with o, and 0 otherwise.

3 Experimental Setup

We consider three object retrieval tasks, with corresponding TREC collections. *Expert finding* uses the test suites of the TREC 2007 and 2008 Enterprise track [1,4]. Objects are experts and each of them is typically associated with multiple documents. *Blog distillation* is based on the TREC 2007 and 2008 Blog track [10,11]. Objects are blogs and documents are posts; each document (post)

Table 3. Object retrieval tasks and collections used in this paper.

Task	Collection (#docs)	Queries
Expert finding	CSIRO (370K)	50 (2007), 77 (2008)
Blog distillation	Blogs06 (3.2M)	50 (2007), 50 (2008)
Vertical ranking	FedWeb13 (1.9M), FedWeb14 (3.6M)	50 (2013), 50 (2014)

belongs to exactly on object (blog). *Vertical ranking* corresponds to the resource selection task of the TREC 2013 and 2014 Federated Search track [5,6]. Objects are verticals (i.e., web sites) and documents are web pages. Table 3 summarizes the data sets used for each task.

For each task, we consider two retrieval models: language models (using Jelinek Mercer Smoothing, $\lambda = 0.1$) and BM25 (with $k_1 = 1.2$ and $b = 0.75$). We further compare two models of document-object associations: binary and uniform.

4   Experimental Results

The results for the expert finding, blog distillation, and vertical ranking tasks are presented in Tables 4, 5, and 6, respectively. Our main observations are the following. First, there is no preferred fusion strategy; early and late fusion both emerge as overall bests in 3–3 cases. While early fusion is clearly preferred for vertical ranking and late fusion is clearly favorable for blog distillation, a mixed picture unfolds for expert finding: early fusion performs better on one query set (2007) while late fusion wins on another (2008). The differences between the corresponding early and late fusion configurations can be substantial. Second, the main difference between binary and uniform associations is that the latter takes into account the number of different documents associated with the object, while the former does not. For expert finding and vertical ranking the binary method is clearly superior. For blog distillation, on the other hand, it is nearly always the uniform method that performs better. The difference between vertical ranking and blog distillation is especially interesting given that these two tasks have essentially identical structure, i.e., each document is associated with exactly one object (see Fig. 1). Third, concerning the choice of retrieval model (LM vs. BM25), we again find that it depends on the task and fusion strategy. BM25 is superior to LM on blog distillation. For expert finding and vertical ranking, LM performs better in case of early fusion, while BM25 is preferable for late fusion.

We also include the TREC best and median results for reference comparison. In most cases, our fusion-based methods perform better than the TREC median, and on one occasion (vertical ranking, 2013) we outperform the best TREC run. Let us emphasize that we did not resort to any task-specific treatment. In the light of this, our results can be considered more than satisfactory and signify the generality of our fusion strategies.

Table 4. Results on the expert finding task. Highest scores are in boldface.

Fusion strategy	Retr. model	Doc-obj. assoc.	2007			2008		
			MAP	MRR	P@10	MAP	MRR	P@10
Early fusion	LM	binary	**0.3607**	**0.4809**	0.1229	0.1927	0.3741	0.1863
	LM	uniform	0.2902	0.3650	0.1083	0.1760	0.3843	0.1725
	BM25	binary	0.2887	0.3654	0.0900	0.1203	0.2599	0.1148
	BM25	uniform	0.1688	0.2159	0.0780	0.0646	0.1517	0.0741
Late fusion	LM	binary	0.3283	0.4730	0.1420	0.2036	0.4342	0.2167
	LM	uniform	0.1978	0.2561	0.0940	0.1146	0.2948	0.1296
	BM25	binary	0.3495	0.4949	**0.1480**	**0.2623**	**0.5048**	**0.2648**
	BM25	uniform	0.2492	0.3065	0.1040	0.1787	0.3988	0.1759
TREC best			**0.4632**			**0.2987**	0.4951	
TREC median			0.3090			0.2606	0.3843	

Table 5. Results on the blog distillation task. Highest scores are in boldface.

Fusion strategy	Retr. model	Doc-obj. assoc.	2007			2008		
			MAP	MRR	P@10	MAP	MRR	P@10
Early fusion	LM	binary	0.2055	0.4660	0.3432	0.1883	0.6996	0.3684
	LM	uniform	0.2479	0.5313	0.3932	0.1897	0.6228	**0.3740**
	BM25	binary	0.2374	0.4773	0.3844	0.1789	0.5731	0.3460
	BM25	uniform	0.2088	0.6316	0.3578	0.1936	0.6180	0.3460
Late fusion	LM	binary	0.1845	0.5349	0.3111	0.1556	0.4755	0.2800
	LM	uniform	0.2605	0.6140	0.4222	0.2040	0.7241	0.3360
	BM25	binary	0.2202	0.5892	0.3489	0.1731	0.5478	0.3140
	BM25	uniform	**0.2987**	**0.7303**	**0.4822**	**0.2245**	**0.7482**	0.3600
TREC best			**0.3695**	**0.8093**	**0.5356**	**0.3015**	**0.8051**	**0.4480**
TREC median			0.2353	0.7425	0.4567	0.2416	0.7167	0.3580

Table 6. Results on the vertical ranking task. Highest scores are in boldface.

Fusion strategy	Retr. model	Doc-obj. assoc.	2013			2014		
			nDCG@20	MAP	P@5	nDCG@20	MAP	P@5
Early fusion	LM	binary	**0.3382**	**0.3656**	**0.4000**	**0.2782**	**0.3052**	**0.4857**
	LM	uniform	0.2271	0.2293	0.3306	0.2184	0.2612	0.3633
	BM25	binary	0.2588	0.2704	0.2500	0.2354	0.2758	0.3920
	BM25	uniform	0.1689	0.1960	0.2612	0.1669	0.2204	0.2960
Late fusion	LM	binary	0.1950	0.1991	0.2163	0.1961	0.2439	0.3000
	LM	uniform	0.1370	0.1641	0.1755	0.1408	0.2094	0.2400
	BM25	binary	0.2373	0.2163	0.2490	0.2220	0.2576	0.3400
	BM25	uniform	0.1548	0.1755	0.1918	0.1658	0.2208	0.3000
TREC best			0.2990			0.3200	**0.7120**		**0.6040**
TREC median			0.1410			0.1850	0.3450		0.2125

5   Conclusions

In this paper we have presented two design patterns, early and late fusion, to the commonly occurring problem of object retrieval. We have demonstrated the generality and reusability of these solutions on three different tasks: expert finding, blog distillation, and vertical ranking. Specifically, we have considered various instantiations of these patterns using (i) language models and BM25 as the underlying retrieval model and (ii) binary and uniform document-object associations. We have found that these strategies are indeed robust and deliver competitive performance using default parameter settings and without resorting to any task-specific treatment. We have also observed that there is no single best configuration; it depends on the task and sometimes even on the particular test query set used for the task. One interesting question for future work, therefore, is how to automatically determine the configuration that should be used for a given task.

References

1. Bailey, P., Craswell, N., de Vries, A.P., Soboroff, I.: Overview of the TREC: enterprise track. In: Proceedings of TREC 2007 (2008)
2. Balog, K., Azzopardi, L., de Rijke, M.: Formal models for expert finding in enterprise corpora. In: Proceedings of SIGIR, pp. 43–50 (2006)
3. Balog, K., de Rijke, M., Weerkamp, W.: Bloggers as experts: feed distillation using expert retrieval models. In: Proceedings of SIGIR, pp. 753–754 (2008)
4. Balog, K., Soboroff, I., Thomas, P., Craswell, N., de Vries, A.P., Bailey, P.: Overview of the TREC: enterprise track. In: Proceedings of TREC 2008 (2009)
5. Demeester, T., Trieschnigg, D., Nguyen, D., Hiemstra, D.: Overview of the TREC: federated web search track. In: Proceedings of TREC 2013 (2014)
6. Demeester, T., Trieschnigg, D., Nguyen, D., Hiemstra, D., Zhou, K.: Overview of the TREC: federated web search track. In: Proceedings of TREC 2014 (2015)
7. Elsas, J.L., Arguello, J., Callan, J., Carbonell, J.G.: Retrieval and feedback models for blog feed search. In: Proceedigs of SIGIR, pp. 347–354 (2008)
8. Fang, H., Zhai, C.X.: Probabilistic models for expert finding. In: Amati, G., Carpineto, C., Romano, G. (eds.) ECIR 2007. LNCS, vol. 4425, pp. 418–430. Springer, Heidelberg (2007). doi:10.1007/978-3-540-71496-5_38
9. Macdonald, C., Ounis, I.: Voting techniques for expert search. Knowl. Inf. Syst. **16**, 259–280 (2008)
10. Macdonald, C., Ounis, I., Soboroff, I.: Overview of the TREC: blog track. In: Proceedings of TREC 2007 (2008)
11. Ounis, I., Macdonald, C., Soboroff, I.: Overview of the TREC-2008 blog track. In: Proceedings of TREC 2008 (2009)
12. Shokouhi, M., Si, L.: Federated search. Found. Trends Inf. Retr. **5**, 1–102 (2011)
13. Weerkamp, W., Balog, K., de Rijke, M.: Blog feed search with a post index. Inf. Retr. **14**, 515–545 (2011)

On the Long-Tail Entities in News

José Esquivel[1,2], Dyaa Albakour[2(✉)], Miguel Martinez[2], David Corney[2], and Samir Moussa[2]

[1] School of Computer Science and Electronic Engineering,
University of Essex, Colchester, UK
[2] Signal Media Ltd., 32-38 Leman Street, London E1 8EW, UK
dyaa.albakour@signalmedia.co, research@signalmedia.co

Abstract. Long-tail entities represent unique challenges for state-of-the-art entity linking systems since they are under-represented in general knowledge bases. This paper studies long-tail entities in news corpora. We conduct experiments on a large news collection of one million articles, where we devise an approach for measuring the volume of such entities in news and we uncover insights on the challenges associated with linking these entities to general knowledge bases.

1 Introduction

In the modern world of fast-flowing news delivery and consumption, searching and filtering documents for entities is becoming a more common information retrieval task. This has been echoed in a number of information retrieval evaluation initiatives such as the TREC KBA track [1] and the NewsIR workshop [2]. Filtering news documents using entities relies on effective Entity Linking (EL) approaches that are capable of identifying mentions of entities in the text and linking them to their entries in knowledge bases (KB)s [3].

State-of-the-art approaches for EL focus on popular entities and rely on general KBs, such as Wikipedia. The success of these approaches depends heavily on the availability of a sufficient quantity of relevant information about the entities in the KB. This includes the textual content of the pages representing the entities from which to learn an appropriate language model that describes them [4]. In addition, the links to the Wikipedia pages representing the entities provide a set of candidate mentions for each entity, as well as the semantic relations between entities in the KB as inferred from the graph of links [5]. In other words, state-of-the-art EL systems rely on general KBs covering popular entities with rich textual content and meta-data about them [6].

Entities which have a less complete profile cannot be easily linked by these approaches [6]. Many less popular or domain-specific entities are under-represented in general KBs such as Wikipedia [7]. We refer to these as *long-tail* entities, and examples of them include small-medium organizations, less popular individuals and rarely-mentioned geographical places. In the literature, long-tail entities have been defined as the large number of entities with relatively few mentions in text corpora [8]. They are characterized as those with limited or no

J.M. Jose et al. (Eds.): ECIR 2017, LNCS 10193, pp. 691–697, 2017.
DOI: 10.1007/978-3-319-56608-5_67

ID: f7ca322d-c3e8-40d2-841f-9d7250ac72ca

Title: Worcester breakfast club for veterans gives hunger its marching orders

VETERANS saluted **Worcester**'s first ever breakfast club for ex-soldiers which won over hearts, minds and bellies. The <u>Worcester Breakfast Club for HM Forces Veterans</u> met at the Postal Order in **Foregate** <u>Street</u> at 10am on Saturday. . . .

Fig. 1. An example from the Signal-1M dataset. **Bold** represents entities identified by the linker, while <u>underlined</u> are entities identified by the NER tagger.

KB profile and sparse or absent resources outside the KB [3]. In this paper, we study long-tail entities in news corpora.

A concrete example of popular and long-tail entities is given in the excerpt from a news article shown in Fig. 1. This shows mentions of two classes of entities. The word "Worcester" is a reference to the town in Worcestershire, England. On the other hand, "Worcester's Breakfast Club for HM Forces and Veterans" is a mention of a specific organization, an entity which does *not* have an entry in Wikipedia and therefore cannot be linked by an off-the-shelf entity linker.

In this paper, we perform an analysis of a large collection of news articles, namely the Signal Media One Million News Articles (Signal-1M) dataset [9], to estimate the volume of long-tail entities which cannot be linked to general KBs. To do this, we compare the entity mentions identified by a Named Entity Recognizer (NER) and the entities linked to a general KB by a state-of-the-art entity linker. Our analysis shows that a large number of entities in news articles are difficult to link as they are either ambiguous or unpopular. Our assumption is that entities that cannot be easily linked are generally long-tail entity mentions, i.e. not well covered in general KBs. Furthermore, we show that even some common entities in the news are not well covered in general KBs.

To summarize, our main contributions are devising an approach for estimating the volume of long-tail entities in the news and uncovering insights into the volume and the types of entities that cannot be easily linked to general KBs.

2   Identifying Long-Tail Entities

To empirically estimate the volume of long-tail entities in a corpus of documents, first we run each document through a NER tagger and an EL tagger separately. The NER tagger identifies mentions of entities in the document along with their types (the NER tag set), while the EL tagger identifies and links entity mentions to their entries in a general KB (the EL tag set). Then, we compute the overlap between these tag sets. We consider this overlap a reasonable proxy for estimating the volume of long-tail entities. In particular, long-tail entities will be typically identified by the NER tagger but not linked by the EL tagger due to their low coverage in the KB. A high overlap indicates a smaller volume of long-tail entities, while a low overlap indicates the opposite. In our approach, we consider two tags as overlapping if either of their start or end offsets is within the other tag's offsets. For example, Fig. 1 shows two cases of overlapping tags.

In the first case, *Worcester* is identified by both taggers. In the second case, *Foregate Street* identified by the NER tagger, whereas the EL tagger marked only *Foregate*.

One limitation of our approach is that it relies on the correctness of the NER tagger. However, we think the resulting NER tag set is an unbiased approximation of the complete set of entities in the corpus. Also, we understand that this is only one possible way of estimating the long-tail entity set. Other approaches, such as getting the least frequent entities in a KB, or the out-of-database entities in the same, should also be explored.

3   Estimating the Long-Tail of Entities in News

3.1   Experimental Setup

To estimate the long-tail of entities in news articles, we applied the procedure described in Sect. 2 on the one million articles in the Signal 1M dataset, originally sourced from tens of thousands of news and blog sources in September 2015. For NER, we used the Stanford tagger [11], and we used DBPedia Spotlight for EL[1], which uses Wikipedia as a KB. When measuring the overlaps of the tagger outputs, we aggregate the results by entity type and by unique entity mentions. The latter is done after normalizing each of the entity mentions by removing any white-space and non-ASCII characters from it, and converting them to their lower case representation. We do this to get a better estimate of the amount of unique entity mentions in the corpus identified by Stanford NER tagger as we decrease the number of duplicate mentions, which only differ in formatting.

Moreover, to further examine the effectiveness of Spotlight in linking long-tail entities, we ran the same procedure described in Sect. 2, but with different subsets of the unique entity mentions identified by the Stanford NER tagger in the corpus. We achieve this by specifying a cut-off point x, at which we consider only the top $x\%$ of normalized unique entity mentions ranked by their frequency in the corpus.

We configured both taggers (Stanford NER Tagger and DBPedia Spotlight) with the recommended parameters according to their documentation. For the Stanford NER Tagger, we used the default English 3-class model trained on news articles without part-of-speech tagging [11]. For DBPedia Spotlight, we used the 'annotate' end-point of the API adjusting the confidence and the support input parameters to 0.4 and 5 respectively as recommended by the API documentation.[2] The API was deployed locally with a Wikipedia dump from July 2013 (two years prior to the dates of the news articles in the dataset). We believe that with this configuration, we may capture newly emerging entities which typically appear in Wikipedia after some lag [10].

Finally, we aggregate entity types into the Stanford types: *PERSON*, *LOCATION*, and *ORGANIZATION*. To do this we map all DBPedia Spotlight types

---

[1] https://github.com/dbpedia-spotlight/.
[2] https://github.com/dbpedia-spotlight/dbpedia-spotlight/wiki/Web-service.

falling under the *Person, Place,* and *Organization* hierarchy to their corresponding Stanford types. We also introduced two other types: (i) any DBPedia type
that does not fall under any of these top-level hierarchies is mapped to *MISC*;
(ii) the DBPedia's default top-level *Thing* type, is mapped to another custom
type *None*.[3]

3.2   Results

Table 1 shows the overlap between the Stanford entities and the Spotlight entities
grouped by type. Overall, we observe that the same-type overlap is relatively
poor across the different types of entities considered. In particular, the same-
type overlap is worst for people (26.59%) and best for locations (64.55%). The
last column "No Overlap" in the table shows the percentage of misses; from this
we can see that the Spotlight linker is not able to provide a link for almost half
of the "people mentions" (more than 4.1 million people mentions in the Signal
1M Dataset). This indicates that there is a large number of people mentioned in
news articles that are hard to link to general KBs. Organizations have a lower
rate of misses, but there is still a large percentage of organizations that are in
the long-tail and hard to link to general KBs. It should be noted that there
are significant cases where Spotlight was able to link the entity but where the
linked entity did not have an identifiable type. This is because there are a large
number of entities in Wikipedia which do not have an explicit type, especially
in the case of organizations, where 27.94% of entity mentions are linked to KB
entities that have no type. This data illustrates that a large number of entities
in news articles are hard to link to general KBs, which is an indication that they
are either not covered in the KB at all or that they are very ambiguous.

Table 1. Overlaps ratio with different types of linked entities by Spotlight. Each row
represents all entity mentions for a certain Stanford type; each column corresponds to
one Spotlight type.

	PERSON	LOCATION	ORG	MISC	None	No Overlap
PERSON (total=7.71M)	26.59%	4.71%	2.52%	1.29%	10.51%	**54.38%**
LOCATION (total=5.52M)	0.65%	**64.55%**	6.43%	1.62%	19.42%	7.33%
ORG. (total=5.37M)	1.49%	11.91%	**39.44%**	4.68%	27.94%	14.54%

We conducted another analysis where we looked at how the overlap changes
for more popular mentions of entities in the corpus. Figure 2 plots the overlap
between Stanford entities and Spotlight entities for different cut-off points of
Stanford entities ranked by their frequency (see Sect. 3.1 for the definition of
cut-off points). Likewise, we plot the misses rate (No Overlaps) in Fig. 3. We
observe that at higher cut-offs, the average same-type overlap increases for all

---

[3] http://mappings.dbpedia.org/server/ontology/classes/.

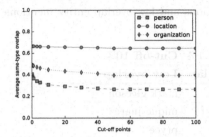

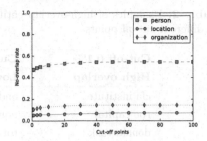

Fig. 2. Same-type overlap between Stanford and Spotlight entities for different cut-off points of Stanford entities ranked by their frequency

Fig. 3. No-overlap rate with Spotlight entities for different cut-off points of Stanford entities ranked by their frequency

entity types, with the largest increase being for people names. Similarly, the Spotlight linker is more successful in finding a link for these mentions, but again the decrease in the misses rate is only marginal. Therefore, even for the very commonly-mentioned entities, the Spotlight linker is still not capable of finding them in Wikipedia.

To examine whether this is due to coverage in the KB or entity ambiguity, we aggregate the overlap per Stanford entity mention at the various cut-off points. The intuition is that understanding the distribution of overlap across entity mentions for different cut-off points (degree of mention popularity) would give more explanation on the effectiveness of Spotlight. For each cut-off point, we present the distribution of overlap percentages per entity mention as a box plot in Fig. 4. For very popular mentions (cut-off point 0.1%) the average overlap is high and the variance is small meaning that the majority of mentions can be linked, in most cases, but there are still hard ones which are never linked to a KB. At higher cut-off points (5%, and 10%), we observe that the average overlap decreases and the variance is very high.

The average lower overlap is expected since less popular entities are less likely to be represented in Wikipedia. However, the high variance indicates that Spotlight is generally either very successful in linking the entity for most of its occurrences or not successful at all. This indicates that the linking is mainly suffering because of the lack of coverage of these entities in Wikipedia.

To further investigate the problem, we manually checked the entity mentions with high mean overlap and with very low mean overlap at

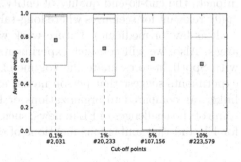

Fig. 4. Average overlap per entity mention at different cut-off points. The number of unique entity mentions is shown below the cut-off point

Table 2. Examples of high-overlap entity mentions and low-overlap entity mentions at different cut-off points.

Cut-off 0.1%	Cut-off 0.1%	Cut-off 10%
High overlap	Low overlap	Low overlap
cfa institute	andy	mark gleeson
rbc capital	nomura	mique juarez
donald tusk	total	pryce
balkans	daesh	amanda sue watson
barclays premier league	diego costa	asigra

the different cut-off points (examples shown in Table 2). As expected, entity mentions with high average overlap are usually referring to popular entities and are not ambiguous, which makes them easy cases for Spotlight. The examples in the table include popular people (Donald Tusk), organizations (CFA institute) and locations (Balkans). On the other hand, very popular mentions with low overlap of linked entities (second column of Table 2) are ambiguous mentions of people or organizations (e.g. Total and Andy) or emerging entities (e.g. Daesh and Diego Costa) that were not well covered in Wikipedia in 2013, the snapshot used in the experiment. Finally for common but less popular mentions in the corpus (cut-off 10%), mentions with low overlap mainly represent people or organizations which are not represented in Wikipedia.

4   Conclusions and Future Work

We have analyzed the overlap between state-of-the-art NER and EL systems and the results show that not only is their overlap relatively poor, but also EL systems clearly under-perform when linking long-tail entities (up to 50% missing rate for people), even for those which are very common in the news. This directly impacts the end-to-end quality of entity linking systems, and it could be especially relevant for scenarios where long-tail entities are common (e.g., niche areas such as law or medicine). Future work will consider other datasets from those areas. Also, we will consider experiments using more recent Wikipedia dumps with Spotlight to estimate the volume of emerging entities. Unsurprisingly, our experiments suggest that person names are the hardest to link by the Spotlight linker, as compared to organizations or locations. Our analysis also highlights some of the challenges of EL in news, such as emerging entities being problematic for EL and that ambiguous mentions of entities are never linked.

References

1. Frank, J.R., Kleiman-Weiner, M., Roberts, D.A., Voorhees, E., Soboroff, I.: TREC KBA overview. In: Proceedings of TREC (2014)
2. Martinez, M., Kruschwitz, U., Kazai, G., Hopfgartner, F., Corney, D., Campos, R., Albakour, D.: Report on the 1st international workshop on recent trends in news information retrieval (NewsIR16). SIGIR Forum **50**(1), 58–67 (2016)

3. Reinanda, R., Meij, E., de Rijke, M.: Document filtering for long-tail entities. In: Proceedings of CIKM (2016)
4. Daiber, J., Jakob, M., Hokamp, C., Mendes, P.N.: Improving efficiency and accuracy in multilingual entity extraction. In: Proceedings of the 9th International Conference on Semantic Systems (I-Semantics) (2013)
5. Ferragina, P., Scaiella, U.: Tagme: on-the-fly annotation of short text fragments (by Wikipedia entities). In: Proceedings of CIKM2010 (2010)
6. van Erp, M., Mendes, P., Paulheim, H., Ilievski, F., Plu, J., Rizzo, G., Waitelonis, J.: Evaluating entity linking: an analysis of current benchmark datasets and a roadmap for doing a better job. In: Proceedings of ELRA (2016)
7. Lin, T., Etzioni, O.: No noun phrase left behind: detecting and typing unlinkable entities. In: Proceedings of EMNLP (2012)
8. Farid, M.H., Ilyas, I.F., Whang, S.E., Yu, C.: LONLIES: estimating property values for long tail entities. In: Proceedings of SIGIR 2016, 1125–1128 (2016)
9. Corney, D., Albakour, D., Martinez, M., Moussa, S.: What do a million news articles look like? In: Proceedings of ECIR NewsIR workshop (2016)
10. Fetahu, B., Anand, A., Anand, A.: How much is wikipedia lagging behind news? In: Proceedings of the ACM Web Science Conference (2015)
11. Finkel, J.R., Grenager, T., Manning, C.: Incorporating non-local information into information extraction systems by Gibbs sampling. In: Proceedings of ACL (2015)

Search Costs vs. User Satisfaction on Mobile

Manisha Verma[✉] and Emine Yilmaz

University College London, London, England
mverma@cs.ucl.ac.uk, emine.yilmaz@ucl.ac.uk

Abstract. Information seeking is an interactive process where users submit search queries, read snippets or click on documents until their information need is satisfied. User cost-benefit models have recently gained popularity to study search behaviour. These models assume that a user gains information at expense of some cost. Primary assumption is that an adept user would maximize gain while minimizing search costs. However, existing work only provides an *estimate* of user cost or benefit per action, it does not explore how these costs are correlated with *user satisfaction*. Moreover, parameters of these models are determined by desktop based observational studies. Whether these parameters vary with device is unknown. In this paper we address both problems by studying how these models correlate with user satisfaction and determine parameters on data collected via mobile based search study. Our experiments indicate that several parameters indeed differ in mobile setting and that existing cost functions, when applied to mobile search, do not highly correlate with user satisfaction.

Keywords: Cost-benefit analysis · Effort · User satisfaction

1 Introduction

Search is an extremely popular means of finding information online. Users repeatedly interact with a search engine to satisfy their information need which makes interactive information retrieval (IIR) an active area of research. Recently, large body of formal models [4,5] have been proposed that capture user cost (or effort) and benefit by incorporating several user actions. Users incur some cost for each of these actions: input a search query, read snippets, click results or scroll up/down search engine result page (SERP). At present, cost of each action is measured in time, keystrokes or number of documents. For instance, query cost can be estimated via $W * c_w$ [4] where W is number of words in query and c_w is the average time it takes a user to type each word. Several models have been proposed [5], simulated [2] or empirically evaluated [3] on real datasets.

However, existing work only provides an *estimate* of user cost or benefit per action, it does not explore how these costs are correlated with user satisfaction. It remains to be seen what cost functions correlate best with user satisfaction. Existing research in IIR is also limited to a desktop setting. User models of search and interaction have been developed for desktop environments and lab

© Springer International Publishing AG 2017
J.M. Jose et al. (Eds.): ECIR 2017, LNCS 10193, pp. 698–704, 2017.
DOI: 10.1007/978-3-319-56608-5_68

studies have been conducted to empirically evaluate and learn these models. However, today users have quick access to information on several devices such as desktops, mobiles and tablets. Whether these models highly correlate with user satisfaction needs to be evaluated on different devices.

In this work we address above mentioned limitations of existing work. We begin by introducing a mobile specific dataset collected during a lab study. We explore different actions and their costs across 25 users and 193 sessions. We also investigate how these cost functions correlate with user satisfaction. Our experiments show that once trained, cost-benefit model parameters are different for mobile search. We also found varied correlation between satisfaction and cost functions proposed in the literature. In following sections, we briefly explain cost functions proposed in literature, followed by examining correlation between user satisfaction and search costs (or effort), benefit (or gain) and profit respectively as proposed in previous work.

2   User Study and Data Statistics

We conduct a small scale search study to collect fine grained user interaction data and explicit labels for satisfaction from some users. We collected data for 10 topics chosen from publicly available dataset [6]. We tailored topic descriptions for mobile search and did not impose any time restrictions for completing these search tasks. We built an Android app[1] for our experimental study. Participants were free to issue as many queries as they liked. Search results were retrieved using Bing Search API[2] with fixed parameters. If two participants issued the same query, we ensured they saw same results by caching results of each query. We customized search interface for image, video and wiki results respectively to reflect existing commercial search engine result pages (SERPs). We logged several interaction signals such as clicks, taps and swipes on SERP. Participants were asked to provide feedback for SERP relevance and satisfaction on Likert scale of 1 (non-relevant/dissatisfied) to 5 (highly-relevant/satisfied). They could begin with any task and perform as many search tasks as they liked.

Participants were recruited via university mailing lists and social media websites. We collected data from 25 participants (7 females and 18 males) for this study whose age lies between 22–55. We asked participants that were familiar with search in mobile browser, to complete the study on their *personal* android phones.

Our data consists of 193 search sessions, 104 unique queries, 161 unique SERP result (URL) clicks and 192 relevance/satisfaction labels for SERPs. The distribution of SERP satisfaction labels is $1 = 13$, $2 = 12$, $3 = 32$, $4 = 54$, $5 = 81$ respectively.

---

[1] Topics, results and app at http://www0.cs.ucl.ac.uk/staff/M.Verma/app.html.
[2] http://datamarket.azure.com/dataset/bing/search.

3 Cost/Benefit vs. Satisfaction Analysis

Cost (or effort) and benefit can be analysed in multiple ways. Existing work [4] investigates user costs on a per-action basis. In this paper, we limit our investigation to two types of costs: query cost and click/scroll cost. Cost of querying solely depends on user's input query i.e. it is directly proportional to query length. However, click/scroll costs are relatively more complex as they depend on factors such as number of snippets read, clicked and number of SERPs examined by the user. We explain different cost/benefit functions, discuss their correlation with SERP satisfaction labels from our study and finally estimate their parameters by optimizing different cost functions in following subsections.

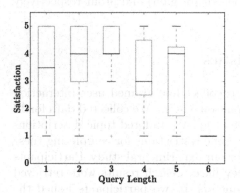
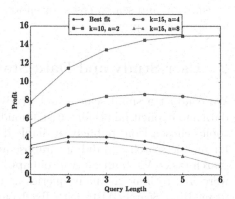

Fig. 1. Satisfaction vs. query length   Fig. 2. Profit curves

3.1 Query Cost-Benefit and User Satisfaction

Users rely on keywords to formulate their information needs. They may incur different costs for issuing query on different mediums. For instance, users can issue a query via keyboard or touch screens on desktop and mobile respectively. Users of our app were required to touch type their queries and we did not provide query auto completion, to ensure that users type all queries explicitly.

Given that a user enters a query with W words and c_w captures the effort required to input each word, we use the model from [4], in Eq. 1, to compute net profit (π), benefit $b(W)$ and cost $c(W)$ for each query:

$$b(W) = k.log_\alpha(W + 1)$$
$$c(W) = W.c_w \qquad\qquad (1)$$
$$\pi(W) = b(W) - c(W)$$

Here, k represents a scaling factor and α captures diminishing returns of typing subsequent words. Distribution of satisfaction labels for queries of varying length is shown in Fig. 1. We use same values for $k \in \{10, 15\}$ and $\alpha \in \{2, 4, 8\}$ as in

Table 1. Pearson ρ b/w query cost and satisfaction.

α	k		
	10	15	20
2	−0.10	−0.14	−0.15
4	**0.312***	−0.009	−0.10
6	0.271*	0.27*	−0.02
8	0.256*	**0.312***	−0.09
10	0.248*	0.295*	0.23

Table 2. Pearson ρ b/w search cost and satisfaction.

β	k			
	2.0	5.0	10.0	16.0
0.03	0.16*	0.14*	0.10*	0.09
0.3	**0.17***	0.13*	0.09*	0.08
0.43	0.16*	0.12	0.08	0.08
1.0	0.11	0.08	0.07	0.06

[4] to compute Pearson correlation (ρ) between query profit and satisfaction. Correlation between satisfaction and profit for each combination of k and α is given in Table 1.

We obtain values of c_w, k, α by optimizing objective function in Eq. 2 which minimizes the difference between user satisfaction ($\hat{\pi}$) and net user profit.

$$\min_{c_w,k,\alpha} \sum_{i=1}^{n} (\hat{\pi} - \pi(W))^2 \tag{2}$$

We can estimate parameters c_w, k and α by minimizing squared loss on satisfaction labels from our study. Parameter values $c_w = 2.18$, $k = 8.5$ and $\alpha = 3.0$ yield best fit on our data. When substituted, net profit has Pearson's ρ of 0.314 (p-val < 0.001) with satisfaction. Profit curves for different parameter settings are shown in Fig. 2. We observe that as the length of query increases, overall profit of user decreases which was also reported in [4]. We also observe a similar trend in our data where profit is highest for three word queries and rapidly drops thereafter. Table 1 shows that higher α yields stronger correlation between satisfaction and user profit which indicates rapid diminishing returns of typing subsequent words. While query cost does not model entire search process, experiments on our data suggest that query costs (in Eq. 2) can affect overall user satisfaction.

3.2   Search Cost-Benefit and User Satisfaction

A user has choice of several actions on submitting any query to the search engine. They can either choose to examine a snippet, click a result, go to the next page or issue a new query. We assume that user submits Q queries, reads S snippets, views V SERP pages per query and reads A clicked documents. If the cost of querying is c_w, the cost of viewing a SERP page is c_v, the cost of reading a snippet is c_s and the cost of reading a clicked document is c_a respectively, we can use cost $c(Q, V, S, A)$ and gain/benefit $b(Q, A)$ function from [2] to compute the net profit π given in Eq. 3. Here, α and β capture user's frequency of issuing multiple queries and reading documents respectively.

$$c(Q, V, S, A) = (c_w + c_v.V + c_s.S + c_a.A).Q$$
$$b(Q, A) = k.Q^\alpha.A^\beta \tag{3}$$
$$\pi = b(Q, A) - c(Q, V, S, A)$$

Distribution of satisfaction with respect to time spent on reading (or examining) A clicked documents, viewing S snippets, cost of reading each snippet (c_s) and clicked document (c_a) is shown in Figs. 3, 4, 5 and 6 respectively. Some users in our study, despite clicking on more than 10 documents for a query, have assigned higher satisfaction grade to SERP. It is worth noting that the median cost of reading a snippet (in milliseconds) is higher on low satisfaction SERPs than on high satisfaction SERPs. However, the trend reverses in the curve depicting examination cost of clicked documents i.e. Fig. 3 where users spend less time reading a document clicked on low satisfaction SERP than on high satisfaction SERP.

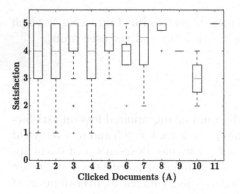

Fig. 3. # clicked documents   **Fig. 4.** # viewed snippets

We optimize the function in Eq. 2 with satisfaction labels and net profit for each SERP. Since our satisfaction labels are per SERP basis, we set $Q = 1$ to compute per SERP cost and benefit function.

We perform optimization similar to Eq. 2 where we minimize the difference between satisfaction labels and benefit obtained from total SERP interaction. We obtained lower value of $k = 2.0$ and $\beta = 0.30$ than previously reported values $k = 5.3$ and $\beta = 0.43$ as given in [1]. Variation in profit curves for different combinations of k and β for clicked documents and viewed snippets is given in Figs. 7 and 8, respectively. Pearson correlation ρ between net profit and satisfaction for different values of k and β is shown in Table 2[3].

Best fit ($k = 2.0$ and $\beta = 0.30$) net profit curve in Fig. 7 shows that *change* in net user gain is highest when only one document is clicked. Net profit gradually increases as more documents are clicked. The kink in curve for two clicked

---

[3] *indicates p-val < 0.05

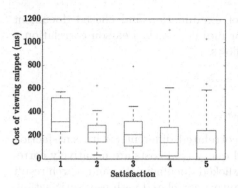

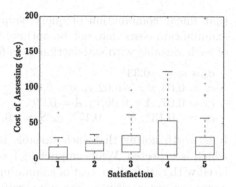

Fig. 5. Cost of reading snippet (c_s) **Fig. 6.** Cost of reading clicked doc (c_a)

documents suggest that other costs dominate cost function, thereby lowering net profit. We did not observe a significant drop in the profit with increase in number of clicked documents. However, net profit when $k = 5.3$ and $\beta = 0.43$ (from [1]) rapidly increases as more documents are clicked. Our data suggests that lower number of clicked documents yield higher user satisfaction on mobile. Profit curves for number of viewed snippets in Fig. 8 shows a different trend. Net gain rapidly increases as users view more snippets but drops significantly when they read between six to eight snippets. Best fit curve shows highest profit when user views four snippets and declines thereafter. Best fit profit curve is similar to curve with $k = 5.3$ and $\beta = 0.43$ (from [1]) when plotted against viewed snippets. Table 2 shows that correlation between satisfaction and net benefit weakens as k and β increase.

Pearson correlation ρ between satisfaction and net search benefit on our data, for parameters obtained by optimizing objective function in Eq. 3 ($k = 2.0$ and $\beta = 0.30$) was significantly low, only 0.17 (p-val < 0.05) which indicates

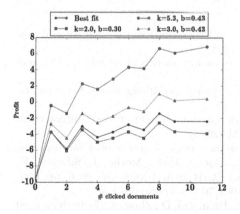

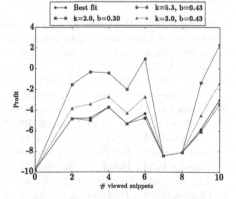

Fig. 7. Clicked doc profit   **Fig. 8.** Viewed snippet profit

that linear combination of query, snippet examination and clicked document examination costs may not be optimal for mobile search. Pearson correlation ρ of each variable with satisfaction is as follows:

- $c_w * w = -0.33^*$
- $c_v = 0.03, v = -0.02, c_v.v = 0.03$
- $c_a = 0.07, A = 0.06, c_a.A = 0.09$
- $c_s = -0.13^*, S = -0.17^*, c_s.S = -0.16^*$

It is worth noting that each variable is correlated differently with satisfaction which is expected. While snippet (c_s) and query (c_w) costs are negatively correlated with satisfaction, cost of examining clicked document (c_a) and search result pages (c_v) are positively (but not significantly) correlated with user satisfaction.

Overall, for both query and search cost-benefit functions, we observed a different optimal value for each parameter on mobile. We observed higher correlation between net query benefit and satisfaction on mobile search data. However, satisfaction correlation with net search benefit was relatively low, which suggests that linear combination of search costs may not be suitable for a mobile setting.

4 Conclusion

Existing models of cost-benefit analysis models in IIR estimate how users maximize their net gain while minimizing search costs. These models do not provide any insight into how these strategies correlate with user satisfaction. Empirical study of these models is also limited to desktop setting. This paper was an investigation of correlation between cost-benefit of querying/searching and user satisfaction in mobile search. We found that optimal parameters of these models differ from desktops. We also found satisfaction to be highly correlated with net query profit but weakly correlated with net search profit. Our study motivates further investigation of non-linear cost models to better capture user behaviour on mobile devices.

References

1. Azzopardi, L.: Economic models of search. In: Proceedings of the 18th Australasian Document Computing Symposium. ACM (2013)
2. Azzopardi, L.: Modelling interaction with economic models of search. In: Proceedings of SIGIR. ACM (2014)
3. Azzopardi, L., Kelly, D., Brennan, K.: How query cost affects search behavior. In: Proceedings of SIGIR. ACM (2013)
4. Azzopardi, L., Zuccon, G.: An analysis of the cost and benefit of search interactions. In: Proceedings of ICTIR, pp. 59–68. ACM (2016)
5. Azzopardi, L., Zuccon, G.: Two scrolls or one click: a cost model for browsing search results. In: Ferro, N., Crestani, F., Moens, M.-F., Mothe, J., Silvestri, F., Nunzio, G.M., Hauff, C., Silvello, G. (eds.) ECIR 2016. LNCS, vol. 9626, pp. 696–702. Springer, Heidelberg (2016). doi:10.1007/978-3-319-30671-1_55
6. Demeester, T., Trieschnigg, D., Nguyen, D., Hiemstra, D., Zhou, K.: Fedweb greatest hits: presenting the new test collection for federated web search. In: Proceedings of WWW. ACM (2015)

On the Efficiency of Selective Search

Fatih Hafizoglu, Emre Can Kucukoglu, and Ismail Sengor Altingovde[✉]

Middle East Technical University, Ankara, Turkey
{fatih.hafizoglu,emre.kucukoglu,altingovde}@ceng.metu.edu.tr

Abstract. Our work shows that the query latency for selective search over a topically partitioned collection can be reduced by up to 55%. We achieve this by physically storing the documents in each topical cluster across all shards and building a cluster-skipping index at each shard. Our approach also achieves uniform load balance among the shards.

1   Introduction

The idea of topically partitioning a large document collection and focusing the search to only those partitions (shards) that are the most relevant to a given query is long known in the IR community, being referred to as *cluster-based retrieval* [1,7] (usually investigated in a centralized setup) or, more recently, *selective search* [6] (within a distributed retrieval setup). By doing so, the goal is only processing the subset of documents that are more likely to be matching the topic of query, which is expected to improve the efficiency, while keeping the result quality the same as (or, maybe better than) searching the entire collection.

Earlier works report that the effectiveness promise of selective search is usually kept, and further, the cost in terms of the *total* work done (i.e., number of documents scored for a query) across the selected shards is lower than searching the full collection, as expected [6]. Yet, efficiency in terms of the *query latency* is rather arguable, as in a parallel setup, the slowest shard determines the latency; and by definition, topical partitioning aims to (ideally) store all documents on a particular topic at a single shard, which may increase the latency. Moreover, since searched topics are not equally popular, certain topical shards are likely to be accessed for most of the queries, which would lead a high load imbalance. We discuss recent studies [4,5] with evidence supporting these concerns in Sect. 2.

As a remedy, we propose to still topically partition the collection, yet store the documents in each topical cluster by distributing to all shards, rather than within a single shard. During search, we employ a cluster-skipping inverted index [1,3] for each shard, which is designed to access only those postings from the required topical clusters while skipping the others. The cluster-skipping index has been shown to yield high savings in processing time in a centralized setup [1,3], and our work here shows that these savings can be transferred to a distributed setup to improve the query latency for selective search. Furthermore, since a slice of each topical cluster is stored in each shard, the load would be uniformly distributed across all shards. Our experiments using TREC and AOL query sets demonstrate

© Springer International Publishing AG 2017
J.M. Jose et al. (Eds.): ECIR 2017, LNCS 10193, pp. 705–712, 2017.
DOI: 10.1007/978-3-319-56608-5_69

that our approach yields up to 55% reduction over traditional selective search in terms of the latency, in return for a slight increase (around 1%) in the total processing cost. Furthermore, for the first time in the literature, our approach achieves uniform load balancing for selective search.

2 Selective Search with Cluster-Skipping Inverted Index

For searching very large document collections (e.g., Web), a firmly established industrial practice (e.g., see [2]) is document-based partitioning of the collection into several (logical) shards, each of which is assigned to a (physical) computing node (machine). The most simple and widely applied document-partitioning approach is *Random*, which assigns each document to a shard uniformly at random [2]. In the literature, the possibility of creating topical shards is also well explored, where document collection can be directly clustered using traditional algorithms like K-means [6] or co-clustered with a set of queries [8]. In this paper, partitioning based on such topical clusters is called *Topic* partitioning.

The search process differs for the systems that involve shards based on *Random* or *Topic* partitioning. As shown in Fig. 1(a), for the former case, a query arriving to a broker is directly forwarded to all shards where local top-k results are computed (using a typical index) and finally merged at the broker to obtain the global top-k result list. Since all shards are fully searched, this is actually an *exhaustive search*. In case of *Topic* partitioning, first the topical shards (i.e., *target clusters*) that are most relevant to a query should be determined (i.e., as in the resource selection phase of federated-search [9]), which usually involves processing of the query over a central sample index (CSI) at the broker. Next, the query is forwarded to only those shards and the final result is constructed by merging the local top-k results only from these shards (see Fig. 1(b) where target cluster is with color "Grey"). This is called *selective search* in [6].

The goal of selective search is generating results that are as good as (or even better) than the exhaustive search, while improving the efficiency, especially for the low-resource environments. Earlier findings reveal that the promise on the effectiveness can be kept (e.g., [6,7]), while the findings regarding the efficiency call for more investigation. In particular, assuming reasonably sized topical-shards and only few of them (typically, less than 10% of all shards [3,6]) will be selected as targets, it is obvious that the total number of documents (i.e., index postings) processed by the selective approach is less than that of processed by the exhaustive search for the same query (and based on the number of target shards, reductions can be more than 95% [6]). However, the former metric serves as an *upper-bound* (see p. 14 in [6]) and what is important for the search system is the query latency, i.e., time to compute the final result. Since the shards are searched in parallel in a distributed setup, the shard with the longest execution time determines the query latency. Earlier works report mixed results regarding the performance of selective search in terms of latency: While Kulkarni et al. again report considerable savings (especially when the number of target shards is small), Kim et al. show that the query processing time per shard is higher for selective search in comparison to the exhaustive search even when dynamic pruning is applied (see Fig. 4 in [4]), and hence, latency of selective search may not always beat that of the exhaustive approach. As a second problem, selective search is vulnerable to load imbalance among shards, as certain topics (say, about celebrities) are known to be queried more often than others. This claim is supported by the recent findings in [5], which shows that certain topical-shards can become a bottleneck and proposes a better shard distribution approach. We will discuss their method and compare to ours in Sect. 4.

As a remedy to these problems, we propose a so-called *Hybrid* approach that again constructs topical clusters but partitions the documents in these clusters to *all* shards in a random fashion (rather than storing them together). To enable efficient search in this setup, we adapt the cluster-skipping index that is shown to improve both the effectiveness and efficiency of retrieval in a single shard setup [1,3]. Here, we show that this data structure can be employed in a distributed retrieval setup to serve us the best of both worlds: The search would be *logically* focused on selected topical clusters, but it will be *physically* conducted on all shards (yet over only those documents from the required topical clusters). Thus, this approach will allow processing shorter lists but over all shards, which would improve both the latency and load balance for selective search.

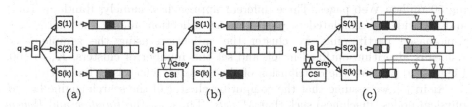

Fig. 1. (a) Exhaustive search (with *Random*), (b) Selective search (with *Topic*), and (c) Selective search (with *Hybrid*). The color of a list element (black, white or grey) represents the cluster in which the document resides. For (b) and (c), target cluster is "Grey". In (c), skipping elements are shown apart from the lists only for visual clarity; o.w., each list is a single entity with its skipping elements and typical postings.

Figure 1(c) shows the use of cluster-skipping index in our scenario. In this index, the documents in a posting list are grouped together based on their cluster membership, and furthermore, for each cluster, there is a skipping element which stores the cluster-id and pointer to the relative start address of this cluster in the list. These skipping elements can be interleaved with other postings [3], or stored at the beginning of the list. During query processing, skipping elements are used to jump to target clusters, and only those documents from the target clusters are processed (i.e., decompressed and scored w.r.t. a retrieval model).

The toy example in Fig. 1 illustrates the benefit of selective search with *Hybrid* partitioning. Assume a query q with term t arrives to the broker B. For *Random*, the query latency is 6, i.e., the size of the longest posting list, which is at shard *S2*. Further assuming that this query best matches to the "Grey" cluster, selective search (with *Topic* partitioning) will forward it to the first shard, *S1*, which would incur a latency of 7. In contrast, our approach will send the query to all shards, where only documents of the "Grey" cluster are processed (plus the skipping elements to guide the search); which adds up to 5 (e.g., 3 skipping elements and 2 "Grey" postings at *S1* in Fig. 1(c)). Note that, due to the overhead of skipping elements, *Hybrid* would seem less efficient if the target was "Black" cluster in this example; but in practical cases (as shown in Sect. 4), the number of clusters is typically a few order of magnitudes smaller than the number of documents, and this overhead turns out to be negligible. Moreover, the inefficiency of *Topic* partitioning for this example may not necessarily be caused by a skew in the cluster sizes (although, less skew is obviously preferable), but also by the fact that *all* documents of a topical cluster are stored at the *same* shard, which may lead to high variation in latency of a query at different shards. Our *Hybrid* partitioning, by definition, eliminates the latter problem.

3   Experimental Setup

Collection and Partitioning. We employ ClueWeb09 Part-B collection including 50 million Web pages. Three different approaches, namely, Random, Topic and Hybrid, are simulated to partition the collection among the shards. For topic-based partitioning, we cluster the collection using the sample-based K-means algorithm outlined in [6], and set the number of clusters, K, to 100. Figure 2(a) presents the distribution of cluster sizes, which is in line with [4].

As in [5], we assume that the computing cluster of the search engine has M physical nodes (machines) such that $M < K$. Thus, for the *Random* and *Hybrid* partitioning, the documents are randomly assigned to one of these M nodes. For *Topic*, we map the topical clusters to physical nodes, so that each physical node stores K/M clusters. The latter assignment process has an effect on the performance of the selective search with *Topic*, which is discussed in detail in the next section. The experimental setup of [6] (see the column T/K in Tables 3, 4 and 5 in [6]) implies that the number of target clusters to be searched, denoted with N, is always set to be equal to number of physical nodes, M. In this paper, assuming a low-resource setup, we experiment for M ($= N$) for the values of 5 and 10.

We employ two query sets. The TREC set includes 200 queries from the TREC Web tracks between 2009 and 2012, while AOL set includes 100 K randomly-sampled distinct queries from the well-known AOL log.

Shard Selection. As in the literature (e.g., [6]), we construct a Central Sample Index (CSI) using a random sample of 1% of documents from each cluster. The well-known ReDDE [10] algorithm is employed to select N topical shards (clusters). For *Topic* partitioning, a given query is forwarded to only those physical nodes that store these N shards. For *Hybrid* partitioning, a query is forwarded to all nodes, but together with the list of N topical clusters to allow skipping during search. For *Random* strategy, naturally, the list of clusters is irrelevant and the query is executed at all the nodes.

Evaluation Metrics. We report early retrieval effectiveness using the Precision, NDCG and ERR metrics. For efficiency, we adopt the total query cost and query latency metrics from [6], as follows: $C_{Total}(q) = |I^q_{CSI}| + \sum_{i=1}^{M} \sum_{c \in T} |I^q_{M[i],c}|$ and $C_{Latency}(q) = \max_{1 \le i \le M}(|I^q_{CSI}| + \sum_{c \in T} |I^q_{M[i],c}|)$, where $|I^q|$ is the sum of the no. of posting elements for each term in q, $|I^q_{CSI}|$ is the query processing cost over CSI, T is the set of target clusters, and $|I^q_{M[i],c}|$ is the cost of processing for a topical cluster $c \in T$ that is stored at machine $M[i]$. We also keep track of the total work done at each node $M[i]$, which is simply $\sum_{c \in T} |I^q_{M[i],c}|$ summed over all the query set. We compute the relative load percentage per machine as the ratio of the former value to the exhaustive processing cost of the query set at a *centralized* setup (i.e., assuming the entire index is stored at a single machine).

4   Experimental Results

In Table 1, we present the effectiveness and efficiency evaluation using TREC queries and for $N = \{5, 10\}$. The effectiveness results are in line with the earlier work in that selecting the most relevant 5 or 10 clusters for selective search yields a comparable effectiveness to exhaustive search (with *Random*). While our actual scores are higher than those in [6], we attribute this to different setup choices (e.g., we do not apply stemming but apply spam filtering).

In terms of the efficiency, we see that the trends for total cost, C_{Total}, is again as expected, and selective search (with *Topic*) yields an improvement of up to 75% over exhaustive search (for $N = 5$). Our approach with *Hybrid*, which does exactly the same work as *Topic* and additionally, processes the cluster-skipping elements in the lists, is only slightly worse than *Topic*, and still provides an improvement of 74% over the exhaustive search. However, in terms of the query latency, we have two striking findings. First, selective search (with *Topic*) yields gains over the exhaustive search over small N, but gains vanish quickly, even for $N = 10$. Note that, for each query, here we made sure that the selected N clusters are located in N different nodes, i.e., reported latency values are *optimal* for selective search (with *Topic*) and may be worse in practice. Second, our approach consistently achieves impressive savings in latency, i.e., 70% (50%) and 40% (55%) over search with *Random* and *Topic* partitioning approaches for $N = 5$ (10), respectively. In Table 2, we provide the efficiency results over the AOL set with 100,000 queries for $K = 100$ and $N = 10$ only (for brevity), which reveals the same trends, indicating the robustness of our findings.

Finally, in Fig. 2(b), for $M = N = 10$, we present the relative load of each physical node to process the larger AOL query set. We find that search over *Random* and *Hybrid* yield almost the same load distribution, so we only plot the latter in Fig. 2(b) for visual clarity. For *Topic*, we apply two methods to distribute $K = 100$ topical clusters to $M = 10$ nodes, as proposed in [5].

Table 1. Results for alternative partitioning & search modes for TREC set. No. of topical clusters (K) is 100, $N = \{5, 10\}$. C_{Total} and $C_{Latency}$ are in million documents.

Partitioning	Search mode	N	P@5	P@10	P@20	NDCG@20	ERR@20	C_{Total}	$C_{Latency}$
Random	Exhaustive	NA	0,34	0,32	0,28	0,19	0,13	4,58	0,93
Topic	Selective	5	0,33	0,32	0,27	0,18	0,12	1,17	0,47
Hybrid	Selective	5	0,33	0,32	0,27	0,18	0,12	1,19	**0,28**
Random	Exhaustive	NA	0,34	0,32	0,28	0,19	0,13	4,58	0,46
Topic	Selective	10	0,33	0,32	0,28	0,18	0,12	1,81	0,51
Hybrid	Selective	10	0,33	0,32	0,28	0,18	0,12	1,83	**0,23**

Table 2. Efficiency of alternative partitioning & search modes for AOL set $(N = 10)$.

Partitioning	Search mode	N	C_{Total}	$C_{Latency}$
Random	Exhaustive	N/A	6,49	0,66
Topic	Selective	10	2,32	0,73
Hybrid	Selective	10	2,35	**0,30**

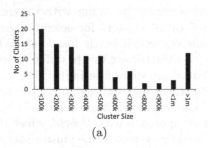

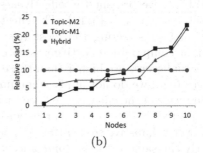

(a)   (b)

Fig. 2. (a) Size distribution of topical clusters, (b) Relative load (%) of nodes with *Topic* and *Hybrid* partitioning. For *Topic*, we employ two methods, *M1* and *M2*, to assign clusters to actual nodes.

The first method, *Topic-M1*, is a simple random assignment, whereas the second one, *Topic-M2*, computes the individual access popularity of each topical cluster (over a training query set) and distributes the clusters to nodes to balance the expected loads as much as possible. Figure 2(b) reveals that the latter method yields better load balance (confirming [5]), but even in this case, there is huge difference among the loads of nodes for selective search (with *Topic*), i.e., the most loaded node works three times more than the one with the least load. In contrary, in our approach, each one of 10 nodes has 1/10 of the exhaustive load, i.e., the load balance is uniform.

5   Discussions and Conclusion

Using an Index per Cluster: Instead of a cluster-skipping index for *Hybrid*, one can construct a separate index for each cluster at each node. This has two disadvantages: First, the bookkeeping and updating of several index files can be a burden. More crucially, if these index files are stored on a traditional HDD, each cluster access would require an extra disk seek, increasing the latency. This can be remedied by using an SSD or storing all index files in the memory, yet this option may be cost prohibitive for a low-resource environment assumed in [6].

Additional Metrics: In [4], the performance of selective search is analyzed through an event-based simulation that also takes into account the inter-query parallelism. While such a setup would not affect the metrics considered here and hence, the validity of our findings, it would allow us computing other important metrics, such as the throughput. Our work towards this direction is underway.

Conclusion. We showed that typical selective search may not always improve query latency over exhaustive search, and suffers from high load imbalance. Our proposed approach distributes a slice of each topical cluster to each shard and constructs a cluster-skipping index per shard. By doing so, we reduce the query latency up to 55% w.r.t. the typical selective search. The proposed approach also guarantees uniform load balancing across the shards for selective search.

Acknowledgements. This work is partially funded by the Ministry of Science, Industry & Technology of Turkey and Huawei Inc. under the grant No. 0441.STZ.2013-2. I.S. Altingovde is also supported by Turkish Academy of Sciences Distinguished Young Scientist Award (TÜBA-GEBİP 2016).

References

1. Altingövde, I.S., Demir, E., Can, F., Ulusoy, Ö.: Incremental cluster-based retrieval using compressed cluster-skipping inverted files. ACM TOIS **26**(3), 15 (2008)
2. Cambazoglu, B.B., Baeza-Yates, R.A.: Scalability Challenges in Web Search Engines. Morgan & Claypool Publishers, San Rafael (2015)
3. Can, F., Altingövde, I.S., Demir, E.: Efficiency and effectiveness of query processing in cluster-based retrieval. Inf. Syst. **29**(8), 697–717 (2004)
4. Kim, Y., Callan, J., Culpepper, J.S., Moffat, A.: Does selective search benefit from WAND optimization? In: Ferro, N., Crestani, F., Moens, M.-F., Mothe, J., Silvestri, F., Nunzio, G.M., Hauff, C., Silvello, G. (eds.) ECIR 2016. LNCS, vol. 9626, pp. 145–158. Springer, Heidelberg (2016). doi:10.1007/978-3-319-30671-1_11
5. Kim, Y., Callan, J., Culpepper, J.S., Moffat, A.: Load-balancing in distributed selective search. In: Proceedings of SIGIR, pp. 905–908 (2016)
6. Kulkarni, A., Callan, J.: Selective search: efficient and effective search of large textual collections. ACM TOIS **33**(4), 17 (2015)
7. Liu, X., Croft, W.B.: Cluster-based retrieval using language models. In: Proceedings of SIGIR, pp. 186–193 (2004)

8. Puppin, D., Silvestri, F., Perego, R., Baeza-Yates, R.A.: Tuning the capacity of search engines: load-driven routing and incremental caching to reduce and balance the load. TOIS **28**(2), 5 (2010)
9. Shokouhi, M., Si, L.: Federated search. Found. Trends Inf. Retrieval **5**(1), 1–102 (2011)
10. Si, L., Callan, J.P.: Relevant document distribution estimation method for resource selection. In: Proceedings of SIGIR, pp. 298–305 (2003)

Does Online Evaluation Correspond to Offline Evaluation in Query Auto Completion?

Alexandros Bampoulidis[1]([envelope]), João Palotti[1], Mihai Lupu[1], Jon Brassey[2], and Allan Hanbury[1]

[1] TU Wien, Favoriten Strasse 9-11/188, Vienna, Austria
{alexandros.bampoulidis,joao.palotti,mihai.lupu,
allan.hanbury}@tuwien.ac.at
[2] Trip Database Ltd., Little Maristow, Glasllwch Lane, Newport, UK
jon.brassey@tripdatabase.com

Abstract. Query Auto Completion is the task of suggesting queries to the users of a search engine while they are typing a query in the search box. Over the recent years there has been a renewed interest in research on improving the quality of this task. The published improvements were assessed by using offline evaluation techniques and metrics. In this paper, we provide a comparison of online and offline assessments for Query Auto Completion. We show that there is a large potential for significant bias if the raw data used in an online experiment is re-used for offline experiments afterwards to evaluate new methods.

1 Introduction

Search logs are the traces that users leave behind when searching for information with a search engine. A number of techniques benefit from analyzing them. Among those, Query Auto Completion (QAC) helps users express their information need by suggesting queries before issuing one. This work is focused on comparing the online and offline evaluation for QAC.

Query Auto Completion is the task of suggesting full queries (*completions*) to the user, which are extensions of what he/she has typed so far (*prefix*). The simplest QAC approach is Most Popular Completion (MPC) [1]. MPC ranks the completions that match the prefix by popularity. More advanced approaches are time-sensitive [3,11] and user-sensitive [1,6], which take into consideration the timeframe and the user's search history, behavior and profile.

Regardless of the method, a significant challenge here is the evaluation of the ranking of the completions. For practical reasons, offline evaluation is the method of choice for researchers in academia.

The central idea of offline QAC evaluation is simulating clicks on completions: Each unique query that the users have issued, as extracted from the search logs, is treated as if they clicked on it as a completion. A list of completions is generated offline, having as prefix various substrings of the query (usually 1–20 first characters) and, given the position of the query in the completions list, a score is calculated for an evaluation metric.

© Springer International Publishing AG 2017
J.M. Jose et al. (Eds.): ECIR 2017, LNCS 10193, pp. 713–719, 2017.
DOI: 10.1007/978-3-319-56608-5_70

Mean Reciprocal Rank (MRR), a precision-oriented metric, is the standard for QAC. Other metrics are weighted MRR [1], which takes into consideration the number of completions available for the given prefix, and Success Rate at top K (SR@K) [6]. Recent studies, where QAC approaches were compared and evaluated offline, were conducted in [2,3,5,6,10,11].

However, none of these studies compare their offline results with online experiments. This is, as we will show, vital, because if we cannot control how the query logs were generated (which QAC method was used in the online system), we will observe misleading results.

To consider online evaluation, we have to draw inspiration from a higher level task: search effectiveness. We have two options: AB testing and Interleaving. AB testing is the standard of online evaluation in IR, used in predicting user satisfaction [8]. It is a controlled experiment where some of the users are exposed to an experimental version of the system. AB testing has low sensitivity due to the high variance of the users and require millions of interactions in order to reach a valid conclusion as to which system is preferred by the users [4].

Interleaving [7] exposes the users to a system which mixes an experimental version and the baseline together in as unbiased a way as possible. Interleaved comparisons have high sensitivity and require much fewer interactions than AB testing [4]. The most widely used algorithm is Team-Draft Interleaving (TDI) [9], which is the one we used in our experiments. User preference is inferred by counting the clicks credited to one version or the other (see Sect. 3).

To the best of our knowledge, no prior work has been published on online QAC evaluation and QAC interleaved comparisons. This fact raises the **research question:** Does online evaluation correspond to offline evaluation in QAC? The answer is *tentatively* no. The results are discussed in Sect. 4.

2   Experimental Methodology

As the core of this work is investigating the link between online and offline evaluation for QAC, we focus only on standard approaches for QAC. While we do not discard the use of more complex methods, such as [1,3,6,10,11] (which are left as future work), we opted for methods that are fast enough to operate in real time, a requirement of production environments. The methods used are:

Most Popular Completion (MPC) is an effective method for QAC [1] often used as a strong baseline when comparing QAC approaches [2,3,5,6,10].

Co-occurrences on Queries (COQ) is a fast method based on an inverted index of past queries. Like MPC, past queries are saved into a database and used to suggest completions. When completions are required, COQ tokenizes the current query, issuing a Boolean request to retrieve all past queries that have all the keywords used in the user query. Then, the most frequent words in the result set are recommended to the user. In the example *asthma children t*, we would filter the past queries first, retaining only those containing both *asthma* and *children*, then we would order the most frequent words that start with *t*.

Co-occurrences on Titles (COT) is similar to COQ, however titles of past clicked documents are used instead of past queries.

In order to implement these methods, we take advantage of the historical clicks of *Trip Database*[1], a commercial medical search engine. We collected a sample of 1.3 million clicks from November 2010 to February 2015, from which the vast majority (around 80%) are recent logs from January 2014 to February 2015. Each entry of these click logs has information regarding: (1) the query issued by a user, (2) the time a user clicked on a document, and (3) the title of the clicked document.

3   Experiments and Results

We trace here parallel experiments using standard online and offline evaluation procedures. Our goal is to understand the insights that each evaluation method would bring us.

During a period of 3 weeks in Sep./Oct. 2016, we collected clicks on query completions while the user was typing the query. In each week, a comparison of two different QAC approaches was done: in the first week we compared MPC and COQ (M-Q), in the second, MPC and COT (M-T), and in the third, COT and COQ (Q-T). Whenever a user clicked on a completion, the interaction was saved into a log file. This data was used for the online evaluation (Sect. 3.1) and posteriori offline evaluation (Sect. 3.3). For the same period of time, we got access to the click logs containing all the clicks made by users (as described in Sect. 2). This data is used for the offline evaluation (Sect. 3.2).

3.1   Online Evaluation

Our online evaluation was done with interleaving. We used a modification of the Team-Draft Interleaving (TDI) algorithm [9], which does not assign a team to the top common elements of both lists [4]. This particular modification of TDI was shown to further increase the sensitivity of the TDI algorithm.

In Fig. 1, we show the results aggregated by experiment day for our three-week comparison between each pair of methods described in Sect. 2. MPC is the dominating method, performing better than COQ and COT across all the days. Note also that COQ outperforms COT.

3.2   Offline Evaluation

At the same period of the online evaluation, we collected 55,805 document clicks made from 27,040 unique queries[2]. These 55k queries were used in the offline evaluation, as described in Sect. 1. The evaluation procedure is the following:

---

[1] https://www.tripdatabase.com.

[2] An average of 2.06 documents were clicked per query. Queries without document clicks were not recorded.

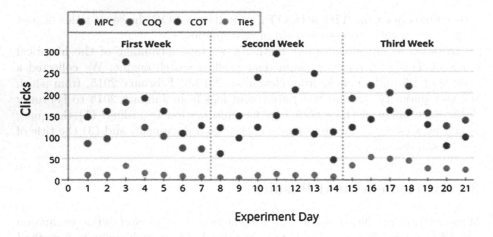

Fig. 1. Online paired interleaving evaluation made in 3 weeks: each week a different pair was compared. *Ties* refer to clicked completions on which no team was assigned.

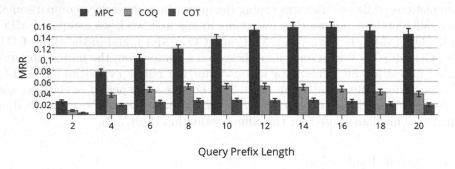

Fig. 2. Offline evaluation: mean reciprocal rank for different prefix lengths. All 3 methods are tested. Error bars show 95% confidence interval.

for each query, a prefix of length L is used to generate completions using MPC, COQ and COT. The Mean Reciprocal Rank (MRR) is then used to evaluate each method. Figure 2 shows the MRR score for each metric with L varying from 2 to 20[3].

We also allocated all 55 K queries into the 3 weeks according to the period in which they were issued. This aims to evaluate the bias regarding the experiment that was in place. For example, in the first week, when MPC was compared to COQ, would it be fair to compare these two methods with another one, such as COT, that was not part of this experiment? Table 1 shows the results of this experiment for different query lengths. Note that, although with varying scores for MRR, MPC consistently outperforms COQ and COT. However, the comparison between COQ and COT, which are similar methods, have a major

---

[3] Note that in our online experiments, the average query length on which the users clicked as a completion is 11 characters.

Table 1. MRR scores using all the data collected during the 3 weeks of our experiments (Sect. 3.2).

	Offline evaluation using all the queries issued															
	Prefix length 4 chars.				Prefix length 10 chars.				Prefix length 16 chars.				Average over all prefixes			
	M-Q	M-T	Q-T	All	M-Q	M-T	Q-T	All	M-Q	M-T	Q-T	All	M-Q	M-T	Q-T	All
MPC	0.077	0.067	0.057	0.066	0.136	0.117	0.097	0.115	0.157	0.134	0.108	0.130	0.118	0.102	0.083	0.099
COQ	0.035	0.025	0.029	0.030	0.052	0.032	0.047	0.043	0.047	0.026	0.042	0.038	0.041	0.026	0.037	0.035
COT	0.017	0.020	0.020	0.019	0.026	0.031	0.035	0.031	0.024	0.029	0.032	0.029	0.021	0.025	0.028	0.025

dependence on the kind of experiment that was in place. When comparing M-Q (data from the first week MPC vs COQ) for any prefix length, we see that COQ clearly outperforms COT, however in M-T weeks we cannot say anymore that COQ outperforms COT.

3.3 Offline Evaluation Using only Online Data

In the last part of our experiments, we explore even further the bias of the offline comparison of different methods. This time, we use only the data produced during our online experiment to perform the offline experiment. This would be the equivalent of using the data generated in campaigns such as CLEF Living Labs[4] and TREC-opensearch[5] to test a new algorithm after these campaigns stop running.

Table 2 shows the result of restricting the offline evaluation to only use the online data produced in the online evaluation. Here we can see how strong the bias towards the methods used in the online data is. For example, consider the week in which COQ and COT were compared (Q-T), if another method such as MPC were tested with the data generated in this week, we would probably say that this is not a good method, as for both small and average prefix length (4 and 10 characters), MPC was outperformed by COQ and COT.

Table 2. MRR scores when only using the queries that received a click as a completion during the 3 weeks of our experiments (Sect. 3.3).

	Offline evaluation using only clicked-completion queries															
	Prefix length 4 chars.				Prefix length 10 chars.				Prefix length 16 chars.				Average over all prefixes			
	M-Q	M-T	Q-T	All	M-Q	M-T	Q-T	All	M-Q	M-T	Q-T	All	M-Q	M-T	Q-T	All
MPC	0.303	0.324	0.100	0.246	0.508	0.476	0.209	0.397	0.611	0.620	0.296	0.500	0.455	0.451	0.200	0.367
COQ	0.172	0.113	0.162	0.148	0.232	0.164	0.254	0.216	0.177	0.113	0.203	0.165	0.182	0.125	0.196	0.168
COT	0.085	0.107	0.128	0.106	0.102	0.164	0.200	0.156	0.069	0.131	0.152	0.119	0.082	0.126	0.151	0.120

---

[4] living-labs.net/clef-lab/.
[5] http://trec-open-search.org/.

4   Discussion and Conclusion

In this paper, we performed and analysed a full online evaluation comparing three query auto completion methods throughout a period of three weeks. Our online experiment collected 6,014 clicks and shows that users systematically prefer the MPC method over COQ and COT, and prefer COQ over COT.

Note that most of the commercial search engines already have a QAC method running. It means that there is always an existing untold bias towards the system in production when data is collected to create query logs for offline evaluation of new QAC methods. We evaluate this bias in two different offline experiments, using all 55k queries issued in the period of the experiment and using only the queries that were clicked on as completion during the period of the experiment. The first offline experiment, performed using a standard approach in the literature (Fig. 2), produces the same results as the online evaluation produces: MPC is the best, followed by COQ, and COT last. However, when breaking this analysis into the 3 different weeks (therefore different QAC methods in the production system), we noticed that similar methods, such as COQ and COT, are harder to tell apart. The data for the weeks in which we were comparing MPC and COT cannot be used to compare COQ as, depending on the prefix length of the queries, COT might outperform COQ, which we know should not happen. The biggest bias is found for comparisons using only the online queries in an offline manner. There we saw that for query lengths of 4 and 10, MPC is the worst method if the data comparing COQ and COT is used. This result is highly undesirable as the development of good methods such as MPC would be impacted by the bias present in the data used.

A major implication of this work is that, although we did not directly use the data created in live campaigns such as CLEF Living and TREC-opensearch, our experiments show that an extra care should be taken when using such data after the evaluation period (in an offline fashion), in order to control the bias towards the methods used in the live system. A way to mitigate this bias is by adding unbiased data, such as additional user clicks, as shown through our experiments.

Acknowledgment. This research was partially supported by the EU Project KConnect (Grant No.: 644753) and the Austrian FWF Project ADmIRE (Project No.: P25905-N23).

References

1. Bar-Yossef, Z., Kraus, N.: Context-sensitive query auto-completion. In: Proceedings of WWW (2011)
2. Cai, F., de Rijke, M.: Selectively personalizing query auto-completion. In: Proceedings of SIGIR (2016)
3. Cai, F., Liang, S., de Rijke, M.: Time-sensitive personalized query auto completion. In: Proceedings of CIKM (2014)
4. Chapelle, O., Joachims, T., Radlinski, F., Yue, Y.: Large-scale validation and analysis of interleaved search evaluation. ACM Trans. Inf. Syst. **30**, 1–41 (2012)

5. Di Santo, G., McCreadie, R., Macdonald, C., Ounis, I.: Comparing approaches for query autocompletion. In: Proceedings of SIGIR (2015)
6. Jiang, J.-Y., Ke, Y.-Y., Chien, P.-Y., Cheng, P.-J.: Learning user reformulation behavior for query auto-completion. In: Proceedings of SIGIR (2014)
7. Joachims, T.: Evaluating retrieval performance using clickthrough data. Text Mining, pp. 79–96 (2003)
8. Kohavi, R., Longbotham, R., Sommerfield, D., Henne, R.M.: Controlled experiments on the web: survey and practical guide. Data Mining Knowl. Discov. **18**, 140–181 (2009)
9. Radlinski, F., Kurup, M., Joachims, T.: How does clickthrough data reflect retrieval quality? In: Proceedings of CIKM (2008)
10. Shokouhi M.: Learning to personalize query auto-completion. In: Proceedings of SIGIR (2013)
11. Shokouhi, M., Radinsky K.: Time-sensitive query auto-completion. In: Proceedings of SIGIR (2012)

A Neural Attention Model for Categorizing Patient Safety Events

Arman Cohan[1]([⊠]), Allan Fong[2], Nazli Goharian[1], and Raj Ratwani[2]

[1] Georgetown University, Washington DC, USA
{arman,nazli}@ir.cs.georgetown.edu
[2] National Center for Human Factors in Healthcare,
MedStar Health, Washington DC, USA
{allan.fong,raj.ratwani}@medicalhfe.org

Abstract. Patient Safety Event reports are narratives describing potential adverse events to the patients and are important in identifying, and preventing medical errors. We present a neural network architecture for identifying the type of safety events which is the first step in understanding these narratives. Our proposed model is based on a soft neural attention model to improve the effectiveness of encoding long sequences. Empirical results on two large-scale real-world datasets of patient safety reports demonstrate the effectiveness of our method with significant improvements over existing methods.

Keywords: Deep learning · Text categorization · Medical text

1 Introduction

In recent years NLP/IR have become increasingly important in understanding, searching, and analyzing medical information [22]. Human or system errors do occur frequently in the health centers, many of which can lead to serious harm to individuals. There are in fact an alarming number of annual death incidents (up to 200K) being reported due to medical errors [1]; medical errors are shown to be the third leading cause of death in the US [14]. Many healthcare centers have deployed patient safety event reporting systems to better identify, mitigate, and prevent errors [5]. Patient safety event reports are narratives describing a safety event and they belong to different safety categories such as "medication", "diagnosis", "treatment", "lab", etc. Recently, due to the importance of patient safety reports, more healthcare centers are enforcing patient safety reporting, resulting in an overwhelming number of daily produced reports. Manual processing of all these reports to identify important cases, trends, or system issues is extremely difficult, inefficient, and expensive. The first step in understanding and analyzing these events is to identify their general categories. This task is challenging because the event descriptions can be very complex; the frontline staff usually focus more on taking care of the patient at the moment than to think through the classification schema when they later write a safety report. For example, an event where a patient fell after being given an incorrect medication might have been classified as

© Springer International Publishing AG 2017
J.M. Jose et al. (Eds.): ECIR 2017, LNCS 10193, pp. 720–726, 2017.
DOI: 10.1007/978-3-319-56608-5_71

"FALL" however, the fall could be due to a mis-medication and therefore belong to the "MEDICATION" safety event. Without the ability to correctly identify the medication category, such problems will not be addressed. Therefore, classifying the patient safety reports not only helps in further search and analytic tasks, but also it contributes to reducing the human reporting errors.

In this paper, we present a method for categorizing the Patient Safety Reports as the first step towards understanding adverse events and the way to prevent them. Traditional approaches of text categorization rely on sparse feature extraction from clinical narratives and then classifying the types of events based on these feature representations. In these conventional methods, complex lexical relations and long-term dependencies of the narratives are not captured. We propose a neural attention architecture for classifying safety events, which performs the feature extraction, and type classification jointly; our proposed architecture is based on a combination of Convolutional Neural Networks (CNNs) and Recurrent Neural Networks (RNNs) with soft attention mechanism. We evaluate our method on two large scale datasets obtained from two large healthcare providers. We demonstrate that our proposed method significantly improves over several traditional baselines, as well as more recent neural network based methods.

2   The Proposed Neural Attention Architecture

Our proposed model for classifying patient safety reports is a neural architecture based on Convolutional Neural Networks (CNN) and Recurrent Neural Networks (RNN) utilizing a soft attention mechanism. Our architecture is partially similar to models by [11,12] in convolutional layers, to [19] in recurrent layer, and to [21] in the document modeling. Our point of departure is that unlike these works which are mainly targeted for sentence and short documents, we utilize a soft neural attention mechanism coupled with CNN and RNN to capture the more salient local features in longer sequences. Below we present the building blocks of our proposed architecture from bottom to the top.

Embedding Layer. Represents a sequence of words $S = \langle w_1; w_2; ...; w_n \rangle$ with an input matrix $\mathbf{x} \in \mathbb{R}^{(n,d)}$ that can be either initialized randomly or by pre-trained word embeddings, and then can be jointly trained with the model.

CNN. CNNs are feed-forward networks which include two main operations: *convolution* and *pooling*. Convolution is an operation on two functions (input and kernel) of real valued arguments [13]. In our context, in layer ℓ in the network, convolution operates on sliding windows of width k_ℓ on the input $\mathbf{x}_{\ell-1}$ and yields a feature map F_ℓ:

$$F_\ell^{(i)} = g(\mathbf{W}_\ell \cdot \mathbf{x}_{\ell-1}^{(i,k_\ell)} + \mathbf{b}_\ell) \tag{1}$$

where $\mathbf{W}_\ell$ and $\mathbf{b}_\ell$ are the shared wights and biases in layer ℓ, g is an activation function, and $\mathbf{x}^{(i,k_\ell)} = \langle x^{i-\frac{(k_\ell-1)}{2}}; ...; x^{i+\frac{(k_\ell-1)}{2}} \rangle$ shows the sliding window of size k_ℓ centered at position i on the input. We use ReLU [6] for the activation function (In our experiments ReLU showed the best results among other activation functions). For pooling, we use "max-pooling" operation whose role is to down-sample the feature map and capture significant local features. Similar to [12], we use filters of sizes from 2 to 6 to capture local features of different granularities. The convolution layer allows the model to learn the salient features that are needed for identifying the type of the safety events.

RNN. Unlike CNNs which are local feature encoders, RNNs can encode large windows of local features and capture long temporal dependencies. Given an input sequence $\mathbf{h} = (x_1, ..., x_T)$ where each $x_t \in \mathbb{R}^d$ is an input word vector of dimension d at time step t, an RNN computes the hidden states $\mathbf{h} = (h_1, ..., h_T)$ and outputs $\mathbf{y} = (y_1, ..., y_T)$ according to the following equations [8]:

$$h_t = g(W^{(hh)}h_{t-1} + W^{(xh)}x_t + b_h) \qquad\qquad y_t = W^{(hy)}h_t + b_y \qquad (2)$$

where W shows the weight matrices for the corresponding input, b denotes the biases, and g is the activation function. RNNs in theory, can capture temporal dependencies of any length. However, training RNNs in their basic form is problematic due to the *vanishing gradient* problem [16]. Long Short-Term Memory (LSTM) [10] is a type of RNN that has several gates controlling the flow of information to be preserved or forgotten, and mitigates the vanishing gradient problem. We use the LSTM formulation as in [9]. We aslo employ bidirectional LSTM to capture both forward and backward temporal dependencies. Using this layer, we capture the dependencies between local features along long sequences.

Neural Attention. The trouble with RNNs for classification is that they encode the entire sequence into the vector at the last temporal step. While the application of RNNs have been successful in encoding sentences or short documents, in longer documents this can result in loss of information [4], and putting more focus on the recent temporal entries [18]. Bidirectional RNNs try to alleviate this problem by considering both the forward and backward context vectors. However, they suffer from the same problem in long sequences.

Inspired by work in machine-translation, to address this problem, we utilize the soft attention mechanism [2]. Neural attention allows the model to decide which parts of the sequence are more important instead of directly considering the context vector output by the RNN. Specifically, instead of considering the final cell state of LSTM for the classification, we allow the model to attend to the important timesteps and build a context vector c as follows:

$$c = \sum_{t=1}^{T} \alpha_t h_t \qquad (3)$$

where α_t are weights computed at each timestep t for the state h_t and are computed as follows:

$$\alpha_t = \frac{\exp(e_t^\top z)}{\sum_{k=1}^{T} \exp(e_k^\top z)} \qquad (4)$$

$$e_t = f_{\text{ATT}}(h_t) \qquad (5)$$

where f_{ATT} is a function whose role is to capture the importance of h_{t_i} and z is a context vector that is learned jointly during training. We use a feed-forward network with "tanh" activation function for f_{ATT}. The context vector c is then fed to a fully-connected and then a softmax layer to perform final classification.

3   Experiments

Setup. We evaluate the effectiveness of our model on two large scale patient safety data obtained from a large healthcare providers in mid-Atlantic US and the Institute for Safe Medication Practices (ISMP). ISMP serves as a safe harbor for all PSE reports from hospitals in Pennsylvania, US. The dataset that was analyzed contains all categories of safety reports (fall, medication, surgery, etc.) and is not limited to medication reports. This study was approved by the Med-Star Health Research Institute Institutional Review Board (protocol 2014-101). The characteristics of the data and the categories are shown in Tables 1 and 2. We split the data with stratified random sampling into 3 sets: train, validation, and test. We tune the parameters of the neural models on the validation set and the test set remains unseen to the models. We compare our results with conventional text classification models (bag of words feature representation with different types of classifiers), as well as related work on neural architectures (CNNs, RNNs and Bidirectional RNNs and their combinations). For space limitation, we do not explain the details of the baselines and refer the reader to the corresponding citations in Table 3. We report accuracy and average F1-score results for the categories which are standard evaluation metrics for this task.

Implementation. We used Keras and TensorFlow for the implementation. We empirically made the following design choices: We used Word2Vec [15] for training the embeddings on both general (Wikipedia) and domain specific corpora

Table 1. Dataset characteristics

	# of reports	# categories	Avg. length (char)	Stdev. length (char)
Dataset 1	82,281	20	410	321
Dataset 2	1,625,512	9	327	174

Table 2. Categories in the larger dataset (dataset 2)

Category	Count	Category	Count
Procedure/treatment/test error	370K	Miscellaneous	140K
Medication error	135K	Adverse drug reaction	34K
Fall	242K	Equipment/supplies/devices	34K
Procedure/treatment/test complication	233K	Transfusion	23K
Skin integrity	234K		

Table 3. Results on the each dataset on both the validation and test sets. Numbers are percentages. Last row shows our method. † (‡) shows statistically significant improvement (McNemar's test) over the next best performance with $p < 0.05$ ($p < 0.01$).

Methods	Dataset 1				Dataset 2			
	Val		Test		Val		Test	
	Acc	F1	Acc	F1	Acc	F1	Acc	F1
SVM [20]	70.7	70.3	70.9	70.6	84.8	84.0	84.7	83.9
MNB [20]	71.2	71.5	71.0	72.3	79.2	79.9	79.0	79.6
XGB [3]	71.4	69.9	72.1	70.8	76.8	75.7	76.7	75.5
cBoW [23]	67.5	62.6	68.0	63.4	84.8	84.2	84.6	84.1
Adaptive cBoW [23]	69.2	63.4	70.6	69.6	83.9	84.3	84.8	84.8
CNN [12]	73.2	70.7	72.2	69.5	83.6	83.1	82.7	83.5
RNN [7]	76.0	74.6	74.5	72.9	84.0	84.2	83.8	83.2
Bi-RNN [7]	76.3	74.5	75.2	73.6	84.7	84.3	84.6	84.5
CNN-BiRNN [19]	77.8	76.9	76.6	76.4	**89.3**	85.9	86.8	84.6
Att-CNN-BiRNN (ours)	**78.3** †	**77.2**	**78.1**‡	**77.3** ‡	89.1	**88.1**‡	**88.9**‡	**88.0**‡

(PubMed), similar to [17]. We used dropout rates of 0.25 for the recurrent and 0.5 for the convolutional layers. We used Adam optimizer with categorical cross entropy loss and early stopping for training.

Results. Table 3 demonstrates our main results. As illustrated, our method (last row) significantly outperforms all other methods in virtually all the datasets. This shows the general effectiveness of our model in comparison with the prior work. We observe that our method's performance improvement is slightly larger in the second (larger) dataset. This is expected since our model can better learn the parameters when trained on larger data. Improvement over RNN and CNN-Bi-RNN baselines shows the effectiveness of the neural soft attention in capturing salient parts of the sequence in comparison with the models without attention.

Error Analysis. While our method effectively outperforms the prior work, we conducted error analysis to better understand the cases that our method fails to correctly perform categorization. In particular, we observed that for both datasets, many wrongly classified samples in the categories were

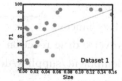 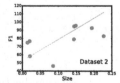

Fig. 1. Performance for each category based on its relative size to the dataset.

misclassified as the "MISCELLANEOUS" category. This pattern was more common for the categories with smaller training samples. This shows that the model learns a broader set of texts for the "MISCELLANEOUS" category, which is expected, given the broad nature of this category. We also observed some misclassified samples in the categories that are closely related together. For example in dataset 1, 32% of the misclassified samples in the "BLOOD-BANK" category were classified as "LAB/SPECIMEN". A similar pattern was observed for the "DIAGNOSIS" and "MEDICATION" safety events. These closely related categories usually have overlaps in terms of training data and this makes it hard for the model to differentiate the edge cases. We furthermore observe that the performance on each category correlates with the number of samples in that category. Figure 1 shows this correlation. We observe that generally, our method performs better with the categories of larger relative size. While the correlation is stronger for dataset 1, both datasets show similar trends. This shows that having more training samples helps our model in better learning the characteristics of that particular category and results in higher performance.

4   Conclusion

We presented a neural network model based on a soft attention mechanism for categorizing patient safety event reports. We demonstrated the effectiveness of our model on two large-scale real-world datasets and we obtained significant improvements over existing methods. The impact of our method and results is substantial on the patient safety and healthcare, as better categorization of events results in better patient management and prevention of harm to the individuals.

Acknowledgments. We thank the 3 anonymous reviewers for their helpful comments. This project was funded under contract/grant number Grant R01 HS023701-02 from the Agency for Healthcare Research and Quality (AHRQ), U.S. Department of Health and Human Services. The opinions expressed in this document are those of the authors and do not necessarily reflect the official position of AHRQ or the U.S. Department of Health and Human Services.

References

1. American Hospital Association: Fast facts on US hospitals (2013)
2. Bahdanau, D., Cho, K., Bengio, Y.: Neural machine translation by jointly learning to align and translate. arXiv:1409.0473 (2014)
3. Chen, T., Guestrin, C.: Xgboost: a scalable tree boosting system. In: KDD (2016)
4. Cho, K., Van Merriënboer, B., Bahdanau, D., Bengio, Y.: On the properties of neural machine translation: encoder-decoder approaches. arXiv:1409.1259 (2014)
5. Clarke, J.R.: How a system for reporting medical errors can and cannot improve patient safety. Am. Surg. **72**(11), 1088–1091 (2006)
6. Dahl, G.E., Sainath, T.N., Hinton, G.E.: Improving deep neural networks for LVCSR using rectified linear units and dropout. In: IEEE ICASSP, pp. 8609–8613 (2013)

7. Dai, A.M., Le, Q.V.: Semi-supervised sequence learning. In: NIPS (2015)
8. Elman, J.L.: Finding structure in time. Cogn. Sci. **14**(2), 179–211 (1990)
9. Graves, A., Jaitly, N.: Towards end-to-end speech recognition with recurrent neural networks. In: ICML, vol. 14, pp. 1764–1772 (2014)
10. Hochreiter, S., Schmidhuber, J.: Long short-term memory. Neural Comput. **9**(8), 1735–1780 (1997)
11. Kalchbrenner, N., Grefenstette, E., Blunsom, P.: A convolutional neural network for modelling sentences. In: ACL, pp. 655–665, June 2014
12. Kim, Y.: Convolutional neural networks for sentence classification. In: EMNLP (2014)
13. Lecun, Y., Bottou, L., Bengio, Y., Haffner, P.: Gradient-based learning applied to document recognition. IEEE **86**(11), 2278–2324 (1998)
14. Makary, M.A., Daniel, M.: Medical error - the third leading cause of death in the US. BMJ **353**, i2139 (2016)
15. Mikolov, T., Sutskever, I., Chen, K., Corrado, G.S., Dean, J.: Distributed representations of words and phrases and their compositionality. In: NIPS (2013)
16. Pascanu, R., Mikolov, T., Bengio, Y.: On the difficulty of training recurrent neural networks. ICML **28**, 1310–1318 (2013)
17. Soldaini, L., Goharian, N.: Learning to rank for consumer health search: a semantic approach. In: Jose, J.M., et al. (eds.) ECIR 2017. LNCS, vol. 10193, pp. 640–646. Springer, Heidelberg (2017)
18. Sutskever, I., Vinyals, O., Le, Q.V.: Sequence to sequence learning with neural networks. In: NIPS, pp. 3104–3112 (2014)
19. Tang, D., Qin, B., Liu, T.: Document modeling with gated recurrent neural network for sentiment classification. In: EMNLP, pp. 1422–1432 (2015)
20. Wang, S., Manning, C.D.: Baselines and bigrams: simple, good sentiment and topic classification. In: ACL, pp. 90–94 (2012)
21. Yang, Z., Yang, D., Dyer, C., He, X., Smola, A., Hovy, E.: Hierarchical attention networks for document classification. In: NAACL-HLT (2016)
22. Yates, A., Goharian, N., Frieder, O.: Extracting adverse drug reactions from social media. In: AAAI, pp. 2460–2467 (2015)
23. Zhao, H., Lu, Z., Poupart, P.: Self-adaptive hierarchical sentence model. In: IJCAI, pp. 4069–4076. AAAI Press (2015)

Promoting Understandability in Consumer Health Information Search

Hua Yang[✉] and Teresa Goncalves

Computer Science Department, University of Evora, Evora, Portugal
huayangchn@gmail.com, tcg@uevora.pt

Abstract. Search engines have become a common way of obtaining health information. Although access mechanism for factual health information search has developed greatly, complex health searches which do not have a single definitive answer still remain indefinable. Answers to a complex health query contain different viewpoints and confuse a non-expert consumer. It is demanding for a consumer with limited medical knowledge background to get a balanced view of the diverse perspectives. This research proposal points out that what consumers need is comprehensive and useful information. To aid consumers get an improved understanding of the retrieved contents, the proposed approach is adding additional information to the retrieved contents. One applicable way is classifying the retrieved contents as support, neutral or oppose. The classification labels serve as the extra information to supplement the retrieved contents. Other potential extra information and ways to incorporate the information into a search engine are to be researched into in our later work. In this proposal, the challenges are narrated and related work are reviewed. Research questions and overall goals are stated. The proposed work is discussed and research outline is depicted.

1 Introduction

Health information search is a domain specific information retrieval (IR) in medical area, which is also known as health information retrieval (HIR).

Search engines have become a common way of obtaining health information; a recent health online report by Fox and Duggan (2013) shows that 35% of American adults have the experience of using Internet as an aid for health decision making.

Nowadays, access mechanism of factual health information search has developed greatly. With a general-purpose search engine, it is easy to get an answer to "what is gout?" or "what are the symptoms of gout?". Nevertheless, it still remains indefinable for complex health searches which do not have a single definitive answer like "does daily aspirin therapy prevent heart attack?". Concerning this kind of searches, not a single answer, but answers of different viewpoints are to be presented. It is demanding for a common user (laypeople without strong medical knowledge background) to get a balanced view of the diverse perspectives. Consumers have difficulty in understanding the answers to a complex query; necessary support information is needed to aid consumers to better

© Springer International Publishing AG 2017
J.M. Jose et al. (Eds.): ECIR 2017, LNCS 10193, pp. 727–734, 2017.
DOI: 10.1007/978-3-319-56608-5_72

understand the retrieved materials. Moreover, consumers vary in medical knowledge and this affects what type of support information are to be returned to them. Goeuriot et al. (2016) point out that development of search and access technologies in this area is still challenging.

My Ph.D work aims to research in health information retrieval and focus on providing support information to improve understandability of the retrieved materials. In the following of this paper, state of the art in this area is discussed in Sect. 2; my proposed research and approach are presented in Sect. 3.

2   State of the Art

There have been abundance of research work in this area. Related work that is relevant to our research includes (1) query expansion techniques applied in HIR; (2) ranking techniques for HIR; (3) study of improving quality of on-line health information. This section briefly describes the related work and concludes with considered state of the art.

2.1   Query Expansion Techniques in HIR

A wide range of query expansion techniques have been applied in HIR. Among them is using thesaurus to find synonymous and related terms serving as extra terms. Widely used thesaurus in HIR include domain specific ones like MeSH[1] and UMLS[2]. Amount of papers have presented the work of using query expansion techniques with domain specific thesaurus. Despite of its effectiveness, papers (Voorhees and Hersh 2012) (Shen and Nie 2015) also have observed that expanding queries with synonyms improved performance for certain instances. It is not always effective of using thesaurus based query expansion techniques and the results can be mixed. Word2vec model (Mikolov et al. 2013) is becoming one of the most efficient approach to learn word embeddings. The work by Wang et al. (2015) present a method of integrating term embedding with medical domain knowledge for healthcare applications.

Medical Subject Headings (MeSH). MeSH is a controlled vocabulary by National Library of Medicine. Synonymous terms are grouped in a concept. One or more concepts closely related to each other in meaning are organized in a MeSH record. Possible relationships between concepts are preferred, related, narrower and broader.

United Medical Language System (UMLS). UMLS brings together many health and biomedical vocabularies and standards to enable interoperability between computer systems. UMLS contains three knowledge sources, Metathesaurus, Semantic Network, and Specialist Lexicon & Tools. These knowledge sources can be used for information retrieval, natural language processing, automated indexing, thesaurus construction, electronic health records and others.

---

[1] https://www.nlm.nih.gov/mesh/.
[2] https://www.nlm.nih.gov/research/umls/.

Metathesaurus clusters terms into concepts and assigns unique identifier to each concept. It contains biomedical and health related concepts, their various names and the conceptual relationship among its source vocabularies. Metathesaurus is not built in one vocabulary. Synonymous terms are clustered into a concept with a unique identifier (CUI). Each term is identified by a unique identifier (LUI). Each term is a normalized name and may have several strings (SUI), which represent the terms lexical variants in the source vocabularies. Each string is associated with one or more atoms (AUI) that represent the concept name.

Word2vec. Word2vec model constructs a vocabulary from training a text corpus and then uses neural networks to learn vector representation of words. The vector representations of words learned by word2vec models have been shown to carry semantic meanings. Word vectors are positioned in the vector space such that words that share common contexts in the corpus are located in close proximity to one another in the space. A specific word2vec model can integrate existing medical knowledge and be trained with large medical corpus. Related words found with the model can be used for expansion.

2.2   Learning to Rank

Learning to rank approaches are based on features vectors and use traditional supervised learning methods. In recent years, learning to rank approaches have been used in HIR. As a supervised learning problem, learning to rank approach needs a training set to be created. The creation of a training set is very similar to the creation of the test set for evaluation. Typically, a training set consists of n training queries $q_i(i = 1, ..., n)$, the associated documents represented by feature vectors X^i, and the corresponding relevance judgements $y^{(i)}$ (Liu 2009). A ranking model is learned by applying a specific learning algorithm. The learned ranking model is then used to predict a new query, sort the documents according to their relevance to the query, and return a corresponding ranked list of the documents as the response to the query. In CLEF eHealth track 2015, most teams used learning to rank approach in their running. These teams employed different strategies of learning to rank and showed that this approach could slightly improve the baseline. Nevertheless, some team results also showed that query expansion techniques indeed obtained higher effectiveness compared to learning to rank alternatives (Palotti et al. 2015). Although learning to rank methods have shown its success in information retrieval, its performance in HIR is not quite clear and not enough research work has proved its usability in this area. No clear performance improvement was reported and its performance compared with other techniques needs to be further explored.

2.3   Search Behaviour Analysis

Web resources provide vast health information for people and many people employ search engines to get health information. Along with it, the issue of the quality of on-line health information arises and is concerned. Mining search logs can provide insights into how people interact with search engine (White and Drucker 2007). Plenty of works research on the analysis of search behaviours and develop techniques to improve the quality of on-line health information. The works (Paparrizos et al. 2016) (Schoenherr and White 2014) (Paul et al. 2015) study the feasibility of learning search behaviours from mining large-scale search log data.

3 Proposed Research

In this section, we first put forward the research questions. Then we discuss our potential contributions to this research area and the goals to achieve. Our proposed research outline is depicted and the techniques to be developed are discussed.

3.1 Research Questions

Techniques for accessing factual health information have matured considerably, but for complex queries they still remain uncertain. So,

1. Users use search engine to get relevant health information. But, how can an search engine aid users to better understand the retrieved materials, which will satisfy users?
2. For non-expert consumers, what kind of support information can be returned to them?

3.2 Research Goals

This work aims to figure out a way to the research queries. The main goal is to not only retrieve relevant materials for users, but also going further into improving understandability of the retrieved materials. To achieve this main goal, we propose to divide our work into the following steps to be accomplished:

1. Study understandability of the retrieved materials; research into the techniques for improving understandability.
2. Research into IR techniques and push the state of the art techniques in HIR area. Improve techniques and models to overcome the shortcomings of existing IR techniques applied in HIR.
3. Integrate support information into a HIR model.
4. Propose effective evaluation metrics. Not only the topical relevance should be evaluated, but also understandability of the information should be taken into account.

3.3   Proposed Research Outline and Techniques

Improve Understandability of Retrieved Materials. Relevant information is not equal to be valuable to users. Valuable information should be relevant, readable, comprehensive and useful to users. In addition to relevance, an ideal search system should also take other elements into account. In our work, we propose to research on improving understandability of the retrieved contents. We define "understandability" from two aspects and our proposed research are discussed respectively. Figure 1 briefly depicts our proposed research.

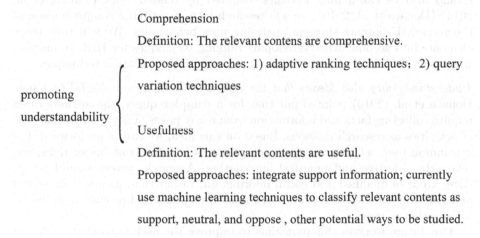

Comprehension

Definition: The relevant contens are comprehensive.

Proposed approaches: 1) adaptive ranking techniques; 2) query

variation techniques

promoting

understandability

Usefulness

Definition: The relevant contents are useful.

Proposed approaches: integrate support information; currently

use machine learning techniques to classify relevant contents as

support, neutral, and oppose , other potential ways to be studied.

Fig. 1. Proposed research outline

Understandability Means that the Relevant Contents are Readable and Comprehensive to Users. With nowadays search engines, after issuing a query to the system, relevant contents are retrieved and ranked. Nevertheless even if a search engine can retrieve relevant contents according to a query raised by a user, but if the user thinks the relevant document is difficult to consume, they tend to give up and move on to another one (Yilmaz et al. 2014). The relevant document is understandable and preferred by some users, but not for other users. When a user can not comprehend the information, even if it is highly relevant, it means nothing to him. Users have different knowledge background, which will affect their comprehension to the retrieved contents. Even for laypeople with the same information need, they may have different choice of reading. We can not ask a user to comprehend all the relevant contents, but we can provide understandable contents to him according to his knowledge background. People with certain medical knowledge may prefer to read technical or professional contents retrieved from professional websites or journal articles. For people with limited medical knowledge, they may enjoy reading more popular contents coming from blogs and forums. To the best of our knowledge, hardly no search system takes this into account; the ranking results are same for a same information need from different level users. For a user, what he needs appears at the first one or two pages; while for another user, the needed content may be ranked at the tenth page. One can be exhausted in finding needed contents, although the contents has been retrieved as relevant.

To make the retrieved contents more readable and comprehensive to users, we are now thinking of two possible ways to do the research. In one way, since users have different preference, we are thinking of employing different ranking strategies to users. Not only the probability of the relevance are considered when ranking retrieved contents, the comprehension of the relevant contents should also be taken into accounts. Inspired by related work (Yilmaz et al. 2014) (Donato et al. 2010), we will research into quantifying comprehension of the retrieved contents through analysing user behaviours. We will take those elements into account when developing ranking techniques for HIR. In another way, we are thinking of research into it through query variation techniques.

Understandability also Means that the Relevant Contents are Useful to Users. Donato et al. (2010) pointed out that for a complex query, the answers often require collecting facts and information from many pages. They refer to this type of activities as research missions. Based on this idea, we can move forward. For a common user, what he needs is not plenty of isolated contents or data, but relatively objective and practical information. A search system should return these kinds of qualified and useful information, rather than plenty of one-sided and lone contents. We can process the related contents and provide more useful information to users.

Our future work in this part aims to improve the usefulness of the relevant contents. Our idea is to draw appropriate summaries from the analysis of the retrieved contents. The work concerns studying on what kinds of retrieved contents can be analysed; what are the potential techniques to do the analysis; what kinds of summaries can be made.

Develop IR Techniques Applied in HIR. General IR techniques have matured considerably. However, due to the characteristics possessed by health information, the techniques that perform well in general IR have obtained limited performance in HIR. Challenges still remain in this area. Our work will explore existing IR techniques applied to HIR. Where do the techniques fail to retrieve relevant contents for HIR? What are the possible improved techniques and models to overcome the shortcomings? What techniques should be developed to achieve our research goal? Our potential contribution in this part includes (1) explore NLP tools for medical texts; (2) developing query expansion technique for HIR; (3) developing ranking techniques to improve the comprehension of the results.

Due to the limitations of applying NLP tools to clinical texts, challenges exist in processing health-related information (Chen 2015). Our future work in this part includes experimenting on existing NLP tools applied in HIR and on the possible combination of the tools.

Despite of its efficiency in HIR, previous work also show that query expansion techniques do not always perform well. So when and how do query expansion techniques fail in HIR? And what should be improved to better solve the problem in certain situations? Or can any alternated approach be developed? Our work will research into this part and developed the techniques. Different ways of query expansion are to be researched in our work. Potential combinations of different query expansion techniques will be experimented and compared. Proposed query expansion techniques include: explore domain specific thesaurus like UMLS and MeSH; pseudo relevant feedback techniques; specific and health-based word2vec models will be trained with applicable medical corpus.

Our proposed work also concerns developing ranking techniques applied in HIR. State of the art ranking techniques are to be explored. We aim to improve readability of the results along with relevance.

Integrate Support Information. As we have talked in the first part, our work aims to improve the understandability of the retrieved contents beyond the relevance. From one part, we propose to develop ranking techniques and query variation techniques to provide more readable and comprehensive results to users; from another part, we propose to develop techniques to make the relevant contents more useful to users. For the second part, our idea is to integrate additional information into retrieved contents as support information. In our proposed work, we deem that a search system can provide more useful information, but not dump a plenty of isolated contents or data to users. For a non yes-no question, a search system can collect data from related web pages, process the data, and return qualified information. Possible ways are to draw objective conclusion based on the analysis of the related contents. In other words, our idea is to develop related techniques to analyse the relevant, which can be done by a search system instead of users themselves. The analysis results are defined as support information.

Different support information can be provided to aid users to better understand the retrieved contents. The kinds of support information may have diverse granularities or dimensionalities. One applicable kind can be the classification of the relevant contents. At present, we are using machine learning techniques to label relevant contents as support, oppose or neutral. Our later work considers including in other kinds of support information. The way to incorporate the support information into a search system will be studied.

4   Discussion

At present, we to use three sorting labels as support information. Sorting schemes may be different, since both user needs and answers to a query vary. Our later work will research into other taxonomic approaches and the applicable way to incorporate support information into the search results. Evaluation metrics for information retrieval or machine learning separately are relatively matured, but are not clear when considering both. Appropriate metrics are needed to evaluate an integrated model. We will go into how to quantify the benefits of the added support information and furthermore how to quantify understandability, which would be helpful in deciding upon the most suitable evaluation metrics. About the learning to rank approaches, should it be better to train different models for each label?

References

Chen, M.W.: Comparison of natural language processing algorithms for medical texts. Ph.D. thesis, Massachusetts Institute of Technology (2015)

Donato, D., Bonchi, F., Chi, T., Maarek, Y.: Identifying research missions in yahoo! search pad. In: Proceedings of the 19th International Conference on World wide web, pp. 321–330. ACM (2010)

Fox, S., Duggan, M.: Health online 2013. Washington, DC: Pew Internet Am. Life Proj. (2013)

Goeuriot, L., Jones, G.J., Kelly, L., Müller, H., Zobel, J.: Medical information retrieval: introduction to the special issue. Inf. Retr. J. 1(19), 1–5 (2016)

Liu, T.-Y.: Learning to rank for information retrieval. Found. Trends Inf. Retr. 3(3), 225–331 (2009)

Mikolov, T., Chen, K., Corrado, G., Dean, J.: Efficient estimation of word representations in vector space (2013). arXiv preprint arXiv:1301.3781

Palotti, J., Zuccon, G., Goeuriot, L., Kelly, L., Hanbury, A., Jones, G., Lupu, M., Pecina, P.: Retrieving information about medical symptoms. In: Proceedings of CLEF (2015)

Paparrizos, J., White, R.W., Horvitz, E.: Detecting devastating diseases in search logs. In: Proceedings of the ACM SIGKDD Conference on Knowledge Discovery and Data Mining (2016)

Paul, M.J., White, R.W., Horvitz, E.: Web search as decision support for cancer. In: Proceedings of the 24th International Conference on World Wide Web, pp. 831–841. ACM (2015)

Schoenherr, G.P., White, R.W.:Interactions between health searchers and search engines. In: Proceedings of the 37th International ACM SIGIR Conference on Research and development in Information Retrieval, pp. 143–152. ACM (2014)

Shen, W., Nie, J.-Y.: Is concept mapping useful for biomedical information retrieval? In: Mothe, J., Savoy, J., Kamps, J., Pinel-Sauvagnat, K., Jones, G.J.F., SanJuan, E., Cappellato, L., Ferro, N. (eds.) CLEF 2015. LNCS, vol. 9283, pp. 281–286. Springer, Heidelberg (2015). doi:10.1007/978-3-319-24027-5_29

Voorhees, E.M., Hersh, W.R.: Overview of the TREC 2012 medical records track. In: TREC (2012)

Wang, C., Cao, L., Zhou, B.: Medical synonym extraction with concept space models (2015). arXiv preprint arXiv:1506.00528

White, R.W., Drucker, S.M.: Investigating behavioral variability in web search. In: Proceedings of the 16th International Conference on World Wide Web, pp. 21–30. ACM (2007)

Yilmaz, E., Verma, M., Craswell, N., Radlinski, F., Bailey, P.: An analysis of document utility. In: Proceedings of the 23rd ACM International Conference on Conference on Information and Knowledge Management, pp. 91–100. ACM (2014)

A Social Framework for Set Recommendation in Group Recommender Systems

Lorena Recalde[✉]

Web Research Group, Department of Information and Communication Technologies,
Universitat Pompeu Fabra, 08018 Barcelona, Spain
lorena.recalde@upf.edu

Abstract. In this research proposal we present a framework that is intended to improve Group Recommender Systems. The framework structure includes a process where an influential group is detected among the target groups of people to recommend to. In order to help the group members agree and make a decision, the visualization of the alternative chosen by the influential group and the reasons why they adopted that recommendation are presented for the target *susceptible group*. (The term susceptibility will be used through the article making reference to the group that is highly perceived as being easily influenced.) Trying to discover influential established groups in a social network and seeing if susceptible groups adopt the recommendations provided by them is considered the main challenge for our future research. Combining this with the kind of item recommendation which involves a sequence of ordered elements will present a novel and original path in Group Recommender Systems design.

Keywords: Influential groups · Group recommender systems · Social factors

1 Introduction

Accessing relevant information stored in the Web is not an easy task. For instance, if a group of people are looking for an enjoyable movie to watch, the searching and choosing processes could both take a long time. That is to say, the quantity of items of interest presented by a search engine and the heterogeneous preferences of the group members are a harmful combination. To address this problem, Group Recommender Systems (*GRSs*) have been implemented last years. A GRS must be able to identify items that the group of users will like so that their needs, explicit or not, are equally satisfied. It can be designed considering three different components: (i) the nature of the target group to recommend to, (ii) the kind of recommendation made (one item, an ordered set of elements, a bunch of items put together), and (iii) external factors that may be involved when formulating the recommendation techniques used to match group - items. Here we introduce an overview of the research motivations, the methodology

© Springer International Publishing AG 2017
J.M. Jose et al. (Eds.): ECIR 2017, LNCS 10193, pp. 735–743, 2017.
DOI: 10.1007/978-3-319-56608-5_73

proposal and expected contributions under the consideration that: (i) the target is a group of users who have a common goal, (ii) the recommendation is made up of an arranged sequence of elements, and (iii) the technique focuses primarily on enabling the target group to know the influential group's choice.

We are exposed to make decisions everyday; and generally, we have to choose one alternative from a wide range of possibilities. Moreover, external factors influence our decisions. As a matter of fact, to make a choice we rely on our friends' judgement and/or opinions, reviews and rates of others. The mentioned behaviour motivates our proposal and makes us think that the same social conduct may be observed when a group is making a decision. In other words, being aware of the options chosen by other groups and their experiences may introduce an appropriate effect and help the given group to make an accurate choice. Consequently, we present a framework where the social context of a specific group is analysed in order to find their corresponding influential group. Once the influencers are detected, their choice is shown to the target group as an alternative option that worked for other people and may be suitable for them as well.

Understanding social structures and mining the knowledge generated in social networks are a matter of interest in GRSs due to the nonexistent isolation of groups. Actually, a GRS is a social platform that might create links that relate a group to others. To the best of our knowledge, this work represents the first attempt to detect influential and susceptible groups in a GRS environment. Accordingly, our scientific contributions are:

- The definition of a social model that detects influential and susceptible groups of users in a social media platform.
- The introduction of how the model can be embedded in a group recommender system to suggest sets of ordered items.

The remainder of the paper is organized as follows. In Sect. 2, the details about the background in sequential elements recommendation, group recommenders and social factors are discussed. Section 3 presents the main idea of the research proposal and its motivations, along with the research questions and methodology. Finally, a discussion and issues about main challenges are explained in Sect. 4.

2   Context and Related Work

In our understanding, no published studies have been done before in implementing, as part of the recommendation/explanation, a system that presents a set of ordered elements already accepted or adopted by an *influential established group of people* to another group in a GRS. Nevertheless, this section presents the previous work on recommendation of sets of elements as a single suggestion, GRSs and social factors in the recommendation process.

2.1 Ordered Sequence of Elements

The state of the art in Recommender Systems is very broad. However, it usually addresses the analysis and improvement of approaches like Collaborative Filtering [1], Content-Based [2], Constraint-Based [3] and Hybrid Recommender Systems [4] considering individual items recommendation to single users. The recommendation of items is generally presented as a ranked list of individual objects and the user can choose one item or another because they are independent (*e.g.* a person to add as friend may be selected from the list of people presented by Facebook friends recommender engine). However, little work has been done when a single recommendation is composed by some ordered units, where their position depends on the user interests or other constraints. In this case, the target user has access to a set of ordered elements that represent the *adopted item*. For example, in [5] the system creates a playlist for the target user, but it is not static so if a new song is added in the system, the playlist is reorganized considering the target user model, and then the recommendation of the new arranged set of songs is made. In [6] another similar system, patented by Amazon Technologies, is detailed. The algorithm output presents three or more items that work well together, so they are recommended as a bundle (without a specific order), refining the idea of *better together*.

2.2 Group Recommender Systems

A GRS supports the recommendation process by using aggregation methods in order to model the preferences of a group of people. This is needed when there is an activity (domain) that can be done or enjoyed in groups [7]. To better understand the difference between designing a GRS and a single-user-oriented recommender, in [8], the authors study how individuals modify their TV viewing habits when they find themselves in group contexts. Accordingly, they present a study of preference aggregation functions to model the group's interests. Jameson and Smyth [9] details the important tasks when designing a Group Recommender System. In [10], Judith Masthoff presents an alternative that models the affective state of the members of the group by combining individual users' models and proposes a framework where an ordered sequence of interactive television programs can be recommended to the group of users. Both authors [9,10] mention some relevant examples of group recommenders and application domains in their research works. They also observe the value of considering influence among the members of the group and its impact in the design of the recommender system.

2.3 Social Factors in Recommender Systems

Social Recommender Systems or Recommenders for the Social Web model the user's preferences by using the information he or she and their friends have published in online social networks [11]. In [12] the authors propose a framework to merge behavioral theory and social recommender systems design. They make their proposal based on the understanding that homophily, tie strength, and trustworthiness leverage the recommendation acceptance (sociological view). The researchers in [13], also model the preferences of the user in a Social Recommender, but they take into account that some of the the the user's friends might have different interests. In their work they represent the diversity of tastes among the user's social connections to improve the accuracy of the recommendation. In [14], the authors propose an approach for Group Recommender Systems by merging Collaborative Filtering and a Genetic Algorithm that learns from known group ratings. The social factor included in the recommender is the preceding interaction among group members reflected in their past ratings. In [15], the authors study individual behaviours, personality and trust relationships among members of the group to make a movie recommendation for them. Other recent works in Group Recommender Systems have tested the way the recommendations are presented in the interface proving that showing members emotions about the item can influence the user adoption.

It is worth noting that social factors are inherent to environments where people interact with each other, whether in online platforms or in group activities. In fact, influence among the group members has been considered in GRSs literature, but there is a lack of research of social factors involved in an intragroup level; for example, influence between groups and its impact.

3 Research Proposal

In the previous section we made evident the effort that has been invested in studying human and social factors to improve Recommender Systems for single users or groups. However, we propose a new recommendation framework aimed at extracting knowledge behind the trust networks [16] to include information about influential groups' decisions (in the preferences model adaptation and in the user interface) in a system where the target is a group of people and the recommendation is a sequence of ordered elements. Questions to guide the research are:

– Can groups of people be influenced by other groups at the moment they are making a decision? If they can, how could this social factor be taken in account in a Group Recommender System?
– Should susceptible groups' preferences be modeled in a different way from influential groups preferences? If so, what preferences aggregation strategy is suitable for each?
– Is it important to let susceptible groups visualize the influential group members, their choice and the reasons why they made that decision? Does it help or manipulate them?

Those research questions are considered as the main components to focus on in the framework outlined in Fig. 1. The components are explained next.

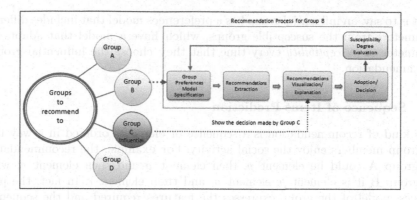

Fig. 1. After identifying the influential group/s and knowing its decision, the recommendation process for the group B will show them the choice of the influential group and why they chose that option. If group B decides to make the same choice, the model of group B preferences will adapt their level of *susceptibility*.

3.1 Influential Group Identification

The detection of the influential group among the target recommender groups needs a methodology which includes the recognition of the groups' members who are known because of their expertise background, good judgment, trust and extroversion and also the application of techniques to track the information diffusion through communities by mining the social network. Strategies to detect communities (hard and soft clustering algorithms) are needed as well as metrics such as betweenness centrality and/or closeness centrality. It would be the starting point to analyze the state of a user, *i.e* creator of relevant information or on the other hand, information consumer (susceptible kind), compared to their social network. Once this information is processed the extraction of main influential groups could be possible.

3.2 Group Preferences Model and Adaptation

What defines a group of people are their similarities. Accordingly, they recognize the social category they belong to, as well as the social categories they do not. A group has its social identity established when the members see themselves as a group. Self categorization theory says that when a person sets the differential parameters with other individuals, he determines his own identity. On the other hand, when he is aware that he has a membership in a group he maximizes perceptually his similarities with the other group members reducing in this way their individual differences [17,18]. This fact will be considered at the moment of formulating the preferences aggregation method. This means that the extraction of individual interests has a lower impact than the rate of items experienced before by the group as a whole, its current expectations, present goals and needs. The model should define the group identity in contrast to other groups [18].

That is to say, an influential group has a preferences model that includes different parameters from the susceptible groups, which have a model that adapts the parameter of *susceptibility* every time that they choose the influential group's recommendation.

3.3 Sequence of Items Prediction

This kind of recommendation is a sequence of elements ordered in a way that all group members enjoy the social activity. For example, the recommendation for group A could be element p, then element q and then element r; while for group B it is element p, element m and then element r. In fact, the preferences model of the group expresses the features required and the sequential integration of the elements recommended in a specific order should match those group needs. Generally, the approaches used depend on the domain of the recommendation: entertainment, content, e-commerce, service or social unit. In the scope of the present research, the recommender system is oriented to suggest leisure activities to enjoy in group. To prepare the ordered sequence of leisure events, the methodology has to incorporate a procedure based not only on the analysis of contextual information, but also on the preferences model and the estimated degree of acceptance of the influential group recommendation. Preliminary empirical experiments may be done by using *Meetup* datasets given the possibility of extracting groups of interest, their members, topics and planned events.

3.4 Recommendations Visualization

The goal of the GRS Interface is to support cooperative work in a way that the members of the group can be aware of each others needs but they still have to see themselves as a whole, that have a common aim. Its design will be centered in characterizing the group interests and offer the option to see why one group they know (the influential one) chose a specific recommendation. As a result, they can trust this is a good recommendation also for them. Figure 2 shows a sketch of the desired elements in the interface. The interface design technique will consider cooperative work and conflict resolution features to help the group, in a non intrusive way, to make a decision. The recommender engine is half of the system. The other half is the interaction that allows groups using the system to find the social activity that best matches their preferences.

4 Challenges and Discussion

In the previous section, the four main components of the social framework were briefly analyzed. Each component faces specific challenges and needs to be implemented by defining its own methodology and techniques. Some of them will have psychological and sociological information as input, others will need to be tested by applying more than one approach and algorithm combinations. In any case,

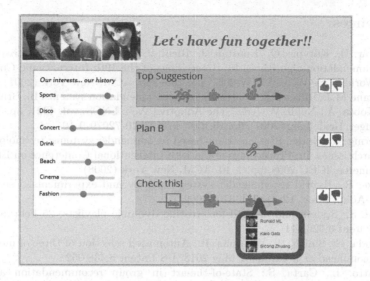

Fig. 2. Components needed in the interface

finding available datasets and/or applying user and field studies [19] to evaluate the results are essential activities. Understanding the nature of a group, their dynamics, how they are formed, size of influential groups and the ways they interact by using online social networks is the first issue to address. Social Web mining results need to be analyzed to figure out online users behavior facing real social interactions. Then, the selection of the approach to compute the recommendations has to be done after evaluating content-based and case-based techniques which are more likely to fit in the entertainment domain. The issues for further discussion are related to the performance of the recommender system. It is to say that the system should find content that the group may be highly interested in, under the condition that without the assistance of the group recommender it would be hard for them to find that item (or ordered set of items). Besides, users have a subjective perception about the recommender system outputs, and properties like diversity and novelty are expected [20]. Then, the framework proposed may enhance the recommender adding diversity when the influential group's adoption is shown. The level of diversity has to be measured in order to evaluate if the social factor actually presents diverse options.

Recommendation visualization, explanation representation and social factors management on collaborative interfaces are important features that require the design of at least two high fidelity prototypes to execute usability tests. Finally, privacy issues are not a fundamental part of the scope; however, levels of intrusion and user information protection need to be considered.

References

1. Sarwar, B., Karypis, G., Konstan, J., Riedl, J.: Item-based collaborative filtering recommendation algorithms. In: Proceedings of the 10th International Conference on World Wide Web, WWW 2001, pp. 285–295. ACM, New York (2001)
2. Pazzani, M.J., Billsus, D.: Content-based recommendation systems. In: Brusilovsky, P., Kobsa, A., Nejdl, W. (eds.) The Adaptive Web. LNCS, vol. 4321, pp. 325–341. Springer, Heidelberg (2007). doi:10.1007/978-3-540-72079-9_10
3. Felfernig, A., Burke, R.: Constraint-based recommender systems: technologies and research issues. In: Proceedings of the 10th International Conference on Electronic Commerce, ICEC 2008, pp. 1–10. ACM, New York (2008)
4. Burke, R.: Hybrid recommender systems: survey and experiments. User Model. User-Adap. Inter. **12**(4), 331–370 (2002)
5. Ward, S.: System and method for creating dynamic playlists, 25 February 2003. US Patent 6,526,411
6. Chanda, G., Smith, B., Whitman, R.: Automated selection of three of more items to recommend as a bundle, 7 May 2013. US Patent 8,438,052
7. Boratto, L., Carta, S.: State-of-the-art in group recommendation and new approaches for automatic identification of groups. In: Soro, A., Vargiu, E., Armano, G., Paddeu, G. (eds.) Information Retrieval and Mining in Distributed Environments. Studies in Computational Intelligence, vol. 324, pp. 1–20. Springer, Heidelberg (2011)
8. Chaney, A.J., Gartrell, M., Hofman, J.M., Guiver, J., Koenigstein, N., Kohli, P., Paquet, U.: A large-scale exploration of group viewing patterns. In: Proceedings of the ACM International Conference on Interactive Experiences for TV and Online Video, TVX 2014, pp. 31–38. ACM, New York (2014)
9. Jameson, A., Smyth, B.: Recommendation to groups. In: Brusilovsky, P., Kobsa, A., Nejdl, W. (eds.) The Adaptive Web. LNCS, vol. 4321, pp. 596–627. Springer, Heidelberg (2007). doi:10.1007/978-3-540-72079-9_20
10. Masthoff, J.: Group recommender systems: combining individual models. In: Ricci, F., Rokach, L., Shapira, B., Kantor, P.B. (eds.) Recommender Systems Handbook, pp. 677–702. Springer, Heidelberg (2011)
11. Guy, I., Carmel, D.: Social recommender systems. In: Proceedings of the 20th International Conference Companion on World Wide Web, WWW 2011, pp. 283–284. ACM, New York (2011)
12. Arazy, O., Kumar, N., Shapira, B.: A theory-driven design framework for social recommender systems. J. Assoc. Inf. Syst. **11**(9), 455–490 (2010)
13. Ma, H., Zhou, D., Liu, C., Lyu, M., King, I.: Recommender systems with social regularization. In: Proceedings of the Fourth ACM International Conference on Web Search and Data Mining, WSDM 2011, New York, USA, pp. 287–296 (2011)
14. Chen, Y., Cheng, L., Chuang, C.: A group recommendation system with consideration of interactions among group members. Expert Syst. Appl. **34**(3), 2082–2090 (2008)
15. Quijano-Sanchez, L., Recio-Garcia, J., Diaz-Agudo, B., Jimenez-Diaz, G.: Social factors in group recommender systems. ACM Trans. Intell. Syst. Technol. **4**, 8:1–8:30 (2013)
16. Jannach, D., Zanker, M., Felfernig, A., Friedrich, G.: Recommender Systems: An Introduction, 1st edn. Cambridge University Press, New York (2010)
17. Turner, J., Oakes, P., Haslam, S., McGarty, C.: Self and collective: cognition and social context. Pers. Soc. Psychol. Bull. **20**(5), 454–463 (1994)

18. Ren, Y., Harper, F.M., Drenner, S., Terveen, L., Kiesler, S., Riedl, J., Kraut, R.E.: Building member attachment in online communities: applying theories of group identity and interpersonal bonds. MIS Q. **36**, 841–864 (2012)
19. Konstan, J., Riedl, J.: Recommender systems: from algorithms to user experience. User Model. User-Adap. Inter. **22**(1–2), 101–123 (2012)
20. Ekstrand, M., Harper, F., Willemsen, M., Konstan, J.: User perception of differences in recommender algorithms. In: Proceedings of the 8th ACM Conference on Recommender Systems, RecSys 2014, pp. 161–168. ACM, New York (2014)

Pyndri: A Python Interface to the Indri Search Engine

Christophe Van Gysel[(✉)], Evangelos Kanoulas, and Maarten de Rijke

University of Amsterdam, Amsterdam, The Netherlands
{cvangysel,e.kanoulas,derijke}@uva.nl

Abstract. We introduce **pyndri**, a Python interface to the Indri search engine. Pyndri allows to access Indri indexes from Python at two levels: (1) dictionary and tokenized document collection, (2) evaluating queries on the index. We hope that with the release of pyndri, we will stimulate reproducible, open and fast-paced IR research.

1 Introduction

Research in Artificial Intelligence progresses at a rate proportional to the time it takes to implement an idea. Therefore, it is natural for researchers to prefer scripting languages (e.g., Python) over conventional programming languages (e.g., C++) as programs implemented using the latter are often up to three factors longer (in lines of code) and require twice as much time to implement [9]. Python, an interactive scripting language that emphasizes readability, has risen in popularity due to its wide range of scientific libraries (e.g., NumPy), built-in data structures and holistic language design [6].

There is still, however, a lack of an integrated Python library dedicated to Information Retrieval (IR) research. Researchers often implement their own procedures to parse common file formats, perform tokenization, token normalization that encompass the overall task of corpus indexing. Uysal and Gunal [11] show that text classification algorithms can perform significantly differently, depending on the level of preprocessing performed. Existing frameworks, such as NLTK [7], are primarily targeted at processing natural language as opposed to retrieving information and do not scale well. At the algorithm level, small implementation differences can have significant differences in retrieval performance due to floating point errors [4]. While this is unavoidable due to the fast-paced nature of research, at least for seminal algorithms and models, standardized implementations are needed.

2 Introducing Pyndri

Fortunately, the IR community has developed a series of indexing frameworks (e.g., Galago, Lucene, Terrier) that correctly implement a wide range of retrieval

---

https://github.com/cvangysel/pyndri.

© Springer International Publishing AG 2017
J.M. Jose et al. (Eds.): ECIR 2017, LNCS 10193, pp. 744–748, 2017.
DOI: 10.1007/978-3-319-56608-5_74

```
index = pyndri.Index('/opt/local/clueweb09')

for int_doc_id in range(index.document_base(),
                        index.maximum_document()):
    ext_doc_id, doc_tokens = index.document(int_doc_id)
```

Code snippet 1: Tokenized documents in the index can be iterated over. The ext_doc_id variable in the inner loop will equal the document identifier (e.g., clueweb09-en0039-05-00000), while the doc_tokens points to a tuple of integers that correspond to the document term identifiers.

```
index = pyndri.Index('/opt/local/clueweb09')
dictionary = pyndri.extract_dictionary(index)

_, int_doc_id = index.document_ids(
    ['clueweb09-en0039-05-00000'])
print([dictionary[token_id]
       for token_id in index.document(int_doc_id)[1]])
```

Code snippet 2: A specific document is retrieved by its external document identifier. The index dictionary can be queried as well. In the above example, a list of token strings corresponding to the document's contents will be printed to stdout.

models. The Indri search engine [10] supports complex queries involving *evidence combination* and the ability to specify a wide variety of constraints involving *proximity, syntax, extracted entities* and *document structure*. Furthermore, the framework has been efficiently implemented using C++ and was designed from the ground up to support *very large databases, optimized query execution* and *fast and concurrent indexing*. A large subset of the retrieval models [1,2,5,12,13] introduced over the course of history can be succinctly formulated as an Indri query. However, to do so in an automated manner, up until now researchers were required to resort to C++, Java or shell scripting. C++ and Java, while excellent for production-style systems, are slow and inflexible for the fast prototyping paradigm used in research. Shell scripting fits better in the research paradigm, but offers poor string processing functionality and can be error-prone. Besides, shell scripting is unsuited if one wants to evaluate a large number of complex queries or wishes to extract documents from the repository as this incurs overhead, causing avoidable slow execution. Existing Python libraries for indexing and searching, such as PyLucene, Whoosh or ElasticSearch, do not support the rich Indri language and functionality required for rapid prototyping.

We fill this gap by introducing pyndri, a lightweight interface to the Indri search engine. Pyndri offers read-only access at two levels in a given Indri index.

2.1 Low-Level Access to Document Repository

First of all, pyndri allows the retrieval of tokenized documents stored in the index repository. This allows researchers to avoid implementing their own format parsing as Indri supports all major formats used in IR, such as the trectext, trecweb, XML documents and Web ARChive (WARC) formats. Furthermore, standardized tokenization and normalization of texts is performed by Indri and

```
index = pyndri.Index('/opt/local/clueweb09')

for int_doc, score in index.query('obama family tree'):
    # Do stuff with the document.
```

Code snippet 3: Simple queries can be fired using a simple interface. Here we query the index for topic wt09-1 from the TREC 2009 Web Track using the Indri defaults (Query Language Model (QLM) with Dirichlet smoothing, $\mu = 2500$).

```
index = pyndri.Index('/opt/local/clueweb09')
query_env = pyndri.QueryEnvironment(
    index, rules=('method:dirichlet,mu:5000',))

results = query_env.query(
    '#weight( 0.70 obama 0.20 family 0.10 tree )',
    document_set=map(
        operator.itemgetter(1),
        index.document_ids([
            'clueweb09-en0003-55-31884',
            'clueweb09-en0006-21-20387',
            'clueweb09-enwp01-75-20596',
            'clueweb09-enwp00-64-03709',
            'clueweb09-en0005-76-03988'
        ])),
    results_requested=3,
    include_snippets=True)

for int_doc_id, score, snippet in results:
    # Do stuff with the document and snippet.
```

Code snippet 4: Advanced querying of topic wt09-1 with custom smoothing rules, using a weighted-QLM. Only a subset of documents is searched and we impose a limit on the size of the returned list. In addition to the document identifiers and their retrieval score, the function now returns snippets of the documents where the query terms match.

is no longer a burden to the researcher. Code snippet 1 shows how a researcher can easily access documents in the index. Lookup of internal document identifiers given their external name is provided by the Index.document_ids function.

The dictionary of the index (Code snippet 2) can be accessed from Python as well. Beyond bi-directional token-to-identifier translation, the dictionary contains corpus statistics such as term and document frequencies as well. The combination of index iteration and dictionary interfacing integrates conveniently with the Gensim[1] package, a collection of topic and latent semantic models such as LSI [3] and word2vec [8]. In particular for word2vec, this allows for the training of word embeddings on a corpus while avoiding the tokenization mismatch between the index and word2vec. In addition to tokenized documents, pyndri also supports retrieving various corpus statistics such as document length and corpus term frequency.

2.2   Querying Indri from Python

Secondly, pyndri allows the execution of Indri queries using the index. Code snippet 3 shows how one would query an index using a topic from the TREC

---

[1] https://radimrehurek.com/gensim.

2009 Web Track using the Indri default retrieval model. Beyond simple terms, the query() function fully supports the Indri Query Language.[2]

In addition, we can specify a subset of documents to query, the number of requested results and whether or not snippets should be returned. In Code snippet 4 we create a QueryEnvironment, with a set of custom smoothing rules. This allows the user to apply fine-grained smoothing settings (i.e., per-field granularity).

3 Conclusions

In this paper we introduced pyndri, a Python interface to the Indri search engine. Pyndri allows researchers to access tokenized documents from Indri using a convenient Python interface. By relying on Indri for tokenization and normalization, IR researchers are no longer burdened by this task. In addition, complex retrieval models can easily be implemented by constructing them in the Indri Query Language in Python and querying the index. This will make it easier for researchers to release their code, as Python is designed to be readable and cross-platform. We hope that with the release of pyndri, we will stimulate **reproducible**, **open** and **fast-paced** IR research. More information regarding the available API and installation instructions can be found on Github.[3]

Acknowledgements. This research was supported by the Google Faculty Research Award program and the Bloomberg Research Grant program. All content represents the opinion of the authors, which is not necessarily shared or endorsed by their respective employers and/or sponsors.

References

1. Balog, K., Azzopardi, L., de Rijke, M.: Formal models for expert finding in enterprise corpora. In: SIGIR, pp. 43–50. ACM (2006)
2. Bendersky, M., Metzler, D., Croft, W.B.: Learning concept importance using a weighted dependence model. In: WSDM, pp. 31–40. ACM (2010)
3. Deerwester, S., Dumais, S.T., Furnas, G.W., Landauer, T.K., Harshman, R.: Indexing by latent semantic analysis. J. Am. Soc. Inf. Sci. **41**(6), 391–407 (1990)
4. Goldberg, D.: What every computer scientist should know about floating-point arithmetic. ACM Comput. Surv. **23**(1), 5–48 (1991)
5. Guan, D., Zhang, S., Yang, H.: Utilizing query change for session search. In: SIGIR, pp. 453–462. ACM (2013)
6. Koepke, H.: Why python rocks for research (2010). https://www.stat.washington.edu/hoytak/_static/papers/why-python.pdf. Accessed 13 Oct 2016
7. Loper, E., Bird, S.: NLTK: the natural language toolkit. In: ACL Workshop on Effective Tools and Methodologies for Teaching NLP and CL, pp. 63–70. Association for Computational Linguistics (2002)

---

[2] http://lemurproject.org/lemur/IndriQueryLanguage.php.
[3] https://github.com/cvangysel/pyndri.

8. Mikolov, T., Corrado, G., Chen, K., Dean, J.: Efficient estimation of word representations in vector space. arXiv:1301.3781 (2013)
9. Prechelt, L.: An empirical comparison of seven programming languages. Computer **33**(10), 23–29 (2000)
10. Strohman, T., Metzler, D., Turtle, H., Croft, W.B.: Indri: a language model-based search engine for complex queries. In: ICIA (2005)
11. Uysal, A.K., Gunal, S.: The impact of preprocessing on text classification. Inf. Process. Manag. **50**(1), 104–112 (2014)
12. Van Gysel, C., Kanoulas, E., de Rijke, M.: Lexical query modeling in session search. In: ICTIR, pp. 69–72. ACM (2016)
13. Zhai, C., Lafferty, J.: A study of smoothing methods for language models applied to ad hoc information retrieval. In: SIGIR, pp. 334–342. ACM (2001)

"Hey, vitrivr!" – A Multimodal UI for Video Retrieval

Prateek Goel*, Ivan Giangreco$^{(\boxtimes)}$, Luca Rossetto, Claudiu Tănase, and Heiko Schuldt

Databases and Information Systems Research Group,
Department of Mathematics and Computer Science,
University of Basel, Basel, Switzerland
{ivan.giangreco,luca.rossetto,c.tanase,heiko.schuldt}@unibas.ch

Abstract. In this paper, we present a multimodal web-based user interface for the vitrivr system. vitrivr is a modern, open-source video retrieval system for searching in large collections of video using a great variety of query modes, including query-by-sketch, query-by-example and query-by-motion. With the multimodal user interface, prospective users benefit from being able to naturally interact with the vitrivr system by using spoken commands and also by applying multimodal commands which combine spoken instructions with manual pointing. While the main strength of the UI is the seamless combination of speech-based and sketch-based interaction for multimedia similarity search, the speech modality has shown to be very effective for retrieval on its own. In particular, it helps overcoming accessibility boundaries and offering retrieval functionality for users with disabilities. Finally, for a holistic natural experience with the vitrivr system, we have integrated a speech synthesis engine that returns spoken answers to the user.

1 Introduction

In recent years, the conventional approach of a user facing a mouse or keyboard for giving inputs to a system has increasingly been challenged by a –more natural– multimodal approach which considers the interaction on both the auditory and visual level. Early work in [1] presents an approach to combining speech and gesture input within a general setting. In the context of search applications, the authors of [2], for instance, show promising results when combining multiple modalities, e.g., the fusion of speech and gesture inputs in the context of a search application.

In this paper, we present a multimodal user interface for the vitrivr [3] video retrieval system which supports both manual pointing and voice commands (alone and in combination) to enhance the user experience. The vitrivr system [3] is a modern open-source video retrieval system which offers users a great variety of query paradigms, including query-by-sketch, query-by-example and

---

* Prateek Goel has been a Google Summer of Code '16 student with the vitrivr project.

© Springer International Publishing AG 2017
J.M. Jose et al. (Eds.): ECIR 2017, LNCS 10193, pp. 749–752, 2017.
DOI: 10.1007/978-3-319-56608-5_75

query-by-motion for searching in large collections of video. It is powered by the ADAM$_{pro}$ database and the Cineast retrieval engine.

The key advantages of the multimodal user interface include a gentle learning curve, an increased efficiency and expressiveness given by the voice interface, which seems to be helpful particularly for novice users, and a natural interaction with the system. Moreover, thanks to the accessibility of the UI, we see a strong use case for users with disabilities, as our approach does not require the explicit use of any pointing device.

2 Multimodal vitrivr UI

The primary goals of the vitrivr UI (see Fig. 1) are to scaffold a query and to present the retrieved results to the user. For the first task, in the vitrivr front-end, queries are specified using one or multiple canvases which can be used to either sketch a query or use an existing input image. The visual information can be further enriched by specifying motion, e.g., to denote the motion of an object within the scene. For presenting results to a user, the vitrivr UI displays a result list of similar shots sorted by relevance.

Speech Interaction: We have carefully hand-crafted an ontology along the lines of the goals of the UI to support in total over 50 actions (e.g., perform a search, choose a specific color, etc.) which can be executed by more than 250 predefined, alternative spoken commands. The matching of spoken text to a command is done either based on exact rule matching (which may contain optional words, however). Furthermore, to allow for a fuzzy matching, for all sentences which could not be exactly matched, we have added a matching strategy based on n-grams; with this, we execute a command although there is no exact matching if the matching score is above a certain threshold.

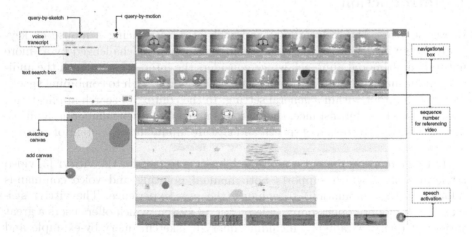

Fig. 1. Screenshot of the multimodal vitrivr UI. The explanations highlighted with a border are relevant for the speech-based UI.

The implemented speech commands allow to specify a query, e.g., by adding keywords to the textual search, by filling the canvas with a specific color, by adjusting the pen size etc. On the other hand, the system provides navigational commands to allow the user to navigate within the result list, e.g., move down the result list, play a specific result elements and hide/highlight certain results, for instance, based on the score. For navigation, after displaying the results to the user, a navigating box is displayed, marking the current video considered, with the most similar result shots within the video being displayed and numerated. With the enumeration, a user can quickly refer to a specific result list element, e.g., by saying "play video of shot number two".

We designed the speech commands to support a natural interaction with the vitrivr system. Follow-up commands, for instance, relate to the previous command, by keeping track of the previously executed command. Consider the action of increasing the pen size: Naturally, a user would first say to the system, e.g. "increase the pen size"; for further increasing the pen size, however, repeating the command would not seem natural. Hence, the UI tracks the executed commands to support follow-up commands, such as "even more" (to adjust the pen size) or "even further" (to move within the results).

Multimodal Interaction: The spoken commands of vitrivr can be enriched by using other modalities, i.e., combining manual pointing and voice commands. We use a simple model for such a multimodal interaction: A spoken query must be followed within a short time frame (e.g., within 5 s) by a pointing action for correctly executing and recognising a multimodal intent. A user can for instance say "play this video" and by that start a timer which expects a click –within the predefined time frame– on a specific result item, which is then played. Similarly, she can add results to relevance feedback by saying "add these videos to positive feedback" and click on multiple videos.

Conversational Feedback: To significantly enhance the natural interaction with the system, a speech synthesis engine in the user interface responds to the commands in a conversational way. For instance, the speech engine can confirm the understanding of a command ("Ok"), give an answer to a specific question as a result of a specific command ("There are hundred results."), or ask to repeat the command in a more specific way if it was not understood ("Did you mean...?"). The latter will be used, if the recognition of the spoken command fails, but the similarity of a command within the ontology and the spoken words is greater than a certain threshold when comparing the n-grams.

3 Implementation

The vitrivr UI is browser-based. The speech recognition is implemented using annyang[1] which works on top of the W3C Web Speech API [4] offering an API for speech recognition in modern browsers. It parses the recognised words to

---

[1] https://www.talater.com/annyang/.

commands using regular expressions. To allow for a fuzzy matching, we have built a matching strategy based on 3-grams on top, which produces a similarity score of the spoken command and a command in the ontology.

The code is available in the vitrivr open-source project.[2]

4   vitrivr in Action

vitrivr uses a large collection of free, creative commons web video. Users are able to search this collection in the context of a known-item search task with the multimodal UI, using a microphone and a mouse/pen input[3].

For this, users are able to browse through a list of present videos in the database, select a target sequence which they would like to re-find using the vitrivr system and apply the search paradigms offered: Starting, for instance, from a hand-drawn sketch by choosing via the speech-interface the brush tool and the color, and use query-by-sketching for retrieving the most similar video snippets in the database. They can, then, for example, choose a result from the result list which appears similar to the query (again using a voice commands) and use it for further searching the system. Finally, the user(s) can navigate using multimodal commands through the results and play video scenes to see if the query scene has been found.

Acknowledgments. This work was partly funded by the Swiss National Science Foundation (SNSF) in the context of the Chist-Era program IMOTION (contract no. 20CH21_151571), and the Google Summer of Code 2016 program.

References

1. Bolt, R.A.: "Put-that-there": voice and gesture at the graphics interface. In: Proceedings of International Conference on Computer Graphics and Interactive Techniques (SIGGRAPH), Seattle, USA, pp. 262–270. ACM (1980)
2. Heck, L., Hakkani-Tür, D., Chinthakunta, M., Tur, G., Iyer, R., Parthasacarthy, P., Stifelman, L., Shriberg, E., Fidler, A.: Multimodal conversational search and browse. In: Proceedings of Workshop on Speech, Language and Audio in Multimedia, Marseille, France. IEEE, August 2013
3. Rossetto, L., Giangreco, I., Tănase, C., Schuldt, H.: vitrivr: a flexible retrieval stack supporting multiple query modes for searching in multimedia collections. In: Proceedings of International Conference on Multimedia (ACM MM 2016), Amsterdam, Netherlands, pp. 1183–1186. ACM, November 2016
4. Shires, G., Wennborg, H.: Web Speech API Specification. W3C community group final report, W3C, October 2012. https://dvcs.w3.org/hg/speech-api/raw-file/tip/speechapi.html. Accessed 11 Jan 2017

---

[2] http://vitrivr-ui.vitrivr.org/tree/natural_language.

[3] A video demoing the system can be found on http://youtu.be/GCqxJ6FMlH0.

FairScholar: Balancing Relevance and Diversity for Scientific Paper Recommendation

Ankesh Anand[1]($\boxtimes$), Tanmoy Chakraborty[2], and Amitava Das[3]

[1] Indian Institute of Technology, Kharagpur, India
ankeshanand@iitkgp.ac.in
[2] University of Maryland, College Park, USA
tanchak@umiacs.umd.edu
[3] Indian Institute of Information Technology Sri City, Sri City, India
amitava.das@iiits.in

Abstract. In this paper, we present FairScholar, a novel scientific paper recommendation system that aims at balancing both *relevance* and *diversity* while searching for research papers in response to keyword queries. Our system performs a *vertex reinforced random-walk*, a time heterogeneous random-walk on the citation graph of papers in order to factor in diversity while serving recommendations. To incorporate semantically similar items in the search results, it uses a query expansion step that finds similar keywords using community detection. An online demo of our search engine is available at http://www.cnergres.iitkgp.ac.in/FairScholar/.

1 Introduction

Given a scientific query topic, the primary goal of any paper recommendation engine is to return a small number of highly relevant articles for that query topic. Most of these paper recommendation systems in the literature utilize PageRank style random-walk based mechanisms as the key criterion to rank the recommended articles. The existing paper recommendation systems primarily aim at suggesting prestigious and well-cited articles for a given query topic. Though it is necessary, we often stand in need of diversified papers for a certain query topic. For instance, diversity helps to improve the coverage of a topic in the recommendation process.

In this paper, we design FairScholar, a novel paper recommendation system, that aims to address this issue by recommending papers in a way that automatically balances prestige and diversity [1]. It is built on an well-established random-walk process, called vertex reinforced random-walk (VRRW) which systematically provides a balance between prestige and diversity in ranking vertices in a network. We experimented on a huge dataset of Computer Science papers. Our empirical results and user study show that FairScholar outperforms other baselines such as Google Scholar, RefSeer [2] in terms of relevance, diversity and user satisfaction.

© Springer International Publishing AG 2017
J.M. Jose et al. (Eds.): ECIR 2017, LNCS 10193, pp. 753–757, 2017.
DOI: 10.1007/978-3-319-56608-5_76

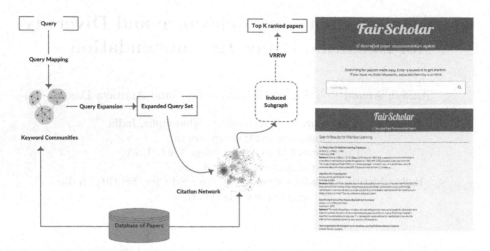

Fig. 1. Left: a schematic diagram of FairScholar's architecture. Top-right: the home page. Bottom-right: the recommendation page of FairScholar.

2   System Description

FairScholar is a web application (see Fig. 1). The homepage provides a search box where the user can enter keyword based queries. Our application then tries to recommend a set of relevant and diverse articles corresponding to the keyword.

The architecture of FairScholar primarily consists of three components: (a) Network Construction, (b) Query Expansion, and (c) Document Retrieval.

The **network construction** component is responsible for building two networks: the citation network and the keyword network. These networks are built on top of a dataset of more than 2 million Computer Science papers crawled from Microsoft Academic Search [3], and subsequently indexed in a MySQL database. Each node in the citation network is a paper and there is a directed edge from paper A to paper B if A cites B. The *Keyword network* is built using the keyword metadata associated with articles in the dataset. We further enrich the keyword set by extracting more keywords from each article using KEA, a keyword extraction tool from text documents [4]. This set of keywords form the nodes in the Keyword network, and two keywords are connected if there is at least one article that contains both the keywords.

The **query expansion** component is responsible for augmenting semantically similar keywords to the search query. A preliminary task in query expansion is to cluster similar keywords from the topological structure of the keyword-keyword network. We utilize the Louvain algorithm [5] to find clusters in the keyword-keyword network. Now given an input query, the system first identifies the community membership of this query, and then all the constituent keywords present in that community are fetched for the next step of the framework.

After expanding the given input query, we obtain an expanded query set Q containing a set of similar keywords. We then collect the set of all articles S

that contain at least one keyword from Q, and construct an induced subgraph $G(V, E)$ of S from the citation network. On this induced subgraph, we run vertex reinforced random-walk (VRRW) [6]. The underlying formulation of VRRW is motivated by this idea – *the transition probability to one vertex from others is reinforced by the number of previous visits to that vertex.*

Vertex Reinforced Random-walk: Let $p_T(u, v)$ be the transition probability from vertex u to vertex v at time T. Then the time-variant random-walk processes can be defined in which $p_T(u, v)$ satisfies the following equation:

$$p_T(u, v) = (1 - \lambda) \cdot p^*(v) + \lambda \cdot \frac{p_0(u, v) \cdot N_T(v)}{D_T(u)} \tag{1}$$

where $D_T(u) = \sum_{v \in V} p_0(u, v) N_T(v)$. Here, $p^*(v)$ is a distribution which represents the prior preference of visiting vertex v. $p_0(u, v)$ is the "organic" transition probability prior to any reinforcement: $p_0(u, v) = \alpha \frac{1}{deg(u)}$, if $u \neq v$; $1 - \alpha$, otherwise, where the parameter α controls the strength of self-links; $deg(u)$ is the out-degree of u; $N_T(v)$ is the number of times the walker has visited v till time T; λ is the damping factor (set to 0.75).

One way to simplify the computation of Eq. 1 is to use $p_t(v)$ directly to approximate $E[N_t(v)]$. Equation 1 is then be simplified as

$$p_T(u, v) = (1 - \lambda) \cdot p^*(v) + \lambda \cdot \frac{p_0(u, v) \cdot p_T(v)}{\sum_{v \in V} p_0(u, v) p_T(v)} \tag{2}$$

When VRRW converges, it will produce a rank-list of vertices (papers). In the **document retrieval** phase, top K papers are returned as a set of final recommendations.

3   Experiments

For evaluating our system, we manually collected a set of survey papers from the Computer Science dataset mentioned earlier. For each survey paper, we extract keywords and use them as query sets. We assume that the references in each survey paper are diverse and serve as the gold-standard for the corresponding query set in our experiment. We use PageRank, Google Scholar and RefSeer[1] [2] as baseline systems. We evaluate the quality of the paper recommendations suggested by the competing baseline methods with two relevance measures – (i) Recall ($R@K$) and (ii) Mean Average Precision ($MAP@K$); and the two graph-based diversity measures – (i) l-hop graph density ($den_l@K$): $den_l(S) = \frac{\sum_{u, v \in S, u \neq v} d_l(u, v)}{|S| \times (|S| - 1)}$, where S is the set of top K results, $d_l(u, v) = 1$ when v is reachable from u within l steps, i.e., $d(u, v) \leq l$; and 0 otherwise; (ii) l-expansion ratio ($\sigma_l(S)$): $\sigma_l(S) = \bigcup_{s \in S} N_l(s)$ where S is the set of top K results and $N_l(s)$ is all the neighbors up to l hops for vertex s. Figure 2 show that FairScholar outperforms other baselines in terms of both relevance and diversity measures with different values of K.

---

[1] http://refseer.ist.psu.edu/.

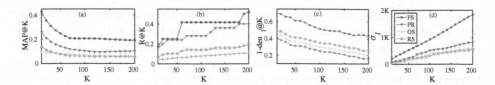

Fig. 2. Comparison of the competing models based on (a)–(b) relevance and (c)–(d) diversity measures (FS: FairScholar, PR: PageRank, GS: Google Scholar, RS: RefSeer).

Table 1. User study: average score and preference (in %) of the competing systems.

		FairScholar	PageRank	Google Scholar	RefSeer
Relevance	Avg. score	**4.21**	1.28	4.00	3.60
	Avg. preference	**48.40**	5.76	31.04	14.80
Diversity	Avg. score	**4.64**	1.48	2.24	1.20
	Avg. preference	**75.56**	10.44	9.80	4.20

User Study: In order to understand the acceptance of FairScholar with the end users, we conducted a focus group user study with 25 members in our department. To generate a query pool, they were asked to recommend different queries in computer science domain, following which total 100 queries were generated (e.g., supervised learning, outlier detection, community analysis etc.). Each member was shown top 10 retrieved papers (as a list of titles and abstracts) of 4 competing systems separately in a single page (note that the systems remained anonymous). For each query, members were asked to score each system (1–5 scale, fractional score was not allowed) separately in terms of relevance (5: highly relevant, 1: not-relevant) and diversity (5: highly-diverse, 1: not-diverse), and to judge which system is preferred most (binary decision). Table 1 shows that users prefer FairScholar most for both relevance and diversity measures.

4   Conclusion

In this paper, we introduced FairScholar which balances diversity and relevance in search results using vertex reinforced random-walk and semantic query expansion. Using empirical experiments and user opinion, we verified and established the efficacy of our system. In future, we would like to expand our system to include other scientific domains.

References

1. Chakraborty, T., Modani, N., Narayanam, R., Nagar, S.: Discern: a diversified citation recommendation system for scientific queries. In: ICDE, pp. 555–566 (2015)
2. Huang, W., Wu, Z., Mitra, P., Giles, C.L.: RefSeer: a citation recommendation system. In: JCDL, London, United Kingdom, pp. 371–374 (2014)

3. Chakrabort, T., Sikdar, S., Tammana, V., Ganguly, N., Mukherjee, A.: Computer science fields as ground-truth communities: their impact, rise and fall. In: ASONAM 2013, Niagara, Ontario, Canada, pp. 426–433 (2013)
4. Witten, I.H., Paynter, G.W., Frank, E., Gutwin, C., Nevill-Manning, C.G.: Kea: practical automatic keyphrase extraction. In: JCDL. ACM, pp. 254–255 (1999)
5. Blondel, V.D., Guillaume, J.L., Lambiotte, R., Lefebvre, E.: Fast unfolding of communities in large networks. JSTAT **2008**(10), P10008 (2008)
6. Mei, Q., Guo, J., Radev, D.: Divrank: the interplay of prestige and diversity in information networks. In: SIGKDD, New York, USA. ACM, pp. 1009–1018 (2010)

The SENSEI Overview of Newspaper Readers' Comments

Adam Funk[(⊠)], Ahmet Aker, Emma Barker, Monica Lestari Paramita,
Mark Hepple, and Robert Gaizauskas

Department of Computer Science, University of Sheffield, Sheffield S1 4DP, UK
a.funk@sheffield.ac.uk

Abstract. Automatic summarization of reader comments in on-line
news is a challenging but clearly useful task. Work to date has produced
extractive summaries using well-known techniques from other areas of
NLP. But do users really want these, and do they support users in realistic tasks? We specify an alternative summary type for reader comments,
based on the notions of issues and viewpoints, and demonstrate our user
interface to present it. An evaluation to assess how well summarization
systems support users in time-limited tasks (identifying issues and characterizing opinions) gives good results for this prototype.

Keywords: User interface · Summarization · Newspaper · Social media

1 Introduction

Many current news websites feature comments, so that readers can engage in
conversations with each other, discussing aspects of a news story and their reactions. But articles can attract hundreds or even thousands of reader comments
within a relatively short time, so users face the problem of making sense of a
sprawling, multi-threaded conversation.

Clearly, it would be useful to have a summary or overview of the conversation
with the option of drilling down for more details. Generating such overviews manually for every news story is obviously impractical, so automatic summarization
is a natural candidate. Several authors have already proposed broadly similar
systems for summarizing reader comments (e.g. [1–3]). Such systems are cluster comments by topic, rank comments within clusters, and finally produce an
extractive summary from selected highly ranked comments. They assume that
topically grouped and ranked comments and extractive summaries are useful to
end users; however, we find no attempt to investigate what the end users really
want in a comment overview. Furthermore, the evaluations generally proposed—
although some have been called user studies—are not task-based evaluations that
demonstrate how well systems are meeting user needs.

This research is supported by the European Union's Seventh Framework Program
project SENSEI (FP7-610916).

2   The Use Case

To help make sense of the sprawling conversations, we apply clustering and summarization techniques and implement a summarization view user interface developed for our use case with *The Guardian* newspaper[1]. This use case aims to give both the general public and news professionals an understanding of the discussion based on the idea of the *town hall meeting summary*: a reporter covering a meeting would summarize it by addressing questions such as the following: what issues were how many people talking about? how did they feel about them? what did they agree or disagree about? how many shared similar views? [4].

The summary parts of the user interface also function as indexes to the underlying comments: they contain links to the relevant comments, with certain parts highlighted, in order to allow the user to drill down to see comments in their original discussion contexts.

3   The SENSEI Prototype

The SENSEI repository [5] is loaded with crawled data, including the content and metadata (username, timestamp, reply-to structure, etc.) of the first 100 comments under the article. (We limited the comment set size for consistency in the evaluation because different articles attract different numbers of comments, but this is not inherent in our system.)

Clustering is carried out using the Markov Clustering Algorithm, which does not require a prespecified number of clusters and allows us to employ a bespoke cluster-cluster similarity measure [6]. The latter is computed using a weighted linear combination of cluster and article features. We experimented with various features, including cosine similarity, word2vec similarity, named entity overlap, reply-to relationship and presence in the same thread. Training data was generated both automatically and using a small set of reference clusters [7]. We extractively summarize the comments from each cluster by ranking sentences within cluster by similarity to cluster centroid and then generate summaries by selecting top-ranked comments from clusters ordered by cluster size. Labels are generated for each cluster by ranking candidate labels (NPs) extracted from the cluster according to a measure trained using sentences in human-authored reference summaries that summarize the cluster [8]. The results are stored as meta-documents in the repository.

The page is generated in PHP beginning with a *summary master* document, which contains cross-references to others: the summary's constituents (clusters, labels, extracts), the article, and the comments. The page contains three columns: a pie chart representing the proportion of comments in each topic; balloons with a selected extract for each topic (colour-coded to match the pie chart); and a column initialized with a brief set of instructions. The pie chart is generated from the clusters and labels with the NVD3 library and has the active features described below. Hovering over a pie wedge causes a pop-up to

---
[1] http://www.theguardian.com/.

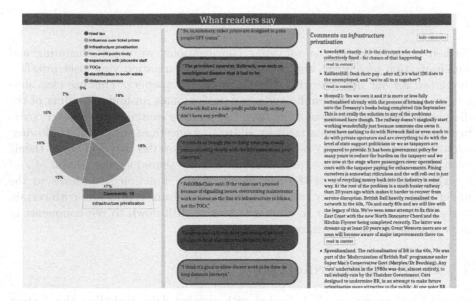

Fig. 1. Screenshot: clicking the pie chart

appear with the cluster label and number of comments it contains. Figure 1 also shows that clicking a wedge emphasizes (in the middle column) the extract from that cluster and shows a list of all the comments from that cluster in the right column. (Hidden content is displayed and then altered with JavaScript and CSS so that interaction takes place quickly in the user's browser.) Clicking an extract emphasizes it and brings up the list of that cluster's comments in the right column, as shown in Fig. 2, with the additional feature of highlighting in purple the comment from which the extract was taken and scrolling the right column so that comment is visible. Every comment also has a *read in context* button, which brings up a pop-up window with the complete set of comments in thread order, and scrolls the window so the selected comment is visible.

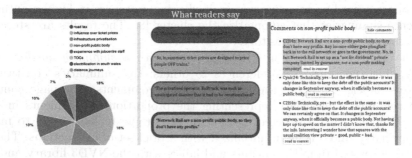

Fig. 2. Cropped screenshot: clicking an extract

The prototype[2] and a video[3] (which was used as training material in the evaluation) are available on-line.

4   Conclusion

We carried out a task-based evaluation with a Latin square design to compare *The Guardian*'s comment tree with our UI. We evaluated how well the users carried out two tasks—*identifying issues* and *characterizing opinion* – as well as their opinions of the two systems on a questionnaire. As reported in detail elsewhere [9,10], our results were good.

Further research will focus on providing more useful and coherent clusters of comments and developing a better UI in response to the evaluation. We hypothesize that with such advances people will do better overall on both tasks and be able to answer questions in less time, since good clusters represent issues and gather together related comments about them.

References

1. Khabiri, E., Caverlee, J., Hsu, C.F.: Summarizing user-contributed comments. In: ICWSM, Barcelona, pp. 534–537 (2011)
2. Ma, Z., Sun, A., Yuan, Q., Cong, G.: Topic-driven reader comments summarization. In: Proceedings of the 21st ACM international Conference on Information and Knowledge Management, CIKM 2012, pp. 265–274 (2012)
3. Llewellyn, C., Grover, C., Oberlander, J.: Summarizing newspaper comments. In: ICWSM, Ann Arbor, June 2014
4. Barker, E., Gaizauskas, R.: Summarizing multi-party argumentative conversations in reader comment on news. In: ACL 3rd Workshop on Argument Mining (2016)
5. Funk, A., Gaizauskas, R., Favre, B.: A document repository for social media and speech conversations. In: LREC (2016)
6. Aker, A., Kurtic, E., Balamurali, A.R., Paramita, M., Barker, E., Hepple, M., Gaizauskas, R.: A graph-based approach to topic clustering for online comments to news. In: Ferro, N., Crestani, F., Moens, M.-F., Mothe, J., Silvestri, F., Nunzio, G.M., Hauff, C., Silvello, G. (eds.) ECIR 2016. LNCS, vol. 9626, pp. 15–29. Springer, Heidelberg (2016). doi:10.1007/978-3-319-30671-1 2
7. Barker, E., Paramita, M., Aker, A., Kurtic, E., Hepple, M., Gaizauskas, R.: The sensei annotated corpus: human summaries of reader comment conversations in on-line news. In: SIGDIAL (2016)
8. Aker, A., Paramita, M., Kurtic, E., Funk, A., Barker, E., Hepple, M., Gaizauskas, R.: Automatic label generation for news comment clusters. In: International Natural Language Generation Conference (INLG), Edinburgh, September 2016
9. Barker, E., Funk, A., Paramita, M., Kurtic, E., Aker, A., Foster, J., Hepple, M., Gaizauskas, R.: What's the issue here? Task-based evaluation of reader comment summarization systems. In: LREC, May 2016
10. Danieli, M., Barker, E., Bechet, F., Caramia, C., Celli, F., Favre, B., Gaizauskas, R., Molinari, L., Palumbo, A., Paramita, M., Riccardi, G., Lanzolla, V.: Final report of prototype evaluation. Deliverable D1.4, SENSEI Consortium (2016)

---

[2] http://sensei.rcweb.dcs.shef.ac.uk/y3C/.

[3] http://sensei.group.shef.ac.uk/sensei/demos.html.

Temporal Semantic Analysis
of Conference Proceedings

Fedelucio Narducci[✉], Pierpaolo Basile, Pasquale Lops,
Marco de Gemmis, and Giovanni Semeraro

Department of Computer Science, University of Bari Aldo Moro, Bari, Italy
{fedelucio.narducci,pierpaolo.basile,pasquale.lops,marco.degemmis,
giovanni.semeraro}@uniba.it

Abstract. T-RecS is a system which implements several computational linguistic techniques for analyzing word usage variations over time periods in a document collection. We analyzed ACM RecSys conference proceedings from the first edition held in 2007, to the one held in 2015. The idea is to identify linguistic phenomena that reflect some interesting variations for the research community, such as a topic shift, or how the correlation between two terms changed over the time, or how the similarity between two authors evolved over time. T-RecS is a web application accessible via http://193.204.187.192/recsys/.

1 Introduction and Related Researches

There are several works in the literature which propose a computational history of science by analyzing corpora of scholarly papers. Generally, these works exploit techniques coming from the natural language processing area and are particularly focused on topic modeling [3,4]. They analyze the trend of research topics over time and study their impact on the evolution of science. In this work we provide a set of techniques with a twofold purpose: first, to perform an author-centered analysis of the topics characterizing the conference, similarly to the work proposed in [1]; second, to examine how the semantics of the topics has evolved over time. The correlation between authors and topics is computed by exploiting two different techniques inspired respectively to the Google Books N-gram Viewer[1] and to the Explicit Semantic Analysis [2]. The topic semantics is studied by means of a framework named Temporal Random Indexing [5] able to outline the evolution of the usage of a particular term over time. Thanks to these techniques we are able to answer to questions like: *What are the authors who studied the influence of emotions in recommender systems?, What was the most used recommendation paradigm in 2007? When did matrix factorization started to be used in collaborative-filtering systems? How did techniques to cope with the cold start problem evolve over time?*

T-RecS can be used for the analysis of any collection of documents and proceedings.

---

[1] https://books.google.com/ngrams.

© Springer International Publishing AG 2017
J.M. Jose et al. (Eds.): ECIR 2017, LNCS 10193, pp. 762–765, 2017.
DOI: 10.1007/978-3-319-56608-5_78

2   Techniques and Methods

In this section we describe the techniques implemented to analyze the RecSys conference proceedings. We just provide an overview of each technique, since more specific technical details are beyond the scope of this paper.

N-gram Analyzer. The N-gram analyzer implemented in this work is inspired to the Google Books N-gram Viewer search engine. Google Books N-gram [6] is a web application that displays through a chart the counting of 1-g, 2-g, 3-g, 4-g, and 5-g from the Google corpus of over 15 million books. In this work we adopted a similar approach and we counted n-grams ($n = 1, \ldots, 5$) in the RecSys proceedings corpus in the time interval from 2007 to 2015. We counted N-grams grouped by author and year. In this way it is possible to show the percentage of N-grams in the corpus for each year, and to split this value by author.

Temporal Random Indexing. In order to analyze how the semantics of a term changes over time, we integrated temporal information in a Distributional Semantic Model (DSM) approach, which consists in representing words as points in a geometric space (*WordSpace*), where two words are similar if represented close in the space. Specifically, given a document collection D annotated with meta-data containing information about the year in which the document was written, we can split the collection in different time periods $T_1, T_2, \ldots, T_p$ we want to analyze. The semantic vector for a word in a given time period is the result of its co-occurrences with other words in the *same* time interval. The ability to compare word vectors in different time periods allows two types of analyses: (1) we can compare the word vector for the word w_i in two different time periods in order to understand if the word changes its semantic over time; (2) we can compute the cosine similarity between the vector representations of word pairs in order to compute their relatedness over time. In our implementation, we adopt the open source version of Temporal RI available on github[2].

Explicit Semantic Analysis. In order to define a correlation between authors and topics of the RecSys conference, we inspired to the Explicit Semantic Analysis (ESA) technique [2]. For this work we build a matrix for each edition of the conference proceedings, in which rows are the terms occurring in the proceedings and columns are the authors, and a correlation score is stored in their intersection. Given a term is thus possible to show the most related authors by extracting the row vector of that term from the matrix. It is also possible to extract the most related authors for a text fragment composed of different terms. In that case the system computes the centroid vector of the row vectors of the terms composing the text fragment.

3   Demonstration Summary

The GUI of T-RecS consists of the following components: a text box to formulate a query, an *analysis* menu that allows the user to select the analysis she desires

---

[2] https://github.com/pippokill/tri.

to perform, a list of retrieved authors that allows to draw a chart tailored on a specific person[3]. The first interaction step is the selection of the analysis to be performed by the corresponding menu. Next, the user selects the time interval for the analysis, and finally she sets the following analysis-dependent parameters:

Topic-Author Correlation. The Topic-Author correlation is powered by the ESA Analyzer. The *author sorting* parameter has two options: total (the ranking considers years with 0 correlation score), and relative (the ranking does not consider years with 0 correlation score).

The *authors list* is shown on the right side of the GUI and contains authors relevant with respect to the query. The user can then select one or more authors and the system shows their contribution on that topic (i.e., the query) for each year.

N-gram Analyzer. The user indicates the n-grams to be searched, separated by commas. The number of n-grams to be retrieved are limited by the *Result Size* parameter. The user can select one or more authors from the *authors list* and the corresponding chart is drawn. For each author, the number of times and the number of papers the input n-grams occur in are reported in brackets. The *with smoothing of* parameter allows the smoothing of the chart by choosing a value from the corresponding menu.

Topic Semantics. The only input required to the user for this analysis is a list of topics (query) whose semantics will be analyzed. Topics must be separated by comma. For each topic (e.g., LDA), the chart shows how its semantics has changed over the years.

TRI Similarity. The input is a pair of topics (query) whose semantic correlation will be analyzed. Topics must be separated by comma (e.g., matrix factorization, collaborative filtering). The query may consist of two or more pairs, sequentially written and separated by commas.

Author Similarities. The user indicates the author name (e.g., Jill Freyne) as query and the system retrieves the most similar authors for each year. This analysis requires to set the neighborhood size, namely the number of similar authors to retrieve, and a flag that allows to exclude co-authors from the computation.

4   Conclusions and Future Applications

In this work we proposed T-RecS, a system which exploits several computational linguistic techniques for analyzing the corpus of the ACM RecSys conference proceedings. The main outcome of the performed analyses is that the implemented techniques are able to discover knowledge generally hidden and not easy to be explicited. Furthermore, the implemented tool is a useful support

---

[3] A video demo is available at https://www.dropbox.com/s/3pyy2ur3lmdkci0/TRecS. mp4?dl=0.

to validate assumptions on the evolution of a topic, the application of a solution for a specific problem, and so on. We suppose that the proposed techniques are exploitable in different tasks and in this section we outline some of possible future applications. The topic-author correlation analysis, for example, might be very useful for supporting the reviewer or meta-reviewer assignment process during the conference management. Indeed, the system is able to retrieve the authors more confident with a specific topic in a given time interval. This tasks can take advantage of the author-similarity function as well. For example, in the reviewer assignment scenario, researchers most similar to the authors of a submission can be selected. Another task this analysis can support is the selection of members for organizing committees (e.g., for workshops), tutorial presenters, invited speakers. The temporal analyses performed by the N-gram analyzer and Temporal Random Indexing framework might help to understand how the topics in the recommender system research area are evolving over time, by identifying existing trends and future research directions. The analysis of the topics that are gaining more attention are very useful in the organization of challenges or in the definition of special tracks. Finally, these analyses, applied to different corpora, can effectively support literature reviews.

References

1. Anderson, A., McFarland, D., Jurafsky, D.: Towards a computational history of the ACL: 1980–2008. In: Proceedings of the ACL-2012 Special Workshop on Rediscovering 50 Years of Discoveries, pp. 13–21. Association for Computational Linguistics (2012)
2. Gabrilovich, E., Markovitch, S.: Wikipedia-based semantic interpretation for natural language processing. J. Artif. Intell. Res. **34**, 443–498 (2009)
3. Griffiths, T.L., Steyvers, M.: Finding scientific topics. Proc. Natl. Acad. Sci. **101**(suppl 1), 5228–5235 (2004)
4. Hall, D., Jurafsky, D., Manning, C.D.: Studying the history of ideas using topic models. In: Proceedings of the Conference on Empirical Methods in Natural Language Processing, pp. 363–371. Association for Computational Linguistics (2008)
5. Jurgens, D., Stevens, K.: Event detection in blogs using temporal random indexing. In: Proceedings of the Workshop on Events in Emerging Text Types, pp. 9–16. Association for Computational Linguistics (2009)
6. Michel, J.-B., Shen, Y.K., Aiden, A.P., Veres, A., Gray, M.K., The Google Book Team, Pickett, J.P., Hoiberg, D., Clancy, D., Norvig, P., Orwant, J., Pinker, S., Nowak, M.A., Aiden, E.L.: Quantitative analysis of culture using millions of digitized books. Science **331**(6014), 176–182 (2011)

QweetFinder: Real-Time Finding and Filtering of Question Tweets

Ameer Albahem[1]([✉]), Maram Hasanain[2], Marwan Torki[2], and Tamer Elsayed[2]

[1] RMIT University, Melbourne, Australia
ameer.albahem@rmit.edu.au
[2] Department of Computer Science and Engineering, Qatar University, Doha, Qatar
{maram.hasanain,mtorki,telsayed}@qu.edu.qa

Abstract. Users continuously ask questions and seek answers in social media platforms such as Twitter. In this demo, we present Qweet-Finder, a Web-based search engine that facilitates finding question tweets (Qweets) in Twitter. QweetFinder listens to Twitter live stream and continuously identifies and indexes tweets that are answer-seeking. Qweet-Finder also allows users to save queries of long-term interest and pushes real-time qweet matches of saved queries to them via e-mail.

1 Introduction

Users turn to social media platforms, such as Twitter, to explore developments of topics and events and to seek opinions of others on matters of interest. Consider a user that is following-up on a running topic like "US election debate". She might be interested in answers of several on-topic questions including subjective ones, e.g., "Who will win the debate?", or more factual ones, e.g., "When is the next debate?". In many cases, users believe that the Web might not have satisfactory answers to some "real-time" questions, however, tweeters might be able to answer them more effectively [4]. Furthermore, the user might not even know what the right questions to ask are, due to lack of full knowledge about the topic. In this work, we propose **QweetFinder**, a system that helps a user find questions on Twitter on a topic of interest. The task of suggesting or retrieving questions given a query is not new, yet the majority of existing systems were designed for community question answering (CQA) platforms [5]; up to our knowledge, only few systems were developed for query-oriented conversation retrieval directly from Twitter (e.g., [2]). Differently from those studies, our system is multilingual as it is currently designed to retrieve Arabic and English questions. In addition, we present an architecture to build such systems using mature open-source technologies.

A large body of literature has showed that users of social media platforms are regularly seeking information from others by posting questions [7]. Several studies focused on understanding the nature and types of such questions or on developing systems to answer questions posted on Twitter. We present an alternative view on how users with questions can interact with Twitter. We

© Springer International Publishing AG 2017
J.M. Jose et al. (Eds.): ECIR 2017, LNCS 10193, pp. 766–769, 2017.
DOI: 10.1007/978-3-319-56608-5_79

propose a ***real-time*** system that suggests questions to users that are relevant to their topics of interest. Studies on CQA suggest that, in many cases, users attempt to find existing questions and answers on topics of interest before posting a new question [5]; we believe Twitter users can benefit from such service as well. The proposed system also enables the user to explore questions of different perspectives she might not have thought of initially. Moreover, a recent study found that, in many cases, users are reluctant to post some kinds of questions on their own social network to avoid disturbing their followers [4]. Our system offers the user the opportunity to anonymously explore existing questions (and their replies) without explicitly asking them; we believe such flexibility will encourage more users to turn to Twitter for answers on questions that are best asked there.

Use cases of QweetFinder are not only limited to those of interest to normal users. For example, QweetFinder can potentially be used by organizations for market research and collection of unsolicited user feedback on a brand or product. In a leading study, Jansen et al. [3] found that out of 2.7K tweets about different brands, 11% were information-seeking tweets. A company can benefit from such user questions for quality management and product improvement.

2   System Description

QweetFinder is a real-time multilingual query-oriented question retrieval and filtering system[1]. It retrieves answer-seeking question tweets (qweets). Qweet-Finder consists of two layers: the back-end engine and the Web-based front-end (i.e., user interface).

2.1   System Back-End Architecture

As shown in Fig. 1, the back-end layer consists of the following components:

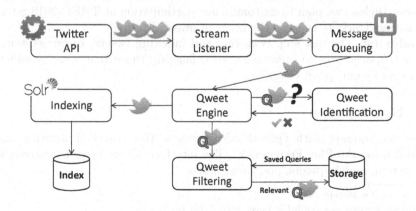

Fig. 1. QweetFinder back-end architecture

---

[1] A prototype of QweetFinder can be found at www.qweetfinder.com.

Tweet Streaming: Twitter provides a streaming API to directly listen to tweets as they are posted. We used Tweepy[2], a Twitter open source streaming library, to access the streaming API. In particular, we subscribed to all tweets that contain Arabic or English question marks, Arabic question phrases or English Wh-question words[3]. In addition, we limit the languages of tracked tweets to only Arabic and English.

Twitter requires streaming clients to process tweets as soon as they arrive and penalizes clients that fall behind. Since we need to perform qweet identification, which adds a time latency on processing the streamed tweets, we used a Message Queue component to tackle the expected latency and store the streamed tweets temporarily until they are consumed by the Tweet Processing Pipeline. We investigated different options such as Apache ActiveMQ and Amazon Kiness, Apache Spark and RabbitMQ. We found RabbitMQ[4] a good option in terms of scalability and learning curve.

Tweet Processing Pipeline: In this component, we developed a consumer that reads tweets from RabbitMQ and processes them through three steps:

1. **Qweet identification**: Not all questions in tweets are answer-seeking [1]. Therefore, we developed Arabic and English qweet classifiers to filter out qweets from a stream of tweets. Arabic qweet filtering was performed using SVM question classifier that leverages groups of features: lexical, structural, question-specific, tweet-specific, and (in)formality aspects of the tweets. The classifier was then trained on a manually annotated tweets collected through crowdsourcing—further details can be found in [1]. We classify English tweets using a Random Forest classifier using n-grams as features.
2. **Qweet Filtering:** Identified qweets are pushed to a filtering service (REST API) that filters them against a list of queries submitted by users. These qweets are then delivered to users via email. Qweet filtering can be done via different filtering mechanisms. In this demo, we used Luwak filtering engine[5]. Nevertheless, we plan to customize our participation at TREC-2016 Summarization Track [6] to suite the needs of qweet filtering.
3. **Indexing:** The last step is storing and indexing tweets. In our system, we used Apache Solr to index all tweets, flagging those that were classified as qweets by our qweet classifier.

2.2 User Interface

Users can interact with QweetFinder using a Web-based application implemented using the Play Framework[6]. The interface allows users to register with our system, search qweets, and save queries.

---

[2] http://www.tweepy.org/.
[3] Arabic phrases are available here: http://bit.ly/2itZDe9.
[4] Version 3.6.5.
[5] https://github.com/flaxsearch/luwak.
[6] https://www.playframework.com.

Fig. 2. A qweet search scenario

Figure 2 shows a user search scenario. An authenticated user issues a search query, then the system returns a list of qweets that match her query. Clicking a qweet will directly allow the user to view it in Twitter, as we assume that users might find answers to the qweet in the replies. The user can then save the query to his question feed. In the back-end, the new saved query is pushed to the filtering service to match incoming qweets against it. The system then pushes the new matched qweets to users via email.

Acknowledgments. This work was made possible by NPRP grant# NPRP 6-1377-1-257 from the Qatar National Research Fund (a member of Qatar Foundation). The statements made herein are solely the responsibility of the authors.

References

1. Hasanain, M., Elsayed, T., Magdy, W.: Identification of answer-seeking questions in Arabic microblogs. In: Proceedings of CIKM 2014 (2014)
2. Herrera, J.M., Parra, D., Poblete, B.: Retrieving relevant conversations for Q&A on Twitter. In: Proceedings of SPS 2015, Co-located at SIGIR 2015 (2015)
3. Jansen, B.J., Zhang, M., Sobel, K., Chowdury, A.: Twitter power: tweets as electronic word of mouth. JASIST **60**(11), 2169–2188 (2009)
4. Oeldorf-Hirsch, A., Hecht, B., Morris, M.R., Teevan, J., Gergle, D.: To search or to ask: the routing of information needs between traditional search engines and social networks. In: Proceedings of CSCW 2014 (2014)
5. Srba, I., Bielikova, M.: A comprehensive survey and classification of approaches for community question answering. ACM Trans. Web **10**(3), 18:1–18:63 (2016)
6. Suwaileh, R., Hasanain, M., Elsayed, T.: Light-weight, conservative, yet effective: scalable real-time tweet summarization. In: TREC 2016 (2016)
7. Zhao, Z., Mei, Q.: Questions about questions: an empirical analysis of information needs on Twitter. In: Proceedings of WWW 2013 (2013)

Integration of the Scientific Recommender System Mr. DLib into the Reference Manager JabRef

Stefan Feyer[1], Sophie Siebert[2], Bela Gipp[1], Akiko Aizawa[3],
and Joeran Beel[3,4(✉)]

[1] University of Konstanz, Konstanz, Germany
{stefan.feyer,bela.gipp}@uni-konstanz.de
[2] Otto-von-Guericke University Magdeburg, Magdeburg, Germany
sophie.siebert@st.ovgu.de
[3] National Institute of Informatics (NII), Tokyo, Japan
{aizawa,beel}@nii.ac.jp
[4] Trinity College Dublin, SCSS, KDEG, ADAPT Centre, Dublin, Ireland
joeran.beel@adaptcentre.ie

Abstract. This paper presents a description of integration of the Mr. DLib scientific recommender system into the JabRef reference manager. Scientific recommender systems help users identify relevant papers out of vast amounts of existing literature. They are particularly useful when used in combination with reference managers. Over 85% of JabRef users stated that they would appreciate the integration of a recommender system. However, the implementation of literature recommender systems requires experience and resources that small companies cannot afford. With the desires of users in mind, we integrated the Mr. DLib scientific recommender system into JabRef. Using Mr. DLib's recommendations-as-a-service, JabRef users can find relevant literature and keep themselves informed about the state of the art in their respective fields.

Keywords: IR · Scientific recommender system · Reference manager · Recommendations-as-a-service

1 Introduction

Scientific recommender systems are used by many academic services, including digital libraries and reference managers. Reference managers are particularly suitable for offering recommendations on related articles because such recommendations enable users to identify relevant information more easily from the large volume of existing reports, articles, and papers. This service not only provides access to local documents; it supports exploration of the relevant documents in external databases such as SowiPort and the ACM digital library.

Although Mendeley, ReadCube, and Docear [4] already have integrated recommender systems, JabRef does not have one. With over 2.5 million downloads

© Springer International Publishing AG 2017
J.M. Jose et al. (Eds.): ECIR 2017, LNCS 10193, pp. 770–774, 2017.
DOI: 10.1007/978-3-319-56608-5_80

over the past 13 years[1], JabRef has become an exceedingly popular reference manager[2]. According to a 2015 survey[3] yielding more than 428 responses, JabRef has users of 29 different languages, mostly German (42%) and English (26%). Respondents stated that they were using JabRef for their professional work (56%), but also for their studies and personal work. The largest field of interest among JabRef users was the natural sciences (36%), but formal sciences (21%), social sciences (10%), and humanities (7%) also commanded interest from users. Over 85% of users stated that a recommender system, if one were integrated into JabRef, would be useful. However, JabRef is developed and run by volunteers with limited time. Additionally, the team has little expertise in recommender systems, and lacks even the limited resources (e.g. servers) necessary to implement a recommender system. Therefore, recommendations-as-a-service, a recommendation service for which a third party is responsible for hosting a recommender system, is particularly attractive for JabRef. In the case of JabRef, it would request recommendations from a third party via an API.

For academic use, BibTip [6] and ExLibris bX[4] provide recommendations based on co-occurrence [3]. The CORE recommender [5] uses collaborative filtering and content-based filtering. Babel [7] was recently developed by DataLab, which is part of the Information School at the University of Washington.

Finally, Mr. DLib[5] (Machine Readable Digital Library) [2] is a free, open-source, RESTful web-service that generates recommendations based on a single document and which offers different recommendation approaches, such as stereotype-based and content-based algorithms with additional re-ranking using bibliometric data[6]. Additionally, Mr. DLib gathers data about the recommendations made. These data are utilized to evaluate the different algorithms, which in the long run will optimize scientific recommender systems. The Mendeley and ReadCube reference managers parse users' libraries and calculate recommendations based on this library. Both are companies with sufficient resources to implement this functionality independently. Docear visualizes the literature in a mind-map, providing several options for displaying additional information, such as annotations and relations to other papers in the mind-map. These mind-maps are utilized for the recommendations using content-based filtering.

The team responsible for JabRef chose to integrate Mr. DLib into the JabRef desktop application because it is open-source and because it includes the CORE dataset, with more than 20 million documents. Another reason Mr. DLib was

---

[1] https://sourceforge.net/projects/jabref/files/jabref/stats/timeline?dates=2003-10-1 2+to+2015-10-01 and https://www.fosshub.com/JabRef.html, accessed on 2016-10-19.

[2] http://www.docear.org/2013/11/11/on-the-popularity-of-reference-managers-and-their-rise-and-fall/, accessed on 2016-10-28.

[3] http://www.jabref.org/surveys/2015/analysis, accessed on 2016-10-21.

[4] http://www.exlibrisgroup.com/category/bXRecommender, accessed on 2016-10-19.

[5] http://mr-dlib.org/, accessed on 2016-10-19.

[6] Detailed information about the specific algorithms and their evaluation will be published in the near future.

chosen is that the recommendation approach of BibTip and ExLibris is mostly used to enrich online catalogues, although it is difficult to integrate.

Our overall goal is to support JabRef users in finding relevant literature and to discover relevant content within the vast amount of existing literature.

2   System Overview and Implementation

JabRef, a free open-source application with modest hardware requirements, is a desktop program available for all major operating systems. The user's literature being managed is presented as a table, as shown in Fig. 1 (left). Every row in the table represents a BibTex item. JabRef provides basic reference management tools such as sorting, filtering, and searching. Selecting an item opens a preview in the bottom area and displays basic information related to the BibTex item including the abstract, if it is given. Double-clicking on an entry gives access to detailed information related to the paper and the BibTex file itself through the so-called *Entry Editor*, which also accommodates editing of the file.

Multiple possibilities presented themselves for displaying the recommendations in the existing JabRef GUI. On the one hand, the area for showing literature recommendations should be highly visible, but on the other hand, it should fit into the existing structure of JabRef. One possible solution was to place recommendations in the preview of a BibTex entry. We considered creating a new section in the preview showing a list of the recommended articles. An alternative was placing a button in this preview, which opens a pop-up containing the recommendations. However, pop-ups would break with the existing structure of JabRefs GUI. In addition, a short evaluation of the user behavior revealed that this preview was not frequently used. Instead, users were using the *Entry Editor*. After considering these alternatives, we chose to add a new tab called *Related articles* in the *Entry Editor*, as shown in Fig. 1 (right). To place it as a new tab in this section fits into the existing structure. It does not feel intrusive and it is easily visible.

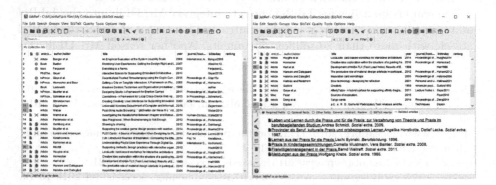

Fig. 1. (left) The JabRef GUI. No entry selected. (right) JabRef GUI with the new *Related articles* tab at the bottom. Six recommendations are displayed.

Whenever a user selects an entry, the title of the entry is transferred to Mr. DLib. Using this title, Mr. DLib calculates related articles and responds with an XML file containing the recommendations. This response is parsed by the JabRef application and is then displayed in the *Related articles* tab, which takes about one second. During this second, a loading animation is displayed in the tab, indicating that a request for the selected entry is in progress.

A click on the link of the recommendation produces a new browser tab, where the PDF can be downloaded. In addition, new recommendations based on the selected article are shown on this site, inviting users to continue browsing.

The application will become available online with the next update of JabRef. The current version is accessible on GitHub.

3   Outlook

The current version of Mr. DLib is specified to deliver recommendations based on a single document. Future versions will make it possible to calculate recommendations using a set of documents and to provide recommendations based on this set. Furthermore, Mr. DLib is constantly extending its database with new documents. The gathered data about the displayed recommendations, e.g. if they were clicked inside the Desktop application of JabRef, are expected to contribute to the improvement of academic literature recommender systems in the long run. The integration of Mr. DLib into JabRef has now enabled the Mr. DLib developers to conduct further research on reproducibility in recommender-systems research [1].

Acknowledgements. This publication has derived from research conducted with the financial support of Science Foundation Ireland (SFI) under Grant Number 13/RC/2106. We are also grateful for support received by Siddharth Dinesh and the JabRef Team: Oliver Kopp, Simon Harrer, Joerg Lenhard, Stefan Kolb, Matthias Geiger, Oscar Gustafsson, Tobias Diez, and Christoph Schwentker.

References

1. Beel, J., Breitinger, C., Langer, S., Lommatzsch, A., Gipp, B.: Towards reproducibility in recommender-systems research. User Model. User-Adap. Inter. (UMUAI) **26**(1), 69–101 (2016)
2. Beel, J., Gipp, B., Langer, S., Genzmehr, M., Wilde, E., Nürnberger, A., Pitman, J.: Introducing Mr. DLib, a machine-readable digital library. In: Proceedings of the 11th Annual International ACM/IEEE Joint Conference on Digital Libraries, pp. 463–464. ACM (2011)
3. Beel, J., Langer, S.: A comparison of offline evaluations, online evaluations, and user studies in the context of research-paper recommender systems. In: Kapidakis, S., Mazurek, C., Werla, M. (eds.) TPDL 2015. LNCS, vol. 9316, pp. 153–168. Springer, Heidelberg (2015). doi:10.1007/978-3-319-24592-8_12
4. Beel, J., Langer, S., Genzmehr, M., Nürnberger, A.: Introducing Docear's research paper recommender system. In: Proceedings of the 13th ACM/IEEE-CS Joint Conference on Digital Libraries - JCDL 2013, p. 459 (2013)

5. Knoth, P.: Linking textual resources to support information discovery. Ph.D. thesis, The Open University (2015)
6. Mönnich, M., Spiering, M.: Erschließung. Einsatz von BibTip als Recommendersystem im Bibliothekskatalog. Bibliotheksdienst **42**(1), 54–59 (2008)
7. Wesley-Smith, I., West, J.D.: Babel: a platform for facilitating research in scholarly article discovery. In: Proceedings of the 25th International Conference Companion on the World Wide Web, pp. 389–393 (2016)

Abstracts of Doctoral Consortium
Papers and Workshops

A Social Framework for Set Recommendation in Group Recommender Systems

Lorena Recalde[✉]

Web Research Group, Department of Information and Communication
Technologies, Universitat Pompeu Fabra, 08018 Barcelona, Spain
lorena.recalde@upf.edu

Abstract. Accessing relevant information stored in the Web is not an easy task. For instance, if a group of people are looking for an enjoyable movie to watch, the searching and choosing processes could both take a long time. That is to say, the quantity of items of interest presented by a search engine and the heterogeneous preferences of the group members are a harmful combination. To address this problem, Group Recommender Systems (*GRSs*) have been implemented last years. A GRS must be able to identify items that the group of users will like so that their needs, explicit or not, are equally satisfied.

As a matter of fact, to make a choice we rely on our friends' judgement and/or opinions, reviews and rates of others. The mentioned behavior motivates our proposal and makes us think that the same social conduct may be observed when a group is making a decision. In other words, being aware of the options chosen by other groups and their experiences may introduce an appropriate effect and help the given group to make an accurate choice. In this research proposal we introduce a framework that is intended to improve Group Recommender Systems and is based on the presented assumption. Consequently, the framework structure includes a process where an influential group is detected among the target groups of people to recommend to. In order to help the group members agree and make a decision, the visualization of the alternative chosen by the influential group and the reasons why they adopted that recommendation are presented for the target *susceptible group*. It is worth mentioning that in the context of the present article, the term susceptibility makes reference to the group that is highly perceived as being easily influenced.

Trying to discover influential established groups in a social network and seeing if susceptible groups adopt the recommendations provided by them is considered the main challenge for our future research. Combining this with the kind of item recommendation which involves a sequence of ordered elements will present a novel and original path in Group Recommender Systems design.

Keywords: Influential groups · Group recommender systems · Social factors

© Springer International Publishing AG 2017
J.M. Jose et al. (Eds.): ECIR 2017, LNCS 10193, p. 777, 2017.
DOI: 10.1007/978-3-319-56608-5

User Behavior Analysis and User Modeling for Complex Search

Jiaxin Mao[✉]

State Key Laboratory of Intelligent Technology and Systems,
Tsinghua National Laboratory for Information Science and Technology,
Department of Computer Science and Technology, Tsinghua University,
Beijing 1000084, China
maojiaxin@gmail.com

Abstract. As the information needs of Web search engine users become more and more diverse, complex search activities, such as exploratory search and multi-step search, have been identified and considered challenging for current search systems. As the user plays a central role in the highly interactive complex search session, user behavior analysis and modeling is vital for making search engines more effective for complex search tasks. To analyze the highly interactive complex search activity, we propose to regard it as a cognitive process that involves dynamic knowledge acquisition and decision making. In the proposed cognitive framework, we assume that user interactively updates his or her cognitive state through reading and then making decisions about next action based on the cognitive state. We further raise four research questions about the proposed framework and discuss the methodology adopted for user behavior analysis. Finally, we report our current progress and summarize empirical findings collected from previous user studies based on the proposed framework.

© Springer International Publishing AG 2017
J.M. Jose et al. (Eds.): ECIR 2017, LNCS 10193, p. 778, 2017.
DOI: 10.1007/978-3-319-56608-5

First International Workshop on Exploitation of Social Media for Emergency Relief and Preparedness (SMERP)

Saptarshi Ghosh[1,2], Kripabandhu Ghosh[3(✉)], Tanmoy Chakraborty[4], Debasis Ganguly[5], Gareth Jones[6], and Marie-Francine Moens[7]

[1] Indian Institute of Technology, Kharagpur, India
saptarshi@cse.iitkgp.ernet.in, sghosh@cs.iiests.ac.in
[2] Indian Institute of Engineering Science and Technology, Shibpur, India
[3] Indian Institute of Technology, Kanpur, India
kripa@iitk.ac.in, kripa@cse.iitk.ac.in, kripa.ghosh@gmail.com
[4] University of Maryland, College Park, USA
tanchak@umiacs.umd.edu
[5] IBM Research Labs, Dublin, Ireland
debasis.ganguly1@ie.ibm.com, debforit@gmail.com
[6] Dublin City University, Dublin, Ireland
Gareth.Jones@computing.dcu.ie
[7] Katholieke Universiteit Leuven, Leuven, Belgium
{sien.moens,marie-francine.moens}@cs.kuleuven.be

Abstract. The enormous growth of the use of Online Social Media (OSM) during emergency situations has led to increasing amounts of information being available to assist emergency relief operations. Additionally, OSM content can also be utilized for emergency preparedness and early warning systems. Effective exploitation of the crowdsourced content posted on OSM requires reliable real-time IR methodologies, and integration of OSM content with other information sources. This workshop will explore the multifarious aspects of effective information extraction and exploitation from social media, for emergency relief as well as emergency preparedness. Along with a peer-reviewed research paper track, the workshop will include a TREC-style data challenge, where we will make available to the research community a large collection of microblogs posted during a recent emergency event, and invite IR methodologies to extract useful information from this dataset.

Keywords: Online social media · Emergency · Disaster · Microblog · Data challenge · Peer-review

1 Background and Motivation

The ever-increasing amounts of user-generated content on online social media (OSM) platforms like Twitter, Facebook, WhatsApp, etc. have become important sources of information about diverse topics and events. Especially, during

© Springer International Publishing AG 2017
J.M. Jose et al. (Eds.): ECIR 2017, LNCS 10193, pp. 779–783, 2017.
DOI: 10.1007/978-3-319-56608-5

emergency events (e.g. natural disasters like earthquakes, cyclones, floods, fire, epidemics, or man-made disasters like terror attacks, riots), various information is posted on OSM, which can contribute significantly to relief operations [1, 3]. Additionally, crowdsourced content from OSM can also be utilized for emergency preparedness, such as for identifying disaster-prone regions and infrastructures, developing early warning systems, developing emergency-resilient communities, and so on.[1]

Given the huge volume of content posted on OSM, and the rapid rates at which content is posted (especially during a disaster event), automated IR techniques need to be developed for extracting, summarizing and presenting the information in a useful way. The proposed workshop aims to provide a forum for researchers working on related fields, to present their results and insights. The workshop will aim to bring together researchers from diverse fields – such as Information Retrieval, Data Mining and Machine Learning, Natural Language Processing, Computational Social Science, Human Computer Interaction, and so on – who can potentially contribute to utilizing social media for emergency relief and preparedness.

Apart from providing a platform for researchers to present their work, the proposed workshop has another motivation. Several recent studies have proposed retrieval methodologies for OSM content posted during emergency events; however, all the studies have used their own datasets, and hence, there has not been any systematic comparison or evaluation of the different algorithms. To bridge this gap, the proposed workshop will include a data challenge, where a large set of microblogs posted during a recent emergency event (the earthquake in Italy in August 2016) will be made available, along with a set of IR tasks that are critical during an emergency (e.g., extracting specific types of information, summarizing the information stream). Interested participants will be invited to develop methodologies for the tasks and submit their results. The organizers will arrange for evaluation of the results, which will enable comparison of the performances of various methodologies.

2   Workshop Objectives

The objectives of the workshop are two-fold, as described below.

2.1   Peer-Review Track

The workshop aims to provide a research platform dedicated to exploring the role of social media in emergency relief and management. The workshop will solicit original research contributions related to the theme, which includes (but is not limited to):

– Information retrieval and extraction from short, noisy content posted on OSM

---

[1] Note that the terms "emergency" and "disaster" have been used inter-changeably.

- Applications of data mining, NLP and machine learning for processing OSM content
- Aggregating information from multiple OSM and online/offline resources
- Addressing the code-mixed and informal vocabulary of OSM content
- Detection of events and emerging themes
- Real-time management and summarization of dynamic content streams
- Detection of rumours, and identification of trustworthy sources and information
- Geo-tagging and geo-localisation of content and sources
- Social network models for information diffusion in emergency situations
- Identifying disaster-prone or accident-prone regions and infrastructures
- Crowdsourcing systems for emergency preparedness and disaster relief
- Mining interactions among emergency preparedness and relief groups

As evident from the list of topics above, the workshop aims to bring together researchers from diverse communities which can contribute to emergency management, including IR, Data Mining, Machine Learning, NLP, and Computational Social Science communities.

2.2 Data Challenge Track

The workshop aims to promote development of IR methodologies for some practical challenges that need to be addressed during an emergency event, along with thorough evaluation and comparison of the methodologies. To fulfil this objective, we include a data challenge, following the style of TREC tracks [2].[2] We will provide a large dataset of microblogs posted during a recent emergency event – the earthquake in Italy in August 2016 – and a set of practical challenges that need to be addressed in such an emergency situation (e.g., real-time summarization, extracting specific types of information). The participants will be invited to submit solutions to the said challenges. We will arrange for evaluation of the submitted results. This process will enable the participants to empirically validate their research methodologies in the ambience of a healthy competition.

The Data Challenge track will offer two problems, viz., Text Retrieval and Text Summarization. A brief description these two problems are given as follows.

Text Retrieval In this problem, the participants will be required to develop methodologies for extracting tweets that are relevant to a set of specified topics. The topics will include practical information needs during an emergency event, such as what resources are needed, what resources are available, what are the casualties and infrastructure damages, and so on.

This problem will have two levels – Level 1 and Level 2. In Level 1, the tweets of first day (24 h) after the earthquake will be provided, and the participants will be asked to extract tweets relevant to each specified topic. The extracted tweets should be submitted, and the organizers will arrange for evaluation of the results

---

[2] http://trec.nist.gov/tracks.html.

of each submission. In Level 2, the tweets collected during the second day (24 h) after the earthquake will be provided. Additionally, some of the tweets that are actually relevant to each topic, from among the tweets used in Level 1, will be indicated. The participants will again be expected to retrieve tweets relevant to each topic from among the tweets posted during the second day.

The purpose of the second level is to provide the participants some idea about which tweets are considered relevant, so that they can apply this knowledge to improve retrieval on an unseen set of tweets. Thus, this exercise will present a platform to evaluate the dynamic reusability and adaptability of the IR system in a practical scenario.

Text Summarization In this problem, the participants will be required to develop summarization techniques tweets that are relevant to each topic. Like the Text Retrieval task, this task will also have two levels to serve the same purpose. That is, in Level 1 the participants will be expected to submit summaries of the tweets of first day (24 h) after the earthquake, on each of the topics and these summaries will be evaluated. In Level 2, the participants will be given the gold standard summaries from Level 1 and the tweets collected during the second day (24 h) after the earthquake on which summaries are to be submitted. Thus, in this task also, the purpose will be to present a platform to evaluate the dynamic reusability and adaptability of the summarization system.

3   Outcomes

The workshop proceedings will be archived online in the CEUR workshop proceedings publication service (http://ceur-ws.org/, ISSN 1613-0073). All papers submitted to the workshop will be peer-reviewed by a program committee, based on technical merit, and the accepted papers will be included in the workshop proceedings.

The methodologies submitted to the data challenge will be evaluated using a procedure similar to what is used by TREC, and the best one or two teams will be invited to submit papers. These papers will be shepherded to ensure that they meet the standards of the papers accepted through the peer-review track, and will then be included in the workshop proceedings. The results of the data challenge will be made public to all participating teams, so that they can calibrate their methodologies.

4   Format/Structure of the Workshop

The full-day workshop will include keynote talks by reputed researchers working on topics related to the theme, reporting and discussion of the results of the data challenge, and presentation of the papers accepted through peer-review. We also plan to have a panel discussion on open and upcoming challenges in utilising social media content for emergency relief and preparedness.

5   Workshop Website

For further details, please refer to the workshop website – http://computing.
dcu.ie/~dganguly/smerp2017/.

References

1. Imran, M., Castillo, C., Diaz, F., Vieweg, S.: Processing social media messages in mass emergency: a survey. ACM Comput. Surv. **47**(4), 67:1–67:38 (2015)
2. Ounis, I., Macdonald, C., Lin, J., Soboroff, I.: Overview of the TREC-2011 microblog track (2011). http://trec.nist.gov/pubs/trec20/papers/MICROBLOG. OVERVIEW.pdf
3. Vieweg, S., Hughes, A.L., Starbird, K., Palen, L.: Microblogging during two natural hazards events: what Twitter may contribute to situational awareness. In: Proc. ACM SIGCHI (2010)

Bibliometric-Enhanced Information Retrieval: 5th International BIR Workshop

Philipp Mayr[1(✉)], Ingo Frommholz[2], and Guillaume Cabanac[3]

[1] GESIS - Leibniz-Institute for the Social Sciences, Cologne, Germany
philipp.mayr@gesis.org
[2] Institute for Research in Applicable Computing,
University of Bedfordshire, Luton, UK
ifrommholz@acm.org
[3] Computer Science Department, University of Toulouse,
IRIT UMR 5505, Toulouse, France
guillaume.cabanac@univ-tlse3.fr

Abstract. Bibliometric-enhanced Information Retrieval (BIR) workshops serve as the annual gathering of IR researchers who address various information-related tasks on scientific corpora and bibliometrics. The workshop features original approaches to search, browse, and discover value-added knowledge from scientific documents and related information networks (e.g., terms, authors, institutions, references). We welcome contributions elaborating on dedicated IR systems, as well as studies revealing original characteristics on how scientific knowledge is created, communicated, and used. In this paper we introduce the BIR workshop series and discuss some selected papers presented at previous BIR workshops.

Keywords: Bibliometrics · Scientometrics · Informetrics · Information retrieval · Digital libraries

1 Introduction

Following the successful workshops at ECIR 2014[1], 2015[2], 2016[3] and JCDL 2016[4], respectively, this workshop is the fifth in a series of events that brought together experts of communities which often have been perceived as different ones: bibliometrics/scientometrics/informetrics on the one hand and information retrieval on the other hand. Our motivation as organizers of the workshop started from the observation that main discourses in both fields are different, that communities are only partly overlapping and from the belief that a knowledge transfer would be profitable for both sides [1, 2]. The need for researchers to

---

[1] http://ceur-ws.org/Vol-1143/.
[2] http://ceur-ws.org/Vol-1344/.
[3] http://ceur-ws.org/Vol-1567/.
[4] http://ceur-ws.org/Vol-1610/.

© Springer International Publishing AG 2017
J.M. Jose et al. (Eds.): ECIR 2017, LNCS 10193, pp. 784–789, 2017.
DOI: 10.1007/978-3-319-56608-5

keep up-to-date with their respective field given the highly increasing number of publications available has led to the establishment of scientific repositories that allow us to use additional evidence coming for instance from citation graphs to satisfy users' information needs.

The first BIR workshops in 2014 and 2015 set the research agenda by introducing each group to the other, illustrating state-of-the-art methods, reporting on current research problems, and brainstorming about common interests. The third workshop in 2016 [3] further elaborated on these themes. For the fourth workshop, co-located with the ACM/IEEE-CS Joint Conference on Digital Libraries (JCDL) 2016, we broadened the workshop scope and interlinked the BIR workshop with the natural language processing (NLP) and computational linguistics field [4]. This 5th full-day BIR workshop at ECIR 2017 aims to foster a common ground for the incorporation of bibliometric-enhanced services (incl. text mining functionality) into scholarly search engine interfaces. In particular we address specific communities, as well as studies on large, cross-domain collections like Mendeley and ResearchGate. This fifth BIR workshop again addresses explicitly both scholarly and industrial researchers.

2   Goals, Objectives and Outcomes

Our workshop aims to engage the IR community with possible links to bibliometrics. Bibliometric techniques are not yet widely used to enhance retrieval processes in digital libraries, yet they offer value-added effects for users [5]. Hence, our objective is to bring together information retrieval, information seeking, science modelling, network analysis, and digital libraries to apply insights from bibliometrics, scientometrics, informetrics and text mining to concrete, practical problems of information retrieval and browsing. We discuss some examples from previous workshops in Sect. 5. More specifically we ask questions like:

- How can we generalize paper tracking on social media?
  a.k.a. altmetrics on steroïds: beyond DOI spotting.
- How can we detect fake reviews [6] to sustain the peer review process?
- How can we improve homonym detection (e.g., Li Li) in bibliographic records [7]?
- To what degree can we automate fact-checking [8, 9] in academic papers?
- How can we support researchers in finding relevant scientific literature, e.g., by integrating ideas from information retrieval, information seeking and searching and bibliometrics [10, 11]?
- How can we build scholarly information systems that explicitly use bibliometric measures at the user interface (e.g. contextual bibliometric-enhanced features [12])?
- How can models of science be interrelated with scholarly, task-oriented searching?
- How can we combine classical IR (with emphasis on recall and weak associations) with more rigid bibliometric recommendations [13, 14]?
- How can we create suitable testbeds (like iSearch corpus) [15]?

3   Format and Structure of the Workshop

The workshop will start with an inspirational keynote "Real-World Recommender Systems for Academia: The Pain and Gain in Developing, Operating, and Researching them" by Joeran Beel (Trinity College Dublin, the School of Computer Science and Statistics) to kick-start thinking and discussion on the workshop topic. This will be followed by paper presentations in a format that we found to be successful at previous BIR workshops: each paper is presented as a 10 min lightning talk and discussed for 20 min in groups among the workshop participants followed by 1-minute pitches from each group on the main issues discussed and lessons learned. The workshop will conclude with a round-robin discussion of how to progress in enhancing IR with bibliometric methods.

4   Audience

The audiences of IR and bibliometrics overlap [1, 2]. Traditional IR serves individual information needs, and is – consequently – embedded in libraries, archives and collections alike. Scientometrics, and with it bibliometric techniques, has a matured serving science policy. We therefore will hold a full-day workshop that brings together IR researchers with those interested in bibliometric-enhanced approaches. Our interests include information retrieval, information seeking, science modelling, network analysis, and digital libraries. The workshop is closely related to the past BIR workshops at ECIR 2014, 2015, 2016 and strives to feature contributions from core bibliometricians and core IR specialists who already operate at the interface between scientometrics and IR. While the past workshops laid the foundations for further work and also made the benefit of bringing information retrieval and bibliometrics together more explicit, there are still many challenges ahead. One of them is to provide infrastructures and testbeds for the evaluation of retrieval approaches that utilise bibliometrics and scientometrics. To this end, a focus of the proposed workshop and the discussion will be on real experimentations (including demos) and industrial participation. This line was started in a related workshop at JCDL (BIRNDL 2016), but with a focus on digital libraries and computational linguistics and not on information retrieval and information seeking and searching.

5   Selected Papers and Past Keynotes

Past BIR workshops had invited talks of several experts working in the field of bibliometrics and information retrieval. Last year, Marijn Koolen gave a keynote on "Bibliometrics in online book discussions: Lessons for complex search tasks" [16]. Koolen explored the potential relationships between book search information needs and bibliometric analysis and introduced the Social Book Search Lab, triggering a discussion on the relationship between book search and bibliometric-enhanced IR. In 2015, the keynote "In Praise of Interdisciplinary Research through Scientometrics" [17] was given by Guillaume Cabanac. Cabanac accentuated the potential of interdisciplinary research at the interface of information retrieval and bibliometrics. He came up with many research questions that lie at the crossroad of scientometrics and other fields, namely information retrieval, digital libraries, psychology and sociology.

Recent examples of BIR workshop publications have shown the potential of informing the information retrieval process with bibliometrics. These examples comprise topics like IR and recommendation tool development, bibliometric IR evaluation and data sets, and the application and analysis of citation contexts for instance for cluster-based search.

As an example of recommendation tool development utilising bibliometrics, Wesley-Smith et al. [18] describe an experimental platform constructed in collaboration with the repository Social Science Research Network (SSRN) in order to test the effectiveness of different approaches for scholarly article recommendations. Jack et al. [19] present a case study on how to increase the number of citations to support claims in Wikipedia. They analyse the distribution of more than 9 million citations in Wikipedia and found that more than 400,000 times an explicit marker for a needed citation is present. To overcome this situation they propose different techniques based on Bradfordizing and popularity number of readers in Mendeley to implement a citation recommending system. The authors conclude that a normal keyword-based search engine like Google Scholar is not sufficient to be used to provide citation recommendation for Wikipedia articles and that altmetrics like readership information can improve retrieval and recommendation performance.

Utilising a collection based on PLOS articles, Bertin and Atanassova [20] try to further unravel the riddle of meaning of citations. The authors analyse the word use in standard parts of articles, such as Introduction, Methods, Results and Discussion, and reveal interesting distributions of the use of verbs for those sections. The authors propose to use this work in future citation classifiers, which in the long-term might also be implemented in citation-based information retrieval.

As an application of citation analysis, Abbasi and Frommholz [21] investigate the benefit of combining polyrepresentation with document clustering, where representations are informed by citation analysis. The evaluation of the proposed model on the basis of the iSearch collection shows some potential of the approach to improve retrieval quality. A further application example reported by Nees Jan van Eck and Ludo Waltman [22] considers the problem of scientific literature search. The authors suggest that citation relations between publications can be a helpful instrument in the systematic retrieval process of scientific literature. They introduce a new software tool called CitNetExplorer that can be used for citation-based scientific literature retrieval. To demonstrate the use of CitNetExplorer, they employ the tool to identify publications dealing with the topic of "community detection in networks". They argue that their approach can be especially helpful in situations in which one needs a comprehensive overview of the literature on a certain research topic, for instance in the preparation of a review article.

Howard D. White proposes an alternative to the well-known bag of words model called *bag of works* [23]. This model can in particular be used for finding similar documents to a given seed one. In the bag of works model, tf and idf measures are re-defined based on (co-)citation counts. The properties of the retrieved documents are discussed and an example is provided.

6   Output

In 2015 we published a first special issue on "Combining Bibliometrics and Information Retrieval" in Scientometrics [1]. A special issue on "Bibliometrics, Information Retrieval and Natural Language Processing in Digital Libraries" is currently under preparation for the International Journal on Digital Libraries. For this year's ECIR workshop we continue the tradition of producing follow-up special issues. Authors of accepted papers at this year's BIR workshop will again be invited to submit extended versions to a special issue on "Bibliometric-enhanced IR" to be published in Scientometrics.

References

1. Mayr, P., Scharnhorst, A.: Scientometrics and information retrieval: weak-links revitalized. Scientometrics **102**(3), 2193–2199 (2015)
2. Wolfram, D.: The symbiotic relationship between information retrieval and informetrics. Scientometrics **102**(3), 2201–2214 (2015)
3. Mayr, P., Frommholz, I., Cabanac, G.: Report on the 3rd international workshop on bibliometric-enhanced information retrieval (BIR 2016). SIGIR Forum **50**(1), 28–34 (2016)
4. Cabanac, G., Chandrasekaran, M.K., Frommholz, I., Jaidka, K., Kan, M.Y., Mayr, P., Wolfram, D.: Report on the joint workshop on bibliometric-enhanced information retrieval and natural language processing for digital libraries (BIRNDL 2016). SIGIR Forum **50**(2), 36–43 (2016)
5. Mutschke, P., Mayr, P., Schaer, P., Sure, Y.: Science models as value-added services for scholarly information systems. Scientometrics **89**(1), 349–364 (2011)
6. Bartoli, A., De Lorenzo, A., Medvet, E., Tarlao, F.: Your paper has been accepted, rejected, or whatever: automatic generation of scientific paper reviews. In: Buccafurri, F., Holzinger, A., Kieseberg, P., Tjoa, AM, Weippl, E. (eds.) CD-ARES 2016. LNCS, vol. 9817, pp. 19–28. Springer, Switzerland (2016)
7. Momeni, F., Mayr, P.: Evaluating co-authorship networks in author name disambiguation for common names. In: Fuhr, N., Kovács, L., Risse, T., Nejdl, W. (eds.) TPDL 2016. LNCS, vol. 9819, pp. 386–391. Springer, Cham (2016). doi:10.1007/978-3-319-43997-6_31
8. Baker, M.: Smart software spots statistical errors in psychology papers. Nature (2015)
9. Ziemann, M., Eren, Y., El-Osta, A.: Gene name errors are widespread in the scientific literature. Genome Biol. **17**(1), 1–3 (2016)

10. Abbasi, M.K., Frommholz, I.: Cluster-based polyrepresentation as science modelling approach for information retrieval. Scientometrics **102**(3), 2301–2322 (2015)
11. Mutschke, P., Mayr, P.: Science models for search. A study on combining scholarly information retrieval and scientometrics. Scientometrics **102**(3), 2323–2345 (2015)
12. Carevic, Z., Mayr, P.: Survey on high-level search activities based on the stratagem level in digital libraries. In: Fuhr, N., Kovács, L., Risse, T., Nejdl, W. (eds.) TPDL 2016. LNCS, vol. 9819, pp. 54–66. Springer, Cham (2016). doi:10.1007/978-3-319-43997-6_5
13. Zitt, M.: Meso-level retrieval: Ir-bibliometrics interplay and hybrid citation-words methods in scientific fields delineation. Scientometrics **102**(3), 2223–2245 (2015)
14. Beel, J., Gipp, B., Langer, S., Breitinger, C.: Research-paper recommender systems: a literature survey. Int. J. Digit. Libr. **17**(4), 305–338 (2016)
15. Larsen, B., Lioma, C.: On the need for and provision for an 'ideal' scholarly information retrieval test collection. In: Proceedings of the Third Workshop on Bibliometric-enhanced Information Retrieval, pp. 73–81(2016)
16. Koolen, M.: Bibliometrics in online book discussions: Lessons for complex search tasks. In: Proceedings of the Third Workshop on Bibliometric-enhanced Information Retrieval co-located with the 38th European Conference on Information Retrieval (ECIR 2016), pp. 5–13, Padova, Italy, 20 March 2016
17. Cabanac, G.: In praise of interdisciplinary research through scientometrics. In: Proceedings of the Second Workshop on Bibliometric-enhanced Information Retrieval, pp. 5–13 (2015)
18. Wesley-Smith, I., Dandrea, R.J., West, J.D.: An experimental platform for scholarly article recommendation. In: Proceedings of the Second Workshop on Bibliometric-enhanced Information Retrieval, pp. 30–39 (2015)
19. Jack, K., López-García, P., Hristakeva, M., Kern, R.: Citation needed: filling in Wikipedia's citation shaped holes. In: Proceedings of the First Workshop on Bibliometric-enhanced Information Retrieval, pp. 45–52 (2014)
20. Bertin, M., Atanassova, I.: A study of lexical distribution in citation contexts through the IMRaD standard. In: Proceedings of the First Workshop on Bibliometric-enhanced Information Retrieval, pp. 5–12 (2014)
21. Abbasi, M.K., Frommholz, I.: Exploiting information needs and bibliographics for polyrepresentative document clustering. In: Proceedings of the First Workshop on Bibliometric-enhanced Information Retrieval, pp. 21–28 (2014)
22. van Eck, N.J., Waltman, L.: Systematic retrieval of scientific literature based on citation relations: introducing the citnetexplorer tool. In: Proceedings of the First Workshop on Bibliometric-enhanced Information Retrieval, pp. 13–20 (2014)
23. White, H.D.: Bag of works retrieval: Tf*idf weighting of co-cited works. In: Proceedings of the Third Workshop on Bibliometric-enhanced Information Retrieval, pp. 63–72 (2016)

OnST'17: The 2nd International Workshop on Online Safety Trust Fraud Prevention

Marco Fisichella[⊠]

Risk, Ident GmbH, Hamburg, Germany
marco@riskident.com

1 Motivation Overview

Almost every aspect of our lives is influenced by the web, such as in entertainment, education, health, commerce, the government, social interaction, and many more. The profound impact these changes have on our lives requires to rethink the way we make decisions in these areas. Besides questions related to cost and benefit, there are important issues raised by users regarding trust and safety. Can I trust this service with my data? Is it safe to use? In many cases, trust-related issues become a deal-breaker for the adoption of online services. We commonly find cases where people avoid online banking or buying products online due to the fear of becoming prey of fraudulent activity. Thus, providing a trustworthy environment for users is of utmost importance.

When online fraud are committed, fraudsters take advantage of gaps allowing them to unjustifiably enrich themselves. However, when fighting fraud, companies face a dilemma, given that no system is perfect: in e-commerce, on the one hand, fraud and its related losses should be reduced; on the other hand, users neither want to be accused of fraud nor treated like criminals. In other areas, such as health, the problems associated with data abuse and security leaks could even result in more severe damage than purely financial matters.

Yet, companies' practical implementations of fraud investigation processes rarely meet scientific standards. Fraud prevention companies offer diverse products on the market, but neither their effectiveness nor their efficiency has been verified scientifically until now. Dealing with these issues, the first step should be a clear definition of what constitutes fraud and how it can be measured. Afterwards, models can be built and tested according to scientific standards of evaluation. On this basis, a careful risk analysis can be conducted in order to weigh pros and cons of pursuing individual suspicious cases.

Companies spend millions to protect themselves from fraudulent activities. One of the most interesting aspects is the fact that fighting fraud is a social interaction that needs constant supervision and improvement: a never ending race between criminals and investigators, in which both actors adjust for the actions of the other.

This workshop aims at bringing together researchers from a wide range of disciplines (mathematics, computer science, economy, philosophy, social science) to (i) understand the cases and motivations of fraudulent activities in online environments, (ii) find solutions to detect and analyze fraud, and (iii) derive means to prevent it.

© Springer International Publishing AG 2017
J.M. Jose et al. (Eds.): ECIR 2017, LNCS 10193, pp. 790–794, 2017.
DOI: 10.1007/978-3-319-56608-5

The OnST workshop focuses on online fraud detection and prevention, but submissions that tackle these challenges in other environments are welcomed as well. Relevant research areas to the workshop are, e.g., "Spam detection", "Trust, authority, reputation, ranking", and "Time series and forecasting" with the goal of anomaly detection. We invite the submission of on-going and mature research work with a particular focus on the following topics:

- Online Safety and Trust
  - User Modeling: personalization of fraudulent and malicious users;
  - Account take-over;
  - Human interactions;
  - IT-forensics investigating a wide variety of crime, including child pornography, fraud, espionage, cyber-stalking, etc.
- Fraud Prevention
  - Features engineering for online detection;
  - Supervised machine learning techniques: fraud rule engines, time series, spatial-based, graph-based, spatio-temporal approaches;
  - Unsupervised machine learning techniques: Outlier and anomaly detection;
  - In Crowdfunding;
  - In Big Streaming Data;
  - Distributed systems;
  - Effective and efficient systems.

The main research questions that the workshop would like to answer are:

(i)   What are the best practices for detecting fraudulent and malicious activities?
(ii) What are the best practices for preventing fraudulent and malicious activities?
(iii) Which is the psychological impact of fraudulent activities on the society?

We invite the submission of original work in these and related areas. Each submission to the workshop will be peer-reviewed by at least two expert reviewers.

2   Workshop Rationale and Significance and Relevance to ECIR

The proposed workshop has a strong relation to most topics of the main program of ECIR, such as "Spam detection", "Trust, authority, reputation, ranking", and "Time series and forecasting" with the goal of anomaly detection. As such, the workshop will be widely accessible to the ECIR community. At the same time, though, this workshop will approach the above ECIR topics from the unique and emerging viewpoint of detecting and preventing malicious activities from a scientific point of view, considering their psychological and, economical impact as well as their risk. This viewpoint and the workshop's focus clearly differentiate the workshop from ECIR's main program and make it an appealing addition to it.

3 Workshop Organizers

Dr. Marco Fisichella is the head of the data science team at Risk Ident (Otto group, Germany) where they devise algorithms in order to detect the online frauds. Before he joined Risk Ident, he was postdoctoral researcher at the L3S Research Center in Hannover, Germany. Until beginning of 2015, he was also lecturer of the Artificial Intelligence course for the Master in Computer Science at the Leibniz University of Hannover. His research interests include data mining, information retrieval, generative model, event detection, clustering methods based on statistical approaches, near duplicate detection. He has worked as project manager in several EU-funded projects including (1) DuraArk-Preservation of architectural building data, and (2) OpenScout - accelerating the use, improvement and distribution of open content in the field of management education and training. He actively participated as proposal consultant and advisory board member in the following accepted proposal: (1) ALEXANDRIA - an ERC Advanced Grant Project on Foundations for Temporal Retrieval, Exploration and Analytics in Web Archives; (2) Zivile Sicherheit - a BMBF German funding program for tracking the Russian flu in U.S. and German medical and popular reports, occurred between 1889 and 1893. He has strong publication records in top-tier conferences, such as CIKM, SPIRE, ECIR and WISE and active professional memberships, i.e., invited reviewer and PC member (e.g., ICDM, WWW, and CIKM) and journal reviewer (e.g., Data & Knowledge Engineering Journal - Elsevier - on the track area about Reasoning Approaches). Finally, he received the best paper award at EC-TEL 2011 for his publication on "Unsupervised Auto-tagging for Learning Object Enrichment".

Prof. Dr. Nattiya Kanhabua is an assistant professor at the Department of Computer Science, Aalborg University, Denmark. Her research interests are information retrieval, data mining, machine learning, and spatial and temporal analytics. She did her PhD at the Department of Computer and Information Science, Norwegian University of Science and Technology (NTNU). She was a postdoctoral researcher at the L3S Research Center Hannover, Germany. At L3S, she worked in several research projects, e.g., (1) EU Project ForgetIT: Concise Preservation by Combining Managed Forgetting and Contextualized Remembering, (2) ALEXANDRIA, an ERC Advanced Grant Project on Foundations for Temporal Retrieval, Exploration and Analytics in Web Archives, and (3) Medical Ecosystem: Personalized Event-based Surveillance. She has published her research work in top-tier conferences, e.g., SIGIR, WSDM, CIKM and ECIR.

Sven Kurras is a senior data scientist at Risk Ident (Otto group, Germany). He focuses on detecting fraudulent behavior from connectivity information by developing scalable techniques for statistical graph-based inference. Beside his work at Risk Ident he finishes his doctoral thesis to the end of 2016 within the DFG Research Unit 1735 "Structural Inference in Statistics: Adaptation and Efficiency". During the preceding four years, he worked full-time as a doctoral researcher in the field of unsupervised machine learning at the machine learning working group of Ulrike von Luxburg at the University of Hamburg. He specialized on theoretical foundations of multi-scale clustering algorithms on random graphs. His results are published at international top-conferences like ICML and AISTATS. Complementary to his theoretical background, he also brings in 15 years of experience as a Java software architect and programmer, recently shifting to Scala and big data architectures such as the SMACK stack.

4 Program Committee

- Prof. Dr. Ismail Sengor Altingovde, Dept. of Computer Engineering Middle East Technical University (METU), Turkey
- Steffen Brauer, Risk.Ident GmbH - Otto Group, Germany
- Andrea Ceroni, L3S Research Center - Leibniz University Hannover, Germany
- Dr. Marco Diciolla, Palantir Technologies, U.K.
- Dr. Marco Fisichella, Risk.Ident GmbH - Otto Group, Germany
- Prof. Dr. Felix Freiling, Friedrich-Alexander-Universität Erlangen-Nürnberg, Germany
- Sergio Govoni, ProQuest, U.S.A.
- Prof. Dr. Rüdiger Grimm, Fellow of the German Informatics Society GI e.V., Germany
- Prof. Dr. Nattiya Kanhabua, Aalborg University, Denmark
- Sven Kurras, Risk.Ident GmbH - Otto Group, Germany
- David Losada, University of Santiago de Compostela, Spain
- Dr. Ida Mele, Faculty of Informatics - Università della Svizzera italiana (USI), Switzerland
- Dr. Katja Niemann, Fraunhofer Institute for Applied Information Technology, Germany
- Simon Schenk, Risk.Ident GmbH - Otto Group, Germany

The entire workshop will gain and leverage from the hard grounded experience of (i) Prof. Dr. Rüdiger Grimm, who was also head of the research group "Security for Virtual Goods" of the Fraunhofer Institute for Digital Media Technology; and (ii) Prof. Dr. Felix Freiling, former advisor of the German constitutional court (Bundesverfassungsgericht) in cases related to data protection and police laws in Germany.

Prof. Dr. Rüdiger Grimm was professor for IT Risk Management at the University Koblenz-Landau since 2005–2015, and he is continuing research and teaching duties in his University after his retirement in October 2015.

During that time, 2002–2005 he was also head of the research group "Security for Virtual Goods" of the Fraunhofer Institute for Digital Media Technology (IDMT) in Ilmenau.

2011–2014 he was elected Dean of the Faculty of Informatics in Koblenz. Since 2010 he is Fellow of the German Informatics Society GI e.V.

Prof. Dr. Felix Freiling is a full professor of computer science at Friedrich-Alexander-Universität Erlangen-Nürnberg (FAU) in Erlangen, Germany. Before joining FAU he held professor positions at RWTH Aachen University and University of

Mannheim. His research interests are in digital forensics and offensive computer security. He is member of the Steering Committee of the International Conference on IT Security Incident Management & IT Forensics (IMF) and was the chair of the TPC of the Digital Forensics Research Conference Europe (DFRWS EU) 2015. He was an advisor of the German constitutional court (Bundesverfassungsgericht) in cases related to data protection and police laws in Germany.

Workshop on Social Media for Personalization and Search (SoMePeAS)

Ludovico Boratto[1]([✉]), Andreas Kaltenbrunner[1], and Giovanni Stilo[2]

[1] Digital Humanities, Eurecat, Av. Diagonal 177, 8th Floor, 08018 Barcelona, Spain
ludovico.boratto@acm.org, kaltenbrunner@gmail.com
[2] Dipartimento di Informatica, Sapienza Università di Roma,
Via Salaria 113, 00198 Rome, Italy
stilo@di.uniroma1.it

Abstract. Social media platforms have become powerful tools to collect the preferences of the users and get to know them more. Indeed, in order to build profiles about what the users like or dislike, a system does not only have to rely on explicitly given preferences (e.g., ratings) or on implicitly collected data (e.g., from the browsing sessions). In the middle, there lie opinions and preferences expressed through likes, textual comments, and posted content. Being able to exploit social media to mine user behavior and extract additional information leads to improvements in the accuracy of personalization and search technologies, and to better targeted services to the users. In this workshop, we aim to collect novel ideas in this field and to provide a common ground for researchers working in this area.

Keywords: Social media · Personalization · Search

1 Introduction

In order to improve the web experience of the users, classic personalization technologies (e.g., recommender systems) and search engines usually rely on static schemes. Indeed, users are allowed to express ratings in a fixed range of values for a given catalogue of products, or to express a query that usually returns the same set of webpages/products for all the users.

With the advent of social media, users have been allowed to create new content and to express opinions and preferences through likes and textual comments. Moreover, the social network itself can provide information on who influences who. Being able to mine usage and collaboration patterns in social media and to analyze the content generated by the users opens new frontiers in the generation of personalization services and in the improvement of search engines. Moreover, recent technological advances, such as deep learning, are able to provide a context to the analyzed data (e.g., Google's *word2vec* provides a vector representation of the words in a corpus, considering the context in which a word has been used).

© Springer International Publishing AG 2017
J.M. Jose et al. (Eds.): ECIR 2017, LNCS 10193, pp. 795–796, 2017.
DOI: 10.1007/978-3-319-56608-5

This workshop solicits contributions in all topics related to employing social media for personalization and search purposes, focused (but not limited) to the following list:

- Recommender systems
- Search and tagging
- Query expansion
- User modeling and profiling
- Advertising and ad targeting
- Content classification, categorization, and clustering
- Using social network features/community detection algorithms for personalization and search purposes

2   Short Biographies of the Organizers

Ludovico Boratto is researcher in the Digital Humanities research group at Eurecat. He previously was a research assistant at the University of Cagliari - Italy, where he also got his Ph.D. in 2012. His main research area is Recommender Systems, with special focus on those that work with groups of users and in social environments. In 2010 and 2014, he spent 10 months at the Yahoo! Lab in Barcelona as a visiting researcher.

Andreas Kaltenbrunner is Head of the Digital Humanities research group at Eurecat. His research is centred on social media and social network analysis. He uses methods from computer science and the study of complex systems to resolve sociological research questions. Dr. Kaltenbrunner obtained his Ph.D. from the University Pompeu Fabra in Computer Science and Digital Communication in 2008.

Giovanni Stilo is an assistant professor at the Department of Computer Science - Sapienza University of Rome. Dr. Stilo obtained his Ph.D. from the University of Aquila in Computer Science and Applications in 2013. He is mainly involved on the study of temporal mining, network analysis over social source, and semantic oriented recommender systems. He was previously contractor with the Web Mining Group of Yahoo! Lab in Barcelona.

Author Index